BORATE GLASSES, CRYSTALS & MELTS 2005

Proceedings of the Fifth International Conference on Borate Glasses, Crystals and Melts: new techniques and practical applications held at University of Trento, Trento, Italy on 10–14 July 2005

Guest Editors
G. Dalba & A. C. Wright

SOCIETY OF GLASS TECHNOLOGY

Borate glasses, crystals & melts 2005

Proceedings of the Fifth International Conference on Borate Glasses, Crystals and Melts: new techniques and practical applications held at University of Trento, Trento, Italy on 10–14 July 2005

A collected volume of papers from the Fifth International Conference, the papers were originally published in *European Journal of Glass Science and Technology*:
Physics and Chemistry of Glasses: European Journal of Glass Science and Technology Part B:
August 2006, pages 313–568; and February 2008, pages 19–25.
Glass Technology: European Journal of Glass Science and Technology Part A:
October 2006, pages 127–156; December 2006, pages 159–176; and February 2007, pages 1–12.

ISBN 978-0-900682-61-2

The objects of the Society of Glass Technology are to encourage and advance the study of the history, art, science, design, manufacture, after treatment, distribution and end use of glass of any and every kind. These aims are furthered by meetings, publications, the maintenance of a library and the promotion of association with other interested persons and organisations.

Society of Glass Technology
Unit 9, Twelve O'clock Court
21 Attercliffe Road
Sheffield S4 7WW, UK
Tel +44(0)114 263 4455
Fax +44(0)114 263 4411
Email info@sgt.org
Web http://www.sgt.org
The Society of Glass Technology is a registered charity no. 237438.

Foreword

The V International Conference on Borate Glasses, Crystals and Melts: New Techniques and Practical Applications was held in Trento (Italy) at the Conference Centre Panorama during the period July 10th to 14th, 2005. This event followed the previous meetings held at Alfred, New York, USA (1977); Abingdon, England (1996); Sofia, Bulgaria (1999) and Cedar Rapids, Iowa, U.S.A. (2002). The fifth conference was dedicated to Professor Dr. Werner Vogel in order to honour his lifetime achievements in Glass Science and, in particular, in Borate Glasses.

In total, 89 scientists from 18 countries gathered in Trento to discuss recent advances in the study of borate glasses, crystals and melts and their applications. The participants came from: Armenia (2), Australia (1), Bulgaria (2), Canada (2), the Czech Republic (2), Egypt (3), France (8), Germany (11), Greece (1), India (2), Italy (8), Japan (9), New Zealand (1), Poland (7), Romania (3), Russia (9), the United Kingdom (6) and the United States of America (12). The Rector of the University of Trento, Professor Davide Bassi, welcomed the participants and officially opened the Conference. The first invited talk was by Steven Feller who summarised the contents of the previous Conference. The lecture in honour of Professor Dr. Werner Vogel, entitled Werner Vogel: Artistry with an Electron Beam, was presented by Doris Ehrt and Adrian C. Wright. The other invited talks were by: B. G. Aitken (USA), A. G. Clare (USA), M. Cutroni (Italy), T. Kloss (Germany), S. Kroeker (Canada), C. Rüssel (Germany), R. N. Sinclair (UK), M. Tatsumisago (Japan), N. Umesaki (Japan), R. Vacher (France), C. Varsamis (Greece) and A. C. Wright (UK).

The conference was structured according to the following sessions: Introduction (Chairman G. Dalba), High Temperature 1 (Y. B. Dimitriev), High Temperature 2 (A. C. Hannon), Spectroscopies and Local Structure (A. Ibanez), Applications 1: Glass Fibres (D. Ehrt), Glass Properties (M. Wasiucionek), AgI-Containing Glasses (P. Fornasini), Structure and Modelling 1 (J. W. Zwanziger), Structure and Modelling 2 (P. Mustarelli), Phase Separation and Thermal Expansion (A. Takada), Anomalies in Thermal Expansion (N. M. Vedishcheva), Thermodynamics and Glass Properties (M. Kodama), Applications 2: Optical properties (M. Affatigato), Applications 3 (F. T. Wallemberger), Local Boron Environment: XAFS and EELS (N. Umesaki). The number of oral presentations was 48 and the number of posters 45.

The success of the conference was due to the efforts and support of many people and institutions. In particular, we are grateful to the University of Trento, Istituto per la Fotonica e le Nanotecnologie del CNR (Trento), Istituto trentino di Cultura and Cassa rurale di Sopramonte (Trento). We thank the members of the International Organising Committee: Adrian C. Wright (United Kingdom), Alexis Clare (USA), Yanko B. Dimitriev (Bulgaria), Doris Ehrt (Germany), Steven A. Feller (USA), Efstratios Kamitsos (Greece), Alex C. Hannon (United Kingdom), Norimasa Umesaki (Japan), Natalia M. Vedishcheva (Russia) and Josef W. Zwanziger (Canada) for their scientific collaboration in the formulation of the Conference. We thank also the members of the local committee, Cristina Armellini, Paolo Fornasini, Roberto Graziola and Rolly Grisenti for the logistic organisation of the Conference. Special mention must be made of the efforts, professionalism and passion of Roberto Graziola in arranging the excursion on the Garda lake, the visit to Malcesine and the conference dinner at Riva del Garda. The careful work of our colleague Rolly Grisenti, as conference accountant, similarly deserves special mention. Many thanks also to Alberta Giacomoni and our PhD students Nasser Afify, Silvia Larcheri, Andrea Sanson and Marco Vaccari, for their praiseworthy collaboration offered to all participants. We express our warm thanks to Paola Bodio and the staff of the Events and Meetings Office of the University of Trento. Finally we want to emphasise the role of Adrian C. Wright and Natalia M. Vedishcheva, as animators of the Conference, and for their work in organising the papers and the manuscripts: to them our sincerest thanks.

The International Organising Committee considered two proposals to hold the next conference and decided that the VI International Conference on Borate Glasses, Crystals and Melts will be held in Japan in 2008. Dr Norimasa Umesaki of the Japan Synchrotron Radiation Research Institute (JASRI) announced his intention to chair the conference and cordially invited all of the participants present in Trento to attend.

Giuseppe Dalba,
Francesco Rocca (Conference Co-chairman)

CONTENTS

FIFTH INTERNATIONAL CONFERENCE ON BORATE GLASSES, CRYSTALS AND MELTS: NEW TECHNIQUES AND APPLICATIONS

Conference Center Panorama, Sardagna – Trento, Italy

10 – 14 July 2005

PROGRAMME

Sunday, July 10th

16-20	*Registration*	
16.30-20	*Welcoming reception*	

Monday, July 11th

Introductory section
Chairman: Dalba G.

8.45-9.10		WELCOME
9.10-9.40 *Invited talk*	**Feller S. A.**	THE FOURTH INTERNATIONAL CONFERENCE ON BORATE GLASSES, CRYSTALS, AND MELTS, 14-18 JULY 2002
9.40 - 10.00 *Invited talk*	**Wright A. C.** Ehrt D.	Lecture in honour of Prof. W. Vogel WERNEL VOGEL: ARTISTRY WITH AN ELECTRON BEAM

Session: High Temperature
Chairman: Dimitriev Y.

10.00-10.30 *Invited talk*	**Aitken, B.G.,** Randall E. Youngman	BOROPHOSPHOSILICATE GLASSES: PROPERTIES AND STRUCTURE
10.30-10.50	**Ehrt D.**	THE EFFECT OF ZnO, PbO, Bi2O3 AND La2O3 ON THE PROPERTIES OF BINARY BORATE GLASSES AND MELTS
10.50-11.10	**Ferlat G.** , Cormier L., Mauri F., Calas G., Charpentier T.	AB-INITIO MOLECULAR DYNAMICS SIMULATIONS OF LIQUID BORON OXIDE
11.10-11.40		***Coffee Break***

Session: High Temperature
Chairman: Hannon A.C.

11.40-12.00	**Herms G.,** Sakowski J.	THE STRUCTURE OF BORATES BETWEEN GLASSY AND LIQUID STATE - RESULTS OF DIFFRACTION METHODS
12.00-12.20	**Anghel E. M.,** Florian P., Bessada C., Massiot D.	HIGH TEMPERATURE AND SOLID STATE NMR SPECTROSCOPY IN $Na_2B_4O_7$-Na_3AlF_6-TiO_2
12.20-12.40	**Cormier L.,** Meneses D. D. S., Neuville D. R., Echegut P., Majérus O., Calas G.	IN SITU EVOLUTION OF THE STRUCTURE OF ALKALI BORATE GLASSES AND MELTS: A STUDY BY INFRARED REFLECTANCE, RAMAN AND NEUTRON DIFFRACTION
12.40-13.00	**Neuville D. R.,** Cormier L.	STRUCTURE OF SODA-LIME BORATE GLASSES AND MELTS BY IN SITU HIGH TEMPERATURE RAMAN SPECTROSCOPY

13.00-14.30		***Lunch Time***

Session: Spectroscopies and Local structure
Chairman: Ibanez A.

14.30-14.50	**Dubiel M.**, Yang X., Ehrt D., Schütz A.	VALENCE STATE AND STRUCTURAL GEOMETRY OF POLYVALENT FE AND TI IONS IN SODIUM BOROSILICATE GLASSES AS A RESULT OF XANES AND EXAFS SPECTROSCOPY
14.50-15.10	**Maia Queiroz L. J.**, Ibanez A., Fick J., Mastelaro V. R., Hernandes A. C.	AMORPHOUS ERBIUM-DOPED BORATES THIN FILMS OBTAINED BY POLYMERIC PRECURSOR METHOD
15.10-15.30	**Shama A.A.A.**, ElBatal H. A.	STRUCTURAL ANALYSIS OF GLASSY LEAD BORATE CONTAINING MoO3 IN RELATION TO ITS PHYSICAL PROPERTIES
15.30-15.50	**Koudelka L.**, Mošner P., Zeyer-Düsterer M., Jäger C.	BORON COORDINATION IN BOROPHOSPHATE GLASSES
15.50-16.10	**Simon S.**	MAGNETIC RESONANCE STUDIES ON AMORPHOUS AND CRYSTALLINE LANTHANUM ALUMINOBORATES
16.10-16.40		***Coffee Break***

Session: Some Applications ...
Chairman: Ehrt D.

16.40-17.10 *Invited talk*	**Clare A.**, Walker J., Raszewski F.	CHARACTERIZATION OF BORATE GLASS AND BORATE GLASS COMPOSITE FIBERS
17.10-17.30	**Jones R. L.**	THE ROLE OF BORON IN THE CORROSION OF E-GLASS FIBRES
17.30-17.50	**Wallenberger F. T.**	THE EFFECT OF B_2O_3 ON THE PROPERTIES OF FIBERGLASS MELTS
17.50-19.00		**Poster Session 1-A**

Tuesday, July 12th

9.00-10.00		**Poster Session 1-B**

Session: Glass properties
Chairman: Wasiucionek M.

10.00-10.30 *Invited talk*	**Cutroni M.**	DYNAMICAL RESPONSE OF BORATE GLASSES DOWN TO LOW TEMPERATURES
10.30-11.00 *Invited talk*	**Vacher R.**	THE VIBRATIONAL MODES OF BORATE GLASSES FROM MEGAHERTZ TO TERAHERTZ FREQUENCIES
11.00-11.30		***Coffee Break***

Session: AgI doped glasses
Chairman: Fornasini P.

11.30-12.00 *Invited talk*	**Varsamis C. P. E.**, Vegiri A., Kamitsos E. I.	STRUCTURE AND DYNAMICS OF IONIC BORATE GLASSES

12.00-12.20	**Hayashi A.**, Ishida T., Minami T., Tatsumisago M.	MECHANOCHEMICAL SYNTHESIS OF SILVER ION CONDUCTING GLASSES IN THE SYSTEM $AgI-Ag_2O-B_2O_3$
12.20-12.40	**Mustarelli P.**, Garcia P. I., Tomasi C., Magistris A.	A NMR AND MOLECULAR DYNAMICS STUDY OF $AgI-Ag_2O-B_2O_3$ GLASSES
12.40-13.00	**Wasiucionek M.**, Foltyn M., Garbarczyk J. E., Nowiński J. L., Gierlotka S. G., Pałosz B.	CONDUCTIVITY AND MICROSTRUCTURE OF COMPOSITES: SILVER BORATE GLASS – ZIRCONIA, PREPARED VIA HIGH PRESSURE ROUTE
13.00-14.30		***Lunch Time***

Session: Structure and Modelling

Chairman: Zwanziger J. W.

14.30-15.00 *Invited talk*	**Sinclair R. N.**, Haworth R., Wright A. C., Taylor J. W., Vedishcheva N. M., Polyakova I. G., Shakhmatkin B. A., Feller S. A., Winslow D. W.	NEUTRON SPECTROSCOPIC STUDIES OF CAESIUM BORATE CRYSTALS AND GLASSES
15.00-15.20	**Hannon A. C.**, Holland D.	FOUR COORDINATED BORON IN BINARY BORATE GLASSES
15.20-15.40	**Takada A.**	MODELING OF B_2O_3 GLASS STRUCTURE BY COUPLED MD/MC SIMULATION
15.50-16.10 *Invited talk*	**Umesaki. N.**, Ohotori N., Handa K.	STRUCTURE OF ALKALINE-EARTH BORATE GLASSES
16.10-16.30	**Wright A. C.**	THE STRUCTURE OF BOROSILICATE GLASSES
16.30-17.00		***Coffee Break***

Session: Structure and Modelling

Chairman: Mustarelli P.

17.00-18.10	**Kroeker S.**, Aguiar P. M., Feller S. A., Affatigato M.	EXPLORING MEDIUM-RANGE ORDER IN BORATE GLASSES: NEW EXPERIMENTAL AND THEORETICAL NMR APPROACHES
17.00-18.10	**Youngman R. E.**	NMR STUDIES OF STRUCTURAL RELAXATION IN SODIUM BOROSILICATE AND RELATED GLASSES
17.00-18.10	**Zwanziger J. W.**, Shaw J. L., Werner-Zwanziger U.	BORATE GLASSES UNDER UNIAXIAL STRESS
18.10-19.00		**Poster Session 2-A**

Wednsday July 13th

Session: Local thermal expansion and phase separation

Chairman: Tanaka A.

9.00–9.30 *Invited talk*	**Rüssel C.**, Tauch D., Garkova R., Woltz S., Völksch G.	PHASE SEPARATION AND CRYSTALLISATION IN BORATE GLASSES

9.30-9.50	**Hovhannisyan R. M.**	LOW THERMAL EXPANSION OF CRYSTALLIZED STOICHIOMETRIC BINARY ALKALINE-EARTH BORATE GLASSES
9.50-10.20	**Dimitriev Y.**, Kashchieva E., Iordanova R.	GLASS FORMATION AND MICROHETEROGENEOUS STRUCTURE IN THE B_2O_3-MoO_3-MenOm (Me=W, Mn, Cu) SYSTEMS

10.20-10.50 ***Coffee Break***

Session: Local thermal expansion and phase separation
Chairman: Vedishcheva N.

10.50-11.10	**Kodama M.**, Ono A., Kojima S., Feller S. A., Affatigato M.	BORATE ANOMALY AND ANHARMONICITY IN SODIUM BORATE GLASSES
11.10-11.30	**Sanson A.**	SHORT RANGE STRUCTURE OF FAST-ION-CONDUCTING AGI DOPED BORATE GLASSES

Session: Thermodynamics and glass properties
Chairman: Kodama M.

11.30-11.50	**Vedishcheva N. M.**, Shakhmatkin B. A., Wright A. C.	THE STRUCTURE-PROPERTY RELATIONSHIP FOR SODIUM BOROSILICATE GLASSES
11.50-12.10	**Markova T. S.**, Yanush O. V., Polyakova I. G., Pevzner B. Z.	GLASS PROPERTIES CALCULATION AND PREDICTION OF NEW COMPOUNDS ON THE BASE OF RAMAN SPECTROSCOPY OF BORATE GLASSES

12.10-14.00 ***Lunch Time***

14.00-24.00 **Excursion to Garda Lake and Banquet**

Thursday, July 14th

9.00-9.30 Poster Session 2-B

Session: ... other applications
Chairman: Affatigato M.

9.30-10.00 *Invited talk*	**Kloss T.**	SPECIALITY BOROSILICATE FLOAT GLASSES – RECENT INNOVATIONS FROM HOME TECH TO HIGH TECH
10.00-10.20	**Kirchhof J.**, Unger S., Aichele C., Grimm S.	BOROSILICATE OPTICAL FIBERS AND PLANAR WAVEGUIDES - TECHNOLOGY AND PROPERTIES
10.20-10.50	**Frischat G. H.**, Helsch G., Krzyzak M., Heide G.	ADHERENT ANTIREFLECTION COATINGS ON BOROSILICATE GLASS FOR SOLAR COLLECTORS

10.50-11.20 ***Coffee Break***

Session: ... other applications
Chairman: Wallenberger F. T.

11.20-11.50 *Invited talk*	**Tatsumisago M.**	ELECTROCHEMICAL PROPERTIES AND LOCAL STRUCTURE OF TIN BORATE GLASSES FOR LITHIUM SECONDARY BATTERIES
11.50-12.10	**Appleby G. A.**, Edgar A., Williams G.V.M.	LITHIUM BORATE GLASS-CERAMICS FOR RADIATION IMAGING
12.10-12.30	**Affatigato M.**, Leipply D., Melhus B., Leonardo M.	STUDIES OF THE ANTIBACTERIAL ACTION OF ALKALI BORATE GLASS SURFACES
12.30-13.00		*... open discussion about applications*
13.00-14.30		***Lunch Time***

Session: Local Boron Environment: XAFS and EELS
Chairman: Umesaki. N.

14.30-14.50 *Invited talk*	**Šipr O.**, Šimůnek A., Vackář J., Dalba G., Rocca F.	STRUCTURAL ASPECTS OF B K-EDGE XANES OF MINERALS
14.50-15.10	**Handa K.**, Ide J., Koganezawa T., Ozutsumi K., Dalba G., Ohtori N., Umesaki N.	XAS STUDY OF BARIUM BORATE GLASSES AND CRYSTALS
15.10-15.30	**Yang G.**, Moebus G., Hand R. J.	EELS & ANALYTICAL STEM STUDIES OF BOROSILICATE GLASSES
15.30-15.50	**Rocca F.**, Armellini C., Boscherini F., Carboni R., Malvestuto M., Šipr O.	XAFS STUDIES AT THE BORON K-EDGE ON AgI:Ag_2O: B_2O_3 GLASSES
15.50-16.00		***Conclusions***

Poster Session 1

Monday 11 July h:17.50-19 and Tuesday 12 h:9-10.00

Heavy Atoms

P1_01
CORROSION BEHAVIOUR OF SOME BINARY AND TERNARY LEAD BORATE GLASSES
El Batal H. A., ElDin F. M. E., Li I. S. A.

P1_02
CHEMICAL DURABILITY AND ELEKTROCHEMICALPROPERTIES OF BORON-ALUMINATE GLASSES CONTAINING FLUORIDES OF ALKALINE METALS
Gasparyan L. A., Muradyan S. G., Kostanyan K. A.

P1_03
XAFS STUDY ON THE LOCAL ORDER OF Pb IN PbO BORATE GLASSES.
Grisenti R., Dalba G., Fornasini P., Rocca F., Afify N., Affatigato M.

P1_04
THE CHARACTER OF THE LIQUATION ALUMINUM BORATE GLASS PHASES' ARRANGEMENT AND OBTAINMENT OF TRANSPARENT GLASS CERAMICS
Haroutunyan N. M., Knyazyan N. B.

P1_05
STUDY OF LOW TEMPERATURE ZINC FLUORID LEAD BORATE GLASSES AND GLASS COMPOSITES ON THEIR BASIS
Hovhannisyan M. R., Knyazyan N.B.

P1_06
MELTING DIAGRAM, CRISTALLINE PHASES AND GLASS FORMATION IN THE $BaO-B_2O_3-TiO_2$ SYSTEM
Hovhannisyan R. M., Grigoryan B. V., Alexanyan H. A., Petrosyan B. V., Toroyan V. P., Abramyan Z. M., Mkhitaryan R. G., Israelyan V. R.

P1_07
XAS STUDY OF BARIUM ALUMINOBORATE GLASSES
Ide J., Ozutsumi K., Handa K., Dalba G., Ohtori N., Umesaki N.

P1_08
INFLUENCE OF THE CATION NATURE ON THE GLASSE PROPERTIES IN THE SYSTEMS RO–Al_2O_3–B_2O_3 (R = Mg, Ca, Sr, Ba)
Klyuev V. P., Pevzner B. Z., Polyakova I. G.

P1_09
GLASSFORMATION, STRUCTURAL FEATURES AND PROPERTIES OF GLASSES ON THE BASIS OF SYSTEM $B_2O_3-V_2O_5-BaF_2$
Kostanyan K. A., Petrosyan M. Z.

P1_10
THE B_2O_3/SrO RATION INFLUENCE ON THE STRUCTURAL AND MAGNETICAL PROPERTIES OF THE $CuO-B_2O_3-SrO$ GLASSES
Lucacel R. C., Ardelean I.

P1_11
AMAZING BARIUM DIBORATE
Pevzner B. Z., Klyuev V. P., Polyakova I. G., Borodzyulya V. F.

P1_12
THE EFFECT OF ASPECT RATIO CHANGES ON THE THERMAL PROPERTIES OF BISMUTH BORATE GLASS FIBERS
Walker J., Raszewski F. C., Clare A. G.

Glass Properties

P1_13
STRUCTURE INTERPRETATION OF WATER CONTENT EFFECT ON THE PROPERTIES OF ONE-ALKALI BORATE GLASSES
Golubeva O. Y., Pavinich V. F., Stolyar S. V.

P1_14
DIFFERENTIAL VOLUME CHANGES IN TRIGONAL AND TETRAHEDRAL BORATE UNITS
Ongie G. J., Feller S. A., Affatigato M., Kodama M.

P1_15
HEAT CAPACITY OF SODA-LIME BORATE AND BOROSILICATE GLASSES AND MELTS
Wahoud F., Roux J., Neuville D. R., Cormier L., Richet P.

Optical Properties

P1_16
OPTICAL CHARACTERISTICS OF FLUORINE CONTAINING BORLANTAN GLASSES BASED ON PbB_2O_4–La_2O_3-LiF SYSTEM
Buniatyan G. L., Knyazyan B., Kostanyan K. A.

P1_17
GAMMA RAY INTERACTION WITH LITHIUM

BORATE SEMICONDUCTOR GLASSES CONTAINING Cr_2O_3
Ewiss M. A. Z., ElBatal H. A., Mohamed S. A., Semary M.A.

P1_18
SYNTHESIS AND THERMAL CHARACTERIZATION OF Er:YAB BY THE Pechini METHOD AND SOL-GEL PROCESS
Ferrari Goncalves R., Maia L. J. Q., Ibanez A., Mastelaro V. R., Hernandes A. C.

P1_20
GAMMA RAY INTERACTION ON THE PROPERTIES OF TRANSPARENT NANOPHASE GLASS-CERAMICS FROM THE SYSTEM BaO-B_2O_3-TiO_2
Ghoneim, N. A., Abdel-Hameed S. A. M., Margha F. H.

P1_21
ERBIUM-DOPED ALUMINOBORATES THIN FILMS BY THE SOL-GEL PROCESS
Ibanez A., Maia L.J.Q.,Fick J.,Mastelaro V.R., Hernandes A. C.

P1_22
FLUORESCENCE SPECTROSCOPY OF Sm_3+: BORATE GLASSES
Jayasankar C. K., Babu S. S., Babu P., Wortmann G., Sievers W.

P1_23
HOLMIUM – DOPED BORATE GLASSES WITH HIGH LEAD OXIDE CONTENT BY SPECTROSCOPIC STUDIES
Pisarska J., Dominiak G., Romanowski D. W., Goryczka T., Pisarski W. A.

P1_24
THERMAL, STRUCTURAL AND OPTICAL PROPERTIES OF ERBIUM DOPED OXYFLUORIDE LEAD BORATE GLASSES
Pisarski W. A., Pisarska J.

P1_25
AN XAFS STUDY OF THE LOCAL STRUCTURE OF EUROPIUM IONS IN CRYSTALLIZED BORO-SILICATE GLASSES
Umesaki, N., Okajima T., Konishi A., Yazawa T., Jin T.

Poster Session 2

Tuesday 12 July - h:18.10-19 and Thursday 14: h:9- 9.30

Preparation and characterization

P2_01
PREPARATION AND APPLICATION OF SOME BOROSILICATE GLASSES CONTAINING MICRONUTRIENTS AS PLANT FERTILIZERS
Azooz M. A., ElBatal H.A., Azooz M. A., ElMoneim M. A., ElAshry S. M.

P2_02
A SOLUTION APPROACH TO HIGH ALKALI CONTENT BORATE GLASSES
Banerjee J., Ongie G., Larson C., Sutton S., Moeller A., Basu A., Edwards T., Affatigato M., Feller S., Kodama M., Kroeker S.

P2_03
STRUCTURE AND PROPERTIES OF IRON BOROPHOSPHATE GLASSES
Bingham P. A., Hand R. J., Forder S. D., Lavaysierre A., Kilcoyne S., Yasin I.

P2_04
INVESTIGATION OF THE INFLUENCE OF BORON-ZIRCONIUM FRIT ON THE PROPERTIES OF PORCELAIN GLAZE FOR SANITARYWARE
Djambazov S. P., Yoleva A. P., Amzov G.

P2_05
SODIUM-SELENOBORATO-CLOSO-DODECABORATES:STRUCTURES AND PROPERTIES
Hammerschmidt A., Döch M., Lindemann A., Pütz S., Krebs B.

P2_06
PREPARATION AND X-RAY CHARACERIZATION OF $(Li_2O.WO_3)_{1-x}(B_2O_3)_x$ GLASSES
Heiba Z. K., El-Sayed K.

P2_07
STRUCTURAL AND DIELECTRIC INVESTIGATION OF ION-EXCHANGED BORATE GLASSES
Kamitsos E. I., Todorov R. , Varsamis C. P. E.

P2_08
BRILLOUIN SCATTERING STUDY OF LITHIUM BORATE AND LITHIUM SILICATE GLASS SYSTEMS
Kojima S., Ike Y., Hushur A., Matsuda Y., Kodama M., Feller S. A., Affatigato M.

P2_09
THERMAL HISTORY OF A LOW ALKALINE BOROSILICATE GLASS PROBED BY IR- AND RAMAN SPECTROSCOPY
Möncke D., Ehrt D., Varsamis C. P. E., Kamitsos E. I., Kalampounias A. G.

P2_10
BROADBAND DIELECTRIC RESPONSE OF SILVER BORATE GLASSES
Raimondo A., Mandanici A., Cutroni M., Federico M., Armellini C., Rocca F.

P2_11
SYNTHESIS AND CHARACTERIZATION OF YAB AND YCOB GLASSES FOR OPTICAL APPLICATIONS
Santos Nascimiento C., Hernandes A. C., Ibanez A.

P2_12
ORDERING EFFECTS IN SPATIAL ARRANGEMENT OF PARTICLES IN PHASE-SEPARATED SODIUM BOROSILICATE GLASS
Shepilov M. P., Kalmykov A. E., Sycheva G.

P2_13
INFLUENCE OF WATER CONTENT ON STRUCTURE AND LINETICS OF STRUCTURAL RELAXATION OF SODIUM BORATE GLASSES
Zhilin A., Shashkin A. V., Shepilov M. P., Golubkov V. V.

Structure and Modelling

P2_14
LOCAL STRUCTURE OF EUROPIUM AND IRON SITES IN ZINC BORATE GLASSES
Congiu F., Concas G., Spano G., Bettinelli M., Speghini A.

P2_15
NEUTRON DIFFRACTION STUDIES OF CAESIUM BORATE CRYSTAL STRUCTURES
Haworth R., Wright A. C., Sinclair R. N., Knight K. S., Vedishcheva N. M., Polyakova I. G., Shakhmatkin B. A., Feller S. A., Winslow D. W.

P2_16
EDGE-SHARING BO_4 TETRAHEDRA AS A

NEW STRUCTURAL FEATURE IN BORATE STRUCTURES
Huppertz H., Emme H.

P2_17
EFFECT OF COBALT IONS ON THE EPR AND DC CONDUCTIVITY IN VANADYL DOPED CoO. $M_2O.B_2O_3$ (M= Li, K) GLASSES
Khasa S.,
Gahlot P. S., Agarwal A., Seth V. P., Gupta S. K.

P2_18
XAS AND XRD STUDY OF LITHIUM BORATE GLASSES AND CRYSTALS
Koganezawa T., Ide J., Ozutsumi K., Kojima K., Handa K., Dalba G., Umesaki N.

P2_19
LOCAL STRUCTURAL DISORDER AND MOBILITY OF Ag^+ IONS IN SILVER BORATE GLASSES CONTAINIG α-AgI STABILIZED AT ROOM TEMPERATURE
Monti F., Dalba G., Fornasini P., Rocca F., Minami T., Tatsumisago M.

P2_20
CRYSTALLIZATION PROCESSES IN RAPID QUENCHED $AgCl-AgI-Ag_2O-B_2O_3$ GLASSES
Nowiński J. L., Garbarczyk J. E., Wasiucionek M., Gierlotka S.

P2_21
MD STUDY OF SODIUM BORATE GLASSES CONTAINING Al_2O_3
Ohtori N., Togashi M., Ogawa M., Takase K., Handa K.

Abstract submitted but not confirmed

Pret_01
SHORT-RANGE STRUCTURE IN $BaO-B_2O_3-TiO_2$ GLASSES ANALYSED BY XAS AND ^{11}B-NMR
Mastelaro V. R., Schneider J. F., Maia L. J. Q., Feitosa C. A. C., Parent P., Laffon C.

Pret_02
THE VISCOSITY OF BISMUTH BORATE GLASS MELTS
Fujino S., Tokunaga H., Hwang C.

Pret_03
OPTICAL PROPERTIES OF $Li_{0.5}Cs_{0.5}B_3O_5$
Adamiv V.T., Burak Ya.V., Teslyuk I.M.

Pret_04
COMPARATIVE STUDIES ON THE ELASTIC PROPERTIES OF SILVER BORATE AND PHOSPHATE GLASS SYSTEM
Aziz S. H.A., Kamari H., Talib Z. A., Shaari A. H., Yusoff W. D. W., Hassan S.

Pret_05
DIELECTRIC RELAXATION IN SINGLE CRYSTAL AND GLASSY $Li_2B_4O_7$:Mn
Rizak V.M., Rizak I. M., Baisa N. D., Bilanych V. S., Stefanovych S. Yu., Bohuslavskii M. B.,Holovey V. M.

Pret_06
SPECTROSCOPY OF CHROMIUM IN BORATE AND BOROSILICATE GLASSES: LOCAL STRUCTURE, REDOX STATES AND THERMAL EXPANSION
Calas G., Galoisy L., Cormier L., Majerus O., Villain O.

Pret_07
LEAD LESS CERAMIC PAINTS GLASS FRITTS ON ZINC-BORATE-PHOSPHATE SYSTEM BASIS
Poghosyan M.

P1_19
NICKEL COORDINATION AS A STRUCTURAL PROBE OF BOROSILICATE GLASS STRUCTURE
Galoisy L., Cormier L., Calas G., Ghaleb D.

Werner Vogel: Artistry with an electron beam

Doris Ehrt

Otto-Schott-Institut für Glaschemie, Friedrich-Schiller-Universität Jena, Fraunhoferstraße 6, D-07743 Jena, Germany

*Adrian C. Wright**

J.J. Thomson Physical Laboratory, University of Reading, Whiteknights, Reading, RG6 6AF, UK

A brief review is given of the life and scientific career of Professor Dr. Werner Vogel, to whom the Fifth International Conference on Borate Glasses, Crystals & Melts is dedicated. Werner Vogel was born, and received his schooling, in the region of the town of Hof and, after World War II, was one of the first students at the newly reopened Friedrich Schiller University in Jena. After leaving the University he joined VEB Jenaer Glaswerk Schott & Gen., as Head of the Chemistry Laboratory, and it was during his time at Jenaer Glaswerk that he began his now famous electron microscopy studies of phase separation in glasses, which included both borate and borosilicate systems. His blossoming international reputation led to his returning to the Friedrich Schiller University, first as a visiting lecturer and eventually, in 1966, to his being invited to establish the Otto Schott Institute, and become its Director, a position from which he retired in 1990. Werner Vogel's great friendship with the late Norbert Kreidl resulted in the continuing series of Otto Schott Colloquia, which initially presented a unique forum for the interaction between glass scientists from Eastern Europe and the former Soviet Union, on the one hand, and Western Europe and the USA on the other. Werner Vogel is thus honoured, both for his seminal work on phase separation in B_2O_3-containing glasses and for his unique contribution to the reunification of the international glass science community.

1. Introduction

To a glass scientist, the name of Werner Vogel immediately brings to mind the superb electron micrographs of glasses, for which he is so justly famous; micrographs that somehow manage to combine great science with true artistry. Even for the non-scientist, they are a fascinating revelation of the (electron) microscopic world of a material that has transformed the lot of mankind, from the humble wine bottle, holding the prime of German winemaking, to the realms of high technology, such as the optical glasses that are so much a part of Jena's vitreous heritage.

The present paper will present a short biography of Werner Vogel and discuss his considerable contributions to glass research in the context both of glass science as a whole and of Jena's special position, particularly in respect of optical glasses. This will require the inclusion of a brief history of the Friedrich Schiller University and Jenaer Glaswerk Schott & Gen., and a short description of the scientific climate pervading during the years of the Cold War and up to German reunification on 3rd October 1990. An account will also be given of Werner Vogel's great efforts, in collaboration with the late Norbert J. Kreidl, to encourage interchange between the glass science community in Eastern Europe and the former Soviet Union and that in the West, as embodied in the famous Otto Schott Colloquia.

This short survey of Werner Vogel's life and achievements will be illustrated by appropriate photographs and electron micrographs, using items taken from his personal archives and from those of the Otto Schott Institute, together with material from his scientific papers and three books:[1–10] *Struktur und Kristallisation der Gläser*[1,2] (*Structure and Crystallisation of Glasses*[3]), *Glaschemie*[4,5,8] (*Chemistry of Glass*;[6] *Glass Chemistry*[10]) and *Glasfehler*[9] (*Imperfections in Glasses*). Much useful information has also been obtained from the short biography of Werner Vogel written by Wolfram Höland,[11] his immediate successor as

Figure 1. Erna and Werner Vogel, ca. 1927

* Corresponding author. Email a.c.wright@rdg.ac.uk

Figure 2. Erna and Werner Vogel, ca. 1928

Figure 4. Werner Vogel during his time at the Volksschule in Sparnberg, ca. 1932

Director of the Otto Schott Institute, on the occasion of Werner Vogel's 65th birthday.

2. Early life

Werner Vogel was born on 21st July 1925 in Sparnberg, Saale (Thuringia), near the town of Hof, the only child of Hermann and Erna Vogel, and early photographs of the young Werner with his mother are reproduced in Figures 1 and 2. A further photograph from his early years is shown in Figure 3.

Werner Vogel received his schooling locally, commencing with the Volksschule (Lower School; 1932–36) in Sparnberg and progressing through the Mittelschule (Middle School; 1936–41) in Hirschberg. In 1941 he entered the Oberrealschule (Upper School) in Hof, from which he graduated in 1943. A photograph of Werner Vogel at age 7, taken during his time at the Volksschule in Sparnberg, may be seen in Figure 4.

After leaving school, during the latter stages of World War II, Werner Vogel was conscripted into the army on 26th August 1943 (Figure 5). He was subsequently sent to the Eastern Front where, on 12th September 1944, he was severely wounded, resulting in the loss of his left leg. He was just 19 years old.

Figure 3. Werner Vogel, ca. 1928

Figure 5. Werner Vogel as a young recruit in 1943

Figure 6. Werner Vogel working in the laboratory as an undergraduate student in 1948

Figure 8. Werner and Käte Vogel (14th July 1951)

3. Friedrich-Schiller-Universität, Jena

After recovering from his injuries and when World War II was over, Werner Vogel was one of the first students to be enrolled at the Friedrich Schiller University in Jena, which reopened on 1st October 1945. He studied chemistry under Professors Ernst Kordes and Franz Hein and received his Diploma on 7th June 1950. The subject of his dissertation was *Gefrierpunktserniedrigung von Salzschmelzen* (*Lowering of the Freezing Point of Salt Melts*) and this work is described in his first scientific paper published in *Zeitschrift für Elektrochemie* in 1951.[12] A photograph of Werner Vogel as an undergraduate student, taken in 1948 during his Diploma years, appears in Figure 6.

Having obtained his Diploma, Werner Vogel remained at the Friedrich Schiller University and began work for his doctorate at the Instutut für Physikalische Chemie (Institute of Physical Chemistry), supervised by Professor Ernst Kordes, the title of his thesis being *Struktur und Eigenschaften von Phosphatgläsern* (*Structure and Properties of Phosphate Glasses*). This work was performed in collaboration with the VEB Jenaer Glaswerk Schott & Gen. The photograph in Figure 7 was taken at the Annual Meeting of Chemists in Leipzig in 1951, during his PhD studies.

The year 1951 was an important one for Werner Vogel in that, on 14th July 1951, he married Käte Oßwald in the church at Sparnberg (Figures 8 to 10). In the following year, he was awarded his PhD. The year 1952 also saw the publication of his second

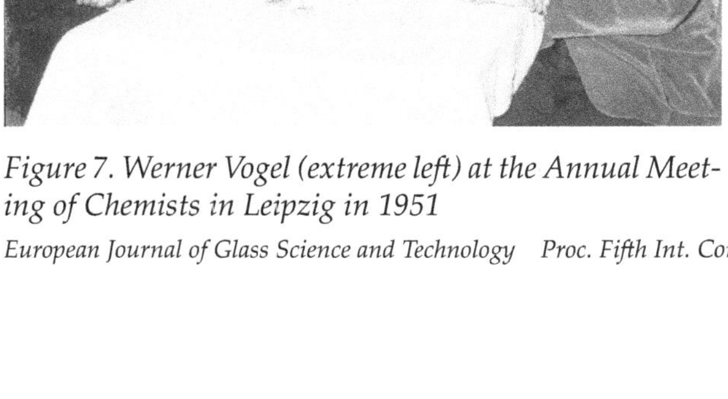

Figure 7. Werner Vogel (extreme left) at the Annual Meeting of Chemists in Leipzig in 1951

Figure 9. Werner and Käte Vogel with flower girls

Figure 10. Werner and Käte Vogel with their wedding group

paper[13] on an apparatus for measuring the refractive index of hygroscopic powders, and his work on the characteristics and fine structures of phosphate glasses is described in a second paper in *Zeitschrift für Elektrochemie* in 1953.[14]

4. Jena heritage

The Friedrich Schiller University in Jena, which has played such an important role in Werner Vogel's life, was founded by Elector Johann Friedrich of Saxonia as an Academisches Gymnasium (Academic School) for theological studies in 1548. It was raised to university status by Kaiser Ferdinand I in 1557 and opened as the University of Jena on 2nd February 1558. Initially it was a centre of protestant scholarship and Lutherian theology and gained an international reputation during the 18th century, when its teaching staff included Johann Gottlieb Fichte, Georg Wilhelm Friedrich Hegel, Friedrich Wilhelm Joseph von Schelling, Friedrich Schlegel and Friedrich Schiller. The first science course to be introduced was in medicine in 1558.

Having been founded as a centre for the new religious opinions of the 16th century, the University of Jena has been one of the most politically radical universities in Germany and hence fell from favour with the German government at the end of the 18th century and the beginning of the 19th century. This continued until until relatively recent times, when the University has benefitted from support from

Figure 11. Coat of arms of the Friedrich-Schiller-Universität, Jena

the Zeiss Corporation. In 1934, the University was renamed the Friedrich-Schiller-Universität, to reflect its association with Friedrich Schiller, and its coat of arms appears in Figure 11.

The origins of Jena's association with optical glasses can be traced back to the beginning of the 19th century, with the arrival of the chemist Johann Wolfgang Döbereiner (1780–1849) in 1810 and that of the fine mechanic instrument maker (Hof- und Universitätsmechanicus) Friedrich Körner (1778–1847) in 1817. With the help of the German poet Johann Wolfgang Goethe, the first chemical institute was established in the castle in 1811. Johann Döbereiner discovered the catalytic action of platinum and, in 1817, proposed his Law of Triads, which was an early step towards the development of the periodic table, reported by Dmitrii Ivanovich Mendeleev in 1869. Johann Döbereiner was appointed Professor of Chemistry at the University of Jena in 1819.

Friedrich Körner received his doctorate from the University in 1818 and gave lectures on levelling technology (Nivellierungstechnik) and the manufacture of chemical, physical and metrological instruments. He had a factory in Jena, with a small glass melting facility on Grietgasse, where he worked on manufacturing optical glasses, with the help of Johann Döbereiner and Carl Dietrich Münchow, and supplied the German court with microscopes and other scientific equipment. Carl Münchow was Professor of Mathematics and Astronomy and the first Director of the first observatory in Jena, which was opened in 1813 in a house that had once belonged to Friedrich Schiller, following the approval of Duke Karl August in 1811.

The ensuing development of optical glasses and instrumentation in Jena is mainly assocated with three persons, Carl Friedrich Zeiss (1816–88), a fine mechanic instrument maker (Hof- und Universitäts-mechanicus), Ernst Abbe (1840–1905), Professor of Physics, and Otto Schott (1851–1935) a glass chemist, yielding a perfect example of productive cooperation between science, technology, and industry. Carl Zeiss worked as an apprentice in the workshop of Friedrich Körner and attended lectures at the University of Jena on a number of subjects, including mathematics, experimental physics and optics. In 1846, Carl Zeiss established a workshop, selling optical and mechanical instruments, and rapidly developed a trade, first in single lens microscopes and then in compound microscopes with two lenses. However, further improvements in design were limited by his lack of a theoretical knowledge of optics. This difficulty was solved in 1866, when Carl Zeiss formed a partnership with Ernst Abbe. Ernst Abbe had also studied mathematics and physics as an undergraduate in Jena and then joined the Jena faculty in 1863 to teach physics. He became Director of the Jena Observatory in 1877.

As a result of his work in designing microscope lenses, Ernst Abbe realised that the major limiting factor was the quality of optical glass and the available range of refractive index and dispersion. These problems were alleviated by Otto Schott, who came from a family of glassmakers and had a doctorate in Physics from the University of Jena. Otto Schott improved the quality of optical glasses and greatly extended the range of optical glass compositions and hence the possible combinations of refractive index and dispersion. He was also responsible for the discovery in 1887 of heat resisting borosilicate glasses, which Werner Vogel was later to investigate so successfully by electron microscopy.

5. Jenaer Glaswerk Schott & Genossen

In 1882, Otto Schott, Ernst Abbe and Carl and Roderich (son of Carl) Zeiss founded the Glastechnische Laboratorium Schott & Genossen (later Jenaer Glaswerk Schott & Genossen) in Jena, and production started in 1884, the original products including borosilicate glasses with special optical dispersion, which reduce chromatic aberration and increase optical resolution. They also have also high thermal and chemical stability and could therefore be used for thermometers, gas cylinders and chemical laboratory and industrial ware. This legacy ensured that, at the beginning of World War I, German optical instrumentation was the best in the world. After World War I, in 1918, the company began producing household glassware bearing the trade name Jenaer Glas which, in the 1930s, came to include products that were visibly influenced by the design concepts of the Bauhaus art movement. With the rise of Adolf Hitler and his Nationalist Party, much of German industry, including Jenaer Glas, was recruited to help in the World War II effort and, as a result, Jenaer Glaswerk was heavily bombed, but continued to operate. After the war, Jena was initially occupied by American troops who, in 1945, took 41 selected managers and specialists from Jena to Western Germany, including Dr. Erich Schott, the son of Otto Schott.[15] These experts were responsible for establishing a new Schott Glas facility in Mainz, which continued the manufacture of Leica and other optical instrumentation. Meanwhile, in 1945, Jena was occupied by Soviet troops and, in 1949, became part of the DDR. Glass production was restarted but then, due to their high scientific and technical standards, about 90% of the production facilities were dismantled and shipped to the Soviet Union, together with 15 further employees and four apprentices.[15]

By March/April 1947, the dismantling was finished and those employees who remained in Jena got production running again and retained the Zeiss name, which they continued to use with great pride. However, almost immediately, the company was expropriated and, in 1948, became VEB (Volkseigenen Betrieben – Nationalised Company) Jenaer Glaswerk Schott & Genossen. As the 1950s progressed, a new generation of specialists became involved in the technical and scientific divisions of the company. Many of these had other professions disrupted by the war. Werner Vogel was the first of the young graduates from the newly founded or reopened universities and colleges and had to instruct his co-workers in glass chemistry. Thus the tradition of optical glass research and manufactuiring continued in Jena, which was noted for its optical glass and instrumentation, especially instruments, such as sextants, that involve fine mechanical engineering and precision optics. Indeed, a colleague of one of the present authors (ACW) at the University of Reading, a keen yatchsman, remarked that by far the best sextants were made in Jena before production there was terminated, following both German and Company reunification and the transfer of all optical glass work to Mainz in 1992. The producion of consumer glassware in Jena ceased in 2005, leaving the company to concentrate on manufacturing high-tech products for the 21st century.

Werner Vogel joined Jenaer Glaswerk on 3rd January 1953, as the head of the chemical laboratory concerned with the development of techniques for the inspection and analysis of raw materials for optical and engineering glasses. The first paper[16] describing his work at Jenaer Glaswerk appeared later the same year and discussed an accurate and rapid method for the determination of fluorine in glasses and raw materials, while subsequent papers considered crystallisation in both silicate[17] and phosphate[18] glasses. It was in 1955, during his time in this laboratory, that Werner Vogel obtained his first electron microscope and began his now famous electron microscopy studies, commencing with fluoroberyllate glasses. These studies will be discussed in Section 7.

In 1958, Werner Vogel was commissioned to set up a new glass melting laboratory for fundamental research and the development of optical glasses using the platinum melting technique. Following in the tradition of Otto Schott, he initiated a new chapter in the basic research and development of optical glasses with the help of a team of young scientists and engineers, many of them his trainees. They were also involved in the solution of new process engineering problems. This was a period of great scientific productivity, resulting in many publications and patents, which established Werner Vogel internationally as an expert in glass structure research.

6. Glass research: divided by ideology

The glass research scene that confronted Werner Vogel in the late 1950s and early 1960s was very different from that which pertains today. The aftermath of the Great October Socialist Revolution of 1917 in Russia, and in particular the replacement of Vladimir Ilyich Lenin, as Head of State, by Josef Stalin in 1924, led to an increasing isolation of Russian/Soviet glass science from that in the West. Thus, while the first papers by Russian[19] (Leningrad) and Western[20] (Washington DC) authors, describing x-ray diffraction measurements on glasses, were published during 1925 in *Zeitschrift für Physik* and the *Journal of the Society of Glass Technology*, respectively, by the 1930s there were two distinct schools of glass science, one in the Soviet Union and the other in the West. As in politics, these schools were separated by opposing ideologies, which continued into the 1970s, no doubt encouraged by the Cold War that followed World War II.

In Russia, the crystallite theory, first suggested by the German scientist Moritz Ludvig Frankenheim (1801–1869) in 1835,[21] was reintroduced by Academician Alexander Alexeevich Lebedev in 1921,[22] to explain sharp changes in the properties of certain silicate glasses at temperatures in the region of the α–β transition of crystalline quartz, and dominated Soviet glass science for the next 50 years although, as Nemilov[23] has pointed out, Academician Lebedev's association of a sudden increase in thermal expansion near 540°C with the α–β transformation of quartz was, in fact, incorrect. The crystallite theory was also initially supported in the West, by the x-ray diffraction measurements of Randall, Rooksby & Cooper,[24–26] but was then abandoned in favour of Zachariasen's random network theory,[27] following further extensive x-ray diffraction studies by Warren & co-workers.[28,29]

As discussed in Ref. 30, the x-ray diffraction studies of Nikolai N. Valenkov & Evgenii Alexandrovich Porai-Koshits,[31,32] which were also published in Western journals, were performed under the supervision of Academician Lebedev to provide evidence for the crystallite theory, and eventually led to what Urnes[33] has termed the modern crystallite theory, according to which there are spatial fluctuations in the degree of intermediate range order within the vitreous network. More highly ordered regions, in which the atomic arrangement approaches that in related crystalline materials (the crystallites), are interconnected by regions where the degree of order is somewhat less. The relative merits of the crystallite vs. the random network theory became a subject of much heated debate during the early All-Union Conferences on the Glassy State in Leningrad, mainly as a result of the views of Evgenii Porai-Koshits, who had rejected the basic postulates of the crystallite theory but, of particular relevance to this biography of Werner Vogel, continued to believe in its prediction that the structures of all binary and multicomponent glasses are chemically heterogeneous.[30] Eventually, it was Werner Vogel's great friend Norbert Kreidl (cf. Section 9) who brought the representatives of the Russian and American schools together for the first time, to discuss the relative merits of the two theories, on a boat on the River Seine during the Fourth International Congress on Glass in Paris in July 1956. A photograph of this event appears in Ref. 34 and Norbert Kreidl remarks that "we found a great difference in the meaning of words in the two languages."

7. Phase Separation in Glasses

The phenomenon of phase separation in glasses has been known for some considerable time and its nature was recognised by Otto Schott:

> "My melting experiments revealed that SiO_2 and B_2O_3 were compatible, while P_2O_5 exhibits hostile behaviour. While SiO_2 + B_2O_3 as a glass forming double acid accept bases in any proportions, the addition of P_2O_5 almost always causes milky opalescence, apparently because phosphates are as insoluble in the melt as oil in water."
> (Otto Schott; quoted by Eberhard Zschimmer[35])

During the 1950s, the prevailing attitude to the concept of phase separation in glasses was very much influenced by the crystallite versus random network debate, in that chemical heterogeneity is predicted by the former for all binary and multicomponent glasses not having crystalline stoichiometry. The Zachariasen–Warren random network theory, on the other hand, predicts a homogeneous structure for such glasses, with the network modifying cations distributed randomly in holes throughout the vitreous network, as in the well known two-dimensional schematic diagram of vitreous Na_2O–SiO_2 by Warren & Biscoe.[36] Indeed, it was this aspect of the crystallite theory that led Evgenii Porai-Koshits to his conclusion that all binary and multicomponent glasses exhibit a heterogeneous structure. Being in the Eastern part of Germany (DDR), under Soviet control, glass research in Jena was strongly influenced by the Russian School, as typified by the crystallite theory. Thus it is not surprising that it was scientists in Jena, and in Leningrad, who took the lead in understanding the phenomenon of phase separation in glasses. As uniquely realised by Werner Vogel, this understanding, together with unambiguous proof of the existence of microheterogeneities in glasses, is inseparably associated with the development of electron microscopy as a structural probe, since it is only the electron microscope that provides a direct image of the resulting microstructure.

8. Electron microscopy

The first (transmission) electron microscopy studies of glasses were performed by Slayter[37] and by Prebus & Michener.[38] Both papers report micrographs for vitreous silica and sodium silicate glasses, together with more complicated multicomponent systems. The earliest (transmission) electron micrographs of glasses from the classic sodium borosilicate (Vycor®) system appear in a paper by the Leningrad group,[39] who combined electron microscopy with a range of other techniques, including small angle x-ray scattering (SAXS)[39] and visible light (Rayleigh) scattering.[40] The replica technique was introduced for glasses by Shelyubsky[41] and by Oberlies,[42] and led to much more easily interpretable micrographs. In particular, Oberlies[42] includes a micrograph of a Vycor® glass ($7Na_2O.25B_2O_3.68SiO_2$) that clearly shows phase separation in the form of connected phases (cf. Section 14).

The results from these early electron microscopy studies of glasses, and their interpretation, were the subject of considerable controversy and there were arguments as to whether the electron micrographs were genuine representations of the glass microstructure, or merely an artifact of the technique. Werner Vogel later comments on this controversy, in the concluding remarks at the end of his first book:[3]

> "But we must not forget to draw attention to a severe criticism of the results obtained. The prerequisite for the correctness of results lies not only in their reproducibility by the same method, but above all the results have to be reproducible by use of various other methods of preparation. It must not be overlooked that one method of investigation by itself – as in our case the electron microscope, be it as sensitive and revealing as it may – will only allow the problem to be viewed from one angle; thus, in the last resort, no comprehensive evidence can be expected. Only a suitable combination with other modern methods of investigation can provide the assurance that the material of glass, brittle in every sense, will yield the secrets of its essence and structure step by step."

In the meantime, after acquiring his first electron microscope (cf. Section 5), Werner Vogel and his young co-worker Klaus Gerth began a systematic investigation of fluoroberyllate glasses, which Goldschmidt[43] had described as "weakened models" of silicate glasses*. The bonding forces in fluoroberyllate glasses are only one quarter of those in the original silicate glasses and therefore phase separation processes proceed to a higher degree and the droplets have a greater diameter. A later electron micrograph[10] showing droplet shaped immiscibility regions in a turbid fluoroberyllate glass fracture plane, obtained via the replica process, is shown in Figure 12. The droplets remaining protruding from the surface, as well as those broken out and leaving a cavity, usually exhibit "fracture flags." The dark shadow lies to the left or right of the droplet. A pair of droplets is cracked right through the middle.

The work on fluoroberyllate glasses led to Werner Vogel's first paper on electron microscopy,[44] pub-

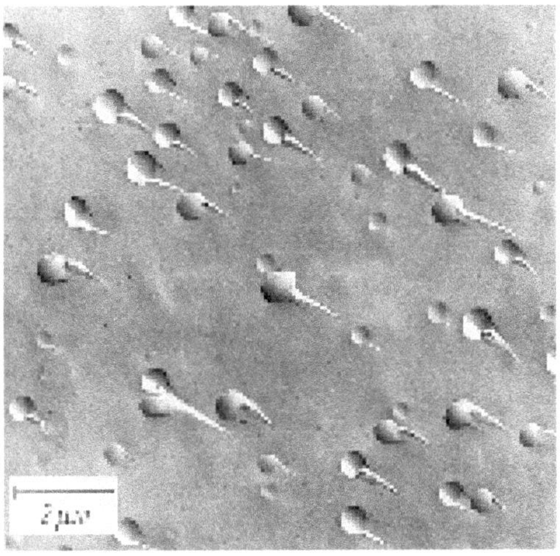

Figure 12. Droplet-shaped immiscibility regions in a turbid fluoroberyllate glass fracture plane (replica process)[10]

THE CELLULAR STRUCTURE OF GLASS

W. Vogel
The Schott Glass Factory, Jena, German Democratic Republic

The Concept of Cellular Glass Structure

When a liquid glass melt is cooled, there is a process of aggregation of like atoms or atomic groups, caused by the same forces as those which lead to the formation of a crystal lattice. The ordering process during glassy solidification terminates at an intermediate stage between ideal disorder of the components in the liquid melt and ideal order in the crystalline state. Cellular aggregation of like ions, which is the initial stage and preliminary condition of crystallization, takes place in definite regions. Owing to the rapid cooling of the melt these regions harden in the glassy form, while because of the considerable energy differences between the regions of honeycomb order and their surroundings they acquire surface boundaries and form drops. The structure of the droplike regions has a higher degree of order than the surrounding glassy phase, and the composition of the drop-like glass phase approximates to the composition of stable crystalline compounds.

Since the droplike segregation regions, detected by means of the electron microscope, are generally very small, no turbidity is observed in the glass. Despite this, most glasses, which appear completely transparent and homogeneous to the naked eye, should be regarded as two- and multiphase systems.

If the cations in the basic glass network have very slight field forces, this leads to very loose cell formation or to total absence of such formation. The structure of such glasses is best described by the random-network theory. If the cations have very high field strengths, the rate of ordering of the cations is high in comparison with the cooling rate of the melt, and may lead to formation of crystallites within the droplike regions. Such glasses are consistent even with the concepts of the original crystallite hypothesis. Most glasses occupy an intermediate position between these two extremes. Since the droplike glassy microphases can be likened to micelles, we shall refer to the "cellular" structure of the majority of glasses.

These characteristics are revealed most clearly in model fluoride glasses. Subsequently it proved possible to extend the experimental investigations to other, more difficult materials — silicate, borate, and phosphate glasses.

Segregation Processes in Glass

We shall illustrate segregation processes in glasses by three examples, which demonstrate the dependence of these processes on: concentration of the cations in the network; cation field strength; heat-treatment of the glass.

Figure 1 shows segregation in lithium beryllium fluoride glasses with increasing concentration of lithium fluoride, established by means of electron micrographs. All the glasses studied were opalescent, with the exception of the glass shown in Fig. 1,e, which was quite clear and free from the Tyndall effect. The formation of droplets and increases of their number and size in this series of glasses are clearly seen.

The effect of cation field strength on segregation is shown in Fig. 2. These micrographs represent binary sodium, potassium, and rubidium beryllium fluoride glasses with equal concentration of alkali fluorides. The tendency to segregation is intensified and the size of the segregation regions increased with increase of the cation field strength. Figure 2,c represents a completely transparent glass which did not exhibit the Tyndall effect.

Figure 13. The opening paragraphs of Werner Vogel's paper[52] at the Second All-Union Conference on the Glassy State, Leningrad, November 1959

lished with Klaus Gerth in *Glastechnische Berichte* in January 1958. Three further papers on fluoroberyllate glasses,[45–47] by the same authors, followed later the same year in *Silikattechnik*, together with Werner Vogels first electron microscopy studies of B_2O_3-containing glasses, viz. sodium borate,[48] sodium borosilicate[48,49] and Pyrex®-type[48] glasses. Thus the year 1958 turned out to be a very important one, with regard to the understanding of phase separation in glasses, especially since it also saw great progress with the complimentary SAXS technique, when Nikolai Sergeevich Andreev & Evgenii Alexandrovich Porai-Koshits[50] succeeded in demonstrating that the SAXS from unleached and leached Vycor® glass was of the same shape and merely differed in absolute magnitude (see Ref. 30) due to the differing scattering contrast.

9. International fame

"He who spoke of microphases in glass ten or twelve years ago was trodden underfoot as a heretic. He who does not concern himself with phenomena of microphase separation in glass today, will always lag behind in glass research." (Norbert Kreidl; quoted by Werner Vogel[2])

Summaries of the work performed by Werner Vogel

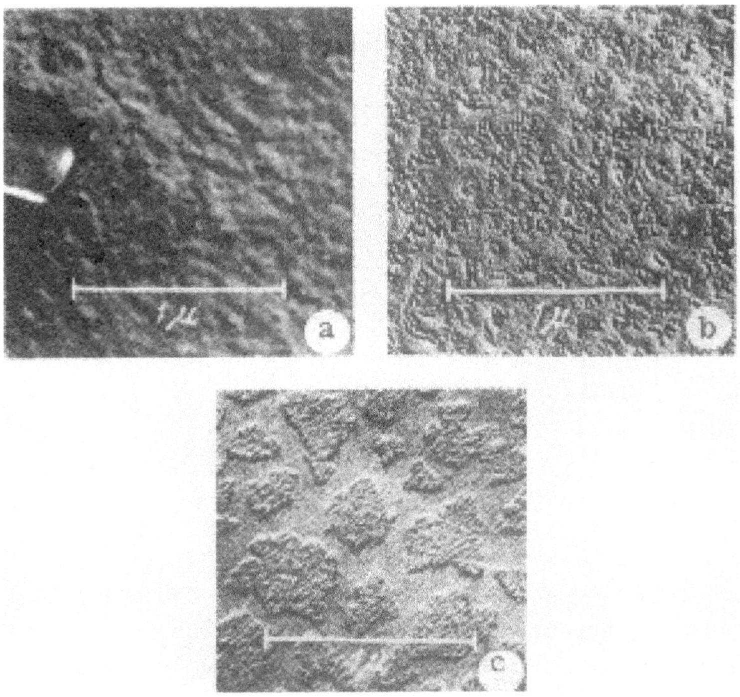

Figure 14. Electron micrographs of transparent sodium borate glasses ($16Na_2O.84B_2O_3$) as a function of preparation conditions and heat treatment:[52] *(a) rapidly cooled from the melt; (b) conventional melt cooling, with the droplets showing a tendency towards aggregation, and (c) after additional heat treatment for 2·5 h at 500°C, clearly showing aggregation of the small droplets. Even glass (c) was completely transparent, with no signs of turbidity*

and his colleagues in Jena were reported at a series of international conferences, commencing with the Symposium sur la Fusion du Verre (Symposium on Glassmelting) in Brussels (October 1958). The title of his paper was *Über Phasentrennung im Glas* (*On Phase Separation in Glass*)[51] and he also participated in the discussions of Work Group III (Le Verre en Tant que Matière).

The following year he presented a paper on *The Cellular Structure of Glass*[52] at the Third All-Union Conference on the Glassy State in Leningrad (November 1959), the opening paragraphs of this paper being reproduced in Figure 13. Results were included on a wide range of different glass forming systems, including lithium and sodium fluoroberyllate, sodium borate (Figure 14), Vycor® type, Pyrex® type, lead silicate, magnesium phosphate, pyroceram and titanium fluorosilicate.

Werner Vogel's first transatlantic visit was to Toronto in April 1961, to participate in the Annual Meeting of the American Ceramic Society, of which he had been a member since 1958. In his paper, *Catalyzed Crystallization in Glass*,[53] he explained the special importance of phase separation for controlled nucleation, crystallisation, and the development of glass-ceramics. It was at this meeting that he first met his great friend Norbert J. Kreidl, who invited him to participate in the 1961 Gordon Research Conference on Glass (Electrical and Relaxational Processes in Glass; Tilton School, August 1961) of which he (Norbert Kreidl) was Vice-Chairman.

The phenomenon of phase heterogeneity in glasses thus changed from a sensation to a firmly accepted aspect of glass structure, a transformation that was elegantly summarised by Norbert Kreidl in the quotation reproduced at the head of this section. The new insights to the structure of binary and multicomponent glasses, afforded by electron microscopy, resulted in a new understanding of these materials. This led to a greatly improved control of the phase separation process and hence to the development and production of new optical and engineering glasses.

10. Back to university

In 1961, as a result of his extremely successful electron microscopy studies of glasses and his rapidly growing international reputation, Werner Vogel was invited to give lectures on the Chemistry of Glass at the Friedrich Schiller University. He submitted his Habilitation (Postdoctoral) thesis, *Neue Erkenntnisse zur Struktur und zum Kristallisationsverhalten der Gläser* (*New Findings on the Structure and Crystallisation Behaviour of Glasses*), in 1963 and was appointed an Honorary Docent in Physical Chemistry in 1964. This was followed on 1st June 1966, by him being offered the newly founded Chair of Glass Chemistry at the Institut für Physikalische Chemie (Institute of Physical Chemistry).

The year 1965 saw the publication of Werner

Figure 15. Jenaer Gespräche über Glas (1973). A question from Alfred R. Cooper. Werner Vogel and Norbert Kreidl are together in the front row and others present include Philip J. Bray and Johannes M. Stevels

Vogel's first book, *Struktur und Kristallisation der Gläser*,[1] which contains many of his beautiful electron micrographs. An English translation[3] (*Structure and Crystallization of Glasses*) appeared in 1971, together with a second (German) edition.[2]

11. Otto-Schott-Institut

On 1st September 1967 Werner Vogel was appointed Director of the newly established Institut für Technische und Glaschemie (Institute of Chemical Engineering and Glass Chemistry), and continued the Jena legacy of Otto Schott, Ernst Abbe and Carl Zeiss by providing a comprehensive programme covering glass chemistry education and research. The lectures included glass chemistry, defects and production engineering, together with practical training, and the Institute began to turn out highly qualified graduates for the glass industry.

As a result of its high academic achievements and quality research, the Institute, which forms part of the Department of Chemistry, was renamed the Otto Schott Institute in 1974, in honour of this famous Jena glass scientist. In the same year, postgraduate courses in glass chemistry were introduced, both for employees of Jenaer Glaswerk and other companies, and for students and scientists from abroad, e.g. Brazil. Thus the scientific basis of glassmaking in the DDR and abroad was raised to a new level. In 1985, the complete Glass Chemistry teaching programme, including the set of scientific instruments, was sold by the University to Brazil.

In this way, under Werner Vogel's direction, an internationally famous school of glass chemistry was established in Jena. Gradually, the conditions for education and research in glass chemistry at the University were improved and, in 1984, the Institute took over a newly reconstucted building and glass melting plant. During Werner Vogel's period as Director, more than one hundred students completed their diploma in glass chemistry and more than thirty graduates received their doctorate, ten of whom qualified for a lectureship (Dr. habil. or Dr. sc.) and five became professors in Germany or Vietnam. At the same time, several Institute scientists became well known and highly respected members of the international glass community.

Figure 16. Jenaer Gespräche über Glas (1973). The conference excursion to Wartburg, Eisennach

12. Jena conferences

The world wide reputation of the Otto Schott Institute as a centre for glass research has been greatly enhanced by the international conferences held in Jena, starting with the Jenaer Gespräche über Glas (Jena Colloquium on Glass) in 1973.

This pioneering meeting, organised by Werner Vogel, together with Norbert Kreidl, was the first important such glass meeting in Jena after the war and brought internationally famous Western glass scientists to Jena. Its objective was to combine the best glass scientists from the East and West. Some of those participating can be seen from Figure 15, which shows Alfred R. Cooper asking a question at one of the sessions. A group of participants is shown in Figure 16 at Wartburg, Eisennach, during the conference excursion, while Figure 17 captures an informal exchange between Werner Vogel and Philip J. Bray.

In 1978, five years after the Jenaer Gespräche

Figure 17. Jenaer Gespräche über Glas (1973). An amusing exchange between Philip Bray and Werner Vogel during the conference trip

über Glas, Wener Vogel was Chairman of the (First) International Otto Schott Colloquium, at the Friedrich Schiller University, which was the first in a series that has been held ever since at four-yearly intervals and will always be associated with his name. The photograph in Figure 18 was taken during a session at the First Colloquium. Before German reunification, the Otto Schott Colloquia provided a unique opportunity for Western scientists to meet their colleagues from Eastern Europe and the Soviet Union and this aspect is still an important factor today. The most recent conference, the eighth, was held in July 2006.

Figure 18. A session at the First International Otto Schott Colloquium. Left to right in the front row are Klaus Kühne, Werner Vogel and Helmut Rötger

Figure 19. Werner Vogel opens the Second International Otto Schott Colloquium

A photograph of Werner Vogel opening the Second International Otto Schott Colloquium in June 1982 is reproduced in Figure 19, and a photograph of a round-table discussion at the same conference, between Peter C. Schultz, Norbert Kreidl, Werner Vogel and Adelbert Feltz appears in Figure 20. Similarly, another account of East–West interaction at an early Otto Schott Colloquium, this time between Evgenii Porai-Koshits and Donald R. Uhlmann, appears in Ref. 30.

One of the enjoyable occasions during the Otto Schott Colloquia is the conference dinner, which is traditionally held at the Fuchsturm (Fox Tower), located at the summit of Hausberg. In former times, this was a meeting place for a student fraternity involved in the sport of duelling. Figure 21 shows Werner Vogel delivering words of welcome during the dinner at the Second International Otto Schott Colloquium, those present comprising (left to right), Doris Ehrt, Norbert Kreidl, Käte Vogel, Werner Vogel and Vera Matioukh, the wife of Evgenii Porai-Koshits, while the ensuing

Figure 20. A round-table discussion at the Second International Otto Schott Colloquium. Left to right: Peter Schultz, Norbert Kreidl, Werner Vogel and Adelbert Feltz

Figure 21. A welcome from Werner Vogel during dinner at the Fuchsturm (Second International Otto Schott Colloquium). Left to right: Doris Ehrt, Norbert Kreidl, Käte Vogel, Werner Vogel and Vera Matioukh (wife of Evgenii Porai-Koshits)

Figure 23. Happy birthday Norbert! (Celebrated during the Second International Otto Schott Colloquium)

good times are illustrated in Figure 22. On this occasion, the opportunity was also taken to celebrate Norbert Kreidl's 78th birthday (Figure 23).

For one of the present authors (ACW), the first invitation (from Werner Vogel) to visit Jena came with the Third International Ottto Schott Colloquium in 1986, when he first had the privilege of meeting not only his co-author but also other scientists from Eastern Europe and the Soviet Union, whose names were familiar from the glass literature, including Adelbert Feltz from the Department of Solid State Chemistry in Jena (cf. Figure 20), Nikolai Andreev and Vladislav Nikolaevich Filipovich, from the Institute of Silicate Chemistry of the Russian Academy of Sciences in Leningrad, and Andrejs R. Silins from the Latvian State University in Riga. Two photographs from this Colloquium appear in Figures 24 and 25.

The Fourth International Otto Schott Colloquium in 1990 honoured Werner Vogel on the occasion of his 65th birthday and retirement as Director of the Otto Schott Institute and opened with a paper by Norbert

Figure 22. Fun at the Fuchsturm during the Second International Otto Schott Colloquium

Kreidl, entitled *Phase Separation in Glasses*, which was "Dedicated to Werner Vogel, Pioneer in Glass Science", while the Fifth was to have continued the celebration of Norbert Kreidl's 90th birthday (3rd July 1994), following the conference held in his honour in Triesenberg (Principality of Liechtenstein; July 3rd to 8th 1994) during the previous week. Sadly, however, Norbert Kreidl passed away in Liechtenstein on the 11th July and did not make it to Jena.

A further important international conference took place in Jena in 1984. This conference, entitled 100 Jahre Jenaer Glas, was held to mark the 100th anniversary of the Glastechnische Laboratorium Schott & Genossen (cf. Section 5). Werner Vogel was one of the scientific organisers and presented an invited paper entitled *On the Microstructure of Glasses*.

13. Scientific career

In addition to the various editions of his three books, during his scientific career Werner Vogel has published over 200 papers and 68 patents. The first (German) edition of his second book, *Glaschemie*, was published in 1979[(4)] and the first English edition (*Chemistry of Glass*) appeared in 1985, translated by Norbert Kreidl.[(6)] This book includes a short section on NMR spectroscopy written by another great friend, Philip J. Bray, to whom the Second International Conference on Borate Glasses, Crystals & Melts was dedicated. The third German edition of 1992[(8)] was also translated into English by Norbert Kreidl, this time with the help of M. Lopes Barotto.[(10)]

An important aspect of Werner Vogel's glass research has been the special attention he has paid to the batch reactions in the melting process and to a comprehensive characterisation of his samples, as discussed in his book *Glasfehler*.[(9)] In addition to his structural studies, he has developed glasses with very low and very high refractive index, new coloured glasses, high performance optical glasses for mi-

Figure 24. Helmut Dislich and Werner Vogel at the Third International Otto Schott Colloquium

crolithographic and laser optics, new photochromic glasses, and glass-ceramics with specific properties (e.g. machinability, bioactivity, magnetism and high strength). The development of many of these glasses has resulted from the understanding of their structure afforded by his electron microscopy studies.

Werner Vogel's prowess in teaching and glass research has led to his being awarded a number of well deserved prizes, including the Verdienter Erfinder (DDR, 1963), the Banner der Arbeit (DDR, 1964) and the Friedrich-Wöhler-Preis der Chemischen Gesellschaft der DDR (1966). In 1970, he received the DDR National Prize for Science and Technology and, in 1990, he was awarded an Honorary Doctorate (h.c. rer. nat.) by the Technische Universität Merseburg. In 1991, he shared (with Hideo Hosono) the Otto Schott

Figure 25. Informal discussion with Ivan Gutzow at the Third International Otto Schott Colloquium. Norbert Kreidl appears in the background

Figure 26. Verner Vogel (left) being awarded an Honorary Doctorate in Engineering (Dr.-Ing. E. h.) from the Technische Universität Clausthal by Günther Frischat (right)

Research Award for his Use of Electron Microscopy to Clarify "Demixing Phenomena" in Glasses and, on 27th June of the same year, he was awarded an Honorary Doctorate in Engineering (Dr.-Ing. E. h.) by the Technische Universität Clausthal. Figure 26 shows him being presented with this Doctorate by Günther Frischat, while the ensuing informal celebrations can be seen in Figure 27. He was made a Fellow of the American Ceramic Society in 1974 and finally, in 1990, an Emeritus Member of the American Ceramic

Figure 27. The celebrations following the award ceremony in Figure 26. Left to right: Rolf Brückner, Werner Vogel, Helmut Schaeffer and Ludwig Merker

Figure 28. Werner Vogel's 60th birthday. A birthday drink with Norbert Kreidl

Society, in recognition of his long membership and outstanding contributions to glass research.

There is no doubt that Werner Vogel's distinguished scientific career has been enriched by his long friendship with Norbert Kreidl, as is amply demonstrated by the pictures from the Jenaer Gespräche über Glas (Figure 15) and various Otto Schott Colloquia (Figures 20, 21, 23 and 25), and that in Figure 28 of them celebrating the former's 60th birthday. Norbert Kreidl's loyal support of both the Otto Schott Institute and Colloquia was rewarded on June 22nd 1984 by his being conferred with an Honorary Doctorate (Dr. h.c.) by the Friedrich Schiller University. Figure 29 shows him on this occasion, with Werner Vogel at the entrance to the Otto Schott Institute.

Figure 29. Norbert Kreidl receives an honorary doctorate (Dr. h.c.) at the Friedrich Schiller University on June 22nd 1984. Left to right: Hans-Joachim Pohl (Director of Jenaer Glaswerk), Werner Vogel, Norbert Kreidl and Egon Uhlig (Dean of the Mathematisch-Naturwissenschaftlich-Technische-Fakultät)

During his time at the Friedrich Schiller University, Werner Vogel participated in many international conferences, including the 1978 Gordon Research Conference on Glass (Figure 30) and the Sixth University Conference on Glass Science (Glass Microstructure - Surface and Bulk; July 29–31 1981) held at Pennsyl-

Figure 30. Official photograph from the 1978 Gordon Research Conference on Glass (Tilton School, August 21st to 25th)

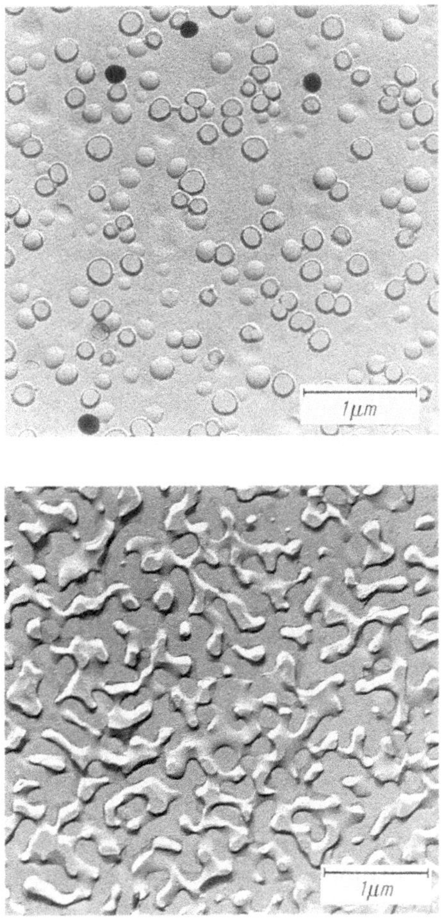

Figure 31. Droplet formation (upper micrograph) and spinodal decomposition (lower micrograph) in sodium borosilicate glasses[10]

vania State University. The latter included an opening session held in memory of the late Frederick William Holder (W.H.) Zachariasen, the first paper being *W.H. Zachariasen – The Melody Lingers On*, by Alfred R. Cooper,[54] and it was on arrival at State College that one of the authors (ACW) was first introduced to Werner Vogel by the Conference Chairman Guy Rindone. Papers were presented in later sessions at this conference by both Werner Vogel, entitled *Electron Microscopical Studies of Glass*,[55] and by Norbert Kreidl, on *Glass Surfaces Then and Now*.[56] During his long and distinguished career, Werner Vogel has been invited to present more than 100 talks/lectures outside the DDR, in Austria, Belgium, Bulgaria, Canada, China, Czechoslovakia, England, the FRG, France, Hungary, Liechtenstein, Poland, the Soviet Union, Sweden, the USA[31] and Yugoslavia. He once remarked (to ACW) that, at one stage, his applications to participate in international conferences in the West were often blocked by the DDR authorities. However, following his refusal to participate in a conference in the Soviet Union, permission was soon restored and he was again allowed to the West.

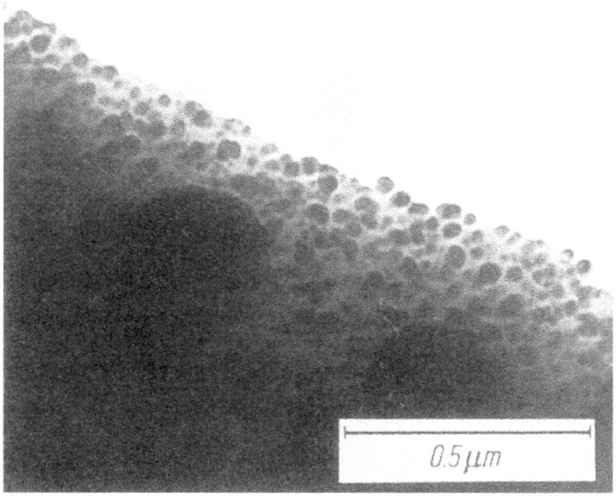

Figure 32. Transmission electron micrograph of the wedge-like edge of a clear binary lithium silicate glass splinter[10]

14. Electron artistry

No account of Werner Vogel's scientific career would be complete without showing a few of his superb electron micrographs. However, with the limited space available, it is only possible to give a few representative examples, which are taken from Werner Vogel's second book, *Glass Chemistry*,[10] and include some of his most famous images.

There are two mechanisms for immiscibility (phase separation) in glasses: Nucleation and growth, and spinodal decomposition. The former leads to a droplet structure, as may be seen from the upper micrograph in Figure 31, and the nucleation stage is governed by an activation energy. This glass is situated in the borate-rich end of the immiscibility region in the Na_2O–B_2O_3–SiO_2 phase diagram, and the structure comprises silica-rich droplets in an alkali borate-rich matrix. The glass is clear when cooled at a conventional rate, but the electron microscope already reveals the presence of droplets of diameter 200–300 Å. On heat treatment, these droplets grow more than 100-fold in diameter, to give the structure in Figure 31, and the glass becomes turbid. Conversely, glasses at the silica-rich end of the immiscibility region yield alkali borate-rich droplets in a silica-rich matrix. Droplet separation is also observed by transmission electron microscopy, which avoids use of the replica technique, as demonstrated by the transmission electron micrograph of the wedge-like edge of a clear binary lithium silicate glass splinter (28·6 mol% Li_2O) in Figure 32,[10] where denser droplet regions appear darker than the matrix because of stronger absorption of electrons.

Spinodal decompostion, on the other hand, can

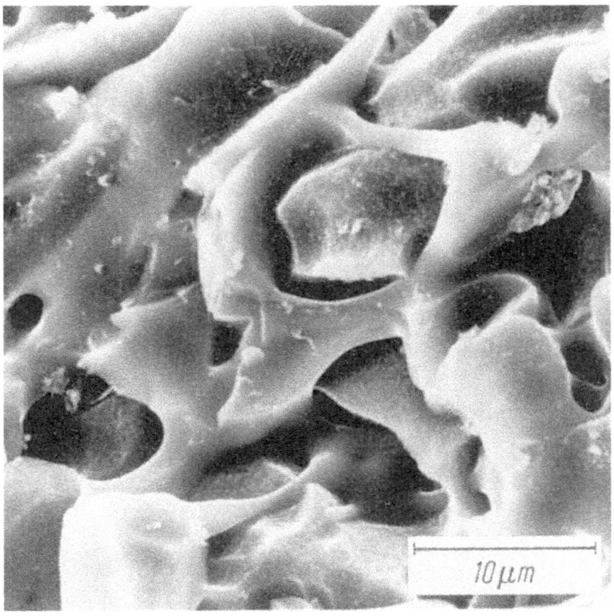

Figure 33. Leached Vycor® glass(10)

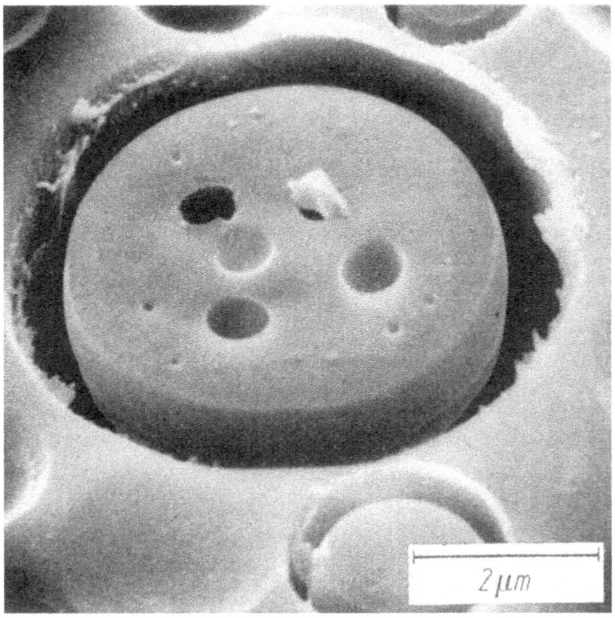

Figure 35. A barium borosilicate glass after HF etching(10)

occur spontaneously, with no energy barrier to nucleation, and leads to a connective microstructure, the glass in the lower micrograph in Figure 31 being from the centre of the immiscibility region. In the case of the sodium borosilicate system, one of the two connective phases is almost pure silica and the other sodium borate rich. The latter can be leached, to give the well known porous (Vycor®) glasses, as shown in Figure 33.

The phase separation described in the previous paragraph is very simple in that only two distinct phases are formed. It frequently occurs that phase separation is very much more complex and involves several stages, as in the case of Werner Vogel's famous micrograph of a barium borosilicate glass in Figure 34. Initially, on cooling the melt, separation occurs to give the largest droplets plus a matrix phase. Each of these then undergoes further phase separation into droplets plus matrix. Finally, both the droplets and the matrix within the largest droplets phase separate, again into droplets plus matrix, thus yielding a total of six distinct phases! The structure can be enhanced by etching, in this case with hydrofluoric acid, as demonstrated in Figure 35.

It is also possible for the droplets formed by phase separation to aggregate into connective phases. During their investigation of bioglass ceramics, Werner Vogel and his colleagues discovered the

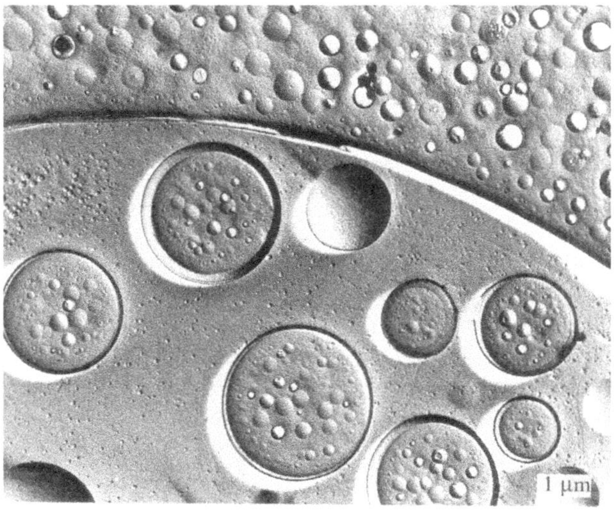

Figure 34. A barium borosilicate glass showing multiple stages of phase separation(10)

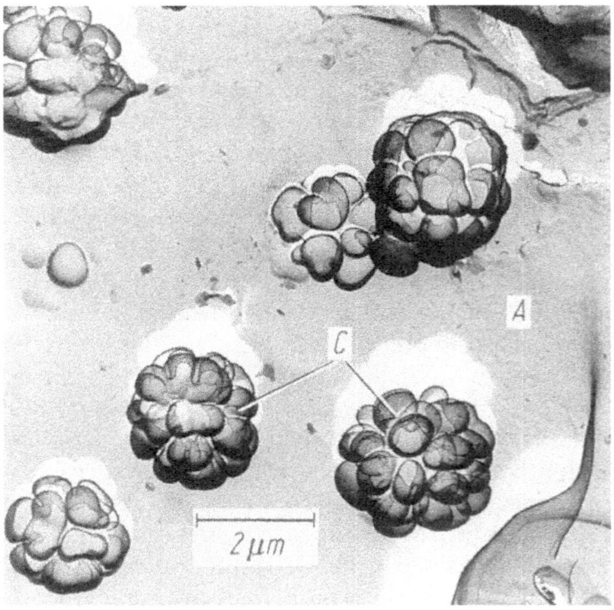

Figure 36. Blackberry-like phase separation in a CaO.SiO_2–$3CaO.P_2O_5$ pseudo-binary glass(10)

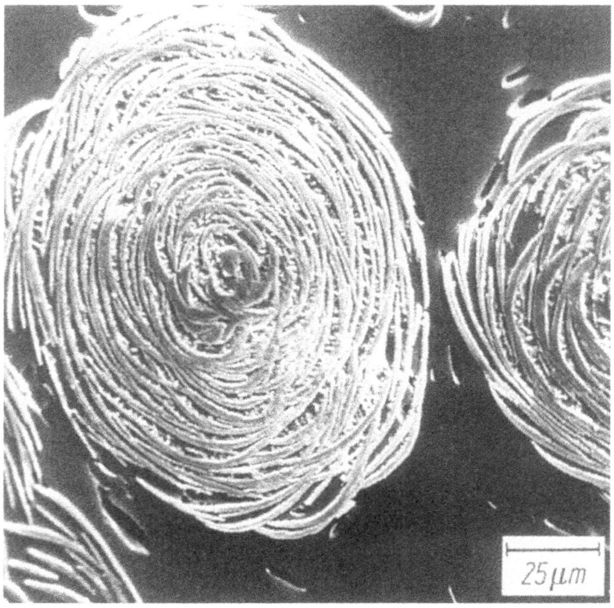

Figure 37. Cross-section through a phlogopite aggregate in a machinable glass ceramic[10]

blackberry-like aggregation of droplets in Figure 36, resulting from secondary phase separation in a $CaO.SiO_2$–$3CaO.P_2O_5$ pseudo-binary glass in the CaO–P_2O_5–SiO_2 system, the glass in question being richer in $3CaO.P_2O_5$. In this system, the calcium metasilicate phase has low solubility in the calcium orthophosphate phase and vice versa. Hence, on cooling the melt, primary phase separation yields silica-rich droplets in a P_2O_5-rich matrix. As the cooling proceeds, secondary immiscibility occurs in both of the primary phases, with P_2O_5-rich droplets forming in the large primary SiO_2-rich droplets and further, smaller SiO_2-rich droplets forming in the primary

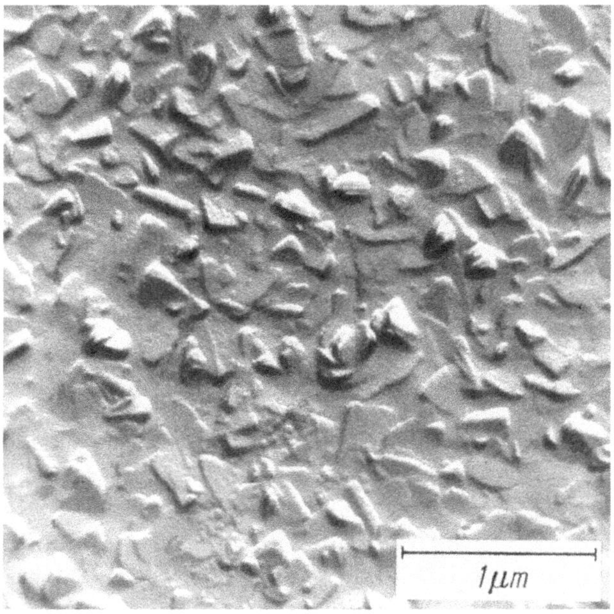

Figure 38. Crystallites in a clear binary magnesium phosphate glass[10]

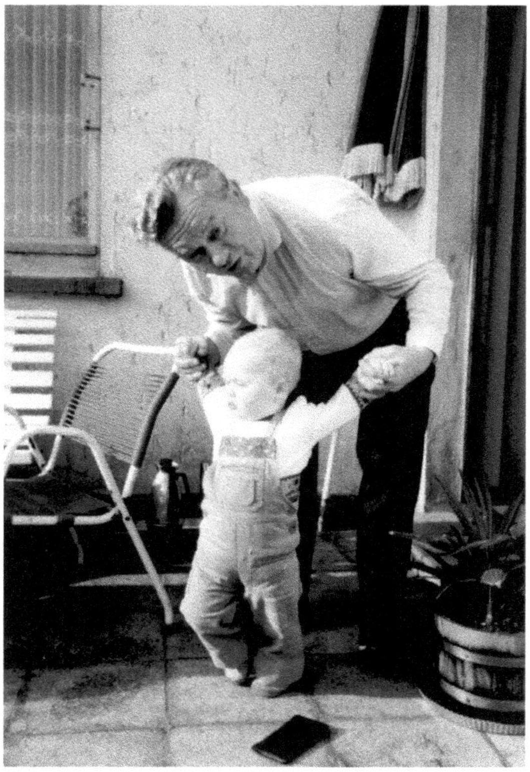

Figure 39. The younger generation in the Vogel line: Albert with his grandfather

matrix phase. In both cases, the droplets separating during the secondary process form blackberry-like aggregations, those in Figure 36 being the silica-rich droplets in the primary matrix phase.

As indicated in Section 9 by Werner Vogel's paper, *Catalyzed Crystallization in Glass*,[53] presented at the American Ceramic Society Annual Meeting in Toronto, electron microscopy is also a powerful tool for investigating crystallisation in glasses. An example is afforded by Figure 37, which shows the cross-section through a phlogopite ($MMg_3AlSi_3O_{10}F_2$; M=Na/K) aggregate in a machinable glass ceramic, comprising a $21{\cdot}2MgO.19{\cdot}5Al_2O_3.59{\cdot}3SiO_2$ base glass doped with F^- (11·5 mol%) and Na_2O/K_2O (6·4 mol%). Werner Vogel likens this to the result of cutting through a cabbage![10]

The question might also be asked as to whether electron microscopy has ever yielded any evidence to support the crystallite theory, which almost certainly contributed to the initial understanding of phase separation phenomena in glasses. There are undoubtedly cases of apparently clear glasses that have a "crystallite" structure, such as the magnesium phosphate "glass" in Figure 38. However, as may be seen from the scale marker, the crystallite dimensions are much (in this case over 100 times) larger than that predicted by the crystallite theory and hence the diffraction pattern would not be that of a glass but exhibit clear polycrystalline Bragg peaks. In this connection,

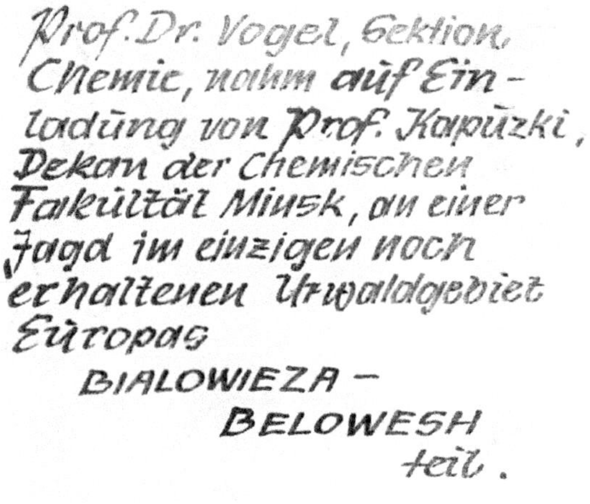
Prof. Dr. Vogel, Sektion,
Chemie, nahm auf Ein-
ladung von Prof. Kapuzki,
Dekan der Chemischen
Fakultät Minsk, an einer
Jagd im einzigen noch
erhaltenen Urwaldgebiet
Europas
BIALOWIEZA –
BELOWESH
teil.

Figure 40. Invitation to the hunt in 1970!

Figure 41. Hunting in the Bialowieza-Belowesh Forest in Belorussia, USSR in 1970

Werner Vogel remarks[10] that it is unfortunate that many early experiments to investigate the validity of the crystallite theory were conducted on vitreous silica prepared by melting polycrystalline α-quartz, in that the preparation of this material requires long melting times at extremely high temperatures and it is difficult to be sure that all "memory" of the starting material has been eradicated.

15. Personal note

Following his retirement as Director of the Otto Schott Institute, Werner Vogel still lives in Jena and participates in the Otto Schott Colloquia, thus maintaining contact with his many scientific friends. He and Käte have two sons, Wilfried and Jürgen, and five grandchildren. His son Wilfried is a veterinary surgeon, while Jürgen was a glass chemist, working in the field of bioglass ceramics at the Otto Schott Institute and later in the Institut für Physikalische Hochtechnologie (Institute for Physical and High Technology) in Jena. Jürgen died in January 2005. Werner Vogel is shown with his grandson Albert in Figure 39.

Throughout his adult life, Werner Vogel has been an enthusiastic and successful hunter, as demonstrated by the many trophies displayed at his house. He has also not only enjoyed hunting in his native Germany but also in other countries, including Czechoslovakia and the former Soviet Union (Figures 40 to 42).

16. Conclusions

"The electron microscope has become a tool of decisive importance not only for the solution of research problems in the sphere of glass, but also for problems of glass manufacture; the working and further processing of new possibilities of application are continuously revealed. This is all the more significant as the material of glass plays an increasingly important role."
(Werner Vogel[3])

Werner Vogel's expertise with the electron microscope led to his being the first scientist who correctly recognised and interpreted phase separation in glasses and his detailed and meticulous studies are a model for students of glass science to follow, particularly in respect of sample preparation and characterisation. The understanding of the phase separation process has resulted in the development and production of many new optical and engineering glasses, both in Jena and worldwide, and he has established electron microscopy as an essential tool in the investigation of glass structure on the scale of ten to several hundred Ångström units.

Following Werner Vogel's retirement and German reunification, the Otto Schott Institute has continued to thrive, in great measure due to the firm foundation that he laid. It continues as one of the world's two leading academic centres for glass research, the other being the New York State College of Ceramics at Alfred University, USA. The Otto Schott Colloquia, too, have become one of the leading international glass conference series, particularly in respect of attracting participants from the former Soviet Union and Eastern Europe.

Many reminders of Jena's optical glass heritage are still visible today. In addition to the revitalised Schott Glas factory, with its modern Borofloat® line described in these proceedings,[57] and the Otto Schott Museum, there is a monument to Ernst Abbe on Otto-Schott-Straße. Several key personages are buried in the historical Johannisfriedhof (Johannis Cemetery) at the Friedenskirche (Church of Peace), close to the Otto Schott Institute, including Johann Wolfgang Döbereiner, Friedrich Körner and Carl Friedrich Zeiss. Thus a visit to Jena is always a stimulating experience, both scientifically and in respect of Jena's unique place in the history of optical glasses.

Figure 42. Hunting in the Bialowieza-Belowesh Forest in Belorussia, USSR in 1970

Werner Vogel's contribution to glass science is immeasurable, especially in the field of phase separation, and his studies have included many B_2O_3-containing systems. Hence it is very fitting that the Fifth International Conference on Borate Glasses, Crystals and Melts is dedicated in his honour, both for his scientific work, following in the famous footsteps of Otto Schott, and for his contribution in bringing together glass scientists separated by the ideologies of the Cold War. It is also highly appropriate that the Conference was held during the month of his 80th birthday. Thus, to combine an old English proverb and a German pun, it is indeed the case that the early bird (in the field of electron microscopy) really did get the (choice) worm, in the form of an understanding of the complex phenomenon of phase separation in glasses and an appreciation of its great industrial potential!

Acknowledgements

The authors would like to thank Werner Vogel and the Otto Schott Institute for access to many of the photographs included in this short biography. The photograph in Figure 18 was kindly provided by Gerhard Herms (Reproduced courtesy of Schott, Jenaer Glas, GmbH) and those in Figures 26 and 27 by Günther Frischat. Thanks are also due to Springer-Verlag and the Consultants Bureau for permission to republish electron micrographs.

References

1. Vogel, W. *Struktur und Kristallisation der Gläser* (VEB Deutscher Verlag für Grundstoffindustrie, Leipzig, 1965).
2. Vogel, W. *Struktur und Kristallisation der Gläser*, 2nd Edn (VEB Deutscher Verlag für Grundstoffindustrie, Leipzig, 1971).
3. Vogel, W. *Structure and Crystallization of Glasses*, 1st English Edn (Pergamon, Oxford, 1971).
4. Vogel, W. *Glaschemie* (VEB Deutscher Verlag für Grundstoffindustrie, Leipzig, 1979).
5. Vogel, W. *Glaschemie*, 2nd German Edn (VEB Deutscher Verlag für Grundstoffindustrie, Leipzig, 1983).
6. Vogel, W. *Chemistry of Glass*, 1st English Edn, tr. N.J. Kreidl (Amer. Ceram. Soc., Columbus, 1985).
7. Vogel, W. *Kemija Stakla* (Croatian tr. of 2nd German Edn by E. Tkalcec & R. Laslo) (Zagreb, 1985).
8. Vogel, W. *Glaschemie*, 3rd German Edn (Springer Verlag, Berlin, 1992).
9. Vogel, W. *Glasfehler* (Springer Verlag, Berlin, 1993).
10. Vogel, W. *Glass Chemistry*, 2nd English Edn (tr. of 3rd 1992 German Edn by M. Lopes Barotto & N.J. Kreidl) (Springer-Verlag, Berlin, 1994).
11. Höland , W. Biography of Prof. Dr.rer.nat.habil. Werner Vogel on his 65th birthday, *J. Non-Cryst. Solids*, 1991, **129**, ix.
12. Kordes, E., Bergmann, W. & Vogel, W. *Z. Elektrochem.*, 1951, **55**, 600.
13. Kordes, E., Vogel, W. & Rackow, B. *Z. Anorg. Allgem. Chem.*, 1952, **268**, 236.
14. Kordes, E., Vogel, W. & Feterowsky, R. *Z. Elektrochem.*, 1953, **57**, 282.
15. Steiner, J. Kappler, D. & Berg, H.-J. *From Jena to Mainz – and back again*, 3rd Edn (Schott Glas, Mainz, 2003).
16. Vogel, W. *Silikattech.*, 1953, **4**, 483.
17. Vogel, W. *Silikattech.*, 1955, **6**, 469.
18. Vogel, W. *Silikattech.*, 1955, **6**, 510.
19. Seljakov, N., Strutinski L. & Krasnikov, A. *Z. Phys.*, 1952, **33**, 53.
20. Wyckoff, R. W. G. & Morey, G. W. *J. Soc. Glass Technol.*, 1925, **9**, 265.
21. Frankenheim, M. L. *Die Lehre von der Cohäsion* (Schulz, Breslau, 1835), p. 389.
22. Lebedev, A. A. *Trudy Gos. Opt. Inst.* (*Transactions of the Optical State Institute*), 1921, **2**, 1.
23. Nemilov, S. V. *Glass Phys. Chem.*, 1995, **21**, 148.
24. Randall, J. T., Rooksby, H. P. & Cooper, B. S. *Z. Kristallogr.*, 1930, **75**, 196.
25. Randall, J. T., Rooksby, H. P. & Cooper, B. S. *J. Soc. Glass Technol.*, 1930, **14**, P219.
26. Randall, J. T., Rooksby, H. P. & Cooper, B. S. *Nature*, 1930, **125**, 458.
27. Zachariasen, W. H. *J. Am. Chem. Soc.*, 1932, **54**, 3841.
28. Warren, B. E. *J. Appl. Phys.*, 1937, **8**, 645.
29. Warren, B. E. *Chem. Rev.*, 1940, **26**, 237.
30. Vedishcheva, N. M. & Wright, A. C. Proc. Third Int. Conf. on Borate Glasses, Crystals & Melts, Eds Y. B. Dimitriev & A. C. Wright (Soc. Glass Technol., Sheffield, 2001), p. xvii.
31. Valenkov, N. N. & Porai-Koshits, E. A. *Z. Kristallogr.*, 1936, **95**, 195.
32. Valenkov, N. N. & Porai-Koshits, E. A. *Nature*, 1936, **137**, 273.
33. Urnes, S. X-Ray Diffraction Studies of Glass, in: *Modern Aspects of the Vitreous State*, Ed. J. D. Mackenzie (Butterworths, London, 1960), pp. 10–37.

34. Anon., 1966 Toledo Award Winner, *Glass Ind.*, 1966, **47**, 15.
35. Zschimmer, E. *Die Glasindustrie in Jena* (Diederichs, Jena, 1912).
36. Warren, B. E. & Biscoe, J. *J. Am. Ceram. Soc.*, 1938, **21**, 49.
37. Slayter, G. *Am. Ceram. Soc. Bull.*, 1952, **31**, 276.
38. Prebus, A. F. & Michener, J. W. *Ind. Eng. Chem.*, 1954, **46**, 147.
39. Zhdanov, S. P., Porai-Koshits, E. A. & Levin, D. I. *Bull. Acad. Sci. USSR Div. Chem. Sci.*, 1955, 177.
40. Levin, D. I., Zhdanov, S. P. & Porai-Koshits, E. A. *Bull. Acad. Sci. USSR Div. Chem. Sci.* 1955, 27.
41. Shelyubsky, V. I. *Steklo Keram.*, 1954, **11** (10), 19.
42. Oberlies, F. *Naturwiss.*, 1956, **43**, 224.
43. Goldschmidt, V. M. *Skr. Nor. Vidensk. Akad. Kl. 1 Mat. Naturvidensk. Kl.*, 1926, **8**, 7.
44. Vogel, W. & Gerth, K. *Glastechn. Ber.*, 1958, **31**, 15.
45. Vogel, W. & Gerth, K. *Silikattech.*, 1958, **9**, 353.
46. Vogel, W. & Gerth, K. *Silikattech.*, 1958, **9**, 495.
47. Vogel, W. & Gerth, K. *Silikattech.*, 1958, **9**, 539.
48. Skatulla, W., Vogel, W. & Wessel, H. *Silikattech.*, 1958, **9**, 51.
49. Vogel, W. Silikattech. 1958, **9**, 323.
50. Andreev, N. S. & Porai-Koshits, E. A. *Dokl. Akad. Nauk. SSSR*, 1958, **118**, 735.
51. Vogel, W. Proc. Symp. sur la Fusion de Verre, Bruxelles, 6-10 Oct. 1958 (L'Union Scientifique Continentale du Verre), p. 741.
52. Vogel, W. The Cellular Structure of Glass, in: *The Structure of Glass*, Vol. 2 (Consultants Bureau, New York, 1960), p. 17.
53. Vogel, W. & Gerth, K. *Am. Ceram. Soc. Bull.*, 1961, **40**, 225.
54. Cooper, Jr., A. R. *J. Non-Cryst. Solids*, 1982, **49**, 1.
55. Vogel, W., Horn, L., Reiss, H. & Völksch, G. *J. Non-Cryst. Solids*, 1982, **49**, 221.
56. Kreidl, N. J. *J. Non-Cryst. Solids*, 1982, **49**, 331.
57. Kloss, T., Schneider, K., Lautenschläger, G. & Fintel, S. This proceedings (Paper I-12).

Structure and properties of iron borophosphate glasses

P. A. Bingham,[1] R. J. Hand

Immobilisation Science Laboratory, Department of Engineering Materials, University of Sheffield, Sheffield, S1 3JD, UK

S. D. Forder, A. Lavaysierre, F. Deloffre

Materials and Engineering Research Institute, Sheffield Hallam University, Howard Street, Sheffield, S1 1WB, UK

S. H. Kilcoyne & I. Yasin

School of Physics and Astronomy, University of Leeds, Leeds LS2 9JT, UK

Thermal stability, structure and aqueous chemical durability of glasses of nominal composition $60P_2O_5$–$40Fe_2O_3$ (mol%) doped with up to 20% B_2O_3 have been investigated. B_2O_3 was substituted for either P_2O_5 or Fe_2O_3, or was added on a pro rata *basis. No substantial melt volatilisation was detected, and refractory corrosion was minimal. Iron cations of both redox states occupied distorted octahedral sites; B_2O_3 additions did not affect iron coordination, although T_g and the $Fe^{2+}/\Sigma Fe$ redox ratio increased. Addition of up to 10% B_2O_3 did not significantly affect durability but it improved thermal stability and increased liquidus temperature. Chemical durability (as measured by the product consistency test, PCT) decreased by approximately one order of magnitude for every further 5% B_2O_3 addition above 10% when substituted for Fe_2O_3. Thermal analysis indicated increasing glass stability with increasing B_2O_3 content. X-ray diffraction showed that one sample of nominal composition $50P_2O_5$–$40Fe_2O_3$–$10B_2O_3$ developed crystalline $B_{0.57}Fe_{0.43}PO_4$ or a similar phase during cooling. There was no evidence that this significantly affected chemical durability. The addition of B_2O_3 at low levels (<10%) may be useful in improving the performance of iron phosphate glasses for applications such as waste immobilisation or sealing glasses.*

1. Introduction

Iron phosphate glasses exhibit high chemical durability, high compositional flexibility and low melting temperatures. This behaviour has prompted substantial research in the past decade aimed at utilising them as hosts for the immobilisation of toxic[1,2] and nuclear[3–7] wastes. Simplified glass compositions such as P_2O_5–Fe_2O_3[8–13] and P_2O_5–Fe_2O_3–R_xO_y where R=Pb,[14] Na and K,[15–17] Zn,[18,19] Ba,[20] and other elements,[21–22] have also been characterised. Many iron phosphate glasses tend to crystallise at relatively low temperatures of 600–650°C.[2,9–11,18] However the addition of boric oxide, B_2O_3, to P_2O_5–BaO,[23] Na_2O–Al_2O_3–P_2O_5,[24] Na_2O–Al_2O_3–P_2O_5–Fe_2O_3[25] and Na_2O–K_2O–BaO–Al_2O_3–P_2O_5[26] glasses is known to improve their thermal stability and crystallisation resistance. B_2O_3 also improves chemical durability in some systems.[27–29] These effects are all desirable for waste immobilisation as crystallisation can increase the aqueous solubility of some phosphate glasses.[3] However glasses in the P_2O_5–Fe_2O_3–B_2O_3 system have received only scant scientific attention.[30–32] In addition, B_2O_3 has thermal neutron absorption and mass absorption coefficients which are ~2 orders of magnitude greater than those of P_2O_5,[33] a fact which may prove useful in the immobilisation of radioactive wastes. The aim of this study was consequently to investigate the possibility of doping iron phosphate glasses with boric oxide to improve their general waste immobilisation performance, however consideration was also given to other potential applications.

2. Experimental procedures

Glasses were prepared from analytical grade >99% $NH_4H_2PO_4$, Fe_2O_3, and H_3BO_3. Batches to produce 200g of glass were placed in preheated mullite crucibles and melted in an electric furnace at 1150 or 1200°C (see Table 1) for 3 h. Melts were stirred for the final 2 h at ~60 rpm using Al_2O_3 stirrers, then were poured into preheated steel moulds, annealed at 450°C for 1 h, and cooled to ambient at 1°C/min. The calculated and analysed glass compositions are given in Table 1. A $60P_2O_5$–$40Fe_2O_3$ (mol%) glass (PFe-1M) was used as the base glass. Boric oxide was substituted for phosphate (B*x*-P glasses), iron oxide (B*x*-Fe glasses) or simply added *pro-rata* (B*x* glasses). Compositions of boron-containing glasses were analysed using inductively-coupled plasma optical emission spectroscopy (ICP-OES) with an ARL 3410 spectrometer. Boron-free samples were analysed using a Link energy dispersive x-ray spectroscopy (EDS) unit fitted to a Philips 500 scanning electron microscope. Densities were measured by the Archimedes method.

Room temperature Mössbauer spectra were

[1] Corresponding author. Email address: p.a.bingham@sheffield.ac.uk

Table 1. Glass compositions and selected property measurements

Sample Name	PFe-1M	B5-P	B10-P	B5	B10	B5-Fe	B10-Fe	B15-Fe	B20-Fe
P_2O_5 mol % (analysed)	60 (58·4)	55 (55·5)	50 (50·9)	57 (58·1)	54 (54·0)	60 (59·5)	60 (59·0)	60 (58·5)	60 (58·6)
Fe_2O_3 mol % (analysed)	40 (41·0)	40 (37·8)	40 (38·0)	38 (35·8)	36 (34·6)	35 (33·9)	30 (28·7)	25 (24·2)	20 (19·0)
B_2O_3 mol % (analysed)	0 (n/m)	5 (5·5)	10 (10·8)	5 (5·3)	10 (10·7)	5 (5·6)	10 (10·8)	15 (15·9)	20 (20·5)
SiO_2 mol % (analysed)	0 (0·3)	0 (0·8)	0 (0·2)	0 (0·2)	0 (0·3)	0 (0·8)	0 (1·1)	0 (0·9)	0 (1·3)
Al_2O_3 mol % (analysed)	0 (0·3)	0 (0·4)	0 (0·2)	0 (0·5)	0 (0·4)	0 (0·3)	0 (0·4)	0 (0·5)	0 (0·6)
Melting Temperature/°C	1150	1150	1150	1150	1150	1150	1150	1150	1200
ρ/g cm^{-3} (±0·005)	2·988	2·982	2·900	2·977	3·014	2·942	2·875	2·842	2·768
Glass Transition T_g/°C	498	503	515	504	511	520	532	547	575
First Crystallisation Tc_1/°C	658	647	679	739	668	856	874	819	815
T_g–Tc_1/°C	160	144	164	235	157	336	342	272	240
Crystalline Phases	-	-	$B_{0·57}Fe_{0·43}PO_4$	-	-	-	-	-	-
P leach rate/g cm^{-2} min^{-1}	2·28 E-11	3·07 E-11	3·76 E-11	3·04 E-11	2·85 E-11	3·60 E-11	4·85 E-11	4·42 E-10	2·52 E-09
Fe leach rate/g cm^{-2} min^{-1}	<1·07 E-11	<1·13 E-11	2·74 E-11	<1·10 E-11	<1·11 E-11	<1·09 E-11	<1·06 E-11	7·15 E-11	2·25 E-10
B leach rate/g cm^{-2} min^{-1}	n/m	<1·13 E-11	<1·07 E-11	<1·10 E-11	<1·11 E-11	<1·09 E-11	<1·06 E-11	6·88 E-11	4·55 E-10
Glass leach rate / g cm^{-2} min^{-1}	<3·34 E-11	<5·33 E-11	<7·57 E-11	<5·24 E-11	<5·08 E-11	<5·79 E-11	<6·97 E-11	5·82 E-10	3·20 E-09
(Fe^{2+}/ΣFe)×100% (±1)	13	22	15	25	35	20	23	32	34
CS (Fe^{2+}) / mm s^{-1} (±0·02)	1·25	1·26	1·11	1·23	1·29	1·28	1·26	1·23	1·22
QS (Fe^{2+}) / mm s^{-1} (±0·02)	2·13	2·09	2·41	2·15	2·21	2·02	1·98	2·10	2·05
LW (Fe^{2+}) / mm s^{-1} (±0·02)	0·15	0·15	0·13	0·17	0·18	0·14	0·16	0·20	0·17
CS (Fe^{3+}) / mm s^{-1} (±0·02)	0·37	0·35	0·35	0·35	0·34	0·37	0·38	0·37	0·38
QS (Fe^{3+}) / mm s^{-1} (±0·02)	0·89	0·94	0·86	0·91	0·95	0·90	0·89	0·89	0·89
LW (Fe^{3+}) / mm s^{-1} (±0·02)	0·15	0·16	0·15	0·15	0·17	0·16	0·15	0·14	0·15

collected over the velocity range ±5 mm s^{-1} using a constant acceleration spectrometer with a 25 mCi source of ^{57}Co in Rh. Eight broadened Lorentzian paramagnetic doublets were fitted to each spectrum, as described elsewhere.[2,8] Figure 1 shows selected fitted spectra. Parameters shown in Table 1 are weighted averages based on doublet areas and are given relative to α-Fe at room temperature. We assumed that the room temperature recoil-free fraction ratio f(Fe^{3+})/f(Fe^{2+})=1·3 in these glasses.[2,8] X-ray diffraction (XRD) spectra were measured using a Philips PW1730/10 goniometer with Cu K_α radiation, at 0·4 °2θ/min. Figure 2 shows XRD spectra for sample B10-P, which was the only sample to exhibit any detectable crystallinity, and sample B15-Fe which typified spectra from all other samples.

Differential thermal analysis (DTA) was performed using a Perkin Elmer DTA 7. Powdered 25 mg samples of particle size < 75 μm were placed in recrystallised Al_2O_3 sample cups and heated at 10°C/min from 20–1150°C. DTA traces are shown in Figure 3. Table 1 shows T_g, the glass transition temperature; T_{c1}, the temperature of maximum crystallisation for the first crystallisation exotherm; and their difference (T_{c1}–T_g). Fourier-Transform infrared (FTIR) spectroscopy was performed using a Perkin-Elmer Spectrum 2000 spectrometer with specular reflectance unit at 16° incidence. Kramers–Kronig transformations were carried out by software. Spectra, shown in Figure 4, were averaged over 20 measurements at 4 cm^{-1} resolution and 1 cm^{-1} data interval.

Chemical durability measurements used product consistency test B (PCT-B) with leachant/sample ratios of 20 ml/g.[34] Tests were carried out in duplicate at 85°C for 7 days. Average values are shown in Table 1 and a plot of log (dissolution rate) for each glass is shown in Figure 5. An oven calibration error led to all measurements unintentionally being made at 85 instead of 90°C as stipulated in the PCT.

3. Results and discussion

Compositional analyses indicated little melt volatilisation. Low levels of SiO_2 and Al_2O_3 were dissolved

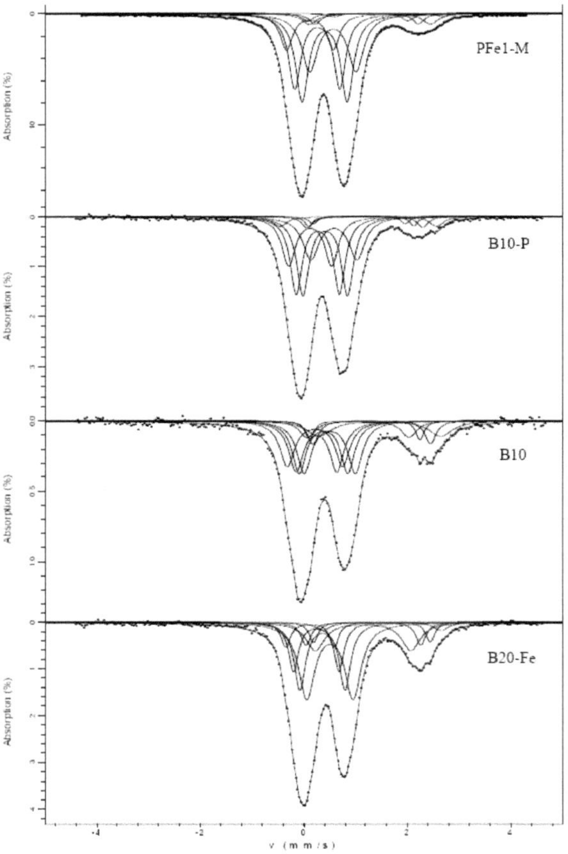

Figure 1. Selected Mössbauer spectra, fitted with eight Lorentzian doublets

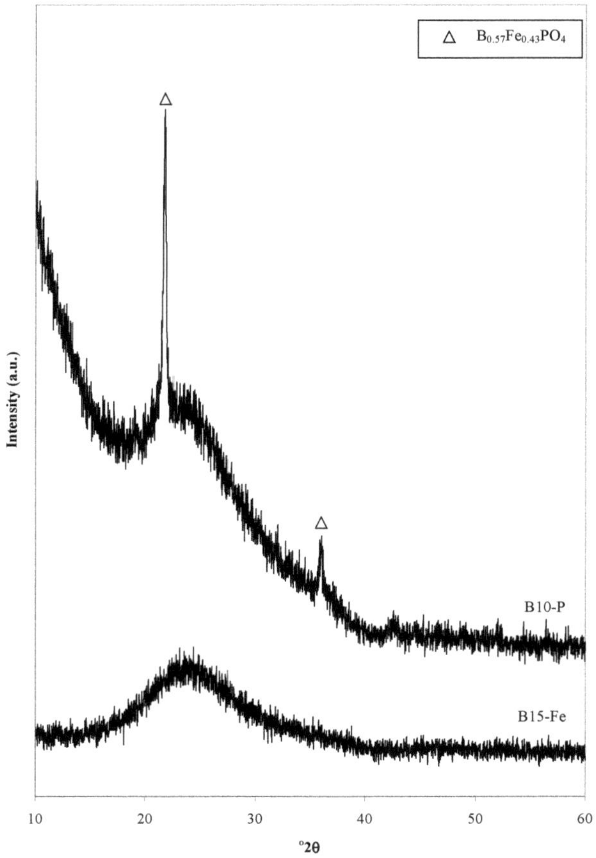

Figure 2. XRD spectra of samples B15-Fe and B10-P

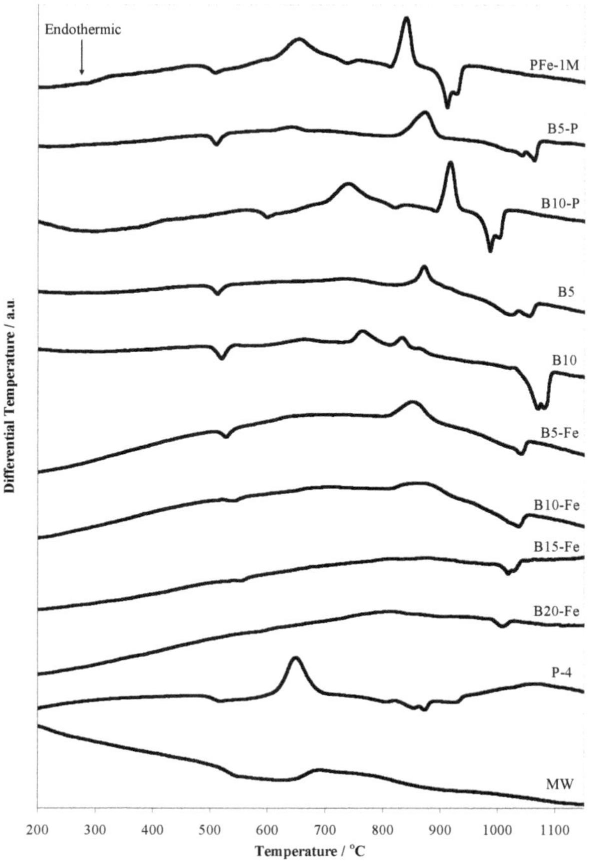

Figure 3. DTA measurements

from crucibles, but compared with similar iron phosphate glasses melted in Al_2O_3 crucibles, the effect of B_2O_3 addition on refractory corrosion was small.[15,16,19] Boric oxide additions increased T_g, consistent with its effects in several other phosphate glasses.[23–28] This behaviour has been attributed by some to an increase in crosslinking and formation of P–O–B bridges.[27–29] Increases in T_g were not attributed to changes in $Fe^{2+}/\Sigma Fe$ ratio because redox does not substantially affect the T_g of iron phosphate glasses.[11]

Mössbauer centre shifts (CS), quadrupole splittings (QS) and linewidths (LW) were consistent with the accepted view that Fe^{2+} and Fe^{3+} ions occupy distorted octahedral sites in iron phosphate glasses.[2,5,7,8,12,14,15,20] Indeed CS, QS and LW were unaffected by the B_2O_3 additions. Therefore the addition of B_2O_3 did not measurably affect iron environments. This was consistent with the results of nuclear magnetic resonance studies of P_2O_5–Fe_2O_3–B_2O_3 glasses.[31] The increase in $Fe^{2+}/\Sigma Fe$ redox ratio with B_2O_3 addition may have been related to changes in glass basicity; similar effects occurred with varying P/Fe ratio in P_2O_5–Fe_2O_3 glasses.[14] However the theoretical optical basicity (Λ_{th}) of glasses B5-P and B10-P was greater than Λ_{th} of glass PFe-1M, based on batch calculations.[35] Increasing Λ_{th} favours Fe^{3+} in binary silicate glasses, although the relationship is more complicated in ternary silicates.[36] Complex relationships may also exist for iron phosphate glasses, possibly due to their very high iron contents.

Only sample B10-P was not fully amorphous, although samples B5-P, B5 and B10 developed weak surface texturing during cooling. This was interpreted as very minor surface crystallisation and was not detected by XRD. XRD confirmed that sample B10-P developed $B_{0.57}Fe_{0.43}PO_4$ or a similar boron iron phosphate phase. This composition was therefore outside the glass formation region, which was at odds with published data.[29] The discrepancy probably lies in melt size, and therefore in sample cooling rate. Glasses in which B_2O_3 replaced Fe_2O_3 showed no tendency for surface crystallisation; this was consistent with their position within the glass-forming region.[29] Imperfect XRD peak matching indicated that the crystal phase formed in sample B10-P may not have had the exact stoichiometry of $B_{0.57}Fe_{0.43}PO_4$. However we believe the peaks were sufficiently close to conclude that the phase formed, if not $B_{0.57}Fe_{0.43}PO_4$, was very similar.

Thermal analysis proved that B_2O_3 additions can improve thermal stability, as shown in Table 1 and Figure 3. In particular the addition of small amounts (5–10%) of B_2O_3 gave some marked improvements in thermal stability. This was demonstrated not only in the increase of (T_{c1}–T_g) from 160°C to as high as ~340°C, but also in the smaller peak areas. Even glass-

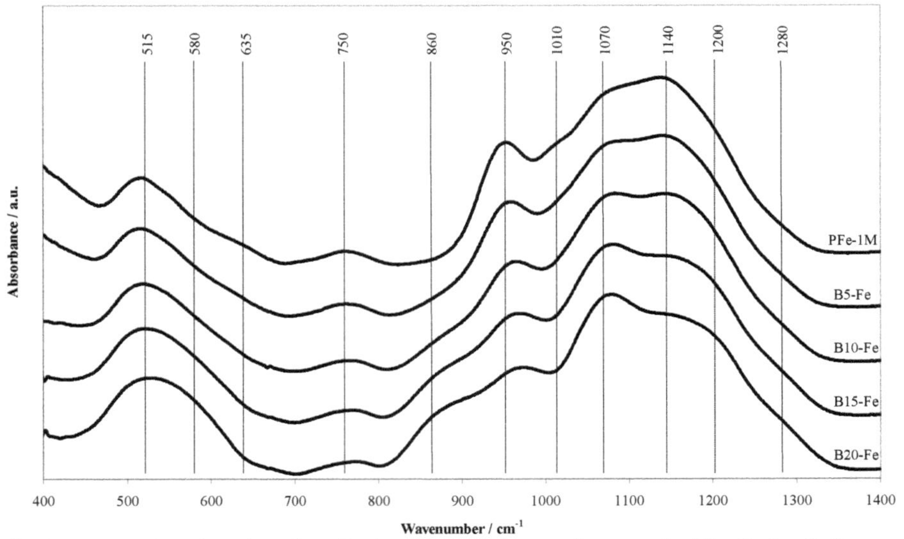

Figure 4. FTIR reflectance spectra showing the effects of increasing replacement of Fe_2O_3 by B_2O_3

es exhibiting surface texturing had (T_{c1}-T_g) values not substantially lower than the base glass, PFe-1M. Glass samples P-4 and MW were included in Figure 3 for comparison. Glass P-4 was a multicomponent iron zinc phosphate glass based on a metal finishing waste(1,2) and MW was a lithium sodium borosilicate glass used to immobilise UK high level nuclear waste.(1,2,7) Glass P-4 performed similarly to PFe-1M. Glass MW was specially developed to have high thermal stability and a low melting temperature,(7) yet its (T_{c1}–T_g) of ~160°C was surpassed by some of the P_2O_5–Fe_2O_3–B_2O_3 glasses. Considering (T_{c1}–T_g), the peak areas and the overall character of the DTA traces, the thermal stabilities of some P_2O_5–Fe_2O_3–B_2O_3 glasses were at least broadly comparable with that of MW and represented substantial improvements over the base glass, PFe-1M.

Figure 4 shows that upon replacement of Fe_2O_3 by B_2O_3, a broad infrared absorption band appeared at ~860 cm^{-1} and the existing band at ~1070 cm^{-1} was strengthened. The band at ~1280 cm^{-1} also increased slightly in intensity. Absorptions at 850–1070 cm^{-1} have been attributed to various BO_4 containing groups,(37–39) therefore we attribute the band at ~860 cm^{-1} to tetrahedral BO_4 groups. Absorptions >1150 cm^{-1} have been attributed to vibrations of trigonal BO_3 groups,(37,38) although bands at 1200–1350 cm^{-1} may also be due to Q^2 and / or Q^3 phosphate groups (where Q=number of bridging oxygen bonds per PO_4 group).(2,5,6,10,11,13,17–20) We believe that absorption in this region probably contains contributions from phosphate groups. Spectral changes at ~750, ~950, 1010, ~1070 and particularly at ~1140 cm^{-1} indicated variations in the distribution of phosphate Q-groups, as were expected due to decreasing Fe_2O_3 concentrations. An increasing P/Fe ratio repolymerises the network and raises the average phosphate Q-value, however borate groups oppose this by crosslinking Q^2 phosphate groups.(28) Absorption at 500–600 cm^{-1} broadened and grew with B_2O_3 replacement of Fe_2O_3. Similar changes occurred in spectra for P_2O_5–Fe_2O_3–ZnO (19) and P_2O_5–Fe_2O_3–BaO(20) glasses with replacement of Fe_2O_3 by RO. Absorptions

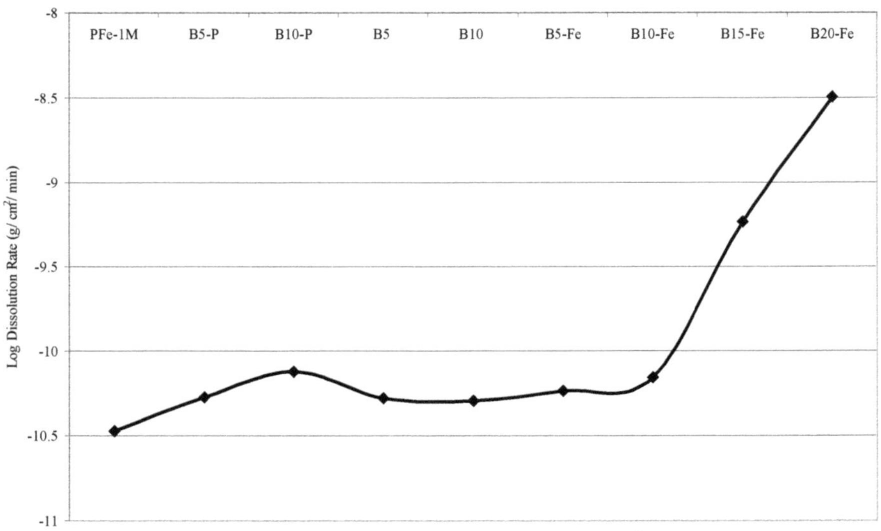

Figure 5. Plot of Log (Leach Rate), averaged over seven days at 85°C, in g/cm^2 min

in the range 400–600 cm^{-1} in iron phosphate glasses have generally been attributed to vibrations of Q^0 and Q^1 phosphate groups with overlapping Fe–O modes,[2,10,11,17–20] although Kim & Day[5] attributed a band at 590 cm^{-1} to vibrations of a $(PO_3)^-$ Q^2 group. Absorptions arising from B–O–B and B–O–Pb bending and borate ring deformation occurred in the range 400-650 cm^{-1} in lead borate glasses,[39] although in iron calcium borate glasses the B–O–B bending mode was attributed to a band at 700 cm^{-1}.[38] It seems likely that the absorption bands at 500–600 cm^{-1} shown in Figure 4 result from a combination of overlapping vibrations of Fe–O bonds, B–O–R bonds and low-Q phosphate groups.

The high chemical durability of iron phosphate glasses has been widely documented.[1,3–6,10,14,15,17–19,22] It is superior to that of borosilicate glass used in nuclear waste immobilisation[1,4,5,22] and to commercial soda–lime–silica glass[6,10,15,18] by up to two or three orders of magnitude. The chemical durability of P_2O_5–Fe_2O_3 glass undergoes its greatest increase between 20 and 30 mol% Fe_2O_3 due to the replacement of P–O–P bonds by P–O–Fe bonds.[15] Further Fe_2O_3 additions above 30 mol% have markedly smaller effects on durability. The same behaviour occurred with the B_2O_3 doped glasses in this study. The PCT-B results shown here in Table 1 demonstrate that addition of B_2O_3 only affects durability at levels of addition above 10 mol%, regardless of the oxide replaced. Plotting our data as log (dissolution rate), as others have done with monolithic test data,[15,17–19] underlines this fact (Figure 5).

4. Conclusions

The addition of up to 10 mol% B_2O_3 to $60P_2O_5$–$40Fe_2O_3$ glasses gave superior thermal stability and had no significant effect on chemical durability compared with the base glass. Greater additions of boric oxide led to decreases in chemical durability. Refractory corrosion, whilst increased by B_2O_3 addition, was low to normal whilst melt volatilisation was very low. Fitted Mössbauer parameters were consistent with published values, and showed that B_2O_3 addition had no effect on the distorted octahedral environments of Fe ions, although the $Fe^{2+}/\Sigma Fe$ ratio increased substantially with B_2O_3 additions. One glass developed $B_{0.57}Fe_{0.43}PO_4$ or a similar crystal phase upon cooling, however this did not substantially affect durability. FTIR spectra indicated the presence four-coordinated B^{3+} ions; other spectral differences were indicative of a changing distribution of phosphate Q-values and the possible presence of B–O–R bonds. Based on the consistency of Mössbauer parameters, it seems more likely that these bonds would be B–O–B or B–O–P rather than B–O–Fe, however further studies would be required to confirm this.

Acknowledgements

The authors wish to thank the UK's Engineering and Physical Sciences Research Council (EPSRC) for funding this work.

References

1. Bingham, P. A. & Hand, R. J. *J. Hazard. Mater.*, 2005, **B119**, 125–33.
2. Bingham, P. A., Hand, R. J. Forder, S. D. & Lavaysierre, A., *J. Hazard. Mater.*, 2005, **B122**, 129–38.
3. Day, D. E., Wu, Z., Ray, C. S. & Hrma, P. *J. Non-Cryst. Solids*, 1998, **241**, 1–12.
4. Mesko, M. G. & Day, D. E. *J. Nucl. Mater.*, 1999, **273**, 27–36.
5. Kim, C. W. & Day, D. E. *J. Non-Cryst. Solids*, 2003, **331**, 20-31.
6. Huang, W., Day, D. E., Ray, C. S., Kim, C. W. & Mogus-Milankovic, A. *J. Nucl. Mater.*, 2004, **327**, 46–57.
7. Donald, I. W., Metcalfe, B. L. & Taylor, R. N. J. *J. Mater. Sci.*, 1997, **32**, 5851–87.
8. Bingham, P. A., Hand, R. J., Forder, S. D., Lavaysierre, A., Kilcoyne, S. H., & Yasin, I. *Mater. Lett.*, 2006, **60**, 844–847.
9. Marasinghe, G. K., Karabulut, M., Ray, C. S., Day, D. E., Shumsky, M. G., Yelon, W. B., Booth, C. H., Allen, P. G. & Shuh, D. K. *J. Non-Cryst. Solids*, 1997, **222**, 144–152.
10. Ray, C. S., Fang, X., Karabulut, M., Marasinghe, G. K. & Day, D. E., *J. Non-Cryst. Solids*, 1999, **249**, 1–16.
11. Fang, X., Ray, C. S., Mogus-Milankovic, A. & Day, D. E. *J. Non-Cryst. Solids*, 2001, **283**, 162–72.
12. Karabulut, M., Marasinghe, G. K., Ray, C. S., Day, D. E., Waddill, G. D., Booth, C. H., Allen, P. G., Bucher, J. J., Caulder, D. L. & Shuh, D. K. *J. Non-Cryst. Solids*, 2002, **306**, 182–92.
13. Karabulut, M., Metwalli, E., Day, D. E. & Brow, R. K. *J. Non-Cryst. Solids*, 2003, **328**, 199–206.
14. Sales, B. C., Abraham, M. M., Bates, J. B. & Boatner, L. A., *J. Non-Cryst. Solids*, 1985, **71**, 103–12.
15. Yu, X., Day, D. E., Long, G. J. & Brow, R. K., *J. Non-Cryst. Solids*, 1997, **215**, 21–31.
16. Mogus-Milankovic, A., Rajic, M., Drasner, A., Trojko, R., & Day, D. E. *Phys. Chem. Glasses*, 1998, **39**, 70–5.
17. Fang, X., Ray, C. S., Marasinghe, G. S. & Day, D. E., *J. Non-Cryst. Solids*, 2000, **263–264**, 293–8.
18. Reis, S. T., Karabulut, M. & Day, D. E., *J. Non-Cryst. Solids*, 2001, **292**, 150–7.
19. Jermoumi, T., Hafid, M., Niegisch, N., Mennig, M., Sabir, A. & Toreis, N. *Mater. Res. Bull.*, 2002, **37**, 49–57.
20. Hafid, M., Jermoumi, T., Niegisch, N. & Mennig, M. *Mater. Res. Bull.*, 2001, **36**, 2375–82.
21. Karabulut, M., Marasinghe, G. K., Ray, C. S., Day, D. E., Ozturk, O. & Waddill, G. D. *J. Non-Cryst. Solids*, 1999, **249**, 106–16.
22. Mesko, M. G., Day, D. E. & Bunker, B. C. *Waste Manage.*, 2000, **20**, 271–8.
23. Harada, T., In, H., Takebe, H. & Morinaga, K., *J. Am. Ceram. Soc.*, 2004, **87**, 408–11.
24. Donald, I. W., Metcalfe, B. L. & Fong, S. K., *Proc. XX Int. Congr. Glass*, Kyoto, Japan, 2004.
25. Donald, I. W. & Metcalfe, B. L., *J. Non-Cryst. Solids*, 2004, **348**, 118–22.
26. Peng, Y. B. & Day, D. E., *Glass Technol.*, 1991, **32**, 166–73.
27. Hadj Youssef, N., Belkhiria, M. S., Videau, J. J. & Ben Amara, M. *Mater. Lett.*, 2000, **44**, 269–74.
28. Koudelka, L. & Mosner, P. *J. Non-Cryst. Solids*, 2001, **293–295**, 635–41.
29. Xu, X. J., Day, D. E., Brow, R. K. & Callahan, P. M. *Phys. Chem. Glasses*, 1995, **36**, 264–71.
30. Sedmale, G. P. *Latvijas Kim. Zurn.*, 1994, **4**, 420–34.
31. Muromtseva, M. E., Ivanova, G. S. & Goland, I. S. *Sov. J. Glass Phys. Chem.*, 1991, **16**, 372–5.
32. Eissa, N. A., Sheta, N. H., El Bahnasawy, H. H., Amer, T. Z. & Sallam, H. A. *J. Radioanal. Nucl. Chem.*, 2001, **247**, 311–16.
33. Bastick, R. E. *J. Soc. Glass Technol., 1958*, **42**, 70–83T.
34. C1285-02, Standard test methods for determining chemical durability of nuclear, hazardous, and mixed waste glasses: the product consistency test (PCT), ASTM, Conshohocken, USA.
35. Duffy, J. A. & Ingram, M. D. *J. Non-Cryst. Solids*, 1976, **21**, 373–410.
36. Bingham, P. A., Parker, J. M., Searle, T., Williams, J. M. & Smith, I., *C. R.. Chimie*, 2002, **5**, 787–96.
37. Efimov, A. M. *J. Non-Cryst. Solids*, 1999, **253**, 95–118.
38. Ardelean, I. & Pascuta, P. *Mater. Lett.*, 2004, **58**, 3499–502.
39. Pisarski, W. A., Pisarska, J. & Ryba-Romanowski, W., *J. Mol. Struct.*, 2005, **744–747**, 515–20.

Proc. Fifth Int. Conf. on Borate Glasses, Crystals and Melts *Phys. Chem. Glasses: Eur. J. Glass Sci. Technol. B*, August 2006, **47** (4), 318–322

Influence of water content on structure and kinetics of structural relaxation of sodium borate glasses

V. Golubkov

Institute of Silicate Chemistry of the Russian Academy of Sciences, ul. Odoevskogo 24-2, St. Petersburg, 199155, Russia

Alexander A. Zhilin,[1] *Alexander V. Shashkin & Michael P. Shepilov*

State Optical Institute, ul. Babushkina 36-1, 192171, St Petersburg, Russia

The influence of the water content on microheterogeneous structure of sodium borate glass containing 2·5 and 5·0 mol% Na_2O and on kinetics of structural relaxation of vitreous boron oxide and sodium borate glasses is examined by small angle x-ray scattering (SAXS) at isothermal heat treatments at various temperatures. An increase in the water content results in an increase of the size of inhomogeneous regions and in a decrease of $<(\Delta\rho)^2>$ of heterogeneous structure. Thus, water promotes aggregation of alkali borate structural elements. Kohlrausch equation describes the kinetics of SAXS by the thermal density fluctuations. In dry glasses β in Kohlrausch's equation is close to 1 and decreases to 0·5 in glasses containing 0·7–1·0 mol% H_2O.

1. Introduction

Water in glasses is known to influence pronouncedly their rheological properties and to lead to decreases in the glass transition temperature, T_g.[1] A clearly expressed microheterogeneous structure has been revealed in low alkali borate glasses.[2–4] Structural inhomogeneities, with regions 10–15 Å in size, were revealed in sodium borate glasses containing up to 10–12 mol% Na_2O. The glasses containing 4·0–4·5 mol% Na_2O demonstrate the greatest degree of inhomogeneity. In glasses with lower Na_2O content, the regions of inhomogeneity are composed of tetraborate groups. Almost all alkali ions are located in the regions of inhomogeneity, *RI*, while the surrounding matrix has a composition of boron oxide.[4] The greatest degree of inhomogeneity is observed at approximately equal concentrations of the high-alkali and the borate components. In glasses containing more than 5·0 mol% Na_2O, the relative content of the high alkali component is above 0·5. Thus the high-alkali *RI* can be considered as the matrix, while the borate phase forms the scattering regions.

A formation of microheterogeneous structure of low alkali borate melts at temperatures below 900–1000°C was found using the SAXS method. Precipitation of alkali borate groups in the form of *RI* stops at 450–500°C. The size of *RI* remains constant at lower temperatures, while the SAXS intensity from a microheterogeneous structure decreases. The temperature where the melt structural change ceases is close to T_g of the sodium tetraborate glass. This cessation may be connected with freezing of a high alkali component so that, at temperatures below T_g, the sample contains nanometre sized regions of high alkali glass in a supercooled borate liquid. Temperature dependences of density and thermal expansion coefficient of boron oxide in the equilibrium state of supercooled liquid at temperatures below glass transition temperature were obtained in Golubkov & Porai-Koshits,[5] on the basis of this mode.

This paper reports results obtained from the SAXS study of the influence of water content on structure and kinetics of structural relaxation for vitreous boron oxide and sodium borate glasses containing 2·5 and 5·0 mol% Na_2O.

2. Experimental

SAXS is a direct structural method for middle range order investigation in disordered systems. It reveals inhomogeneities several Angstrom to several hundreds Angstroms in size.[6] The SAXS intensity of one-component homogeneous and supercooled liquids is due to their inhomogeneity which is related to thermal density fluctuations (TDF).

The SAXS intensity from the TDF at near to zero angles, $I_\rho(0)$, is expressed by the equation

$$I_\rho(0)=\rho^2 kT\chi_T V \quad (1)$$

where ρ is the mean electron density, k is Boltzmann's constant, T is the absolute temperature, χ_T is the iso-

[1] Corresponding author. Email address: ntiom@goi.ru

thermal compressibility and V is the volume.

This intensity is independent of the scattering angle. However, due to the influence of light scattering at large angles, an increase in I_ρ is observed with increasing the scattering angle. The angular dependence of the SAXS intensity can be expressed as

$$I(s)=I(0)\exp(Ks^2)$$

$$s=4\pi\sin(\varphi/2)/\lambda \qquad (2)$$

where φ is the scattering angle, λ is a wavelength of x-ray irradiation and K is a constant which depends on the glass type.

In binary and multicomponent systems light scattering due to the concentration fluctuations, I_c, can pronouncedly influence the SAXS intensity, especially when some components contain elements with large numbers.[4] In the general case, the scattering by a homogeneous single phase liquid is a sum of the SAXS scattering by thermal density and concentration fluctuations, $I_{\rho c}=I_\rho I_c$.

Liquids and glasses, whose inhomogeneity originates only from the thermal density and concentration fluctuations, can be regarded as homogeneous media. Glasses, liquids, and melts are regarded as microheterogeneous media if they contain *RI*, regions with electron density different from the density of the surrounding matrix. Appearance of *RI* can be connected with phase separation, crystallisation and clustering of different origin. Their presence can be easily proved by the characteristic angular dependence of the SAXS scattering. In the case of spherical regions with identical composition, the intensity of SAXS from the structure, I_{is}, may be presented as

$$I_{is}(s)=Nn^2\exp(-s^2R^2/5)$$

$$=N(4\pi/3)^2R^6(\rho_1-\rho_2)^2\exp(-s^2R^2/5)$$

$$\text{at } sR\ll 1 \qquad (3)$$

where N is the number of *RI*, n is the number of scattering electrons in a region, R is a radius of the spherical region with a constant volume density, ρ_1 and ρ_2 are the mean electron densities of the *RI* and the surrounding matrix.

The important characteristics of structure is the mean square of the electron densities difference, $<(\Delta\rho)^2>$. In the case of a two-phase structure

$$<(\Delta\rho)^2>=(\rho_1-\rho_2)^2w_1w_2 \qquad (4)$$

where ρ_1 and ρ_2 are electron densities of phases, and w_1 and w_2 are the respective volumes of the phases. The $<(\Delta\rho)^2>$ value can be determined from Equation (5)

$$<(\Delta\rho)^2>=\frac{1}{4\pi^2 v}\int_0^\infty sI(s)\,ds \qquad (5)$$

where $I(s)$ is the SAXS intensity in e.u. measured with an 'infinitely' high primary beam (e.u. is the scattering intensity for one electron).

It should be noted that in the range of small angles $\sin\varphi=\varphi$, thus for the calculation of $<(\Delta\rho)^2>$ by Equation (4) one can use the dependence $\varphi I(\varphi)$.

If Equation (3) holds true, there is a maximum in the curve of the angular dependence of $sI(s)$ and $\varphi I(\varphi)$. The radius of the particle and the values of s or φ corresponding to the maximum in the curve and denoted as s_0 or φ_0 are connected by the simple correlation

$$R=(5/2)^{1/2}s_0^{-1}\cong 1{\cdot}33\times10^3\,\varphi_0^{-1} \qquad 6)$$

where φ_0 is a scattering angle measured in angular minutes.

CuK_α radiation was used in this study. The samples were glass plates prepared by seizing the melt drop between two metal plates. The samples were heat treated in a high temperature attachment of a SAXS instrument.[7] By this means scattering from the same sample volume was registered during all the experiment. The ultimate temperature was close to the softening point, which is the temperature at which the sample became deformed under the action of its own weight. When studying the kinetics of structural relaxation, I_ρ was measured at a fixed scattering angle of 350°. At this angle I_ρ from the *RI* becomes negligible so that the measured I_ρ can be considered equal to the intensity of scattering from the TDF. The impact of scattering at large angles can be neglected, as it is independent of temperature. Thus all changes in $I(350°)$ are due to change in $I_{\rho c}$. Angular dependences of I_ρ were measured at different temperatures and at different relaxation stages.

Monochromatisation of the primary beam by a total external reflection allowed measurement of the intensity of the primary beam after attenuation by a standard absorber. Intensity of the primary beam was measured before every measurement of I_ρ. Statistical mean square error of the intensity measurements was ±0·5–1%; when studying samples in the equilibrium state or at low temperatures (long relaxation times) it was 0·2% at most. At above T_g temperature dependences of I_ρ were measured at both increasing and decreasing the temperature at every 10–20°C. The intensity of SAXS in absolute units was obtained by a comparison with a scattering for a standard sample of quartz glass which was taken as 0·55 e.u./Å^3 for the unit volume.

Glass samples with different water content were prepared from the melt held on air at different temperatures and after additional holding in vacuum. We did not determine the water content in glasses directly; it was estimated from the T_g values with the use of data presented in Visser & Stevels.[1]

2.1. Determination of the glass transition temperature and the fictive temperature

According to Equation (1), the intensity of scattering from TDF of liquid is temperature dependent. Density and isothermal compressibility depend on temperature as well. According to our findings, in all glass forming systems under study I_ρ from supercooled liquids linearly depends on temperature in a wide temperature range from T_g and up to a liquidus temperature. Linear dependence is also revealed in the case of prolonged heat treatments in some temperature range below T_g, which implies that the sample can be in the supercooled liquid state above the temperature of transition to a solid non-crystalline state, T_s. For boron oxide T_s is 50°C below T_g. A linearity of the temperature dependence of the SAXS intensity allows determining the fictive temperature, T_f, and T_g with high accuracy. The fictive temperature is the temperature at which the SAXS intensity from the TDF of the sample in the equilibrium state is equal to the measured SAXS intensity. The glass transition temperature is the fictive temperature of the sample cooled to a temperature at which the relaxation time of the structure significantly exceeds the time of the measurement of the SAXS intensity. Usually the temperature of the observation is 60–70°C below T_g. In our experiments the error in determining the T_g and T_f does not exceed ±1°C.

3. Results

3.1. Influence of water on structure of sodium borate glasses

The dependence on syntheses conditions of the SAXS intensity at a scattering angle of 350° is presented in Figure 1 for the glass samples containing 5·0 mol% Na_2O. The SAXS intensity in the supercooled liquid state increases and T_g decreases with increasing water content in glass. The similar behaviour of the temperature dependences is characteristic of glasses containing 2·5 mol% Na_2O. However, for this glass the dependence of the SAXS intensity on the water content is less pronounced than for the glass containing 5·0 mol% Na_2O. The SAXS intensity for a boron oxide in the supercooled state is independent of the water content. The increase of the water content leads only to decreasing the glass transition temperature.

The angular dependence of $\varphi I(\varphi)$ for glass containing 2·5 mol% Na_2O prepared at different synthesis conditions is presented in Figure 2. With decreasing the water content, a maximum on the curve of angular dependences of $\varphi I(\varphi)$ moves to larger angles which manifests decreasing the sizes of the scattering regions. The square under the curve of $\varphi I(\varphi)$ increases which according to Equation (5) means the increase of $<(\Delta\rho)^2>$.

Table 1 presents the values of T_g determined from the SAXS data for the boron oxide and sodium borate glasses, the size of the *RI* and $<(\Delta\rho)^2>$ for sodium borate glasses prepared under different synthesis conditions. For all the glasses, the size of the *RI* and $<(\Delta\rho)^2>$ are obtained at 360°C. The sizes of the *RI* and $<(\Delta\rho)^2>$ depend on the water content in glass (Table 1). When it decreases, the size of the *RI* is decreased and $<(\Delta\rho)^2>$ is increased. In alkali borate glasses, the sizes of the *RI*, the $<(\Delta\rho)^2>$ and T_g depend on Na_2O content as well. For the compositions under study, an increase of the Na_2O content results in diminishing the sizes of the *RI* and in increasing T_g. However taking into account that after additional heat treatment in air the decrease of T_g is accompanied by the decrease of $<(\Delta\rho)^2>$ and the increase of the size of the *RI*, one can conclude that the change in sizes and

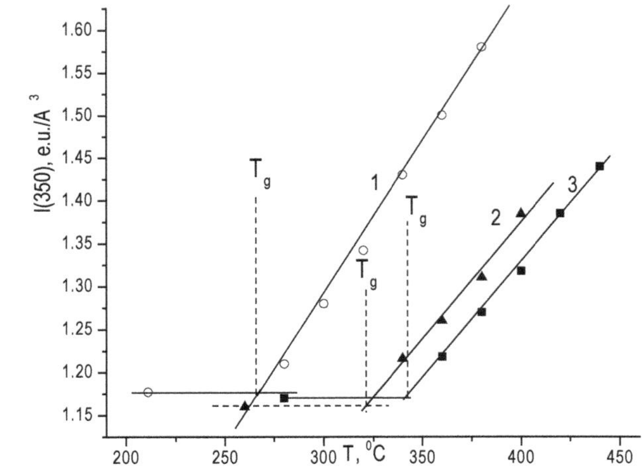

Figure 1. Temperature dependence of the SAXS intensity at a scattering angle of 350° for the sodium borate glass containing 5 mol% Na_2O and prepared at: (1) 850°C for 1 h (more than 1 mol% H_2O); (2) 1350°C for 1 h (0·5 mol% H_2O); (3) 900°C for 1 h in vacuum (without water)

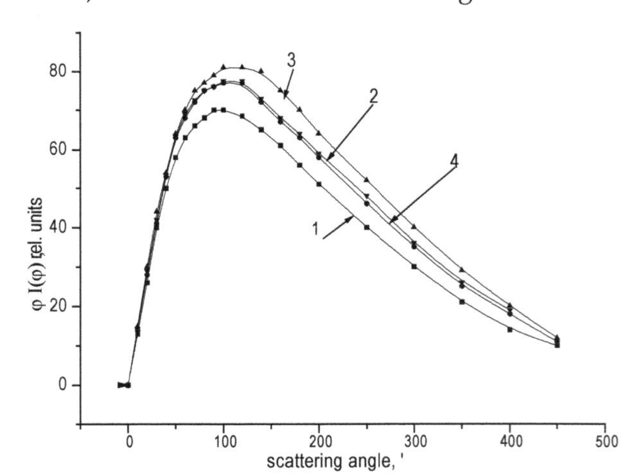

Figure 2. Angular dependence of φI(φ) for borate glass containing 2·5 mol% Na_2O and prepared at (1) 900°C for 1 h (0·8 mol% H_2O); (2) 1300°C for 1 h (0·35 mol% H_2O); (3) 900°C for 1 h in vacuum (without water); (4) additional holding at 800°C for 3 h (0·3 mol% H_2O)

Table 1. The influence of the water content (synthesis conditions) in glasses on glass transition temperature, T_g, for the boron oxide and sodium borate glasses, sizes (R) of the RI and $<(\Delta\rho)^2>$ for sodium borate glasses

Content of Na_2O (mol%)	Synthesis conditions T (°C)	Time (h)	T_g (°C)**	Content of H_2O (mol%)	R (Å)	$<(\Delta\rho)2>$ 10^4 $(e/Å^3)^2$
0	900	1	250	0·65	-	-
	1250	2	275	0·20	-	-
	900 (vacuum)	1	290	0	-	-
2·5	900	1	275	0·80	13·5	4·70
	1300	3	300	0·35	11·5	5·00
	900 (vacuum)	1	315	0	10·5	5·60
	800*	3	295	0·30	12	4·90
5·0	850	1	280	>1	11·5	4·50
	1350	1	320	0·50	9·0	4·70
	900 (vacuum)	1	340	0	7·8	5·05

* After holding in vacuum the melt was held at 800°C for 3 h in air.
** The samples were cooled through the glass transition region with the rate of 5 K/min.

$<(\Delta\rho)^2>$ originate solely from the change of the water content in glasses.

3.2. Influence of the water content on kinetics of structural relaxation

The kinetics of structural relaxation were studied from the change of the SAXS intensity by the TDF with heating time. The sample was cooled from the high temperature to 60–70°C below T_g.

Dependence of the SAXS intensity at scattering angle of 350° on heating time at temperatures 60°C below T_g for a vitreous boron oxide with different water contents (Figure 3) is satisfactorily described by the Kohlrausch's equation

$$I(t)-I_\infty=(I(0)-I_\infty)\exp(-(t/\tau)^\beta) \qquad (7)$$

where $I(t)$, I_∞ and $I(0)$ are the SAXS intensities at the instant t, in the equilibrium state and at the initial time, respectively, τ is an effective relaxation time and β is a coefficient. Dotted curves are obtained for $I(t)$ for the same values of I_∞, $I(0)$ and τ at β=1. For the water-containing glass β=0·5, for the dry glass the curve is of exponential character, β is near one. The similar are dependences for the dry and water-containing sodium borate glass.

4. Discussion

4.1. Influence of water on glass structure

Water in glasses changes their structure (Table 1, Figures 1 and 2). For the one-component vitreous boron oxide, T_g decreases with increasing the water content. The SAXS intensity for the samples in the supercooled state is independent of the water content. According to Equation (1), the product $\rho^2\chi_T$ is independent of the water content.

In sodium borate glasses, water leads to increasing the sizes of *RI* and decreasing the $<(\Delta\rho)^2>$. In glasses under study, the regions of inhomogeneity are composed of high alkali tetraborate groups. Hence water promotes aggregation of alkali borate structural groups. As precipitation of these groups is completed at 450–500°C and their size remain unchanged at cooling down the sample, water influences the structure of melts catalysing aggregation of alkali borate structurals groups.

Decreasing $<(\Delta\rho)^2>$ can be a result of equalisation of densities of the high alkali and high borate components of the glass. It becomes possible if either the density of the high borate component increases due to increased content of alkali borate groups in it or if the density of high alkali regions decreases due to the presence of water. An increase of $<(\Delta\rho)^2>$ can follow

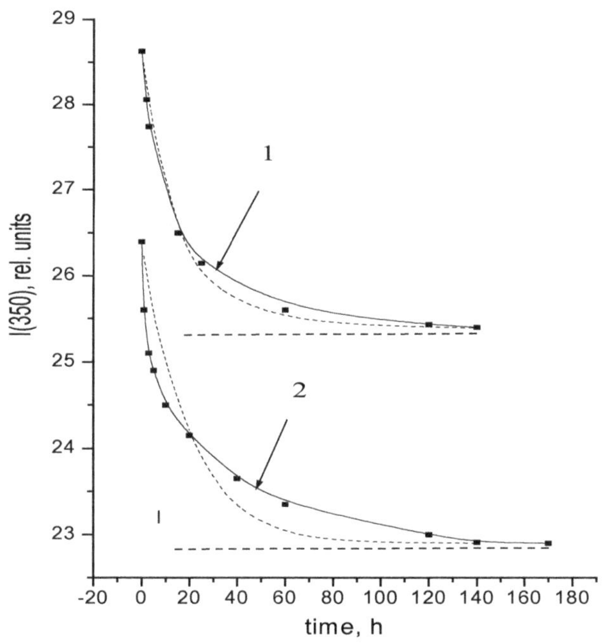

Figure 3. Dependence of the SAXS intensity at a scattering angle of 350° on heating time at temperature 60°C below T_g for a vitreous boron oxide containing different water contents: (1) dry glass, (2) glass containing 0·5 mol% H_2O. Dash lines are exponential dependences for the same values of I_∞, $I(0)$, and τ, short dot lines correspond to I_∞

from the appearance of additional number of small groups. It is equivalent to an increase of the SAXS intensity by concentration fluctuations and can be an origin of the increase of I(350°) upon decreasing the water content in sodium borate glass (Figure 1).

4.2. Influence of water on relaxation kinetics

The dependence of the SAXS intensity on heating time for water containing glasses can be regarded as a superposition of fast and slow relaxations. It implies that these glasses contain regions with different relaxation times probably due to different water contents in these regions. In the regions with increased water content relaxation times are shorter and consequently relaxation rates are higher. Probably the glass transition range should also depend on the water content.

The regions with fast and slow relaxations would be expected to differ in density, which should result in additional small angle scattering. However, this difference in densities is probably very small and cannot be detected even for a one-component vitreous boron oxide.

In spite of the fact that there are no direct structural confirmations of water distribution in glass, the character of the SAXS intensity dependence on heating time is a convincing prove of existence of microinhomogeneities caused by an irregular water distribution in a glass volume, i.e., with appearance of regions with different water contents.

5. Conclusions

For boron oxide containing from 0–1 mol% H_2O the SAXS intensity in the supercooled liquid state is independent of the water content. This means that the level of TDF is independent of the water content for the viscosity change more than two orders of magnitude.

In microheterogeneous sodium borate glasses an increase of the water content leads to enlargement of the *RI* and to increasing the SAXS intensity by the thermal density and concentration fluctuations, and to decreasing $<(\Delta\rho)^2>$ of microheterogeneous structure.

Kohlrausch's equation describes the kinetics of SAXS intensity by the TDF upon isothermal heating. In dry glasses β in this equation is close to one and decreases up to 0·5 in glasses containing 0·7–1·0 mol% H_2O both in one component boron oxide and in sodium borate glasses.

A decrease of β can be explained by an appearance of the regions with pronouncedly different effective relaxation times, which can be a result of different water contents in these regions. Consequently, in water containing glasses additional microheterogeneous structure appears. However we failed to estimate the structural parameters, particularly, the sizes of *RI*, probably due to the fact that water weakly influences the glass density.

Acknowledgments

This work was supported by the Russian Foundation for Basic Research (projects no. 03-03-32269, 04-03-32657 and 05-03-08013-ofi_p), and by Grant NSH-1405.2003.3 of President of Russian Federation.

References

1. Visser, T. & Stevels, J. M. *J. Non-Cryst. Solids*, 1972, **7**, 394.
2. Golubkov, V. V. J. *Non-Cryst. Solids*, 1995, **192–193**, 463.
3. Golubkov, V. V. In *Borate Glasses, Crystals and Melts*. Edited by A. C. Wright, S. A. Feller and A. C. Hannon. The Society of Glass Technology, Sheffield, 1997, P. 88–98.
4. Golubkov, V. V. *Glass Phys. Chem.*, 1998, **24** (3) 196.
5. Golubkov, V. V. & Porai-Koshits, E. A. *Fiz. Khim. Stekla*, 1991, **17** (5), 796.
6. Andreev, N. S., Mazurin, O. V., Porai-Koshits, E. A., Roskova, G. P. & Filipovich, V. N. *Phase separation in glasses*. Edited by O. V. Mazurin and E.A. Porai-Koshits, North-Holland; M. M. Shultz, Nauka, Leningrad, 1974.
7. Golubkov, V. V., Titov, A. P., Porai-Koshits, E. A. *Prib. Techn. Eksp*. 1975, **215** (1).

MD study of sodium borate glasses containing Al_2O_3

Norikazu Ohtori,[1] *Masanao Togashi, Keiichi Takase*

Graduate School of Science and Technology, Niigata University, Niigata 950-2181, Japan

Katsumi Handa, Junko Ide

Ritsumeikan University, Shiga 525-8577, Japan

Efstratios I. Kamitsos

Theoretical and Physical Chemistry Institute, NHRF, 48, Vass, Constantinou Ave., 116 35 Athens, Greece

K. Itoh, T. Fukunaga

Research Reactor Institute, Kyoto University, Osaka, 590-0494, Japan

Norimasa Umesaki

Japan Synchrotron Radiation Research Institute, Hyogo 679-5198, Japan

The short range structure of $0{\cdot}05Al_2O_3.xNa_2O.(0{\cdot}95-x)B_2O_3$ glasses ($0{\cdot}10<x<0{\cdot}45$) has been investigated using molecular dynamics calculations employing a fully ionic model. Regardless of x, the coordination number of oxygen atoms around the Al atom was approximately four. It was found that these Al atoms are present as network formers of AlO_4 units in the glasses in the same manner as are BO_4 units, and that they cause a decrease of the coordination number of O around Na. It is suggested that the exclusively formed AlO_4 units may suppress the conversion of the BO_3 unit to the BO_4 unit caused by an increase of Na_2O. The decrease of the fraction of four coordinated boron atoms, N_4, due to the addition of Al_2O_3 is supported by the results obtained from preliminary neutron diffraction measurements.

1. Introduction

The compositions of alkali and alkaline earth borate mixtures have limited glass forming ranges. Their behaviour is somewhat complicated: it is known that certain alkaline earth borate mixtures have non-glass forming ranges at a low alkaline earth metal oxide content.[(1)] Aluminium oxide, Al_2O_3, has been empirically known as an effective additive for the extension of the glass forming range of these glasses. Although this means that Al ions should effectively enhance the stability of the vitreous state of alkali and alkaline earth borates, the change of the glass structure and the role of Al ions have not yet been elucidated from the viewpoint of the stability of the glasses.

So far, through the use of molecular dynamics (MD) and neutron diffraction methods, we have focused on the constitutional units in the network to investigate the short range structure of a series of alkaline earth borate glasses in the entire glass-forming range, compared with that for sodium and potassium borate glasses.[(2–4)] The results showed the presence of well defined triangular and tetrahedral constitutional units, BO_3 and BO_4, respectively, in these glasses. As well, the behaviour of the fraction of BO_4 unit of all boron atoms, N_4, was elucidated. It follows $x/(1-x)$ at a low metal oxide content, where the x indicates the molar fraction of metal oxide, but tends to deviate from this, as the radius of metal cation decreases in the higher x range.

On the basis of these findings, therefore, it is important to investigate the changes in short range structures such as the constitutional units (changes which are caused by small amounts of additives such as Al_2O_3) and to clarify the role of a modifier such as Al in the network, in regard to the stability of the glass structure. For this purpose, in the present study we have performed MD simulation for sodium borate glasses containing 5 mol% Al_2O_3 using fully ionic potentials, and have investigated the short range structure around B, Al, and Na. Furthermore, we present some results from our current neutron diffraction experiments, and compare results of simulation and experiment, focusing

[1] Corresponding author. Email address: ohtori@chem.sc.niigata-u.ac.jp

on $5Al_2O_3.29Na_2O.66B_2O_3$ glass. A comprehensive report on the neutron diffraction experiments will be presented elsewhere.

2. Calculation and experimental procedure

2.1. MD calculation

The simulated compositions of the alumina sodium borate glass system were $0{\cdot}05Al_2O_3.xNa_2O.(0{\cdot}95-x)B_2O_3$ with x=0·10, 0·19, 0·29, 0·38, and 0·45, where x denotes the mole fraction of Na_2O. All MD calculations were carried out using the *NPT* ensemble. The total number of ions, *N*, was chosen to be slightly over 1000. The pressure, *P*, during the calculations had been adjusted using the Andersen method[5] so as to maintain the density evaluated from the experimentally determined values for each real glass, using a pycnometer method, within ±0·01 g/cm^3. The temperature was controlled using the Nose–Hoover thermostat.[5] The function used for pair potential energy, U_{ij}, between the ions *i* and *j* was a form of the Busing approximation of the Born–Mayer–Huggins' type without the dispersion terms. The function is given as

$$U_{ij}\left(r_{ij}\right)=\frac{e^2}{4\pi\varepsilon_0}\frac{Z_iZ_j}{r_{ij}}+f_0\left(b_i+b_j\right)\exp\left(\frac{a_i+a_j-r_{ij}}{b_i+b_j}\right) \quad (1)$$

where *Z* is the ionic charge, *e* is the unit charge, *r* is the interionic distance, and f_0 is a constant (=6·9472×10^{-11} N). Since the model used in this study is fully ionic, Z_B=+3, Z_O=−2, Z_{Na}=+1, and Z_{Al}=+3. The parameters, *a* and *b*, are the values related to the radius and the compressibility of each ion, respectively, and their values for each ion were those proposed by Xu *et al*,[6] as shown in Table 1. The Coulomb interactions were treated using the Ewald method[7] under periodic boundary conditions.

The initial atomic configuration was generated randomly and the initial atomic velocities were also provided randomly in accordance with the Maxwell distribution of velocities. The equations of motion were integrated using the RESPA algorithm.[8] Initially, the system was kept at 5000 K for 303 ps with a calculation time step of 1 fs in order to obtain a fully disordered configuration in the equilibrium state. The molten state was confirmed by checking the mean square displacement of the atoms. Subsequently, it was gradually cooled to 300 K for vitrifying at a rate of 50 K/ps; the time step was 1 fs at above 3000 K, and was 2 fs below 3000 K. In the process of cooling, the system was kept at 3000, 2000, 1000 and 300 K for 40, 40, 40, 40, and 800 ps, respectively, to relax the structure. The trajectories of the positions for the subsequent 100 ps were analysed to obtain the structural information. On the basis of the obtained pair distribution functions, $g_{ij}(r)$, the partial static structure factor, $S_{ij}(Q)$, in the Faber–Ziman form[9] was evaluated for each pair of ions *i* and *j*, as follows

Table 1. Potential function parameters used in the MD calculations

Ion	Z	*a* (nm)	*b* (nm)
Na	1	0·1260	0·0080
Al	3	0·1064	0·0080
B	3	0·0720	0·0080
O	−2	0·1629	0·0085

$$S_{ij}(Q)=1+4\pi\rho_0\int_0^{r_{max}} r^2\left[g_{ij}(r)-1\right]\frac{\sin(rQ)}{rQ}dr \quad (2)$$

where ρ_0 is the average number density of atoms and r_{max} is the cutoff value of *r* for the Fourier transformation. The neutron weighted static structure factor, *S*(*Q*), is given by

$$S(Q)=\frac{\sum_i\sum_j x_ix_jb_ib_jS_{ij}(Q)}{\left(\sum_k x_kb_k\right)^2} \quad (3)$$

where b_i is coherent scattering length of the *i*th nuclear species and x_i is the atomic fraction. The total structure function, $Q{\cdot}i(Q)$, is defined as follows :

$$Q{\cdot}i(Q)=Q(S(Q)-1) \quad (4)$$

The neutron weighted total pair distribution function, $g^N(r)$, is expressed by

$$g^N(r)=\sum_i\sum_j x_ix_jb_ib_jg_{ij}(r)/\left(\sum_k x_kb_k\right)^2 \quad (5)$$

The coordination numbers, n_{BO} and n_{OB}, for O around B and B around O were estimated from the integration of $g_{BO}(r)$ using the upper integration limit, r_{cutoff}, of 0·200 nm. In the same way, the n_{AlO} and n_{NaO} for O around Al and Na were obtained from corresponding $g_{ij}(r)$ with a r_{cutoff} of 0·200 and 0·300 nm, respectively.

2.2. Sample preparation

The compositions of the glass samples used in the present neutron diffraction study were $0{\cdot}05Al_2O_3.xNa_2O.(0{\cdot}95-x)B_2O_3$ with x=0·29, 0·38, and 0·45. These glasses were made from analytical reagent powders of H_3BO_3, Na_2CO_3, and Al_2O_3. The H_3BO_3 was enriched with 99·27 at% ^{11}B to avoid neutron absorption by ^{10}B. The details of sample preparation are described in a previous paper.[4]

2.3. Neutron diffraction

Pulsed neutron diffraction experiments were carried out using a time-of-flight method employing the High Intensity Total scattering spectrometer (HIT-II) at the KENS Neutron Scattering Facility of the High Energy Accelerator Research Organization (KEK), Tsukuba, Japan. Each sample was contained in a thin walled vanadium cell that was 8 mm in inner diameter and 25 μm in thickness. The neutrons scattered from each sample filled in the cell at 298 K were detected at nominal angles 2θ of 15, 20, 25, 30, 50, 90, and 150°, which covered a region of momentum transfer, $hQ/2\pi$, where $Q=(4\pi/\lambda)\sin\theta$ and ranged from 1–500 nm^{-1}. The observed intensity was corrected for the cell, background, absorption, and multiple and incoherent scattering. The coherent scattering intensity, $I(Q)$, was normalised with respect to the scattering intensity from a vanadium rod used as a standard. The total structure factor, $S(Q)$, in the Faber–Ziman form was derived from the $I(Q)$, and the neutron weighted total pair distribution function, $g^N(r)$, was obtained from the $S(Q)$. The average coordination number of O around B, n_{BO}, was based on the simple integration of $g^N(r)$ for the B–O correlation. The upper limit of integration, r_{cutoff}, was fixed at 0·170 nm for the glasses containing the additive in order to avoid the Al–O correlation, while the n_{BO} was evaluated both at r_{cutoff}=0·170 and 0·200 for the glasses not containing the additive for comparison with the MD results. The experimental error of n_{BO} was estimated to be ±0·05 from the reproducibility. The value of N_4 can be determined experimentally from n_{BO} using the relation:

$$n_{BO}=3(1-N_4)+4\,N_4=N_4+3 \qquad (6)$$

The details of the analytical procedure are provided in a previous paper.[4]

3. Results and discussion

The pair distribution functions, $g(r)$, obtained from MD calculation are shown in Figure 1(a) for $5Al_2O_3.29Na_2O.66B_2O_3$ glass. Although the first peak position in all these functions is reasonably close to the approximate sum of ionic radii of each pair, the peak height is approximately twice the experimental value in the neutron weighted radial distribution function, $g^N(r)$, as shown in Figure 1(b), owing to the use of a fully ionic model. However, since this is due simply to a strong ionic interaction that should suppress excessively interionic thermal fluctuations, these disagreements are not problems critical to discussion in this paper.[2–4] Actually, the first peak positions in the simulation and experiment are in close agreement, as shown in Figure 1(b). Furthermore, as shown in Figure 2, the neutron-weighted total structure function from MD simulation is in good agreement with that derived from the experiment, although strong oscillations observed up to the high Q range in the MD result are also caused by the intense first B–O and O–O peaks in the $g^N(r)$, as shown in Figure 1(b).

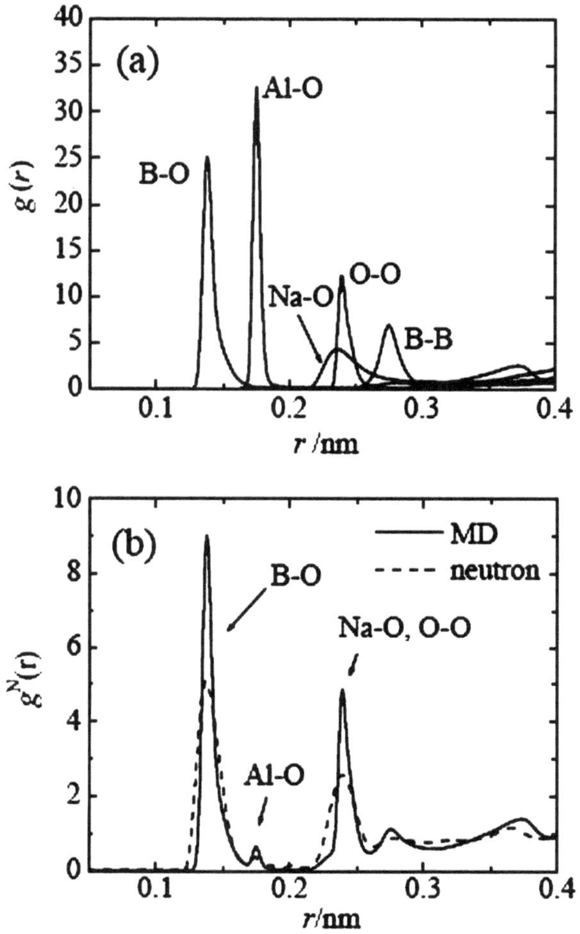

Figure 1. (a) Pair distribution functions obtained from MD calculation for $5Al_2O_3.29Na_2O.66B_2O_3$ glass. (b) Neutron weighted radial distribution functions, $g^N(r)$, from MD calculation for $5Al_2O_3.29Na_2O.66B_2O_3$ glass. The dashed line denotes $g^N(r)$ from the neutron diffraction experiment

Table 2 shows the coordination numbers evaluated from the B–O, Al–O, and Na–O pair distribution functions. The coordination number of O around Al, n_{AlO}, was found to be approximately four, regardless of x. This shows that Al atoms are usually present as AlO_4 units in these glasses, which was confirmed also by direct evaluation of coordination number distribution of O around Al. Theses results are consistent with those from NMR studies.[10,11] Both the n_{BO} and n_{OB} for the glasses added with Al_2O_3 are smaller than those without Al_2O_3. As a result, the fraction of four-coordinated B, N_4, decreases on addition of Al_2O_3 for the entire simulated x range, as is shown in Figure 3(a).

On the other hand, the fraction of four-coordinated B and Al atoms, N_4, was evaluated without distinction between B and Al, and shows the same quantitative dependence on x as that for the glasses without Al_2O_3, except for the high x range, as isshown in Figure 3(b).

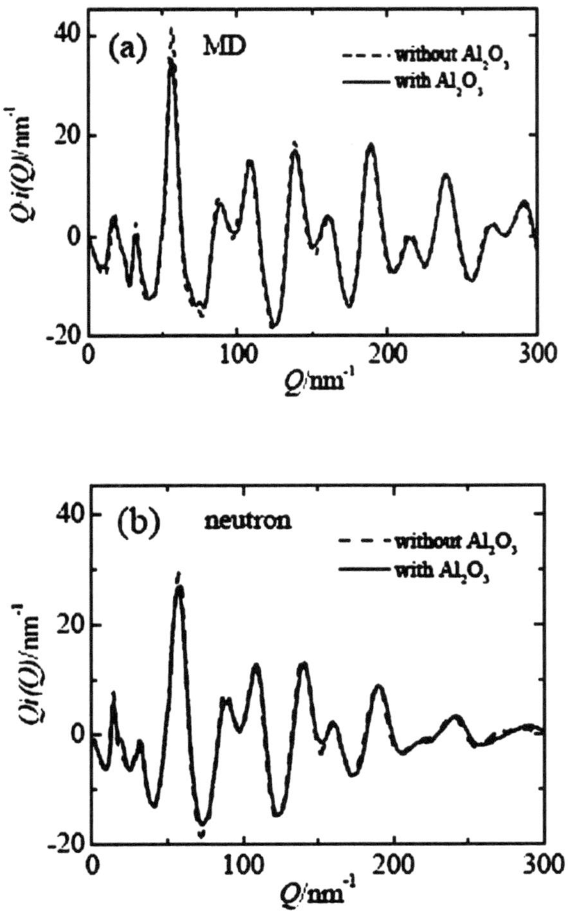

Figure 2. Comparison of structure functions $Q{\cdot}i(Q)$ between (a) MD calculation and (b) neutron diffraction for $5Al_2O_3.29Na_2O.66B_2O_3$ glass. The dashed line denotes $Q{\cdot}i(Q)$ for $30Na_2O.70B_2O_3$ glass

Furthermore, in the same way as described above, the coordination numbers of B and Al around O, n_{OBAl}, obtained without distinction between B and Al, also show the same dependence as the n_{OB}. That is, the fraction of nonbridging oxygen (NBO) evaluated without distinction between B and Al shows the same x-dependence quantitatively as that for the glasses without Al_2O_3, as is shown in Figure 4, except for

Table 2. Average coordination numbers from MD calculation and neutron diffraction

Composition	$n_{BO}^{(a)}$	$n_{BO}^{(b)}$	$n_{OB}^{(a)}$	$n_{OB}^{(b)}$	$n_{OBAl}^{(a)}$	$n_{AlO}^{(a)}$	$n_{NaO}^{(a)}$
B_2O_3	3·00	-	2·00	-	-	-	-
$5Al_2O_3.10Na_2O.85B_2O_3$	3·07	-	1·86	-	2·00	4·05	6·30
$5Al_2O_3.19Na_2O.76B_2O_3$	3·18	-	1·83	-	1·99	4·01	6·19
$5Al_2O_3.29Na_2O.66B_2O_3$	3·33	3·22	1·82	1·76	1·98	4·05	6·38
$5Al_2O_3.38Na_2O.57B_2O_3$	3·44	3·24	1·77	1·67	1·93	4·03	6·07
$5Al_2O_3.45Na_2O.50B_2O_3$	3·46	3·21	1·66	1·54	1·84	4·00	5·70
$10Na_2O.90B_2O_3$	3·11	3·11	2·00	2·00	-	-	6·41
$20Na_2O.80B_2O_3$	3·24	3·23	1·99	1·98	-	-	6·55
$30Na_2O.70B_2O_3$	3·38	3·37	1·97	1·96	-	-	6·43
$40Na_2O.60B_2O_3$	3·53	-	1·93	-	-	-	6·63
$50Na_2O.50B_2O_3$	3·61	-	1·80	-	-	-	6·42

(a) MD calculation, (b) Neutron diffraction experiment

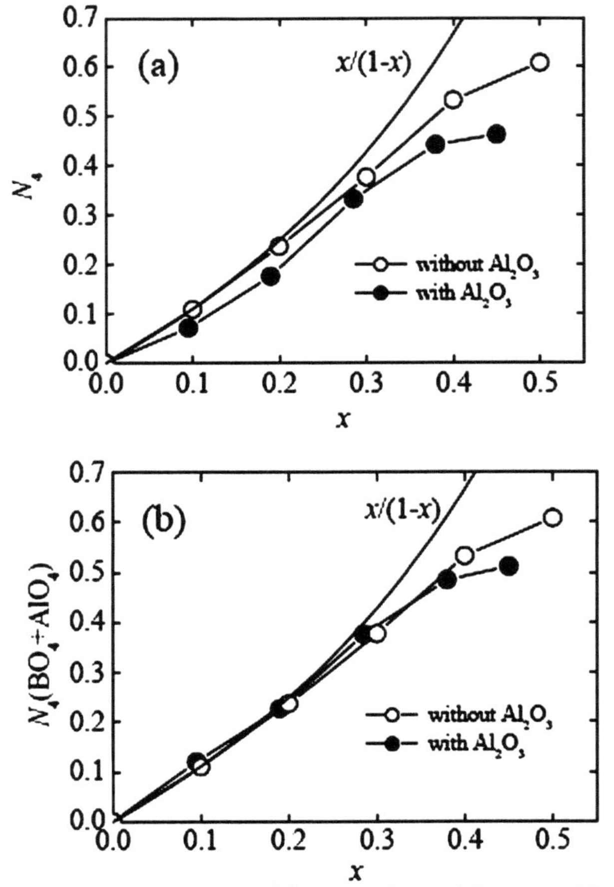

Figure 3. (a) Fraction of four-coordinated B atoms, N_4, and (b) that of four-coordinated B and Al atoms, N_4 (BO_4+AlO_4), as a function of the mole fraction of alkali oxide content

the high x range. These results show that Al atoms participate in the B–O glass network as glass-forming AlO_4 units.

If all the B atoms are three-coordinated by O in pure B_2O_3 glass, and all the O atoms added as Na_2O into B_2O_3 participate in the formation of four-coordinated B atoms by bridging two boron atoms, then N_4 follows $x/(1-x)$, as shown in the low x region in Figure 3. Actually, O atoms supplied by a small addition of metal oxide to borate glass convert $BØ_3$ to $BØ_4^-$ units, where Ø indicates the bridging oxygen atom. However, as the added amount increases, the formation of units containing NBO such as $BØ_2O^-$ could occur even at low Na_2O content as shown in Figure 4,[12,13] (where O denotes NBO), and N_4 then deviates downward from $x/(1-x)$. Therefore, the results described above suggest that the exclusive formation of AlO_4 units from Al atoms could suppress the conversion from $BØ_3$ to $BØ_4^-$ units, compared with the glasses without Al_2O_3. It is concluded that Al_2O_3 has an effect of suppressing the formation of the BO_4 unit in the case of sodium borate glasses with a composition of x~0·4.

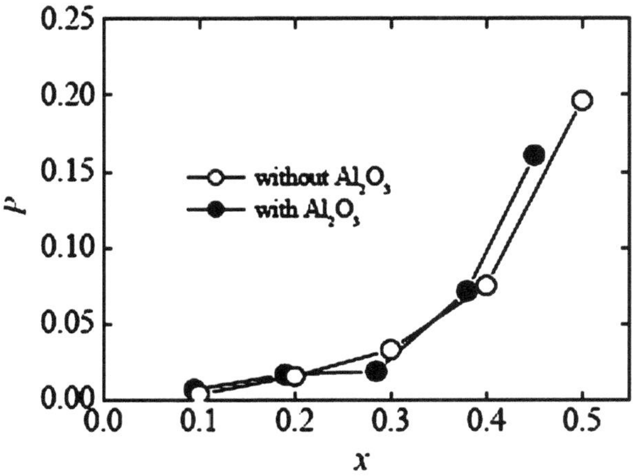

Figure 4. Fraction of nonbridging oxygen, p, evaluated without distinction between B and Al as a function of the mole fraction, x, of alkali oxide content

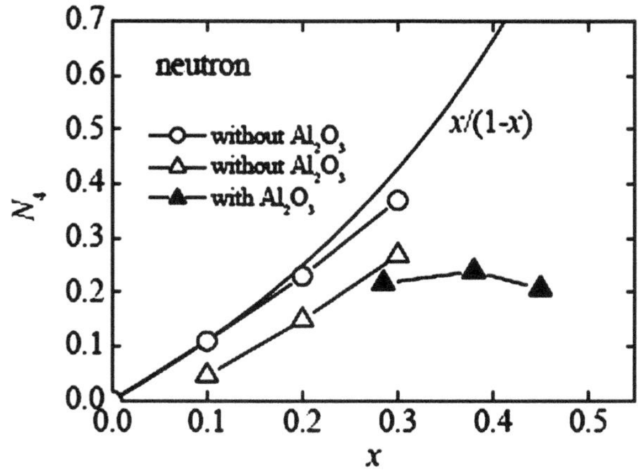

Figure 5. Fraction of four-coordinated B, N_4, obtained from neutron diffraction as a function of the mole fraction of alkali oxide content. The values evaluated using the upper integration limit of 0·170 and 0.200 nm are indicated by triangles and circles, respectively

Although the coordination number of O around Na, n_{NaO}, could not be definitively estimated since the coordination shell is somewhat ambiguous, the values were estimated with a given r_{cutoff}=0·30 nm. The n_{NaO} for the glasses with Al_2O_3 shows a decrease by a few percent on average from those for the glasses without Al_2O_3. This result suggests that the additive Al_2O_3 inclines to draw O atoms around Na and to form AlO_4 units.

Figure 5 shows the N_4 obtained from neutron diffraction experiments as a function of x. Although the r_{cutoff} was always 0·200 nm in MD simulation, it was 0·170 nm for the experimental results of glasses with Al_2O_3 and 0·200 nm for those without Al_2O_3, since the B–O correlation cannot be separated from the Al–O correlation in the former case. The N_4 was then also evaluated using a r_{cutoff} of 0·170 nm in the experimental results of the glasses without Al_2O_3. In both cases, the decrease in N_4 resulting from the addition of Al_2O_3 is seen in a comparison of N_4 between the glasses with and those without Al_2O_3. Therefore, both the simulation and experimental results show similar behaviour regarding the decrease in N_4 due to the addition of Al_2O_3.

4. Conclusions

The short range structure of $0{\cdot}05Al_2O_3.xNa_2O.(0{\cdot}95-x)B_2O_3$ glasses (0·10<x<0·45) has been investigated using MD calculations. The n_{AlO} was approximately four, regardless of x. On the basis of the analysis of other coordination numbers and the fraction of four-coordinated B and Al, it was concluded that these Al atoms are present as network-formers of AlO_4 units in the glasses in the same manner as the BO_4 unit and causes a decrease of the coordination number of O around Na. This summary suggests that the exclusively formed AlO_4 units could suppress the conversion of the BO_3 unit to the BO_4 unit caused by an increase of Na_2O. The decrease of the fraction of four-coordinated boron atoms, N_4, due to the addition of Al_2O_3 was supported by the results obtained from preliminary neutron diffraction measurements.

Acknowledgements

The authors thank Professors M. Misawa & K. Maruyama and Dr E. M. Anghel for their enlightening discussion. This work was supported by a Grant-in-Aid for Scientific Research from the Ministry of Education, Science, Sports and Culture of Japan (No. 16560599).

References

1. Yiannopoulos, Y. D., Kamitsos, E. I., Chryssikos, G. D. & Kapoutsis, J. A. *Proc Second Int. Conf. on Borate Glasses, Crystal and Melts*, 1997, 514.
2. Ohtori, N., Takase, K., Akiyama, I., Handa, K., Iwadate, Y. & Umesaki, N. *Phys. Chem. Glasses*, 2000, **41** (6), 369–72.
3. Ohtori, N., Takase, K., Akiyama, I., Suzuki, Y., Handa, K., Sakai, I., Iwadate, Y., Fukunaga, T. & Umesaki, N. *J. Non-Cryst. Solids*, 2001, **293–295**, 136.
4. Suzuki, Y., Ohtori, N., Takase, K., Handa, K., Itoh, K., Fukunaga, T. & Umesaki, N. *Phys. Chem. Glasses*, 2003, **44** (2), 150–4.
5. Martyna, G. J., Toblas, D. J. & Klein M. L. *J. Chem. Phys.*, 1994, **101**, 4177.
6. Xu, Q., Kawamura, K. & Yokokawa, T. *J. Non-Cryst Solids*, 1988, **104**, 261.
7. Allen, M. P. & Tildesley, D. J. *Computer simulation of liquids*. Oxford University Press, Oxford, 1987, Pp 156–182.
8. Martyna, G. J., Tuckerman, M. E., Toblas, D. J. & Klein, M. L. *Mol. Phys.*, 1996, **87**, 1117.
9. Waseda, Y. *The Structure of Non-Crystalline Materials*. McGraw-Hill, 1986, Pp 8–13.
10. Bertmer, M., Züchner, L., Chan, J. C. C. & Eckert, H. *J. Phys. Chem. B*, 2000, **104**, 6541.
11. Du, L.-S. & Stebbins, J. F. *Solid State Nucl. Magn. Reson.*, 2005, **27**, 37.
12. Varsamis, C.-P. E., Vegiri, A. & Kamitsos, E. I. *Phys. Rev. B*, 2002, **65**, 104203.
13. Vegiri, A., Varsamis, C.-P. E. & Kamitsos, E. I. *J. Chem. Phys.*, 2005, **123**, 14508.

Proc. Fifth Int. Conf. on Borate Glasses, Crystals and Melts Phys. Chem. Glasses: Eur. J. Glass Sci. Technol. B, August 2006, **47** (4), 328–331

A solution approach to high alkali content borate glasses

Joy Banerjee,[1] *Greg Ongie, Jacob Harder, Trenton Edwards, Chris Larson, Scott Sutton, Anthony Moeller, Abhirup Basu, Mario Affatigato, Steve Feller*

Physics Department, Coe College, Cedar Rapids, IA 52402, USA

Masao Kodama

Department of Applied Chemistry, Sojo University, 4-22-1 Ikeda, Kumamoto, 860-0082, Japan

The glass forming range of alkali borates has been extended to R=5·0 (83 mol% alkali oxide) using the solution method devised by Kodama. This method involves the reaction between solutions of boric acid (H_3BO_3) and alkali hydroxide (MOH). The densities, molar volumes, packing fractions, and glass transition temperatures (T_g) were determined. We have obtained results for the entire alkali borate system including lithium, sodium, potassium, rubidium and cesium. This method has the advantage of forming high alkali content borate glasses which are practically devoid of carbon dioxide retention.

1. Introduction

Alkali borate glasses [$R(M_2O).B_2O_3$, where M is the alkali metal] are generally made by fusing alkali carbonates and boric acid or boron oxide. The physical properties and glass formation ranges of these glasses have been determined, even though the alkali content in the glass was limited due to carbon dioxide retention.(1) Using the solution method,(2) we made glasses in the alkali borate system with *R*-values ranging from 0·2 to 5·0, where *R* is the molar ratio of alkali oxide to boron oxide in the glass, and we determined their densities and glass transition temperatures. Molar volumes and packing fractions of these glasses were then ascertained using the density data and ionic radii.

This is a continuation and completion of the study of potassium borates by Moeller *et al.*(3) Their research was used to produce potassium borates and study their physical properties using the solution method. Their data showed that the molar volumes of the glasses were consistent with the change in the coordination number of boron from three to four and back to three with the increase in alkali content. Carbon dioxide retention from alkali carbonates can change the structure and give inconsistent results for density and glass transformation temperatures. However, due to the absence of carbonates in the solution method, there is no carbon dioxide retention unless it is due to absorption from air. The solution method is not the first attempt to bypass the carbon dioxide retention in alkali borate glasses. Royle *et al*(4) prepared alkali borate glasses using alkali oxides which allowed for extension of the composition of alkali borate glasses beyond those previously achieved using alkali carbonates. Royle made alkali borates up to *R*=3·0 except for sodium borate which was prepared to *R*=3·5. However, this method was cumbersome as the starting materials were very hygroscopic and the glass samples needed to be prepared under an inert environment. Another problem that they encountered was the loss of 10–15% of the initial alkali content when heated to 1000°C before quenching. Finally, not all the oxides were readily available. The solution method, on the other hand, needs an inert environment only in the last heating before quenching. These advantages of using alkali hydroxides instead of alkali oxides makes the solution method is an ideal successor to the preparation of high alkali content borate glasses.

2. Experimental procedure

The starting materials for $R(M_2O).B_2O_3$ are H_3BO_3 (ACS reagent, 99·5%+, Aldrich) and MOH (50% solution, Aldrich). While making the H_3BO_3 solution (1g H_3BO_3/100 g water), it was heated for 15 min during solution stirring as this facilitated quicker dissolving. After mixing the boric acid solution with alkali hydroxide in a Teflon beaker, we dried them at 130°C for periods ranging from a day to a week. The high alkali content borate precipitate generally took 4–5 days to dry. A slight deviation in procedure occurred

[1] Corresponding author

in lithium borates because LiOH (98%+ powder, Aldrich) is only available in powder and has to be made into a solution to be used efficiently. Lithium hydroxide required careful handling due to the fine nature of the powder.

The validity of this method was verified by producing low alkali content borate glasses to compare their properties against glasses produced using the conventional alkali carbonates and traditional melts. The low alkali content borate glasses were made in platinum crucibles by heating the precipitates two or three times in an electric muffle furnace at 1000°C for 15 min. These melts were then plate quenched to form bulk, clear, colourless glass. Physical properties of these glasses were comparable to literature data.[(5)]

Using earlier procedures for the preparation of glasses using alkali hydroxides,[(3)] glasses of $R>1{\cdot}0$ were made in vitreous carbon crucibles as the sample started reacting noticeably with the platinum crucible. We used about 10–12 g of precipitate to yield sufficient glass for physical characterisation tests. For cesium, potassium and rubidium samples, the first heating was done in air at 650°C for 10 min in an electric muffle furnace to remove most of the water from the precipitate. The second and third heatings were done in an inert N_2 environment at 900°C for maximal expulsion of moisture from the sample. These melts were then roller quenched to make bulk, clear, and colourless glass as plate quenching did not offer the rate of cooling needed to make these glasses. Since high alkali content borate glasses were very hygroscopic, we performed several heatings and then observed that most of the water was driven out in the first two heatings at 900°C. The water expulsion in the third and fourth heatings were negligible. Lithium and sodium samples followed a slightly different sequence of heatings. The first heating was done in air at 650°C and the other heating was done at 900°C in an inert N_2 environment for 20 min. Since these glasses were easier to make than the heavier alkali borates, the melts were able to be plate quenched up to R=3·0. The glasses beyond this molar ratio were roller quenched to make samples that contained more glass than crystal. Using the solution method and available means of rapid quenching, the glass formation for alkali borates ended at R=5·0.

The densities of the glasses were measured using a Quantachrome Ultrapycnometer 1000 and an analytical balance. Multiple runs were averaged to obtain a reliable density. The sample pycnometer runs were sandwiched between two high purity aluminum runs for calibration purposes. The error is estimated to be ±1%. The T_gs were obtained using a Perkin Elmer Differential Scanning Calorimeter 7 (DSC). The samples were heated at 40°C/min from 50°C to 600°C and the T_g was found using the onset method. The error achieved was ±5°C.

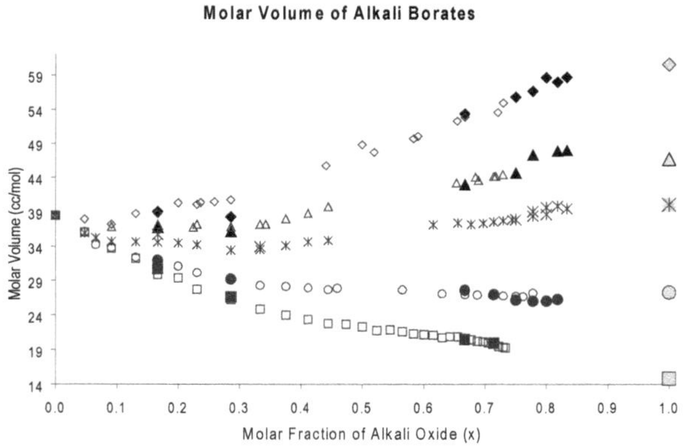

Figure 1. Molar volume of alkali borates in cc/mole given as a function of the molar fraction of alkali oxide (x), where the small open symbols are the literature values[5] for the system, and the filled large points are the current solution method experimental data. ◆ represent the cesium system, ▲ represent the rubidium system, ✱ represents the potassium system, ● represents the sodium system, and ■ represents the lithium system. Also, the large symbols with grey background at x=1 are values deduced from alkali oxide crystals. The error in the data points are comparable to the size of the data points

3. Results

In this study, we obtained glasses and densities for low R ($0{\cdot}0\leq R\leq 0{\cdot}5$) and high R ($2{\cdot}0\leq R\leq 5{\cdot}0$) alkali borates, from which we calculated the molar volume (Figure 1) and packing fraction (Figure 2). In addition to density, we also measured T_gs for the low R, and some high R alkali borate glasses in order to compare our results to studies using carbonate and oxide starting materials.[(5)] These results, as a function of R, are listed in Table 1.

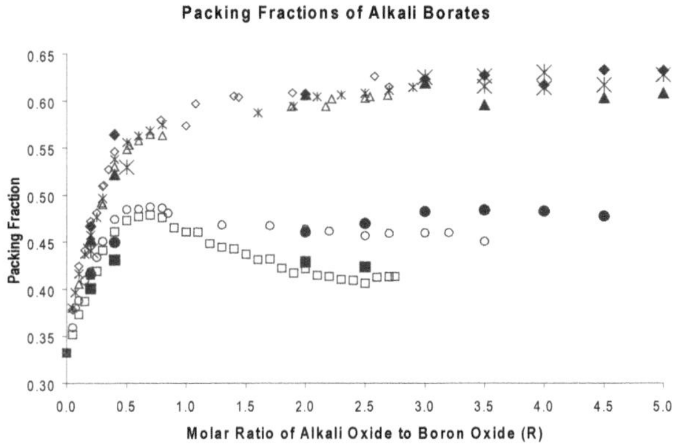

Figure 2. The packing fractions of the alkali borates given as a function of the molar ratio of alkali oxide to boron oxide (R), where the small open symbols are the literature values for the system, and the filled large points are the current solution method experimental data. ◆ represents the cesium system, ▲ represents the rubidium system, ✱ represents the potassium system, ● represents the sodium system, and ■ represent the lithium system. The error in data is comparable to the size of the data points

Table 1. The experimental data and calculated values for the alkali borate systems found from literature values[3,5] and the present study including density, T_g, molar volume, and the packing fraction. Literature values presented are in the neighbourhood of current experimental data

R Value Value	Source	Molar fraction alkali oxide (x)	ρ (in g/cc) Error: ±1%)	T_g (in K) (Error: ±5 K)	Packing fraction
Lithium					
0·2	E	0·17	2·05	708	0·40
0·2	L	0·17	2·11	707	0·41
0·4	E	0·29	2·19	784	0·45
0·4	L	0·29	2·22	769	0·46
2·0	E	0·67	2·11	572	0·42
2·0	L	0·67	2·12	554	0·42
2·5	E	0·71	2·07	543	0·41
2·5	L	0·71	2·07	-	0·41
2·8	L	0·74	-	546	-
3·0	E	0·75	-	540	-
5·0	E	0·83	-	453	-
Sodium					
0·20	E	0·17	2·14	-	0·42
0·20	L	0·17	2·14	702	0·42
0·40	E	0·29	2·31	753	0·47
0·40	L	0·29	2·32	-	0·47
2·00	E	0·67	2·34	-	0·45
2·00	L	0·67	2·39	530	0·46
2·50	E	0·71	2·38	508	0·46
2·50	L	0·71	2·38	516	0·46
3·00	E	0·75	2·44	-	0·47
3·00	L	0·75	2·39	505	0·46
3·50	E	0·78	2·45	-	0·47
3·50	L	0·78	2·34	-	0·45
4·00	E	0·80	2·44	-	0·47
4·50	E	0·82	2·41	-	0·47
Potassium					
0·20	E	0·17	2·06	675	0·44
0·20	L	0·17	2·13	675	0·46
0·50	E	0·33	2·29	707	0·55
0·50	L	0·33	2·32	-	0·56
2·90	L	0·74	2·32	-	0·61
3·00	E	0·75	2·32	-	0·62
3·50	M	0·78	2·31	-	0·62
3·50	M	0·78	2·27	-	0·61
3·50	M	0·78	2·31	-	0·62
4·00	M	0·80	2·31	-	0·62
4·00	M	0·80	2·25	-	0·61
4·00	M	0·80	2·31	-	0·62
4·50	M	0·82	2·25	-	0·61
4·50	M	0·82	2·25	-	0·61
5·00	M	0·83	2·28	-	0·62
Rubidium					
0·20	E	0·17	2·43	670	0·46
0·20	L	0·17	2·40	-	0·45
0·40	E	0·29	2·85	697	0·54
0·40	L	0·29	2·80	723	0·53
1·88	L·O	0·65	3·38	-	0·59
2·00	E	0·67	3·44	-	0·60
2·17	L·O	0·68	3·40	-	0·60
3·00	E	0·75	3·53	-	0·61
3·15	L·O	0·76	-	495	-
3·50	E	0·78	3·40	-	0·59
4·50	E	0·82	3·45	-	0·60
5·00	E	0·83	3·48	-	0·60
Cesium					
0·20	E	0·17	2·69	630	0·47
0·20	L	0·17	2·71	648	0·47
0·40	E	0·29	3·40	657	0·58
0·40	L	0·29	3·20	703	0·55
2·00	E	0·67	3·95	-	0·60
2·00	L·O	0·67	3·99	-	0·61
2·85	L·O	0·74	-	501	-
3·00	E	0·75	4·10	-	0·62
3·21	L·O	0·76	-	492	-
3·28	L·O	0·77	-	480	-
3·50	E	0·78	4·14	-	0·62
3·62	L·O	0·78	-	483	-
4·00	E	0·80	4·08	-	0·61
4·50	E	0·82	4·19	-	0·63
5·00	E	0·83	4·20	-	0·63

Symbols:
E - Present experimental data
L·O - Literature data (Alkali oxide)[5]
L - Literature data (Alkali carbonates)[5]
M - Experimental data by A. Moeller[3]

We speculate that the differences between our density results and the literature values from carbonates may be due to the absence of CO_2 retention in our glasses. However, the differences between literature and current experimental values were close to the experimental error in our measurements.

The T_g data of the alkali borates show consistency with the literature values[5] of glasses prepared with both carbonate and oxide starting materials. For each system, the T_g trends show a maximum near R=0·4, mirroring the abundance of the four-coordinated borons in the glass.[6]

4. Discussion

To further analyse the structural changes the glass undergoes, we determined molar volume, V_{mol}, using the following formula[7]

$$V_{mol} = \frac{MolWt\left[R(M_2O)\cdot B_2O_3\right]}{(1+R)(\rho)} \quad (1)$$

where *MolWt* represents the molecular weight of the glass, and ρ represents the experimental density. Note that the factor (1 + R) is used in the denominator due to the fact there are R moles of alkali oxide and one mole of B_2O_3.

In Figure 1, lithium borate and sodium borate glasses show a monotonic decrease in the molar volume due to the formation of tetrahedral units which are more compact than three-coordinated trigonal planar units until x=0·5 where x is the molar fraction of alkali oxide in the glass. Beyond x=0·5, although four-coordinated borate units convert to three-coordinated borate units with one nonbridging oxygen, three coordinated borate units with two nonbridging oxygens, and three coordinated borate units with three nonbridging oxygens, there are increasing amounts of lithium and sodium ions in the glass structure per mole, and since these ions are small and interstitial, the molar volume keeps decreasing, but at a lesser rate. However, in cesium, rubidium, and potassium borate glasses, there is an initial nearly constant molar volume in the range of tetrahedral boron formation followed by a monotonic increase because the larger size of the ions dominates the volume with increase in cesium, rubidium and potassium content.

Another indicator of structural information is the packing fraction, defined as the volume of the ions present divided by the total volume of the glass, including empty space. The packing fraction, p, is calculated using[7]

$$p = density \times \sum \frac{\left(\frac{4}{3}\pi r_i^3 n_i\right)}{MolWt} \qquad (2)$$

where r_i is the radius of the ith ion, and n_i is the amount of that ion present in the structural formula. In our calculations, Shannon radii[8] were used for the alkali ions, while some other radii were obtained from x-ray and neutron scattering results.[7] The values of these radii are listed in Table 2.

The packing fractions of the alkali borates are listed in Table 1 and graphed as a function of R in Figure 2. The packing fractions of the alkali borates follow either 'ionic' packing or 'covalent' packing, the former being dominated by the alkali ion present and the latter being dominated by the oxygen network.[7] Cesium, rubidium, and potassium borates, having alkali with Shannon radii larger than the oxygen radius, pack ionically; sodium and lithium borates, having alkali with Shannon radii smaller than the oxygen radius, pack covalently.

The ionic packing dominated by alkali ions (see Figure 2) have a much higher packing fraction than the covalent packing oxygen network, approaching the packing of random spheres (≈64%). In the region, $0 \leq R \leq 0{\cdot}4$, each packing trend shows a sharp increase as the triangular planar units convert into compact tetrahedral units. Following this region, the packing divides into the two kinds of packing given above. An observation in sodium borate glasses is the apparent difference in the packing fractions obtained from current experimental solution method data and literature values. At present, no explanation is given.

A tangible method for examining our high-R data is to look at packing of alkali oxide crystals, since approximately 75–85% of these samples are alkali oxide. The packing fractions of high alkali content borate glasses from the current experimental solution method data approach asymptotically the packing fractions of alkali oxide crystals.

Table 2. The Shannon radii[8] and other ionic radii[7] with their calculated volumes

	Li_4^{+1}	Na_6^{+1}	K_8^{+1}	Rb_9^{+1}	Cs_{10}^{+1}	B_3	B_4	$O_{2\ or\ 1}$
Radius (in Å)	0·73	1·16	1·65	1·77	1·95	0·15	0·25	1·21
Volume (in Å^3)	1·63	6·54	18·82	23·23	31·06	0·014	0·075	7·42

Subscripts refer to the coordination number

5. Conclusion

The formation of alkali borate glasses has been extended to R=5·0 (83 mol% alkali oxide) using a solution method employing alkali hydroxides. The densities and T_gs of these glasses were measured, and molar volumes and packing fractions were deduced from our measurements. The packing fractions of the alkali borates are split into two separate groups based on their bonding, one being an ionically dominated group (K, Rb, Cs) and other being covalently dominated (Li, Na). Our experimental data follow literature values.[5]

The solution method is similar to using alkali oxides though the solution method utilises alkali hydroxides instead. Solutions of alkali hydroxides are easier to use as they are stable in air as opposed to the hygroscopic nature of alkali oxides. Since the methods are similar and produce glasses without CO_2 retention, their data can be compared to draw conclusions about glass structure for binary alkali borates.

Acknowledgments

The United States NSF is acknowledged for funding under grants numbered DMR 0211718 and DMR 0502051. Coe College is thanked for housing and other support.

References

1. Zhang, H., Koritala, S., Farooqui, K., Boekenhauer, R., Bain, D., Kambeyanda, S. & Feller, S.A. A model for carbon dioxide retention in alkali borate and borosilicate systems. *Phys. Chem. Glasses*, 1991, **32** (5), 185–7.
2. Kodama, M. & Kodama, S. Velocity of sound and elastic properties of alkali borate glasses. *Proc. Second Int. Conf. on Borate Glasses, Crystals, and Melts*. Edited by A. C. Wright, S. A. Feller, & A. C.Hannon. Society of Glass Technology, Sheffield. 1996, 181–8.
3. Moeller, A. R., Olesiak, M. D., Jian Gao, Giri, S. K., Affatigato, M., Feller, S. A., & Kodama, M. A solution approach to high potassium content borate glasses. *Phys. Chem. Glasses*, 2003, **44** (3), 237–40.
4. Royle, M., Sharma, M., Feller, S., MacKenzie, J. & Nijhawan, S. Densities and structural models resulting from extremely modified cesium and rubidium borate glasses. *Phys. Chem. Glasses*, 1993, **34** (4), 149–52.
5. Feller, S. A., Kottke, J., Welter, J., Nijhhawan, S., Boekenhauer, R., Zhang, H., Feil, D., Parameswar, C., Budhwani, K., Affatigato, M., Bhatnagar, A., Bhasin, G., Bhowmik, S., Mackenzie, J., Royle, M., Kambeyanda, S., Pandikuthira, P., & Sharma, M. Physical properties of alkali borosilicate glasses. *Proc. Second Int. Conf. On Borate Glasses, Crystals, and Melts*. Edited by A. C. Wright, S. A. Feller & A. C. Hannon. Society of Glass Technology, Sheffield. 1996, 246–53).
6. Shelby, J. E. *Introduction to Glass Science and Technology*. RCS, Cambridge, UK. 1997, 90.
7. Giri, S., Gaebler, C., Helmus, J., Affatigato, M., Feller, S. & Kodama, M. A general study of packing in oxide glass systems containing alkali." *J. Non-Cryst. Solids*. 2004, **347**, 87.
8. Shannon, R. D. *Acta. Crystallogr.*, 1976, **A32**, 751.

Preparation and x-ray characterisation of $(Li_2O.WO_3)_{1-x}(B_2O_3)_x$ glasses

Z. K. Heiba,* A. Abou Shama & Karimat El-Sayed

Physics Department, Faculty of Science, Ain Shams University, 11566 Abbassia, Cairo, Egypt

Glasses in the system $(Li_2O.WO_3)_{1-x}(B_2O_3)_x$ were synthesised with equal molar ratios in Li_2O and WO_3 and 0·0≤x≤0·65 by the usual rapid quenching technique. From the analysis of x-ray diffraction data, it was found that for 0·0≤x≤0·25 the materials formed are crystalline, while for x≥0·30, glasses are obtained. The Rietveld method is used to characterise the crystalline systems obtained. On the other hand, the glassy phases are investigated by analysing the experimental radial distribution function RDF. The first observed real space correlations for the sample x=0·35 were located at 1·454 Å, 1·736 Å and 2·538 Å which are attributable to B–O, W–O and B–B atomic pairs, respectively. The corresponding atomic correlations for x=0·55 were at 1·479 Å, 1·960 Å and 2·289 Å. Concerning the boroxol group, in x=0·55 it is found to be in form of BO_3 connected units having both nonbridging oxygen (NBO) and bridging oxygen (BO) types. In case of x=0·35, the units BO_4 are formed indicating the change in the formed network units in medium range order MRO as a composition dependent.

Introduction

Coloured glasses can generally be divided into two classes according to whether the generation of colour centres is reversible or irreversible. Both classes have specific applications, the most prominent being the reversible darkening of sunglasses[1,2] and the irreversible coloring of utility glasses.[3,4] Lithium borate tungstate glasses also have a colour centre of both types; a reversible photochromic blue coloration has been observed in the presence of hydrogen and was later attributed to the formation of tungsten bronzes H_xWO_3.[5] A widespread set of very different borate glasses with optical, superionic conductivity, and other technologically interesting properties are currently produced.[6] The optimisation of such properties as a function of composition and other preparation parameters requires a good knowledge of the microscopic glassy structure. A precise knowledge of the structure and properties of B_2O_3 based glasses and melts is increasingly required from both fundamental and industrial points of view and, in particular the structure of borate glasses at high temperature which is still controversial.[7] Among the binary borate glasses, alkali borate glasses have long been paid considerable attention by many researchers because of the "borate anomaly" in the compositional dependence of some physical properties. The structure of alkali borate glasses have been extensively studied by XRD, NMR, Raman spectroscopy, etc.[7] Also the introduction of WO_3 in the alkali borate glasses opened a new broad range of technological interests for more industrial findings for these amorphous materials and also their crystalline phases.

Experimental and analytical procedure

The system $(Li_2O.WO_3)_{1-x}(B_2O_3)_x$ with 0·0≤x≤0·65 has been prepared by the usual rapid quenching technique. X-ray diffraction data were collected with a Philips X'pert diffractometer applying Cu K_α radiation for the crystalline phases and Mo K_α for amorphous ones, with a graphite monochromator in the diffracted beam.

Qualitative phase analysis of obtained diffraction patterns was carried out using the ICDD diffraction file and the search-match program from Philips software. Quantitative phase analysis was carried out applying the MAUD 1.999 software[8] (Lutterotti, 2002) based on Rietveld method. The RDF analysis for the two samples x=0·35 and 0·55 was carried out applying the Fourier transform. The simulation of the observed RDF peaks was carried out using a programs package (implemented in the storage ring laboratory at Orsay-Lure-France). The simulated radial distance R, the coordination number N, and the disordering parameter σ, are determined for the best fittimg between the observed and calculated RDF. The Gaussian form of the RDF peaks was the usual applied form.[7,9]

Results and discussion

Figure 1 shows the x-ray diffraction patterns of the system $(Li_2O.WO_3)_{1-x}(B_2O_3)_x$ with 0·05≤x≤0·65. As

[1] Corresponding author. Email Zein_Kh@gawab.com

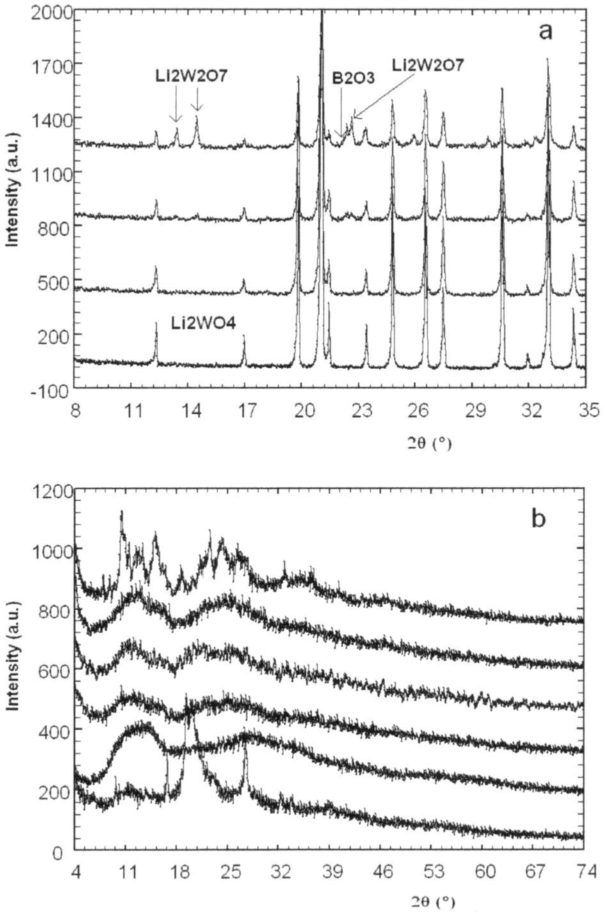

Figure 1. The diffraction patterns obtained for the system $(Li_2O.WO_3)_{1-x}(B_2O_3)_x$; (a) x=0, 0·15, 0·20, 0·25 and (b) x=0·30, 0·35, 0·40, 0·50, 0·55, 0·65 from bottom to top, respectively

shown in Figure 1(a), for low content of the glass former B_2O_3, up to x=0·25, the phases formed are well crystalline while for x≥0·30 the samples obtained are amorphous. Applying the Search-match program for crystalline phases, it is found that for x<0·20, the patterns are found to be a single phase which is Li_2WO_4, while for x=0·2 and 0·25 three phases could be identified Li_2WO_4, $Li_2W_2O_7$ and B_2O_3. The quantitative phase analysis gave the weight percentages 94·3, 3·9, 1·8 for x=0·2 and 73·0, 25·0, 2·0 for x=0·25 for the three phases, respectively. The profile fitting resulting from Rietveld analysis is depicted in Figure 2. Comparing these values with those initially intended during preparation, we found that the B_2O_3 weight content (1·8, 2·0%) is far less than should be present (12, 15%). We may safely claim that B is incorporated as substitutional or interstitial in Li_2WO_4 lattice. However, due to the small x-ray scattering power of boron atoms, attempts to accommodate B interstitially or substitutionally during Rietveld refinement were failed. Further investigation using neutron diffraction may be needed.

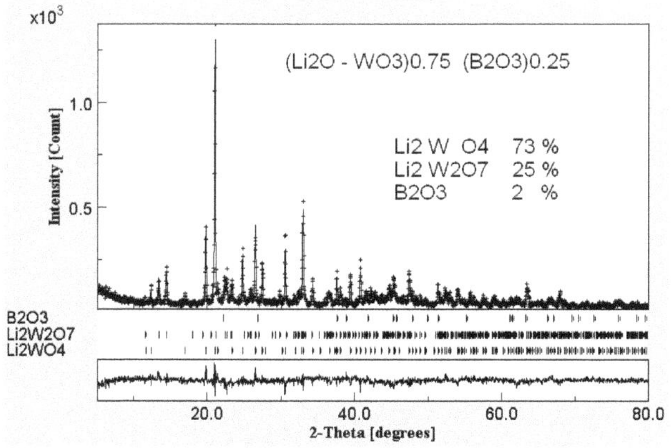

Figure 2. The profile fitting of the sample $(Li_2O.WO_3)_{0·75}(B_2O_3)_{0·25}$ resulting from Rietveld analysis

Inspecting Figure 1(b), one can observe that for x=0·30 as well as x=0·65 a degree of crystallisation is highly declared. For the other mid curves x=0·35, 0·45, 0·50 and 0·55, amorphous phases are well formed with a distinct two broad humps in the small 2-theta range.

The reduced RDF, $G(R)$, versus R is depicted in Figure 3, for x=0·35 and 0·55. The Li–O pairs are not observed in the RDF results because the scattering power of Li is much weaker than that of other atoms in the matrix.[7] Moreover, Li^+ ions are embedded in the amorphous matrix of the borate tungstate within the hexagonal tunnels of the connected W–O tissue and can be cycled among W–O and B–O pairs.[10] In the SRO, the main dominant peak for x=0·35 at 1·736 Å is due to W–O atomic pairs, and this peak hides the expected B–O correlated pairs at ~1·45 Å. The simulation of this peak gives a coordination number ~5·0 for W–O and ~3·0 for B–O pairs. The asymmetric tail detected at longer R-side indicates that the connected W–O pairs are

Figure 3. The reduced RDF versus R (Å) for the system $(Li_2O.WO_3)_{1-x}(B_2O_3)_x$, x=0·35 and 0·55

Table 1. The first coordination shells with their coordination numbers for the two samples x=0·35 and 0·55

x	R_1	N_1	R_2	N_2	R_3	N_3	R_4	N_4	R_5	N_5	R_6	N_6
0·35	1·454	~2·60	1·736	~5·0	--	--	2·538	3·150	3·410	4·969	3·584	3·100
0·55	1·479	~2·63	1·960	2·500	2·289	2·458	2·756	3·512	3·200	2·559	3·658	7·360

$R\pm0{\cdot}0449$ Å, $N_i\pm0{\cdot}150$

arranged in highly distorted octahedral units. The well resolved peak at 2·538 Å with a lower intensity is due to correlated B–B pairs. For the x=0·55 sample, the first peak is of composite structure and it is resolved to more than one peak with a pronounced shoulder at about 1·479 Å, which revealing the connected B–O pairs. The next peaks at 1·960 Å and 2·289 Å could be attributed to W–O pairs with coordination number 2·5 and 2·46, respectively. This splitting in the main W–O peak and the shift to a longer R side in x=0·55 compared with x=0·35 reveals more distortion in the W–O octahedral units. Also, the peak of B–B pairs appeared at ~2·54 Å for x=0·35 is shifted to ~2·75 Å in x=0·55.

In the MRO, the observed two maxima at 3·41 Å and 3·584 Å in x=0·35 represent the correlated B–O pairs. The first B–O pairs represents nonbridging oxygen type (NBO) and the other one, of longer distance, represents bridging oxygen type (BO). These B–O pairs are evidenced in the sample x=0·55 as two unresolved shoulders at 3·20 Å and 3·658 Å. The observed trough around 3·4 Å in x=0·55 may refer to abnormal change which may be evolved in the MRO in the network when x changes from 0·35 to 0·55. Inspecting the corresponding coordination numbers in Table 1 in the MRO at R_5, an interesting observation could be extracted concerning the conversion of the boroxol group BO_3 in x=0·55 to BO_4 in x=0·35 which reflects the role of Li_2O as an alkali modifier in the given amorphous matrix. This finding is confirmed by the obtained optical properties for x=0·35 and 0·55 to be published later.

Conclusions

1. The system $(Li_2O.WO_3)_{1-x}(B_2O_3)_x$ is formed in a pure crystalline phase for $0{\cdot}0\leq x\leq0{\cdot}25$, while for $0{\cdot}30\leq x\leq0{\cdot}65$ the obtained phases are amorphous.
2. For x<0·20, a single phase which is Li_2WO_4, while for $0{\cdot}2<x<0{\cdot}3$, the phases formed are Li_2WO_4, $Li_2W_2O_7$ and B_2O_3.
3. In the SRO, the W–O correlated pairs are in form of distorted octaheda.
4. The correlated B–O pairs in the MRO are found to be of both BO and NBO types. In x=0·55, the boroxol group is formed of BO_3 units and in x=0·35, the conversion of BO_3 to BO_4 units is a composition dependent.

References

1. Wollenhaupt, M., Ahrens, H., Frobel, P., Barner, K., Giessinger, E. R. & Braunstein, R. *J. Non-Cryst. Solids*, 1996, **194**, 191.
2. Deb, S. K. *Phil. Mag.*, 1973, **22**, 801.
3. Braunstein, R. & Barner, K. *Solid State Commun.*, 1978, **28**, 847.
4. Braunstein, R. & Barner, K. *Solid State Commun.*, 1980, **33**, 941.
5. Faughnan, B. W., Crandall, R. S. & Heyman, P. M. *RCA Review*, 1975, **36**, 177.
6. Ruf, Ch., Barner, K. & Braunstein, R. *Solid State Commun.*, 1985, **54**, 11.
7. Umesaki, N., Kita, Y., Iida, T., Handa, K., Kohara, S., Suzuya, K., Fukunaga, T. & Misawa, M. *Phys. Chem. Glasses*, 2000, **41**, 304.
8. Lutterotti, L. 2003 Maud 1.999, http://www.mg.unitn.it/~Luttero/maud.
9. Abou Shama, A. *Egypt. J. Solids*, 2005, **28** (2), 337.
10. Senthil Murugan G. & Varma, K. B. R. *J. Non-Cryst. Solids*, 2001, **279**, 1.

Elastic and thermal properties of lithium borate glass system

Yuji Ike,[1,a] *Yu Matsuda,*[a] *Chihiro Matsui,*[c] *Masao Kodama*[b] *& Seiji Kojima*[a, c]

[a] *Graduate School of Pure and Applied Sciences, University of Tsukuba, Tsukuba, Ibaraki 305-8573, Japan*
[b] *Department of Applied Chemistry, Sojo University, Ikeda 4-22-1, Kumamoto 860-0082, Japan*
[c] *Engineering Sciences, University of Tsukuba, Tsukuba, Ibaraki 305-8573, Japan.*

Elastic and thermal properties of lithium borate glasses xLi_2O–$(1-x)B_2O_3$ *(x=0·06–0·28) were investigated by means of Brillouin scattering spectroscopy and modulated differential scanning calorimetry (MDSC). The temperature dependences of elastic modulus and absorption coefficient of lithium borate glasses were measured from 25 to 600°C. The glass system showed a stiffening tendency and a higher onset temperature of softening with increasing mole fraction of* Li_2O*. We successfully observed the change in the elastic modulus at the glass transition temperature* (T_g)*. The elastic parameters showed larger temperature dependences above* T_g*, while they change little below* T_g*. From MDSC measurement,* T_g *and the profile of heat capacity exhibited significant composition dependence. With increasing concentration of* Li_2O*,* T_g *increased and the temperature range of* T_g *became narrower. These results might be due to the change in some amount of the crosslinking borate units from triangle* BO_3 *to tetrahedral* BO_4*. The Cole–Cole plot of complex heat capacity was examined and well fitted by Davidson–Cole function. The parameter derived from the function was determined and β was 0·58 at x=0·18. This non-Debye nature is explained by the variation of intermediate range structures of borate glass.*

Introduction

Since physical properties of alkali metal borate glasses exhibit maxima or minima against their compositions called borate anomalies, the composition dependence of structure and physical properties of alkali borate glasses are interesting. The random network of B_2O_3 glass is mostly constructed by BO_3 planar triangle units. The addition of alkali metal oxide to the B_2O_3 glass causes the formation of BO_4 tetrahedral groups by crosslink between the BO_3 group units to build the glass network. Such properties have been studied by several experimental techniques (nuclear magnetic resonance, Raman spectroscopy, etc.).[(1–4)]

The concentration of BO_4 group units increases up to around x=0·25 and the rate is closely expressed by $x/1-x$.[(5)] This configuration change affects the B_2O_3 glass network. It causes a stiffening of the structure and this fact is proved by the increase in the elastic moduli.[(6)] Regarding to the ultrasonic study of lithium borate glasses, the measurement of elastic property plays a significant role in understanding the structural characteristics of glass network. The elastic parameters related to density are sensitive and indicative about the changes occurring in the structure of glass network. However, the measurement of elastic properties in the high temperature region is scarce. Therefore, the elastic properties such as velocity, elastic modulus and attenuation of composition and temperature dependencies are great interest. Especially, these properties at the glass transition temperature have attracted much attention.

Brillouin scattering is known to be a powerful tool to examine condensed materials, not only to probe the high frequency elastic properties, but also as a method for exploring the elastic properties in an extremely wide temperature range. Another technique for studying the glass transition phenomenon is calorimetry. The most common calorimetric tool for identifying a glass transition temperature (T_g) has been differential scanning calorimetry (DSC). Furthermore, recent improvement has been achieved in the measurement of a frequency dependent specific heat, and it can be used to demonstrate both a glass transition and relaxation processes.[(7)] The method of temperature modulated differential scanning calorimetry (MDSC) involves superimposing a small sinusoidal temperature perturbation onto the linear temperature variation, which is used in the conventional DSC technique. In this study, the elastic and thermal properties of several compositions of lithium borate glass have been investigated by using the Brillouin scattering and MDSC in the hope of better understanding the glass transition phenomena.

Experimental

The composition formula of lithium borate glass is denoted by $x\mathrm{Li_2O}.(1-x)\mathrm{B_2O_3}$, where x indicates

[1] Corresponding author

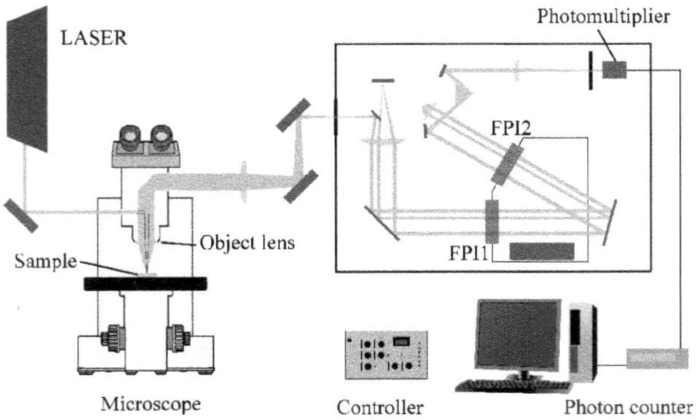

Figure 1. Schematic diagram of the micro-Brillouin setup. The objective lens focuses the incident laser beam on to a sample and also corrects the scattered light from a sample. The collimated scattered light passes through two Fabry–Perot interferometers six times. The transmitted light is detected by a photomultiplier tube and acquired a Brillouin spectrum

the mole fraction of Li_2O (x=0·06–0·28). All glasses were prepared with high homogeneity in order to investigate the inherent nature of the binary glass. The starting materials were analytical reagent grade $LiOH.H_2O$ and H_3BO_3. To achieve the high homogeneity, the starting materials were made to react first in an aqueous solution. The aqueous solution was prepared in a beaker of poly(tetrafluoroethylene) and a chemically synthesised lithium borate powder was obtained after complete evaporation of water in a dry box. Details of the high homogeneity glass preparation method can be seen elsewhere.[6]

The experimental setup of a micro-Brillouin scattering apparatus is shown in Figure 1. The Brillouin scattering spectra were measured at a backward scattering geometry. The light source was the diode-pumped solid-state laser (DPSS532, Coherent), operating in a single mode at 532 nm with 150 mW. A conventional Sandercock-type 3+3 pass tandem Fabry-Perot Interferometer (FPI) was operated to acquire the spectra of the scattered light. The free spectral range of 60 GHz was applied to measure the Brillouin spectra, and applied scan range was between ±50 GHz. A cryostat cell (THMS 600, LINKAM) was used to control the temperature from 25°C to 600°C. Temperature stability was within ±0·1°C below 500°C.

The MDSC measurements were performed by a DSC 2920, TA Instruments and provided glass transition temperatures of different concentrations of lithium borate glass. The sample was powdered by a mortar and poured in a standard aluminum pan. The mass of the samples weighed approximately 20 mg. The sample was then put into the sample chamber at room temperature. Dry nitrogen gas was used as a purge gas with a flow rate of 20 ml/min, and flows through the sample chamber during the measurement. In order to obtain accurate measurements, calibrations were performed by examining the heat of fusion of highly pure metals of indium, tin and zinc. Also the heat capacity calibration was accomplished by the use of a sapphire disc as standard. The temperature control procedure for measurements was programmed to heat from 200°C through a glass transition temperature up to 550°C. The heating rate was 5°C/min with the modulation amplitude of ±1°C every 100 s. From the modulated heat flow signals one can obtain the complex heat capacity $C_p^*=C_p'-iC_p''$ which separates into the real (C_p') and imaginary (C_p'') parts.

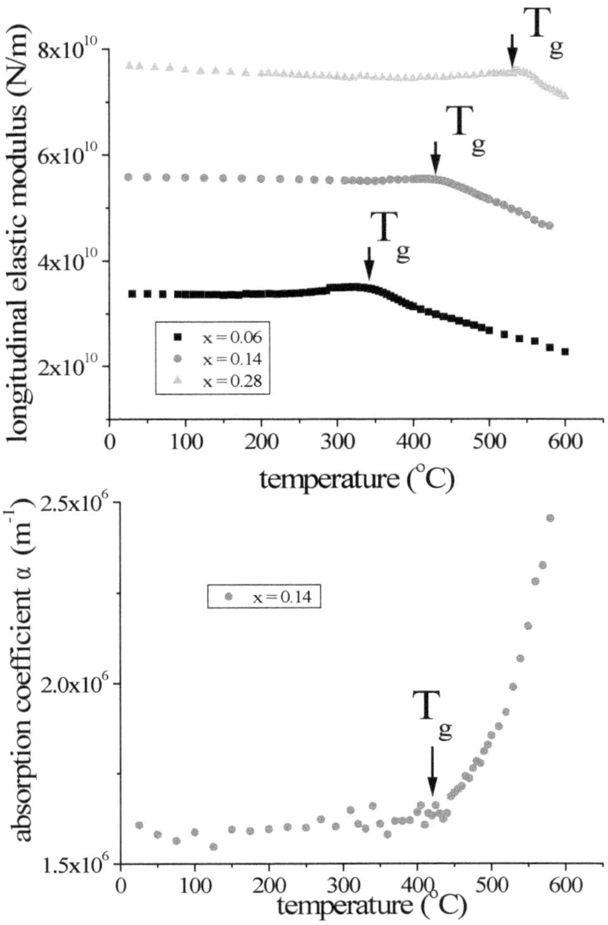

Figure 2. The temperature dependence of derived longitudinal elastic modulus and absorption coefficient of the LA mode in lithium borate glasses

Results

The Brillouin spectra acquired for the lithium borate glasses showed a longitudinal acoustic (LA) mode. The Brillouin shifts and the full width at half maximum (FWHM) have been obtained from the spectra by using the fitting function which takes in consideration the instrumental function. The longitudinal velocity of sound V_l was obtained from the Brillouin shift ν (=$\mathbf{q}V_l/2\pi$) and the scattering wave vector $\mathbf{q}$

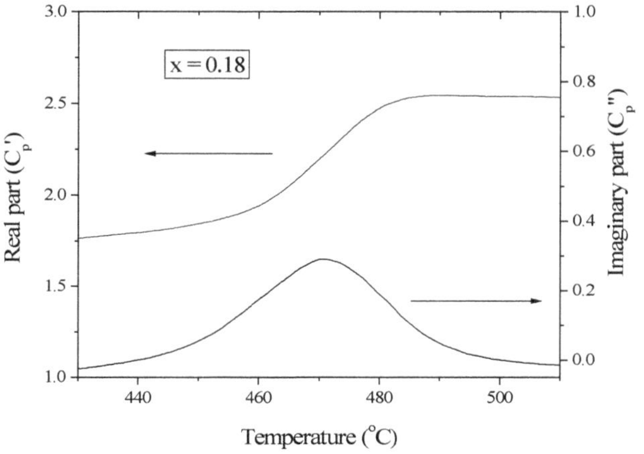

Figure 3. The real part of heat capacity (C_p') and imaginary part of heat capacity (C_p'') obtained from lithium borate glass at x=0·18

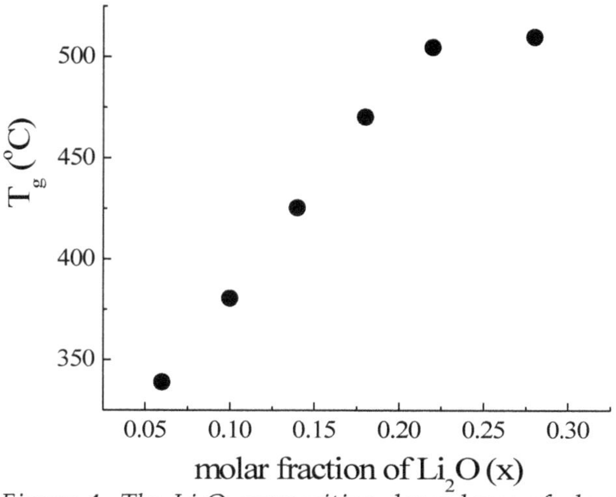

Figure 4. The Li_2O composition dependence of glass transition temperatures which determined from the peak temperature of C_p''

as being (=$2\pi n\sin(\theta/2)/\lambda$), where n, λ and θ are the refractive index of the sample, the wavelength of the laser and scattering angle, respectively. The longitudinal elastic modulus M is derived by $M=\rho V_l^2$, where ρ is the density. The absorption coefficient of LA mode α can be determined by the FWHM of Brillouin component using the equation, $\alpha=\pi\Gamma/V_l$ where Γ is FWHM. The temperature dependence of derived elastic modulus of lithium borate glasses (x=0·06, 0·14 and 0·28) are plotted in Figure 2.

The complex heat capacity was decomposed into the real and imaginary parts. Figure 3 represents the real part (C_p') and imaginary part (C_p'') of the lithium borate glass at x=0·18. The curve of C_p' shows a step-like increase and C_p'' has a peak around T_g which is the characteristic feature of a glass transition phenomenon. T_g was derived from the centre of peak temperature in C_p''. Figure 4 shows the glass transition temperatures as a function of the composition of lithium borate glasses. These characteristic temperatures change linearly with increasing the Li_2O fraction and have a maximum at Li_2O fraction around 0·28. Another parameter, defining the glass transition temperature region, (ΔT), is defined as a temperature between the onset and the end of a glass transition. These points were determined by extrapolations of the line in C_p' shown in Figure 3.

Discussion

Figure 2 indicates the elastic modulus becomes larger with increasing mole fraction of Li_2O, which leads to the stiffening of the borate glass network. This might be due to the change in some amount of the cross-linking borate units from triangle BO_3 to tetrahedral BO_4 on addition of Li_2O to the borate glass matrix. This transformation of the glass network leads to a more rigid glass. Also we clearly observed the curve kinked in the present temperature range, the kinked point being a glass transition temperature, T_g. The point of inflection, T_g, is shifted to higher temperatures with increasing the Li_2O content. This tendency is also due to variations in the glass network, and a higher T_g describes the tightening process of the borate glass network. Above T_g, the slope of curves become steeper. These glasses are softened by increasing temperature while the dependence below T_g is constant. Another elastic parameter, α exhibits the relation to the elastic modulus and has a huge increase beyond T_g which is in a supercooled temperature region. We observed this behavior is in common with all different concentrations of lithium borate glass.

From the results of thermal measurements, the composition dependence of ΔT was obtained which shown in Figure 5 and it becomes narrower with increasing the lithium fraction. The additive effect in borate system is related to the change in fragility.[8]

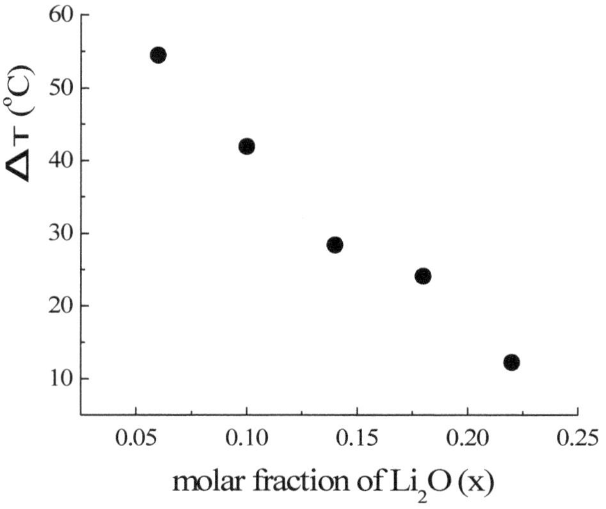

Figure 5. The glass transition temperature range ΔT as a function of mole fraction of Li_2O

Further analysis was examined to identify the distribution function in the vicinity of T_g in lithium borate glass. Figure 6 shows the Cole–Cole plot of lithium borate glass with the composition of x=0·18 and it is noted that the shape of the curve shows a distorted semicircle. This distributed relaxation profile is well expressed by Davidson–Cole equation[9]

$$C_p = C_{p\infty} + \frac{C_{p0} - C_{p\infty}}{(1+i\omega\tau)^{\beta}} \quad (1<\beta<1) \qquad (1)$$

The distributed Davidson–Cole curve well describes the experiment data, and the parameter obtained was β=0·58. The result indicates well the feature of glass shows a dispersion from a single relaxation time. This non-Debye nature can originate from the variation of structural units in the random glass network structure.

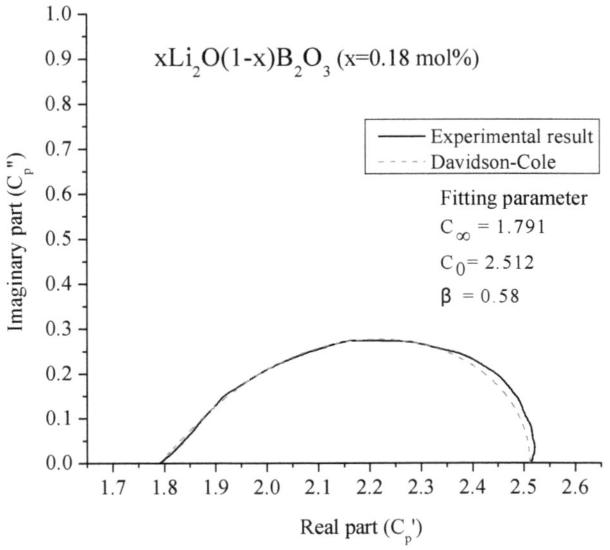

Figure 6. Cole–Cole plot of complex heat capacity of lithium borate glass (x=0·18)

Summary

In the present study, we have investigated the elastic and thermal properties of lithium borate glasses over a wide temperature range. At the glass transition temperature, the elastic modulus and absorption coefficient of the LA mode show an obvious change in its behaviour between that above and below the transition temperature. The composition dependence in the thermal properties also show a remarkable difference at the glass transition temperature and in its transition regime. These results might be due to the change in crosslinking borate units from triangle BO_3 to tetrahedral BO_4 by adding the Li_2O. This transformation of borate units leads to the glass more rigid and fragile compared to the strong pure borate glass. By using the analogous analysis to the dielectric relaxation, the Cole–Cole plot was examined and was well fitted by the Davidson–Cole function. The distribution parameter was determined, and the β was 0·58. This non-Debye property may come from the diversity of structural units of boron atoms. The additive effect of lithium oxide is significant, thus the remarkable composition dependence of the distribution function is expected. Further analysis of the composition and frequency dependence in lithium borate glass is in progress.

References

1. Carini, G., Cutroni, M., Fontana, A., Mariotto, G. & Rocca, F. *Phys. Rev. B*, 1984, **29**, 3567.
2. Bray, P. J. *Phys. Chem. Glasses*, 1963, **4**, 37.
3. Zhong, J. & Bray, P. J. *J. Non-Cryst. Solids*, 1989, **111**, 67.
4. Chryssikos, G. D. & Kamitsos, E. I. *Proc. Second Int. Conf. on Borates Glasses, Crystals and Melts*, 1997, 128.
5. Zhong, J. & Bray, P. J. *J. Non-Cryst. Solids*, 1989, **111**, 67.
6. Kodama, M., Matsushita, T. & Kojima, S. *Jpn. J. Appl. Phys.*, 1994, **34**, 2570.
7. Birge, N. O. & Nagel, S. R. *Phys. Rev. Lett.*, 1985, **54**, 2674.
8. Kojima, S., Novikov, V. N. & Kodama, M. *J. Chem. Phys.*, 2000, **113**, 6344.
9. Davidson, H. W. & Cole, R. H. *J. Chem. Phys.*, 1951, **19**, 1484.

Ordering effects in spatial arrangement of particles in phase separated sodium borosilicate glass

M. P. Shepilov[1]

Vavilov State Optical Institute, ul. Babushkina 36/1, Saint-Petersburg, 192171, Russia

A. E. Kalmykov

A.F. Ioffe Physico-Technical Institute, ul. Politechnicheskaya 26, Saint-Petersburg, 194021, Russia

G. A. Sycheva

Institute of Silicate Chemistry of the Russian Academy of Sciences, ul. Odoevskogo 24-2, Saint-Petersburg, 199155, Russia

A structure of the 13·9Na_2O.36·0B_2O_3.50·1SiO_2 (mol%) glass heat treated at 610°C for 5 h (glass G1) and 10 h (glass G2) is studied by a stereological method on the basis of electron microscopy data obtained from planar sections of samples. It has been found that liquid phase separation leads to formation of spherical particles with the volume fractions v_1=0·047 (G1) and v_2=0·065 (G2). The distributions of particle radii r are characterised by the average values r_{pa1}=21 nm (G1) and r_{pa2} = 34 nm (G2), and by standard deviations σ_{p1}=4 nm (G1) and σ_{p2}=6 nm (G2). The analyses of pair correlation functions of particle traces on a secant plane indicate ordering effects in mutual arrangement of particles for the both glasses G1 and G2. Since the volume fraction of particles for the glass G1 does not reach the equilibrium value (the latter is unknown, but not smaller than v_2 = 0·065), we can conclude that the ordering effects take place even at a growth stage of phase separation. It is also noted that these effects are crucial for a turbidity and other of physical properties of material.

1. Introduction

A quantitative study of structure of many phase materials formed in the course of phase transformations in glasses is interesting from both fundamental and practical points of view. In general, the problem is very difficult. One of the simplest objects for such a study is a two phase material consisting of spherical particles embedded in a matrix. Materials of this type may be formed in the course of nucleation-and-growth phase separation.(1) Using stereological method, Barnett & Douglas in their classical work(2) determined a volume fraction, number density and mean radius of particles in the course of phase separation in the soda–lime–silica glasses. Additional information on the structure is given by size distribution of particles.(3)

However, the structure of the system of spherical particles embedded in a matrix and the properties of the material are determined not only by the characteristics of individual particles but also by their mutual arrangement. For the description of mutual arrangement, the pair correlation function (PCF)(4) of particles is of special importance.

PCFs of several model types have been used in the literature to describe the effects of interparticle interference in small angle x-ray and neutron scattering and in scattering of visible light by phase separated glasses and glass-ceramics.(1,5–9) These effects are caused by elements of ordering in the mutual spatial arrangement of particles, and the PCFs used in cited papers correspond to more or less ordered systems. Note that only for two types of PCFs, the ordering is sufficient for the explanation of interference effects under consideration. The PCF of the first type(5) is purely model PCF and, in our opinion,(9) cannot be observed for particles in glasses. The PCF of the second type used in Refs 6–8 was calculated by Hayter & Penfold(10) for a system of monodisperse spherical particles with the repulsive screened Coulomb pair potential of interaction between particles. This PCF seems to be applicable for the description of the structure of glasses and glass-ceramics examined in Refs 6–8. Nevertheless, such an application leads to some questions. For example, it is noticed(7, p.114) that the best fit of experimental data on scattering is often achieved for a particle volume fraction less

[1] Corresponding author. Email address shep@ms3992.spb.edu

than one half of that expected by the authors of Ref. 7. Besides, the use of scattering data for estimation of PCF needs a number of unjustified assumptions, such as assumption about monodispersity and sphericity of particles.

Thus the question about PCF of particles in glasses requires further study. Recently we used a stereological method for the determination of the structure of phase separated sodium borosilicate glass (of the glass G2 – see below).[11,12] The preliminary data on PCF of particle traces on a secant plane showed that a mutual arrangement of particles is characterised by elements of ordering. In the present work, we carry out a detailed stereological study of the structures of two glasses of the same composition heat treated for different times at the same temperature.

2. Experimental

The initial glass with the composition $13{\cdot}9Na_2O.36{\cdot}0B_2O_3.50{\cdot}1SiO_2$ (mol%, as-analysed) was prepared by melting a 180 litre batch at 1250°C. The melt was then cooled down to 500°C for 70 h, held at this temperature for 8 h, and cooled to room temperature at a rate of 4 K/h. The samples of the initial glass were heat treated at temperature (610±1)°C for 5 h (glass G1) and 10 h (glass G2).

For the preparation of carbon replicas, the surfaces of freshly fractured samples were etched with a solution of hydrofluoric acid. The replicas were examined in an EM-125K electron microscope at 50 kV with 0·3 nm point-to-point resolution. Seven and six micrographs were prepared for the glasses G1 and G2, respectively. The micrographs were scanned and files of electronic images were obtained in JPEG format for computer treatment.

Particle traces in the form of circles were observed in the micrographs for both glasses G1 and G2. On this basis, it was concluded that the particles of a precipitated phase are spherical in shape.[2] The coordinates of centres of traces and their radii were determined as a result of computer treatment of electronic images of micrographs. These data were used as a basis for all subsequent calculations.

A value for the area fraction of traces was calculated for each micrograph. A sample of these values for a given glass was used for the calculation of sample average v_t and its statistical error for a confidence level 0·95. Following a stereological approach,[14] a sample average v_t was taken to be an estimate for volume fraction v of particles.

Data on radii of traces for a given glass were used for the calculation of particle radii distribution by the method proposed by Filipovich *et al.*[15,16] As a result of these calculations, a volume fraction of particles and their number density were also determined.

Coordinates of centres of traces form the basis for the calculation of PCF of traces $g(r)=dn/(n_0 2\pi r dr)$. Here n_0 is the average number of trace centres per unit area of the section and dn is the average number of trace centres within the ring $[r, r+dr]$ circumscribed around the centre of a given trace. The $g(r)$ dependence was calculated as a histogram with the step Δr in the range $r \in [0, r_1]$. Note that the PCF was averaged over the ensemble of traces, whose distances from the edge of the micrograph exceed r_1, because, only for these traces, the circle with the radius r_1 (centred at the chosen trace) falls within the field of vision. The statistical errors of histogram evaluation were estimated for a confidence level 0·95.

Systematic errors connected with etching[14, p.57] and with inaccuracy in determination of magnification were not considered.

3. Results and discussion

Some results obtained from the micrographs are given in Table 1. First of all, it is seen that the volume fraction v_1 of precipitated phase in the glass G1 does not reach an equilibrium value v_e. Really, although v_e is unknown, it cannot be smaller than the volume fraction v_2 of particles in the glass G2, the latter being appreciably greater than v_1. Thus, we can conclude that the process of transformation in the glass G1 is related to the growth stage.

The number density of particles in the glass G2 is smaller than that for the glass G1. This means that Ostwald ripening stage is realised in the glass G2.

The calculated probability density of distribution of particle radii is almost symmetrical with respect to the mean radius r_{pa}, and especially for the glass G2. It should be noted that the probability density for the glass G2 differs significantly from the asymptotic function of Lifshitz & Slyozov.[17,18] However, it is necessary to keep in mind that etching effects[14, p.57] decrease the number of small traces and small particles, so that the true probability density would be somewhat greater in the region of small radii.

The histograms of PCF of traces, $g(\rho)$, for the glasses G1 and G2 are presented in Figure 1. In the Figure, the dimensionless intertrace centre-to-centre distance ρ is measured in the units of the mean diameter of particles ($\rho=r/(2r_{pa})$). Analyses of the obtained histograms of the PCFs for traces lead to the conclusion that the arrangement of particles exhibits spatial correlations. Actually, it can be shown that, when nonoverlapping polydisperse particles with a small volume fraction (<0·1) are randomly arranged, the PCF of traces should satisfy the condition $g(\rho)\approx 1$ at $\rho > r_{pmax}/r_{pa} \approx 1{\cdot}3$ where the maximum radius of par-

[2] Using the data on tie-line directions in the region of phase separation for the $Na_2O\text{–}B_2O_3\text{–}SiO_2$ system,[13, p.158] we estimated compositions and volume fractions of coexisting phases and found that the droplet phase is enriched in silica.

Table 1. Some details and results of stereological analysis of phase-separated glasses G1 and G2

	G1	G2
Number of micrographs	7	6
Number of traces analysed	4758	2576
Volume fraction of particles		
(a) as area fraction of traces	0·047±0·004	0·065±0·002
(b) as a result of calculation of radius distribution of particle	0·048	0·066
Number density of traces n_t, cm^{-2}	(0·44±0·01)×10^{10}	(0·232±0·005)×10^{10}
Characteristics of radii of traces:		
sample average r_{ta}, nm	17·7±0·2	28·8±0·3
sample standard deviation σ_t, nm	5·0	7·7
σ_t/r_{ta}	0·28	0·27
maximum radius r_{tmax}, nm	28	44
Number density of particles n_p, cm^{-3}	1·2×10^{15}	0·38×10^{15}
Characteristics of radius distribution of particles:		
mean value r_{pa}, nm	21	34
standard deviation σ_p, nm	4·0	6·0
σ_p/r_{pa}	0·19	0·18
Number of traces in ensemble for averaging of PCF histogram	4064	2218

ticles r_{pmax} is estimated as a maximum radius of traces (see Table 1). As can be seen from Figure 1, this condition is not satisfied for the both glasses G1 and G2; i.e. the particle arrangement cannot be random. A rapid decrease in $g(\rho)$ with a decrease in ρ at ρ<2·5 (G1) and 2·0 (G2) suggests the spatial separation of particles (to put it figuratively, the particles are 'repelled' from each other). In our opinion, this fact is quite sufficient in order to draw inference about the presence of correlations in the particle arrangement.

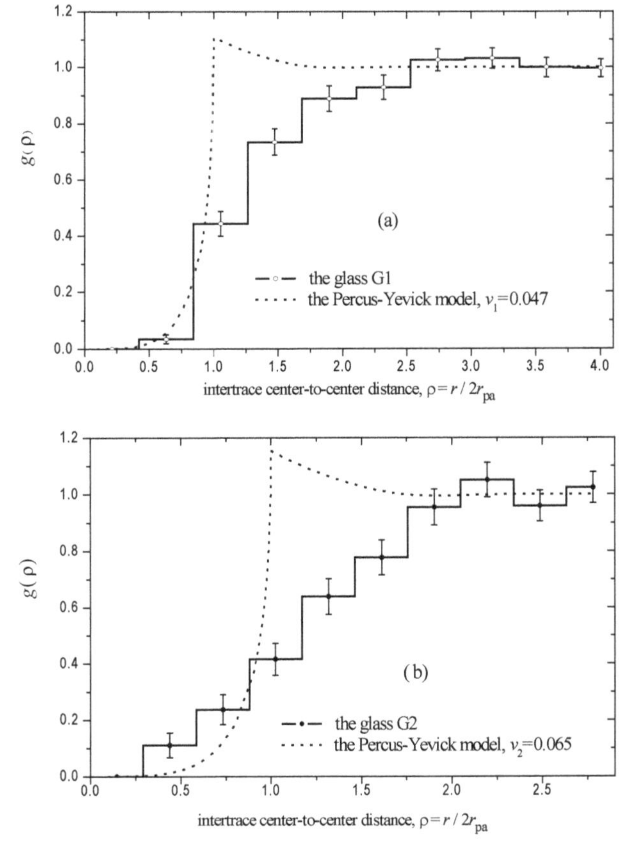

Figure 1. The pair correlation function of traces calculated on the basis of stereological data and for the Percus–Yevick model for the glass G1 (a) and G2 (b)

It should be mentioned that the small maximum $g(\rho)\cong1{\cdot}05$ at $\rho\cong2{\cdot}2$ for the glass G2 and the broader maximum $g(\rho)\cong1{\cdot}03$ at $\rho\cong3$ for the glass G1 also indicate the presence of correlations (weakly pronounced predominant interparticle distances). However, the fact of the spatial separation of particles, first, is revealed with a higher degree of confidence, and, second, is more important for the structural characterization of the system (as shown in the preliminary study[11,12] of the glass G2).

To reinforce the conclusion about the correlations in particles arrangement, it is interesting to compare the observed PCF of traces with those in a system characterized by random particle arrangement. In this work we consider the simplest example of such a type – the system of nonpenetrating monodisperse spherical particles in the Percus–Yevick approximation.[19] In this case the PCF $G_{PY}(r)$ of particles with a diameter D is expressed in analytical form and, as a function of dimensionless distance $\rho=r/D$, is determined by the only parameter – by the volume fraction of particles. The PCF of traces $g_{PY}(r)$ is related with the PCF of particles by the equation[4,20]

$$g_{PY}(r) = 2D^{-2}\int_{r}^{\sqrt{r^2+D^2}} R\,G_{PY}(R)\left(\frac{D}{\sqrt{R^2-r^2}}-1\right)\mathrm{d}R$$

which holds for any system of monodisperse spherical particles (in Stoyan[4] this equation is presented in another but equivalent form). Using the values of particle volume fraction for the glasses G1 and G2 (see Table 1), we calculated $g_{PY}(r)$ by this equation. Comparing the results of the calculations with the observed PCF (Figure 1), we consider that the radius of particles in the Percus–Yevick model is equal to the mean radius of particles in corresponding glass. Figure 1 shows that the observed and calculated PCFs differ significantly.

Earlier,[12] to show quantitatively that the particle arrangement in the glass G2 is not random, we carried out computer simulation of the PCF of particle traces in a plane section of the sample for a model of random arrangement of spherical particles. In the model, the volume fraction of particles and their size distribution correspond to those calculated from experimental data for the glass G2. For a random arrangement of particles, the procedure of "random sequential adsorption"(21) generalised to the case of polydisperse particles was used. In this procedure, spheres were placed sequentially and randomly in a given volume. If a new sphere was placed in the volume that it intersected a sphere already existing then the position of the new sphere was rejected and a new one was generated. The simulated PCF of traces $g_{CS}(\rho)\approx 1$ at $\rho>1{\cdot}1$ and differs significantly from the PCF of traces observed for the glass G2. The difference between $g_{CS}(\rho)$ and $g_{PY}(r)$ is not so essential: the positions of maxima are slightly different (ρ_{max}=1·2 and 1·0 for $g_{CS}(\rho)$ and $g_{PY}(r)$, respectively) and the maximum of $g_{CS}(\rho)$ is not so sharp as in the case of $g_{PY}(r)$.

It is interesting to compare the PCFs of traces for the glasses G1 and G2 in dimensionless and usual units (Figure 2). It should be noted that in the region of intermediate values of ρ (ρ=0·8–1·6) the PCFs for the both glasses are close to each other. The distinction in the region of small values of ρ may be partly connected with etching effects which must be greater for the glass G1 where particles are smaller. The weak maximum of PCF is more pronounced for the glass G2.

Using the results of preliminary study of the PCF of traces for glass G2, the PCF of particles and structure factor were calculated in the approximation of a monodisperse system.[11,12] The behaviour of structure factor shows that the correlations in particles arrangement are very important for light scattering of the material.[9] Namely, the interparticle interference effects can decrease a scattering coefficient by an order of magnitude or more in comparison with the case of randomly arranged particles. It will be interesting to carry out similar calculations for the glass G1.

In our opinion, the correlations in spatial arrangement of particles are formed as a result of nucleation process. Volmer[22] considered a nucleation of particles of a new phase in solution and noted that diffusion zones depleted of precipitating component are formed around growing particles where the nucleation of new particles is suppressed (the zones of exclusion for nucleation). Therefore, the probability of small distances between particles is low. Using the notion of Volmer and the diffusion growth law obtained by Zener,[23] a simple model for calculation of kinetics of nucleation-and-growth stage of diffusion limited phase separation was proposed[24] and described in detail.[25] The particle PCF calculated in this model for the time of completion of nucleation[24] shows the existence of short range order in spatial arrangement of particles (the spatial separation of particles and weakly pronounced predominant interparticle distances). After completion of nucleation and until the end of growth stage, the PCF of particles in this model does not change. We assume that at the Ostwald ripening stage, the order in particle arrangement increases since the diffusive interactions (i.e. dissolution and growth processes) are most intensive in pairs of closely spaced particles. Besides, in such pairs, a motion of particles to each other and their coalescence are experimentally observed.[26]

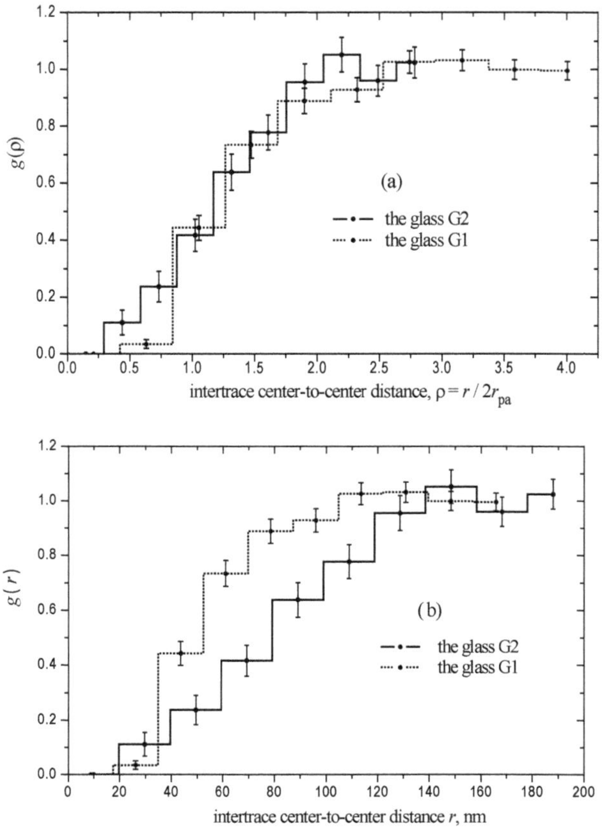

Figure 2. Comparison of pair correlation functions of the glasses G1 and G2 with use of dimensionless (a) and usual (b) units for distances

There are a number of papers where models for calculation of phase separation kinetics are described (see, for example, Refs 27–29) which are more rigorous than the model mentioned above.[24,25] In these models, correlations in spatial arrangement of particles are taken into account. The existence of such correlations manifests itself as a maximum in structure function (small angle scattering) curve calculated for these models (see Figure 11 of Ref. 27 and Figures 8 and 12 of Ref. 28). To our regret, these papers do not contain any data on PCF of particles in explicit form.

4. Conclusions

The correlations in arrangement of particles in phase-separated glass were observed by a stereological method for both growth and Ostwald ripening stages. A strong effect of these correlations on physical properties of the material is to be expected.

Acknowledgments

We are very grateful to G. P. Roskova who kindly supplied the initial glass samples.

This work was supported by the Russian Foundation for Basic Research (projects no. 04-03-32657 and ofi-p-05-03-08013), by Grant RI-112/001/192 of President of Russian Federation and by International Science and Technology Centre (project no. 2696).

References

1. Andreev, N. S., Filipovich, V. N., Mazurin, O. V., Porai-Koshits, E. A. & Roskova, G. P. *Phase Separation in Glass*. Elsevier Science Publishers B.V., Amsterdam, North-Holland, 1984.
2. Barnett, D. G. & Douglas, R. W. Liquid–liquid phase separation in the soda–lime–silica system. *Phys. Chem. Glasses*, 1970, **11** (5), 125–35.
3. Zarzycki, J. & Naudin, F. A study of kinetics of the metastable phase separation in the $PbO–B_2O_3$ system by small-angle scattering of x-rays. *Phys. Chem. Glasses*, 1967, **8** (1), 11–18.
4. Stoyan, D., Kendall, W. S. & Mecke, J. *Stochastic Geometry and its Applications*. Wiley, Chichester, 1995.
5. Andreev, N. S., Aver'yanov, V. I. & Voishvillo, N. A. Structural interpretation of anomalous scattering of visible light in sodium-borosilicate glasses. *Fiz. Tverd. Tela*, 1960, **2** (5), 1011–21. [*Sov. Phys. Solid State*, 1960, **2** (5), 916–25].
6. Wright, A. F., McMillan, P. W. & Brett N. H. Establishment of short-range order during nucleation and growth of precipitates within amorphous solids. *Proc. 2nd Int. Conf. Structure of non-crystalline materials*. Edited by P. H. Gaskell, J. M. Parker & E. A. Davies. Taylor & Francis, London. Pp. 569–81.
7. Wright, A. F., Fitch, A. N., Hayter, J. B. & Fender, B. E. F. Nucleation and crystallisation of cordierite-TiO_2 glass-ceramic. Part 1. Small angle neutron scattering measurements and simulations. *Phys. Chem. Glasses*, 1985, **26** (4), 113–18.
8. Labarbe, P. D., Lin, J. S. & Wright A. F. Small angle neutron scattering study of phase separation in the binary glass system $TiO_2–SiO_2$. *Phys. Chem. Glasses*, 1988, **29** (3), 91–9.
9. Shepilov, M. P. On the problem of theoretical description of anomalous light scattering by phase-separated glasses. *Proc. Seventh ESG Conf. on Glass Sci. and Technol.*, 2004. *Phys. Chem. Glasses*, 2005, **46** (2), 173–7.
10. Hayter, J. B. & Penfold, J. An analytic structure factor for macroion solutions. *Mol. Phys.*, 1981, **42** (1), 109–18.
11. Kalmykov, A. E., Shepilov, M. P. & Sycheva, G. A. Electron microscopic investigation of spatial ordering of particles formed in the course of liquid phase separation in sodium borosilicate glass. *Glass Phys. Chem.*, 2000, **26** (3), 307–9.
12. Shepilov, M. P. & Kalmykov, A. E. The observation of correlations in mutual spatial arrangement of phase-separated particles in glass by stereological method. *Glass Sci. Technol.*, 2002, **75** (C2), 458–61.
13. Mazurin, O. V., Roskova, G. P., Aver'yanov, V. I. & Antropova T.V. *Two-Phase Glasses: Structure, Properties, Application*. Nauka, Leningrad, 1991. (in Russian).
14. *Quantitative Microscopy*. Edited by R. T. DeHoff & F. N. Rhines, McGraw-Hill Book Company, New York, 1968.
15. Filipovich, V. N. & Kalinina, A. M. Determination of nucleation rate for a new phase in glasses at the initial stage of phase transformation. In: *Likvatsionnye Yavleniya v Steklakh* (Phase Separation in Glasses) (in Russian). Nauka, Leningrad, 1969, Pp.40–4.
16. Morozova, E. V., Kalinina, A. M. & Filipovich, V. N. A quantitative analysis of the kinetics of droplet phase separation by the example of sodium borosilicate glass undoped and doped with ZrO_2. *Fiz. Khim. Stekla*, 1986, **12** (4), 396–405.
17. Lifshitz, I. M. & Slyozov, V. V. On the kinetics of diffusional decomposition of supersaturated solid solutions. *Zh. Eksp. Theor. Fiz.*, 1958, **35** (2), 479–92.
18. Lifshitz, I. M. & Slyozov, V. V. The kinetics of precipitation from supersaturated solid solutions. *J. Phys. Chem. Solids*, 1961, **19** (1–2), 35–50.
19. Percus, J. K. & Yevick, G. J. Analysis of classical statistical mechanics by means of collective coordinates. *Phys. Rev.*, 1958, **110** (1), 1–13.
20. Kalmykov, A. E. & Shepilov, M. P. A method for direct determination of the pair correlation function for particles formed in the course of phase transformation. *Glass Phys. Chem.*, 1997, **23** (3), 190–7].
21. Evans, J. V. Random and cooperative adsorption. *Rev. Modern Phys.*, 1993, **65**, 1281–329.
22. Volmer, M. *Kinetik der Phasenbildung*. Th. Steinkopf, Dresden, 1939. Ch. 6.
23. Zener, C. Theory of growth of spherical precipitates from solid solution. *J. Appl. Phys.*, 1949, **20** (10), 950–3.
24. Shepilov, M. P. & Bochkariov, V. B. The calculation of phase separation kinetics in metastable systems for two models of particle nucleation. *Proc. XVI Int. Congr. on Glass, Madrid, Bol. Soc. Esp. Ceram. Vidr.*, 1992, **31-C**, **Vol. 2**, 199–204.
25. Shepilov, M. P. A model for calculation of isothermal kinetics of the nucleation-and-growth type phase separation in the course of one-step heat treatment. *J. Non-Cryst. Solids*, 1996, **208** (1–2), 64–80.
26. Aver'yanov, V. I. & Kalmykov, A. E. Coalescence and motion of particles in liquid phase separation. *Sov. J. Glass Phys. Chem.*, 1992, **18** (4), 288–92].
27. Marder, M. Correlations and Ostwald ripening. *Phys. Rev. A*, 1987, **36** (2), 858–74.
28. Tokuyama, M. & Enomoto, Y. Theory of phase-separation dynamics in quenched binary mixtures. *Phys. Rev. E*, 1993, **47** (2), 1156–79.
29. Tokuyama, M. & Enomoto, Y. Effects of encounters on phase separation dynamics from diffusion-controlled growth to coarsening. *Physica A*, 1995, **220**, 261–76.

Crystallisation processes in rapid quenched $AgCl–AgI–Ag_2O–B_2O_3$ glasses

J. L. Nowińskia,[a] J. E. Garbarczyka,[a] M. Wasiucioneka[a] & S. Gierlotkab[b]

[a] Faculty of Physics, Warsaw University of Technology, Koszykowa 75, 00-662 Warszawa, Poland
[b] Institute of High Pressure Physics, Polish Academy of Sciences, Sokołowska 29/39, 01-142 Warszawa, Poland

The influence of silver chloride on formation of the crystalline precipitates in borate glasses obtained in the system $AgCl–AgI–Ag_2O–B_2O_3$ by rapid quenching was investigated. X-ray diffraction and field emission scanning microscopy methods. The glasses selected for the detailed investigations were related to the glass of the $78AgI.16{\cdot}5Ag_2O.5{\cdot}5B_2O_3$ composition, which demonstrates some interesting properties. Rapid quenching of the $78[xAgI.(1-x)AgCl].16{\cdot}5Ag_2O.5{\cdot}5B_2O_3$, $0\leq x\leq 1$ glass melt produces glass in which crystalline inclusions are embedded. Two types of the inclusions were identified. When $0\leq x<0{\cdot}70$, the $AgCl_yI_{1-y}$ particles of fcc cubic AgCl structure are formed in the glass host. For the $0{\cdot}7\leq x$ composition range, the α-AgI nanoparticles of bcc cubic structure, stable at room temperature, precipitate in the glass host. The dependence of a lattice parameter a of an inclusion crystal structure as a function of silver chloride contents x in the glass was found out.

1. Introduction

A successful stabilisation of crystallites α-AgI phase down to room temperature in ultrafast quenched silver oxysalt glasses, reported in 1991 by Tatsumisago *et al* in their seminal paper,(1) started a series of studies on crystallisation and stability issues in AgI-rich systems.(2–6) Because of unique properties of AgI and, especially, the high conductivity of its α-phase, these studies almost solely have mainly focused on systems with AgI dopant. It seems however that a second co-dopant, accompanying AgI, could participate in formation of crystalline precipitates, and possibly it could affect the course of crystallisation. Recognition of such processes, in consequence, could enable preparation of novel glass composites, and should provide an opportunity to modify some properties of the well known materials. Such modifications could be interesting also for applications.

One of the most natural candidates for such a co-dopant, which easily dissolves and co-exists with silver iodide in the glass host is AgCl. The compound was itself tested as a single dopant in silver ion conducting glasses. However a poor ion conductivity of the resulting materials was considered discouraging.(7) Interest in the application of AgCl was renewed in 2004, when Agraval *et al* demonstrated that a AgCl–AgI mixture used as a co-dopant led to a significant enhancement of the ionic conductivity of phosphate glasses.(8) The same group earlier reported that the [0·75AgI.0·25AgCl] applied as host in composite electrolytes, in which a second, usually non-ion conducting phase, such as for example Al_2O_3 or TiO_2 was dispersed, exhibited improved ionic conductivity.(9,10)

In this work, we aim to study the influence of silver chloride on the formation of the crystalline precipitates in borate glasses obtained in the system $AgCl–AgI–Ag_2O–B_2O_3$ by rapid quenching. Such an investigation, according to our knowledge, has not been reported in the literature. Glasses selected for the detailed investigations were related to the glass of the $78AgI.16{\cdot}5Ag_2O.5{\cdot}5B_2O_3$ composition, which demonstrates some interesting properties described elsewhere.(11) The molar ratio $[Ag_2O]/[B_2O_3]$ was kept constant and equal to 3, which means that basic borate network was strongly disrupted by the Ag_2O modifier.

2. Experimental

Glasses of composition described by the formula $78[xAgI.(1-x)AgCl].16{\cdot}5Ag_2O.5{\cdot}5B_2O_3$, $0\leq x\leq 1$ were prepared by a ultrafast quenching from reagent grade AgI, AgCl, $AgNO_3$ and H_3BO_3 chemicals. For each composition, the weighed amounts of starting chemicals were ground in a mortar and the mixture, placed in a crucible, was calcined at 400°C for ~30 min to remove water. Then the temperature was raised to 650°C and the molten mixture was held for approximately 20 min to complete the evaporation of nitric oxides and remaining water. Finally, the melts were poured into the rotating twin-roller (1000 rpm) to produce thin flakes, about 0·05 mm thick.

X-ray diffraction patterns (XRD) were collected at

[1] Corresponding author. Email nowin@if.pw.edu.pl

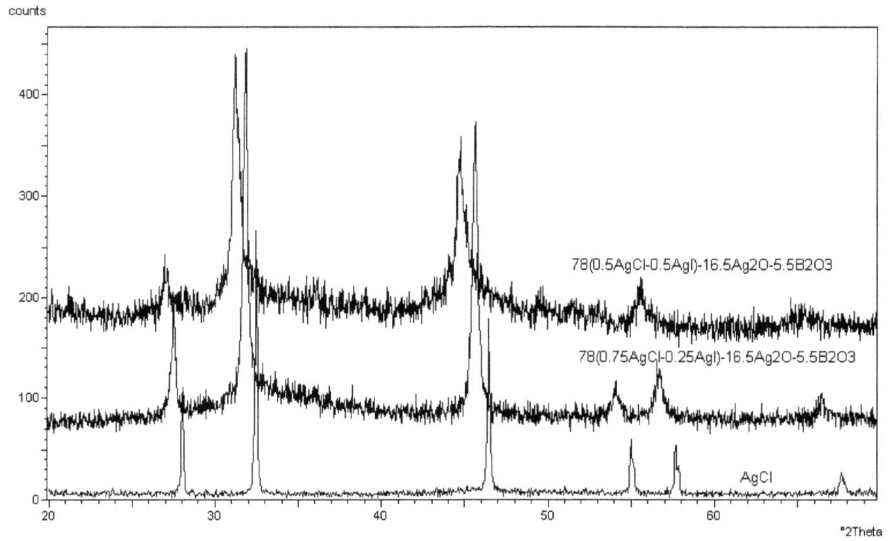

Figure 1. XRD patterns of pure AgCl and the 78[xAgI.(1−x)AgCl].16·5Ag$_2$O.5·5B$_2$O$_3$ glasses described by x=0·25 and x=0·5

room temperature using Cu K_α radiation on a Philips X'Pert Pro diffractometer set in the Bragg–Brentano geometry. Unit cell refinements were carried out employing Philips X'Pert High Score and Philips X'Pert Plus programs. Phase identification analysis was performed by means of PCPDFWIN program (PDF-2 version of ICDD software).[15]

A LEO 1530 Field Emission Scanning Electron Microscope (FE-SEM) was used in microstructure investigations.

3. Results and discussion

XRD patterns of the obtained samples contain diffraction lines due to crystalline phases together with a wide halo typical for amorphous systems. The former observation should have been expected for materials with such high content (78 mol%) of dopants. A phase identification analysis of the diffraction data for the material containing only AgCl (x=0) indicates that the peaks correspond to fcc cubic AgCl, (JCPDS No 31-1238).[15] Similarly, for the AgI dopant (x=1) all detected lines agree with pattern of the α-AgI.[1,11] In the case of the other glasses studied, the reflection positions 2θ on the patterns varied with chemical composition, though demonstrating some regularities. Thus, for the glasses characterised by $0\leq x<0{\cdot}70$ the peaks on the patterns were regularly shifted in respect to those of x=0 (Figure 1) suggesting the same crystal structure type of the precipitates as pure AgCl, although with different lattice parameters.

Based on this observation, it was assumed that only one crystalline phase precipitates in glass host when $0\leq x<0{\cdot}70$ and, furthermore, the crystal structure

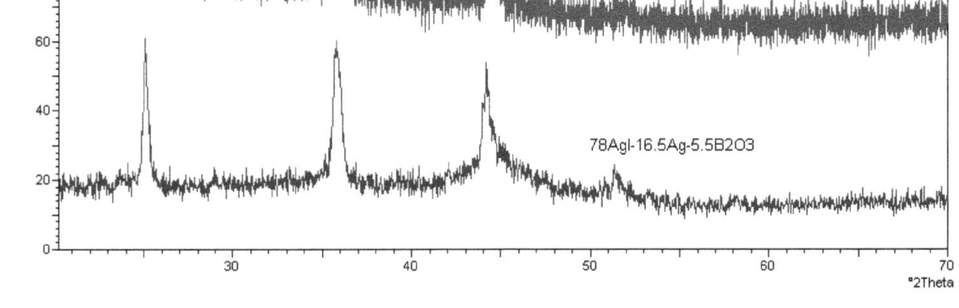

Figure 2. XRD patterns of the 78[xAgI.(1−x)AgCl].16·5Ag$_2$O.5·5B$_2$O$_3$ glasses described by x=0·87 and x=1

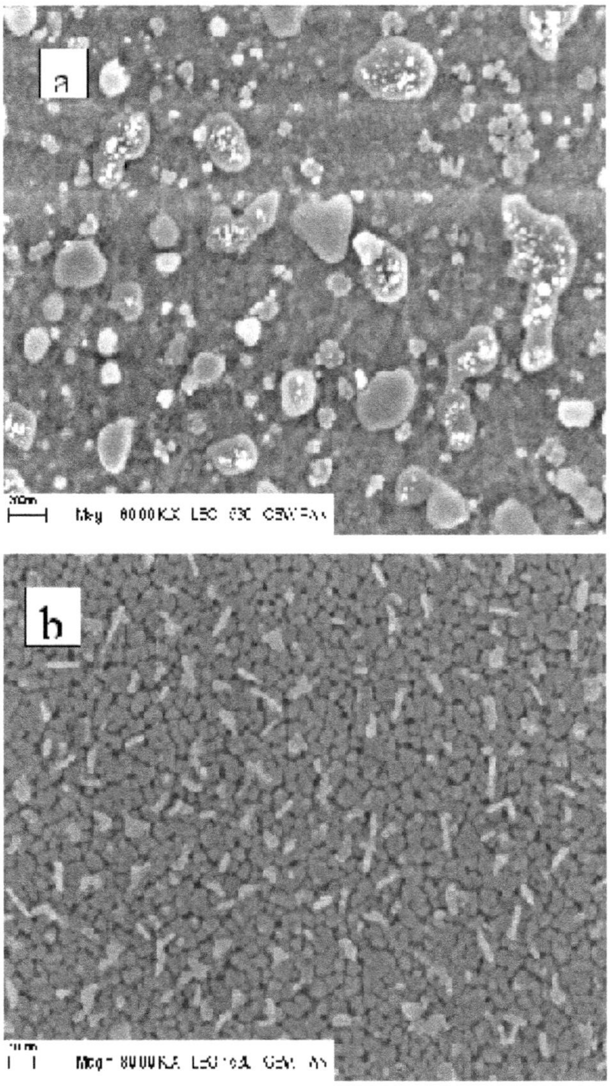

Figure 3. FE-SEM micrographs of the $78AgI.16{\cdot}5Ag_2O.5{\cdot}5B_2O_3$ glass (a) and the $78[0{\cdot}25AgI.0{\cdot}75AgCl].16{\cdot}5Ag_2O.5{\cdot}5B_2O_3$ glass (b)

of that phase was related to that of AgCl. On the other hand, the patterns recorded for materials described by $0{\cdot}70 \leq x$ appeared to be very similar in respect of the peak positions to that of the α-AgI one (Figure 2). Consequently, also these materials were considered as the composites comprising a single crystalline phase only. A preliminary analysis of the x-ray data suggested that the materials investigated could be classified, taking into account the structure type of the precipitated phases, into two groups: those with the AgCl related structure and those which adopted the α-AgI one.

Further investigations were conducted by means of FE-SEM method. Obtained images revealed the presence of elongated particles uniformly dispersed in the glass host for materials characterised by $0 \leq x < 0{\cdot}70$, whereas the images recorded for the glasses of compositions $0{\cdot}70 \leq x$ presented oval shape objects,

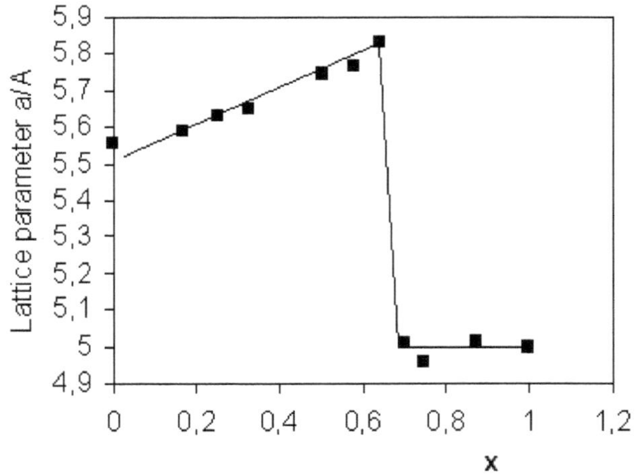

Figure 4. Dependence of the lattice parameter *a* versus index x for the $78[xAgI.(1-x)AgCl].16{\cdot}5Ag_2O.5{\cdot}5B_2O_3$ glass

about 100–200 nm in diameter, in which small inclusions ~20–45 nm were embedded (see Figure 3). Similar oval objects were observed by Tatsumisago *et al*, reporting formation of the α-AgI room temperature stable phase.[1–5] Thus, the FE-SEM investigations confirmed the earlier suggestions of the formation of the two groups of the materials.

The conclusion makes feasible a unit cell search of an inclusion crystal structure. The refinements found that the cubic structure could be assigned to the crystal structure of the precipitates formed for the $0 \leq x < 0{\cdot}70$ glasses. The lattice parameter *a* appeared to be chemical composition dependent, increasing almost linear by with index *x* from a=5·55 Å – the value characteristic for pure AgCl to x=5·83Å (see Figure 4). The dependence satisfies the Vegard's law,[14] characteristic for many solid solutions, where it is observed for the compounds in which some atoms in the structure are substituted by other ones with a different atom radius. The increase of the lattice parameter a could then be explained as the results of partial substitution of chlorines by iodine atoms in the fcc cubic AgCl structure of a precipitate. This leads to the conclusion that chemical composition of a precipitate for the $0 \leq x < 0{\cdot}70$ glasses should be expressed by the formula $AgCl_yI_{1-y}$.

For higher nominal contents of silver iodide in the material, the lattice parameter abruptly decreases to about 5·0 Å and then it remains constant up to x=1 (Figure 4). Since the x-ray diffraction profile shapes are also almost identical in respect to the diffraction line positions within this composition range, it seems scarcely likely that the chlorine atom participates in the formation of the crystalline inclusions. All the evidence strongly suggests that, in this case, the precipitated crystalline particles contain silver iodide only, which adopts the α-AgI crystal structure. It is worth noticing, that for geometric reasons, the lattice parameter of the fcc cubic AgCl structure, despite of a smaller chlorine atom size in respect to iodine one, appears to be bigger than that of the bcc cubic (α-AgI). Therefore, we observe a sudden decrease in the parameter a.

The crystallisation in borate glasses doped with AgCl–AgI, promoted by ultrafast quenching, demonstrates some fundamental differences when compared with the equilibrium crystallisation in the system of pure AgCl–AgI. As Lauer & Maier[12] have found, the approximate miscibility gap for the silver chloride/silver iodide mixture varies from $AgCl_{0\cdot90}I_{0\cdot10}$ to $AgCl_{0\cdot04}I_{0\cdot96}$. Slightly different values were reported by Cornwell & Dyson,[13] nevertheless both studies agreed that the immiscibility range for the AgCl–AgI system was quite wide. In contradiction of this, our observations indicate a wide range of miscibility of silver iodide in AgCl, and they seem to suggest continuous solubility of AgCl in AgI. However one should remember that the comparison relates to dissimilar thermodynamics conditions, i.e. to non-equilibrium and equilibrium materials respectively.

Non-equilibrium crystallisation demonstrates peculiar feature. Each glass melt just before quenching is kept at about 650°C, the temperature at which all glass constituents, AgCl and AgI included, are molten. In consequence, they are microscopically well mixed. This rapid quenching should freeze, in some respects, the local structure of the melt. Therefore one should expect a similar or simple proportionality in the AgI/AgCl molar ratio in a melt and in the relevant precipitate. As discussed earlier, we observe an almost linear increase of the lattice parameter with increasing total silver iodide content in the material this supporting such a relation to be valid.

Other surprising experimental results concern the crystallisation process which occurs when silver iodide concentration exceeds a threshold value of x=0·70. Then silver chloride stops participating in crystallisation, as all our experimental evidences indicates, and the chemical composition of a precipitate does not relate to the chemical composition of the melt.

Conclusions

Rapid quenching of the $78[x\text{AgI}.(1-x)\text{AgCl}].16{\cdot}5Ag_2O.5{\cdot}5B_2O_3$, $0\leq x\leq 1$ glass melt produces glass in which crystalline inclusions are embedded. The crystal structure and chemical compositions of the inclusions are determined by the value of the x index. When $0\leq x<0{\cdot}70$, the $AgCl_yI_{1-y}$ particles of fcc cubic AgCl structure are formed in the glass host. For the $0{\cdot}70\leq x$ composition range, the α-AgI nanoparticles of bcc cubic structure, stable at room temperature, precipitate in the glass host.

Acknowledgements

This work was financed by EU CEPHOMA Centre (Contract ENK5-CT-2002-80666).

References

1. Tatsumisago, M., Shinkuma, Y. & Minami, T. *Nature*, 1991, **354**, 217.
2. Tatsumisago, M., Shinkuma, Y., Saito, T. & T. Minami, T. *Solid State Ionics*, 1992, **50**, 273.
3. Tatsumisago, M., Taniguchi, A. & Minami, T. *J. Am. Ceram. Soc.*, 1993, **76**, 235.
4. Itakura, N., Tatsumisago, M. & Minami, T. *J. Am. Ceram Soc.*, 1997, **80**, 3209.
5. Tatsumisago, M., Okuda, K., Itakura, N. & Minami, T. *Solid State Ionics*, 1999, **121**, 193.
6. Minami, T., Shimizu, T. & Tanaka, J. *Solid State Ionics*, 1983, **9/10**, 577.
7. Agraval, R. C., Gupta, R. K., Bhatt, A., Verma, M. L. & Chandra, A. *Ionics*, 2004, **10**, 126.
8. Agraval, R. C., Gupta, R. K., Kumar, R. & Kumar, A. *J. Mater. Sci.*, 1994, **29**, 3673.
9. Agraval, R. C., Kumar, R., Gupta, R. K. & Saleem, M. *J. Non-Cryst. Solids*, 1995, **181**, 110.
10. Nowiński, J. L., Lewandowski, P. A., Garbarczyk, J. E., Wasiucionek, M. & Gierlotka, S. Submitted to *Solid State Ionics*, 2005.
11. Nowiński, J. L., Lewandowski, P. A., Garbarczyk, J. E., Wasiucionek, M. & Gierlotka, S. *SSI-15 International Conference on Solid State Ionics*, 17–22 July 2005, Baden-Baden/Germany, Book of Abstracts, P308.
12. Lauer, U. & Maier, J. *Ber. Bunsenges. Phys. Chem.*, 1992, **98**, 111.
13. Cornwell, K. & Dyson, R. W. *J. Phys. D*, 1969, **2**, 305.
14. Denton, A. R. & Ashcroft, N. W. *Phys. Rev. A*, 1991, **43**, 3161.
15. PCPDFWIN Version 1.0, ICPDS-ICDD licence.

Local structure of europium and iron sites in zinc borate glasses

G. Concas, F. Congiu[1] & G. Spano

Dipartimento di Fisica, Università di Cagliari, S.P. Monserrato-Sestu km 0.700, I-09042 Monserrato (CA), Italy

M. Bettinelli & A. Speghini

Dipartimento Scientifico e Tecnologico, Università di Verona and INSTM, UdR Verona, Ca' Vignal, Strada le Grazie 15, I-37134 Verona, Italy

Samples of zinc borate glasses, doped with europium or iron ions, were prepared and examined by ^{151}Eu and ^{57}Fe Mössbauer spectroscopy. The structural features of the rare earth and transition metal ion sites have been investigated with particular attention to the coordination number, distortion and degree of disorder. The relative concentrations of the different oxidation states have been also determined. The disorder of the sites was investigated by using the convolution method for the analysis of the spectra of glasses. The resulting distibution of the quadrupole interaction parameter is due to the structural disorder of the dopant ion sites and its standard deviation gives an evaluation of the degree of disorder. The results for zinc borate glasses are compared with the ones found for iron doped phosphate glasses and europium metaphosphate glass.

1. Introduction

Borate glasses containing rare earth or transition metal ions have been extensively investigated because they are of interest for applications in optical and magnetic devices;[1,2] ^{151}Eu and ^{57}Fe Mössbauer spectroscopy is a powerful technique which provides information about the symmetry and disorder of the dopant sites, and the oxidation states of the dopant ions.[3–5]

In this paper, samples of zinc borate glasses, doped with europium or iron ions, were prepared and examined by ^{151}Eu and ^{57}Fe Mössbauer spectroscopy. In order to clarify the characteristics of iron and europium sites in borate glasses, we studied two glasses with composition $4ZnO.3B_2O_3$ doped with 2·5 mol% of Fe or Eu (substituting for Zn) called FZBO and EZBO respectively. The experimental data have been fitted with computer programs that account for the disorder typical of a glass structure, and compute a probability distribution of the Mössbauer parameters.

2. Experimental procedure

Zinc borate glasses of molar composition $4ZnO.3B_2O_3$ doped with 2·5 mol% of ^{57}Fe or 2·5 mol% Eu with respect to Zn were prepared by melting a 0·8 g batch composed of analytical grade ZnO, H_3BO_3 and ^{57}Fe enriched Fe_2O_3 or Eu_2O_3 in a platinum crucible with a tightly fitting lid for 4 h at 1250°C. The melts were then quenched in air in a brass mould. The samples were annealed for 12 h at 500°C. Glass samples with excellent optical quality were thus obtained. Losses due to volatilisation from the glass melt were found to be lower than 5% of the total mass of the batch.

The Mössbauer absorption spectrum of the FZBO glass was obtained in a standard transmission geometry, using a source of ^{57}Co in rhodium (37 MBq). The calibration has been performed using a 6 μm thick natural iron foil. Measurements were carried out at room temperature on powder samples with absorber thickness of 5·6 mg/cm^2 of glass. The values of isomer shift are referred to α-Fe.

The spectrum of the EZBO glass was obtained using a $^{151}SmF_3$ source of activity 3·7 GBq; the absorber thickness is 3·8 mg/cm^2 of Eu. The values of the isomer shift are referred to the anhydrous fluoride EuF_3. The full width at half maximum of the crystalline absorption peak, determined with our source, is 1·76±0·01 mm/s and has been measured using $Cs_2NaEuCl_6$, which contains Eu^{3+} in a cubic site.[6] The width was measured for an absorber thickness of 3·8 mg/cm^2 of Eu; this value corresponds to an effective thickness t=1, when calculated using the recoilless fraction of the source, f=0·6.[5]

The absorption spectra have been tentatively examined using a program that fits the data by single Lorentzian line shapes, but the quality of the fit was poor. These trial fits showed that it was necessary to fit the absorption spectrum of the FZBO sample us-

[1] Corresponding author. Email address: franco.congiu@dsf.unica.it

ing a method proposed by Hesse & Rubartsch[7] and improved by Wivel & Mørup.[8] Amorphous materials do not show simple Lorentzian line shapes, owing to variations of Fe–O bond lengths in the glass network. This method makes use of a set of Lorentzian quadrupole doublets with fixed width and splitting, and computes the contribution of each curve to the absorption spectra by a least square fitting procedure, allowing for the empirical linear relation between isomer shift δ and quadrupole splitting Δ of each doublet[7]

$$\delta=\xi+\alpha\Delta \quad (1)$$

The program gives the best fit values of the free parameters ξ and α and the probability distribution of the doublets versus quadrupole splitting. Following Wivel & Mørup[8] minimisation is performed using the Lagrangean multipliers γ and β, which control the smoothing and the end point behaviour of the distribution profiles, respectively.

The spectrum of the EZBO glass has been analysed using a modified method derived from that used for iron;[9] this method uses a convolution of Lorentzian quadrupole multiplets as required by the ^{151}Eu spectroscopy. In this case, the γ-ray is emitted during a transition from an excited state with spin 7/2 to the ground state with spin 5/2 and the quadrupole splitting of the lines is determined by the quadrupole interaction parameter eQV_{zz} and by the asymmetry parameter η. The parameter eQV_{zz} replaces Δ in Equation (1). If there is no threefold or fourfold axis of symmetry passing through the nucleus, the asymmetry parameter is not zero; there are 12 allowed transitions which give 12 absorption lines. Each multiplet of the set is characterised by a value of the quadrupole interaction parameter and the probability distribution of this parameter has a fixed Gaussian shape. The width of the Lorentzian components is fixed at the crystalline width. The analysis with the convolution gives the standard deviation of the Gaussian distribution, which is an evaluation of the structural disorder in the glass.

3. Results

In Figure 1 we show the spectrum of the FZBO sample. A first standard fit with two Lorentzian doublets shows that the spectrum can be interpreted as the superposition of two components, with relative area of 55% (component I) and 45% (component II); the more intense can be assigned to ferric iron and the less intense to ferrous iron.[4] The ratio of the areas corresponds to ratio of the ion concentration if the recoil free fractions of Fe^{3+} and Fe^{2+} are supposed to be equal. We used a set of 50 Lorentzian doublets, 25 to fit the contribution to the absorption spectra due to Fe^{3+} and 25 for Fe^{2+}. The Lorentzian full width has

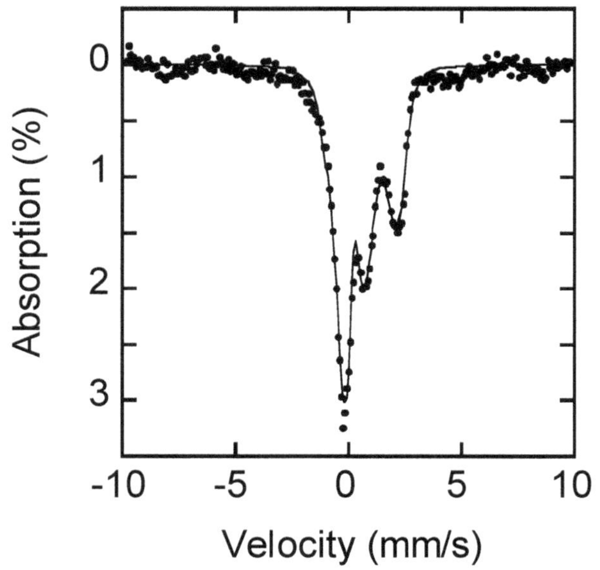

Figure 1. Absorption spectrum of the FZBO sample. The experimental data are reported as dots; the full line shows the fit

been fixed to 0·24 mm/s, typical of a spectrum of thin α-iron foil.[8] The quadrupole splitting ranges from 0–4 mm/s for Fe^{3+} and from 0–6 mm/s for Fe^{2+}. Some trial fits led to the use of the values γ=100 and β's=10 for the Lagrangean multipliers.

The spectrum also shows a small sextet component due to a magnetically ordered phase with external peaks at −7·8 mm/s and +8·4 mm/s, corresponding to isomer shift of about 0·4 mm/s (referred to α-Fe) and hyperfine field of about 51 T; these values identify a residual of α-Fe_2O_3 used for the preparation.[10] This component amounts to less than 2% of iron atoms and has a very small effect on the fit results presented here because only the two less intense internal peaks of the sextet are overlapped to the Fe^{2+} and Fe^{3+} doublets.

In Table 1 we present the results of the fit of the spectrum. The fit parameters ξ and α, the most probable value of quadrupole splitting (*QI*), the most probable value of isomer shift (*IS*) and the standard deviation (σ_{QI}) of the probability distribution of the splitting are given. The most probable value of isomer shift is computed by means of Equation (1).

In Figure 2 we show the probability distribution of the 25 Lorentzian doublets versus quadrupole splitting for Fe^{3+}, which gives the contribution of each doublet to the total resonant absorption. For high values of quadrupole splitting, a shoulder can be observed, which is probably due to a superpara-

Table 1. Mössbauer parameters as obtained by fitting the absorption spectra

Sample	Component	ξ (mm/s)	α	IS (mm/s)	*QI* (mm/s)	σ_{QI} (mm/s)
FZBO	I	0·35	0·09	0·42	0·84	0·77
	II	1·24	−0·10	0·98	2·51	0·99
EZBO		0·86	0·06	0·6	−6·9	1·1

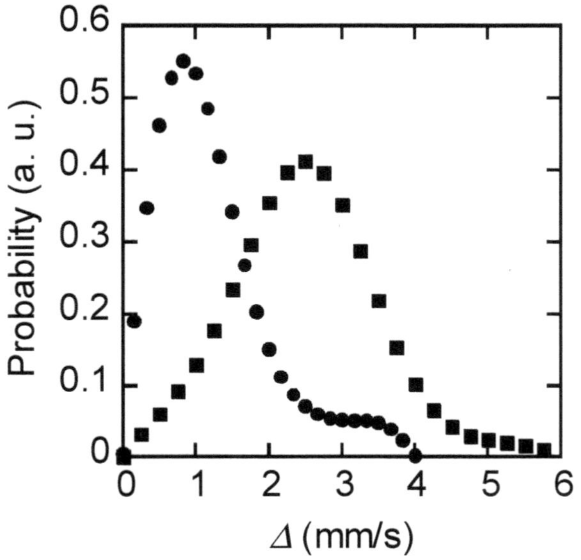

Figure 2. Probability distribution of the quadrupole splitting (Δ) for Fe^{3+} (circles) and Fe^{2+} (squares) in the FZBO sample

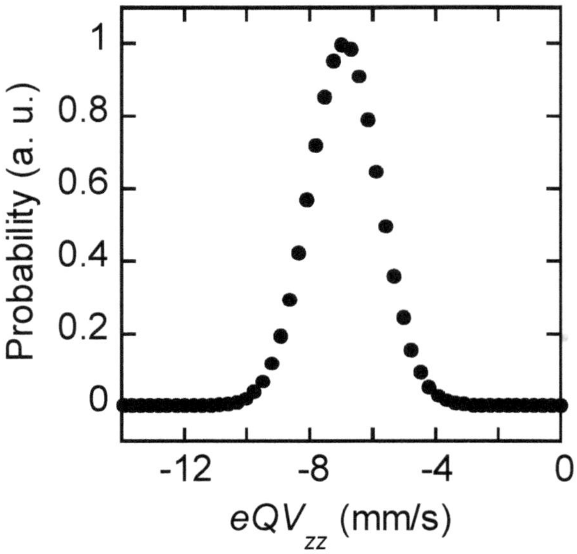

Figure 4. Probability distribution of the quadrupole interaction parameter (eQV_{zz}) for Eu^{3+} in the EZBO sample

magnetic phase.[11] The probability distribution of the 25 Lorentzian doublets versus quadrupole splitting for Fe^{2+} is also shown in Figure 2.

The absorption spectrum of the EZBO sample is shown in Figure 3. In this case, a good fit can only be obtained by making allowance for a quadrupole interaction with a non-zero asymmetry parameter. We used a set of 51 Lorentzian multiplets; the Lorentzian full width has been fixed to 1·76 mm/s. The multiplets of the set have values of the quadrupole interaction parameter ranging from eQV_{zz}=0 to eQV_{zz}=−14 mm/s (about twice the most probable value of the parameter). Table 1 gives the most probable value (QI=−6·9±0·1 mm/s) and the standard deviation of

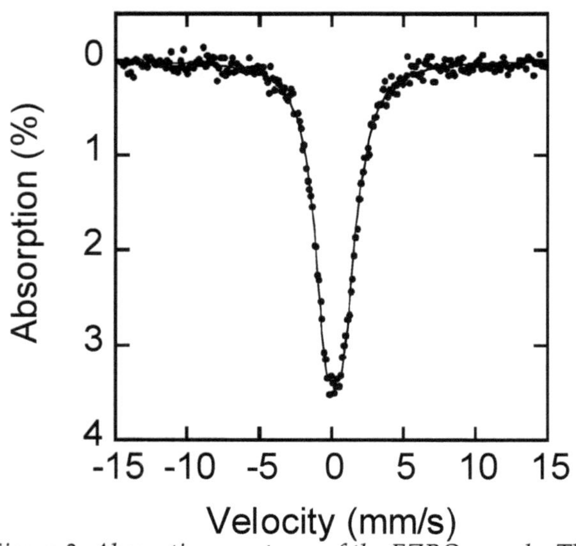

Figure 3: Absorption spectrum of the EZBO sample. The experimental data are reported as dots; the full line shows the fit

the distribution (σ_{QI}=1·1±0·1 mm/s) of the quadrupole interaction parameter, obtained by the analysis with the convolution of multiplets, and the asymmetry parameter η. The asymmetry parameter has been fixed to the value η=0·80±0·06 found by a fit with the standard method.[12] The distibution of the quadrupole interaction parameter is shown in Figure 4.

4. Discussion

In FZBO sample the most probable values of isomer shift and quadrupole splitting are typical of high spin Fe^{3+} and high spin Fe^{2+}.[3] Various review papers show that there is a relation between the isomer shift and the coordination number of iron atoms in crystalline and amorphous materials.[4,13–16] The most probable values of isomer shift and quadrupole splitting obtained for both ferric and ferrous iron are consistent with the octahedral coordination.[14,15] The non-zero value of the quadrupole splitting points out that the iron site is distorted in comparison with a regular octahedron.[16]

The variation of the quadrupole splitting is due to the distribution of the Fe–O bond lengths in the glass and gives an evaluation of the disorder of the iron sites; it can be well reproduced using the convolution method. The standard deviation of the distribution of the quadrupole splitting directly measures this disorder; in the FZBO glass, the standard deviation is equal to 92 and 39% of the most probable value for Fe^{3+} and Fe^{2+}, respectively. These values are larger than those found in sodium phosphate glasses for ferric and ferrous iron sites.[17]

The spectrum of the EZBO sample shows only one resonance peak with value of the isomer shift corresponding to an oxidation number +3. There is no detectable presence of Eu^{2+}. The quadrupole interaction indicates that there is a deviation of the Eu^{3+} sites from cubic symmetry. Moreover, the non-zero asymmetry parameter indicates that the Eu^{3+} site has a low symmetry (no threefold or fourfold axis passing through the Eu atom). The quadrupole interaction parameter is negative, as usual in the trivalent europium oxide compounds. In general, values are reported for most of the compounds ranging from −4–−8 mm/s. The values of the asymmetry parameter are in the range 0·8–1·0, as found in some crystalline oxide compounds[5] and oxide glasses.[18]

In ^{151}Eu spectroscopy, the variation of the quadrupole interaction parameter, reproduced by the convolution method, gives an evaluation of the disorder of the europium sites.[19] The standard deviation of the distribution of the quadrupole interaction parameter, in the EZBO glass, is equal to 16% of the most probable value; this value is smaller than in europium metaphosphate glass ($Eu(PO_3)_3$).[9]

5. Conclusions

In the zinc borate glass under investigation, we found that iron is present as ferric and ferrous ions in octahedral coordination with sites distorted compared to a regular geometry. The convolution method applied to spectroscopy of ^{57}Fe and ^{151}Eu gives the variation of the hyperfine parameters due to the structural disorder of iron and europium sites in glasses. In zinc borate glasses, the structural disorder of the iron sites is larger than in sodium phosphate glasses while the disorder of the europium sites is smaller than in europium metaphosphate glass.

References

1. Pozza, G., Ajò, D., Bettinelli, M., Speghini, A. & Casarin, M. *Solid State Comm.*, 1996, **97**, 521.
2. Kumar Mendiratta, S. & Almeida Valente, M. *Phys. B*, 1995, **211**, 338.
3. Gutlich, P., Link, R. & Trautwein, A. *Mössbauer Spectroscopy and Transition Metal Chemistry*. Springer-Verlag, Berlin, 1978, P. 56.
4. Dyar, M. D. *Am. Miner.*, 1985, **70**, 304.
5. Grandjean, F. & Long, G. J. *Mössbauer Spectroscopy Applied to Inorganic Chemistry*. Plenum Press, New York, 1989, **Vol. 3**, P. 513.
6. Bettinelli, M. & Flint, C. D. *J. Phys., Condens. Matter,* 1991, **3**, 4433.
7. Hesse, J. & Rubartsch, A. *J. Phys. E*, 1974, **7**, 526.
8. Viwel, C. & Mørup, S. *J. Phys. E*, 1981, **14**, 605.
9. Concas, G., Congiu, F., Spano, G., Speghini, A. & Gatterer, K. *J. Non-Cryst. Solids*, 1998, **232–234**, 341.
10. Murad, E. & Johnston, J. H. *Mössbauer Spectroscopy Applied to Inorganic Chemistry*. Edited by G. J. Long. Plenum Press, New York.1987, **Vol. 2**, P. 507.
11. Wickman, H. H. *Mössbauer Effect Methodology*. Edited by I. J. Gruverman. Plenum Press, New York, 1966, Vol. 2, P. 39.
12. Shenoy, G. K. & Dunlap, B. D. *Nucl. Instrum. Methods*, 1969, **71**, 285.
13. Kurkjian, C. R. & Sigety, E. A. *Phys. Chem. Glasses*, 1968, 9 (3), 73–83.
14. Dyar, M. D. *J. Am. Ceram .Soc.*, 1986, **69**, C-160.
15. Coey, J. M. D. *J. Phys.*, 1974, **C-6**, 89.
16. Dyar, M. D. *Am. Miner.*, 1984, **69**, 1127.
17. Concas, G., Congiu, F., Muntoni, C. & Pinna, G. *J. Phys. Chem. Solids*, 1995, **56**, 877.
18. Concas, G., Congiu, F., Muntoni, C., Bettinelli, M. & Speghini, A. *Phys. Rev. B*, 1996, **53**, 6197.
19. Medda, M. P., Piccaluga, G., Pinna, G., Bettinelli, M. & Cormier, G. *Z. Naturforsch.*, 1994, **49a**, 977.

The polymorphs of crystalline caesium enneaborate

Richard Haworth, Adrian C. Wright,* Roger N. Sinclair

J.J. Thomson Physical Laboratory, University of Reading, Whiteknights, Reading, RG6 6AF, UK

Kevin S. Knight

ISIS Pulsed Neutron & Muon Source, Rutherford Appleton Laboratory, Chilton, Didcot, OX11 0QX, UK

Natalia M. Vedishcheva, Irina G. Polyakova & Boris A. Shakhmatkin

Institute of Silicate Chemistry of the Russian Academy of Sciences, Ul. Odoevskogo 24, Korp. 2, St. Petersburg, 199155, Russia

The high temperature crystal structure of caesium enneaborate, γ-$Cs_2O.9B_2O_3$, has been refined using the Rietveld technique, from neutron powder diffraction data obtained with the high resolution powder diffractometer (HRPD) at the ISIS spallation pulsed neutron source. γ-$Cs_2O.9B_2O_3$ is tetragonal, space group $P4_122$, with unit cell dimensions a_0=8·77646±0·00016 and c_0=15·94929±0·00030 Å. This polymorph was formed by devitrification at high temperature, before undergoing a displacive transition to the orthorhombic polymorph, β-$Cs_2O.9B_2O_3$, detailed in earlier neutron studies, upon cooling below 316 K. The topology of the borate network remains unchanged, comprising two independent, interpenetrating borate networks of boroxol and triborate groups in the ratio 2:1. An analysis is presented of the detailed geometry of the borate units, and the probable existence of a third, low temperature, polymorph is discussed.

1. Introduction

The structure of crystalline caesium enneaborate, $Cs_2O.9B_2O_3$, is not only interesting in its own right, but it is also highly relevant concerning the role of boroxol (B_3O_6) groups in that of vitreous B_2O_3, since it is the only crystalline polymorph known to have a high proportion of its constituent boron atoms in boroxol groups. As indicated by the x-ray diffraction study of Krogh-Moe & Ihara,[1] the structure of caesium enneaborate comprises two independent interpenetrating borate networks made up from boroxol and triborate groups in the ratio 2:1, with no independent BO_3 triangles or BO_4 tetrahedra, and hence establishes a tendency for boroxol groups to form at low alkali contents. The Gmelin handbook[2] also lists crystalline enneaborate phases in the Na_2O–B_2O_3 (α, β, and γ polymorphs), K_2O–B_2O_3, Rb_2O–B_2O_3, Ag_2O–B_2O_3 and Tl_2O–B_2O_3 systems, but in no case has the crystal structure been determined.

A previous investigation of the structure of polycrystalline $Cs_2O.9B_2O_3$ by the current authors,[3] at ambient temperature (β-$Cs_2O.9B_2O_3$) and using high resolution diffraction data, yielded an accurate refinement in the orthorhombic space group $P222_1$.[4] This structure is qualitatively similar to that of Krogh-Moe & Ihara[1] (with the latter being refined in the higher symmetry tetragonal space group $P4_122$[4]). However, the lower symmetry orthorhombic space group means that there are now two symmetrically distinct boroxol groups in the asymmetric unit. For each of these, the bonds within the boroxol ring are longer than those outside, indicating that the origin of any stabilisation energy leading to the formation of these groups in the crystal (and so in the glass) is unlikely to be delocalised π-bonding.

The departure from tetragonal symmetry is slight, and suggested that a higher temperature tetragonal polymorph (γ-$Cs_2O.9B_2O_3$) may form initially on devitrification, before transforming to the orthorhombic polymorph observed at ambient temperatures. The objective of the present study, therefore, is to obtain accurate quantitative structural information concerning the borate network in $Cs_2O.9B_2O_3$ from ambient to high temperatures, and to chart any changes in symmetry that occur.

2. Experimental procedures

2.1. Sample

The sample of polycrystalline $Cs_2O.9B_2O_3$ was prepared at the Institute of Silicate Chemistry by devitrification from a glass of the same composition,

[1] Corresponding author. Email a.c.wright@rdg.ac.uk

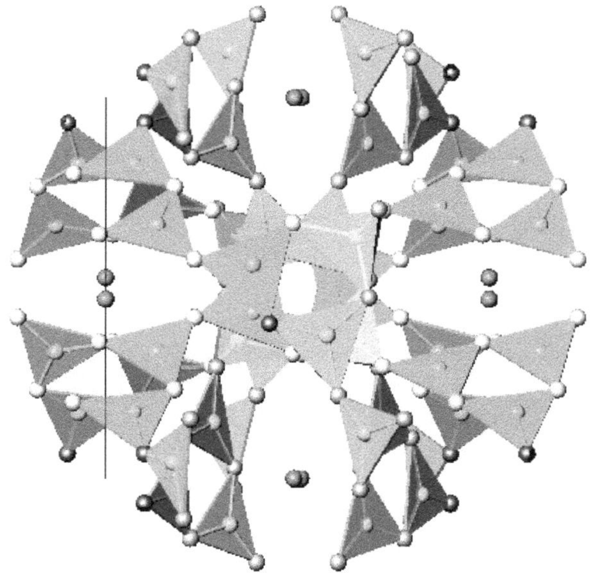

Figure 1. The unit cell of β-$Cs_2O.9B_2O_3$ (orthorhombic polymorph). In the electronic version of this paper, the non-symmetric BO_3 units are blue and green, the BO_4 basic structural units are yellow, and the caesium ions are red

using high purity Cs_2CO_3 and boric acid enriched in ^{11}B (99·27% ^{11}B), to avoid the high neutron absorption by ^{10}B. The Cs_2CO_3 was calcined at 500°C for 120 h before use. The glass was crystallised at 490°C for 10 days, followed by a slow increase to 560°C during 5 days and an anneal at this temperature for a further 7 days. The phase purity of the sample was investigated by x-ray diffraction and by determining the melting point by differential thermal analysis (DTA).

2.2. Neutron diffraction measurements

The neutron diffraction measurements employed the high resolution powder diffractometer (HRPD) at the ISIS spallation pulsed neutron source.[5] For the ambient[3] and low temperature experiments, powdered cæsium enneaborate was sealed into a 5 mm thick aluminium sample can with slab geometry and thin vanadium windows, through which the neutron beam passed. A gadolinium neutron absorbing mask was attached to the aluminium frame of the sample can facing the incident beam and back scattering detectors, to prevent contaminant Bragg peaks arising from the body of the sample can and the stainless steel frames surrounding the vanadium windows. For the high temperature experiment, the powder was sealed into a silica tube instead of the aluminium sample can. The back scattering detectors were focussed to a mean flight path of 95·8924 m and a scattering angle, 2θ, of 168·329°. Diffraction data were collected at incremental temperatures; between 4·2 and 200 K in the low temperature experiment, and between 183 and 560 K in the higher temperature experiment. The

Table 1. Unit cell parameters and R-factor

Parameter	*β-$Cs_2O.9B_2O_3$*[3]	*γ-$Cs_2O.9B_2O_3$*
Space group	$P222_1$	$P4_122$
a_0	8·80577(8) Å	8·77646(16) Å
b_0	8·75298(8) Å	8·77646(16) Å
c_0	15·79161(16) Å	15·94929(30) Å
ρ_0	1·64316(3)×10^{-3} c.u. Å^{-3}	1·62798(10)×10^{-3} c.u. Å^{-3}
R factor	R_p=0·054, R_{wp}=0·051	R_p=0·031, R_{wp}=0·032

resulting data were corrected for neutron absorption, normalised to the incident flux distribution via an upstream beam monitor and corrected for detector efficiency with a vanadium standard. Data in two time-of-flight windows, 30–130 and 100–200 ms were collected and data in the ranges 35–120 and 105–187 ms were binned logarithmically as $\Delta t/t$=3×10^{-4} for use in profile refinement. The crystalline unit cell, atomic coordinates and isotropic displacement parameters were extracted by Rietveld refinement,[6] in the tetragonal space group $P4_122$ (high temperature data) and the monoclinic space group P2 (low temperature data) using the GSAS suite of programs.[7] The full crystal structure was determined at 473·2 and 4·2 K, whilst only the unit cell parameters were found for the incremental data.

3. Results

Two distinct polymorphs are seen at high (>316 K) and ambient temperatures. The lattice parameters from the Rietveld refinement (at 473·2 K) are compared with those of our previous study[3] in Table 1 and the atomic positions and displacement parameters (again, at 473·2 K) are given in Table 2. The γ-$Cs_2O.9B_2O_3$ unit cell contains two $Cs_2O.9B_2O_3$ composition units (c.u.). The final structure is qualitatively very similar to that of the ambient temperature polymorph, β-$Cs_2O.9B_2O_3$, but differs in detail due to the higher symmetry space group ($P4_122$, cf. $P222_1$), as may be seen from Figures 1 and 2. In particular, there is now only one symmetrically distinct boroxol group instead of two. The atomic numbering in the asymmetric unit of the β-phase follows that of Krogh-Moe

Table 2. Atomic positions and isotropic displacement parameters for γ-$Cs_2O.9B_2O_3$

	Position[4]	x_i	y_i	z_i	u_{11}×100 *(Å^2)*
Cs	2c	0	0·4893(5)	0	4·7(25)
B1	4e	0·3254(4)	0·13433(35)	0·05895(19)	1·7(12)
B2	4e	0·43723(33)	0·38328(32)	0·11092(13)	0·8(12)
B3	4e	0·28238(28)	0·86609(28)	0·02817(16)	0·5(12)
B4	4e	0·17746(35)	0·0588(4)	0·94222(22)	2·7(12)
B5	4e	0·35067(30)	0·35067(30)	0·3750000(0)	0·3(10)
O1	4e	0·2314(4)	0·1678(4)	0·99307(23)	5·3(14)
O2	4e	0·29000(34)	0·71961(33)	0·05131(18)	4·3(13)
O3	4e	0·3187(4)	0·48874(34)	0·11693(18)	3·4(11)
O4	4e	0·40467(35)	0·2296(4)	0·10206(21)	5·0(14)
O5	4e	0·3462(4)	0·9788(4)	0·07556(20)	6·2(17)
O6	4e	0·2001(4)	0·9081(4)	0·95701(19)	5·4(15)
O7	4e	0·0903(5)	0·0903(5)	0·8750000(0)	11·1(20)
O8	4e	0·5758(4)	0·5758(4)	0·3750000(0)	8·3(17)

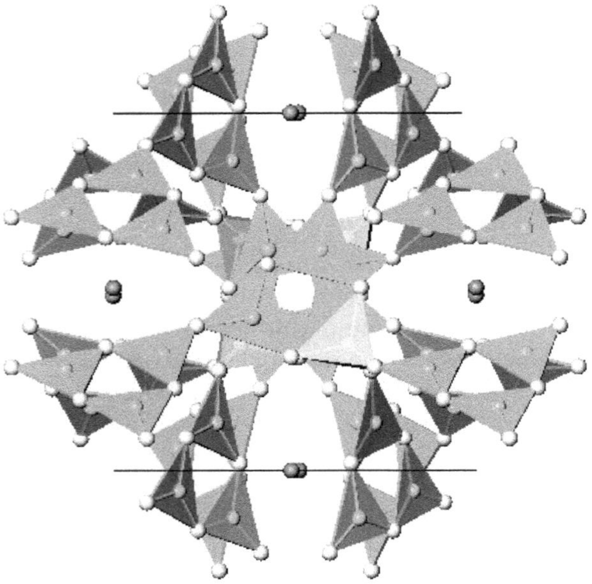

Figure 2. The unit cell of γ-$Cs_2O.9B_2O_3$ (tetragonal polymorph). In the electronic version, BO_3 and BO_4 basic structural units are green and yellow, respectively, whilst the caesium ions are red

& Ihara[1] with a prime to indicate the second atom of each of those pairs separated by the lower symmetry, and for ease of comparison with the corresponding, equivalent, pairs in the γ-phase. Interatomic distances and bond angles were calculated with the DISAGL program[7] and are given in Tables 3 to 8. Further lowering of symmetry, with lower temperatures, was suspected and a subsequent experiment using the same sample, at nine temperatures between 4·2 and 200 K, revealed a change to monoclinic symmetry below 145 K.

4. Discussion

4.1. Structure of β-caesium enneaborate

The topology of the borate network is unchanged from Krogh-Moe & Ihara[1] and the authors' previous paper.[3] Another structural study[8] of the ambient temperature β-$Cs_2O.9B_2O_3$, by N. Penin *et al*, was published shortly after that in Ref. 3, using a single crystal and x-ray diffraction, also with the same topology. The triborate ring is linked only to (four) boroxol groups, by sharing oxygen atoms O2, O2′, O4 and O4′, while the two boroxol groups in the asymmetric unit are interconnected via O7. The bond lengths and angles determined in the latter two studies are in good agreement, with the greatest differences found in the average B–O distances. In this system, the x-ray diffraction pattern will be dominated by the caesium scattering and the contrast between the boron and oxygen scattering is low, leading to less accurate measurements of the B–O distances, and conversely, the Cs–O distances may be more accurate than those obtained by neutron diffraction. Surprisingly, Penin *et al* did not predict a higher symmetry polymorph to exist, even though the observation is made that the departure from tetragonal symmetry is very small.[8]

Table 3. Boroxol ring bond lengths for β- and γ-$Cs_2O.9B_2O_3$

B–O bond	*β Ring 1*	*β Ring 2(′)*	*γ Ring*
Internal bonds			
B1–O1	1·407(6) Å	1·378(7) Å	1·3685(5) Å
B1–O5	1·377(7)	1·402(7)	1·402(4)
B3–O5	1·362(7)	1·362(7)	1·3653(34)
B3–O6	1·416(6)	1·383(7)	1·3947(31)
B4–O1	1·301(6)	1·386(8)	1·341(4)
B4–O6	1·397(7)	1·369(7)	1·358(4)
Average	1·377(3)	1·380(3)	1·3716(13)
External bonds			
B1–O4	1·332(7)	1·311(6)	1·287(4)
B3–O2	1·307(6)	1·336(6)	1·3392(32)
B4–O7	1·361(8)	1·358(8)	1·346(4)
Average	1·333(4)	1·335(4)	1·3240(22)

Table 4. Boroxol ring bond angles for β- and γ-$Cs_2O.9B_2O_3$

	β Ring 1	*β Ring 2(′)*	*γ Ring*
Internal B–Ô–B angles			
B1–O1–B4	118·3(4)°	118·2(5)°	121·62(31)°
B1–O5–B3	125·1(5)	123·2(5)	123·19(3)
B3–O6–B4	117·5(5)	120·7(5)	118·31(31)
Average	120·3(3)	120·7(3)	121·04(18)
Internal O–B–O angles			
O1–B1–O5	117·4(5)	117·4(5)	115·64(3)
O5–B3–O6	115·8(4)	117·1(4)	118·14(28)
O1–B4–O6	125·2(5)	121·1(5)	122·58(32)
Average	119·5(3)	118·5(3)	118·79(17)
External O–B–O angles			
O1–B1–O4	124·8(5)	125·0(5)	126·64(32)
O4–B1–O5	117·3(5)	117·5(5)	117·47(3)
O2–B3–O5	123·9(5)	124·2(4)	121·51(28)
O2–B3–O6	120·2(5)	118·7(4)	120·24(29)
O1–B4–O7	122·9(5)	123·0(5)	122·42(34)
O6–B4–O7	111·7(5)	115·8(5)	114·92(33)
Average	120·1(2)	120·7(2)	120·53(13)

Table 5. Triborate ring bond lengths for β- and γ-$Cs_2O.9B_2O_3$

	β	*γ*
Internal B^{III}–O bonds		
B2–O3	1·333(6) Å	1·3315(34) Å
B2–O8	1·431(7)	1·3710(33)
B2′–O8	1·378(8)	-
B2′–O3′	1·301(6)	-
Average	1·361(4)	1·3513(24)
External B^{III}–O bonds		
B2–O4	1·368(7)	1·3701(34)
B2′–O4′	1·392(6)	-
Average	1·380(5)	1·3701(34)
Internal B^{IV}–O bonds		
B5–O3	1·475(6)	1·4428(33)
B5–O3′	1·478(6)	-
Average	1·476(4)	1·4428(33)
External B^{IV}–O bonds		
B5–O2	1·462(7)	1·4301(30)
B5–O2′	1·460(7)	-
Average	1·461(5)	1·4301(30)

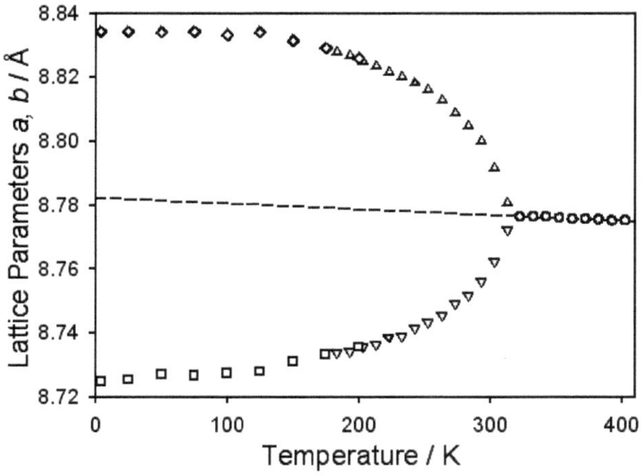

Figure 3. The lattice parameters change with temperature, as the symmetry of the structure changes from monoclinic to orthorhombic, and finally to tetragonal at high temperature

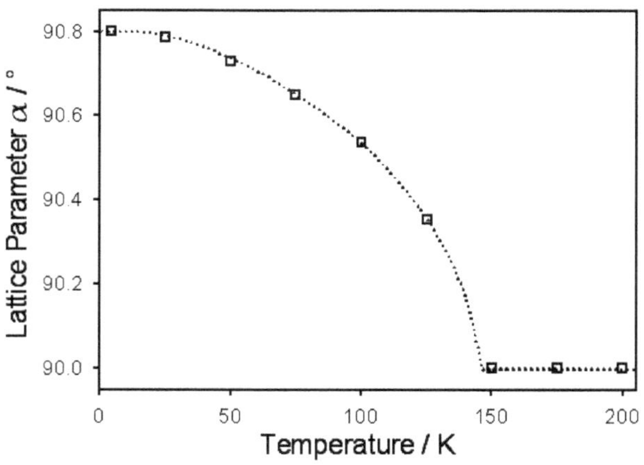

Figure 4. The orthorhombic to monoclinic transition occurs at 145 K as the angle, α, between a and b, distorts beyond 90°

4.2. Structure of γ-caesium enneaborate

In the current study of γ-$Cs_2O.9B_2O_3$, the average B^{III}–O and B^{IV}–O bond lengths (1·356±0·001 and 1·436±0·002 Å, respectively) are in good agreement with those found in other alkali borate structures. The caesium atom has ten oxygen atom neighbours between 3·14 and 3·48 Å, the mean Cs–O bond length being 3·294±0·002 Å. The distribution of Cs–O distances is very broad, the root mean square (RMS) bond length variation being 0·114 Å. In comparison with the β-phase, the two boroxol rings are no longer non-equivalent, as the relative tilt between them is absent due to the equivalence of what were the O2 and O2′ oxygen atoms in the BO_4 tetrahedra. The average

Table 6. Triborate ring bond angles for β- and γ-$Cs_2O.9B_2O_3$. Note that the prime indicates non-equivalent sites in the β-phase only

Internal B^{III}–Ô–B^{III} angle	*β*	*γ*
B2–O8–B2′	116·2(5)°	121·7(5)°
Internal B^{III}–Ô–B^{IV} angles		
B2–O3–B5	121·5(4)	123·13(31)
B2′–O3′–B5	120·1(4)	-
Average	120·8(3)	123·13(31)
Internal O–B^{III}–O angles		
O3–B2–O8	121·6(5)	119·1(4)
O8–B2′–O3′	125·9(5)	-
Average	123·8(4)	119·1(4)
External O–B^{III}–O angles		
O3–B2–O4	127·2(5)	126·03(32)
O4–B2–O8	111·2(5)	114·34(33)
O8–B2′–O4′	111·7(5)	-
O3′–B2′–O4′	122·3(5)	-
Average	118·1(3)	120·19(23)
Internal O–B^{IV}–O angle		
O3–B5–O3′	114·1(4)	112·8(4)
External O–B^{IV}–O angles		
O2–B5–O3	104·7(4)	106·01(16)
O2–B5–O2′	111·4(4)	110·74(32)
O2–B5–O3′	112·9(4)	110·68(16)
O3–B5–O2′	109·3(4)	110·68(16)
O2′–B5–O3′	104·5(4)	106·01(16)
Average	108·6(2)	108·82(9)

internal B–O bond lengths in the boroxol rings are in good agreement between the two phases, as are the average internal angle(s). This confirms the prediction that the boroxol rings, to first approximation, behave as rigid structural units during phase transitions,[(3)] and so it is not unreasonable to expect them to retain this property during the glass transition. The average B^{III}–O bond length, in the β- and γ-phases of $Cs_2O.9B_2O_3$, when compared to that in vitreous B_2O_3, also supports this (1·364±0·002, 1·356±0·001 and 1·365±0·001 Å, respectively). As previously discussed in Ref. 3, the origin of a stabilisation energy could not be due to delocalised π-bonding, as the external B–O bond lengths were clearly less than the internal bond lengths (not the reverse, as required by delocalised π-bonding). This remains true for the B–O bond lengths in the γ-phase also (1·372±0·001 compared with 1·324±0·002 Å for the external bonds).

The triborate group does not remain unchanged between the two phases, with considerable differences between most of the bond lengths and angles in the two cases (Tables 5 and 6). The only value to remain similar is the average of the external O–B^{IV}–O angles. However, the RMS bond length variation for all of the bonds in BO_3 triangles (0·029 Å) is much greater than that for the BO_4 tetrahedron (0·001 Å), with much of this variation resulting from the boroxol

Table 7. Oxygen bond angles between superstructural units for β- and γ-$Cs_2O.9B_2O_3$. Note that the prime indicates non-equivalent sites in the β-phase only

B^{III}–Ô–B^{III} angle (between 2 boroxol groups)	*β*	*γ*
B4–O7–B4′	143·3(6)°	150·2(5)°
B^{III}–Ô–B^{III} angles (between boroxol and triborate groups)		
1–O4–B2	138·8(5)°	140·61(35)°
B1′–O4′–B2′	143·4(5 °	-
Average	141·1(4)°	140·61(35)°
B^{III}–Ô–B^{IV} angles (between boroxol and triborate groups)		
B3–O2–B5	128·4(5)°	131·24(27)°
B3′–O2′–B5	128·8(4)°	-
Average	128·6(3)°	131·24(27)°

Table 8. Cs–O bond lengths for β- and γ-$Cs_2O.9B_2O_3$ (two neighbours at each distance). Note that the prime indicates non-equivalent sites in the β-phase only

Oxygen	*β Cs*	*β Cs′*	*γ Cs*
1 or 1′	3·422(8) Å	3·410(8) Å	3·478(5) Å
2 or 2′	3·374(7)	3·339(7)	3·352(4)
3 or 3′	3·425(5)	3·294(5)	3·262(3)
4 or 4′	3·233(6)	3·188(6)	3·240(4)
5 or 5′	3·162(8)	3·151(8)	3·140(4)
Average	3·323(3)	3·276(3)	3·367(2)

group, as in the β-phase.

The obvious difference between the phase structures is manifest in the bridging oxygen bond angles between the superstructural units (Table 7), with those found in the γ-phase all systematically larger than in the β-phase. It is the opening out of these bond angles which results in the reduction of the external B–O bond lengths in both the boroxol and triborate groups (Tables 1 and 3).

4.3. The phase transitions

The phase transitions of caesium enneaborate are a fine example of a classic displacive process. From the tetragonal phase, symmetry is broken as the lattice parameters a_0 and b_0 start to diverge at 316±3 K (from 8·77646±0·00016 to 8·80577±0·00008 Å, and from 8·77646±0·00016 to 8·75298±0·00008 Å, respectively) into the orthorhombic phase (Figure 3). The orthorhombic to monoclinic transition then occurs at 145±2 K as the angle α, between b and c distorts beyond 90° to 90·8° (Figure 4). Note that the monoclinic unit cell is quoted in a tetragonal setting with $a_0 \approx b_0 \neq c_0$. It might be expected that the monoclinic distortion would be between a and b, i.e. with $\gamma \neq 90°$ and space group $P2_1$, but in fact the shear develops between one of the short axes and the long axis, with space group P2.

5. Conclusions

The high temperature polymorph, γ-$Cs_2O.9B_2O_3$, remains topologically unchanged from the ambient temperature structure, but undergoes a displacive transition at 316±3 K from tetragonal to orthorhombic symmetry upon cooling. As with β-$Cs_2O.9B_2O_3$, the bonds within the boroxol rings are longer than those outside indicating that delocalised π-bonding still has no part in the stabilisation energy of these groups, even upon their formation at high temperatures. The boroxol groups within these phases remain structurally very similar, with the major structural differences arising in the B–O–B angles between superstructural units, and in the triborate group to a lesser extent. A lower temperature, monoclinic phase, α-$Cs_2O.9B_2O_3$, has also been observed below 150 K, but problems with the refinement have so far prevented a complete structural analysis beyond the crystal classification.

Acknowledgements

The authors are grateful for technical and scientific support received from the staff at the ISIS spallation pulsed neutron source (Rutherford Appleton Laboratory, Chilton, UK) and to the EPSRC for financial support.

References

1. Krogh-Moe, J & Ihara, M. *Acta Crystallogr.*, 1967, **23**, 427.
2. Heller, G. *Gmelin Handbuch der Anorganischen Chemie, Vol. 28, Borverbindungen Part 7*, Springer-Verlag, Berlin, 1975.
3. Wright, A. C., Sinclair, R. N., Stone, C. E., Knight, K. S., Polyakova, I. G., Vedishcheva, N. M. & Shakhmatkin, B. A. *Phys. Chem. Glasses*, 2003, **44** (3), 197.
4. *International Tables for X-Ray Crystallography, Vol. 1*, Eds Henry, N. F. M. & Lonsdale, K. Kynoch, Birmingham, 1969, 103.
5. http://www.isis.rl.ac.uk/crystallography/hrpd
6. Rietveld, H. M. *Acta Crystallogr.*, 1967, **22**, 151.
7. Larson, A. C. & Von Dreele, R. B. *GSAS (General Structure Analysis System) LAUR 86-748*, Los Alamos Nat. Lab., 1994.
8. Penin, N., Touboul, M. & Nowogrocki, G. *J. Solid State Chem.*, 2003, **175**, 348.

Proc. Fifth Int. Conf. on Borate Glasses, Crystals and Melts Phys. Chem. Glasses: Eur. J. Glass Sci. Technol. B, August 2006, 47 (4), 357–363

Local structural disorder and mobility of Ag^+ ions in silver borate glasses containing α-AgI stabilised at room temperature

G. Dalba, P. Fornasini

Dipartimento di Fisica, Universit`a defli Studi di Trento, Via Sommarive 14, I-38050 Povo, Italy

F. Rocca

IFN-CNR, Istituto di Fotonica e Nanotecnologie - Sezione "ITC-CeFSA" di Trento, Via Sommarive 18, I-38050 Povo, Italy

F. Monti[1]

Dipartimento di Informatica Facolt`a di Scienze Matematiche Fisiche e Naturali, Universit`a di Verona C`a Vignal 2, Strada le Grazie, I-37134 Verona, Italy

T. Minami & M. Tatsumisago

Department of Applied Sciences, Osaka Prefecture University, Sakai 599, Japan

EXAFS temperature dependent measurements at the I K-edge have been done on AgI rich silver borate glasses containing crystallites of AgI stabilised in the α-phase at room temperature, as obtained by fast roller quenching. Data analysis was performed in the cumulant expansion approximation, to distinguish between structural and vibrational disorder, and the first shell I–Ag radial distribution functions were determined. The thermal behaviour of this composite will be compared to that of a homogeneous AgI–AgO–B_2O_3 glassy systems. The relation between temperature dependent local disorder and mobility of Ag^+ ions in both systems will be discussed also in the light of previous results obtained by x-ray diffraction, conductivity and heat capacity measurements.

1. Introduction

Among typical ionic conductors, AgI and AgI related compounds show very large ionic conductivities. It is well known that the high conductivity (>10 S cm^{-1}) α-phase of AgI transforms into the low conductivity β-phase (10^{-5} S cm^{-1} at room temperature) below 147°C. To improve the ionic conductivity at room temperature (RT), the most promising AgI-based compounds were produced by Minami and co-workers, who succeeded in stabilising the α-AgI phase at RT in a silver borate glass. Such composites were obtained by quenching a melt of AgI–Ag_2O–B_2O_3 with a high content of AgI,[1,2] and show a RT ionic conductivity comparable to that of α-AgI. In particular, they demonstrated that crystallites of a strained α-AgI phase are present in the rapidly quenched systems AgI–Ag_2O–B_2O_3 having a ratio Ag_2O/B_2O_3=3 (orthoborate composition) with an AgI content higher than 75% mol (composite): a typical composition for this composite is 82AgI.13·5Ag_2O.4·5B_2O_3. When the AgI concentration is greater than 85% mol, β-AgI precipitation occurs; at 75% mol of AgI and below, no crystalline phase of AgI is observed by diffraction (pure glass).

X-ray diffraction (XRD), field emission scanning electron microscopy (FE-SEM), differential scanning calorimetry (DSC) and conductivity measurements were performed by Minami and co-workers to understand formation, microstructure, stability and temperature behaviour of these glassy systems[3–5] above and below RT.

In the high temperature range, at about T=350 K, crystallisation of β-AgI has been observed in the pure glass of composition 75AgI.18·75Ag_2O.6·25B_2O_3.[4] At the same temperature, in the composite 82AgI.13·5Ag_2O.4·5B_2O_3 there is a steep decrease in the strain parameter; above it, growth and aggregation of larger particles of α-AgI appears.[4] As a result, on cooling the composite down to RT, the expected α to β transformation occurs. On the contrary, α-AgI cannot transform to β-AgI if the maximum temperature reached on heating is lower than T=350 K. In any case, it must be noted also that the α to β transition takes place at a lower temperature than the phase

[1] Corresponding author.

transition temperature of the AgI crystal (420 K), since the highly viscous glassy matrix depresses the α to β phase transformation: in fact, AgI in the crystalline β-phase needs a larger volume than in the α-phase, while the crystallites are expected to be compressed by the matrix glass.[2,4]

In the low temperature range (below RT), the composite shows an unexpected anomalous behaviour in conductivity, heat capacity and crystal lattice strain: at about 260 K, on heating, the lattice strain and the activation energy decrease steeply, while the heat capacity shows an endothermic peak.[4]

The aim of this work is to gain new insights into the temperature behaviour of these AgI rich silver borate glasses from a local structure point of view by EXAFS. EXAFS is one of the most powerful techniques able to provide detailed information on the local structure and dynamics around a selected atomic species in disordered systems. In this work, the local structure around I atoms is investigated in the composite and in the pure glass as a function of temperature. In order to characterise structural and vibrational disorder, highly accurate EXAFS measurements have been carried out and the cumulant expansion method has been applied in the analysis of the first I–Ag shell. Particular care has been devoted to the temperature behaviour of the composite around 260 K where the cited anomalous behaviour occurs. The pure glass has been measured to compare its temperature behaviour to that of the composite and to isolate the temperature behaviour of the α-AgI phase in the crystallites by assuming that the amorphous AgI phase present in the matrix glass is structurally equivalent to the pure glass one. The results of the present EXAFS investigation will be discussed in connection with the experimental data obtained by Minami *et al.*[3–5]

2. Experimental

We have investigated three samples: (1) crystalline β-AgI, which was also used as the reference sample for extracting EXAFS phases and amplitudes; (2) the composite, $82AgI.13{\cdot}5Ag_2O.4{\cdot}5B_2O_3$, containing the α-AgI phase stabilised at RT; (3) the pure glass, $75AgI.18{\cdot}75Ag_2O.6{\cdot}25B_2O_3$. The last two compounds were prepared[1,2] by twin-roller quenching a melt of $AgI–Ag_2O–B_2O_3$ containing different amounts of AgI and the same proportion of Ag_2O with respect to B_2O_3 (3:1). FE-SEM and XRD measurements[1,2] show that, at RT, the pure glass contains fine particles of amorphous AgI (40–60 nm in diameter), while the composite contains finer particles (crystallites of 20–30 nm in diameter) homogeneously dispersed in a matrix glass of $AgI–Ag_2O–B_2O_3$. These very fine particles in the composite are grains of a strained crystalline α-phase of AgI. From calorimetric measurements (DSC), described in detail in Ref. 6, the amount of α-AgI crystallites inside the composite has been estimated to be about 70%, while the remaining 30% of AgI is present in the matrix glass in the same amorphous phase as in the pure glass.

EXAFS measurements described in the present work were performed at the I K-edge in transmission mode at the BM29 beamline of ESRF as a function of temperature, from 15–380 K. The crystalline β-AgI sample was measured on heating from 15 up to 350 K. The composite was measured starting from RT, cooling it down to 15 K and again heating it up to 380 K. The pure glass was first measured at RT, then directly cooled down to 15 K and measured on heating up to RT. Two independent measurements were made for each temperature on all the samples.

3. Data analysis

The EXAFS characteristic functions $\chi(k)$, where k is the photoelectron wave vector, were extracted from the experimental absorption spectra.[7,8] The $\chi(k)$ of the α-AgI crystallites were isolated at each available temperature (below 300 K) by subtracting from the characteristic function of the composite the known contribution of the amorphous phase of AgI present in the matrix glass, assumed to be equal to the pure glass one. All the $\chi(k)$ functions were k^2 weighted and Fourier transformed in the range k=2·57–14 Å^{-1} using a Gaussian window.

The first shell contribution to the EXAFS characteristic function was singled out by backtransfoming the corresponding peak and the data analysis was performed using the phase difference and amplitude ratio method,[8] with the crystalline β-AgI sample measured at 15 K as reference, in the cumulant expansion approximation. This method, as it is fully described elsewhere,[9] allows to reconstruct in a model independent way the radial distribution function (RDF) by determining some meaningful parameters which characterise the distribution – the cumulants C_n^*. In fact, the single shell EXAFS $\chi(k)$ is sensitive to the cumulants C_n of the so called effective distribution of distances, which is strictly related to the RDF, so that only the first cumulant requires corrections when passing from the experimental effective distribution C_1 to the RDF C_1^*.[9] The number of cumulants which have to be determined from the analysis depends on the behaviour of phases and amplitudes and on the convergence requirements of the characteristic function. C_1^* is the mean value and C_2 is the variance of the RDF. C_3 and higher order odd cumulants are related to the asymmetry of the distribution, while C_4 and higher order even cumulants give the symmetric deviation from a Gaussian shape.

The knowledge of the RDF cumulants in temperature dependent measurements is very useful since their temperature behaviour allows, in principle, to separate the structural from the vibrational (or thermal) contribution to disorder. This can be done because the temperature dependence of the cumulants due to thermal disorder is known,[10,11] as it is derived in the classical approximation from the anharmonic mean force effective pair potential. In amorphous systems, the contribution of local structural disorder at low temperatures can be estimated by comparison with the corresponding crystalline phase, assuming a similar local vibrational disorder; conversely, any deviation from the expected behaviour of the cumulants at increasing temperature can be attributed to temperature dependent structural changes.[12]

In our case, the phases and amplitudes of the first coordination shell were analysed by the cumulant expansion method up to the sixth cumulant C_6. The need for the sixth cumulant in the amplitude analysis of all the samples was justified by the temperature behaviour of the even cumulants and, more important, by the convergence requirements of the characteristic functions. In fact, it is well established and it has been verified in the case of our analysis, that the characteristic function does not converge if the cumulant expansion is truncated to the fourth cumulant when $C_4>0$. Due to the presence of a beating at $k\approx6$ Å reflecting the resonant behaviour of the silver backscattering amplitude, the phase differences were fitted in the range 6·5–11·5 Å^{-1}. The logarithms of amplitude ratios were first fitted up to C_4 in the range 2·6–14 Å^{-1} taking the ratio of the coordination numbers N_s/N_r as a free parameter. Then, a second fit was performed up to C_6 in the range 6·5–11·5 Å^{-1} keeping the coordination number fixed. The absolute values of the first two cumulants were calculated assuming for the first shell of the reference β-AgI sample a Gaussian shape with C_1=2·81 Å and C_2=0·00305 Å^2, determined by fitting an Einstein model to the C_2 temperature behaviour. Checks on the determined coordination numbers and cumulants (and consequent error bar determination) were made on all the samples at any temperature by choosing different backtransform ranges, as well as taking the sample itself at T=15 K as the reference compound.

4. Results

The first shell coordination number around iodine is lower than in β-AgI both in the crystallites and in the pure glass, while in the composite it is midway between them (Table 1). The reason for these lower coordination numbers can be attributed either to structural disorder (the lower the coordination number the higher the structural disorder) or to a higher mobility of Ag^+ ions (the lower the coordination number the higher the mobility). From conductivity measurements,[3–5] we know that the mobility of Ag^+ ions in the composite is higher than in β-AgI

Table 1. I–Ag coordination numbers and mean values (in Å) of the RDFs at 15 and 300 K

	N	C1* (15 K)	C1* (300 K)
Crystallites	3·7±0·1	2·825±0·001	2·823± 0·001
Composite	3·5±0·1	2·833±0·001	2·829±0·001
Pure glass	3·2±0·1	2·865±0·002	2·838±0·003

(and the coordination number results lower), while in the pure glass it is lower than in the composite: this means that the lower coordination number in the pure glass with respect to the composite and to the crystallites cannot be attributed to a higher mobility of Ag^+ ions, but must be due to structural disorder, i.e. in the amorphous phase I ions are on average coordinated to less than 4 Ag ions. Then, as regards the coordination number in the composite, the contribution from the matrix glass is only related to structural disorder, while the presence of the α-AgI crystallites adds a contribution which could be due to the higher mobility of Ag^+ ions (as well as, partly, also to the crystallites straining).

The values of the first four cumulants for the composite, measured on heating from 15 to 380 K, are plotted in Figure 1 (solid triangles) together with that of the α-AgI crystallites (open triangles), of the

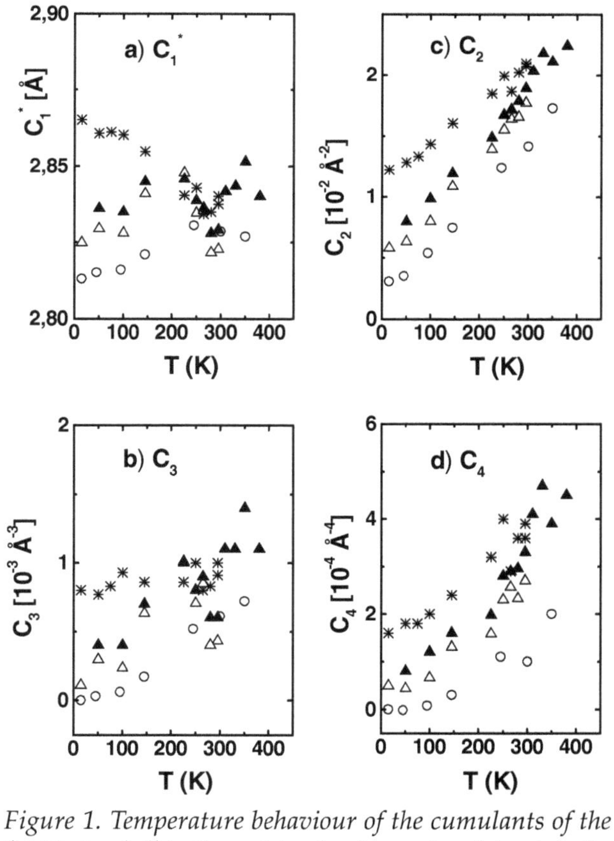

Figure 1. Temperature behaviour of the cumulants of the first I–Ag shell in the matrix glass/pure glass (stars), in the composite (solid triangles), in the α-AgI crystallites (open triangles) and in β-AgI (circles). Error bars are within the size of the symbols

pure glass (stars) and of β-AgI (circles). As regards the composite, three regions can be distinguished: (1) below 225 K, the temperature dependence of all the cumulants is similar to that of the β-AgI crystal, with only an overall average constant contribution of structural disorder (Table 2), which affects the I–Ag distance (ΔC_1^*) and the asymmetry of the I–Ag RDF (ΔC_3); (2) between 225 and 300 K, an anomalous steep reduction appears both in C_1^* and in C_3, until they reach the same values of β-AgI at RT; correspondingly, an anomalous slight step appears in C_2 and, more pronounced, in C_4. Anyway, contrary to C_1^* and C_3, neither C_2 nor C_4 become equal to the corresponding values of β-AgI at T=300 K. In the centre of this region, at 260 K, the anomaly cited in the introduction was observed; (3) above RT, only a few data points were collected, which anyway allow to recognise some trends: all the cumulants grow with a higher slope than below 225 K; moreover, an irregular behaviour can be seen at around T=350 K.

Table 2. Structural disorder contribution relative to β-AgI in the first three cumulants at T=15 K

T = 15 K	ΔC_1^* (Å)	ΔC_2 (10^{-2} Å^{-2})	ΔC_3 (10^{-3} Å^{-3})
Crystallites	0·010±0·001	0·23±0·01	0·11±0·02
Composite	0·018±0·001	0·44±0·02	0·25±0·03
Pure glass	0·050±0·002	0·92±0·03	0·80±0·05

At this temperature (350 K) FE-SEM and diffraction measurements made on heating indicate the growth and aggregation of bigger particles of α-AgI in the composite[13] and crystallisation of β-AgI in the pure glass.[4]

As regards the crystallites, the temperature behaviour of the cumulants is very similar to the composite, with only an overall reduction of structural disorder (Table 2), suggesting that the temperature behaviour of the cumulants found in the composite should be mainly attributed to the presence of the crystallites themselves. These considerations are supported by the quite different behaviour shown by the cumulants of the pure glass, which was largely discussed in a previous paper.[12] The low T values of all the cumulants are higher than in β-AgI and in the crystallites, but with increasing temperature to 300 K they tend to reach those of β-AgI.

The interplay between the temperature behaviour of the odd and even cumulants is summarised by the corresponding RDF. The first shell I–Ag RDF below and above 225 K (up to 300 K) normalised to unity are shown in Figure 2 for β-AgI, for AgI in the crystallites (strained α-phase) and for AgI in the pure glass/matrix glass (amorphous AgI). Below 225 K, the α-AgI RDF shows a higher degree of disorder with respect to the β-AgI RDF, but the broadening

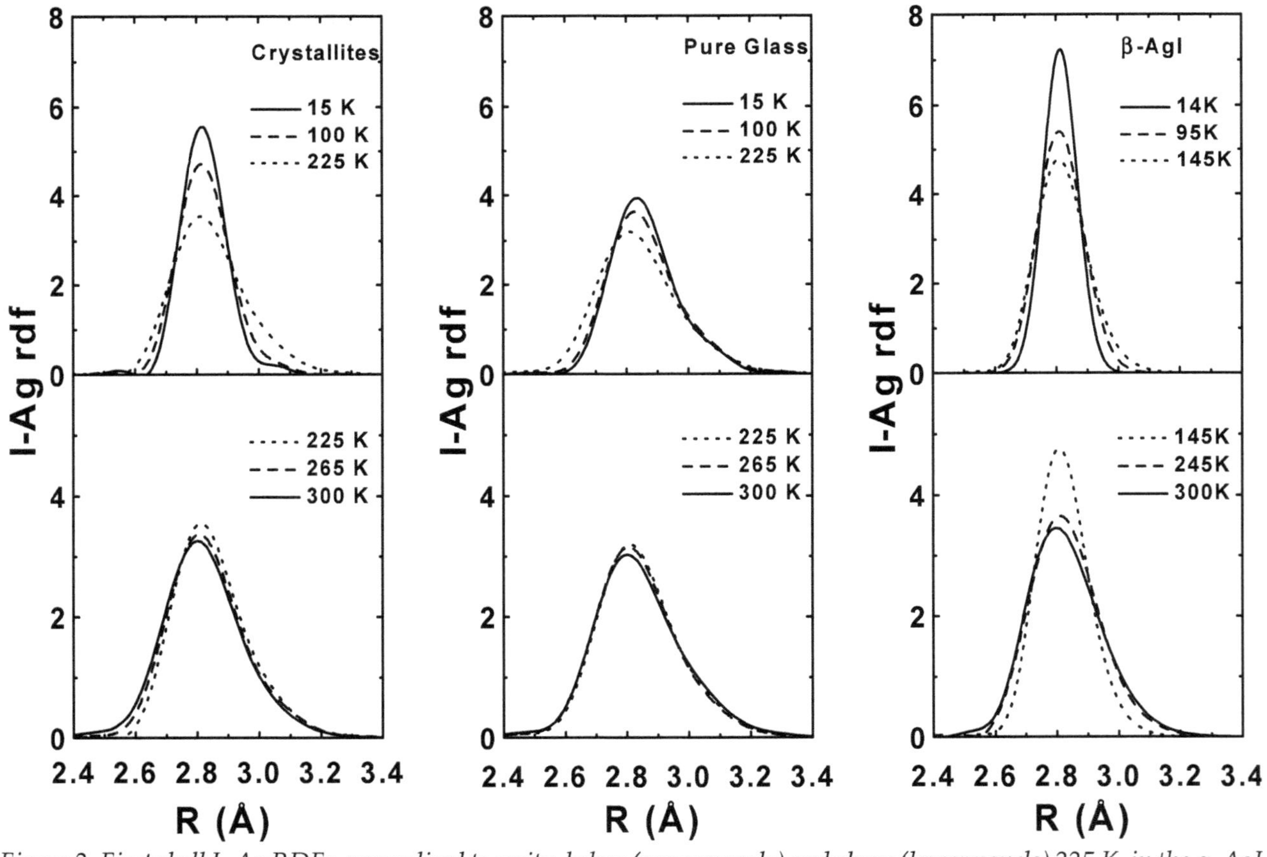

Figure 2. First shell I–Ag RDFs, normalised to unity, below (upper panels) and above (lower panels) 225 K, in the α-AgI crystallites, in the matrix glass/pure glass and in β-AgI

both in the low-*R* and in the high-*R* side at increasing temperature appears to be similar for the two phases. On the contrary, the RDF in the pure glass broadens at the short distances, while the tail at the long distance side remains the same. Above 225 K up to 300 K, the I-Ag RDF of the α-AgI phase shows a net overall shift towards lower distances, while in the pure glass it undergoes only slight changes both in the low-*R* side and in the high-*R* side. At the same time, the I–Ag RDF of the β-phase broadens on the high-*R* side. As a result, the β-phase and the α-phase RDF become more and more similar in shape on the high-*R* side and tend to slightly differ on the low-*R* side. As regards the composite (Figure 3), below RT the behaviour of the I–Ag RDF is very similar to that of the α-AgI crystallites. Above RT, a broadening is seen only in the high-*R* tail of the RDF, suggesting a higher mobility of Ag^+ ions in the high temperature range.

5. Discussion

The thermal behaviour of the pure glass was discussed in a previous paper by Dalba *et al.*[12] This glass is characterised by a first-shell mean I–Ag distance higher than β-AgI, and exhibits a contraction in the range 15–300 K as shown by C_1^* in Figure 1 (stars). By observing that the first three cumulants of the glass tend to assume the same values of the corresponding cumulants of β-AgI, the negative thermal expansion was ascribed to a local relaxation of the amorphous AgI network inside the borate matrix. In other words, the I–Ag distance distribution at low temperature is continuously modified with increasing temperature till the same first mean I–Ag distance of β-AgI is reached.

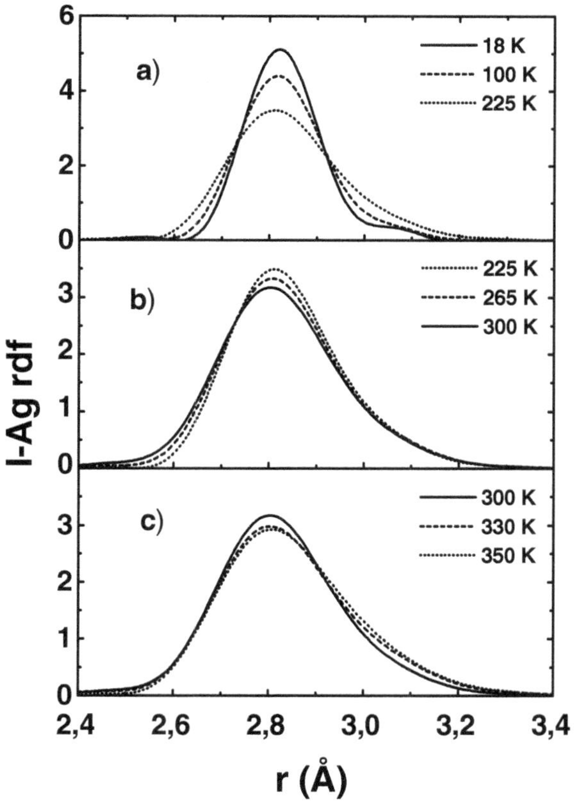

Figure 3. First shell I–Ag RDFs, normalised to unity, in the composite below 225 K (a), between 225K and 300 K (b) and above 300 K (c)

As regards the cumulants and the RDF of the composite, on heating we have observed a decrease in the odd terms between 225 and 300 K, particularly steep at about 260 K (Figure 1). This suggests that a local structural reorganisation occurs around iodine, as also shown by the RDF, which becomes more similar to the crystalline β-AgI above 260 K, due to the pronounced broadening at the low-*R* side with respect to the high-*R* side (Figures 2 and 3). On the other hand, the change in slope of the temperature dependence of the cumulants (Figure 1) above RT (reflected as a strong broadening at the high-*R* side in the corresponding RDF, see Figures 2 and 3) with respect to their temperature dependence below 225 K (Figure 1) (reflected as a broadening both at the low-*R* side and at the high-*R* side in the RDF, see Figures 2 and 3), indicates that the resulting average RDF is characterised by an increased thermal disorder. This process may be related to a higher mobility of Ag^+ ions above RT. This interpretation is supported by Minami and co-workers findings: the endothermic peak in the heat capacity at 260 K can be explained by a disorder–order transition; the steep decrease of the strain parameter at 260 K suggests a relaxation of the glassy matrix, which allows the disorder–order transition to take place, while the steep decrease in the activation energy indicates an increase in the mobility of the Ag^+ ions. The fact that the ordering of the local structure around Iodine is accompanied by an increased mobility of the Ag^+ ions is consistent with the Ag^+ ions being less mobile in the amorphous phase than in the crystalline α-AgI phase, as shown by conductivity measurements. Moreover, since the thermal behaviour in the composite is very similar to the α-AgI crystallites one, the local structural ordering at T=260 K seen by EXAFS in the composite should be related to the presence of the crystalline grains. It could then be explained as an ordering of the α-AgI phase inside the crystallites. These considerations are also consistent with the fact that the above discussed anomalous behaviour seen by Minami and co-workers at T=260 K on heating the composite disappeared if it had been previously heated above 350 K and then cooled down before heating again, so that it had undergone the α to β transition during cooling.

We can conclude that at T=260 K there is a local structural transition temperature: on heating the composite, at this temperature the matrix glass relaxes and ordering of the α-AgI phase inside the crystallites occurs, with an increased mobility of the Ag^+ ions.

Finally, it is interesting to compare the above discussed temperature behaviour in the composite around T=260 K during the heating process to what happens during the cooling process. It is important to note that the composite was first cooled down starting from RT and then heated up to 380 K, so that in the cooling process the α to β transition did not occur. Temperature dependent diffraction measurements[14] show that the α-phase peaks broaden on cooling from RT to 86 K, due to the crystallites of α-AgI becoming more and more strained; on heating again up to RT, the peaks become sharper again until at RT the same diffraction pattern as before cooling is seen: nonetheless, the relaxation of the glassy matrix seems not to be a reversible process. From present EXAFS measurements, by comparing the behaviour of the cumulants and of the first shell RDF in the composite during the cooling and during the heating processes, we saw that the same RDF was reached at 300 K after cooling and heating again (Figure 4) and that the rate of the transition between 300 and 225 K on cooling is higher than on heating (Figure 5): on cooling, the structural disordering of the I–Ag distribution accompanied by the lowering of mobility of the Ag^+ ions seems to be more rapid than the inverse structural ordering process accompanied by an increased mobility of Ag^+ ions on heating. This hysteresis can be clearly seen by comparing the obtained I–Ag RDF at T=300 and 225 K (Figure 4): the RDF obtained after heating is the same as before cooling, while the RDF at 225 K on cooling and on heating are different. This behaviour indicates that the rigidity of the silver-borate glassy network is such that the local ordering process at T=260 K during heating is slower than the local disordering transition during cooling.

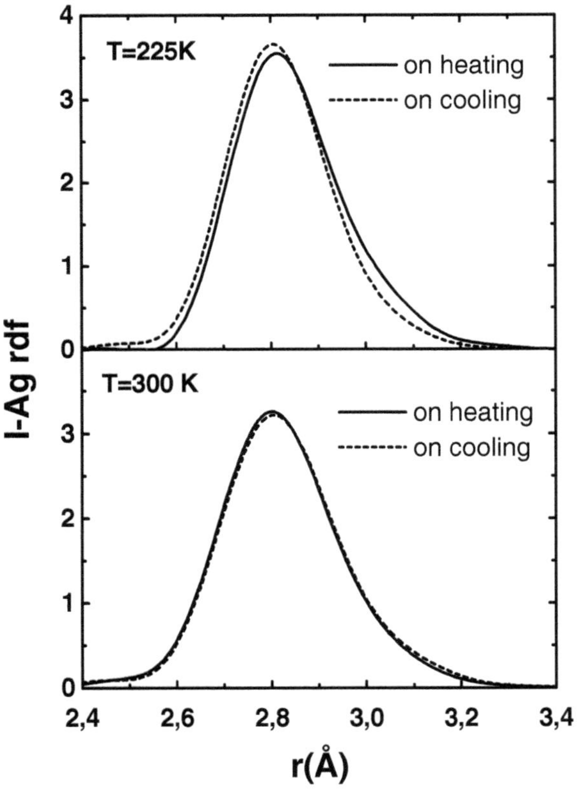

Figure 4. First shell I–Ag RDFs, normalised to unity, in the composite measured on heating (solid lines) and on cooling (dashed lines), at T=225 K (upper panel) and at 300 K (lower panel)

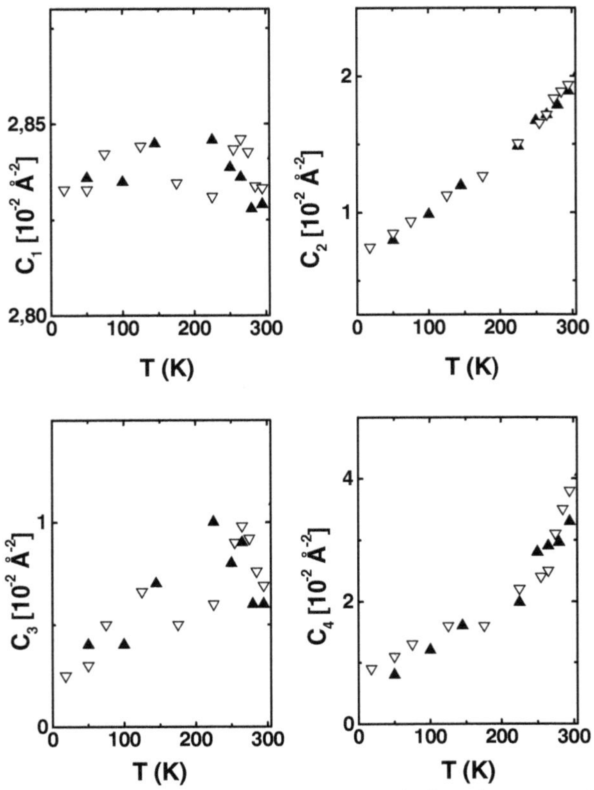

Figure 5. Cumulants of the first I–Ag shell in the composite measured on cooling from 300 to 15 K (open triangles) and on heating up to 300 K (solid triangles). Error bars are within the size of the symbols

6. Conclusions

The I–Ag coordination in AgI rich silver borate glasses – the composite and the pure glass – has been studied by temperature dependent EXAFS measurements below and above room temperature. The temperature behaviour of the first shell I–Ag RDF and of the corresponding first four cumulants of the composite and of the α-AgI crystallites, as extracted from the composite, has been discussed in the light of previous results obtained by other techniques. Particular care has been devoted to the temperature behaviour around 260 K where an unexpected anomaly was seen in the composite: the decrease in the cumulants on heating from 225 K to RT together with the RDF becoming more similar to β-AgI has been associated to a local structure ordering process, while the higher slope in the temperature dependence of the cumulants above room temperature together with the strong broadening in the high distance side of the RDF has been associated to an increase in thermal disorder. By comparing the temperature behaviour of the composite and of the α-AgI crystallites, we suggest that this local structure ordering accompanied by an increased mobility of Ag^+ ions is strictly related to the presence of the crystalline grains inside the glass where the strained α-phase of AgI is present. We can conclude that, on heating the composite, between 225 K and RT the matrix glass relaxes and an ordering of the strained α-AgI phase inside the crystallites occurs.

Finally, we have compared the above discussed behaviour of the composite on heating to what happens in the inverse cooling process: similarly to previous diffraction measurements, we recognise that, although the same local structure is reached at 300 K after cooling and heating again, the relaxation of the glassy matrix at around 260 K is not reversible and the corresponding local ordering of the strained α-phase in the crystallites on heating is a slower process than the inverse local disordering on cooling.

References

1. Tatsumisago, M., Shinkuma, Y. & Minami, T. *Nature*, 1991, **354**, 217.
2. Tatsumisago, M., Shinkuma, Y., Saito, T. & Minami, T. *Solid State Ionics*, 1992, **50**, 273.
3. Tatsumisago, M. & Minami, T. *Thermochim. Acta*, 1996, **280–281**, 333–41.
4. Tatsumisago, M. & Minami, T. *Proc. Sixth Asian Conf. on Solid State Ionics*, 1998 81–90.
5. Tatsumisago, M., Okuda, K., Itakura, N. & Minami, T. *Solid State Ionics*, 1999, **121**, 193–200.
6. Saito, T., Tatsumisago, M. & Minami, T. *Solid State Ionics*, 1993, **61**, 285–91.
7. Crozier, E. D. *et al. X-Ray Absorption: Principle and Application Techniques of EXAFS, SEXAFS and XANES*. Edited by D. C. Konigsberger & R. Prins, Wiley, 1988.
8. Dalba, G., Fornasini, P. & Rocca, F. *Phys. Rev. B*, 1993, **47**, 8502.
9. Fornasini, P., Monti, F. & Sanson, A. *J. Syncrotron. Radiat.*, 2001, **8**, 1214–20.
10. Trabquada J. M. & Ingallsa, R. *Phys. Rev. B*, 1983, **28**, 3520.
11. Stern, E., Livins, A. & Zhang, Z. *Phys. Rev. B, 1991*, 43, 8850–60.
12. Dalba, G., Fornasini, P., Rocca, F. & Monti, F. *J. Non-Cryst, Solids*, 2001, **293–295**, 93–8.
13. Torata, N., Saito, T., Tatsumisago, M. & Minami, T. *J. Mater. Sci. Lett.*, 1997, **16**, 1012–16.
14. Tatsumisago, M., Saito, T., Minami, T., Hanaya, M. & Oguni, M. *J. Phys. Chem.*, 1994, **98**, 2005–7.

Edge sharing BO_4 tetrahedra as a new structural feature in borate structures

*Holger Emme & Hubert Huppertz**

Department Chemie und Biochemie, Ludwig-Maximilians-Universität München, Butenandtstraße 5-13 (Haus D), D-81377 München, Germany

High pressure/high temperature conditions are used to synthesise new rare earth oxoborates $Ln_4B_6O_{15}$ (Ln=Dy, Ho; 8·0 GPa, 1000°C) and α-$Ln_2B_4O_9$ (Ln=Sm–Ho; 10·0 GPa, 1150°C) in a Walker-type multianvil apparatus. These new compositions, which till now were unattainable under normal pressure conditions, exhibit the new structural motif of edge sharing BO_4 tetrahedra, leading to a structural extension of oxoborate chemistry.

1. Introduction

For the class of oxoborates, the numbers of synthetic studies, structural characterisations, and materials processing studies have increased continuously during the past decade.[1–4] The structural chemistry of oxoborates exhibits a considerable diversity that results from the ability of boron to bind to three or four oxygen atoms, forming BO_3 or BO_4 groups. These groups can occur isolated or linked to groups, chains, bands, sheets, or complex networks.[5,6] In principle, the structural characteristics are analoguous to those of the silicates with the triangular BO_3 group as an additional structural component. As a common feature it can be pointed out that the BO_3 triangles and the BO_4 tetrahedra link to each other only via common corners (oxygen atoms) and not via common edges or faces. This characteristic was postulated already in 1967 by Ross & Edwards, and was still valid until a short while ago.[7,8]

In contrast to ambient pressure investigations on borates, high pressure investigations are rare. The major part of the high pressure results comes from the area of geological research, investigating borate containing minerals. Several studies were performed by Werding & Schreyer, collecting data in the pressure range from 3 to 5 GPa.[9] To investigate the synthesis and stability of boron bearing systems at elevated temperatures and pressures, they used hydrothermal bombs up to about 0·6 GPa, 650 to 800°C; internally heated gas pressure vessels up to about 0·8 GPa, 1000°C; and several types of piston cylinder apparatus up to 6 GPa, 1100°C. Because of demanding experimental requirements, only a few experiments were performed at higher pressures (>6 GPa) with a multianvil apparatus at the Bayerisches Geoinstitut, Bayreuth, Germany.[9] In our group we were able to use a Walker multianvil device to perform experiments in oxoborate chemistry routinely up to a pressure of 15 GPa.[10] Therefore, we expect to obtain new insights in the structural behaviour of these compounds at elevated pressures.

Corresponding to the pressure coordination rule,[11] an increase of pressure can lead to a higher coordination number of the boron atoms. For examples in the normal pressure modification of boron oxide B_2O_3-I, all boron atoms have the coordination number three,[12] whereas in the high pressure modification B_2O_3-II, all boron atoms are coordinated tetrahedrally.[13]

Recently, we found that the use of high pressure during synthesis extends the compositional and structural scope. Known compositions in the system Ln_2O_3/B_2O_3 under ambient conditions are Ln_3BO_6 (3:1) (which can be regarded as orthoborates $((LnO)_3BO_3)$), the orthoborates $LnBO_3$ (1:1), and the metaborates LnB_3O_6 (1:3) $(Ln(BO_2)_3)$.[14,15] While the composition of the orthoborates $REBO_3$[16–33] and the meta-borates REB_3O_6[34–40] is beyond doubt, there exist several uncertainties concerning the rare earth borates RE_3BO_6.[14,41] Lin *et al* established the crystal structures of "La_3BO_6",[42] "Y_3BO_6",[43] and "Gd_3BO_6"[44] where they found that the composition of the lanthanum compound is $La_{26}(BO_3)_8O_{27}$, showing a slight La_2O_3 excess relative to La_3BO_6 ($8La_3BO_6.La_2O_3$). In contrast to these results, the structure of the yttrium compound $Y_{17\cdot33}(BO_3)_4(B_2O_5)_2O_{16}$ and the gadolinium compound $Gd_{17\cdot33}(BO_3)_4(B_2O_5)_2O_{16}$, determined by x-ray powder diffraction data, were found to have an excess of borate. Although the Rietveld refinements resulted in reasonable residuals, some uncertainties remained. In contrast to these results, Cohen-Adad and coworkers confirmed the compositions RE_3BO_6 (RE=Y, Gd).[27]

In the system Ln_2O_3–B_2O_3 we were able to realise

[1] Corresponding author. Email huh@cup.uni-muenchen.de

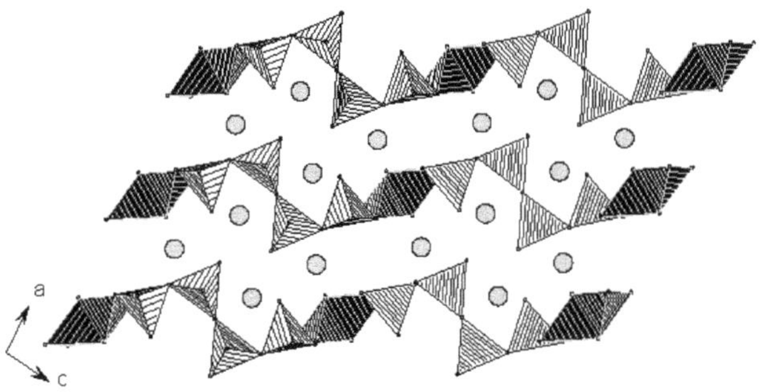

Figure 1. Crystal structure of $Ln_4B_6O_{15}$ (Ln=Dy, Ho), view along [010]. The structure is built up from corrugated layers of corner (light) and edge sharing (dark) BO_4 tetrahedra. The Ln^{3+} ions are positioned between the layers

the new compositions $Ln_4B_6O_{15}$ (Ln=Dy, Ho)[15,45] and α-$Ln_2B_4O_9$ (Ln=Sm–Ho)[46–48] under high pressure/high temperature conditions (>6 GPa). In contrast to all nearly 500 structurally characterised oxoborates, in which the linkage of BO_3- and BO_4-units occurs exclusively via corners, these new oxoborates are the first examples exhibiting edge sharing BO_4 tetrahedra next to corner sharing BO_4 tetrahedra.

2. Results and discussion

$Ln_4B_6O_{15}$ (Ln=Dy, Ho)

The structure refinements for the mentioned compositions were performed using single crystals for all compounds. The structure of monoclinic $Ln_4B_6O_{15}$ (Ln=Dy, Ho) (a=1167·7(2), b=437·7(1), c=1892·3(3) pm, β=96·72(2)° for $Dy_4B_6O_{15}$ and a=1163·8(3), b=436·5(1), c=1888·6(4) pm, β=96·77(3)° for $Ho_4B_6O_{15}$) is built up from corrugated layers of linked BO_4 tetrahedra (Figure 1). The Ln^{3+} ions are positioned between the layers. The linkage of the BO_4 tetrahedra inside the layers is achieved via common corners as well as common edges. In detail, two pairs of edge sharing tetrahedra (dark polyhedra) are linked via two additional corner sharing BO_4 tetrahedra (light polyhedra) to form six-membered rings. The linkage of these rings by further corner sharing BO_4 tetrahedra leads to rings consisting of ten BO_4 tetrahedra forming corrugated layers. More details concerning $Ln_4B_6O_{15}$ (Ln=Dy, Ho) with the introduction of a new descriptor "⧅" for edge-sharing tetrahedra in the concept of fundamental building blocks can be found in Ref. 15.

α-$Ln_2B_4O_9$ (Ln=Sm–Ho)

Figure 2 gives a view of the crystal structure of α-$Ln_2B_4O_9$ (Ln=Sm–Ho) along [010]. The monoclinic structure (e.g. α-$Sm_2B_4O_9$: a=2557·2(5) b=445·16(9), c=2496·7(5) pm, β=99·85(3)°) exhibits a complex network of linked BO_4 tetrahedra. After $Ln_4B_6O_{15}$ (Ln=Dy, Ho), this new structure of α-$Ln_2B_4O_9$ (Ln=Sm–Ho) is the second example for an oxoborate, wherein the linkage of BO_4 tetrahedra is realised via common corners as well as via common edges. For a clearer representation, the corner sharing tetrahedra are drawn as light polyhedra and the edge sharing tetrahedra as black polyhedra. For a more detailed discussion of the structure see Ref. 47.

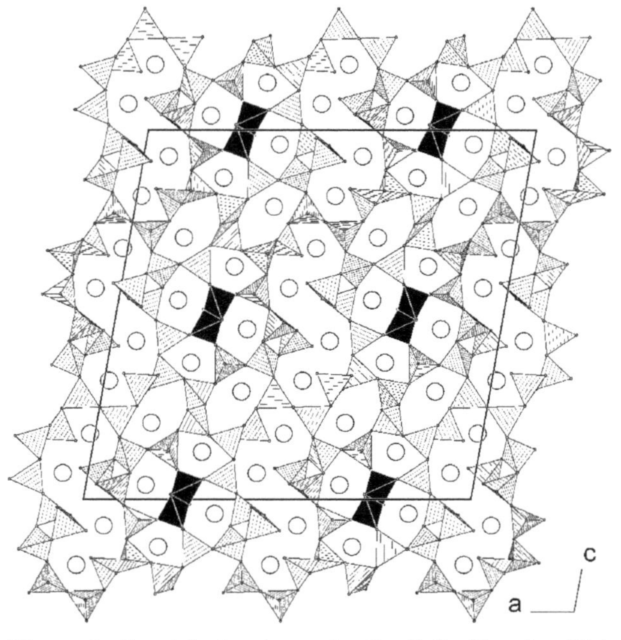

Figure 2. Crystal structure of α-$Ln_2B_4O_9$ (Ln=Sm–Ho), view along [010]

Comparison of the edge sharing BO_4 tetrahedra

The B–O distances and angles in the edge sharing BO_4 tetrahedra of α-$Ln_2B_4O_9$ (Ln=Sm–Ho) in comparison to the data of $Ln_4B_6O_{15}$ (Ln = Dy, Ho) are of special interest. The B–B distances in the edge sharing tetrahedra are 207·1(9) in α-$Sm_2B_4O_9$, 204(2) pm in α-$Gd_2B_4O_9$, 205·3(9) pm in α-$Eu_2B_4O_9$, 205·9(9) pm in α-$Tb_2B_4O_9$, and 204(3) in α-$Sm_2B_4O_9$. These are even shorter than in $Ln_4B_6O_{15}$ (Ln=Dy, Ho) (207·2(8) and 207(1) pm). The angles in the planar B_2O_2-ring exhibit average values of 86·5° and 93·5°. Conclusively, the bond distances and angles inside the edge sharing BO_4 tetrahedra of α-$Ln_2B_4O_9$ (Ln=Sm–Ho) and $Ln_4B_6O_{15}$ (Ln=Dy, Ho) are similar.

3. Conclusion

By the high pressure/high temperature syntheses of the compounds $Ln_4B_6O_{15}$ (Ln=Dy, Ho)(15,45) and α-$Ln_2B_4O_9$ (Ln=Sm–Ho)(46-48) it was possible to identify edge sharing of BO_4 tetrahedra for the first time in the structural chemistry of oxoborates. Besides the new structural motif, these compounds exhibit new compositions, which up to now were unreachable under normal pressure conditions.

Acknowledgements

The authors gratefully acknowledge the continuous support of these investigations by Prof. Dr. W. Schnick, Department Chemie and Biochemie of the University of Munich (LMU). Special thanks go to the Deutsche Forschungsgemeinschaft for financial support of this work.

References

1. Becker, P. *Adv. Mater.*, 1998, **10**, 979.
2. Sasaki, T., Mori, Y., Yoshimura, M., Yap, Y. K. & Kamimura, T. *Mater. Sci. Eng.*, 2000, **30**, 1.
3. Keszler, D. A. *Curr. Opin. Solid State Mater. Sci.*, 1996, **1**, 204.
4. Keszler, D. A. *Curr. Opin. Solid State Mater. Sci.*, 1999, **4**, 155.
5. Heller, G. "A Survey of Structural Types of Borates and Polyborates" in *Topics in Current Chemistry*, 1986. Springer Verlag, Berlin.
6. Christ, C. L. & Clark, J. R. *Phys. Chem. Miner.*, 1977, **2**, 59.
7. Ross, V. F. & Edwards, J. O. "The structural chemistry of the borates" in *The Chemistry of Boron and its Compounds*, 1967. Wiley, New York.
8. Filatov, S. K. & Bubnova, R. S. *Phys. Chem. Glasses*, 2000, **41**, 216.
9. Werding, G. & Schreyer, W. Experimental studies on borosilicates and selected borates; In: *Boron, Mineralogy, Petrology and Geochemistry.* (Eds. E. S. Grew, L.M. Anovitz), Reviews in Mineralogy 33, 1996. Mineralogical Society of America, Washington.
10. Huppertz, H. *Z. Kristall.*, 2004 , **219**, 330.
11. Neuhaus, A. *Chimia*, 1964, **18**, 93.
12. Gurr, G. E., Montgomery, P. W., Knutson, C. D. & Gorres, B. T. *Acta Cryst.*, 1970, **B26**, 906.
13. Prewitt, C. T. & Shannon, R. D. *Acta Cryst.*, 1968, **B24**, 869.
14. *Gmelin Handbook of Inorganic and Organometallic Chemistry*, C11b, Eighth edition, 1991. Springer Verlag, Berlin.
15. Huppertz, H. *Z. Naturforsch.*, 2003, **58b**, 278.
16. Levin, E. M., Roth, R. S. & Martin, J. B. *Am. Miner.*, 1961, **46**, 1030.
17. Morgan, P. E. D., Carroll, P. J. & Lange, F. F. *Mater. Res. Bull.*, 1977, **12**, 251.
18. Bradley, W. F., Graf, D. L., & Roth, R. S. *Acta Cryst.*, 1966, **20**, 283.
19. Henry, J.-Y. *Mater. Res. Bull.*, 1976, **11**, 577.
20. Roth, R. S., Waring, J. L. & Levin, E. M. *Proc. Third Conf. Rare Earth Res.*, Clearwater, Fla., 1963, 153.
21. Chadeyron, G., El-Ghozzi, M., Mahiou, R., Arbus, A. & Cousseins, J. C. *J. Solid State Chem.*, 1997, **128**, 261.
22. Ren, M., Lin, J. H., Dong, Y., Yang, L. Q., Su, M. Z. & You, L. P. *Chem. Mater.*, 1999, **11**, 1576.
23. Palkina, K. K., Kuznetsov, V. G., Butman, L. A., & Dzhurinskii, B. F. *Acad. Sci. USSR*, 1976, **2**, 286.
24. Meyer, H. J. *Naturwissenschaften*, 1969, **56**, 458.
25. Meyer, H. J. & Skokan, A. *Naturwissenschaften*, 1971, **58**, 566.
26. Meyer, H. J. *Naturwissenschaften*, 1972, **59**, 215.
27. Cohen-Adad, M. Th., Aloui-Lebbou, O., Goutaudier, C., Panczer, G., Dujardin, C., Pedrini, C., Florian, P., Massiot, D., Gerard, F. & Kappenstein, Ch. *J. Solid State Chem.*, 2000, **154**, 204.
28. Keszler, D. A. & Sun, H. *Acta Crystall.*, 1988, **C44**, 1505.
29. Abrahams, S. C., Bernstein, J. L. & Keve, E. T. *J. Appl. Crystall.*, 1971, **4**, 284.
30. Huppertz, H. *Z. Naturforsch.*, 2001, **56b**, 697.
31. Böhlhoff, R., Bambauer, H. U. & Hoffmann, W. *Z. Kristall.*, 1971, **133**, 386.
32. Lemanceau, S., Bertrand-Chadeyron, G., Mahiou, R., El-Ghozzi, M., Cousseins, J. C., Conflant, P. & Vannier, R. N. *J. Solid State Chem.*, 1999, **148**, 229.
33. Weidelt, J. & Bambauer, H. U. *Naturwissenschaften*, 1968, **55**, 342.
34. Canneri, G. *Gazz. Chim. Ital.*, 1926, **56**, 450.
35. Weidelt, J. *Z. Anorg. Allg. Chem.*, 1970, **374**, 26.
36. Tananaev, I. V., Dzhurinskii, B. F. & Belyakov, I. M. *Izv. Akad. Nauk SSSR Neorg. Mater.*, 1966, **2**, 1791.
37. Bambauer, H. U., Weidelt, J. & Ysker, J. St. *Z. Kristall.*, 1969, **130**, 207.
38. Tananaev, I. V., Dzhurinskii, B. F. & Chistova, B. F. *Izv. Akad. Nauk SSSR Neorg. Mater.*, 1975, **11**, 165.
39. Abdullaev, G. D., Mamedov, Kh. S. & Dzhafarov, G. D. *Sov. Phys. Cryst.*, 1975, **20**, 161.
40. Sieke, C., Nikelski, T. & Schleid, Th. *Z. Anorg. Allg. Chem.*, 2002, **628**, 819.
41. Leskelä, M. & Niinistö, L. *Handbook on the Physics and Chemistry of Rare-Earth*, (K. A. Gschneider, Jr., L. Eyring, Ed.) 1986. Elsevier Science, Amsterdam P. 203.
42. Lin, J. H., Su, M. Z., Wurst, K. & Schweda, E. *J. Solid State Chem.*, 1996, **126**, 287.
43. Lin, J. H., Zhou, S., Yang, L. Q., Yao, G. Q. & Su, M. Z. *J. Solid State Chem.*, 1997, **134**, 158.
44. Lin, J. H., You, L. P., Lu, G. X., Yang, L. Q. & Su, M. Z. *J. Mater. Chem.*, 1998, **8**, 1051.
45. Huppertz, H. & von der Eltz, B. *J. Am. Chem. Soc.*, 2002, **124**, 9376.
46. Emme, H. & Huppertz, H. *Z. Anorg. Allg. Chem.*, 2002, **628**, 2165.
47. Emme, H. & Huppertz, H. *Chem. Eur. J.*, 2003, **9**, 3623.
48. Emme, H. & Huppertz, H. *Acta Cryst.*, 2005, **C61**, i29.

Broadband dielectric response of binary borate glasses

A. Mandanici,[1] M. Cutroni,[1,*] A. Raimondo,[1] M. Federico,[1] F. Rocca[2] & C. Armellini[2]

[1] Dipartimento di Fisica – Università degli Studi di Messina, Italy
[2] IFN, Istituto di Fotonica e Nanotecnologie del Consiglio Nazionale delle Ricerche, Sezione di Trento-38050 Povo (Trento), Italy

The dielectric response of borate glasses $Ag_2O.nB_2O_3$ has been investigated as a function of the composition using a broadband measurement technique in the frequency range 10^{-3} Hz–10^9 Hz. This approach reveals a d.c. plateau at low frequencies, a power law dispersive behaviour at intermediate frequencies, and a nearly constant loss regime in the high frequency limit, already at room temperature. The log of the d.c. conductivity linearly increases with increasing Ag_2O molar fraction in the compositional range investigated. Analysing the experimental data either in terms of frequency dependent conductivity or in terms of complex dielectric modulus the results indicate that the shapes of the electrical dispersive behaviour in the $Ag_2O.nB_2O_3$ borate glasses, as well as their dielectric response in the nearly constant loss regime are influenced by the composition.

Introduction

A main challenge in the field of condensed matter physics is the structural study of ionic conducting glasses. Understanding the microscopic structure and the link with the macroscopic physical properties as for example the high ionic conductivity, is critical from the point of view of applications. With this purpose in mind, we start to study the dielectric properties of borate glasses doped with various amount of silver oxide. For many years the experimental information on binary borate glasses doped with silver oxide has been derived from acoustic relaxation processes and from d.c. conductivity.[1–2] More recently numerous studies have been conducted on borate network glass former doped with the addition of a different network modifier (e.g. Li_2O, Na_2O, K_2O, Cs_2O, Ag_2O). Different experimental techniques, such as electrical conductivity, infrared spectroscopy, neutron diffraction etc,[3–8] seem to indicate that the basic structural units BO_3 are more or less converted in BO_4 structural units, depending on the mole fraction and on the nature of the network modifier. The structure of the glassy matrix and the ionic interactions determine the dynamical properties. A suitable approach to gaining useful information about the ion kinetics is the study of the dielectric response of the ion conducting glasses in wide frequency and temperature ranges.

A common feature which characterises the imaginary part, $\varepsilon''(\omega)$, of the frequency dependent permittivity of ionic glasses at sufficiently low temperatures/high frequencies is the emergence of a region where ε'' becomes nearly constant after a normal dispersive behaviour.[4–18] This is the so called "Nearly Constant Loss" regime (NCL). Therefore, to account for the dielectric permittivity behaviour in a wide frequency range, the most used representation has been in terms of conductivity

$$\sigma_{ac}(\omega)=\sigma_0+B\omega_p+A\omega$$

where σ_0 gives an evaluation for the dc conductivity, while the power law ω_p and the linear ω contributions, respectively, account for the dispersive and for the constant loss regions.

Recently it has been proposed[14,17] that a correlation could exist between the activation energy characterising the d.c. conductivity and the NCL region which appears at high frequencies/low temperatures. In this sense, the origin of the NCL behaviour could be related to the motion of the same ions which are responsible for the long range ionic conduction in the material. Furthermore it must be underlined that the NCL behaviour appears also in some glasses which have no free mobile charges,[15] so that it could also arise from a relaxational feature of the glassy matrix. As a consequence it has been hypothesised that in general the NCL effect could originate from different kinds of processes.

In the present paper we show that the binary silver borate glasses $Ag_2O.nB_2O_3$ with n=2, 3, 4 exhibit a constant loss behaviour at room temperature in the mHz–GHz frequency region, and we study how the dynamical response on very different time scales is influenced by the composition. We point out that several dynamical observables (d.c. conductivity value,

* Corresponding author. Email andrea.mandanici@unime.it

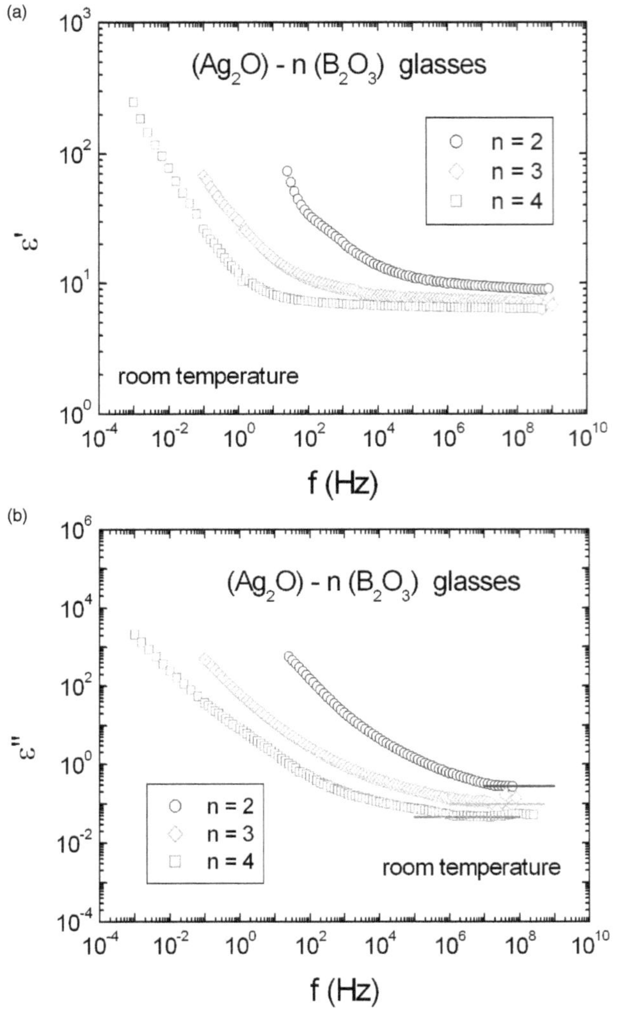

Figure 1. Real (a) and imaginary part (b) of the dielectric permittivity of $Ag_2O.nB_2O_3$ glasses (n=2, 3, 4) measured as a function of frequency in the mHz–GHz frequency range at 295 K

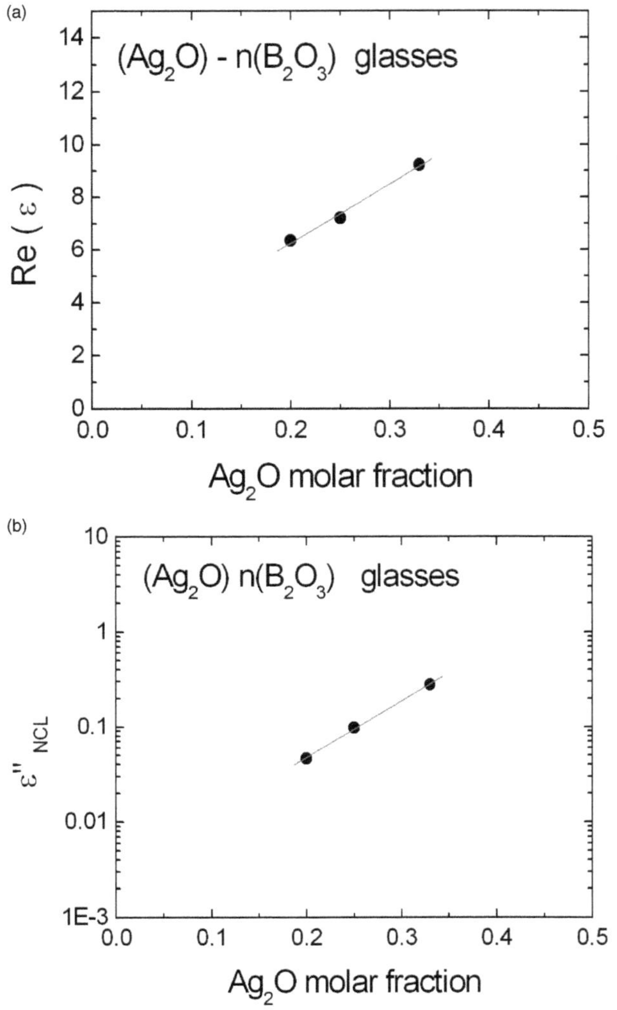

Figure 2. High frequency permittivity (a) and constant loss (b) of $Ag_2O.nB_2O_3$ glasses (n=2, 3, 4) at 295 K as a function of the Ag_2O molar fraction

σ_{dc}, high frequency constant loss, ε''_{NCL}, high frequency permittivity, $\varepsilon'(\infty)$) are correlated in a simple way to the amount of silver oxide in the glasses.

Experimental

Silver borate glasses $Ag_2O.nB_2O_3$ where n=2, 3, 4, were prepared by the usual melt quenching technique. The stoichiometric mixture of dried powders Ag_2O, B_2O_3 (Aldrich, 99% pure) was melted in an alumina crucible kept for 2 h at constant temperature in the range 850°C–900°C depending on the composition. The melts were equilibrated for about 1 h at 350°C and then quenched in brass moulds. The glassy nature of all the samples was confirmed by x-ray diffraction analysis. The samples (13 mm diameter, ~3 mm thickness) were polished in order to obtain flat and parallel surfaces, on which silver paint electrodes were carefully applied. The complex dielectric permittivity has been measured as a function of frequency using a broadband measurement method(19) in the wide range between 10^{-3} Hz and 10^9 Hz. The home made measurement set up has been used in connection with different instruments working on complementary and partially overlapping frequency ranges: Solartron 1260 impedance/gain analyser equipped with a SI1296 Dielectric Interface (1 mHz–100 kHz), HP4284A precision LCR meter (20 Hz – 1 MHz), Agilent 4395A Vector Network Analyzer (10 Hz–500 MHz), Agilent 8753E Vector Network Analyzer (30 kHz–3 GHz). A calibration based on the measured response of two known standards was used(19) for the removal of systematic errors.

Results and discussion

The real $\varepsilon'(\omega)$ and imaginary $\varepsilon''(\omega)$ part of the dielectric permittivity of binary glasses $Ag_2O.nB_2O_3$ (n=2,3,4) at room temperature are reported in Figure 1. The real part of the permittivity $\varepsilon'(\omega)$ approaches a

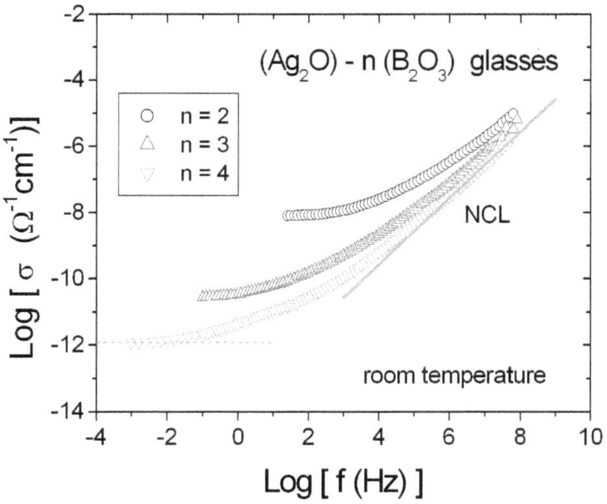

Figure 3. Log–log plot of the a.c. conductivity of $Ag_2O.nB_2O_3$ glasses (n=2, 3, 4) at 295 K as a function of frequency in the wide frequency range from 10^{-3} Hz up to microwave frequencies. The straight line with a slope 1 indicates the nearly constant loss regime of the conductivity for the composition $Ag_2O.4B_2O_3$. The flat line in the low frequency region indicates the d.c. plateau region

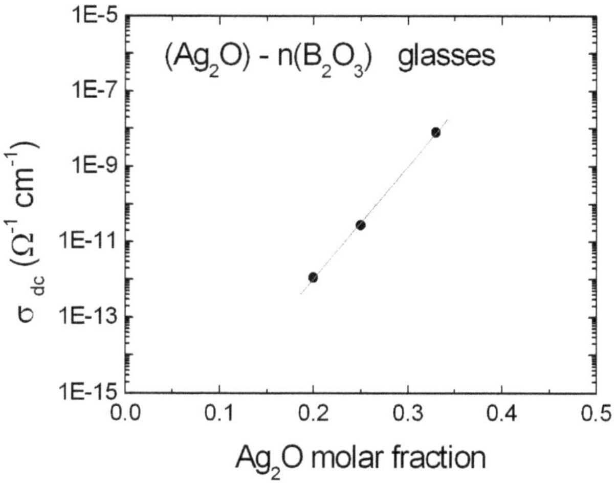

Figure 4. Dependence of the d.c. conductivity value on the Ag_2O content. The log of σ_{dc} linearly increases at increasing silver oxide molar fraction

constant value $\varepsilon'(\infty)$ at high frequencies. The imaginary part of the permittivity, $\varepsilon''(\omega)$, tends to increase with increasing as $1/f$ at low frequencies, as a result of the d.c. ionic conductivity. It is evident that at increasing frequency the slope of the permittivity $\varepsilon''(\omega)$ decreases and finally a nearly constant loss plateau value ε''_{NCL} is achieved.

The value of the high frequency real permittivity $\varepsilon'(\infty)$as well as the value ε''_{NCL} of the nearly constant loss contribution in the $Ag_2O.nB_2O_3$ glasses depend on the composition. This $\varepsilon'(\infty)$ linearly increases as a function of the Ag_2O molar fraction (Figure 2(a)). Indeed the relation between ε''_{NCL} and the Ag_2O content appears to be linear on a semi-logarithmic scale (Figure 2(b)). Also, the onset of the constant value of $\varepsilon''(\omega)$ shifts towards higher frequencies at increasing silver oxide content (see Figure 1(b)).

This effect, as indicated out in the raw $\varepsilon''(\omega)$ data at MHz frequencies and room temperature can be checked also in the conductivity and/or modulus representation of the same data. The frequency dependent behaviour of the ac conductivity is shown in Figure 3. At sufficiently low frequency the conductivity assumes a d.c. plateau value, it then increases as a function of frequency until a $\sigma \propto \omega$ behaviour is reached, corresponding to the region of nearly constant loss.

The composition of the glasses influences the d.c. conductivity value as illustrated in Figure 4 thus the log of σ_{dc} linearly increases at increasing silver oxide molar fraction. This behaviour is similar to that observed in AgI doped oxide glasses such as the $(AgI)_x[(Ag_2O).n(B_2O_3)]_{1-x}$ glasses,[13,20] where the increase in conductivity typically corresponds to a decrease in the activation energy associated with the d.c. conductivity. However the composition does not only affect the d.c. conductivity value. In fact, at increasing Ag_2O content, the σ_{ac} dispersive region as well as the outset of the linear dependence are shifted towards higher frequencies (see Figure 3). Also the shape of the frequency dependent conductivity spectra seems to depend on the composition: the scaling technique adopted in Figure 5, which exploits the d.c. conductivity value (Summerfeld scaling) of each glass, helps to point out the differing behaviours of the dispersive ac conductivity for different compositions.

Analysing the data in terms of the modulus formalism, the electrical response appears very similar

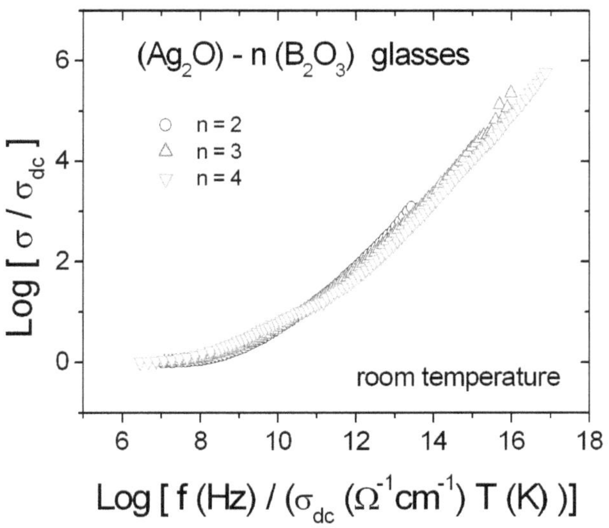

Figure 5. Scaling of a.c. conductivity of $Ag_2O.nB_2O_3$ glasses (n=2, 3, 4) at 295 K, based on the d.c. plateau value (Summerfeld scaling)

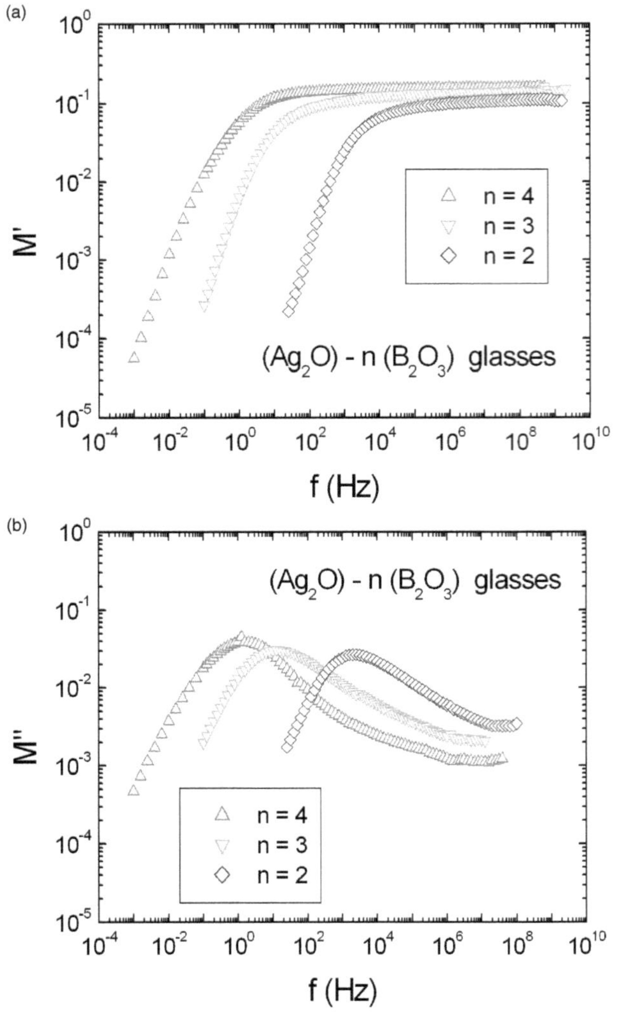

Figure 6. Frequency dependent electric modulus of Ag_2O. nB_2O_3 glasses (n=2, 3, 4) at 295 K: Real part (a); imaginary part (b)

to the mechanical response of glass forming materials close to their T_g.[21] The real part of the modulus (Figure 6(a)) increases up to a high frequency plateau value, while the imaginary part (Figure 6(b)) exhibits a broad and asymmetric relaxation peak. The region where the real modulus *M'* becomes constant (Figure 6(a)) shifts towards higher frequencies at increasing Ag_2O contents. Correspondingly (Figure 6(b)) the relaxation peak in *M''* shifts towards high frequencies and becomes broader at increasing silver oxide content. At high frequencies the NCL regime can be observed. However analysing the shape of the *M''* curve between the main peak and the NCL region, especially in the glass with lowest Ag_2O content, the existence of another relaxational contribution lying under the high frequency flank of the main peak could be hypothesised. If this is true, in a time temperature transposition, a secondary relaxation process should come out in the low temperature part of the isochronal spectra. On this basis, interesting outcomes are expected from the study of the dielectric properties of the $Ag_2O.nB_2O_3$ glasses down to low temperatures.

Conclusions

The analysis of the broadband dielectric permittivity data on the binary borate glasses $Ag_2O.nB_2O_3$ (with n=2, 3, 4) shows the existence of a "Nearly Constant Loss" regime at room temperature in the MHz–GHz frequency range. The value of the constant loss increases regularly at increasing Ag_2O contents within the compositional range investigated. It has been shown that many aspects of the electrical response, corresponding to different timescales, appear to be simply related to the Ag_2O molar fraction. Because of their properties, silver borate glasses are very interesting materials to study the features of the NCL down to low temperatures and their possible connection to the microscopic structure.

Acknowledgments

This work has been realised in the framework of the PRIN 2004-2006 programme "Correlazione tra struttura a corto e medio raggio e dinamica ionica in sistemi disordinati" coordinated by prof. M. Cutroni and financially supported by MIUR.

References

1. Carini, G., Cutroni, M., Federico, M., Galli, G. & Tripodo, G. *Phil. Mag.*, 1989, **B59**, 43.
2. Chiodelli, G., Magistris, A., Villa, M. & Bjorkstam, J. *J. Non-Cryst. Solids*, 1983, 51, 142; Avogadro, A., Mancini, S. & Villa, M. *Phys. Rev. Lett.*, 1980, **44**, 256.
3. Wong J. & Angell, C. A. *Glass Structure by Spectroscopy*, Dekker, New York, 1976.
4. Lee, W. K., Liu, J. F. & Nowick, A. S. *Phys. Rev. Lett.*, 1991, **67**, 1559.
5. Lu, X. & Jain, H. J. *Phys. Chem. Solids*, 1994, **55**, 1433.
6. Cramer, C., Funke, K. & Saatkamp, T. *Philos. Mag.*, 1995, **B71**, 701.
7. Sidebottom, D. L., Green, P. F. & Brow, R. K. *Phys. Rev. Lett.*, 1995, **74**, 5068.
8. Ngai, K. L., Jain, H. & Kanert, O. *J. Non-Cryst. Solids*, 1997, **222**, 383.
9. Nowick, A. S., Vaysleb, A. V. & Liu, W. *Solid State Ionics*, 1998, **105**, 121.
10. Ngai, K. L. & Moynihan, C. T. *Mater. Res. Bull.*, 1998, **23**, 51.
11. Ngai, K. L. J. *Chem. Phys.*, 1999, **110**, 10576.
12. Ngai, K. L. *J. Non-Cyst. Solids*, 1999, **248**, 194.
13. Cutroni, M. & Mandanici, A. *Solid State Ionics*, 1998, **105**, 149 and references cited therein.
14. Ngai, K. L. & Leon, C. *Phys. Rev.*, 2002, **B66**, 64308.
15. Sidebottom, D. L. & Murray-Krezan, C. M. *Phys. Rev. Lett.*, 2002, **89**, 195901.
16. Ngai, K. L. & Casalini, R. *Phys. Rev.*, 2002, **B66**, 132205.
17. Leon, C., Ngai, K. L. & Rivera, A. *Phys. Rev.*, 2004, **B69**,134303.
18. Funke, K., Banhatti R. D. & Cramer, C. *Phys. Chem. Chem. Phys.*, 2005 **7**, 157.
19. Pelster, R. *IEEE Trans. Microwave Theory Tech.*, 1995, **43**, 1494.
20. Cutroni, M., Mandanici, A., Raimondo, A. & Sanson, A. *Phys. Chem. Glasses*, submitted
21. Mandanici, A., Shi, X., McKenna, G. B. & Cutroni, M. *J. Chem. Phys.*, 2005, **122**, 114501.

Effect of cobalt ions on the EPR and d.c. conductivity in vanadyl doped $CoO–M_2O–B_2O_3$ (M=Li, K) glasses

S. Khasa,[1] V. P. Seth,[2] P. S. Gahlot,[3] A. Agarwal[2] & S. K. Gupta[4]*

[1] *Physics Department, Govt. College for Women Rohtak- 124001, India*
[2] *Applied Physics Department, Guru Jambheshwar University, Hisar-125001, India*
[3] *Radiology Department, PGIMS, Rohtak - 124001, India*
[4] *EPR Group, National Physical Laboratory, New Delhi - 110012, India*

Glasses having the composition $xCoO.(0{\cdot}30-x)M_2O.0{\cdot}70B_2O_3$ (M=Li, K) have been prepared over the range $0{\cdot}00 \le x \le 0{\cdot}15$ containing 1·0 and 2·0 mol% of V_2O_5. Spin-Hamiltonian parameters, dipolar hyperfine coupling parameter, P, and Fermi contact interaction term, K, of the VO^{2+} ions have been calculated. It is found the V^{4+} ions in these glasses exist as VO^{2+} in octahedral coordination with a tetragonal distortion. Results suggest that hyperfine spectra on vanadium are not observed when the amount of CoO exceeds a certain value, which depends on the type of alkali and the amount of V_2O_5. The 3dxy orbital in the vanadium ion expands and the tetragonal distortion around vanadium ion increases with the increase in the $CoO.M_2O$ ratio. There is virtually no change in the microscopic structure around vanadium due to annealing or when Li_2O is replaced by K_2O. The d.c. conductivity decreases when Li_2O is replaced by K_2O, keeping the concentration of CoO and B_2O_3 constant. Addition of CoO to the sample shows a "blocking effect" on the overall mobility of alkali ions. As a result the conductivity of the sample decreases with increase in $CoO:M_2O$ ratio.

Introduction

Many authors have reported[1–9] detailed information regarding structural phenomena and to identify the site symmetry of sites around the transition metal (TM) ion in glasses by using electron paramagnetic resonance (EPR) spectroscopy as an experimental technique.[1–5] However, a very small number of papers have been devoted to borate glasses with mixed TM ions.

It is well known[9] that in the first row of transition elements the Co^{2+} ion relaxes fast and its electron paramagnetic resonance (EPR) lines are too broad to be observed at temperature greater than 77 K. The EPR spectra of Co^{2+} ions in oxide glasses could only be detected at 4·2 K. It has also been noted[8] that when Co^{2+} is introduced into oxide glasses containing π ions (where π=Cu^{2+}, VO^{2+}, Mn^{2+} and W^{5+}) the EPR spectra of these ions attenuates. In earlier studies,[10–12] it was found that in ternary borate glasses ($Li_2O–CdO–B_2O_3$ and $Li_2O–BaO–B_2O_3$) containing 2·0 mol% of V_2O_5 variation in the Li_2O:CdO or Li_2O:BaO ratio for constant B_2O_3 has no effect on the vanadium site. In these glasses Li_2O, BaO and CdO are diamagnetic. However, the octahedral site symmetry on the V^{4+} site is improved[11] when a diamagnetic modifier, Ba^{2+}, is replaced by the paramagnetic ion, Co^{2+}, keeping B_2O_3 constant in $CoO–BaO–B_2O_3$ glasses containing 2·0 mol% of V_2O_5.

It was, therefore, of interest to study the effect of Co^{2+} ions on the features of EPR spectra of VO^{2+} doped in other oxide glasses. Structural studies of alkali borate glasses presently attract a significant interest because of their ionic conductivity and potential usages as solid electrolytes in various electro-chemical devices such as solid state batteries, fuel cells and chemical sensors.[13–15] Alkali and alkaline earth cations incorporated in the structure of glasses are responsible for ionic conduction in the materials.[16–17] Study of d.c. conductivity is also interesting because vanadium ions can exist as V^{4+} and V^{5+} species depending by the matrix composition and the preparation conditions.

In the present paper reports EPR and d.c. conductivity of the xCoO.(0·30–x)M_2O.0·70B_2O_3 (M=Li or K) ($0{\cdot}0 \le x \le 0{\cdot}15$) doped with 1·0 and 2·0 mol% of V_2O_5.

Glass preparation

The starting materials used to prepare the glasses were the Analar grade reagents of Li_2CO_3, K_2CO_3, $CoCO_3$, H_3BO_3 and V_2O_5 which were mixed in appropriate proportions. To each batch 1·0 and 2·0 mol% of V_2O_5 was added. Melting was done in porcelain crucibles at approximately 1273 K in an electrical muffle for 1 h. When all bubbles disappeared, the melts were quickly poured onto a carbon plate and

[1] Corresponding author. Email skhasa@rediff.com

pressed with another plate. To free the glass from internal stress a part of each sample was annealed at 473 K for 2 h.

Experimental

1. Electron Paramagnetic Resonance
First derivative EPR spectra of annealed and unannealed of the crushed samples were recorded at room temperature (300 K) in the X-band frequency on EPR spectrometers Varian E-112 as discussed in earlier paper.[12]
2. D.C. conductivity
To measure the d.c. conductivity glasses in the form of slices of nearly 1mm uniform thickness were chosen. Colloidal silver paint was used as an electrode material. D.C. conductivity was measured with the help of Kiethley programmable electrometer/source. A constant voltage of 20 V was applied across the sample.[12] Measurements were made at different temperatures between 373 K and 533 K, first by increasing, and then by decreasing the temperature.

Results

1. EPR

Figures 1 and 2 illustrate hyperfine spectra of VO^{2+} ion in unannealed $CoO–M_2O–B_2O_3$ (M=Li or K) samples at 300 K. There is no change in the EPR spectra on annealing. The spectra have structures which are characteristic of a hyperfine interaction arising from an unpaired electron with the ^{51}V nucleus whose spin is 7/2 and which is present in 99·75% abundance.[18] These spectra were analysed by assuming[5,10,19] that vanadium is present as a vanadyl ion in a ligand field of C4V symmetry.

The spin Hamiltonian used is of the form[20]

$$\mathbf{H}=\beta g_{\parallel}B_zS_z+\beta g_{\perp}(B_xS_x+B_yS_y)+A_{\parallel}S_zI_z+A_{\perp}(S_xI_x+S_yI_y) \quad (1)$$

where the symbols have their usual meaning. Quadrupole and nuclear Zeeman interaction terms are ignored. The solutions[21] of the spin-Hamiltonian[1] are given in Equations (2) and (3) for the parallel and perpendicular orientation, respectively

$$B_{\parallel}(m) = B_{\parallel}(o)-mA_{\parallel}-\{63/4-m^2\}A^2_{\perp}/2B_{\parallel}(o) \quad (2)$$

$$B_{\perp}(m) = B_{\perp}(o)-mA_{\perp}-\{63/4-m^2\}(A^2_{\parallel}+A^2_{\perp})/4B_{\perp}(o) \quad (3)$$

where m is the magnetic quantum number of the vanadium nucleus having values ±7/2, ±5/2, ±3/2 and ±1/2

$$B_{\parallel}(o)=hv/g_{\parallel}\beta \text{ and } B_{\perp}(o)=hv/g_{\perp}\beta$$

with h as the Planck constant, v as the frequency of the spectrometer and β as the Bohr magneton. Measurements for the $B_{\parallel}$ position were taken which correspond to a maximum in the first derivative curve

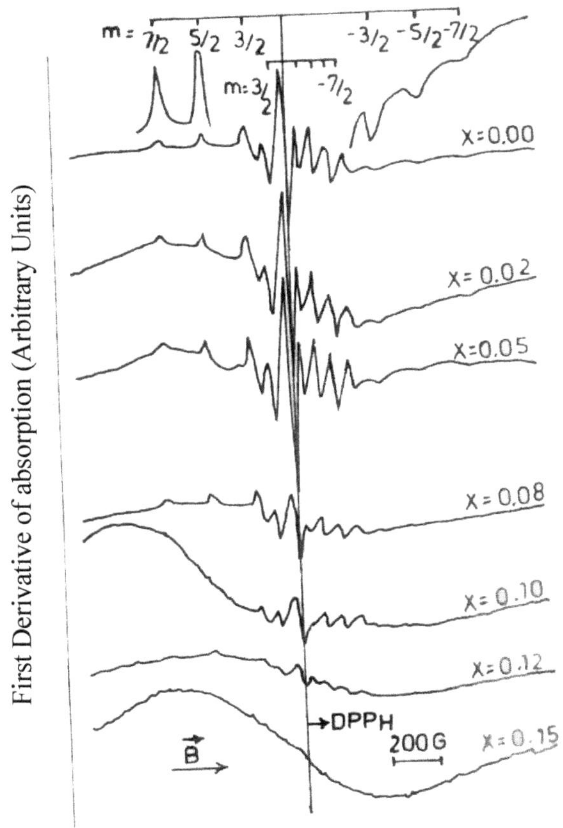

Figure 1. The EPR spectra of VO^{2+} ions in $xCoO–(0{\cdot}30–x)Li_2O–0{\cdot}70\,B_2O_3$ ($0{\cdot}00\leq x\leq 0{\cdot}15$) glasses containing 2·0 mol% of V_2O_5 in the X-band (v ~9·3 GHz) at 300 K

of the parallel hfs component for a given m value, whereas the $B_{\perp}$ position is enclosed between the first derivative perpendicular peak and its "zero".[20] Spin-Hamiltonian parameters of VO^{2+} ion determined from the observed positions of spectral lines and using Equations (2) and (3) are given in Tables 1 and 2. From the values of these parameters the dipolar hyperfine coupling parameter, $P=2\gamma\beta_e\beta_N<r^{-3}>$ and the Fermi contact interaction term, K, are evaluated by using

Table 1. Spin Hamiltonian parameters# and Λ_{th} of VO^{2+} in unannealed $xCoO–(0{\cdot}3–x)\,Li_2O–0{\cdot}7B_2O_3$ glasses at room temperature

Glass No.	x	V_2O_5 (mol%)	$g_{\parallel}$ (±0·001)	$g_{\perp}$ (±0·001)	$\lvert A_{\parallel}\rvert$ ($10^{-4}cm^{-1}$) (±1·0)	$\lvert A_{\perp}\rvert$ ($10^{-4}cm^{-1}$) (±1·0)	Λ_{th}
L1	0·00	1·0	1·9421	1·9741	170·1	58·8	0·4940
L2	0·02	1·0	1·9422	1·9743	170·1	58·6	0·4897
L3	0·04	1·0	1·9414	1·9744	170·1	59·2	0·4854
L4	0·05	1·0	1·9415	1·9752	169·9	59·2	0·4831
L5	0·06	1·0	1·9404	1·9760	169·9	60·0	0·4809
L6	0·08	1·0	-	-	-	-	0·4767
L7	0·00	2·0	1·9412	1·9721	169·5	58·8	0·4940
L8	0·02	2·0	1·9413	1·9745	169·6	59·2	0·4897
L9	0·05	2·0	1·9401	1·9752	169·7	59·2	0·4831
L10	0·08	2·0	1·9404	1·9752	169·8	59·2	0·4767
L11	0·10	2·0	1·9404	1·9764	169·8	59·7	0·4724
L12	0·12	2·0	-	-	-	-	0·4681
L13	0·15	2·0	-	-	-	-	0·4616

$A_{\parallel}$ and $A_{\perp}$ are negative

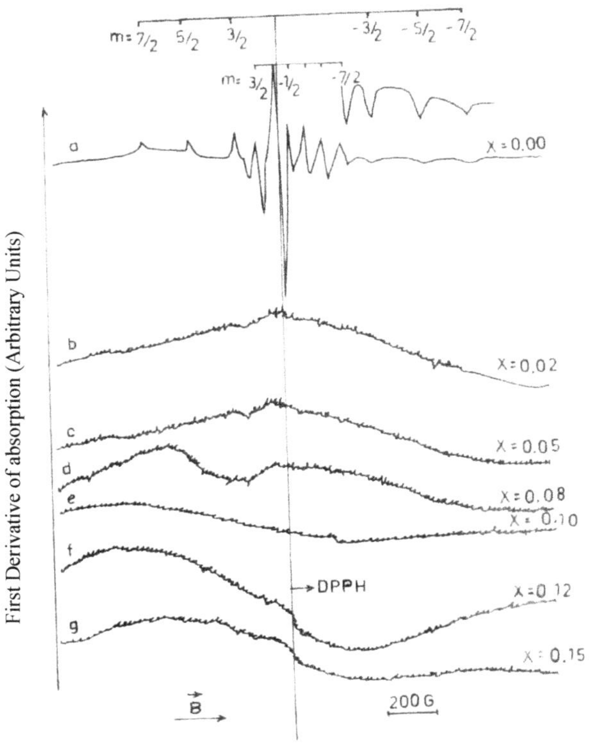

Figure 2. The EPR spectra of VO^{2+} ions in $xCoO-(0{\cdot}30-x)K_2O-0{\cdot}70B_2O_3$ $(0{\cdot}00\leq x\leq 0{\cdot}15)$ glasses containing 2·0 mol% of V_2O_5 in the X-band (υ ~9·3 GHz) at 300 K

the expressions developed by Kivelson & Lee[22]

$$A_{\parallel}=P[K+4/7-\Delta g_{\parallel}-(3/7)\Delta g_{\perp}] \quad (4)$$

$$A_{\perp}=P[K-2/7-(11/14)\Delta g_{\perp}] \quad (5)$$

where $\Delta g_{\parallel}=g_{\parallel}-g_e$; $\Delta g_{\perp}=g_{\perp}-g_e$ and g_e(=2·0023) is the g factor of free electrons. Both $A_{\parallel}$ and $A_{\perp}$ are found to be negative by the method proposed by Muncaster & Parke.[18] The term $-PK$ in Equations (4) and (5) is due to s-character of the magnetic spin of the vanadium. This s-character arises due to partial unpairing or polarisation of the s electrons as a result of an interaction with the unpaired d electrons.[23] The effect of polarisation on the hyperfine coupling was determined by Heine[23] and is included as $-PK$ in the expression for hyperfine coupling. For transition metal ions K is found to be positive.[24] From the molecular orbital theory it can also be shown[7] that the components $A_{\parallel}$ and $A_{\perp}$ consist of the contributions $A'_{\parallel}$ and $A'_{\perp}$ of the electron to the hyperfine structure and the PK term arising due to the anomalous contribution of the s-electrons. Equations (4) and (5) can be rewritten in the following form

$$A_{\parallel}=-PK-P[4/7-\Delta g_{\parallel}-(3/7)\Delta g_{\perp}]$$
$$=-PK+A'_{\parallel} \quad (6)$$

$$A_{\perp}=-PK+P[2/7+(11/14)\Delta g_{\perp}]$$
$$=-PK+A'_{\perp} \quad (7)$$

Calculated values of $A'_{\parallel}$ and $A'_{\perp}$ are given in Tables 3 and 4.

Table 2. Spin Hamiltonian parameters# and Λ_{th} of VO^{2+} in unannealed $xCoO-(0{\cdot}3-x)K_2O-0{\cdot}7B_2O_3$ glasses at room temperature

Glass No.	x	V_2O_5 (mol%)	$g_{\parallel}$ (±0·001)	$g_{\perp}$ (±0·001) (±1·0)	$\lvert A_{\parallel}\rvert$ ($10^{-4}cm^{-1}$) (±1·0)	$\lvert A_{\perp}\rvert$ ($10^{-4}cm^{-1}$)	Λ_{th}
K1	0·00	1·0	1·9422	1·9751	168·9	59·0	0·5394
K2	0·02	1·0	-	-	-	-	0·5325
K3	0·04	1·0	-	-	-	-	0·5256
K4	0·05	1·0	-	-	-	-	0·5214
K5	0·06	1·0	-	-	-	-	0·5176
K6	0·08	1·0	-	-	-	-	0·5104
K7	0·00	2·0	1·9422	1·9734	168·5	57·9	0·5394
K8	0·02	2·0	-	-	-	-	0·5325
K9	0·05	2·0	-	-	-	-	0·5214
K10	0·08	2·0	-	-	-	-	0·5031
K11	0·10	2·0	-	-	-	-	0·4957
K12	0·12	2·0	-	-	-	-	0·4846
K13	0·15	2·0					

$A_{\parallel}$ and $A_{\perp}$ are negative

2. Theoretical optical basicity

Theoretical optical basicities, Λ_{th} have also been calculated using the expression given by Duffy & Ingram[25] and are included in Table 1.

3. D.C. conductivity

Figures 3(a) and (b) and 4(a) and (b) show the variation of logσ (σ is the electrical conductivity) against 1000/T for these glasses. In general the conductivity increases with the temperature and the variation of the conductivity of the samples over an accessible temperature range of about 100 K is about two orders of magnitude in all these glasses. From these figures it clear that logσ against 1000/T plots are linear for temperature higher than 400 K. These linear curves satisfy the relation

$$\sigma=\sigma_o\exp(-W/kT) \quad (10)$$

where σ_o is a constant for a given glass, k is the Boltzmann constant and W is the activation energy

Table 3. P, K, $\lvert A'_{\parallel}\rvert$, $\lvert A'_{\perp}\rvert$ and $\Delta g_{\parallel}/\Delta g_{\perp}$ of VO^{2+} in unannealed $xCoO-(0{\cdot}3-x)Li_2O-0{\cdot}7B_2O_3$ glasses at room temperature

Glass No.	P (10^{-4} cm^{-1})	K	$\lvert A'_{\parallel}\rvert$ (10^{-4} cm^{-1})	$\lvert A'_{\perp}\rvert$ (10^{-4} cm^{-1})	$\Delta g_{\parallel}/\Delta g_{\perp}$
L1	122·7	0·743	79·0	32·3	2·1491
L2	122·7	0·741	79·0	32·3	2·1532
L3	122·2	0·748	78·7	32·2	2·1860
L4	121·7	0·751	78·5	32·2	2·2492
L5	120·8	0·762	77·9	32·0	2·3383
L6	-	-	-	-	-
L7	122·9	0·745	78·7	32·0	2·0472
L8	121·6	0·750	78·4	32·0	2·1471
L9	121·3	0·752	78·4	32·0	2·2552
L10	121·4	0·752	78·4	32·1	2·3161
L11	120·9	0·761	78·9	32·0	2·3513
L12 –L13	-	-	-	-	-

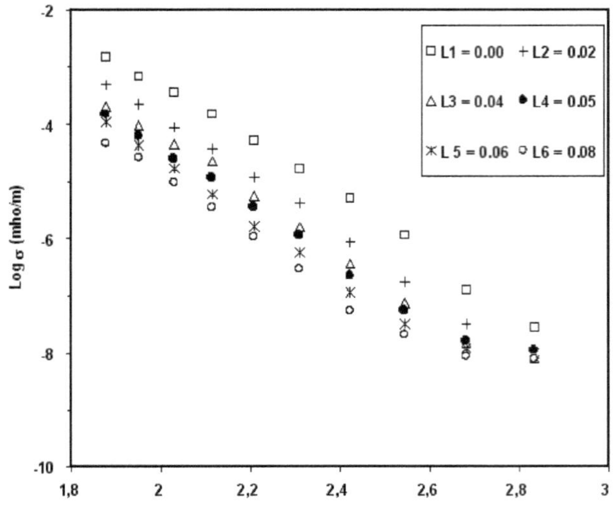

Figure 3(A)

Figure 3(a). Variation of log σ as a function of 1000/T for $xCoO–(0·30–x)Li_2O–0·70B_2O_3$ (0·00≤x≤0·08) glasses containing 1·0 mol% of V_2O_5

for conduction. The activation energy, *W* and pre-exponential term, σ_o, were determined by using the least square fitting of the experimental data with the relation

$$\log\sigma = \log\sigma_o - (W/1000k)(1000/T) \qquad (11)$$

The calculated values of activation energy, pre-exponential term and measured value of conductivity are given in Tables 5 and 6.

Discussion

1. EPR

The value of $\Delta g_{\parallel}/\Delta g_{\perp}$, which measures the tetragonality of the vanadium site[7] is also included in Tables 3, 4. EPR spectra (Figures 1 and 2) reveal that the hyperfine spectra of VO^{2+} ion in $CoO–Li_2O–B_2O_3$ samples is not observed when 1·0 mol% of V_2O_5 along with ≥8·0 mol% of $CoCO_3$ is added in the samples (whereas the hyperfine spectra of VO^{2+} ion in these samples is not observed when 2·0 mol% of V_2O_5 with more than 12·0 mol% of $CoCO_3$ are added in the samples). In $CoO–K_2O–B_2O_3$ samples with 1·0 mol% or 2·0 mol% of V_2O_5 the hyperfine spectra is not observed even when only 2·0 mol% of $CoCO_3$ is added in the sample. The V^{4+} ions presences are not influenced not only by the cobalt ones but also by the other components. At so small V_2O_5 content of vanadium must be present in V^{4+} and V^{5+} ions but as V^{5+} is diamagnetic so its contribution towards EPR spectra is absent. Absence of the hyperfine spectra may be due to the absence of V^{4+} ions or presence of cobalt ions as discussed below.

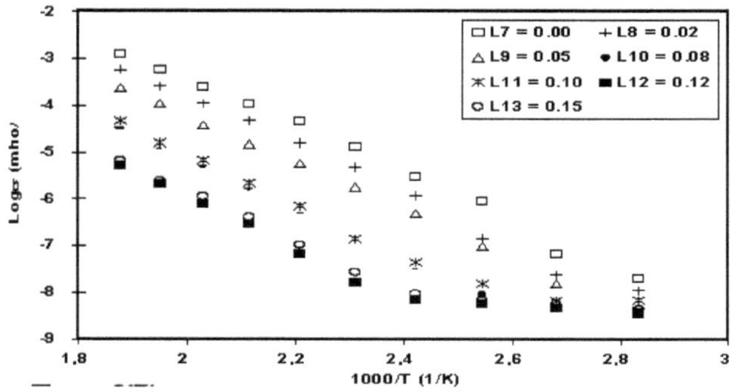

Figure 3(b). Variation of log σ as a function of 1000/T for $xCoO–(0·30–x)Li_2O–0·70B_2O_3$ (0·00≤x≤0·15) glasses containing 2·0 mol% of V_2O_5

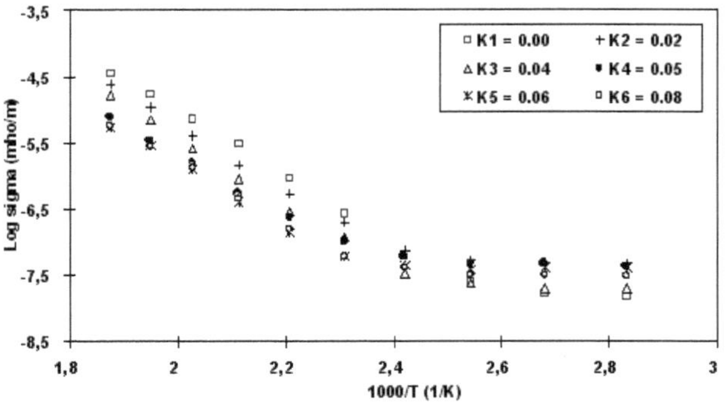

Figure 4(a). Variation of log σ as a function of 1000/T for $xCoO–(0·30–x)K_2O–0·70B_2O_3$ (0·00≤x≤0·08) glasses containing 1·0 mol% of V_2O_5

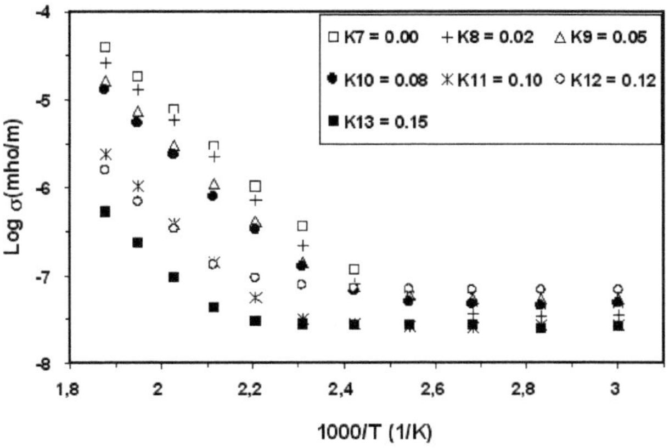

Figure 4(b). Variation of log σ as a function of 1000/T for $xCoO–(0·30–x)K_2O–0·70B_2O_3$ (0·00≤x≤0·15) glasses containing 2·0 mol% of V_2O_5

Following hypothesis may be given to try to explain the absence of hyperfine spectra in these samples. Co^{2+} has very short spin lattice relaxation time, as a result its EPR in borate glasses is not observed down to liquid nitrogen temperature.[26] On the other hand,

Table. 4 P, K, $|A'_{\parallel}|$, $|A'_{\perp}|$ and $\Delta g_{\parallel}/\Delta g_{\perp}$ of VO^{2+} in unannealed $xCoO–(0·3–x)K_2O–0·7B_2O_3$ glasses at room temperature

Glass No.	P (10^{-4} cm^{-1})	K	$\lvert A'_{\parallel}\rvert$ (10^{-4} cm^{-1})	$\lvert A'_{\perp}\rvert$ (10^{-4} cm^{-1})	$\Delta g_{\parallel}/\Delta g_{\perp}$
K1	121·2	0·751	78·0	32·0	2·1701
K2-K6	-	-	-	-	-
K7	121·2	0·738	78·5	32·0	2·0902
K8-K13	-	-	-	-	-

Table 5. D.C. conductivity, σ activation energy, W and pre-exponential term, log σ₀ of $xCoO$–$(0·3-x)Li_2O$–$0·70B_2O_3$ *glasses*

Glass No.	x	V_2O_5 (mol%)	$\sigma_{at\ 433\ K}$ ($\Omega^{-1}m^{-1}$)	$\sigma_{at\ 533\ K}$ ($\Omega^{-1}m^{-1}$)	$\log\sigma_o$	W (eV)
L1	0·00	1·0	$1·7\times10^{-5}$	$1·5\times10^{-3}$	5·68	0·90
L2	0·02	1·0	$4·0\times10^{-6}$	$4·9\times10^{-4}$	6·04	0·98
L3	0·04	1·0	$1·0\times10^{-6}$	$2·0\times10^{-4}$	5·80	0·99
L4	0·05	1·0	$1·0\times10^{-6}$	$1·5\times10^{-4}$	5·73	1·01
L5	0·06	1·0	$5·8\times10^{-7}$	$1·1\times10^{-4}$	6·14	1·06
L6	0·08	1·0	$2·9\times10^{-7}$	$4·5\times10^{-5}$	5·93	1·07
L7	0·00	2·0	$1·2\times10^{-5}$	$1·3\times10^{-3}$	5·96	0·93
L8	0·02	2·0	$4·0\times10^{-6}$	$5·9\times10^{-4}$	5·99	0·97
L9	0·05	2·0	$1·0\times10^{-6}$	$2·4\times10^{-4}$	5·77	0·99
L10	0·08	2·0	$1·4\times10^{-7}$	$4·5\times10^{-5}$	5·82	1·08
L11	0·10	2·0	$1·4\times10^{-7}$	$4·6\times10^{-5}$	5·88	1·08
L12	0·12	2·0	$1·7\times10^{-8}$	$5·0\times10^{-6}$	4·91	1·08
L13	0·15	2·0	$2·8\times10^{-8}$	$6·0\times10^{-6}$	5·02	1·08

Table 6. D.C. conductivity, σ activation energy, W and pre-exponential term, log σ₀ of $xCoO$–$(0·3-x)K_2O$–$0·70B_2O_3$ *glasses*

Glass No.	x	V_2O_5 (mol%)	$\sigma_{at\ 433\ K}$ ($\Omega^{-1}m^{-1}$)	$\sigma_{at\ 533\ K}$ ($\Omega^{-1}m^{-1}$)	$\log\sigma_o$	W (eV)
K1	0·00	1·0	$2·7\times10^{-7}$	$3·5\times10^{-5}$	4·71	0·96
K2	0·02	1·0	$2·0\times10^{-7}$	$2·3\times10^{-5}$	4·47	0·96
K3	0·04	1·0	$1·2\times10^{-7}$	$1·6\times10^{-5}$	4·46	0·98
K4	0·05	1·0	$1·0\times10^{-8}$	$7·0\times10^{-6}$	3·92	0·95
K5	0·06	1·0	$6·2\times10^{-8}$	$5·0\times10^{-6}$	3·89	0·96
K6	0·08	1·0	$6·0\times10^{-8}$	$5·0\times10^{-6}$	3·69	0·94
K7	0·00	2·0	$3·6\times10^{-7}$	$4·0\times10^{-5}$	4·41	0·93
K8	0·02	2·0	$2·2\times10^{-7}$	$2·7\times10^{-5}$	4·38	0·94
K9	0·05	2·0	$1·4\times10^{-7}$	$1·6\times10^{-5}$	4·20	0·95
K10	0·08	2·0	$1·3\times10^{-7}$	$1·3\times10^{-5}$	4·20	0·96
K11	0·10	2·0	$3·2\times10^{-8}$	$2·4\times10^{-6}$	3·64	0·98
K12	0·12	2·0	$7·7\times10^{-8}$	$1·6\times10^{-6}$	2·56	0·88
K13	0·15	2·0	$2·8\times10^{-8}$	$5·2\times10^{-7}$	2·30	0·91

VO^{2+} has long spin lattice relaxation time. We assume that samples containing VO^{2+} and Co^{2+} can be treated as two spin systems having quite different spin lattice relaxation times. According to this assumption there is a difference in the spin temperatures of these systems. When there is thermal contact between these system (realised through dipole-dipole interaction) equalisation of their temperature can take place (cross relaxation).[8] Although the overlap regions for the Co^{2+} and VO^{2+} spectra are insignificant but for sufficiently high Co^{2+} concentration (12·0 mol% in case of CoO–Li_2O–B_2O_3 sample, more than 2·0 mol% in case of CoO–K_2O–B_2O_3 sample), there is sharp decrease in the spin lattice relaxation time for all VO^{2+} ions regardless of where this ion makes its contribution to the spectrum. We can, therefore, conclude that the presence of the medium of rapidly relaxing Co^{2+} ions leads to the rapid relaxation of adjacent VO^{2+} ions so that the latter do not make any contribution to the net intensity of EPR signal. Thus, hyperfine spectra of VO^{2+} in these compositions are not observed.

Tables 1 to 4 indicate that the values of $g_\perp$, $A_\perp$, K and $\Delta g_\parallel/\Delta g_\perp$ increase and the values of $A_\parallel$, P, $|A'_\parallel|$ and $|A'_\perp|$ decrease when M_2O is replace by CoO. As suggested by Kivelson & Lee,[22] we assume the increased in the values of K with increase in the CoO:M_2O ratio as due[27–30] to an increase in the tetragonal nature of the $V^{4+}O_6$ complex because of a strongly bonded oxygen atom at the V^{4+} ion in the site opposite to the vanadyl oxygen. The increase in $\Delta g_\parallel/\Delta g_\perp$ also shows that the octahedral symmetry is reduced when M_2O is replaced by CoO. The decrease of the anisotropic contribution (i.e. $|A'_\parallel|$ and $|A'_\perp|$) of the 3dxy electron to the hyperfine splitting is brought about by increasing[7] screening of the 3dxy orbital from its nucleus through overlap of the electron orbits of the surrounding oxygen ligands. This screening produces an expansion of the 3dxy orbital, resulting in a decreased interaction between this magnetic electron with the vanadium nucleus. The decreasing value of P also supports the argument that 3dxy orbit expands with increase in CoO:M_2O ratio.

2. Theoretical optical basicity

Theoretical optical basicity serves in the first approximation as a measure[3] of the ability of oxygen to donate a negative charge in the glass. In other words, the optical basicity reflects the Lewis basicity of oxide glasses. As the ability of the equatorial ligands to donate the electron (i.e. Lewis basicity) decreases, π bonding between V^{4+} and the ligands reduces.[4] This reduction, in turn increases the positive charge on V^{4+} and increases the π bonding between V^{4+} and vanadyl oxygen. This increase, decreases bond length of V^{4+} (vanadyl oxygen). Consequently, the tetragonal nature of the $V^{4+}O_6$ complex is enhanced. In the present system , the theoretical optical basicity, Λ_{th}, of the host glass decreases when M_2O is replaced by CoO.

3. D.C. conductivity

From Tables 5 and 6 following observations can be made:

(1) The d.c. conductivity increases with increase in temperature.

(2) The d.c. conductivity decreases when M_2O (M=Li or K) is replaced by CoO.

(3) The d.c. conductivity decreases when Li_2O is replaced by or K_2O keeping mol% of CoO and B_2O_3 constants.

(4) The d.c. conductivity increases when V_2O_5 is changed from 1·0 mol% to 2·0 mol% in cobalt alkali borate glasses.

Similar results have been reported by others.[31–35] Figures 3(a) and (b) and 4(a) and (b) depict that log Conductivity of CoO–M_2O–B_2O_3 samples from 310 K to 410 K ($1000/T\geq2·4$)is almost constant. It is reported by Angel *et al*[35] in potassium silicate glass that for alkali concentration greater than 7·0 mol% occupation of available nearest neighbour sites begins to increase. In order for an alkali jump to a new site, a nearest neighbour alkali atom may also have to jump to open an adjacent site. Because of the high site occupancy, the probability of an alkali jumping back to its original site is also much higher. In the present system M_2O is equal to 15·0 mol%. As a result at temperature below 400 K the probability of an alkali ion jumping back to its original site may be more than the probability of its jumping to its adjacent site and observed variation of conductivity in this region is very small. The decrease in d.c. conductivity by replacing Li_2O with K_2O suggests that conductivity has contribution from ionic conductivity because K^+ ion is a heavy particle as compared to Li^+. The d.c. conductivity of lithium glass is maximum due to the fact Li^+ ions are lightest as compared to K^+ ions. As Vanadium ions can be present in more than one valence states, i.e. V^{4+} and V^{5+}, conduction could also take place by transfer of electrons from low to high valence states. Decrease in conductivity with time reveals that the electrical conductivity is mainly due to alkali ions and it may have only negligible contribution of electronic conductivity. S_o, Electronic conductivity due to hopping of electrons between V^{4+} and V^{5+} does not play an important role in total conductivity in the present glass system.

Decrease in d.c. conductivity and increase in activation energy by increasing the ratio of CoO: M_2O (M=Li or K) indicate that CoO in the sample shows "blocking effect" by cobalt ions on the overall mobility of alkali ions. A similar argument was considered by Khawaja *et al*[34] for the decrease in the conductivity of sodium germanate glass when copper is added to it.

Conclusions

From the present study of EPR and d.c. conductivity in alkali borate glasses, the following conclusions can be made:

(1) Hyperfine spectra of vanadium are not observed when the amount of CoO exceeds certain values which depends on the type of alkali and amount of V_2O_5.

(2) By keeping B_2O_3 constant in CoO–M_2O–B_2O_3 (M=Li or K) samples, the increase in CoO:M_2O ratio may result in an increase in the tetragonal nature of the $V^{4+}O_6$ complex.

(3) Increase in the CoO:M_2O ratio may result in the expansion of 3dxy orbit of the unpaired electron in the vanadium ion in CoO–M_2O–B_2O_3.

(4) There is virtually no change in the microscopic structure around the vanadium ion due to annealing in CoO–M_2O–B_2O_3 (M=Li or K) samples There is virtually no change in the microscopic structure around the vanadium ion when Li_2O is replaced by K_2O in the 0·3Li_2O.0·7B_2O_3 sample.

(5) The theoretical optical basicity of the CoO–M_2O–B_2O_3 samples decreases when CoO is substituted for M_2O predicting that the tetragonal distortion of $V^{4+}O_6$ should increase with increase in the concentration of CoO. It also predicts that the bond length of V^{4+} and vanadyl oxygen should increase with increase in mol% of CoO. These predictions are consistent with the experimental findings.

(6) In CoO–M_2O–B_2O_3 glasses, the d.c. conductivity increase with increase in temperature.

(7) Electrical conductivity decreases when Li_2O is replaced by K_2O, keeping the concentration of CoO and B_2O_3 constant.

(8) Addition of CoO in the sample shows a "blocking effect" on the overall mobility of alkali ions. As a result the conductivity of the sample decreases with increase in CoO:M_2O ratio.

References

1. Rama Kumar, R., Bhatnagar, A.K. & Rao, J. L. *Mater. Lett.*, 2002, **57**, 178.
2. Kreidl, N. J. *J. Non-Cryst. Solids*, 1990, **123**, 377.
3. Klonkowski, A. *Phys. Chem. Glasses*, 1985, **26**, 11.
4. Hosono, H., Kawazoe H. & Kanazawa, T. *J. Non-Cryst. Solids*, 1974, **33**, 125.
5. Toyuki, H. & Akagi, S. *Phys. Chem. Glasses*, 1974, **15**, 1.
6. Bogomolova, L. D., Filatova, I. V., Glassova, M. P. & Spasibkina, S.N. *J. Non-Cryst. Solids*, 1987, **90**, 625.
7. Bandopadhayay, A. K. *J. Mater. Sci.*, 1981, **16**, 189.
8. Bogomolova, L. D., Lazukin, V. N. & Petrovykh, N. V. *Sov. Phys. Doklady*, 1969, **14**, 243.
9. Bogomolova, L. D. & Dolgolenko, T. F. *J. Non-Cryst. Solids*, 1982, **47**, 427.
10. Seth, V. P., Gupta, S., Jindal, A. & Gupta, S. K. *J. Non-Cryst. Solids*, 1993, **162**, 263.
11. Seth, V. P., Yadav, Satya Pal & Gupta, S. K. *Radiat. Effects Defects Solids*, 1994, **132**, 187.
12. Khasa, S., Seth, V. P., Prakash, D. & Chand, Prem *Radiat. Effects Defects Solids*, 1997, **140**, 197.
13. Minami, T. & Yamani, A. *Chem. Express*, 1986, **1**, 567.
14. Angell, C. A. *Ann. Rev. Phys. Chem.*, 1992, **23**, 693.
15. Schiraldi, A. & Pezzat, E. *Mater. Chem. Phys.*, 1989, **23**, 75.
16. Baskaran, N. *J. Appl. Phys.*, 2002, **92**, 825.
17. Cormack, A. N., Du, J. & Zeitler, T. R. *Phys. Chem. Chem. Phys.*, 2002, **4**, 3193.
18. Muncaster, R. & Parke, S. *J. Non-Cryst. Solids*, 1977, **24**, 399.
19. Hochstrasser, G. *Phys. Chem. Glasses*, 1966, **7**, 178.
20. Hecht H. G. & Johnston, T. S. *J. Chem. Phys.*, 1967, **46**, 23.
21. Bleaney, B., Bowers, K. D. & Pryce, M. H. L. *Proc. R. Soc. (London)*, 1955, **A228**, 147.
22. Kivelson D. & Lee, S. *J. Chem. Phys.*, 1964, **41**, 1896.
23. Heine, V. *Phys. Rev.*, 1957, **107**, 1002.
24. Mcgarvey B. R. in *Transition Metal Chemistry*, Vol 3, Ed. R. L. Carlin, Marcel Dekker, New York, p.115.
25. Duffy, J. A. & Ingram, M. D. *J. Inorg. Nucl. Chem.*, 1975, **37**, 1203.
26. Kojima, K., Yano, H. & Matsuda, J. *J. Am. Ceram. Soc.*, 1990, **73**, 2134.
27. Hosono, H., Kawazoe, H. & Kanazawa, T. *J. Non-Cryst. Solids*, 1980, **37**, 427.
28. Ardelean, I., Cozar, O. & Iloica, Gh. *J. Non-Cryst. Solids*, 1984, **33**, 68.
29. Seth, V. P. & Yadav, A.Phys. Chem. Glasses, 1986, 28, 109.
30. Yadav, A., Seth, V. P., Chand, I. & Chand, Prem *Phys. Chem. Glasses*, 1998, **39**, 133.
31. Feridoonian, S. & Hogarth, C. A. *J. Mater. Sci.*, 1991, **26**, 3481.
32. Seth, V. P., Yadav, A. & Gupta, S. K. *J. Mater. Sci.*, 1988, **23**, 3495.
33. Sayer, M.& Man Singh, A. *Phys. Rev.*, 1972, **B6**, 4629.
34. Khawaja, E. E., Khan, M. A., Khan, M. N., Li, A. S. W. & Hwang, J. S. *J. Mater. Sci. Lett.*, 1984, **3**, 593.
35. Angel, P. W., Hann, R. E. & Cooper, A. R. *J. Non-Cryst. Solids*, 1995, **183**, 277.

Sodium selenoborato-*closo*-dodecaborates: structures and properties

Adrienne Hammerschmidt, Matthias Döch, Arno Lindemann, Steffen Pütz & Bernt Krebs[1]

University of Münster, Institute of Inorganic and Analytical Chemistry, 48149 Münster, Germany

Systematic studies on selenoborates containing a B_{12} cluster entity and alkali metal cations led to the new crystalline phases $Na_8[B_{18}Se_{18}]$, $Na_6[B_{18}Se_{17}]$ and $Na_2[B_{18}Se_{16}]$ which were prepared in solid state reactions from sodium selenide, amorphous boron and selenium in evacuated carbon coated silica tubes. Monoclinic $Na_8[B_{18}Se_{18}]$ (space group $P2_1/c$) consists of icosahedral B_{12} clusters completely substituted with trigonal-planar BSe_3 units and sodium counter ions. $Na_6[B_{18}Se_{17}]$ is also monoclinic, space group C2/c. Icosahedral B_{12} cluster entities as found in $Na_8[B_{18}Se_{18}]$ are connected in one direction via exocyclic selenium atoms forming the infinite chain anion $([B_{18}Se_{16}Se_{2/2}]^{6-})_\infty$. In $Na_2[B_{18}Se_{16}]$ (space group $P\bar{3}$) the icosahedral B_{12} cluster entities are connected to each other via seleno and perseleno bridges. Thus a three-dimensional network results with $[B_{18}Se_{12}Se_{6/2}]^{2-}$- and $[B_{18}Se_{15}Se_{6/2}]^{2-}$ anions in a ratio of 2:1.

1. Introduction

Numerous thio- and selenoborates have been synthesized and characterised in recent years due to improved preparation techniques.[1,2] Binary boron sulfides and selenides[1,3–5] as well as ternary and quaternary thio- and selenoborates contain boron atoms in a trigonal-planar coordination, for which various novel types of chalcogenoborate anions have been observed (Figure 1 illustrates the diversity of hitherto known chalcogenoborates). Typical examples are the small, highly charged anion entities BS_3^{3-},[6–8] BSe_3^{3-},[9] $B_2S_4^{2-}$,[10] $B_2S_5^{2-}$[11] and $B_3S_6^{3-}$,[8,12–14] which are characteristic structural features in nonoxide chalcogenoborates of alkali and alkaline earth metals. In contrast, examples for boron atoms in a trigonal planar coordination with selenium atoms are scarce, whereas tetrahedral chalcogen coordination which is found not only in thio- but also extensively in selenoborates.[15–22] Apart from the BSe_3^{3-} anion present in the thallium compound Tl_3BSe_3[9] as well as in $Ba_7(BSe_3)_4Se$,[23] such a boron–selenium coordination sphere is also observed in the recently discovered cluster phases of general formulae $M_8[B_{12}(BSe_3)_6]$ (M=K, Rb, Cs) and $M_4Hg_2[B_{12}(BSe_3)_6]$ (M=Rb, Cs).[24–27] With $M_8[B_{12}(BS_3)_6]$ (M=Rb, Cs) the first thioborate-*closo*-dodecaborates could be synthesised and characterised.[28] In all these structures $[B_{12}(BQ_3)_6]^{8-}$ anions (*Q*=S, Se) occur which are formed by B_{12}-*closo*-clusters which are completely substituted with six BQ_3 units.

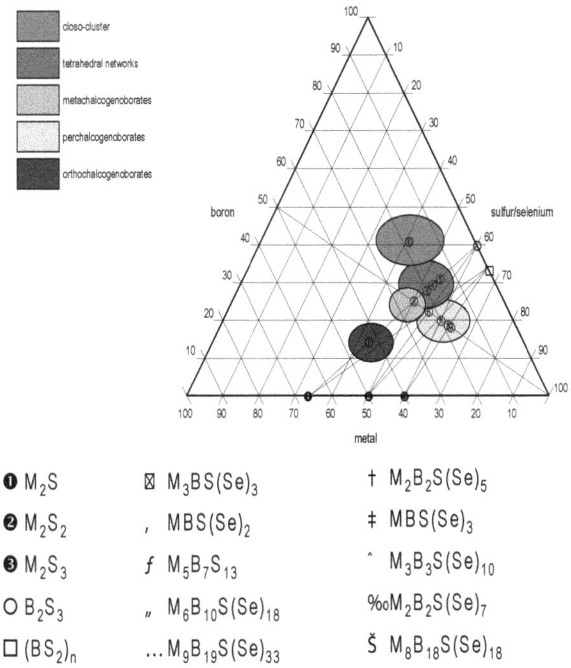

Figure 1. Structurally characterised chalcogenoborates in the M^I/B/S(Se) system

2. Experimental

2.1. Synthesis

The synthesis of well defined and highly pure boron chalcogen compounds is rather difficult because of the high reactivity of *in situ* built boron chalcogenides towards a variety of container materials at elevated temperatures. The fused silica tubes usually employed for solid state reactions are attacked by boron chalcogenides at temperatures above 400°C forming silicon chalcogen compounds by B–Si exchange at the surface of the ampoules. For the synthesis of pure samples the reaction vessel must either be made of boron nitride or graphite, or silica tubes coated with glassy carbon must be used. The latter are prepared by slowly turning a silica ampoule filled with acetone vapour through the flame of an oxygen–hydrogen

[1] Corresponding author

operated welding torch at about 1000°C. In some cases, especially when longer annealing is necessary, the former type of crucibles are employed. To protect them against oxidation they are encapsulated in steel or tantalum ampoules under an argon atmosphere, and these again are enclosed in evacuated silica tubes.

As starting materials the following products were used: sodium selenide (prepared following a method by Thiele *et al*[29]), amorphous boron (Alfa, amorphous powder, 95%), and selenium (Strem, powder, 99·5%).

For the syntheses of $Na_8[B_{18}Se_{18}]$, $Na_6[B_{18}Se_{17}]$ and $Na_2[B_{18}Se_{16}]$ appropriate amounts of the starting compounds were mixed up and filled into a carbon-coated silica tube which was thereafter sealed at a pressure of 6 Pa and inserted into a horizontal one-zone furnace. Heating and cooling procedures were performed as follows:

$$Na_8[B_{18}Se_{18}]: RT \xrightarrow{8\,h} 850°C(6\,h) \xrightarrow{4\,h} 700°C \xrightarrow{150\,h} 300°C \xrightarrow{10\,h} RT$$

$$Na_6[B_{18}Se_{17}]: RT \xrightarrow{4\,h} 850°C(6\,h) \xrightarrow{4\,h} 700°C \xrightarrow{150\,h} 300°C \xrightarrow{4\,h} RT$$

$$Na_2[B_{18}Se_{16}]: RT \xrightarrow{8\,h} 900°C(4\,h) \xrightarrow{20\,h} 500°C \xrightarrow{100\,h} 300°C \xrightarrow{10\,h} RT$$

$Na_2[B_{18}Se_{16}]$ gave bright orange plate shaped crystals, colourless crystals of $Na_8[B_{12}(BSe_3)_6]$ and $Na_6[B_{18}Se_{17}]$ suitable for single crystal x-ray diffraction were obtained. The products are air and moisture sensitive and were therefore handled under dry argon in a glove box.

2.2. Single crystal structure analysis

For the data collection single crystals were sealed in Mark capillaries under an argon atmosphere. The x-ray diffraction data for $Na_8[B_{18}Se_{18}]$ and $Na_2[B_{18}Se_{16}]$ were collected on a Bruker AXS Smart CCD diffractometer, in case of $Na_6[B_{18}Se_{17}]$ we used a STOE IPDS. The structure solutions were achieved by applying direct statistical methods of phase determination using the SHELXTL PLUS program,[30] and full-matrix least-squares refinements were performed using the SHELXL-97 software programs.[31] The complete data collection parameters and details of the structure solutions and refinements are given in Table 1.

3. Results and discussion

Starting from the compound $Na_8[B_{18}Se_{18}]$ with monomeric molecular $[B_{18}Se_{18}]^{8-}$ entities a formal decrease of Na_2Se content leads to the 1D-polymeric

Table 1. Crystallographic data for $Na_8B_{18}Se_{18}$, $Na_6B_{18}Se_{17}$ and $Na_2B_{18}Se_{16}$

Formula	$Na_8B_{18}Se_{18}$	$Na_6B_{18}Se_{17}$	$Na_2B_{18}Se_{16}$
Crystal system	monoclinic	monoclinic	trigonal
Space group	$P2_1/c$ (no. 14)	C2/c (no. 15)	$P\bar{3}$ (no. 147)
Lattice parameters	a=10·990(1) Å	a=18·005(4) Å	a=17·021(1) Å
	b=16·279(1) Å	b=16·549(3) Å	
	c=11·490(1) Å	c=11·245(2) Å	c=8·903(1) Å
	β=117·82(1)°	β=91·35(3)°	
Cell volume	1817·9(8) $Å^3$	3349·7(12) $Å^3$	2233·9(2) $Å^3$
Z	2	4	3
Density (calculated)	3·288 g/cm^{-3}	3·321 g/cm^{-3}	3·354 g/cm^{-3}
Diffractometer	BRUKER APEX	STOE IPDS	BRUKER APEX
Measurement temperature	298(2) K	293(2) K	298(2) K
Radiation	Mo-K_α, λ=0·71073 Å	Mo-K_α, λ=0·71073 Å	Mo-K_α, λ=0·71073 Å
Scan mode	φ-ω-scan; $\Delta\varphi$=0·3°	φ-scan, $\Delta\varphi$=1·0°	φ-ω-scan; $\Delta\varphi$=0·3°
Range in θ	2·1–28·02°	5·16–28·21°	2·40–27·99°
Absorption correction	Sadabs[32]	Abscor[32]	Sadabs[32]
Absorption coefficient	18·156 mm^{-1}	18·6 mm^{-1}	19·621 mm^{-1}
Reflections observed	18340	14821	22937
Independent reflections (m)	4405 [R(int)=0·045]	4060 [R(int)=0·0659]	3613 [R(int)=0·0881]
Data/Restraints/Parameters (n)	4405/0/199	4060/0/193	3613/0/164
Goodness-of-fit on F^2	1·024	1·057	0·754
Residuals	R_1=0·0248	R_1=0·0401	R_1=0·0323
[reflections with $I>2\sigma(I)$]	wR_2=0·0507	wR_2=0·1021	wR_2=0·0584
Residuals	R_1=0·0328	R_1=0·0563	R_1=0·0454
[all reflections]	wR_2=0·0529	wR_2=0·1112	wR_2=0·0598
Weighting factor $-g, h$	0·6095	0·0554; 42·7045	0·0212
Largest peak and hole	0·74/–0·55 $eÅ^{-3}$	1·07/–1·44 $eÅ^{-3}$	1·635/–1·469 $eÅ^{-3}$
Depository number	CSD-414010	CSD-411729	CSD-414456
Ø exocyclic B–Se bond length	1·910(4) Å	1·899(4) Å	
Ø μ_2 B–Se bond length (BSe_3 unit)	1·993(4) Å	1·973(4) Å	1·924(4) Å
Ø μ_2 B–Se bond length (B12 unit)	1·972(4) Å	1·996(4) Å	1·993(4) Å
Ø μ_2 B–Se bond length (seleno bridge)		1·962(4) Å	1·955(4) Å
Se–Se bond length (perseleno bridge)			2·333(2) Å
Ø Na···Se distance	3·141(2) Å	3·200(2) Å	3·306(2) Å
Ø B–B bond length	1·780(5) Å	1·780(5) Å	1·781(5) Å

compound $Na_6[B_{18}Se_{17}]$, a further decrease suggests the hypothetical formulae '$Na_4[B_{18}Se_{16}]$' (2D-polymeric) and '$Na_2[B_{18}Se_{15}]$' (3D-polymeric). Actually the formation of a three-dimensional network succeeded within the system Na/B/Se with the synthesis of (3D)$Na_2[B_{18}Se_{16}]$ which consists of $[B_{18}Se_{12}Se_{6/2}]^{2-}$ and $[B_{18}Se_{15}Se_{6/2}]^{2-}$ anions connected via seleno- and perseleno bridges, respectively.

B_{12} cluster entities are also observed in elementary boron as well as in various *closo*-boranes. Examples of completely substituted *closo*-dodecaboranes apart from halide or hydroxide substituted B_{12} clusters[33–37] are rare. Although a hydrogen–selenium exchange succeeded in the formation of the cluster anion $[B_6(SeCN)_6]^{2-}$ [38] without any lower substitution side products, only monosubstitution by a selenocyanate ligand is observed for a B_{12} cluster even under basic conditions.[39] Another example of selenium bonding to a B_{12} icosahedron is the binary boron selenide $B_{12}Se_{2-x}B_x$.[40] Since it crystallises in the B_6P-type structure no close structural resemblance to the herein described selenoborates is present. The significant stability of the B_{12} icosahedron is based on the ability of boron to form multicentre bonds. Wade's rules[41] apply for the $[B_{12}(BQ_3)_6]^{n-}$ entities in the herein reported selenoborate-*closo*-dodecaborates as well, for an aesthetic mathematical treatment of such polyhedra in crystal structures see Andersson & Jacob.[42]

As illustrated in Figure 2 the sodium selenoborato-*closo*-dodecaborate $Na_8[B_{18}Se_{18}]$ contains isolated B_{12} cluster entities saturated with BSe_3-ligands. Due to the perpendicular arrangement of the bidentate BSe_3-ligands the $[B_{18}Se_{18}]^{8-}$ entities show nearly D_{2h} symmetry. The monomeric $[B_{12}(BSe_3)_6]^{8-}$ unit exhibits two types of endocyclic B–Se bond lengths in the chelate-type B_3Se_2 rings: B–Se bonds to boron atoms in the cluster core range from 1·967(4)–2·021(4) Å and B–Se bonds within the trigonal planar BSe_3 units vary from 1·911(4)–1·989(4) Å. In contrast, the B–Se bonds to the exocyclic selenium atoms are significantly shorter with 1·927(4) and 1·908(4) Å. Strikingly, the heavier homologues with potassium, rubidium and cesium exhibit different substitution patterns[24–27] with a lower symmetry. The B–B bonds in the B_{12} cluster entity show only slight deviations from the mean value for all distances (computable to 1·78 Å). Therefore the boron icosahedra are nearly ideal platonic solids with a mean antipodal B···B distance of 3·38 Å. These values are in very good agreement with those found for the aforementioned isolated cluster compounds. Four crystallographically distinguishable sodium cations are located between the isolated anion entities. The sodium cations are in a six-fold and seven-fold selenium coordination, respectively. Selenium atoms forming the coordination polyhedra of these Na sites are found at distance of 2·774(2)–3·798(3) Å

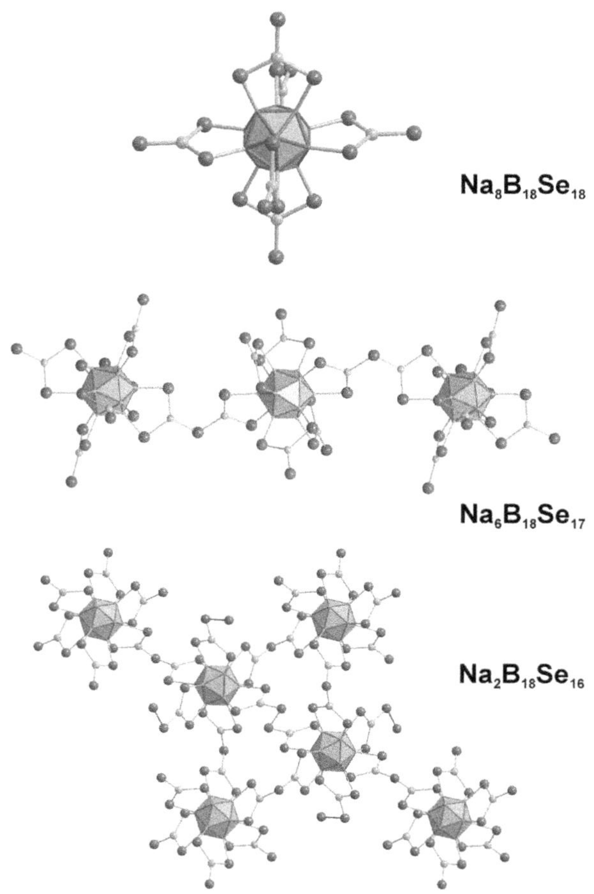

Figure 2. From monomeric to polymeric selenoborato-closo-dodecaborates

(please see Table 1 for a comparing presentation for all three selenoborato-*closo*-dodecaborates).

In the sodium selenoborato-*closo*-dodecaborate $Na_6[B_{18}Se_{17}]$ neighbouring B_{12} cluster entities are connected in one direction along the a-axis, via exocyclic selenium atoms of the BSe_3 chelate bonding ligands. This leads to the formation of infinite chain anions of formula $([B_{18}Se_{17}]^{6-})_\infty$. Due to the inversion symmetry in the substituted B_{12} icosahedron, the B_2Se_3 bridges point alternately in two directions. Neighbouring chain anions are linked by electrostatic interaction with the sodium cations. The molecular $[B_{12}(BSe_3)_4(BSe_2Se_{1/2})_2]^{6-}$ unit exhibits two types of endocyclic B–Se bond lengths in the chelate-type B_3Se_2 rings: B–Se bonds to boron atoms in the cluster core range from 1·985(7)–2·005(8) Å and B–Se bonds within the trigonal planar BSe_3 units vary from 1·943(8)–1·990(9) Å. In contrast, the B–Se bonds to the exocyclic selenium atoms are significantly shorter with 1·897(9) and 1·901(9) Å. The B–Se bond length in the boron selenium bridge between the cluster entities is slightly longer with 1·964(8) Å. As expected the B–B bonds in the B_{12} cluster entity exhibit only slight deviations from the mean value for all distances (computable to 1·78 Å). For $Na_6[B_{18}Se_{17}]$ a perpendicular arrangement with nearly D_{2h} symmetry of the six chelate type BSe_3

ligands is obvious as found in $Na_8[B_{18}Se_{18}]$. Four crystallographically distinguishable sodium cations are located between the infinite anion chains. The sodium cations are in a fivefold and sixfold selenium coordination, respectively. Selenium atoms forming the coordination polyhedra of these Na sites are found at distances of 2·813(12) to 3·586(6) Å.

In the sodium selenoborato-*closo*-dodecaborate $Na_2[B_{18}Se_{16}]$ B_{12} cluster entities substituted with BSe_3-ligands occur. In contrast to the perpendicular arrangement of the bidentate BSe_3-ligands in $Na_8[B_{18}Se_{18}]$ and $Na_6[B_{18}Se_{17}]$ the $[B_{18}Se_{12}Se_{6/2}]^{2-}$ and $[B_{18}Se_{15}Se_{6/2}]^{2-}$ anions present in $Na_2[B_{18}Se_{16}]$ exhibit D_{6h} and D_{3h} symmetry, respectively. The $[B_{18}Se_{12}Se_{6/2}]^{2-}$ and $[B_{18}Se_{15}Se_{6/2}]^{2-}$ anions form layers running parallel to (010) which are connected via perseleno bridges. Since all recent characterized perselenoborates exhibit BSe_4-tetrahedra connected to perseleno bridges $Na_2[B_{18}Se_{16}]$ is the first chalcogenoborate with trigonal planar BSe_3 entities linked to a Se–Se bond. The mean B–Se bond length (1·957(4) Å) agrees very well with the ones found in other selenoborates. Here two types of B–Se bonds are distinguishable. The average B_{ico}–Se bond length (icosahedral boron atom) is calculated to 1·993(4) Å, the B_{trig}–Se bond length (boron atom within the BSe_3-ligand) is found to be 1·950(4) Å. The B–B bonds in the B_{12} cluster entities show only slight deviations from the mean value for all distances (computable to 1·78 Å). These values are in very good agreement with those found for the aforementioned isolated cluster compounds. In contrast to the hitherto known perselenoborates the Se–Se bond length (2·333(2) Å) is sligthly shorter. The Na–Se distances vary in a range from 3·115(2)–3·543(2) Å (mean distance 3·306(2) Å). The structure contains one sodium atom in a seven-fold selenium coordination, the coordination polyhedron can be regarded as a threefold capped tetrahedron.

4. Conclusion

Since polymerisation of selenoborato-*closo*-dodecaborates has proven to be possible in one dimension with the synthesis of (1D)$Na_6[B_{18}Se_{17}]$ the formation of a three-dimensional network succeeded even within the system Na/B/Se with the synthesis of (3D)$Na_2[B_{18}Se_{16}]$. Further studies will concentrate on the question whether it is possible to synthesise a compound without perselenobridges of formula '(3D)$Na_2[B_{18}Se_{15}]$' or a 2D- polymeric compound '$Na_4[B_{18}Se_{16}]$'.

Acknowledgements

We would like to thank the Deutsche Forschungsgemeinschaft (SFB 458 'Ionic Motion in Materials with Disordered Structures – From Elementary Steps to Macroscopic Transport') and the Fonds der Chemischen Industrie for substantial support of this work.

References

1. Krebs, B. *Angew. Chem.*, 1983, **95**, 113; *Angew. Chem. Int. Ed. Engl.*, 1983, **22**, 113.
2. Conrad, O., Jansen, C. & Krebs, B. *Angew. Chem.*, 1998, **110**, 3396; *Angew. Chem. Int. Ed.*, 1998, **37**, 3208.
3. zum Hebel, P., Diercks, H. & Krebs, B. *Z. Kristallogr.*, 1988, **185**, 401.
4. Krebs, B. & Hürter, H.-U. *Acta Crystallogr.*, 1981, **A37**, C163.
5. Krebs, B. & Hürter, H.-U. *Angew. Chem.*, 1980, **92**, 479; *Angew. Chem. Int. Ed. Engl.*, 1980, **19**, 481.
6. Vinatier, P., Gravereau, P., Ménétrier, M., Trut, L. & Levasseur, A. *Acta Crystallogr.*, 1994, **C50**, 1180.
7. Hiltmann, F., Jansen, C. & Krebs, B. *Z. Anorg. Allg. Chem.*, 1996, **622**, 1508.
8. Hiltmann, F. & Krebs, B. *Z. Anorg. Allg. Chem.*, 1995, **621**, 424.
9. Krebs, B. & Hamann, W. *J. Less-Common Met.*, 1988, **137**, 143.
10. Hammerschmidt, A., Jansen, C., Küper, J., Püttmann, C. & Krebs, B. *Z. Anorg. Allg. Chem.*, 1995, **621**, 1330.
11. Jansen, C., Küper, J. & Krebs, B. *Z. Anorg. Allg. Chem.*, 1995, **621**, 1322.
12. Püttmann, C. Diercks, H. & Krebs, B. *Phosphorus, Sulfur, Silicon*, 1992, **65**, 1.
13. Chopin, F. & Hardy, A. *C. R. Acad. Sci.*, 1965, **261**, 142.
14. Chopin, F. & Turrell, G. *J. Mol. Struct.*, 1969, **3**, 57.
15. Hammerschmidt, A., Küper, J., Stork, L. & Krebs, B. *Z. Anorg. Allg. Chem.*, 1994, **620**, 1898.
16. Püttmann, C., Hiltmann, F., Hamann, W., Brendel, C. & Krebs, B. *Z. Anorg. Allg. Chem.*, 1993, **619**, 109.
17. Chopin, F. & Capdepuy, B. *Bull. Soc. Chim. Fr.*, 1970, **1**, 505.
18. Püttmann, C., Hamann, W. & Krebs, B. *Eur. J. Solid State Inorg. Chem.*, 1992, **29**, 857.
19. zum Hebel, P., Krebs, B., Grüne, M. & Müller-Warmuth, W. *Solid State Ionics* 1990, **43**, 133.
20. Krebs, B. & Diercks, H. *Z. Anorg. Allg. Chem.*, 1984, **518**, 101.
21. Hammerschmidt, A., zum Hebel, P., Hiltmann, F. & Krebs, B. *Z. Anorg. Allg. Chem.*, 1996, **622**, 76.
22. Lindemann, A., Küper, J., Hamann, W., Kuchinke, J., Köster, C. & Krebs, B. *J. Solid State Chem.*, 2001, **157**, 206.
23. Hammerschmidt, A., Lindemann, A., Küper, J., Köster, C. & Krebs, B. *Z. Anorg. Allg. Chem.*, 2001, **627**, 1253.
24. Küper, J., Conrad, O. & Krebs, B. *Angew. Chem.* 1997, **109**, 1995; *Angew. Chem. Int. Ed. Engl.*, 1997, **36**, 1903.
25. Lindemann, A., Kuchinke, J. & Krebs, B. *Z. Anorg. Allg. Chem.*, 1999, **625**, 1165.
26. Hammerschmidt, A., Lindemann, A., Döch, M. & Krebs, B. *Z. Anorg. Allg. Chem.*, 2003, **629**, 1249.
27. Hammerschmidt, A., Döch, M., Pütz, S. & Krebs, B. *Z. Anorg. Allg. Chem.*, 2004, **630**, 2299.
28. Kuchinke, J., Hammerschmidt, A. & Krebs, B. Solid State Sci., 2003, **5**, 189.
29. Thiele, K.-H., Steinicke, A., Dümichen, U. & Neumüller, B. *Z. Anorg. Allg. Chem.*, 1996, **622**, 231.
30. Sheldrick, G. M., SHELXTL PLUS, Siemens Analytical X-Ray Instruments, 1990.
31. Sheldrick, G. M. *SHELXL-97, Program for the refinement of crystal structures*, University of Göttingen 1997.
32. Saint Plus Version 6.02, Bruker Axs Inc., Madison, WI, 1999; Stoe IPDS Software, Version 2.75, Stoe & Cie GmbH, 1996.
33. Solntsev, K. A., Mebel, A. M., Votinova, N. A., Kuznetsov, N. T. & Charkin, O. P. *Koord. Khim.*, 1992, **18**, 340.
34. Kuznetsov, N. T., Zmeskova, L. A. & Ippolitov, E. G. *Zh. Neorgan. Khim.*,1981, **26**, 1862; *Russ. J. Inorg. Chem.*, 1981, **26**, 1003.
35. Kuznetsov, N. T., Zmeskova, L. A., Alikanova, Z. M. & Ippolitov, E. G. *Zh. Neorgan. Khim.*, 1981, **26**, 1331; *Russ. J. Inorg. Chem.*, 1981, **26**, 715.
36. Srebny, H.-G., Preetz, W. & Marsmann, H. C. *Z. Naturforsch. B*, 1984, **39**, 189.
37. Miller, H. C. & Muetterties, E. L. US-A 3551120 (1970), *Chem. Abstr.* 1971, **74**, 55815c.
38. Preetz, W. & Hake, M. G. *Z. Naturforsch. B*, 1992, **47**,1119.
39. Nachtigal, C., Steuer, B. & Preetz, W. *Chem. Ber.*, 1996, **129**, 1531.
40. Bolmgren, H. & Lundström, T. *J. Alloys Comp.*, 1993, **202**, 73.
41. Wade, K. *Advan. Inorg. Radiochem.*, 1976, **18**, 1.
42. Andersson, S. & Jacob, M. *The Mathematics of Structures.* Oldenbourg Verlag, München, 1997.

Borophosphosilicate glasses: properties and structure

B. G. Aitken & R. E. Youngman

SP-FR-05, Corning Incorporated, Corning, NY 14831, USA

The properties and structure of xB_2O_3:30–xP_2O_5:70SiO_2 *glasses, where* x=0, 2·5, 5, 7·5, 10, 15, 20, 22·5, 25, 27·5 *and* 30 *are examined. In order to minimise P volatilisation losses, ternary glasses were synthesised from mixtures of previously prepared 30P_2O_5:70SiO_2 and 30B_2O_3:70SiO_2 glasses. Chemical analysis confirmed that analysed compositions were within 0·5% of nominal values. Although the glass transition temperature (T_g) of the binary P, B silicate glass is 600, 440°C, respectively, T_g of ternary glasses attains a maximum value of 695°C for intermediate compositions containing ~12·5% B_2O_3. The compositional dependence of the thermal expansion coefficient and refractive index of these glasses is also discontinuous at B/P~1. Structural characterisation was carried out using wideline, magic angle spinning (MAS) and triple quantum MAS NMR spectroscopic techniques. A very small percentage (≤3%) of the total Si is present as both 5- and 6-fold coordinated species in the most P-rich glasses. ^{31}P wideline NMR spectra of the same glasses show an asymmetric lineshape characteristic of an axially symmetric O=$PO_{3/2}$ species, whereas the spectra of B-rich glasses show only the Gaussian resonance typical of symmetric $PO_{4/2}$ tetrahedra. ^{11}B results indicate that, as B is progressively replaced by P, B coordination changes monotonically from fully trigonal to dominantly tetrahedral, with the tetrahedral B fraction ranging from 80 to 95% for the most P-rich glasses. Overall the results are consistent with a structural scenario of associated $BO_{4/2}$ and $PO_{4/2}$ tetrahedra comprising a fully polymerised silica-like network for the glasses of intermediate composition. The greater degree of network connectivity caused by the preferential association of B and P to form BPO_4-like groups is the likely origin of the T_g maximum.*

1. Introduction

Thermally stable, high strain point (≥650°C) glasses that are expansion matched to Si have recently been of interest as substrate materials for the fabrication of active matrix liquid crystal displays (LCDs). Current commercial glasses are typically alkaline earth boroaluminosilicates,[1] but forming these materials can sometimes be problematic due to the crystallisation of the aluminosilicate phase mullite. Al-free glasses in the B_2O_3–P_2O_5–SiO_2 system have sufficiently high strain point, especially for the more silica-rich compositions. Moreover, due to the fact that BP silicate glasses contain no network modifying components, it is likely that they may have the high liquidus viscosity (typically >300,000 poise) required for forming by the so-called "overflow downdraw" or "fusion" sheet manufacturing process.[2] As such, B_2O_3–P_2O_5–SiO_2 glasses are worthy of examination as alternate candidate materials for LCD applications. In addition, glasses from this chemical system are ideal materials with which to study the degree to which network-forming cations, in the absence of classical network modifiers, tend to either associate or form random distributions. An earlier Raman and NMR investigation of the structural effect of P_2O_5 additions to a 40B_2O_3:60SiO_2 glass showed evidence for the formation of BPO_4-like species.[3] That result is consistent with the well documented, strong tendency for BPO_4 to crystallize from borophosphosilicate gels and glasses.[4–6] A more recent NMR study by Youngman *et al* examined a wider compositional range of ternary glasses and glass ceramics.[7] Their results also indicated the presence of a BPO_4-like network in the glasses, as well as little connectivity between SiO_2 and the other glass forming oxides, suggesting the existence of SiO_2-rich domains.

Previous work in the B_2O_3–P_2O_5–SiO_2 system has outlined the regions of glass formation using conventional crucible melting procedures.[5,8,9] Glasses with SiO_2 content ≤ 50% were characterised by poor chemical durability, although this was improved by about three orders of magnitude on heat treating the glasses at 900–1000°C to form glass ceramics containing crystalline BPO_4 plus a silica rich residual glass.[5,6] The latter result suggests that BP silicate glasses with higher SiO_2 concentration may well have sufficient chemical durability for practical applications. Thus, the possibility of high strain point as well as improved chemical durability at higher silica concentration for potential applications in the field of LCD substrates

[1] Corresponding author

prompted us to investigate the physical properties and structure of borophosphosilicate glasses with a constant SiO_2 content of 70%.

2. Experimental

xB_2O_3:30–xP_2O_5:70SiO_2 glasses, with x ranging from 0 to 30 in increments of 2·5, were synthesised from mixtures of previously prepared 30P_2O_5:70SiO_2 and 30B_2O_3:70SiO_2 glasses. 500 g batches of the latter were melted in Pt crucibles for 2 h at 1500–1550°C. After casting onto a steel plate, the glasses were annealed by heating near the glass transition temperature (T_g) followed by slow cooling to room temperature. The chemical composition of all samples was determined by inductively coupled plasma emission spectroscopy. T_g was determined to ±5°C by differential scanning calorimetry using a heating rate of 10°C/min. Strain (T_{str}) and annealing points (T_a), corresponding to viscosities of ~10^{14}, 10^{13} poise, respectively, were measured by beam bending viscometry on 2″×1/8″×1/8″ bars. The softening point (T_s), corresponding to a viscosity of $10^{7·6}$ poise, was measured by the parallel plate technique using 5 mm diameter × 5 mm thick discs. T_{str}, T_a and T_s are believed to be accurate to ±2°C. The thermal expansion coefficient (α) was determined to ±0·1 ppm/°C by dilatometry relative to a fused silica reference. Refractive index was measured on glass powder using the Becke line technique.

NMR data were collected using commercial instruments and probes, in conjunction with 11·7 and 4·7 T superconducting magnets. ^{11}B MAS and triple quantum NMR were carried out at 11·7 T (160·3 MHz resonance frequency) using a 3·2 mm MAS (magic angle spinning) NMR probe and sample spinning of 20 kHz. To obtain quantitatively accurate values of the tetrahedral B fraction, very short radio-frequency pulse widths (typically <π/12) were used along with 2 s recycle delays. ^{11}B NMR data were processed without additional line broadening and are referenced against BF_3–Et_2O. ^{31}P wideline NMR spectra were measured at 4·7 T (80·9 MHz resonance frequency) using a 9·5 mm MAS NMR probe without sample spinning. Short pulse widths (1·7 μs ~ π/6) and 30 s pulse delays were used to collect these spectra. Data collected with longer pulse delays were essentially unchanged, so the above conditions were deemed adequate for obtaining accurate spectra, which were processed without line broadening and referenced to an external 85% H_3PO_4 standard. ^{29}Si MAS NMR spectra were obtained at 4·7 T (39·7 MHz resonance frequency) using a 9·5 mm MAS NMR probe with 5 kHz sample spinning. Short pulse widths (~π/6) and 90 s recycle delays were chosen to maximise the signal to noise. These data were processed with 25 Hz Gaussian line broadening and plotted against tetramethylsilane.

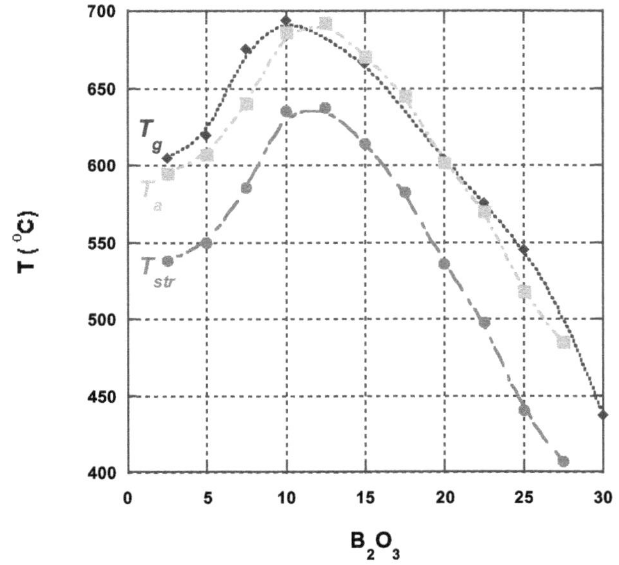

Figure 1. Variation of the characteristic temperatures T_{str}, T_a and T_g of 70% SiO_2 borophosphosilicate glasses with B_2O_3 concentration

3. Results

3.1 Chemical composition and physical properties

Clear, colorless glasses were formed from all investigated compositions. The results of chemical analysis and selected physical property measurements on the studied borophosphosilicate glasses are reported

Table 1. Composition and properties of xB_2O_3–(30–x)P_2O_5–70SiO_2 glasses

% B_2O_3 (nom.)	% B_2O_3 (anal.)	% P_2O_5 (anal.)	T_g (°C)	T_{str} (°C)	T_a (°C)	T_s (°C)	α (ppm/°C)	n
0	n.a.	30·16				801	4·91	1·504
2·5	2·48	27·41	605	538	595	814	4·40	1·502
5	4·88	25·44	619	555	607	835	4·30	1·499
7·5	7·35	22·83	675	585	640	880	4·22	1·497
10	10·3	19·85	694	635	686	912	4·40	1·495
12·5				637	692		4·04	
15	14·47	15·31	666	614	671	1018	3·61	1·492
17·5				582	645		3·65	
20	19·61	10·03	604	536	602	960	3·60	1·480
22·5	21·46	7·48	575	498	570	939	3·45	1·474
25	24·53	4·99	545	440	518	908	3·54	1·468
27·5	26·58	2·45		407	485	878	3·54	1·464
30	29·66	n.a.						

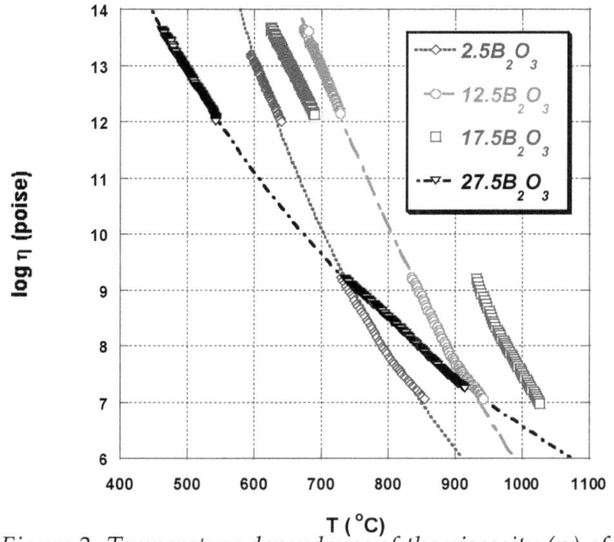

Figure 2. Temperature dependence of the viscosity (η) of xB_2O_3:30−xP_2O_5:70SiO_2 glasses for x= 2·5, 12·5, 17·5 and 27·5. Dashed curves are Fulcher fits to the experimental data

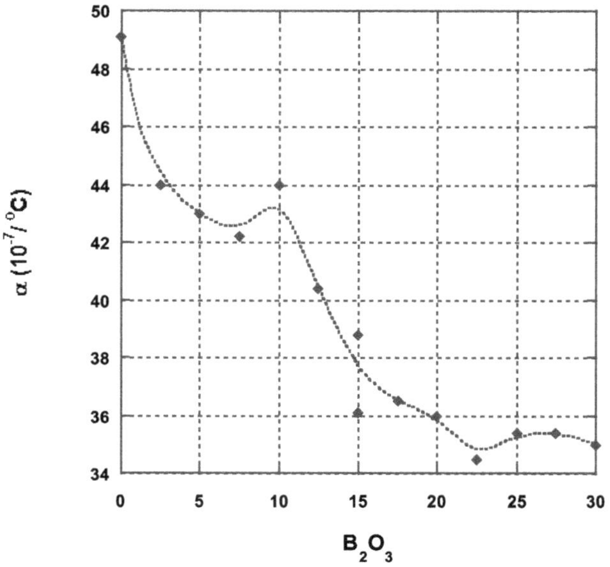

Figure 3. Compositional dependence of α of xB_2O_3:30−xP_2O_5:70SiO_2 glasses; dashed line is a smooth curve through the data

in Table 1. The chemical data indicate minor loss of B_2O_3 or P_2O_5 during glass preparation and therefore, in the following, glass samples are referred to using the nominal B_2O_3 and/or P_2O_5 contents.

The compositional dependence of the characteristic temperatures of the studied samples is illustrated in Figure 1. Starting from a B-poor composition, T_a increases monotonically with rising B_2O_3 content until it attains a maximum value of 692°C at 12·5% B_2O_3. Thereafter, T_a drops steadily with further increase in B_2O_3 concentration. T_{str} and T_g exhibit analogous compositional trends, with both properties passing through a maximum at intermediate compositions.

The temperature dependence of viscosity for glasses containing 2·5, 12·5, 17·5 and 27·5% B_2O_3 is displayed in Figure 2. Glasses with B_2O_3 content ranging from 5 to 10% have viscosity curves (not shown) that nest at progressively higher temperatures between the curves for the 2·5 and 12·5% glasses; the viscosity data for all glasses with B_2O_3 content ranging 2·5–15% B_2O_3 are well fit by the Fulcher expression

$$\log \eta = A + B/(T - T_0)$$

The viscosity curve of the 27·5% B_2O_3 glass is also well fit by the Fulcher expression, but is seen to have a much shallower slope than that of the P-rich glasses, intersecting the latter in the viscosity range of 10^7–10^9 poise. The viscosity of glasses with intermediate B_2O_3 concentration, as exemplified by the illustrated data for the 17·5% B_2O_3 glass, diverges to higher values from a smooth Fulcher curve in the softening range. These results are consistent with the qualitative observations made during glass preparation that B_2O_3-rich melts are much more viscous than those with B/P<1.

α of the studied glasses ranges from 3·5 to 4·9 ppm/°C. As shown by the data in Table 1 and Figure 3, α drops from 4·9 to 3·7 ppm/°C as the B_2O_3 content increases from 2·5 to 17·5%, but remains essentially constant with further increase in B_2O_3 concentration.

The refractive index of xB_2O_3:30−xP_2O_5:70SiO_2 glasses decreases from 1·505 to 1·460 with rising B_2O_3 content over the studied range. However, as shown in Figure 4, there is a break in the slope of this trend at ~15% B_2O_3, with the trend of decreasing index with increasing B_2O_3 being steeper for B-rich glasses.

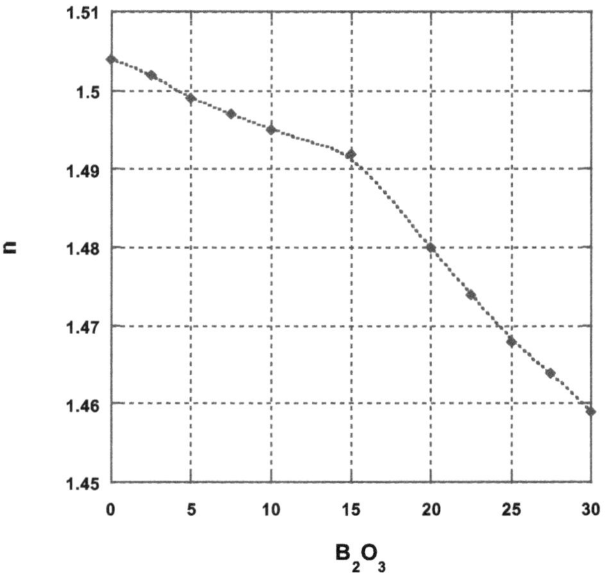

Figure 4. Refractive index (n) of 70% SiO_2 borophosphosilicate glasses as a function of B_2O_3 concentration; dashed line is a guide to the eye

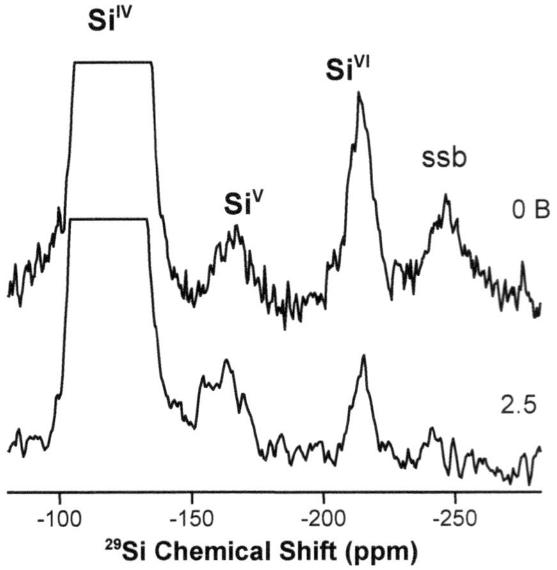

Figure 5. ^{29}Si MAS NMR spectra of xB_2O_3:30–xP_2O_5:70SiO_2 glasses for x=0 and 2·5. Spinning side band denoted by "ssb"

3.2 NMR spectroscopy

The ^{29}Si MAS NMR spectrum of the B-free phosphosilicate glass used as a starting material, as well as that of a 2·5% B_2O_3 glass, are shown in Figure 5. Both spectra are dominated by an intense resonance with a chemical shift of −120 ppm. However, both spectra also contain additional weaker peaks at −160 and −210 ppm, the intensities of which are reduced as the B_2O_3 concentration increases to 2·5%.

^{31}P wideline NMR spectra of ternary glasses are illustrated in Figure 6. The spectra of B-poor glasses containing ≤15% B_2O_3 consist of a superposition of uniaxial and Gaussian lineshapes, the latter centered around −30 ppm. The uniaxial powder pattern contains several singularities, including the peak at 30 ppm and a pronounced shoulder near −160 ppm, and basically extends over a chemical shift range of approximately 300 ppm. The relative intensity of this uniaxial component decreases with rising B_2O_3

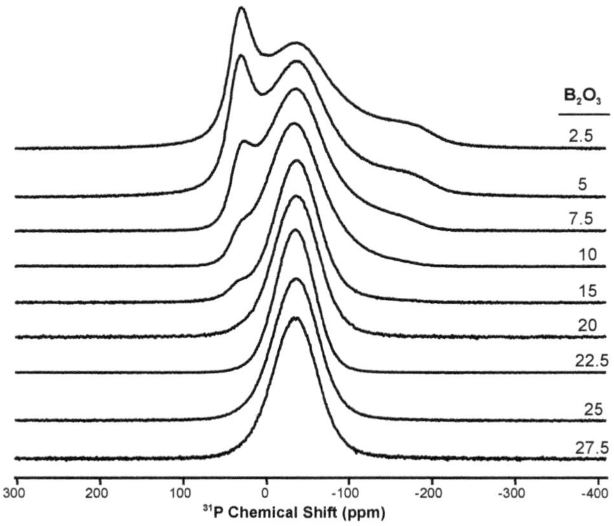

Figure 6. ^{31}P wideline NMR spectra of xB_2O_3:30–xP_2O_5:70SiO_2 glasses, with x ranging from 2·5 to 27·5

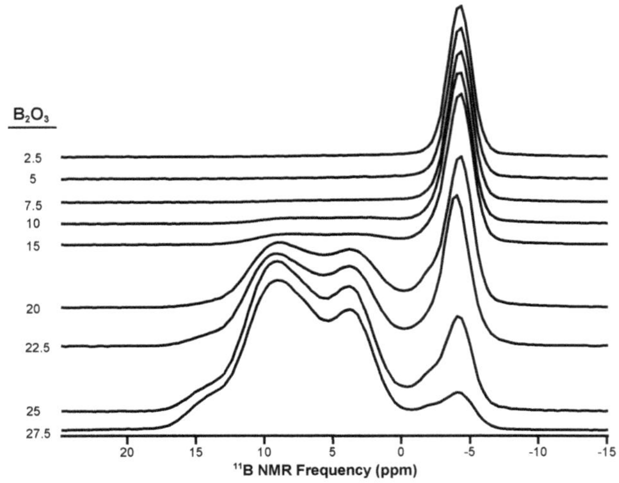

Figure 7. ^{11}B MAS NMR spectra of xB_2O_3:30–xP_2O_5:70SiO_2 glasses, with x ranging from 2.5 to 27.5

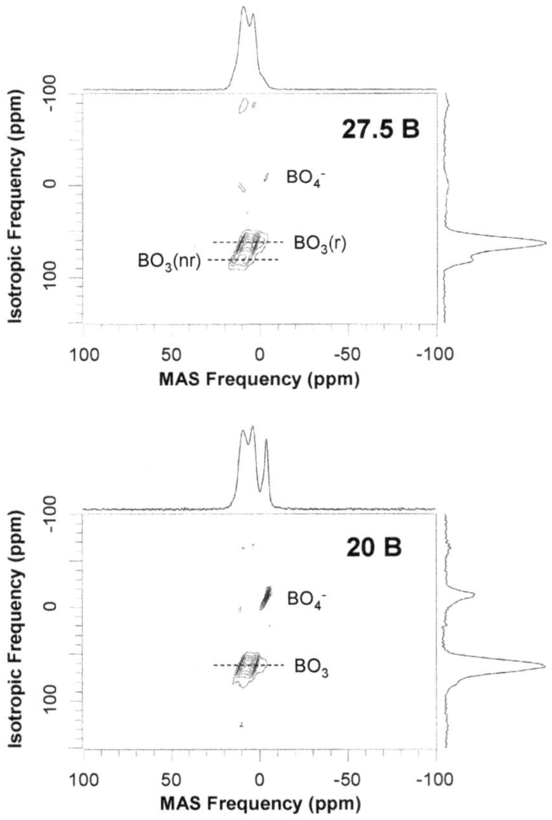

Figure 8. Triple quantum ^{11}B NMR spectra of xB_2O_3:30–xP_2O_5:70SiO_2 glasses for x=27·5 and 20

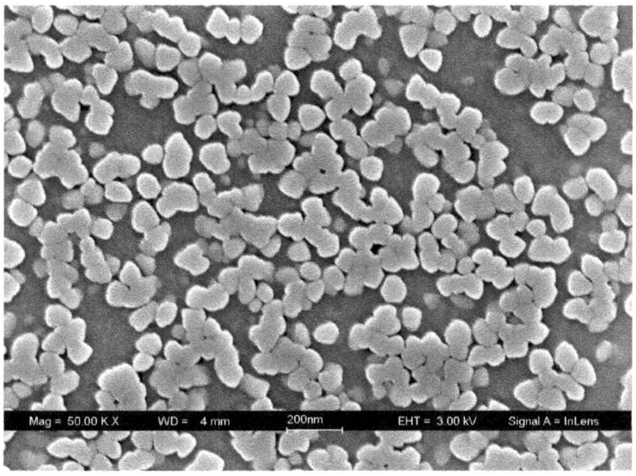

Figure 9. SEM photomicrograph of a sample of $15B_2O_3$:$15P_2O_5$:$70SiO_2$ glass after heating to 1050°C during parallel plate viscometry. White scale bar measures 200 nm

concentration. Only the −30 ppm Gaussian powder pattern is present in the spectra of B-rich glasses containing ≥15% B_2O_3.

^{11}B MAS NMR spectra of xB_2O_3:30−xP_2O_5:$70SiO_2$ glasses with x ranging from 2·5 to 27·5 are presented in Figure 7. A broad resonance at 0–15 ppm dominates the spectrum of the most B-rich glass, where it is accompanied by a weaker and narrower peak centered at −4·2 ppm. As the B_2O_3 concentration decreases, the relative intensity of the broad peak decreases, such that this feature is apparently absent in glasses containing ≤7·5% B_2O_3. Conversely, the relative intensity of the narrow −4·2 ppm resonance increases monotonically with decreasing B_2O_3 concentration. In the most B-rich glasses, the latter feature is seen to be composed of a dominant peak at −4·2 ppm and a weaker shoulder at −2 ppm. High resolution ^{11}B triple quantum (3Q) MAS NMR spectra of glasses containing 20 and 27·5% B_2O_3 are shown in Figure 8. In the spectrum of the latter sample, the dominant 0–15 ppm resonance is seen to be comprised of two components with isotropic frequencies of 60 and 85 ppm. However, as the B_2O_3 concentration drops to 20%, the latter resonance is comprised of only the 60 ppm component. In addition, the narrow −4·2 ppm MAS resonance is clearly resolved.

Table 2. B and P speciation of xB_2O_3–(30−x)P_2O_5–$70SiO_2$ glasses

% B_2O_3 (nominal)	% BIV	% $PO_{4/2}$
0		0
2·5	98·9	21
5	97·9	32
7·5	96·7	43
10	89·4	76
12·5	80·4	92
15	67·2	100
17·5	52·5	100
20	40·7	100
22·5	32·2	100
25	17·9	100
27·5	9·1	100
30	0	

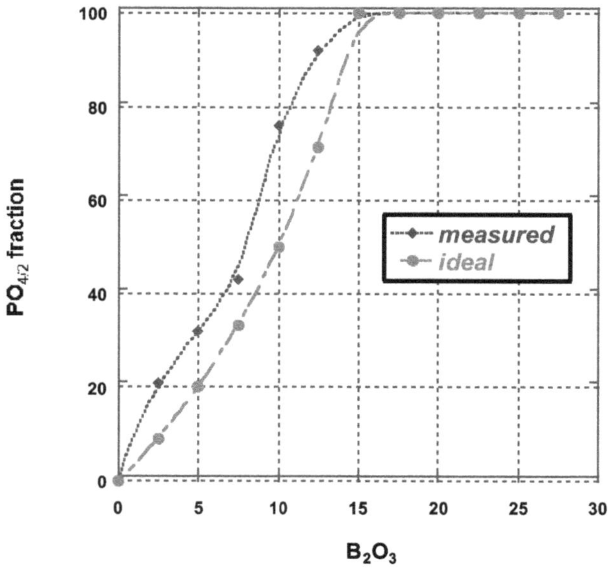

Figure 10. $PO_{4/2}$ or Q^4 P fraction estimated from ^{31}P wideline spectra (◆) as a function of B_2O_3 concentration. "Ideal" (●) curve is calculated assuming a 1:1 conversion of Q^3 to Q^4 P by each added B

Discussion

The outstanding feature of the compositional dependence of the physical properties of xB_2O_3:30−xP_2O_5:$70SiO_2$ glasses is the observed discontinuity at B/P~1. In the case of the characteristic temperatures of these materials, this discontinuity is manifested as a maximum in T_{str}, T_a and T_g at ~12·5% B_2O_3. For α and refractive index, the discontinuity is seen as a change in the slope of the property's compositional dependence. These features strongly suggest that a significant change in the structural evolution of these glasses occurs at B/P ~1 in response to the progressive replacement of P by B. Given that this structural change gives rise to a maximum in characteristic temperatures, it is likely that it is associated with an increase in the degree of polymerisation or network connectivity of these glasses. One possible explanation for an increase in overall network connectivity is that B and P associate to form BPO_4-like species in xB_2O_3:30−xP_2O_5:$70SiO_2$ glasses with intermediate composition. As BPO_4 is a silica-like material in which alternating corner-shared $BO_{4/2}$ and $PO_{4/2}$ tetrahedral groups are linked in three dimensions, the formation of similar species as B is substituted for P in P-rich BP silicate glasses or as P replaces B in B-rich glasses would be expected to further polymerise the glassy network and, hence, increase the characteristic temperatures of these materials.[10] This interpretation is consistent with the observed

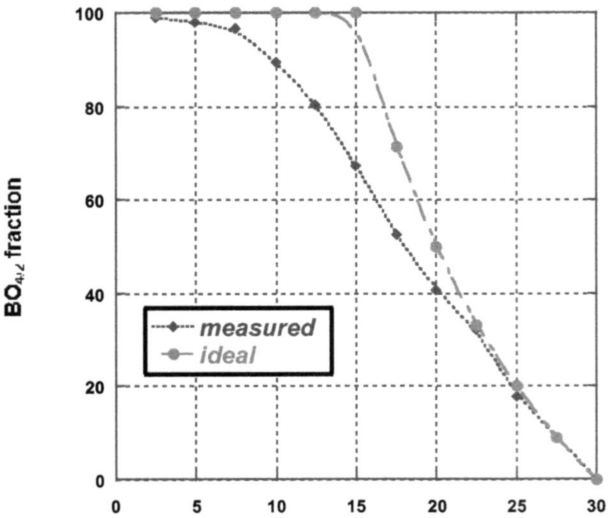

Figure 11. $BO_{4/2}$ fraction determined from ^{11}B MAS spectra (◆) as a function of B_2O_3 concentration. "Ideal" (•) curve is calculated assuming a 1:1 conversion of trigonal to tetrahedral B by each added P

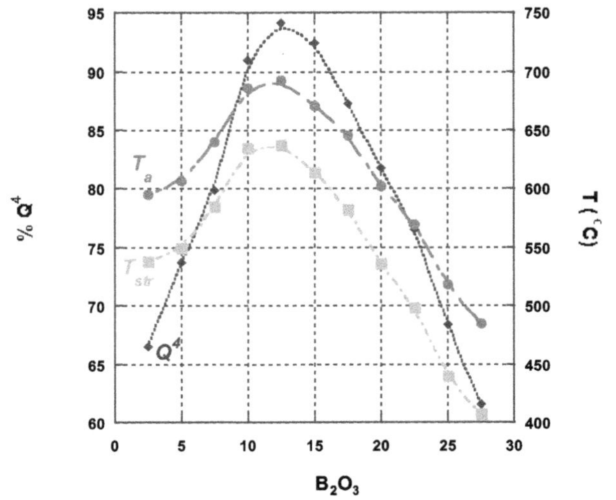

Figure 12. Total Q^4 species concentration (◆) as a function of B_2O_3 content; for comparative purposes, also plotted are the corresponding data for T_a (•) and T_{str} (■) taken from Figure 1

divergence to higher temperature of the softening regime viscosity of glasses with intermediate B_2O_3 concentration (cf. Figure 2). Following viscometry, the anomalously increased viscosity of such glasses at higher temperature is seen to have been due to the precipitation of ~100 nm diameter spherulites of BPO_4 nanocrystals (cf. Figure 9).

The NMR results reported above are in agreement with this structural scenario. Considering first the ^{31}P wideline spectra, these show clear evidence for increasing network connectivity with increasing B_2O_3 concentration, at least up to ~15% B_2O_3. The asymmetric component of the NMR lineshape dominates the spectra of B-poor xB_2O_3:30-xP_2O_5:70SiO_2 glasses and is reasonably attributed to Q^3 $O{=}PO_{3/2}$ tetrahedral species with their characteristic uniaxial symmetry.[11] With increasing B_2O_3 content, the intensity of the uniaxial component and, hence, the concentration of $O{=}PO_{3/2}$ groups decreases, whereas the Gaussian component becomes correspondingly predominant. The latter feature can be assigned to an isotropic $PO_{4/2}$ tetrahedral species, analogous to the Q^4 $PO_{4/2}$ tetrahedra present in crystalline BPO_4.[12] The dependence of the $PO_{4/2}$ fraction on B_2O_3 concentration is illustrated in Figure 10; also plotted in Figure 10 is the expected compositional dependence of the $PO_{4/2}$ fraction assuming that each added B converts a Q^3 $O{=}PO_{3/2}$ species into a Q^4 group. The agreement between the observed and predicted trends is striking.

The ^{11}B MAS spectral results are also consistent with the progressive formation of BPO_4-like species as intermediate compositions are approached. As P replaces B in B-rich glasses, there is a decrease in the relative intensity of the broad resonance at 0–15 ppm feature that is characteristic of 3-coordinated or trigonal B.[13] In addition, the 3Q MAS NMR spectra illustrated in Figure 8 show that there is a change in the trigonal B distribution from a mixture of ring, including boroxol groups, and non-ring species in the most B-rich glasses to essentially only non-ring species in glasses containing ≤20% B_2O_3.[14,15] Conversely, there is a monotonic increase in the relative intensity of the narrower −4·2 ppm resonance that is assigned to 4-coordinated or tetrahedral B.[13] It should be noted that the chemical shift of this resonance is in good agreement with that observed for tetrahedral B in crystalline BPO_4[16] and, therefore, is indicative of not only tetrahedral B but also of tetrahedral B with P next nearest neighbours. It is only in the most B-rich glasses containing ≥25% B_2O_3 that a second resonance at ~−2 ppm, attributable to tetrahedral B with B or Si next nearest neighbors, can be clearly resolved (cf. Figure 7).[13,17] The pronounced tendency of B and P to associate to form BPO_4-like units, also observed by Dickinson *et al* in relatively SiO_2-poor glasses,[3] implies that there is little structural connectivity between $BO_{4/2}$, $PO_{4/2}$ tetrahedra and the silicate component of these glasses. The latter suggests that xB_2O_3:30−xP_2O_5:70SiO_2 glasses with intermediate composition may comprise BPO_4-rich and SiO_2-rich subnetworks, which in turn may explain the observed internal nucleation of BPO_4 nanocrystals during high temperature viscometry (cf. Figure 9).

The compositional dependence of the observed tetrahedral B fraction is plotted in Figure 11, along with the trend expected for 1:1 conversion of trigonal to tetrahedral B for each added P. Although the agreement between observed and predicted trends for xB_2O_3:30−xP_2O_5:70SiO_2 glasses is good, especially at low and high x, the reaction between trigonal B and

Q^3 P species to form BPO_4-like groups, namely

$$BO_{3/2} + O{=}PO_{3/2} \rightarrow BO_{4/2} + PO_{4/2}$$

does not proceed to completion in glasses with intermediate composition.

The ^{29}Si MAS NMR spectra show that the overwhelmingly dominant Si species is the Q^4 $SiO_{4/2}$ tetrahedral group that is typical of most silica-rich glasses. The chemical shift of these Q^4 groups, near −120 ppm, is shifted to slightly lower values (cf. −112 ppm in pure silica) due to the effect of P second neighbors.(18) It is interesting to note, however, that a small fraction of higher coordinated Si is present in the most P-rich glasses. As shown in Figure 5, there are weak resonances at −160 and −210 ppm in the spectra of these glasses that can be assigned to Si in 5- and 6-fold coordination.(19) Nevertheless, these species comprise only 3 and 2·5% of the total Si in the 0 and 2·5% B_2O_3 glasses, respectively. It is expected that all Si in the glasses containing ≥5% B_2O_3 is present solely as Q^4 species, although the ^{29}Si MAS NMR spectra of these samples were not recorded.

Overall, the structural evolution of xB_2O_3:30−xP_2O_5:70SiO_2 glasses, as determined by NMR spectroscopy, can be described as follows. In P-rich glasses, P is present principally in Q^3 $O{=}PO_{3/2}$ groups that link either with Q^4 $SiO_{4/2}$ tetrahedra or with other $O{=}PO_{3/2}$ groups. As B replaces P, Q^3 $O{=}PO_{3/2}$ groups are converted into Q^4 $PO_{4/2}$ tetrahedra, with B initially assuming tetrahedral coordination and associating with the latter species to form the B–O–P linkages of BPO_4-like groups. The latter conversion continues as the B_2O_3 concentration approaches 12·5–15%. At this point, all P is present as Q^4 $PO_{4/2}$ tetrahedra and, therefore, any additional increase in the B_2O_3 content results in the formation of trigonal $BO_{3/2}$ groups. With further increase in the B_2O_3 concentration, the decreasing amount of P continues to occur as Q^4 $PO_{4/2}$ tetrahedra with an equally decreasing population of associated $BO_{4/2}$ tetrahedra, whereas the B in excess of P forms an increasing number of $BO_{3/2}$ species that ultimately associate to form ring structures, including boroxol groups, at the highest B_2O_3 concentrations.

The structural evolution of xB_2O_3:30−xP_2O_5:70SiO_2 glasses described above is therefore one in which network polymerisation or connectivity initially increases as x increases from 0 to ~15. At the latter point, there is a maximum in the concentration of Q^4 P and B species and, hence, a maximum in network connectivity, as further increase in x results in the formation of less connected trigonal B species. One can estimate the total Q^4 species concentration in these glasses from the reported $PO_{4/2}$ and $BO_{4/2}$ fractions (cf. Figures 10, 11) and assuming that all Si is Q^4. The latter assumption results in a small overestimation of total Q^4 only for glasses containing < 5% B_2O_3, which contain a small fraction of higher coordinated Si. This quantity is plotted in Figure 12 as a function of the B_2O_3 content, along with the measured T_a and T_{str}. The similar compositional dependences of characteristic temperatures and the total Q^4 fraction strongly suggest that the former is governed principally by the variation in overall network connectivity.

Conclusions

The compositional dependence of the physical properties of xB_2O_3:30−xP_2O_5:70SiO_2 glasses is discontinuous at intermediate compositions. In particular, the characteristic temperatures attain a maximum near B/P~1. NMR spectroscopy demonstrates that, in the same glasses, BPO_4-like species consisting of associated Q^4 $BO_{4/2}$ and $PO_{4/2}$ tetrahedra also reach their highest concentration. The resultant maximum in network connectivity is believed to be the underlying structural basis of the observed property variation.

References

1. Dumbaugh, W. H., Lapp, J. C. & Moffat, D. M. US Pat. 5,374,595, 1994.
2. Dockerty, S. M. US Pat. 3,338,696, 1967.
3. Dickinson, J. E., de Jong, B. H. W. S. & Schramm, C. M. *J. Non-Cryst. Solids*, 1988, **102**, 196.
4. Woignier, T., Phalippou, J. & Zarzycki, J. *J. Non-Cryst. Solids*, 1984, **63**, 117.
5. MacDowell, J. F. *Proc. XV Int. Congr. Glass*, Leningrad, 1989, 90.
6. MacDowell, J. F. US Pat. 4,576,920, 1986.
7. Youngman, R. E., Aitken, B. G. & Dickinson, J. E. *J. Non-Cryst. Solids*, 2000, **263&264**, 111.
8. Horn, W. F. & Hummel, F. A. *J. Soc. Glass Tech.*, 1955, **39**, 113.
9. Englert, W. J. & Hummel, F. A. *J. Soc. Glass Tech.*, 1955, **39**, 121.
10. Shulce, G. E. *Z. Phys. Chem.*, 1934, **B24**, 215.
11. Eckert, H. *Prog. NMR Spectroscopy*, 1992, **24**, 159-293.
12. Turner, G. L., Smith, K. A., Kirkpatrick, R. J. & Oldfield, E. *J. Magn. Reson.*, 1986, **70**, 408.
13. Turner, G. L., Smith, K. A., Kirkpatrick, R. J. & Oldfield, E. *J. Magn. Reson.*, 1986, **67**, 544.
14. Youngman, R. E., Werner-Zwanziger, U. & Zwanziger, J. W. *Z. Naturforsch.*, 1996, **51a**, 321.
15. Du, L.-S. & Stebbins, J. F. *J. Phys. Chem.*, 2003, **B107**, 10063.
16. Grimmer, A. R., Müller, D., Gözel, G. & Kniep, R. *Fres. J. Anal. Chem.*, 1997, **357**, 485.
17. Du, L.-S. & Stebbins, J. F. *J. Non-Cryst. Solids*, 2003, **315**, 239.
18. Dupree, R., Holland, D., Mortuza, M. G., Collins, J. A. & Lockyer, M. W. G. *J. Non-Cryst. Solids*, 1989, **112**, 111.
19. Dupree, R., Holland, D. & Mortuza, M. G. *Nature*, 1987, **328**, 416.

Dynamical response of borate glasses down to low temperatures

M. Cutroni,[1] A. Mandanici,[1,*] A. Raimondo[1] & A. Sanson[2]

[1] Dipartimento di Fisica, Università di Messina, Salita Sperone 31, 98100 Messina, Italy
[2] IFN, Istituto di Fotonica e Nanotecnologie del Consiglio Nazionale delle Ricerche, Sezione CeFSA di Trento - 38050 Povo (Trento), Italy

Relaxation dynamics of silver ions in the $(AgI)_x.(Ag_2O.2B_2O_3)_{1-x}$ superionic glasses have been studied as a function of temperature between 300 and 90 K in the frequency range from 10^{-3} to 10^9 Hz. The frequency dependent dielectric permittivity has been analysed in terms of conductivity as well as in the formalism of the electric modulus. From the conductivity spectra the Arrhenius dependence of d.c. conductivity as well as the power law exponent characterising the dispersive behaviour at high frequencies have been studied for samples with different AgI content. The analysis of the same data, at fixed frequencies and variable temperatures, in terms of the electrical modulus shows an impressive agreement between the dielectric and the mechanical relaxation process on the $(AgI)_{0.6}.(Ag_2O.2B_2O_3)_{0.4}$ glass.

Introduction

It has been stressed by all researchers in this field that the understanding of the ion dynamics in the conducting glasses is very important both from the basic and applied research point of view. The most studied fast ion conducting glasses are the silver-based glasses and particularly the silver borate glasses in that they exhibit a high ionic conductivity at room temperature, can be prepared in a wide glass-forming region and are stable. It is remarkable but also well known that universal model to account for the dispersive behaviour of the conductivity does not exist, not with standing the considerable amount of conductivity data on different kind of ionic glasses.

Several transport models, such as the weak electrolyte model,[1] the random site model,[2] the diffusion pathway model,[3] migration model,[4] cluster model and cluster bypass model[5,6] and so on, have been proposed to account for the ionic conductivity behaviour. The main unsolved problem is the lack of a link between the microscopic structure and the fast ionic conductivity. It has been established from the microscopic point of view that, in the most extensively studied borate glasses, the addition of metal oxide modifies the B_2O_3 structure in such a way that the number of the tetrahedral BO_4 units increases while the three-coordinated boron oxide, typical of pure B_2O_3, decreases. On the other hand, the room temperature conductivity at increasing AgI salt dopant drastically increases. In the past, Raman[7] and NMR investigations[8] as well as high-Q neutron scattering indicated[9,10] that the local structure of the binary borate glasses is not affected by dopant salt, so that the AgI silver ions could be diluted in the interstices and could weakly interact with the glass network. More recently, neutron and x-ray diffraction experiments[11,12] have revealed the existence of a considerable intermediate range order, which increases at increasing AgI content, and it has been observed that the host borate glassy network considerably expands by about 60% at highest AgI salt content, while its short range order is still preserved. Also in the $(AgI)_{0.6}.(Ag_2O.2B_2O_3)_{0.4}$ glass, it has been found that most Ag+ ions can be coordinated to both I^- ions and BO^{-4} units.[12] The existence of mixed oxy-iodide environments, in addition to oxide sites and iodide sites for silver ions in AgI-doped diborate glasses, has been pointed out also on the basis of infrared measurements.[13,14] Nevertheless, the conclusions drawn from the infrared studies indicate a systematic influence of AgI on the borate short range order,[13,14] so that a unified picture of the microscopic structure has not yet been achieved. However, concerning the ion conduction mechanism, from the results of the microscopic structural investigations, Swenson *et al*[11] hypothesises that the high ionic conductivity can be attributed to the formation of the new conduction pathways rather than the AgI cluster as proposed in the past from macroscopic mechanical studies.[15]

Experiment

Glass samples of compositions $(AgI)_x.(Ag_2O.nB_2O_3)_{1-x}$ where x=0, 0·25, 0·4, 0·6 and n=2 were prepared by quenching the melts. The appropriate amounts of the dried powders Ag_2O, B_2O_3 and AgI pure 99%

* Corresponding author. Email andrea.mandanici@unime.it

furnished by Aldrich were mixed and then melted for two hours in an alumina crucible in the temperature range from 850°C to 900°C depending on the composition. The melts were equilibrated for about 1 h at about 350°C and then quenched. The XRD diffraction analysis confirmed the glassy nature of all samples whose colour is transparent at low AgI content and becomes red at increasing AgI. After the exposure to the air, a silver diffusion on the disk surfaces (13 mm diameter and 2–3 mm thickness) can be observed. The electrical measurements were carried out using a home made apparatus based on a broadband dielectric measurement method[16] which gives us the possibility to obtain the real (ε′) and the (ε″) imaginary part of the dielectric permittivity in the frequency range 1 mHz–1 GHz using vectorial network analysers HP8753E and Agilent 4395A at high frequencies, a HP4284A precision LCR meter in the range 20 Hz–1 MHz, and a Solartron SI1260 impedance/gain analyser coupled to a SI1296 dielectric interface at frequencies down to 1 mHz. The measurements acquired automatically were carried out in the temperature range 90–300 K using an Oxford Instruments continuous flow cryostat. Silver paint electrodes were applied on the surfaces of the samples, previously polished in order to obtain flat and parallel surfaces. The samples were normally kept under vacuum in a dry darkened dessiccator. Electrode polarisation effects were observed in all the samples, so the usual approach[17] based on the complex impedance plots was used in order to separate the bulk electrical response of the samples from the interfacial effects.

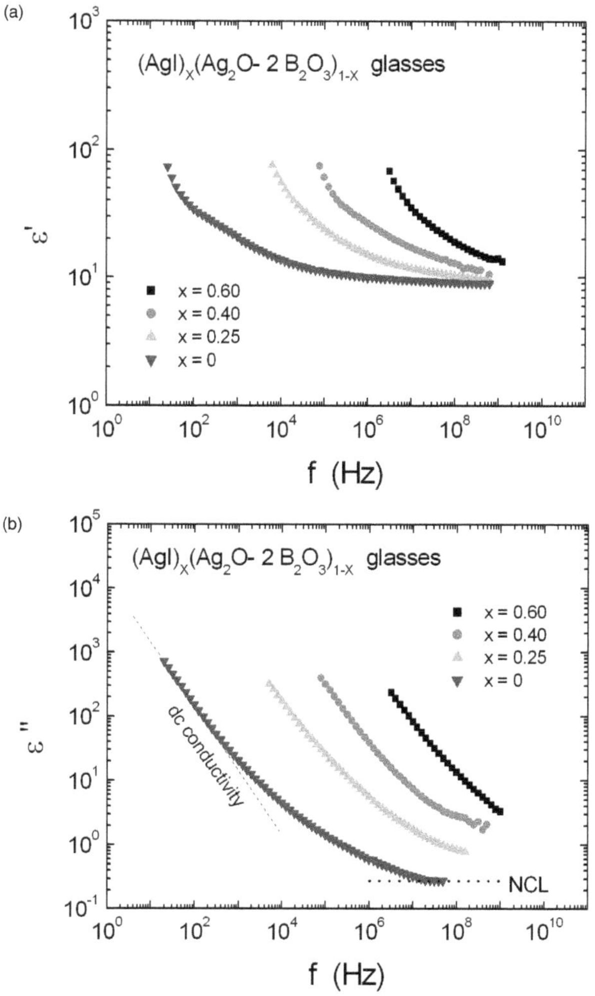

Figure 1. Real (a) and imaginary part (b) of the dielectric permittivity of $(AgI)_x.(Ag_2O.2B_2O_3)_{1-x}$ glasses measured as a function of frequency in the mHz–GHz frequency range at 295 K

Results and discussion

The dielectric response of $(AgI)_x.(Ag_2O.2B_2O_3)_{1-x}$ glasses with x=0, 0·25, 0·4, 0·6, as a function of frequency at room temperature is shown in Figure 1(a,b). With increasing frequency, the real part of permittivity regularly decreases towards a high frequency limit ε′(∞), whose value changes as a function of the AgI content. The imaginary permittivity (Figure 1(b)), measured on a broad frequency range from mHz up to microwave frequencies, also depends on the composition. The low frequency behaviour corresponds to a dc conduction process, while at higher frequencies a dispersive region is observed in which the slope of the log ε″ versus log f curve progressively decreases. The same experimental results for $(AgI)_x.(Ag_2O.2B_2O_3)_{1-x}$ glasses at room temperature are reported in Figure 2, in terms of the frequency dependent conductivity. The low frequency plateau value σ_0 corresponds to the d.c. conductivity. At higher frequencies the conductivity increases above the d.c. plateau value as a function of frequency (dispersive region). Evidently, doping with AgI allows one to enhance the dc conductivity of $(AgI)_x.(Ag_2O.2B_2O_3)_{1-x}$ glasses by several orders of magnitude. As shown in Figure 3, the log of the σ_{dc} values extracted from the broadband experimental data at room temperature increase almost linearly as a function of the dopant salt concentration. The increase of the d.c. conductivity at increasing AgI content corresponds to a decrease in the activation energy E_{dc} associated with the ionic diffusive motion (Figure 4). The dopant salt concentration also influences the onset of the dispersive region (Figure 2), above the σ_{dc} plateau value. In fact, the onset is shifted towards lower frequencies as the AgI content decreases, so that a larger part of the conductivity dispersive behaviour lies within the frequency range experimentally available, and a region of linearly frequency dependent conductivity $\sigma(\omega) \propto \omega^1$ appears on the high frequency side of the experimental spectra. The same effect is observed considering the frequency dependent conductivity of a given glass

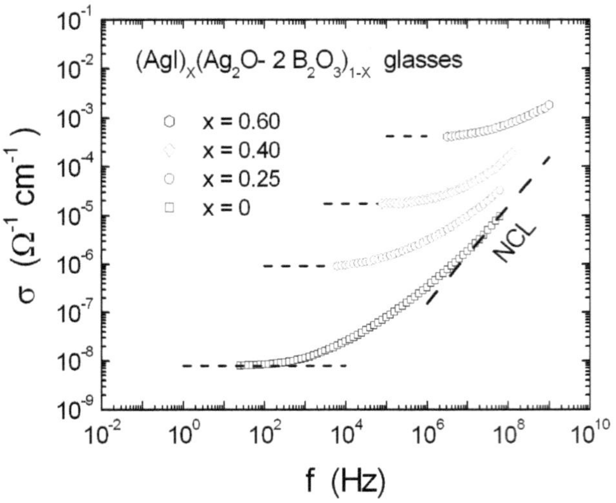

Figure 2. Frequency dependent conductivity of the $(AgI)_x.(Ag_2O.2B_2O_3)_{1-x}$ glasses with x=0, 0·25, 0·4, 0·6 at 295 K in the range from 10^{-3} Hz up to microwave frequencies

at decreasing temperatures (Figure 5). Usually,[18,19] all the ionic glasses show this kind of behaviour that may be described by a power law

$$\sigma_{ac}(\omega)=\sigma_0+B\omega^p \quad (1)$$

Some researchers[18-23] have proposed to add a linear contribution in order to account for the frequency linear dependence of the conductivity

$$\sigma_{ac}(\omega)=\sigma_0+B\omega^p+A\omega \quad (2)$$

The $\sigma(\omega)\propto\omega$ term that reflects a nearly constant value on the $\varepsilon''(\omega)$ imaginary part of the permittivity (nearly constant loss or NCL) is usually found at low temperatures /high frequencies. On the other hand, other researchers[24] have shown that the loss contributions due to the cooperative ion hopping and the

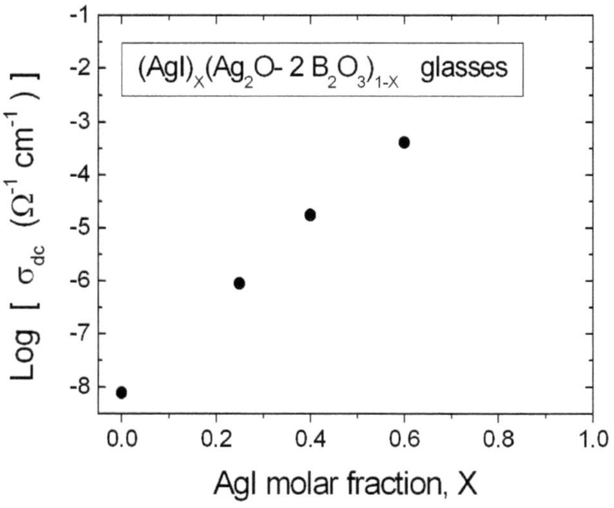

Figure 3. Effect of the dopant salt concentration on the dc conductivity of $(AgI)_x.(Ag_2O.2B_2O_3)_{1-x}$ glasses at room temperature

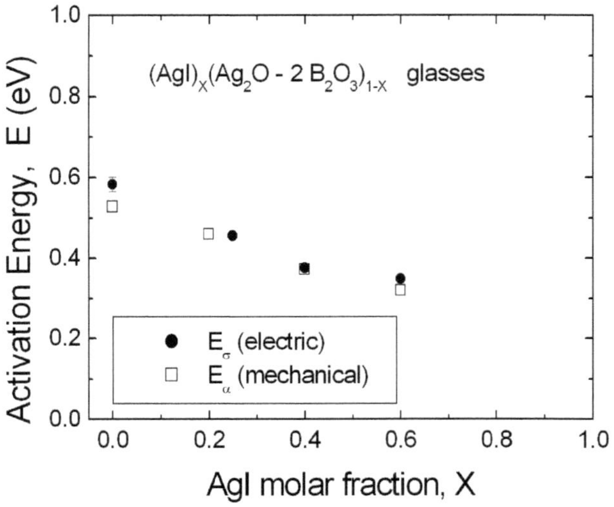

Figure 4. Activation energy of $(AgI)_x.(Ag_2O.2B_2O_3)_{1-x}$ glasses from electric and mechanical measurements. The values of E_σ have been obtained from the Arrhenius fit of the dc conductivity values (present work) according to the equation $\sigma_{dc}T=K\exp(-E_{dc}/k_BT)$. The values of E_σ have been obtained from the analysis of ultrasonic attenuation data[26]

NCL are not additive. However from the frequency dependence of a.c. conductivity experimental data (Figure 2), we find that in the glasses with lowest dopant salt content, at sufficiently high frequencies the conductivity assumes a linear frequency dependence and also at room temperature. Therefore, as recently hypothesised by Ngai[25] the $\sigma(\omega)$ linear regime might be directly related to the more or less disordered nature of glasses. However the properties and the origin of the nearly constant loss behaviour of glasses still needs to be fully clarified.

The phenomenological Equation (2) allows one to describe the $\sigma_{ac}(\omega)$ experimental data on

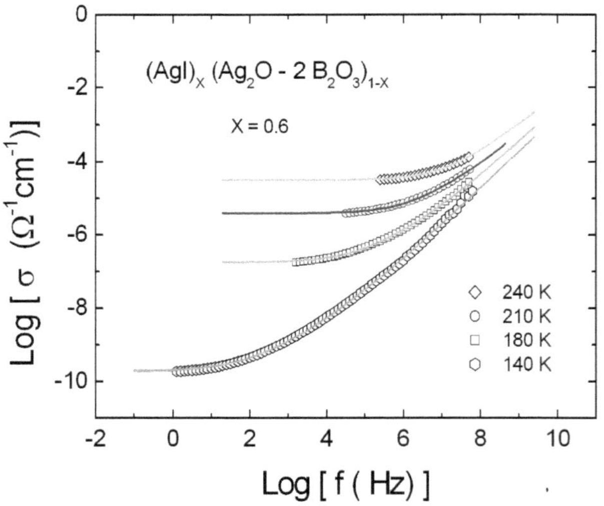

Figure 5. Log–log plot of the frequency dependent conductivity for the $(AgI)_{0\cdot6}.(Ag_2O.2B_2O_3)_{0\cdot4}$ glass at selected temperatures. Solid lines have been obtained from the best fit of experimental data to the phenomenological Equation (2)

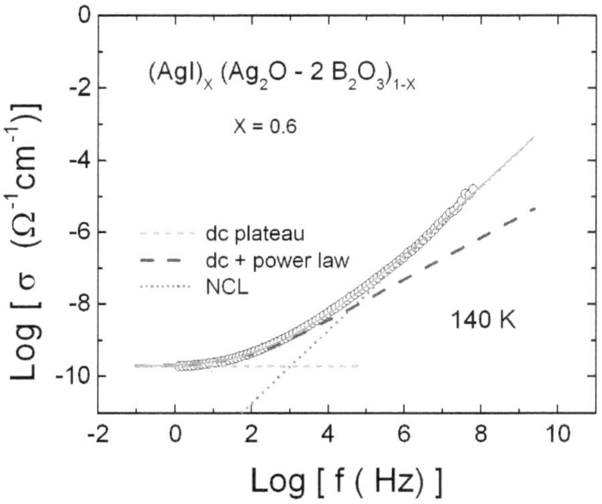

Figure 6. Frequency dependent conductivity of the $(AgI)_{0{\cdot}6}.(Ag_2O.2B_2O_3)_{0{\cdot}4}$ glass at 140 K. The solid line, corresponding to the best fit parameters of the experimental data to the phenomenological Equation (2) is shown along with the separate contributions: d.c. conductivity, power law term, and nearly constant loss behaviour

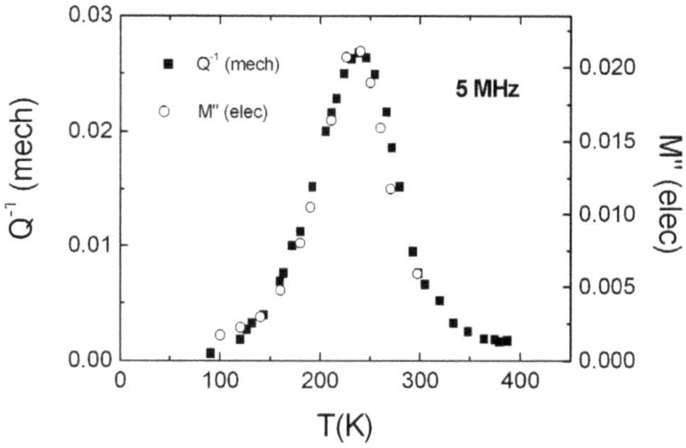

Figure 7. Direct comparison between electric relaxation data (present work) and mechanical relaxation data (Ref. 26) on the $(AgI)_{0{\cdot}6}.(Ag_2O.2B_2O_3)_{0{\cdot}4}$ glass at the same frequency (5 MHz) as a function of temperature

$(AgI)_x.(Ag_2O.2B_2O_3)_{1-x}$ glasses at all the temperatures in the wide range up to the microwave frequency region: the relevance of the different contributions at different frequencies is shown in Figure 6 for the $(AgI)_{0{\cdot}6}.(Ag_2O.2B_2O_3)_{0{\cdot}4}$ glass at 140 K. The values of the parameter p obtained from the best fit of all the compositions investigated, $p\approx0{\cdot}5$–$0{\cdot}7$, are close to those found for several different ionic glasses. The variation of the power law exponent p with temperature and AgI content has been intensively studied as it was considered in the past a good parameter to account for the dispersive behaviour of the conductivity[19,20] in ionic glasses as well as in crystalline conductors. In the present case, slight variations of the exponent p have been obtained as a function of temperature. However it has to be pointed out that the value of p deduced from the best fit procedure is sensitive to the nearly constant loss value ε''_{NCL}, which can be clearly estimated only if a sufficiently large part of the dispersive region lies in the frequency range investigated. In this sense, broadband measurements techniques represent a very useful tool to study the dynamical response of disordered systems.

It must be underlined that silver borate glasses with the same composition have been intensively studied in the past from mechanical point of view. The acoustic attenuation in these silver borate glasses at fixed MHz frequency and variable temperature shows a very high peak whose position shifts towards lower temperatures and whose height increases with increasing AgI content.[26] All the revealed mechanical relaxation peaks are broader than a Debye process, so they have been connected with a distribution of thermally activated relaxation processes involving silver ions. The observed dissipative processes have been described in terms of inelasticity theory.[27] As it is well known, an intense debate has taken place and still continues between the researchers who choose the electric modulus[28–30] and those who prefer the conductivity representation.[31] Leaving aside the division between the alternative representations, which in our opinion can be equally valid, we would like to do a comparison between mechanical and dielectric response on the same glass in order to find useful information on ionic dynamics. So in order to extract more information it could be useful to compare at fixed MHz frequency the mechanical and dielectric response in the same temperature range and in glasses with the same composition. The activation energies (see Figure 4) characterising the mechanical and dielectric relaxation processes show a very good agreement. The lack of the permittivity data at highest temperatures in glasses $(AgI)_x.(Ag_2O.2B_2O_3)_{1-x}$ where x=0, 0·2, 0·4, prevents us from performing a direct comparison between mechanical and dielectric relaxation processes in all the compounds. Therefore, going back to the original dielectric permittivity data on the $(AgI)_{0{\cdot}6}.(Ag_2O.2B_2O_3)_{0{\cdot}4}$ glass, we analyse these data in terms of electric modulus $M^*=1/\varepsilon^*$ and directly compare them with the mechanical response in terms of internal friction Q^{-1} (Figure 7). Notwithstanding the glasses have been prepared by different researchers in different periods and also that they could have different thermal histories the agreement between the mechanical and electrical relaxation data at 5 MHz is very impressive. Other investigations have been planned in order to see if this condition holds for all the silver borate glasses.

Conclusions

A broadband dielectric technique has been used to investigate the dynamical response of $(AgI)_x$.

$(Ag_2O.2B_2O_3)_{1-x}$ glasses in a wide frequency range from mHz to GHz frequencies. The d.c. conductivity increases at increasing silver iodide content while the activation energy associated with the ionic conduction decreases. The frequency dependent conductivity shows a typical power law behaviour, and a nearly constant loss regime, which can be observed also close to room temperature in the glasses with low dopant concentration. The activation energies obtained from the temperature dependence of the d.c. conductivity are in good agreement with the activation energies associated with the mechanical relaxation process in the same glasses. Furthermore the direct comparison between the mechanical and the dielectric relaxation data at fixed ultrasonic frequency in the temperature range between 90 K and 300 K show very good agreement, suggesting that the same ionic motion could be responsible for the mechanical as well as for the dielectric relaxations.

Acknowledgments

This work has been realised in the framework of the PRIN 2004-2006 programme "Correlazione tra struttura a corto e medio raggio e dinamica ionica in sistemi disordinati" coordinated by Prof. M. Cutroni and financially supported by MIUR. The authors thank Cristina Armellini, from the IFN-ITC Trento, for the preparation of the samples and their x-rays analysis.

References

1. Ravaine, D. *J. Non -Cryst. Solids*, 1985, **73**, 287.
2. Glass, A. M. & Nassau, K. *J. Appl. Phys.*, 1980, **51**, 3756.
3. Minami, T. J. Non-Cryst. Solids, 1985, 73, 273; Ingram, M. D. *Phil. Mag. B*, 1989, **60**, 729.
4. Funke , K. Banhatti, R. D. & Cramer, C. *Phys. Chem. Chem. Phys.*, 2005, **7**, 157 and references cited therein.
5. Tachez, M., Mercier, R., Malugani J. P. & Dianoux, A. J. *Solid State Ionics*, 1986, **20**, 93; Malugani, J. P., Tachez, M., Mercier, R., Dianoux, A. J. & Chieux, P. *Solid State Ionics*, 1986, **44**, 151.
6. Rousselot, C., Malugani, J. P., Mercier, R., Tachez, M., Chieux, P., Pappin, A. J. & Ingram, M. D. *Solid State Ionics*, 1995, **78**, 211.
7. Carini, G., Cutroni, M., Fontana, A., Mariotto G. & Rocca, F. *Phys. Rev. B*, 1984, **29**, 3567.
8. Chiodelli, C., Magistris, A., Villa, M. & Bjorksta, J. L. *J. Non-Cryst. Solids*, 1982, **51**, 143.
9. Borjesson, L., Torrel, L. M. & Howells, W. S. *Phil. Mag. B*, 1989, **59**, 105.
10. Borjesson, L., Torrel, L. M., Dahlborg, U. & Howells, W. S. *Phys. Rev B*, 1989, **39**, 3404.
11. Swenson, J., Borjesson,L. McGreevy R. L. & Howells, W. S. *Phys. Rev B*, 1996, **55**, 11236.
12. Cervinka, L., Rocca, F., Fornasini, P.& Dalba, G. *J. Non-Cryst. Solids*, 1992, **150**, 140.
13. Kamitsos, E. I., Kapoutsis, J. A., Chryssikos, G. D., Hutchinson, J. M., Pappin, A. J., Ingram, M. D. & Duffy, A. J. *Phys. Chem. Glasses*, 1995, **36**, 141.
14. Varsamis, C. P., Kamitsos, E. I. & Chryssikos, G. D. *Phys. Rev. B*, 1999, **60**, 3885.
15. Carini, G., Cutroni, M., Federico,M., Galli, G.& Tripodo, G. *Phys. Rev B*, 1984, **30**, 7219.
16. Pelster, R. *IEEE Trans. Microwave Theory Tech.*, 1995, **43**, 1494.
17. Cutroni, M., Mandanici, A., Piccolo, A., Fanggao, C., Saunders, G. A. & Mustarelli, P. *Phil. Mag. B*, 1996, **73**, 349.
18. Schroder, T. B. & Dyre, J. C. *Phys. Rev. Lett.*, 2000, **84**, 310.
19. Jonscher, A. K. *Nature (London)*, 1977, **267**, 673.
20. Almond, D. P. & West, A. R. *Nature (London)*, 1983, **306**, 456.
21. Sidebottom, D. L. *Phys. Rev. B*, 1999, **61**, 1450.
22. Roling, B., Happe, A., Funke K. & Ingram, M. D. *Phys. Rev. Lett.*, 1997, **78**, 2160.
23. Sidebottom, D. L. *Phys. Rev. Lett.*, 1999, **82**, 3653.
24. Ngai K. L.& Casalini, R. *Phys. Rev. B*, 2002, **66**, 132205.
25. Ngai K. L.& Leon, C. *Phys. Rev. B*, 2002, **66**, 064308.
26. Carini, G., Cutroni, M., Federico, M., Galli, G. & Tripodo, G. *J. Non-Cryst. Solids*, 1983, **56**, 393.
27. Nowick A. S. & Berry, B. S. in *Anelastic Relaxation in Crystalline Solids*, Academic, New York, 1972.
28. Moynihan, C. T. *J. Non-Cryst. Solids*, 1996, **203**, 359.
29. Moynihan, C. T. *Solid State Ionics*, 1998, **105**,175.
30. Hodge, I. M., Ngai, K. L. & Moynihan, C. T. *J. Non-Cryst. Solids*, 2005, **351**, 104 and references cited therein.
31. Sidebottom, D. L., Roling, B. & Funke, K. *Phys. Rev. B*, 2000, **6**, 024301.

Proc. Fifth Int. Conf. on Borate Glasses, Crystals and Melts Phys. Chem. Glasses: Eur. J. Glass Sci. Technol. B, August 2006, 47 (4), 393–396

Alkali dependence of tetrahedral boron in alkali borate glasses

Scott Kroeker, Pedro M. Aguiar, Anthony Cerquiera, Joy Okoro*

Department of Chemistry, University of Manitoba, Winnipeg, Manitoba, R3T 2N2, Canada

Warren Clarida, Jesse Doerr, Martin Olesiuk, Greg Ongie, Mario Affatigato & Steven A. Feller

Department of Physics, Coe College, Cedar Rapids, Iowa, 52402, USA

Previous ^{11}B magic-angle spinning NMR work establishing the presence of multiple distinct tetrahedral boron units in cesium borate glasses has been extended to five alkali borate glass series. The heavier alkali borates (Cs, Rb, K) exhibit similar behaviour across a wide range of glass compositions, while the lighter alkali borates (Li, Na) behave differently. Namely, the former exhibit two tetrahedral boron peaks, whereas the latter possess only one. Other differences include a lower four-coordinate boron fraction at high alkali loadings for the heavier alkalis relative to the lighter alkalis, and composition independent peak widths for the heavier alkalis contrasted with variable peak widths for the lighter alkali borate glasses. Similar differences between the heavier and lighter alkali borate glasses have been previously observed in some bulk properties. The spectroscopic measurements presented here suggest that the structural origin of these effects may lie in intermediate range order.

Introduction

Binary alkali borate glasses have been studied by many physical means for decades. Detailed information about short range order is well established and the transformation of three- to four-coordinate boron with alkali content explains the main features of many bulk properties. Much work has also been done to elaborate on the intermediate range order in borate glasses. The dominant discourse on this topic invokes the presence of discrete multipolyhedral anionic network clusters – sometimes called "superstructural units" – which are thought to occur in borate glasses at concentrations much higher than expected from a strictly random network model.(1) The structural characteristics of these units are predicated upon the structures of known "fundamental building blocks" in borate minerals and crystalline compounds.(2) In spite of extensive research using a variety of analytical methods, detailed information about the identities, abundance, and significance of such units in glasses remains difficult to obtain.

Recent nuclear magnetic resonance (NMR) spectroscopic work established a quantitative handle on ordering beyond the first coordination sphere in cesium borates, through the observation of multiple four-coordinate boron signals.(3) Attempts to assign the observed peaks by comparison with crystalline compounds,(4) *ab initio* calculations of chemical shifts and electric field gradients,(5) rotational echo double resonance NMR experiments(6) and Raman spectroscopy(5) have been moderately successful, but have raised questions about whether the nature of intermediate range order is uniform across the alkali series. In this paper, we report a comprehensive ^{11}B magic angle spinning (MAS) NMR analysis of five alkali borate glass series over a wide range of glass-forming compositions, and find that the behaviour of the lighter alkali borate glasses differs markedly from that of the heavier alkalis.

* Corresponding author. Email Scott_Kroeker@UManitoba.ca

Experimental procedure

Sample preparation

Glasses were prepared from alkali carbonates and B_2O_3 by the conventional melt-quench approach, or by a solution based precursor synthesis method.(7) The NMR spectra of selected glasses prepared by both methods yielded indistinguishable spectra. The solution approach to glass synthesis is intended to extend the glass forming region and enhance compositional precision and final product purity.(7) Solutions containing the dissolved or suspended precursors were mixed and dried to a powder at 145°C, and then heated to a melt in a Pt crucible for one hour. The liquids were cooled to a glass on a metal plate. Glass transition temperatures measured by differential scanning calorimetry differed by less than 7°C from previously reported values.(8,9)

Nuclear Magnetic Resonance

Boron-11 magic-angle spinning (MAS) NMR spectra were recorded at 192·4 MHz on a Varian Inova 600 spectrometer with a 3·2 mm double resonance probe. Glass samples were crushed to a fine powder and packed into a rotor; hygroscopic compositions were handled in a glove box under nitrogen. Sample sizes ranged from 15 to 35 mg. A typical experiment involved quantitative and homogeneous excitation of the nuclear spins in a sample rotating at 16 kHz by a 0·8 μs radio frequency pulse, which corresponds to about 15% of the pulse duration required to reach maximum signal intensity for a solution sample. Signal averaging of 64 transients, collected with a relaxation delay ranging from 6 to 60 s, produced the time domain data which were Fourier transformed to yield spectra exemplified by Figure 1. ^{11}B chemical shifts are reported relative to 0·1 M aqueous boric acid, which is observed at +19·6 ppm with respect to $BF_3.(OEt_2)$.[10]

To obtain quantitative populations from MAS NMR spectra with substantially different quadrupolar interactions, the effects of finite spinning speeds and applied magnetic fields must be accounted for. Hence, this correction results in augmentation of the measured integrated intensity when significant central transition (−1/2, +1/2) signal is distributed into spinning sidebands, as is the case for sites with larger C_Q values. The correction diminishes the measured value when significant intensity from satellite transitions (3/2, 1/2) contributes to the central peak. Greater detail on this straightforward procedure can be found in Massiot *et al.*[11]

Results

Figure 1 illustrates the quality of spectra obtained for different alkali borate glasses with the same composition. It is immediately apparent that the four-coordinate boron region between 0–3 ppm is occupied by two distinct signals for the heavier alkali borate glasses: potassium, rubidium and cesium. By contrast, the lighter two alkali modified borate glasses exhibit only a single symmetrical peak. With little exception, this trend persists across the compositional range studied: lithium and sodium borate glasses possess a single peak, whereas potassium, rubidium and cesium borate glasses show evidence of at least two distinct signals in the four-coordinate region (Table 1). Of the lithium and sodium series, only the *R*=0·2 sodium borate glass has a hint of asymmetry in the lineshape; of the heavier alkalis, only the *R*=0·7 potassium borate glass lacks clear evidence of multiple tetrahedral boron peaks.

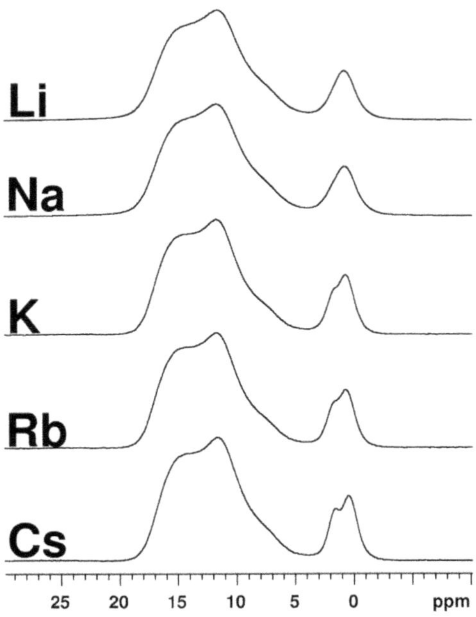

Figure 1. ^{11}B MAS NMR spectra of alkali borate glasses (R=0·1) collected at 14·1 T with spinning speeds of 16 kHz

The three-coordinate boron region of the spectra also appears to exhibit more definition for the heavier alkali borates, however the effect is subtle and no attempt to quantify the ring and non-ring contributions were made.

The total $^{[4]}$B fractions (N_4) were evaluated for each

Table 1. Four-coordinate boron fractions for alkali borate glasses, with contributions of high and low frequency components, where observed; estimated errors ±0·01

R	N_4 *Li*	*Na*	*K*			*Rb*			*Cs*		
	Total	*Total*	*Total*	*Low*	*High*	*Total*	*Low*	*High*	*Total*	*Low*	*High*
0·03	0·03	0·04	0·05	0·03	0·02	0·05	0·03	0·02	0·06	0·03	0·03
0·10	0·12	0·12	0·14	0·09	0·05	0·15	0·09	0·06	0·13	0·07	0·06
0·20	0·21	0·22	0·22	0·11	0·11	0·21	0·13	0·08	0·22	0·11	0·11
0·30	0·31	0·31	0·32	0·22	0·10	0·33	0·16	0·17	0·32	0·19	0·13
0·32	0·37	–	0·35	0·22	0·13	0·36	0·12	0·24	0·33	0·20	0·13
0·40	0·37	0·38	0·38	0·20	0·18	0·39	0·17	0·22	0·38	0·19	0·19
0·48	–	0·39	0·43	0·23	0·20	0·42	0·19	0·23	0·40	0·16	0·24
0·50	0·41	0·42	0·43	0·22	0·21	0·41	0·19	0·22	0·39	0·12	0·27
0·54	–	–	0·43	0·18	0·25	0·43	0·18	0·25	0·38	0·13	0·25
0·60	–	–	0·38	*	*	0·37	0·12	0·25	0·36	0·12	0·24
0·70	0·42	0·39	0·31	*	*	0·34	0·11	0·23	0·34	0·11	0·23

* Insufficient lineshape resolution for accurate fitting

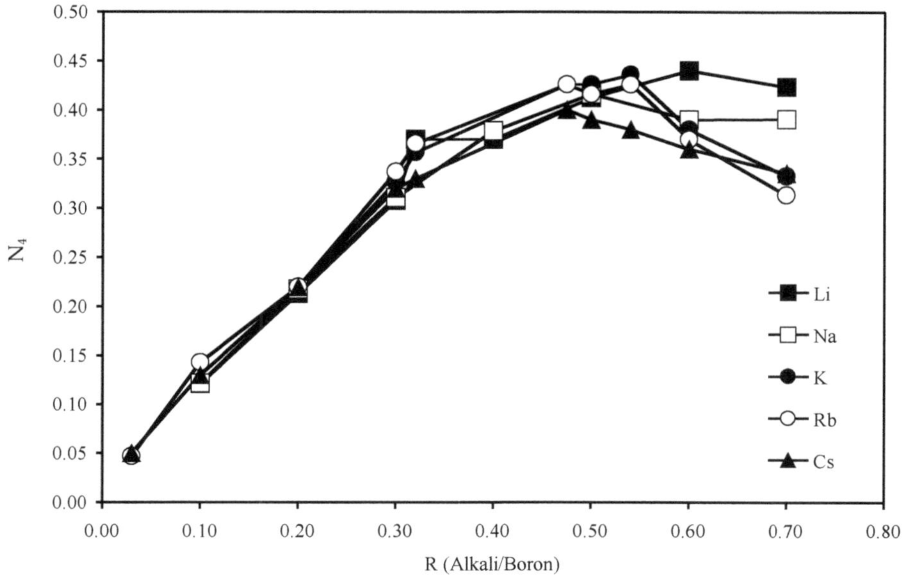

Figure 2. Four-coordinate boron fraction as a function of composition, for alkali borates

of the glasses. Since the three- and four-coordinate boron signals are completely resolved at this magnetic field (14·1 T), no lineshape fitting is necessary to quantify the short range order and the reported values are correspondingly very precise (Table 1, Figure 2). As thoroughly documented in the literature,[3,12–16] the fraction of borons in four-coordination increases approximately linearly with R at low alkali loadings, before deviating negatively around R=0·3, when nonbridging oxygens on $^{[3]}$B units begin to form in competition with $^{[4]}$B tetrahedra. Above this alkali concentration the $^{[4]}$B fraction in the heavier alkali borates drops more rapidly than in the lighter alkali borates (Figure 2). This dispersion in the $^{[4]}$B fraction has not been previously measured to our knowledge, and indicates different behaviour between the lighter and heavier alkali modified borate glasses.

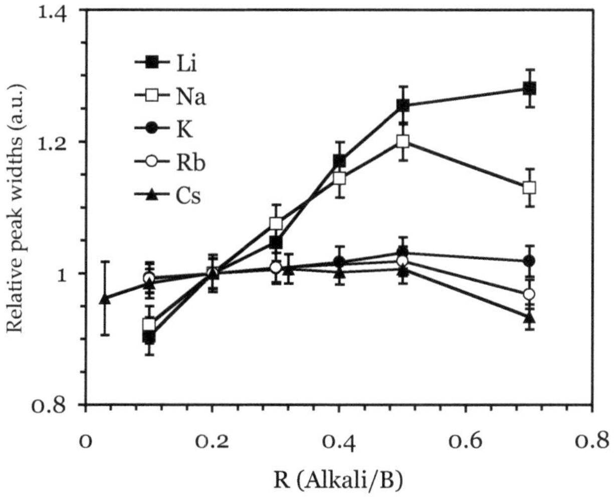

Figure 3. Relative ^{11}B MAS NMR total peak width for four-coordinate boron peaks for alkali borate glasses as a function of composition

The high and low frequency contributions to the total $^{[4]}$B fractions – evaluated by fitting the composite experimental peaks by two peaks of fixed width and Gaussian/Lorentzian character – are reported in Table 1. Common to all three series possessing multiple four-coordinate peaks is that the two components increase in approximately equal measure up to R=0·2. Above that composition, one or the other component predominates, with the precise behaviour differing for each series.

Another comparison amongst the alkali borates is the total width of the four-coordinate boron signal. While the absolute values are difficult to compare, the relative values – with R=0·2 as the reference – are depicted in Figure 3. Once again, the potassium, rubidium and cesium borates behave qualitatively differently than the lithium and sodium borates: the peak widths of the heavier alkali borates are essentially invariant to composition, while those of the lighter alkali borates display a steady increase in width with R. Only above R=0·45 do the lithium and sodium borate widths begin to level off or decrease.

Discussion

Two distinct environments are observed for the tetrahedral boron in all heavier alkali glasses, the positions of which change very little from low to high alkali content. This suggests that whatever structural feature gives rise to these peaks persists across a wide compositional range. Raman spectroscopy[5] and chemical common sense militate against the occurrence of the same set of fundamental building blocks being present at low and high alkali. However, *ab initio* calculations[5] predict distinct chemical shifts for four-coordinate boron participating in one ring versus two rings simultaneously. Consistent with these results is a model in which all four-coordinate borons are present in single three-membered rings with two three-coordinate borons ("triborate units") and linking two similar three-membered rings ("pentaborate units") at low alkali loadings (see Ref. 6 for depictions). At higher loadings (R>0·4) tetrahedral boron is present in single three-member rings containing another four-coordinate and a three-coordinate boron ("di-triborate") and a double ring structure containing two four-coordinate borons and two three-coordinate borons ("diborate"). That this appears to occur only in the heavier alkalis may indicate that the large, polarisable alkali cations exert very little structural control over network formation, acting strictly in a charge balancing capacity.

The lithium and sodium borate series, while displaying no distinction between individual signals, nevertheless possess peak breadths comparable to those of the heavier alkalis, signifying a similar distribution of local boron environments. This may be related to the smaller size of these alkali ions which permit stronger interactions with the network oxygen ions, and hence a broader range of local structural distortions. Supporting this hypothesis is that increasing alkali content results in a broader distribution of local boron environments, as evidenced by the increase in relative linewidths for lithium and sodium only (Figure 3). If this interpretation is correct, it implies that the lighter alkalis may stabilize four-coordinate borons with respect to nonbridging oxygens on three-coordinate borons, as observed by the higher N_4 for lithium and sodium borates above R=0·5, relative to potassium, rubidium and cesium borates which act more passively to balance charge.

Finally, it should be noted that the spectroscopic measurements differentiating the lighter and heavier alkali borates are reflected by bulk measurements of some physical properties. For example, stiffness measurements[17] reveal steadily increasing values for the lithium and sodium borates up to R=0·4, whereas potassium, rubidium and cesium borate exhibit more complex, but uniform, behaviour. Moreover, the inflection points in the latter curves show tantalizing correspondence with the variations in individual tetrahedral boron contributions measured by NMR. These results suggest that it is necessary in some cases to consider the influence of intermediate range ordering on bulk properties.

Conclusions

This study represents a systematic ^{11}B MAS NMR analysis of five series of alkali borate glasses. While some of these compositions exist in the literature, a direct comparison is facilitated by examining all glasses under the same experimental conditions. This approach reveals interesting distinctions between the lighter (Li and Na) and the heavier (K, Rb and Cs) alkali borate glasses. In particular, spectroscopic signatures of distinct types of intermediate range order are observed for the latter, but not the former. The fact that these distinctions are mirrored in the elastic properties suggests that intermediate range order may play a decisive role in governing some bulk properties of interest for applications. Furthermore, the extension to higher alkali compositions uncovers a significant difference in the fraction of tetrahedral boron for different alkali. Greater clarity about the peak assignments may provide insight into the roles of lighter and heavier alkali ions in the stabilisation of tetrahedral boron versus nonbridging oxygens.

Acknowledgements

SK thanks the Natural Sciences and Engineering Research Council (NSERC) of Canada and the Canada Foundation for Innovation (CFI) for operating and infrastructure support. AC is grateful to the University of Manitoba Faculty of Science for scholarships. This research is funded, in part, by the US National Science Foundation, through grants to SAF and MA (DMR 0211718, DMR 0502051, INT 0089510).

References

1. Bray, P. J. *Inorg. Chim. Acta*, 1999, **289**, 158–173.
2. Burns, P. C., Grice, J. D. & Hawthorne, F. C. *Can. Miner.*, 1995, **33**, 1131.
3. Kroeker, S., Feller, S. A., Affatigato, M., O'Brien, C. P., Clarida, W. J. & Kodama, M. *Phys. Chem. Glasses*, 2003, **44**, 54–58.
4. Kroeker, S., Aguiar, P. M., Feller, S. A. & Affatigato, M. In *227th American Chemical Society National Meeting*, Anaheim, California, 2004, p INOR-82.
5. Aguiar, P. M. & Kroeker, S. In *88th Canadian Chemistry Conference*, Saskatoon, Saskatchewan, 2005, p 684.
6. Aguiar, P. M. & Kroeker, S. *Solid State NMR*, 2005, **27**, 10–15.
7. Kodama, M. & Kojima, S. In *Second International Conference on Borate Glasses, Crystals and Melts*; Wright, A. C., Feller, S. A., Hannon, A. C., Eds.; Society of Glass Technology: Sheffield, UK, 1997, p 181–188.
8. Berryman, J. R., Feller, S. A., Affatigato, M., Kodama, M., Meyer, B. M., Martin, S. W., Borsa, F. & Kroeker, S. *J. Non-Cryst. Solids*, 2001, **293–295**, 483–489.
9. Feller, S. et al. In *Second International Conference on Borate Glasses, Crystals and Melts*; Wright, A. C., Feller, S. A., Hannon, A. C., Eds.; Society of Glass Technology: Abingdon, UK, 1996, p 246–253.
10. Dewar, M. J. S. & Jones, R. *J. Am. Chem. Soc.*, 1967, **89**, 2408–2410.
11. Massiot, D., Bessada, C., Coutures, J. P. & Taulelle, F. *J. Magn. Reson.*, 1990, **90**, 231–242.
12. Bray, P. J. & O'Keefe, J. G. *Phys. Chem. Glasses*, 1963, **4**, 37–46.
13. Greenblatt, S. & Bray, P. J. *Phys. Chem. Glasses*, 1967, **8**, 213–217.
14. Yun, Y. H. & Bray, P. J. *J. Non-Cryst. Solids*, 1981, **44**, 227–237.
15. Youngman, R. E. & Zwanziger, J. W. *J. Am. Chem. Soc.*, 1995, **117**, 1397–1402.
16. Youngman, R. E. & Zwanziger, J. W. *J. Phys. Chem.*, 1996, **100**, 16720–16728.
17. Feller, S., Affatigato, M., Giri, S. K., Basu, A. & Kodama, M. *Phys. Chem. Glasses*, 2003, **44**, 117–120.

Proc. Fifth Int. Conf. on Borate Glasses, Crystals and Melts Phys. Chem. Glasses: Eur. J. Glass Sci. Technol. B, August 2006, 47 (4), 397–404

Phase separation and crystallisation in borate glasses

Christian Rüssel, Diana Tauch, Ralitsa Garkova, Sebastian Woltz & Günter Völksch

Otto-Schott-Institut, Jena University, Fraunhoferstr. 6, 07743 Jena, Germany

Although phase separation and crystallisation in borate glasses are mainly of fundamental interest, valuable preparation routes especially for the formation of nanocrystalline oxides are offered. In this paper, the formation of Fe_3O_4, SnO_2 and In_2O_3 crystals with mean crystallite sizes in the nanometre range is described. After thermal treatment the glassy matrices ($Na_2O/CaO/Al_2O_3/B_2O_3$) were dissolved and the crystals isolated. The effects of chemical compositions and thermal treatment on the crystal sizes and morphologies are described. The crystalline phase exhibits a negative thermal expansion coefficient in the direction of the crystallographic c*-axis while it is approximately zero in the direction of the crystallographic* a*-axis up to a temperature of 150°C and then gets positive. Glass ceramics produced from a glass with the same stoichiometry exhibit a thermal expansion of around zero, or a highly negative thermal expansion depending on the preparation route. The effect of the latter as well as of the chemical composition on the microstructure and the thermal expansion is discussed.*

1. Introduction

Phase separation and crystallisation in borate glasses are well known phenomena, since the pioneer works of Werner Vogel.[1–6] In the past few years, the crystallisation of borate glasses proved to be a versatile route for the preparation of nanocrystalline powders of various compositions. Using these routes, borate glasses are melted, and crystallised by subsequent thermal treatment.[7–14] The borate glass matrix is then dissolved in water or acid, and the nanocrystalline powder is separated by filtration or dialysis.

The preparation of ferromagnetic powders[7–14] for many applications, such as ferro-fluids[12,13] and also for sintering of compact magnetoceramics should be mentioned. Other applications of ferrite powders with mean particle sizes in the range from 10 to 50 nm are in cancer diagnostics and cancer therapy.[15–20] Super paramagnetic powders, which do not show a permanent magnetic moment due to their small particle sizes, are coated, e.g. by polysaccharides. These coated particles are preferentially incorporated into cancer cells. This can be utilised for the *in vitro* separation of cancer cells by strong magnetic fields and hence for diagnostics. Alternatively, a strong alternating magnetic field is supplied *in vivo*, which leads to an increase in temperature and to a destruction of cancer cells (hyperthermia).

Other powders which can be prepared via the crystallisation of borate glasses are semi conductive oxide powders.[21–23] In the past few years, the preparation of indium and tin oxide has been reported in the literature. One possible application of these powders is in the preparation of polymer matrix composites. Due to the optical properties of the nano particle, these do not show transparency in the near infrared.

The crystallisation of borate glasses is not only a valuable tool for the preparation of powders, but some borate glass ceramics also show some notable physical properties. The preparation of glass ceramics with zero or even negative thermal expansion coefficients should especially be mentioned.[24–26] Among the systems which show this effect, these are glass ceramics in the system $BaO/Al_2O_3/B_2O_3$. After appropriate thermal treatment, the phase $BaAl_2B_2O_7$ is crystallised which shows a strongly anisotropic and negative thermal expansion coefficient. Although, the chemical durability is much less than that of glass ceramics based on lithium aluminosilicates and the temperature range where they can be applied is not as wide, potential applications are fairly numerous. Among these, especially applications in the field of micromechanics are to be mentioned.

This paper reports some examples for the preparation of nanocrystalline oxide powders as well as on preparation and properties of glass ceramics in the system $BaO/Al_2O_3/B_2O_3$.

2. Experimental procedure

A glass with the base composition of $BaO.Al_2O_3.B_2O_3$ was melted from reagent grades $BaCO_3$, $Al(OH)_3$ and $B(OH)_3$ at 1500°C in a platinum crucible. A part of the glass was crushed into pieces < 1 mm. Then 0·01 wt% platinum as $PtCl_4$/acetone solution was added to the

[1] Corresponding author

crushed glass. After homogenising and subsequently drying, the mixture was transferred to a platinum crucible, remelted at 1480°C, and stirred at 1450°C for 30 min. The melt was then cast on a copper block and transferred to a furnace preheated to 500°C, which was subsequently switched off to allow the sample to cool. For the preparation of sintered glass ceramics the glass was milled and sieved to obtain grain sizes < 25 μm. Then 1% polyvinyl alcohol (PVA) solutions were added to the milled glass and the mixture was homogenised. After drying, pellets with a diameter of 40 mm and a thickness of 10 mm were prepared by uniaxial pressing. The bulk glass samples were thermally treated at 720°C for 24 h and 780°C for 8 h using heating and cooling rates of 10 K/min. The powder sample was thermally treated at 600°C (burn out of PVA) for 90 min and at 850°C for 40 min.

For the preparation of nano crystalline In_2O_3 and SnO_2, glasses were prepared from reagent grade $B(OH)_3$, Na_2CO_3, $Al(OH)_3$, SnO_2 and/or In_2O_3. The batch was melted in a platinum crucible at 1250 to 1500°C in a resistance heated furnace, soaked for 15 min to 1 h. The liquid was subsequently quenched on a copper block. The glasses were then heat treated at temperatures in the range from 500 to 700°C for 20 min to 4 h.

Glasses for the preparation of nanocrystalline magnetite with the mol% composition $33{\cdot}3CaO.4{\cdot}9Na_2O.17{\cdot}1Fe_2O_3.44{\cdot}7B_2O_3$ were melted in 100 g batches from reagent grade Na_2CO_3, $CaCO_3$, Fe_2O_3 and $B(OH)_3$ in a platinum crucible at 1200°C and poured on a copper plate; 20 g of the received sample were put in another platinum crucible equipped with a platinum down pipe (inner diameter: 2 mm) and located inside an alumina tube with a water cooled flange. After remelting at 1200°C, the alumina tube with the crucible was flushed with nitrogen for 20 to 360 min in order to adjust the Fe^{2+}/Fe^{3+} ratio. The melt was then rapidly quenched by twin rollers (copper). The flakes obtained were around 150 μm thick. The glass was thermally treated at temperatures in the range from 550 to 620°C.

The glasses were characterised by differential thermal analysis (DTA). Crystallised samples were studied by room temperature and, in the case of $BaO.Al_2O_3.B_2O_3$ glasses by high temperature x-ray diffraction (Siemens D5000 with a PAR HTK 10 heating chamber and Cu K_α radiation, in the case of iron containing samples with Mo K_α radiation). The thermal expansion coefficient was measured by dilatometry (Netzsch 402 E) using a heating rate of 5 K/min. The crystallised samples were further studied by scanning electron microscopy (SEM, Zeiss DSM 940 A). Densities were measured using a helium pycnometer (Accupyc 1330). In order to dissolve the glass matrix of samples in the systems $Na_2O/CaO/B_2O_3/Fe_2O_3/FeO$ and $Na_2O/Al_2O_3/B_2O_3/SnO_2/In_2O_3$, 1 g powdered glass was stirred with 50 ml water for 6 d at 40°C. The crystalline residue was separated by filtration and the obtained powder characterised by TEM using an acceleration voltage of 200 kV.

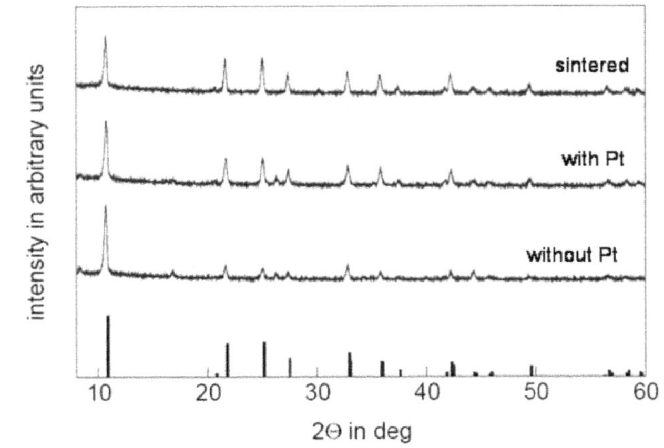

Figure 1. XRD patterns of samples with and without platinum after thermal treatment and the sintered sample. Bottom: schematic diffraction pattern of $BaAl_2B_2O_7$ (JCPDS Nr. 86-2168)

3. Results and Discussion

3.1. Glass ceramics in the system $BaO/Al_2O_3/B_2O_3$

The transformation temperature of the glasses prepared with and without the addition of platinum were 585°C and the softening point 660°C, as shown by dilatometry. The glass was x-ray amorphous, any hint of crystallisation was not observed.

As shown in Figure 1, the XRD patterns of all crystallised samples prepared exhibit solely lines attributable to rhombohedral $BaAl_2B_2O_7$ (JCPDS no. 86-2168). The relative intensities of the respective XRD lines, however, are different in the bulk glass sample without platinum, the intensity of the line at 10·8° ((003)-peak) is larger in comparison to those of the samples with platinum and the sintered sample. Figure 2 shows micrographs of the samples with and without platinum. The sample without platinum exhibits a surface layer which is approximately 500 to 700 μm thick. In the attributed XRD pattern of this layer (not shown), it is seen that the crystals are arranged with their crystallographic c-axis perpendicular to the surface. In the bulk, large spheroidal crystals with mean diameters in the 500 to 700 μm range are observed. In the sample with platinum, a surface layer with a structure different from the bulk is not observed. This sample exhibits much smaller crystals.

In Figure 3, results from dilatometric measurements are shown for the crystallised sample without platinum, after mechanical removal of the surface layer. It exhibits a negative thermal expansion coefficient in the temperature range from 25 to 310°C. At

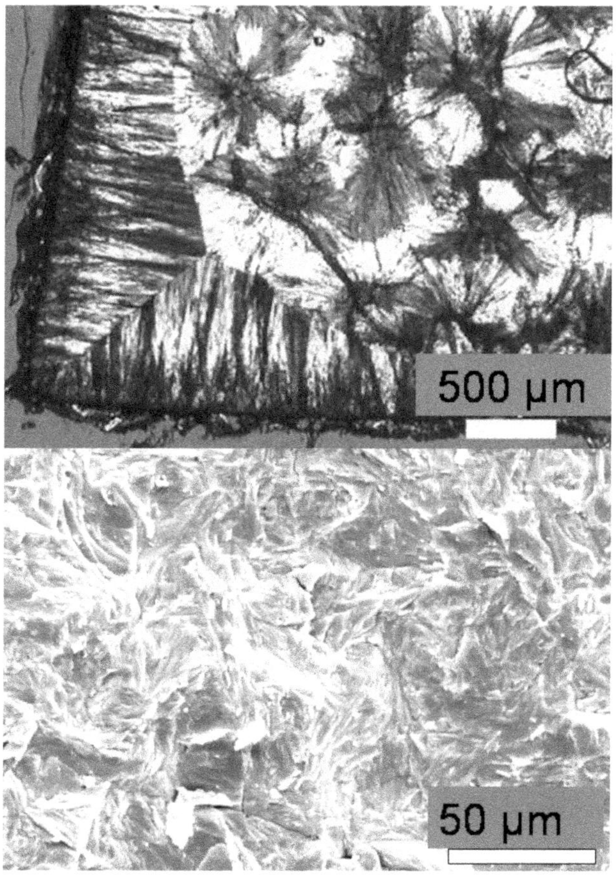

Figure 2. Micrographs of bulk samples with (bottom) and without the addition of platinum (top). The sample without platinum exhibits a surface layer with a thickness of around 500–700 µm

around 390°C, it possesses approximately the same length as at room temperature. The thermal expansion coefficient of the sample with platinum, however, is negative up to a temperature of around 75°C. The thermal expansion coefficient of the sintered sample is approximately zero up to around 80°C. At higher temperatures, the expansion coefficients are positive. The expansion coefficients of the respective samples are notably different depending on the preparation procedure applied, the type of the crystalline phase, however, is the same in all cases.

Figure 3 also includes results from high temperature XRD studies, illustrated by means of the relative changes of the crystallographic *a* and *c* axes and the changes in the mean length of the unit cell ($l=\sqrt[3]{V}$, V=volume of the unit cell) with temperature. The relative length of the *a*-axis remains approximately constant up to 150°C and then increases. At 400°C, $\Delta a/a$ reaches a value of 2×10^{-4}. The relative length of the *c*-axis decreases continuously; at 400°C,[26] a value of $\Delta c/c$ is around -2×10^{-3}. By analogy, the relative mean length of the unit cell $\Delta l/l$ is negative within the entire temperature range studied. At a temperature of 400°C, $\Delta l/l$ reaches a value of -7×10^{-4}. It is seen that the thermal expansion of the sample without platinum is close to $\Delta l/l$ up to a temperature of 300°C. This should, in a first approximation, be expected from a sample with statistically arranged crystals. As shown in Figure 2, this samples exhibits a surface layer of crystals aligned with their *c*-axis perpendicular to the surface. This layer is most probably formed by surface nucleation and subsequent crystal selection. Since the layer was mechanical removed before recording the dilatometric curves, it did not affect the thermal expansion. Large spheroidal crystals occur in the bulk. Here, from the centre, the crystals start to grow with the *c*-axis as the fast growing axis.

The samples with platinum and the sintered sample exhibit thermal expansions close to $\Delta a/a$ up to a temperature of around 250°C. The expansion in the crystallographic *a*-axis seems to determine the

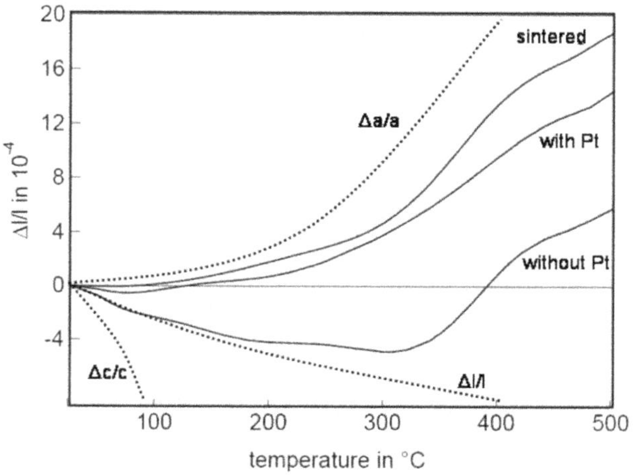

Figure 3. Dilatometric curves of samples with and without platinum and the sintered sample in comparison to the relative length change of a- (Δa/a) and c-axes (Δc/c) as well as of the mean length of the unit cell (Δl/l)

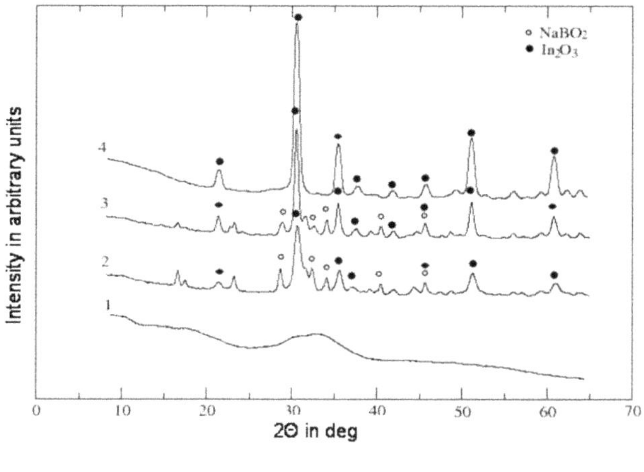

Figure 4. XRD-patterns of a sample with the composition $40B_2O_3.45Na_2O.10Al_2O_3.5In_2O_3$. Curve 1: without thermal treatment, 2: thermally treated at 560°C for 80 min, 3: thermally treated at 700°C for 60 min, 4: thermally treated at 650°C for 60 min after dissolution of the glass matrix

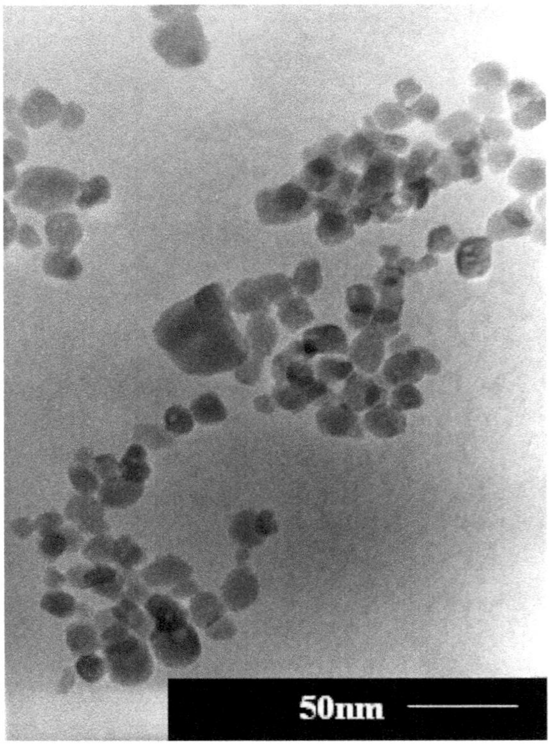

Figure 5. TEM micrograph of a sample with the composition 45Na_2O.40B_2O_3.10Al_2O_3.5In_2O_3 tempered at 580°C for 1 h

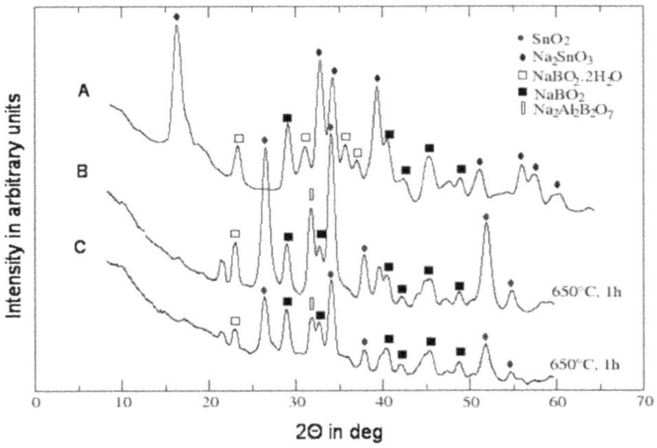

Figure 6. XRD patterns of A: a sample with the composition 40·20B_2O_3.55·28Na_2O.4·48SnO_2, B: 40·24B_2O_3. 40·2 4Na_2O.9·47Al_2O_3.10·05SnO_2 and C: 42B_2O_3.38·48Na_2O. 9·47Al_2O_3.10·05SnO_2

macroscopic expansion. In these samples, the crystals are much smaller and hence the nucleation rate must have been much higher. In the case of the sample with platinum most probably heterogeneous nucleation occurred and platinum acts as nucleating agent. This is a typical prerequisite for a statistical orientation of crystals. In the case of the sintered sample, most probably, nucleation occurs at the surface of each glass particle already before the sintering process is completed. This is a well known mechanism during the preparation of sinter glass ceramics, which has already been described in the literature (see e.g. Refs 27, 28). Both mechanisms obviously result in a more statistical orientation of crystals as in the sample A without platinum. Obviously the microstructure is widely responsible for the thermal expansion of the resulting glass ceramics.

During crystallisation, the densities of the glass ceramic samples with and without platinum decreased from 3·45 g/cm^3 to 3·07 g/cm^3, while that of the sintered sample was 2·53 g/cm^3. It should be noted, however, that the latter exhibits some pores which easily explains the smaller density. A glass with the composition BaO.Al_2O_3.B_2O_3 exhibits a higher density than the crystal with the same chemical composition and hence during crystallisation, a volume expansion takes place. It should be noted that, for most glass compositions, volume contraction is observed during crystallisation. Although this behaviour is unusual, it is not unique, and e.g. also observed during crystallisation of cordierite glass.[29]

3.2. Precipitation of SnO_2 and In_2O_3 from glasses in the system $Na_2O/Al_2O_3/B_2O_3/SnO_2/In_2O_3$

Glasses in the system $Na_2O/Al_2O_3/B_2O_3/SnO_2/In_2O_3$ exhibit a fairly wide glass forming range. They can be prepared up to In_2O_3 concentrations of around 7%. Since the dissolution of SnO_2 in the borate melts is fairly slow, the SnO_2 raw material was prepared *in situ* from an aqueous $SnCl_2.2H_2O$ solution by adding H_2O_2 and NH_4OH to the raw materials. Using this route, also in the case of the $Na_2O/Al_2O_3/B_2O_3/SnO_2$ system, glasses with SnO_2 concentrations of up to 10 mol% could be obtained. At higher SnO_2 concentrations homogeneous glasses could not be obtained. In the $Na_2O/Al_2O_3/B_2O_3/In_2O_3$ system, the alumina concentration of the glass is decisive for the type of crystalline phase formed. From glasses in the ternary system $Na_2O/B_2O_3/In_2O_3$, a controlled crystallisation of In_2O_3 could not be achieved. In the sample with the largest B_2O_3 concentration (36·1Na_2O.58·9B_2O_3.5In_2O_3), only the crystallisation of sodium borate was observed. In a sample with the composition 55Na_2O.40B_2O_3.5In_2O_3, thermal treatment led to the crystallisation of $NaInO_2$, while in samples with Na_2O concentrations of 50 and 45 mol%, spontaneous crystallisation of In_2O_3 and $NaBO_2$, respectively, occurred. The introduction of Al_2O_3 into the glass gave rise to an extended glass forming range. Whereas during tempering in the B_2O_3-rich glasses (35Na_2O.50B_2O_3.10Al_2O_3.5In_2O_3), indium containing phases did not crystallise, in the Na_2O-rich compositions, In_2O_3 precipitated. At a [Na_2O]/[B_2O_3] ratio of 1·11, and an Al_2O_3 concentration of 4·8 mol%, a transparent glass is obtained. By further adding alumina to a composition with the same [Na_2O]/[B_2O_3] ratio, amorphous samples are obtained up to an Al_2O_3 concentration of 12 mol%. It

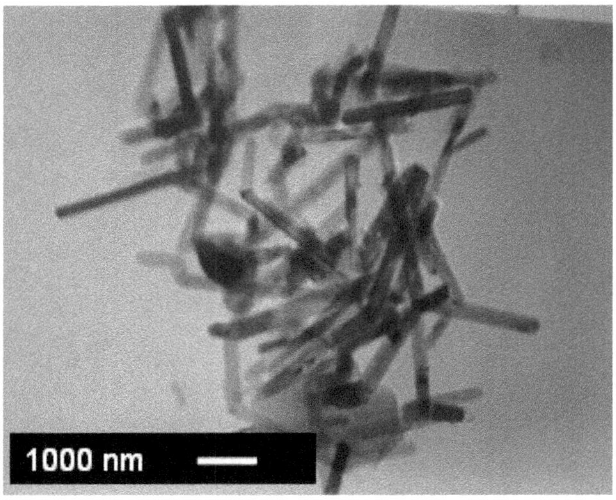

Figure 7. TEM micrograph of a sample with the composition 42·22B_2O_3.47·5Na_2O.?35?Al_2O_3.5·27SnO_2, tempered at 700°C for 18 h

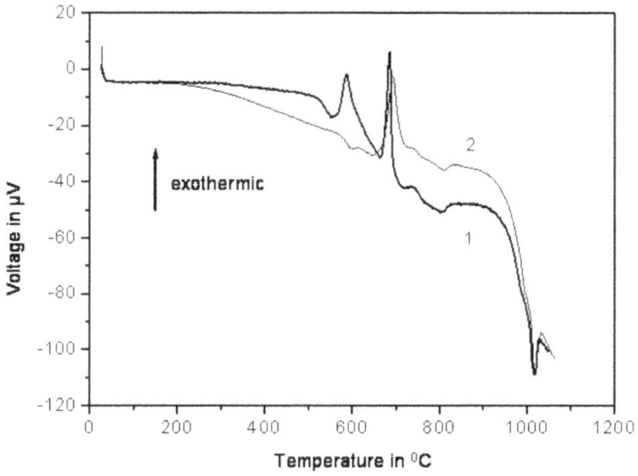

Figure 8. DTA profile of 1: a quenched glass and 2: a glass thermally treated at 570°C for 7 h

should be noted that a similar effect of Al_2O_3 on the crystallisation of In_2O_3 containing compounds has also been observed for silicate glasses.[21]

Figure 4 shows XRD-patterns of a sample with the composition 40B_2O_3.45Na_2O.10Al_2O_3.5In_2O_3. While in the sample without thermal treatment, no hint of crystallisation is obtained, the patterns of samples thermally treated at temperatures in the range from 560 to 700°C show distinct lines attributable to crystalline In_2O_3 (JCPDS-file Nr.: 6-0416) and $NaBO_2$ (JCPDS-file Nr.: 32-1046). In Figure 4, curve 4, a pattern of a crystallised sample in which the glass matrix has been dissolved is shown. Only lines attributed to In_2O_3 are visible. Figure 5 presents a TEM micrograph of a powder sample obtained from a glass with the composition $Na_2O/Al_2O_3/B_2O_3/In_2O_3$. Here approximately spherical particles can be seen, which possess sizes in the range from 15 to 30 nm. The crystal size distribution is fairly narrow and in agreement with crystallite sizes calculated by SCHERRER equation from XRD line broadening.

The system $Na_2O/B_2O_3/Al_2O_3/SnO_2$ exhibits a fairly large glass forming range. Amorphous products were obtained at $[Na_2O]/[B_2O_3]$ ratios in the range of 0·364 to 1·723. At smaller $[Na_2O]/[B_2O_3]$ ratios, spontaneous crystallisation of SnO_2 occurred. In contrast to the $Na_2O/B_2O_3/Al_2O_3/In_2O_3$ system, the Al_2O_3 concentration has neither an obvious effect on the glass forming range nor on the phase formed during thermal treatment. Tempering of the amorphous samples led to the crystallisation of tin containing phases. From samples with large $[Na_2O]/[B_2O_3]$ ratios (see samples A and B), Na_2SnO_3 was precipitated. At smaller $[Na_2O]/[B_2O_3]$ ratios, the more Sn-rich phase $Na_2Sn_2O_5$ crystallised (samples C to F) while SnO_2 was precipitated from samples with a $[Na_2O]/[B_2O_3]$ ratio ≤1·125. In summary, the larger the $[Na_2O]/[B_2O_3]$ ratio, the larger the Na_2O concentration of the crystallised phase. The ratio $[Na_2O]/[B_2O_3]$=1·125 seems to be a limiting value. At all glasses studied, additionally $NaBO_2$ and in the case of glasses containing Al_2O_3, also $Na_2Al_2B_2O_7$ crystallises. As shown in Figure 6, the XRD lines attributed to the latter two phases, are not as broadened as those of SnO_2. After dialysis in a cellulose membrane tube, the XRD lines attributed to borate phases are no longer detected. Hence, the salts resulting from the dissolution of the glass and crystal phases are supposed to have diffused through the membrane. The crystallite sizes calculated using Scherrer's equation increase with increasing temperature and were in the range from 18 to 60 nm. It should be noted that tin should predominately occur as Sn^{4+}, as also known from equilibrium studies and high temperature electrochemical studies.[30,31] Compounds of Sn^{2+} were not observed.

The crystal morphology was studied by transmission electron microscopy. The SnO_2 crystals exhibit needle-like morphologies. In a sample with the composition 47·5B_2O_3.42·2Na_2O.5Al_2O_3.5·3SnO_2 already at a temperature of 600°C, needle-like SnO_2 crystals were formed. They possess thicknesses in the range from 4–7 nm and were 25 to 100 nm long. The aspect ratios (length/thickness) were hence in the range from 6 to 14. In Figure 7, a TEM micrograph of a samples tempered at 700°C is shown. Here, the crystals were up to 2 µm long and exhibit smaller aspect ratios. The thicknesses obtained from the micrographs were in good agreement with the respective mean crystallite sizes as determined from XRD line broadening using the Scherrer equation. It should be assumed that the crystal morphology is caused by kinetic reasons, e.g. crystal growth along screw dislocations. During thermal treatment at constant temperature, the aspect ratio of the crystals decreases, supposedly they approach their thermodynamic equilibrium

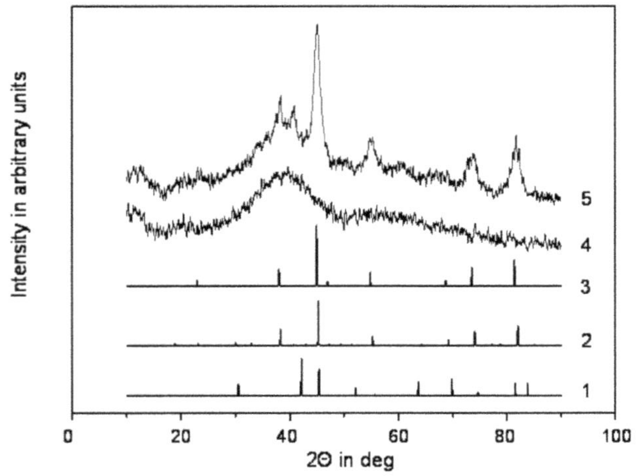

Figure 9. XRD patterns of: 4: the quenched glass; 5: the sample tempered at 570°C; 1: pattern of hematite; 2: pattern of maghemite; 3: pattern of magnetite

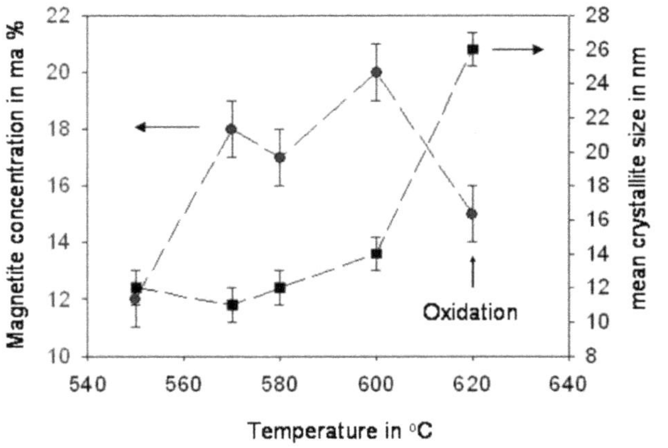

Figure 10. Magnetite concentration and mean crystallite size, calculated by SCHERRER equation from XRD line broadening as a function of the temperature of thermal treatment

shape, while only slight increase in the volume of each crystal was observed. This behaviour is known as "crystallisation pendulum".[32]

3.3. Crystallisation of magnetite from glasses with the composition 33·3CaO.4·9Na_2O.17·1Fe_2O_3.44·7B_2O_3

In Figure 8, graph 1, a DTA-profile of a roller quenched glassy sample is shown. The glass transition temperature, T_g is around 520°C. At 550 and 685°C, sharp exothermic peaks are observed. As proved by XRD of samples tempered at 550°C, the first peak is due to the crystallisation of magnetite. The second peak at 685°C is due to the crystallisation of calcium borates and possibly also to the oxidation of magnetite. After tempering the sample at 570°C for 7 h, another DTA profile was recorded (see curve 2). Here T_g is observed at around 570°C, i.e. at a temperature 50 K higher than of the untempered sample. During further heating, an exothermic peak is observed at a temperature of 690°C. Again the exothermic peak is caused by crystallisation as proved by XRD. Figure 9 presents XRD patterns of the roller quenched (curve 4) and the sample subsequently tempered at 570°C (see curve 5). While the quenched sample is amorphous without any hint at the occurrence of a crystalline phase, curve 5 shows distinct lines all attributable to crystalline magnetic (see curve 3). Lines attributed to crystalline hematite are not seen. Hence, within the limits of error of XRD, the only phase formed is magnetite. However, the lines are notably broadened indicating the occurrence of nanocrystals. Using Scherrer equation, a mean crystallite size of 10 nm is calculated.

One of the major problems associated with the formation of magnetite from glasses is the adjustment of the Fe^{2+}/Fe^{3+} redox ratio. At high temperatures, these redox species form an equilibrium with the physically dissolved oxygen of the melt. This equilibrium depends on temperature; at higher temperatures, the reduced state, Fe^{2+}, becomes more and more favoured. For low iron concentrations (<2 mol%), the thermodynamics of redox equilibria can easily be measured using electrochemical techniques (see e.g. Ref. 33). At iron concentrations as large as considered in this paper, these methods unfortunately cannot be applied. Hence, the redox ratio can hardly be predicted and has to be adjusted empirically. In contrast to silicate melts which usually exhibit much higher viscosities, borate melts can much faster be reduced by attaching an atmosphere with small oxygen fugacity. While the reduction of few cm^3 of a silicate melt usually requires many hours, 20 g of the borate melt used in this paper were reduced within 6 h. This is due to the low viscosity at melting temperature which gives raise to enhanced convection. Possibly, also the electronic conductivity resulting from electron hopping plays an important part.

Figure 10 presents the quantity of crystalline magnetite as well as the mean crystallite size calculated from SCHERRER equation as a function of the temperature of thermal treatment. With increasing temperature, the quantity of formed crystalline magnetite increases. At a temperature of 620°C, magnetite oxidises partially to hematite and hence the quantity of magnetite decreases again. The mean crystallite size remains the same within the limits of error (11–14 nm), after tempering at temperatures in the range from 550 to 600°C. In Figures 11 and 12, the effect of the time of thermal treatment on the formed quantity of crystalline magnetite and the mean crystallite size is shown for the temperatures of 550 and 570°C, respectively. After tempering at 550°C, the mean crystallite size is in the range from 10 to 12 nm, and does not increase with time. The quantity of magnetite formed increases with time

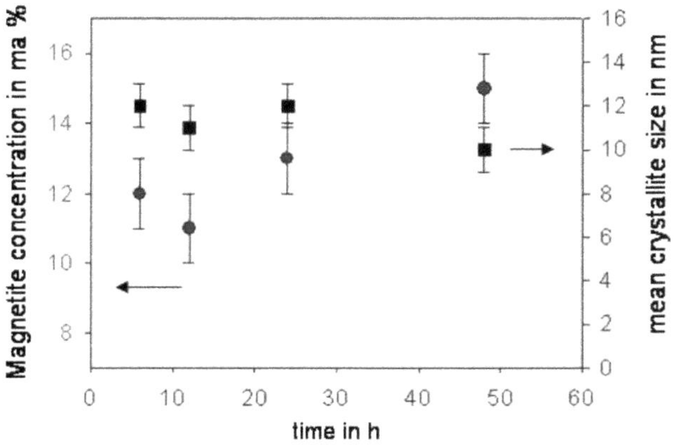

Figure 11. Magnetite concentration and mean crystallite size, calculated by SCHERRER equation from XRD line broadening as a function of the time at a temperature of 550°C

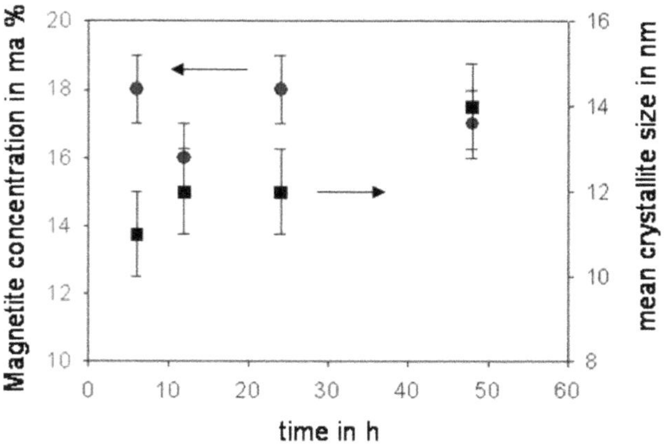

Figure 12. Magnetite concentration and mean crystallite size, calculated by SCHERRER equation from XRD line broadening as a function of the time at a temperature of 570°C

from around 11 to 15%. By contrast after tempering at 570°C, the mean crystallite size increases from 11 to 14 nm, while the quantity of formed crystalline magnetite is around 17%.

Obviously at 550°C the formed crystallites do not grow further while new nuclei are formed and grow to a size equal to that of the crystallites previously formed. At 570°C, a slight growth of the crystals is observed and the quantity of magnetite remains constant. Hence, the number of crystals should slightly decrease with time. Qualitatively, the behaviour at 570°C can be explained as "Ostwald ripening". The thermodynamic driving force is the minimisation of the total glass/crystal interface energy. The process is achieved by diffusion, i.e. by the concentration gradient in the glassy phase between a large and a small crystal. The smaller crystals disappear with time and the large crystals grow with time according to $d{\sim}t^{1/3}$.[20] As described above, the growth with time is much smaller (570°C) or negligible (550°C) as in the case of Ostwald ripening. If the crystals grew according to $d{\sim}t^{1/3}$, they should be twice as large after tempering for 48 h as after thermal treatment for 6 h. As shown in Figure 8, the glass transformation temperature increases during tempering. It should be noted that the T_g values are solely related to the residual glassy phase. While T_g of the untempered sample is 510°C, those of samples tempered at 550, 570 and 580°C are 557, 570 and 582°C, respectively. Obviously, T_g of the tempered samples is approximately equal to the temperature at which the samples are tempered. This effect is caused by an increasing quantity of magnetite formed, i.e. T_g increases with decreasing iron concentration of the glass.

During crystallisation of magnetite, the composition of the glassy phase near the crystals changes, the iron concentration decreases and the viscosity increases. The crystals grow and a further decrease in the iron concentration near the crystals occurs which leads to a further increase in viscosity. Since diffusion and hence crystal growth are strongly affected by the viscosity, the change in the chemical composition decelerates the further crystal growth. When the viscosity reaches a value of around 10^{13} dPas, T_g of the glassy phase around the crystal is equal to the respective temperature supplied. Then further crystal growth does no longer occur. However, nucleation still takes place in between the crystals, where the composition is still the same as that of the initial glass. These processes continue until the whole glassy phase has a viscosity of around 10^{13} dPas. Then further nucleation and crystal growth gets more and more negligible and the crystallisation process is frozen. The crystallite size and the quantity of formed magnetite approach a constant value. If, however, higher temperatures are supplied, a viscosity of 10^{13} dPas is attributed to a smaller iron concentration in the glassy phase and hence to larger quantities of crystalline magnetite. Hence, the quantity of formed magnetite increases with increasing temperature.

Crystallisation of magnetite from a rapidly quenched $Na_2O/CaO/Fe_2O_3/B_2O_3$ glass is an example for a self organised process. During nucleation and crystal growth, the growth kinetics change and control the size of the crystals.

Conclusions

The crystallisation of borate glasses enables the preparation of materials with challenging properties.

Among these are oxide nanoparticles, such as SnO_2, In_2O_3 or Fe_3O_4. The particles possess sizes in the nanometre range and a fairly narrow size distribution. In the case of the $Na_2O/CaO/Fe_2O_3/B_2O_3$ system, the viscosity near the crystals increases during nucleation and crystal growth, which decreases the crystal growth velocity. If the viscosity reaches a value of 10^{13} d Pas, the crystal growth process is frozen.

Borate glass ceramics of high application potential are found in the $BaO/Al_2O_3/B_2O_3$ system. $BaAl_2B_2O_7$ crystals exhibit strongly negative thermal expansion up to temperatures of 500°C. The thermal expansions of glass ceramics containing this crystal phase, however, strongly depend on the preparation route. Most negative thermal expansions are observed in the pure stoichiometric system with large spheroidal $BaAl_2B_2O_7$ crystals, while glass ceramics prepared from platinum doped glasses or via a sintering route exhibit a thermal expansion of around zero near room temperature.

References

1. Vogel, W., Schmidt, W. & Horn, L. *Z. Chem.*, 1969, **9**, 401.
2. Vogel, W., Schmidt, W. & Horn, L. *Z. Chem.*, 1969, **9**, 276.
3. Vogel, W., Schmidt, W. & Horn, L. *Silikattechnik*, 1972, **23**, 112.
4. Ehrt, D., Reiß, H. & Vogel, W. *Silikattechnik*, 1976, **27**, 304.
5. Heidenreich, E., Ehrt, R., Doenitz, F.-D. & Vogel, W. *Silikattechnik*, 1976, **27**, 402.
6. Ehrt, D., Reiß, H. & Vogel, W. *Silikattechnik*, 1977, **28**, 359.
7. Woltz, S. & Rüssel, C. *J. Non-Cryst. Solids*, 2004, **337**, 226.
8. Müller, R., Hiergeist, R., Steinmetz, H., Aqoub, N., Fujisaki, M. & Schüppel, W. *J. Magn. Magn. Mater.*, 1999, **201**, 34.
9. Lee, Y.-K. & Choi, S. Y. *J. Am. Ceram. Soc.*, 1967, **9**, 992.
10. Blackburn, W. J. S. & Tilley, B. P. *J. Mater. Sci.*, 1974, **9**, 1265.
11. Klupsch, Th., Steinbeiss, E., Müller, R., Ulbricht, C., Schüppel, W., Steinmetz, H. & Höche, Th. *J. Magn. Magn. Mater.*, 1999, **196–197**, 264.
12. Shirk, B. T. & Buesseln, W. R. *J. Am. Ceram. Soc.*, 1970, **53**, 192.
13. Görnert, P., Sinn, E., Schüppel, W., Pfeiffer, H., Rösler, M., Schubert, T., Jurisch, M. & Sellger, R. *IEEE Trans. Magn.*, 1990, **26**, 12.
14. Rezlescu, N., Rezlescu, E., Pasnicu, P. & Crans, M. L. *J. Magn. Magn. Mater.*, 1994, **131**, 273.
15. Hergt, R., Hiergeist, R., Hilger, I. & Kaiser, W. A. *Recent Res. Devel. Mater. Sci.*, 2002, **3**, 723.
16. Hilger, I., Andrä, W., Hergt, R., Hiergeist, R. & Kaiser, W. A. *Radiology*, 2002, **218**, 570.
17. Chan, D. C. F., Kirpotin, D. B. & Bunn Jr., P. A. *J. Magn. Magn. Mater.*, 1993, **122**, 374.
18. Jordan, A., Scholz, R., Wurst, P., Schirra, H., Schiestel, T., Schmidt, H. & Felix, R. *J. Magn. Magn. Mater.*, 1999, **194**, 185.
19. Hiergeist, R., Andrä, W., Buske, N., Hergt, R., Hilger, I., Richter, U. & Kaiser, W. *J. Magn. Magn. Mater.*, 1999, **201**, 420.
20. Woltz, S., Hiergeist, R., Görnert, P. & Rüssel, C. *J. Magn. Magn. Mater.*, 2006, **298**, 7.
21. Löser, C. & Rüssel, C. *J. Non-Cryst. Solids*, 2001, **282**, 228.
22. Garkova, R., Gugov, I. & Rüssel, C. *J. Non-Cryst. Solids*, 2003, **320**, 291.
23. Garkova, R., Völksch, G. & Rüssel, C. *J. Non-Cryst. Solids*, 2005, **351**, 2287.
24. Hovhannisyan, R. M. *Glass Technol.*, 2003, **44**, 96.
25. Keding, R., Tauch, D. & Rüssel, C. *J. Non-Cryst. Solids*, 2004, **348**, 123.
26. Tauch, D. & Rüssel, C. *J. Non-Cryst. Solids*, 2005, **351**, 2294.
27. Lira, C., de Oliveira, A. P. N. & Alarcon, O. E. *Glass Technol.*, 2004, **45**, 43.
28. Szabo, I., Nagy, B., Völksch, G. & Höland, W. *J. Non-Cryst. Solids*, 2000, **272**, 191.
29. Müller, R., Naumann, R. & Reinsch, S. *Thermochim. Acta*, 1996, **280 & 281**, 119.
30. Rüssel, C. *Glastech. Ber.*, 1993, **66**, 68.
31. Xiang, Z. D. & Cable, M. *Phys. Chem. Glasses*, 1997, **38**, 167.
32. Avramov, A., Hoeche, T. & Rüssel, C. *J. Chem. Phys.*, 1999, **110**, 8676.
33. Wiedenroth, A. & Rüssel, C. *J. Non-Cryst. Solids*, 2002, **297**, 173.

Neutron spectroscopic studies of caesium borate crystals and glasses

Roger N. Sinclair, Richard Haworth, Adrian C. Wright*

J.J. Thomson Physical Laboratory, University of Reading, Whiteknights, Reading, RG6 6AF, UK

Ben G. Parkinson, Diane Holland

Physics Department, Warwick University, Coventry CV14 7AL, UK

Jon W. Taylor

ISIS Pulsed Neutron & Muon Source, Rutherford Appleton Laboratory, Chilton, Didcot, OX11 0QX, UK

Natalia M. Vedishcheva, Irina G. Polyakova, Boris A. Shakhmatkin

Institute of Silicate Chemistry of the Russian Academy of Sciences, Ul. Odoevskogo 24, Korp. 2, St. Petersburg, 199155, Russia

Steven A. Feller, B. Rijal & T. Edwards

Department of Physics, Coe College, Cedar Rapids, IA 52402-5092, USA

Inelastic neutron scattering data have been measured for the vitreous forms of caesium enneaborate and pentaborate along with polycrystalline caesium enneaborate, pentaborate and triborate. The known crystal structures allow associated superstructural unit modes in the neutron weighted vibrational density of states (VDOS) of the glass forms to be identified. The widths of the features in the VDOS give additional information on the dispersive nature of the modes and hence the molecular character. To assist in the identification of larger structural units, new Raman measurements have been completed for a range of five glasses and four crystalline compositions in the same system. It is concluded that boroxol units are only formed in vitreous samples containing alkali up to the tetraborate composition. A glass with the enneaborate composition contains both boroxol and triborate units, while glasses of pentaborate composition appear to contain pentaborate and possibly triborate groups.

Introduction

The use of spectroscopic techniques to investigate the superstructural unit concentrations in borate glasses began with the detailed study of the Raman spectra of vitreous B_2O_3, where there is an extremely sharp line at 100meV (803 cm^{-1}),[1,2] due to the breathing mode of a $B_3Ø_6$ boroxol ring, where Ø represents a bridging oxygen atom. Earlier structural work on borate glasses[3–5] had provoked controversy when Hägg[6] challenged Zachariasen's continuous random network hypothesis and proposed groupings in glasses larger than the basic structural units. Krogh-Moe[7] and Bray and co-workers[8,9] both confirmed the existence of these larger structural groupings in borate glasses.

Raman spectroscopy[10] has proved valuable in identifying these larger groupings by comparisons with results for crystalline systems where peak assignments can be made reliably using the results of crystal structure determinations. Whilst there have been many crystal structure studies of complex synthetic borates and borate minerals, there is a surprising lack of structural data for the simple anhydrous binary borates required for the interpretation of both neutron and NMR data for the corresponding glasses. For example, in the case of the $Na_2O–B_2O_3$ system, a full crystal structure determination has only been performed for seven out of a total of 22 polymorphs listed in the Gmelin Handbook.[11] It is important to note that the crystalline borate networks necessary for the interpretation of data for related glasses should ideally be formed by a single superstructural unit species, with no independent BO_3 triangles or BO_4 tetrahedra. Table 1[12] lists such crystalline compounds or, for superstructural units where no known crystalline compound exists, one or more compounds containing the given superstructural unit, together with a list of the other units present. In Table 1, compounds of the form $mCs_2O.nB_2O_3$ are denoted by the abbreviation mCs.nB, NBO indicates a nonbridging oxygen atom.

[1] Corresponding author. Email r.n.sinclair@rdg.ac.uk

Table 1. Borate crystalline compounds containing superstructural units

Unit	*Formula*	*Crystalline compound(s)* *Single unit*	*Multiple Units*
Structural units			
BO_3 triangle	$B\varnothing_3$	B-I	-
BO_4 tetrahedron	$B\varnothing_4^-$	γ-Li.B	-
BO_3 triangle with NBO	$B\varnothing_2O^-$	Li.B, Ca.B	-
Superstructural units			
Boroxol	$B_3\varnothing_6$	-	Cs.9B ($+B_3\varnothing_7^-$)
Pentaborate	$B_5\varnothing_{10}^-$	K.5B, Rb.5B, Cs.5B	-
Triborate	$B_3\varnothing_7^-$	Li.3B, Cs.3B	-
Di-pentaborate	$B_5\varnothing_{11}^{2-}$	-	Na.2B ($+B_3\varnothing_6O^{2-}$), Ba.2B ($+B_3\varnothing_8^{2-}$)
Diborate	$B_4\varnothing_9^{2-}$	Li.2B, Zn.2B, Mg.2B	-
Tri-pentaborate	$B_5\varnothing_{12}^{3-}$	-	2Ca.3B ($+B\varnothing_4^-$)
Di-triborate	$B_3\varnothing_8^{2-}$	-	3Li.7B ($+B_3\varnothing_7^-+B\varnothing_3$), K.2B & Rb.2B ($+B_4\varnothing_9^{2-}+B\varnothing_3$), Ba.2B ($+B_5\varnothing_{11}^{2-}$)
Triborate with NBO	$B_3\varnothing_6O^{2-}$	-	Na.2B ($+B_5\varnothing_{11}^{2-}$)
Metaborate	$B_3\varnothing_9^{3-}$	Ca.B-IV, Sr.B-IV, Cu.B	-
Anions			
Cyclic metaborate	$B_3\varnothing_3O_3^{3-}$	Na.B, K.B, α-Rb.B, Cs.B, Ba.B	-
Pyroborate ($2\times B\varnothing O_2^{2-}$)	$B_2\varnothing O_4^{4-}$	2Na.B, 2Mg.B, 2Ca.B, 2Sr.B	-
Orthoborate	BO_3^{3-}	α-3Li.B, 3Na.B, 3Ag.B-I, 3Tl.B, 3Zn.B, 3Mg.B, 3Ca.B, 3Sr.B	

Greek letters are used for polymorphs that are stable at different temperatures and Roman numerals for those formed at different pressures.

Unfortunately, Raman spectroscopy cannot be employed to determine the absolute numbers of superstructural units in borate glasses, because of unknown (and large) matrix element effects. This is not the case for the neutron inelastic scattering technique, which was used[13,14] to derive an effective VDOS for vitreous B_2O_3. The relative size of the symmetric breathing mode of a $B_3\varnothing_6$ boroxol ring showed that the fraction of the boron atoms in such rings (*f*) is indeed very close to that given by neutron diffraction experiments, 0·80±0·05.[15] A similar study of the VDOS of a glass of composition $Rb_2O.0{\cdot}33B_2O_3$[14] agreed with Raman data[16] that boroxol rings are not present in the structure of this glass. However, a sharp peak observed at 610 cm^{-1} (75·6 meV) in the Raman spectrum attributed to the stretching mode of a $B_3\varnothing_3O_3^{3-}$ cyclic metaborate ion was not visible in the neutron data.

More recently, spectroscopic measurements using the neutron inelastic scattering method have begun[17,18] for the caesium borate glass system, which has the advantage that crystalline polymorphs contain known fractions of a variety of superstructural units that may also appear in the equivalent glasses. Data were presented for the neutron weighted VDOS for samples of polycrystalline caesium pentaborate ($Cs_2O.5B_2O_3$), which contains only pentaborate groups (so *f* is equal to zero), for polycrystalline caesium triborate ($Cs_2O.3B_2O_3$; only triborate groups) and vitreous caesium enneaborate (v-$Cs_2O.9B_2O_3$). The boroxol ring breathing mode was also observed in vitreous caesium enneaborate (at 99·9 meV) and, as expected, in the VDOS of polycrystalline caesium enneaborate. The triborate ring breathing mode was also identified from Raman data, and characterised in the neutron weighted VDOS for polycrystalline caesium enneaborate and polycrystalline caesium triborate. The sharpness of the split peak and it's position (at 95·5 meV and 98·6 meV) in the polycrystalline caesium pentaborate effective VDOS, enabled its association with vibrational modes of the pentaborate double ring unit.[7]

This paper reports and discusses a neutron scattering measurement of the VDOS for a vitreous sample with the caesium pentaborate composition, together with Raman measurements for a range of glasses and crystals in the same system.

2. Inelastic neutron scattering technique

The inelastic neutron scattering technique[19] can have an energy transfer resolution comparable to that available optically. A dynamical structure factor, $S(Q,E)$, may be measured with the optimum resolution in scattering vector and energy transfer for a variety of different types of excitation. Q is the magnitude of the scattering vector and E is the energy transfer ($E=E_0-E'$, where E_0 and E' are the initial and final neutron energies). Statistical accuracy remains a significant problem for detailed studies due to the limited source intensities available at present. For borate glasses local atomic configurations in the short range order result in localised vibrations (e.g. the breathing modes of superstructural units) revealed as sharp features in the effective vibrational density of states (VDOS).

The dynamical structure factor, $S(Q,E)$, may be derived from the measured quantity in an inelastic neutron scattering experiment which is the double differential scattering cross section.[20] If the scattering function is averaged over a wide range of the scattering vector Q to eliminate coherent scattering effects, an effective neutron weighted VDOS, $G(E)$,

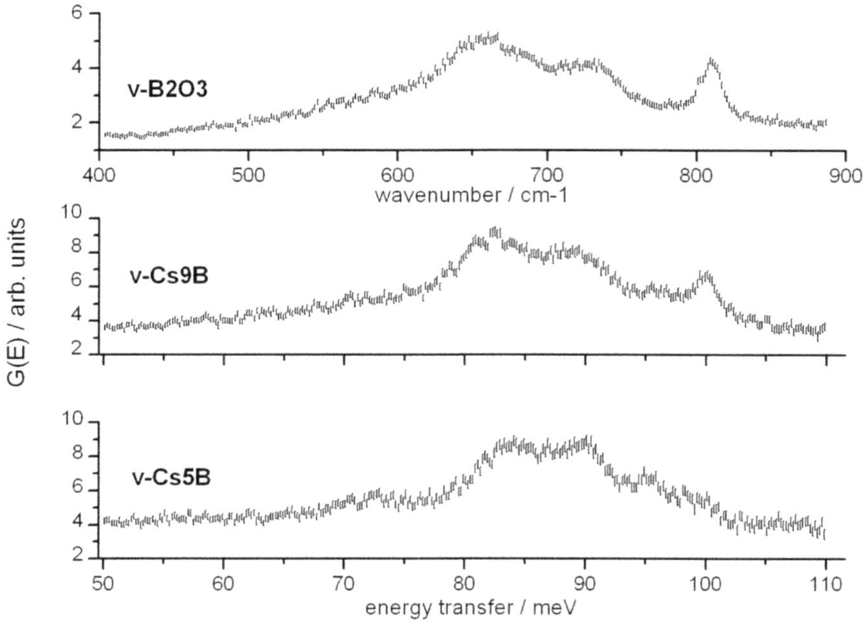

Figure 1. The neutron weighted VDOS, G(E), for samples of vitreous B_2O_3, Cs_2O $9B_2O_3$ and Cs_2O $5B_2O_3$, measured using neutrons with 135 meV incident energy

can be derived from the one phonon creation self scattering function in the incoherent approximation as described in Ref. 20. A more complete description of the theory is given in Ref. 18.

3. Experimental details and results

3.1. Sample preparation

All of the samples for this series of neutron measurements were prepared using boric acid enriched to 99·27% ^{11}B to minimise neutron absorption due to the ^{10}B isotope. A vitreous caesium pentaborate sample was prepared in Cedar Rapids. Boric acid was converted to B_2O_3, crushed, mixed with high purity Cs_2CO_3, and then heated (at 1100°C) in a platinum crucible under N_2 before pouring onto a stainless steel plate (still under N_2). The preparation conditions for the samples of polycrystalline caesium enneaborate, caesium pentaborate and caesium triborate compounds from glasses of the correct composition have been described in Refs 17 and 18. The phase purity of the crystalline samples was investigated

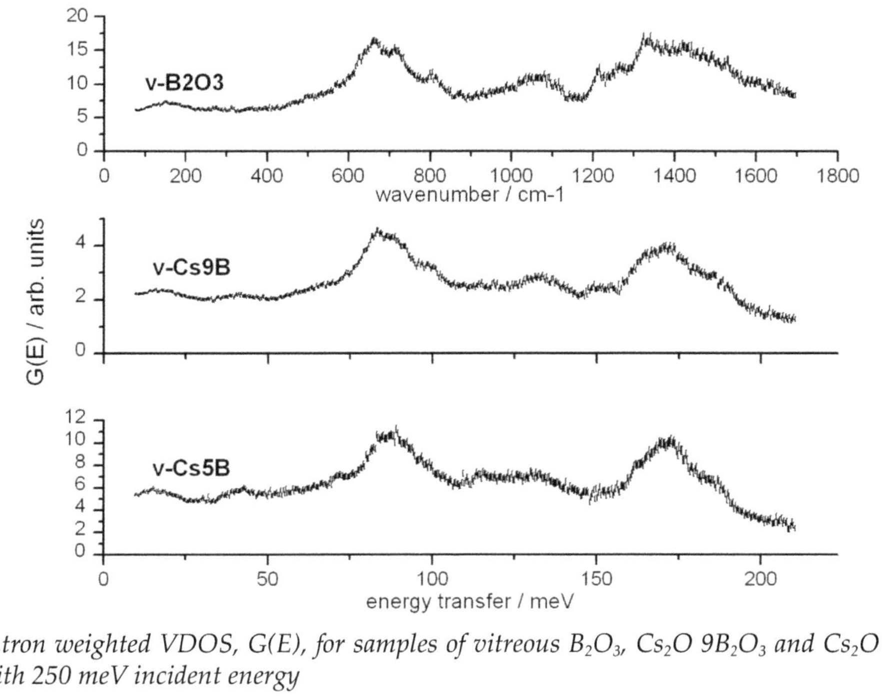

Figure 2. The neutron weighted VDOS, G(E), for samples of vitreous B_2O_3, Cs_2O $9B_2O_3$ and Cs_2O $5B_2O_3$, measured using neutrons with 250 meV incident energy

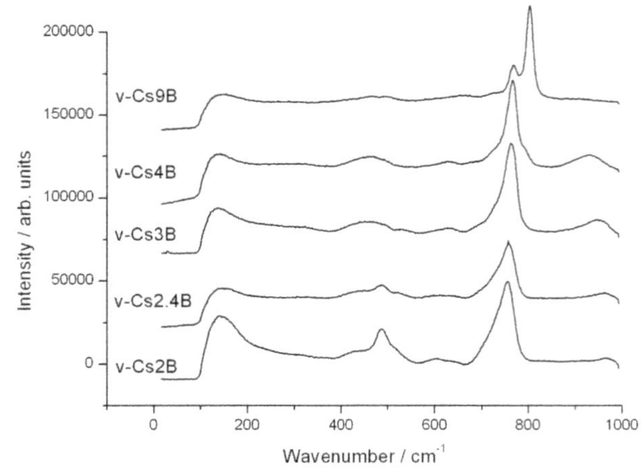

Figure 3. Raman spectra for five glasses in the caesium borate system

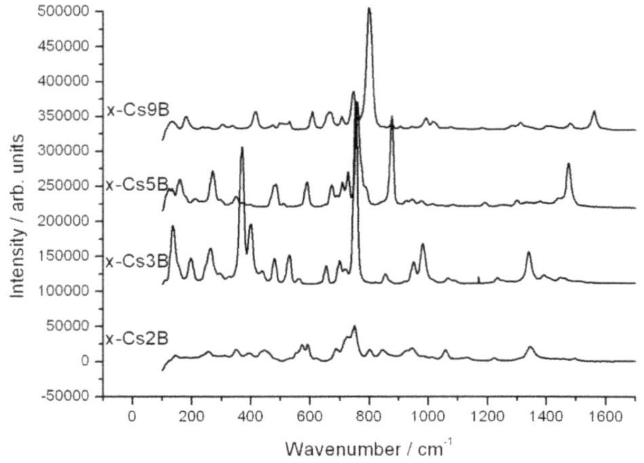

Figure 4. Raman spectra for four crystalline compositions in the caesium borate system

by x-ray diffraction and by DTA determination of the melting point.

3.2. Inelastic neutron scattering

The phased rotor inelastic scattering spectrometer MARI[21] uses the time of flight method at the ISIS UK pulsed neutron source. Measurements were performed at two incident neutron energies, 250 meV to investigate the full VDOS and 135 meV to obtain the best energy transfer resolution in the region of sharp superstructural breathing modes. In each case, the scattering was recorded as a function of the scattered neutron time of flight in 922 detectors located at scattering angles between 3° and 135° at an evacuated path length of 4·0 m from the sample. A powdered sample of vitreous caesium pentaborate weighing 53 g was held at 15 K in a thin walled aluminium alloy container in a closed circuit refrigerator. The neutron weighted VDOS ($G(E)$) for this sample (Figures 1 and 2) was obtained from data corrected for instrumental background and container scattering by then integrating the 135 meV scattering function between Q values of 4 Å^{-1} and 12 Å^{-1} and the 250 meV data between 7 Å^{-1} and 15 Å^{-1} as discussed in section 2. The spectrometer resolution at 100 meV (800 cm^{-1}) energy transfer for the 135 meV data was 1·6 meV (12 cm^{-1}).

3.3. Raman spectroscopy

Raman spectra were obtained at Warwick University with a Renishaw Invia Raman spectrometer equipped with a 20 mW Argon Laser operating at 514 nm. The measurements reported here were acquired at room temperature with 10 mW incident laser power and a ×50 objective. The spectrometer resolution was approximately 2 cm^{-1}. Measurements for five glass compositions in the caesium borate system are shown in Figure 3. Raman spectra for the same polycrystalline samples whose VDOS was reported in Ref. 18 were also recorded (Figure 4). Peak positions obtained by Gaussian fitting techniques are given Table 2 for the glass samples and in Table 3 for the polycrystalline samples.

4. Discussion

As the alkali (Cs_2O) content is increased, the fraction of four-fold coordinated boron atoms, x_4, increases linearly up to a Cs_2O/B_2O_3 ratio, R, of about 0·4. x_4 is equal to R in the linear region. The pentaborate group requires the least alkali content to form fol-

Table 2. Peak positions (cm^{-1}) in Raman spectra for caesium borate glasses and vitreous boron oxide

$Cs_2O.2B_2O_3$-v Raman	$Cs_2O.2{\cdot}4B_2O_3$-v Raman	$Cs_2O.3B_2O_3$-v Raman	$Cs_2O.4B_2O_3$-v Raman	$Cs_2O.9B_2O_3$-v Raman	B_2O_3-v Raman
432	447	440	442		
465				467	
486	487	483	475		471
510	523	534	500	498	501
589					
606					601
635		625	627		
				662	665
710	714	704	696		
729	732	736	747	729	737
755	759	764	767	771	
			794	803	808
				883	

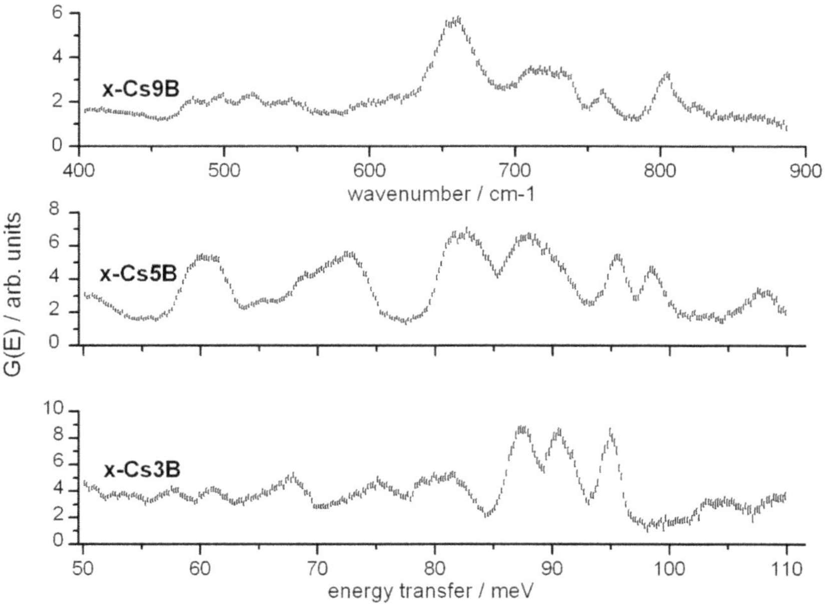

Figure 5. The neutron weighted VDOS, G(E), for samples of polycrystalline $Cs_2O\ 9B_2O_3$, $Cs_2O\ 5B_2O$ and $Cs_2O\ 3B_2O$, obtained using neutrons with 135 meV incident energy

lowed by the triborate unit. As these groups increase in concentration, the fraction of boroxol rings should decrease. To attempt to detect the formation of these superstructural units all the VDOS and Raman spectra were fitted using a Gaussian line shape to extract frequencies, intensities and line widths wherever this was possible. By comparison to the result for vitreous B_2O_3, signatures of 3-fold boron coordination in all the measured VDOS (Figures 1, 2, 5 and 6) for the other crystals and glasses are provided by broad features at 661 and 730 cm^{-1} with a further unresolved feature at about 1335 cm^{-1}. These features correspond to peaks 4, 5 and 6 in the numbering scheme used by Galeener & Thorpe[1] for the Raman spectrum. The lower two modes arise from the symmetric stretch and out-of-plane deformation of BO_3 triangles while the highest at 1335 cm^{-1} has been found to be a signature of 3-fold coordinated boron.[22]

4.1. Vitreous and crystalline caesium pentaborate

The crystal structure of $Cs_2O.5B_2O_3$[23] is based on zizag boron oxygen layers built up from rigid pentaborate groups and the measured Raman spectrum (Figure 4) is dominated by sharp modes at 589, 760,

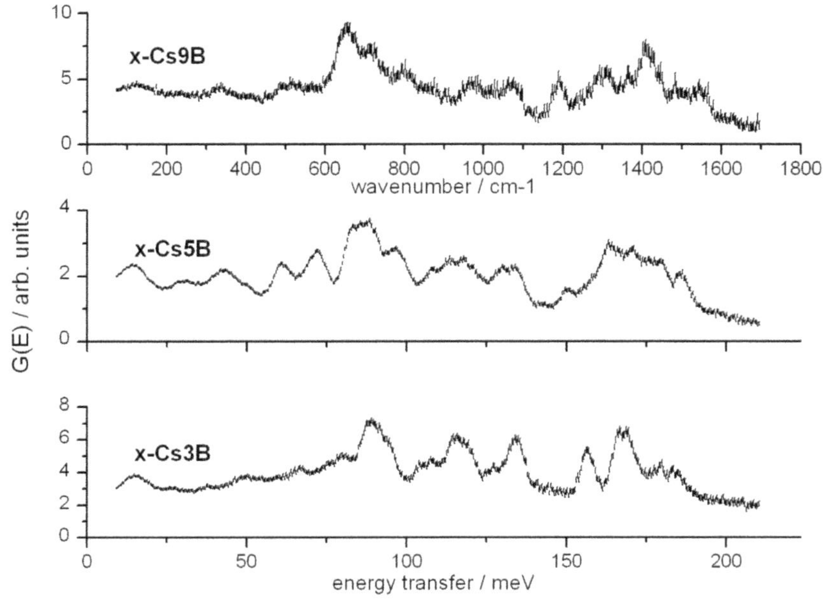

Figure 6. The neutron weighted VDOS, G(E), for samples of polycrystalline $Cs_2O\ 9B_2O_3$, $Cs_2O\ 5B_2O$ and $Cs_2O\ 3B_2O$, obtained using neutrons with 250 meV incident energy

786, 876 and 1476 cm^{-1} with FWHM values of 15·7, 17·4, 18·0, 13·7 and 18·9 cm^{-1}, respectively (Table 3). The Raman spectrum for a glass of composition $Cs_2O.4B_2O_3$ (Figure 3, Table 2) is also dominated by an equally narrow mode at 767 cm^{-1} (FWHM 22·1) with a nearby small unresolved feature at 794 cm^{-1} (FWHM 15·3). The Raman results are inconclusive but suggest that the pentaborate group is present in this glass. There are no indications of a significant fraction of boroxol rings.

$G(E)$ derived from $S(Q,E)$ measured for the caesium pentaborate crystalline sample (Figures 5 and 6) has a broad feature at 588 cm^{-1}, along with distinctive narrow peaks 769, 794 and 1205 cm^{-1} associated with the pentaborate grouping. The result of an identical experiment for a vitreous sample with the caesium pentaborate composition $Cs_2O.5B_2O_3$ (Figures 1 and 2) show distinguishing weak peaks at 589, 764 and 796 cm^{-1} thus confirming the presence of pentaborate groups in the glass. As the key signature mode for the triborate group is at 764 cm^{-1} (see section 4.2 below) the presence of this grouping cannot be eliminated by these data. A strong feature in the Raman spectrum for the crystal at 1205 cm^{-1} was not present in the glass and is shown later to be indicative of the fraction of 4-fold coordinated boron atoms, x_4. Again the data do not indicate any boroxol rings in the glass form.

Table 3. Peak positions (cm^{-1}) in Raman spectra for crystals in the caesium borate system

$Cs_2O.9B_2O_3$	$Cs_2O.5B_2O_3$	$Cs_2O.3B_2O_3$	$Cs_2O.2B_2O_3$
415			
		433	447
474			
497	480	480	
512	510		
531		529	531
	572	562	553
	589		573
			593
607			618
656		653	
669	672		
	690	698	688
707	709	720	723
745	724	753	
	760		750
	774		
801	786		801
			846
	861	855	865
876	876	874	
905	931		926
	949	950	949
975	977	981	977
992		1001	
1018	1016		1017
1030			1057
1077	1084	1072	
			1131
1182	1192	1171	
1286	1251	1233	1224
		1244	
1312	1299		1310
	1334	1340	1345
	1379	1391	1373
		1406	
1414		1417	
	1445	1445	1449
1481	1476	1462	
1561			

4.2. Vitreous and crystalline caesium triborate

Crystalline caesium triborate also contains one type of borate grouping, in this case the triborate group.[24] The measured Raman spectrum (Figure 4, Table 3) for this compound has strong narrow peaks at 753 and 1340 cm^{-1} (FWHM 15·9 and 20·8 cm^{-1}) with weaker narrow features at 653, 950 and 981 cm^{-1} (FWHM 14·6, 20 and 19 cm^{-1}). The results for a vitreous sample with the same composition (Figure 3, Table 2) also show a strong narrow feature at 764 cm^{-1} (FWHM 28·4 cm^{-1}), which has been assigned to a ring breathing mode of six atom borate rings in the triborate grouping.[24]

The VDOS derived from neutron scattering measurements from the polycrystalline sample (Figures 5 and 6) should also indicate features which can provide indications of the presence of triborate groups. Strong features are found at 765, 730, 704 and 1353 cm^{-1}. The first and last of these peaks provide a signature of the group in both Raman and neutron spectra. A broad feature centred at about 600 cm^{-1} is also found in the VDOS of the pentaborate glass.

4.3. Vitreous and crystalline caesium enneaborate

The crystal structure of all phases of caesium enneaborate ($Cs_2O.9B_2O_3$) consists of two separate interpenetrating boron–oxygen networks.[25] Each network is made up of planar boroxol groups ($B_3Ø_6$) and triborate groups, ($B_3Ø_7$). The number of boron atoms associated with each group is in the ratio 2:1 respectively and thus f is equal to 0·67.

The Raman spectrum (Figure 4, Table 3) for a polycrystalline sample has features at 669 and 800 cm^{-1} (FWHM 15·7 and 22·8 cm^{-1}), which characterise the boroxol groups in the structure.[2] Peaks at 656 and 745 cm^{-1} (FWHM 15·6 and 14·5) identify the triborate groups, as found for the caesium triborate crystal in section 4.2. For the glass with the enneaborate composition, the Raman spectrum (Figure 3, Table 2) exhibits a peak at 803 cm^{-1} (FWHM 18·2 cm^{-1}) arising from the symmetric stretching mode of the boroxol ring which is the most intense peak. A second broad peak at 662 cm^{-1} may contain components from the 669 and 656 cm^{-1} excitations of boroxol and triborate groups, respectively, as observed for the crystal form. The strong peak at 771 cm^{-1} also provides evidence of the presence of the triborate group in this glass but may prevent observation of pentaborate modes at 769 and 794 cm^{-1}.

In the effective VDOS for crystalline caesium

enneaborate (Figures 5 and 6) a sharp feature at 803 cm^{-1}, as in the VDOS for vitreous B_2O_3 (Figure 1), corresponds to the boroxol ring symmetric stretch mode in which two thirds of the boron atoms participate It has a weighting similar to that at 808 cm^{-1} in the VDOS for vitreous B_2O_3. Thus f for vitreous B_2O_3 cannot be significantly less than that found in this crystalline structure (0·67). A sharp feature observed at 761 cm^{-1} is associated with the peak at 771 cm^{-1} in the equivalent Raman spectrum and confirms the assignment as a stretching mode of the triborate unit. There is no indication of the other triborate signature mode at 1353 cm^{-1} although a complex feature is observed at 1306 cm^{-1}. The signature of the boroxol ring at 806 cm^{-1} is also observed for the sample of vitreous caesium enneaborate (Figures 1 and 2) with a similar weighting to the same mode in the VDOS for vitreous B_2O_3. A complex small feature centred at 770 cm^{-1} is unfortunately not well defined statistically but does suggest that pentaborate and triborate groups may be present in the structure of this glass. In contrast, the Raman spectrum (Figure 3, Table 2) would indicate that only triborate groups are present.

5. Conclusions

The experiments described above demonstrate that neutron scattering studies, with high energy transfer resolution, can provide important extra information about the structure of borate glasses when used in conjunction with measured Raman spectra. Narrow lines arising from localised vibrational modes of boroxol, triborate and pentaborate units have been identified by studying the VDOS and Raman spectra of polycrystalline caesium borates containing structural units which may occur in the glass.

The size of the symmetric breathing mode features in the VDOS of the vitreous and crystalline samples provides evidence that the fraction of boron atoms in boroxol rings in vitreous B_2O_3 is close to that given by the diffraction and NMR experiments.

For the caesium borate system, boroxol units are only formed in vitreous samples containing alkali up to the tetraborate composition. A glass with the enneaborate composition contains both boroxol and triborate units, while a glass of pentaborate composition definitely contains pentaborate groups and possibly triborate units. It is not possible to deduce the fractions of these more complex superstructural units from the results presented here. These can only be determined by comparing the measured VDOS with those derived from dynamical calculations that are based on realistic model structures containing predicted populations of superstructural units.

6. Acknowledgements

Support both financial and technical from ISIS, RAL CLRC, EPSRC and the J. J. Thomson Physical Laboratory are gratefully acknowledged. This work was also supported by the United States National Science Foundation under grant DMR 0502051.

7. References

1. Galeener F. L. & Thorpe, M. F. *Phys. Rev.*, 1983, **B28**, 5802.
2. Galeener, F. L., Lucovski G. & Mikkelsen, J. C. Jnr. *Phys. Rev.*, 1980, **B22**, 3983.
3. Zachariasen W. H. *J. Am. Chem. Soc.*, 1932, **54**, 3841.
4. Warren, B. E., Krutter, H. & Morningstar, O. *J. Am. Ceram. Soc.*, 1936, **19**, 202.
5. Biscoe, J. & Warren, B. E. *J. Am. Ceram. Soc.*, 1938, **21**, 287.
6. Hägg, G. *J. Chem. Phys.*, 1935, **3**, 42.
7. Krogh-Moe, J. *Phys. Chem. Glasses*, 1965, **6**, 46.
8. Jellison, G. E. Jnr. & Bray, P. J. *J. Non-Cryst. Solids*, 1978, **29**, 187.
9. Feller, S. A, Dell , W. J. & Bray, P. J. *J. Non-Cryst. Solids*, 1982, **51**, 21.
10. Meera, B. N. & Ramakrishna, J. *J. Non-Cryst. Solids*, 1993, **159**, 1.
11. G. Heller, *Gmelin Handbuch der Anorganischen Chemie*, Vol. 28, Borverbindungen Part 7 (Springer-Verlag, Berlin, 1975).
12. Wright, A. C. Private communication.
13. Hannon, A. C., Sinclair, R. N., Blackman, J. A., Wright, A. C. & Galeener, F. L. *J. Non-Cryst. Solids*, 1988, **106**, 116.
14. Sinclair, R. N., Wright, A. C., Wanless, A. J., Hannon, A. C., Feller, S. A., Mayhew, M. T., Meyer, B. M., Royle, M. L., Wilkerson, D. L., Williams, R. B. & Johanson, B. C. In *Borate Glasses, Crystals & Melts*. 1997. Edited by Wright, A. C., Feller, S. A. & Hannon, A. C. Society of Glass Technology, Sheffield. P 140.
15. Hannon, A. C., Grimley, D. I., Hulme, R. A., Wright A. C. & Sinclair, R. N. *J. Non-Cryst. Solids*, 1994, **177**, 299.
16. Kamitsos, E. I., Karakassides, M. A. & Chryssikos, G. D. *Phys. Chem. Glasses*, 1989, **30**, 2290.
17. Sinclair, R. N., Stone, C. E., Wright, A. C., Polyakova, I. G., Vedishcheva, N. M., Shakhmatkin, B. A., Feller, S. A., Johanson, B. C., Venhuizen, P., Williams, R. B. & Hannon, A. C. *Phys. Chem. Glasses*, 2000, **41**, 286
18. Haworth, R., Shaw, J. L., Wright, A. C., Sinclair, R. N., Knight, K. S., Taylor, J. W., Vedishcheva, N. M., Polyakova, I. G., Shakhmatkin, B. A., Feller, S. A., Affatigato, M. & Winslow, D. W. *Phys. Chem. Glasses*, 2005, **46**, 476.
19. Sinclair, R. N. *J. Non-Cryst. Solids*, 1985, **76**, 61.
20. Price D. L. & Carpenter, J. M. *J. Non-Cryst. Solids*, 1987, **92**, 153.
21. Arai, M., Taylor, A. D., Bennington, S. M. & Bowden, Z. A. In *Proc. Int. Sym. Structure and Dynamics of Liquids and Gases*. 1992. Adam Hilger, Bristol. F 321.
22. Kamitsos, E. I., Karakassides, M. A. & Chryssikos, G. D. *Phys. Chem. Glasses*, 1989, **30**, 229.
23. Filatov,. S, Bubnova.,R., Shepelev, Yu., Anderson, J. & Smolin, Yu. *Cryst. Res. Technol.*, 2005, **40**, 65
24. Krogh-Moe, J. *Acta Cryst.*, 1974, **B30**, 1178
25. Krogh-Moe, J. & Ihara, M. *Acta Cryst.*, 1967, **23**, 427.

Connection between spectral features of B *K* edge XANES of minerals and the local structure

O. Šipr,[1] A. Šimůnek, J. Vackář

Institute of Physics, Academy of Sciences of the Czech Republic, Cukrovarnická 10, 162 53 Prague, Czech Republic

G. Dalba

INFM and Dipartimento di Fisica, Universitá di Trento, Via Sommarive 14, I-38050 Povo, Trento, Italy

F. Rocca

IFN-CNR, Istituto di Fotonica e Nanotecnologie del Consiglio Nazionale delle Ricerche, Sezione "CeFSA" di Trento, Via Sommarive 18, I-38050 Povo Trento, Italy

The potential of XANES spectroscopy for studying the structure of boron-containing compounds is assessed by theoretical modelling of B K edge spectra of minerals. Spectra arising from three-fold coordinated boron sites differ considerably from spectra arising from four-fold coordinated sites. The characteristic shape of the spectra is governed first of all by short range order and, to a lesser extent, by medium range order. Areas of main spectral peaks depend only on the coordination of the photoabsorbing site and can thus be used for quantifying the portion of boron atoms in BO_3 and in BO_4 units, in minerals as well as in glasses. The width of the main peak of spectrum at a four-fold coordinated boron site generally increases with the the spread of the B–O distances but deviations from this correlation may be substantial. In contrast to spectra of minerals, XANES of BPO_4 crystal is governed by long range order and is thus not a suitable model for studying XANES of borate glasses.

1. Introduction

Knowing the structure of borates glasses is crucial for a deeper understanding of their physical properties. This work concentrates on studying the local neighborhood around B atoms. Boron atoms participate in creating the B–O network which can be seen as a basic matrix onto which the whole complex structure of binary and ternary glasses is hooked. In a B_2O_3 glass, a B atom is coordinated by three O atoms.[(1)] In binary and ternary glasses, boron atoms occur both in BO_3 and BO_4 structural units. Nuclear magnetic resonance (NMR) spectra of ^{11}B proved to be very useful in determining the ratio of B atoms occurring in BO_3 and BO_4 environments.[(2)] It would be useful, nevertheless, to have an independent method of measuring this quantity. Moreover, NMR cannot be used in situations where the content of boron is very small, such as in thin films or in materials where boron is the dopant. Therefore, there is a need for alternative ways of determining the ratio of threefold and four-fold coordinated B atoms. Also, it would be very useful to have a technique which might be able to probe not only the short range order but the medium range order as well, informing thus about participation of B atoms in superstructural units such as boroxol rings.

X-ray absorption spectroscopy has established itself as a method for studying structure of materials which often cannot be investigated by other means. It offers a view on the local geometry around a chemically specific atom. The extended x-ray absorption fine structure (EXAFS) spectrum can usually be analysed quantitatively and information about nearest-neighbour distances and about coordination numbers can be obtained in a more or less straightforward way.[(3)] However, measuring EXAFS oscillations is often not possible in amorphous systems, due to their strong disorderrelated damping. X-ray absorption near edge structure (XANES) signal can be measured under much broader circumstances. Unfortunately, it is not possible to extract structural information from XANES in a straightforward way. Typically, one relies on comparing XANES of a compound with unknown structure with spectra of compounds where the structure is known. Employing this technique bears several dangers because the correspondence between the shape of XANES spectrum and the local structure is not unique.[(4)] Therefore, a thorough theoretical investigation of how relevant spectral features arise should be done wherever possible.

Recently, B *K* edge XANES spectra of several glasses were measured.[(5–8)] In order to use these data

[1] Corresponding author. Email address: sipr@fzu.cz

efficiently, one needs to understand the mechanism by which the XANES peaks are generated in such compounds. This can be achieved by studying theoretically XANES spectra of boron containing minerals. The local neighborhood of B atoms in these systems is, namely, similar as in borate glasses; they are coordinated by three or four O atoms. Various features in B *K* edge XANES spectra of these minerals were identified as markers of presence of either BO_3 or BO_4 units and even were employed in studying the structure of borosilicate glasses in the past.[9–11] However, these assignments were done mostly on a semi-empirical basis. Theoretical investigations were limited only to small model clusters of four or five atoms.[10] No calculations of XANES involving full geometric structure of boron containing minerals have been performed so far.

The purpose of this work is thus to explore theoretically spectra of boron containing minerals, to find out what structural elements give rise to characteristic spectral features, to compare the roles of short, medium and long range order in generating the XANES spectra and to assess the applicability of the XANES technique in studying the structure of borate glasses.

2. Theoretical method

XANES was calculated *ab initio* via a real space full multiple scattering technique,[12,13] which means that infinite crystals were substituted by finite clusters centred on the photoabsorbing B atom. Spectra presented in this work were obtained for clusters containing around 120 atoms, unless explicitly stated otherwise. We checked that this cluster size is fully sufficient – adding more atoms does not change the resulting spectra in a considerable way.

A non-self-consistent muffin tin potential constructed according to the Mattheiss prescription was mostly used. The 1s core hole was taken into account by relying on the final state rule ('relaxed and screened' approximation).[12,13] For several compounds we obtained also selfconsistent scattering potentials by performing SCF-Xα molecular calculations for clusters of about 20 atoms embedded in a crystal.[14,15] We found out that the general shape of theoretical XANES spectra does not change if selfconsistent instead of non-selfconsistent potentials are used.

3. Minerals containing only BO_3 or only BO_4 units

3.1. Comparison between theory and experiment

In order to verify the applicability of our theoretical scheme, we start by comparing theoretical and experimental XANES for minerals containing only one type of B atoms. In that way, possible ambiguity connected with superposing spectra arising from nonequivalent sites can be avoided. Figure 1 summarises our results for minerals containing boron in a BO_3 coordination (calciborite, ludwigite, vonsenite) and in a BO_4 coordination (danburite, datolite, sinhalite). The experimental spectra were taken from

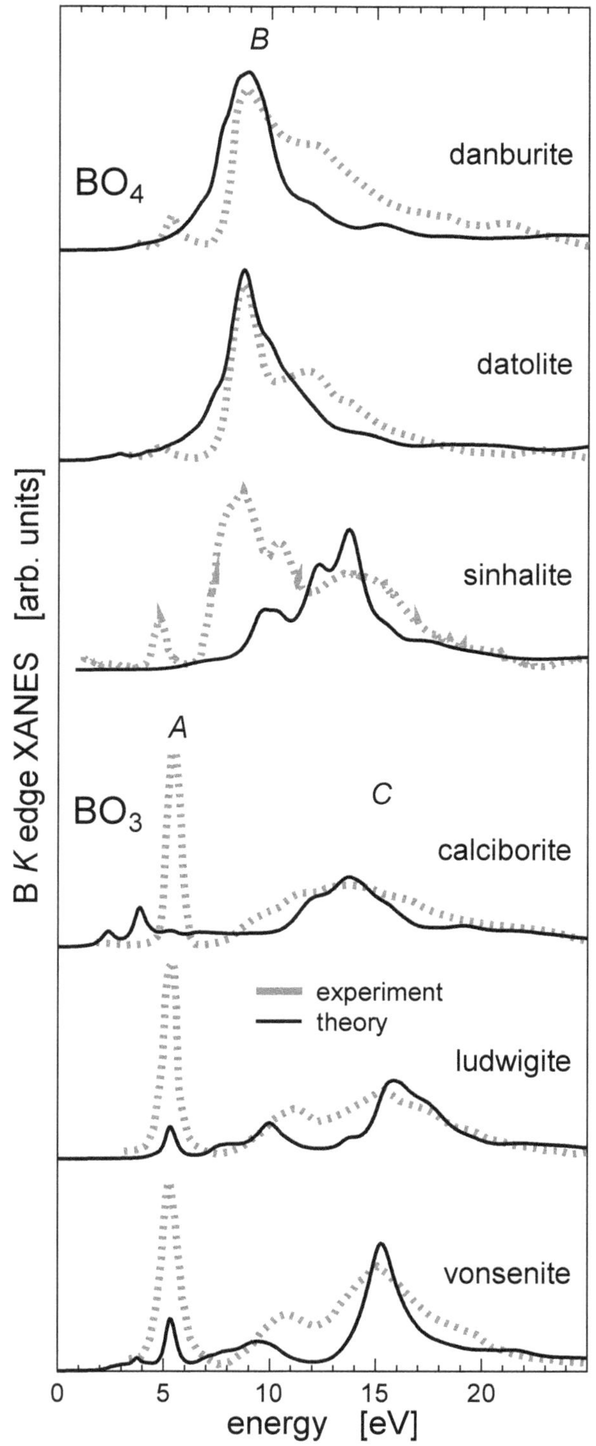

Figure 1. Theoretical (full lines) and experimental spectra (dashed lines) for minerals containing only BO_3 or only BO_4 units. The experiment is taken from Fleet et al[16,17]

earlier fluorescence yield measurements of Fleet and coworkers.[16,17] The alignment between theoretical and experimental spectra of the same compound was done so that the best agreement between them is achieved; spectra of different compounds were aligned so that corresponding experimental peak positions match. The origin of the energy scale is arbitrary.

The first thing to notice in Figure 1 is that the spectra split into two groups, according to whether the compound contains boron in BO_3 or in BO_4 units. This distinction appears both in experimental and in theoretical spectra. Spectra of minerals with BO_3 units are characterised by an intensive prepeak just before the onset of continuum states (around 5 eV in the scale of Figure 1) and by a main peak around 15 eV. Minerals with BO_4 units give rise to spectra with one single main peak at ~9 eV. We denote these peaks as A, B, and C (see Figure 1). Spectrum of sinhalite does not quite fit into this scheme; this can be attributed to the fact that coordination of boron in sinhalite is 'three plus one' rather than four (three B–O distances are 1·46 Å while the fourth distance is 1·58 Å). The intensity of the theoretical prepeak in BO_3-containing compounds is significantly smaller than the experimental prepeak. This difference probably stems from our static treatment of the core hole. We found, namely, that the calculated intensity of this prepeak and even its very occurrence depend quite a lot on the scheme used for accounting of the core-hole (see Šipr *et al*[14] and Brydson *et al*[18] for more technical details). This indicates that the prepeak is closely related to the way in which other electrons react to the presence of the core hole and its accurate description is thus not possible within a static scheme.[12,13]

Experimental spectra of BO_4-containing minerals also exhibit a small prepeak. However, it is much less intensive than in the case of BO_3-containing materials and it does not occur in theoretical spectra. This prepeak probably arises due to the presence of a small amount of threefold coordinated B atoms, which appear in these compounds as a result of radiation damage.[16,17] Note also that the samples used in the experiments were natural minerals, so they could have contained a significant amount of impurities even before that start of the measurements.

The shoulders at the high energy side of the main peak in danburite and in datolite (around ~13 eV) are not reproduced by our theory very accurately; we checked that employing selfconsistent potentials does not improve this situation. Spectra calculated for fourfold coordinated boron sites in minerals containing both BO_3 and BO_4 units do not display such features either (see below). This suggests that the highenergy shoulder observed in experimental XANES of BO_4-containing minerals could also contain a significant contribution arising from structural defects, similarly as it is the case with the pre-edge in these compounds. Deficiencies of our theoretical scheme, such as muffin tin approximation to the potential and/or one electron treatment of electron excitations, could be another source of the disagreement between theory and experiment. These effects are usually not important but their precise quantification is very difficult.[13]

3.2. Cluster size effect

Knowing the size of the region of the crystal within which the observed spectral structure is generated is essential for any physical interpretation of XANES. Therefore, spectra of all minerals displayed in Figure 1 were calculated for several cluster sizes. Representative results for datolite and for ludwigite are shown in Figure 2. One can see immediately that the basic features of XANES spectra, i.e. the single peak B for a BO_4-containing mineral datolite and the prepeak *A* and the main peak *C* for a BO_3-containing mineral ludwigite, are reproduced even for the smallest clusters employed. The fine structure between the

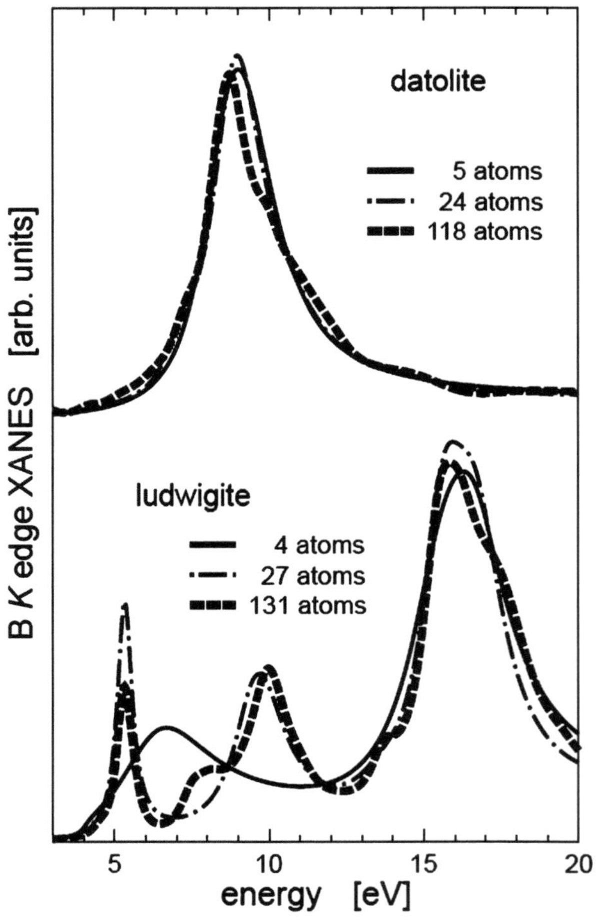

Figure 2. Spectra of datolite and ludwigite calculated for clusters of different sizes. Number of atoms in the cluster is indicated for each line type in the plot

peaks *A* and *C* in XANES of ludwigite is reproduced if around 30 or more atoms are taken into account. Because a similar behaviour with respect to cluster size can be observed for spectra of the other minerals as well, one can draw a general conclusion that the characteristic shape of XANES of minerals is generated within the basic BO_3 or BO_4 units. The spectrum is thus dominated by short range order effects. The largest influence of medium range order can be observed in the fine structure between the peaks A and C in spectra of BO_3-containing minerals.

3.3. XANES of BPO_4

Boron phosphate BPO_4 has a simple crystal structure, with boron coordinated by four oxygens. Therefore it was frequently used as a model system for studying spectra of compounds with four-fold coordinated boron.[16,17] In order to gain a better intuitive understanding of XANES of BPO_4, we calculated the B *K* edge XANES of BPO_4 and observed how it changes if the cluster size is varied. The results are summarised in Figure 3.

One can see that, unlike in the case of boron-containing minerals, XANES of BPO_4 is governed by long range order – even 17 atoms are not enough to reproduce the basic shape of experimental spectrum. The reason behind different behaviour of BPO_4 and of minerals may rest in the different symmetry of their crystal structures. The BPO_4 crystal is highly symmetrical and well defined coordination shells around boron can be resolved even at large distances. The scattering of excited photoelectron by more distant shells can thus contribute to the formation of XANES structure. On the other hand, the symmetry of boron-containing minerals is very low and the distribution of more distant atoms around boron thus resembles

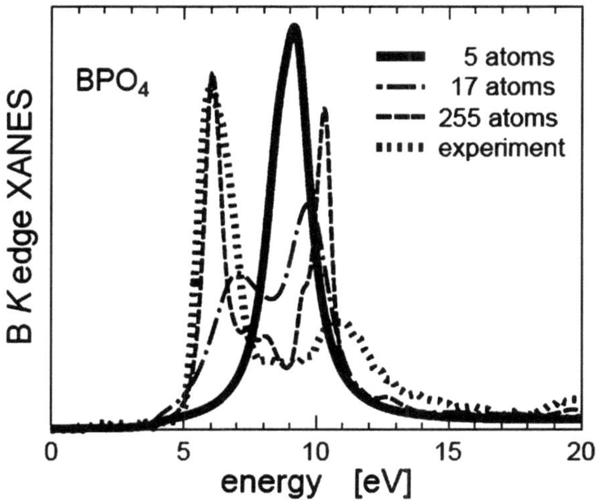

Figure 3. Spectra of BPO_4 crystal calculated for different cluster sizes (as indicated for each line type in the plot). The experiment is taken from Rocca et al[8]

rather a continuum than individual coordination shells. The phases of photoelectrons scattered by these more distant atoms are consequently nearly random and the corresponding contributions cancel each other.

Another interesting aspect is that the theoretical spectrum of the 5 atom cluster in BPO_4 looks similar to spectrum calculated for the BO_4 unit in datolite (Figure 2). That suggests that, on the short range, spectra of BPO_4 and of BO_4-containing minerals are similar. The difference between them stems from the long range order effects. The dominance of long range order over the short range order in generating spectrum of BPO_4 implies that BPO_4 is not a good model example for studying XANES of glasses.

4. Minerals containing both BO_3 and BO_4 units

X-ray absorption spectrum of any material is a superposition of spectra generated at edges of all the inequivalent photoabsorbing sites. It is quite difficult to disentangle these individual contributions from experimental XANES. In a theoretical study, however, one calculates the spectra for each of the sites separately, making it thus possible to search for features characteristic for each individual coordination.

Our study is aimed at obtaining knowledge which could be useful in studying structure of borate glasses, so a significant statistical disorder has to be anticipated. Therefore we focus not on a detailed analysis of XANES of a single compound but rather on observing spectral features which would be common to all boron sites with threefold or fourfold coordinations. That requires an analysis of a large number of spectra. Figure 4 displays theoretical spectra of ten minerals, listed in Table 1. The spectra are shown for each boron site, meaning that we have 13 curves corresponding to threefold coordinated boron and 26 curves corresponding to fourfold coordinated boron. As an aid to an eye, spectra averaged over all sites in BO_3 units and in BO_4 units are shown via thick lines. The horizontal alignment of spectra generated at different sites of the same compound was provided by the theory itself, spectra of different compounds were aligned so that positions of the peak B at ~9 eV approximately coincide. Similarly, the vertical scale for spectra calculated at different sites of the same mineral is fixed by the theory, while mutual alignment of spectra of different compounds was done so that their intensities at high energies ($E \geq 20$ eV) would coincide.

Two basic patterns can be distinguished in the spectra shown in Figure 4, according to the coordination of the photoabsorbing site (similarly as in Figure 1). Characteristic shapes of spectra generated at BO_3 and BO_4 sites are thus well separated. This clearly indicates that the gross shape of B *K* edge XANES of

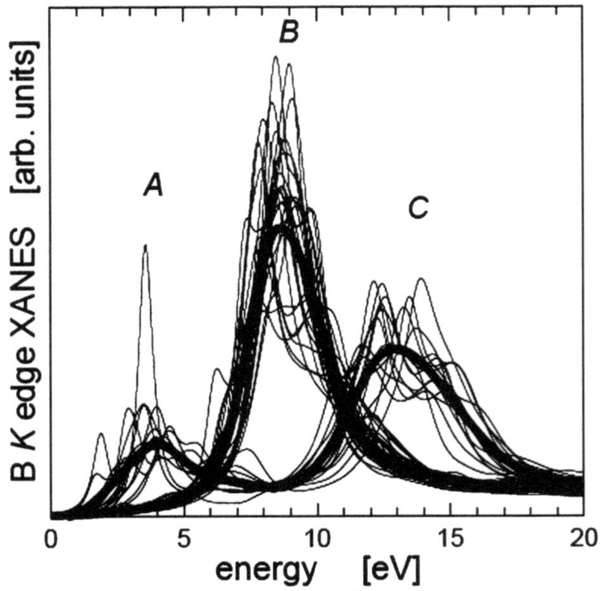

Figure 4. Spectra calculated at all boron sites in minerals listed in Table 1. Thin lines represent spectra at individual sites, thick lines show averages over all spectra at sites with BO_3 or with BO_4 coordination

these compounds is dominated by short range order. The medium and long range order, which differs from compound to compound and from site to site, affects the 'fine structure' of peaks A, B, and C but has little influence on their very existence.

It follows, therefore, that peaks B and C could be involved in studying the short range order in borate compounds. For a quantitative treatment, assessing peak areas seems to be a natural choice. We evaluated areas of peaks B and C by subtracting from the spectra the high energy background and integrating the intensities over a 10 eV energy range. It turns out that all spectra connected with a particular boron coordination, either BO_3 or BO_4, yield very similar areas of the peaks: for each of the peaks B and C, individual areas deviate from the respective average by less then 5%. However, the peak area depends on whether the spectrum relates to boron in BO_3 or in BO_4 coordination. The average area of the peak C in spectra at threefold coordinated boron sites constitute only ~75% of the average area of the peak B in spectra at fourfold coordinated boron sites. We checked that this conclusion remains valid even if various 'technical parameters' are varied (such as the scattering potential or the energy range over which the peak area is integrated).

Table 1. Minerals involved in this study that contain both three-fold and four-fold coordinated B atoms. For each compound, numbers of boron sites in BO_3 and in BO_4 coordinations are shown in the second and in the third columns. Theoretical XANES spectra of these minerals are displayed in Figure 4

Name	BO_3	BO_4
boracite	1	6
borax	1	1
colemanite	1	2
howlite	1	4
kernite	2	2
kurnakovite	1	2
meyerhofferite	1	2
probertite	2	3
tincalconite	1	1
ulexite	2	3

It appears therefore that the ratio of areas of peaks C and B could serve as a criterion for estimating the ratio of BO_3 and BO_4 units in borate minerals and in borate glasses. However, before this criterion is applied, one would have to account for the effect of structural changes due to radiation damage and due to surface treatment. There is a strong evidence that these effects may 'artificially' increase the portion of threefold coordinated boron.[16,17]

Contrary to the case of peaks B and C, the core-hole-related prepeak A depends quite a lot on the electronic structure of the material. One can expect that its intensity and area will be affected by complex many body processes, meaning that the direct inuence of the local geometry will be obscured. Therefore, although the prepeak A could serve as a marker of presence of BO_3 units, its usability in quantifying the ratio of threefold and fourfold coordinated B atoms seems to be limited.

5. Peak width and distortion of BO_4 tetrahedra

It follows from the intuitive picture of generating x-ray absorption spectra that positions of XANES peaks depend on bond lengths.[12] Hence, presence of several slightly different B–O distances ought to give rise to broader 'smeared' peaks in comparison with situation when all the B–O distances are equal. Fleet & Liu[17] suggested that the width of the peak B could serve as an indicator of the spread of the B–O distances in BO_4 tetrahedra. We can test this conjecture against our theoretical spectra. Figure 5 shows the dependence of the width of the peak B on the spread of B–O distances for all the spectra at fourfold coordinated boron sites displayed in Figures 1 and 4 and, additionally, for spectrum of axinite (which contains BO_4 units only). The width of the peaks was evaluated at the height which corresponds to 40% of the maximum peak height (after subtracting the background), the spread of the B–O distances is defined simply as the difference between the longest and the shortest bond in the BO_4 tetrahedron as obtained from the crystallographic data. One can see that sites with larger spread of B–O distances generally give rise to broader peaks. However, this correlation is not very strong. For most typical spreads of B–O distances, the widths of peak B appear to be distributed quite uniformly between 3 and 5 eV. It seems, therefore, that in most situations the main peak width is not a

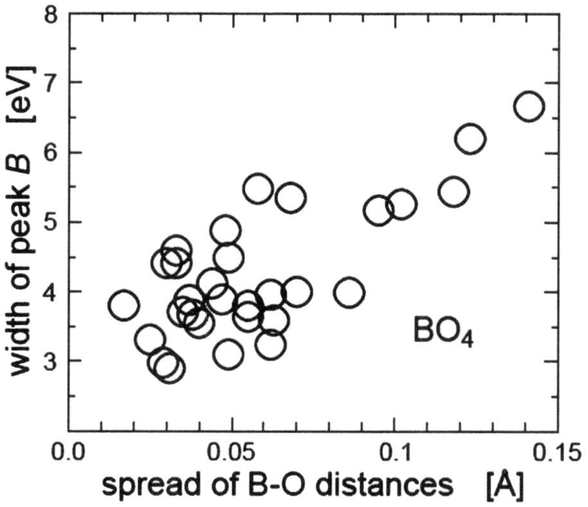

Figure 5. Dependence of the width of the main peak B of spectra at four-fold coordinated boron sites on the spread of the B–O distances in the BO_4 tetrahedron

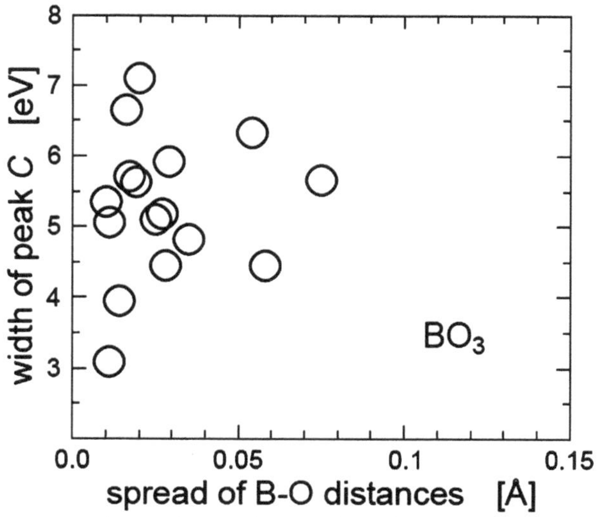

Figure 6. Dependence of the width of the main peak C of spectra at three-fold coordinated boron sites on the spread of the B–O distances in the BO_3 unit

reliable indicator of the distortion of the BO_4 tetrahedron. Rather, one could assume that exceptionally large distortions get revealed through exceptionally broad peaks.

Because of the irregular shape of the B peak, the height in which the peak width is measured may affect the outcome. The overall picture offered by Figure 5 remains essentially unchanged as long as the peak width is evaluated between 30–45% of the maximum peak height. Beyond these limits, the correlation between the peak width and the spread of the B–O distances gradually fades away.

A similar analysis was done also for the widths of the main peak C in spectra generated at the BO_3 sites (Figure 6). In that case, however, no manifest relation between the peak widths and the spread of B–O distances in the BO_3 units can be found.

6. Discussion

Our main conclusion, namely, that the B *K* edge XANES in minerals is dominated by short range order effects, was obtained under broad circumstances. It remains valid even in case that the disagreement between our calculations and the experiment at the high energy side of the peak B in XANES of danburite and datolite is caused by deficiencies of the theory (see Figure 1 and accompanying text). Namely, factors not described by our scheme, such as non-muffin tin effects or multielectron excitations, would influence the calculated spectra always in the same way, independently on the size of the cluster involved in the calculation. Therefore, it is fair to assume that the trends in cluster size effect, evident from Figure 2, would not change if a more accurate theory was used.

Our study involves boron in many crystallographic positions, in various compounds. Because of this variety of nearest boron neighborhoods, our results can be transferred from crystals to glasses. An argument in support of a local character of XANES spectra of a B_2O_3 glass was recently presented by Preobrajenski *et al*, who analysed resonant Auger spectra with photon energies covering the B *K* absorption edge.[19]

In our theoretical study, we did not address the problem of isolating the peaks B and C from the experimental spectrum of a compound with boron atoms in BO_3 as well as in BO_4 units. Such an effort will have to cope with the fact that the high energy shoulder of peak B partially overlaps with peak C. At least part of this shoulder comes from radiation damage – a process too complicated to be described by *ab initio* approaches. Therefore, in order to identify peaks B and C in experimental data, one should also rely on comparisons with experimental spectra of materials with different BO_3 and BO_4 content, recorded under similar circumstances.

Relation between the width of the peak B and the spread of B–O distances was also investigated by Fleet & Muthupari.[16] Their results based on analysing experimental spectra seem to suggest that a stronger correlation exists between these two quantities than what was found in our study (Figure 9b of Fleet & Muthupari[16] exhibits less scatter of data points than our Figure 5). However, the widths of peak B in the experimental spectra were estimated from fitting these spectra by several gaussians,[16] which involves some ambiguity. Such a procedure was obviously not needed in our theoretical study. Apart from that, our Figure 5 includes nearly twice as many data points as the corresponding plot of Fleet & Muthupari. Therefore, it seems that our study describes the relation between the width of the peak

B and the spread of B–O distances more reliably than the earlier work.

7. Conclusions

B *K* edge XANES spectra of minerals arising from sites in BO_3 units differ considerably from spectra arising from sites in BO_4 units; the characteristic shape of these spectra is governed by short range order. The prepeak appearing in spectra of three-fold coordinated boron is essentially affected by the core hole. Areas of main XANES peaks depend only on whether the photoabsorbing atom is in a BO_3 or in a BO_4 unit and can thus be used for studying structure of materials. The width of the main peak of spectra at four-fold coordinated boron sites generally increases with the spread of the B–O distances but deviations from this correlation may be substantial. In contrast to spectra of minerals, XANES of a BPO_4 crystal is governed by long-range order and is thus not a suitable model for studying XANES of borate glasses.

Acknowledgement

This work was supported by the project 202/04/1440 of the Grant Agency of the Czech Republic and by the CNR-AV CR Common Research Project 'Interplay between structural and electronic properties in nanostructures'. The research in the Institute of Physics AS CR was supported by the project AV0Z-10100521 of AS CR.

References

1. Hannon, A. C., Grimley, D. I., Hulme, R. A., Wright, A. C., & Sinclair, R. N. *J. Non-Cryst. Solids*, 1994, **177**, 299.
2. Bray, P. J. *Borate glasses, crystals and melts*. Edited by A. C. Wright, S. A. Feller & A. C. Hannon. The Society of Glass Technology, Sheffield. 1992, Pp 1–20.
3. Teo, B. K. *EXAFS: Basic Principles and Data Analysis*. Springer, Berlin, 1986.
4. Šipr, O., Dalba, G. & Rocca, F. *Phys. Rev. B*, 2004, **69**, 134201.
5. Carboni, R., Pacchioni, G., Fanciulli, M., Giglia, A., Mahne, N., Pedio, M., Nannarone, S. & Boscherini, F. *Appl. Phys. Lett.*, 2003, **83**, 4312.
6. Handa, K., Ide, J., Koganezawa, T., Ozutsumi, K., Dalba, G., Norikazu, O. & Umesaki, N. *Phys. Chem. Glasses*, 2006, **47**, 445–448.
7. Yang, G., Moebus, G. & Hand, R. J. *Phys. Chem. Glasses*, 2006, **47**, 507–512.
8. Rocca, F., Armellini, C., Boscherini, F., Carboni, R., Malvestuto, M. & Šipr, O. *Fifth Int. Conf. on Borate Glasses, Crystals and Melts*, unpublished.
9. Sauer, H., Brydson, R., Rowley, P. N., Engel, W. & Thomas, J. M. *Ultramicroscopy*, 1993, **49**, 198.
10. Garvie, L. A., Craven, A. J. & Brydson, R. *Am. Miner.*, 1995, **80**, 1132.
11. Fleet, M. E. & Muthupari, S. *J. Non-Cryst. Solids*, 1999, **255**, 233.
12. Vvedensky, D. D. *Unoccupied Electronic States*. Edited by J. C. Fuggle & J. E. Inglesfield. Springer, Berlin, 1992, Pp 139–76.
13. Natoli, C. R., Benfatto, M., Della Longa, S. & Hatada, K. *J. Synchrotron. Radiat.*, 2002, **10**, 26.
14. Šipr, O., Rocca, F. & Dalba, G. *J. Synchrotron. Radiat.*, 1999, **6**, 770.
15. Šipr, O. & Šimůnek, A. *J. Phys., Condens. Matter*, 2001, **13**, 8519.
16. Fleet, M. E. & Muthupari, S. *Am. Miner.*, 2000, **85**, 1009.
17. Fleet, M. E. & Liu, X. *Phys. Chem. Miner.*, 2001, **28**, 421.
18. Brydson, R., Vvedensky, D. D., Engel, W., Sauer, H., Williams, B. G., Zeitler, E. & Thomas, J. M. *J. Phys. Chem.*, 1988, **92**, 962.
19. Preobrajenski, A. B., Vinogradov, A. S., Kleimenov, E., Knop-Gericke, A., Krasnikov, S. A., Szargan, R. & Mårtensson, N. *Phys. Script.*, 2005, **T115**, 1071.

Proc. Fifth Int. Conf. on Borate Glasses, Crystals and Melts Phys. Chem. Glasses: Eur. J. Glass Sci. Technol. B, August 2006, 47 (4), 419–429

Structure and dynamics of ionic borate glasses

C. P. E. Varsamis,[1] A. Vegiri & E. I. Kamitsos

Theoretical and Physical Chemistry Institute, The National Hellenic Research Foundation, 48 Vass. Constantinou Ave., 116 35 Athens, Greece

The structural and dynamic properties of ionic borate glasses with composition $xM_2O–(1-x)B_2O_3$ *(M=Li, Cs) are revisited in view of recent results of molecular dynamics simulations. Simulations were carried out in collections of ~512 atoms in the canonical ensemble for compositions* $0·20 \leq x \leq 0·50$ *and temperatures 300, 600 and 1250 K. The short range order structure of these glasses was found to depend on temperature and on the alkali oxide modifier content, as well as on the nature of the alkali ion. The structural trends resulted from simulations are compared with those derived from different experimental techniques. At the nanoscale size, it was possible to distinguish long lived and well distinguished network sites hosting alkali metal ions. The distribution of cations in the glassy network was found to be inhomogeneous and consistent with the predictions of the modified random network model. Cation dynamics were evaluated in both short and long time scales. In particular, the vibrational properties of cations were found to depend strongly on the nature of their hosting sites, whereas their diffusivity is facilitated by the presence of nonbridging oxygen (NBO) atoms. The detailed analysis of the ion transport mechanism revealed discrete hopping of ions mostly between similar neighbouring sites. The calculated cluster size distribution curves for cations and NBO atoms were found to obey the predictions of percolation theory. The formation of percolating clusters above a certain composition threshold supports strongly the existence of a percolating diffusion mechanism for ionic borate glasses.*

Introduction

Ionic borate glasses are characterised by the continuous non-monotonic variation of their structural, dynamic and macroscopic properties with increasing the modifier oxide content. This phenomenon, known as the "borate anomaly" effect, has been the subject of numerous investigations.(1) The detailed knowledge of the dependence of their structural and dynamic aspects on composition is highly desirable in order to correlate them with their macroscopic properties. The establishment of such correlations is of great importance since these materials have potential applications in electrochemical devices.(2)

The structural properties of ionic borate glasses have been extensively investigated over the years. Early experimental studies including infrared,(3) NMR(4–6) and Raman(7) investigations concluded that the structure of vitreous B_2O_3 consists of neutral $BØ_3$ triangles found in random configurations and in B_3O_6 boroxol ring formations (Ø denotes an oxygen atom bridging two boron centres). Addition of alkali oxide content to vitreous B_2O_3 results in the progressive transformation of borate triangles into charged tetrahedral entities, $BØ_4^-$.(3–7) It was further suggested that in alkali borate glasses, $xM_2O–(1-x)B_2O_3$ (M=alkali ion), for compositions with $x<0·30$ the fraction N_4 of tetrahedral units follows the $N_4=x/(1-x)$ law.(4–6) For higher compositions, there is the progressive depolymerisation of the borate network by the formation of charged triangles with nonbridging oxygen (NBO) atoms, like metaborate, $BØ_2O^-$, pyroborate, $BØO_2^{2-}$, and orthoborate, BO_3^{3-} triangles.

Despite the fact that the aforementioned structural mechanisms in ionic borate glasses are widely accepted, thus providing the basis for the explanation of the "borate anomaly", the dependence of the short range order (SRO) structure on the nature of the modifier ion is still a controversial issue. On one hand, early NMR studies concluded that the SRO structure has no dependence on the alkali type(4–6) while on the other, a plethora of early experimental measurements including viscosity,(8) absorption edge(9) optical basicity,(10,11) glass transition temperature, thermal expansion coefficient,(12,13) as well as sound velocity and elastic constant measurements(14,15) showed the clear dependence of the SRO structure on the nature of the alkali modifier. Along the same lines, spectroscopic investigations employing ^{11}B NMR,(16) Raman,(17) infrared,(18) DAS NMR(19) and neutron scattering, as well as molecular dynamics simulations(21–23) provided also to the dependence of the SRO structure on the type of the alkali ion. The controversy is still going on, since recent ^{11}B MAS-NMR investigations concluded that the type of alkali metal does not affect the network structure,(24,25) thus leaving this issue open for further investigations.

Besides the SRO network structure, an equally important structural aspect is the identification of

[1] Corresponding author. Email: cvars@eie.gr

the local structures hosting the metal cations and the knowledge of cation site interactions in borate glasses. Early ideas, based on the continuous random network model proposed by Zachariasen[26] and applied to alkali silicate glasses by Warren & Biscoe,[27,28] suggested the homogeneous distribution of alkali metal ions in voids or sites of the silicate network. On the contrary, on the basis of EXAFS studies in sodium silicate glasses, Greaves proposed the inhomogeneous distribution of metal ions in the silicate network, i.e. the so-called modified random network model.[29]

In ionic borate glasses, early studies suggested that metal ions are not homogeneously distributed in the glassy network.[30–33] Further evidence for structural inhomogeneity at the nanoscale size was provided by far infrared spectroscopy since vibrations of metal ions in their hosting sites are active in the far infrared region.[34–36] In borate glasses, as well as in most ionic glasses, the asymmetric far infrared profiles showed the existence of at least two distinct environments of cation hosting sites.[37–41] The so-called "two site" model has been initially attributed to the ability of the borate network to provide cation hosting sites with variable charge density and coordination number.[39] In a different viewpoint, however, the far infrared absorption of sodium borate glasses was assigned to vibrational modes of Na ions in a single type of hosting sites.[42] Differently again, Wright and coworkers concluded that the rigidity of borate superstructural units along with the lack of periodicity in the glass structure should lead normally to a considerable variation of cation hosting sites in terms of site distortion and coordination numbers.[43]

It is clear from the aforementioned discussion that the microstructure of borate glasses and the nature and distribution of metal ion hosting sites as well as their spatial distribution constitute an issue of open discussion. At the same time, this information is particularly valuable towards a better understanding of transport properties in ionic glassy materials. In fact, different models for ionic transport in glasses proposed so far involve, in general, the nature and distribution of metal ion hosting sites. For example, theoretical studies of the ionic conductivity mechanism were based on the existence of distinct cation sites and preferred pathways for ion transport,[44] microsegregation of cations in channels suitable for ion migration,[29] or cation site memory effects.[45] Moreover, recent approaches of the ion transport mechanism in single[46–48] or mixed[49] alkali glasses invoked the 'two site' model.

In the present manuscript we review recent results of molecular dynamics simulations in Li- and Cs-containing borate glasses. Emphasis is given on the dependence of the SRO structure on the composition, temperature and nature of the metal ion and the

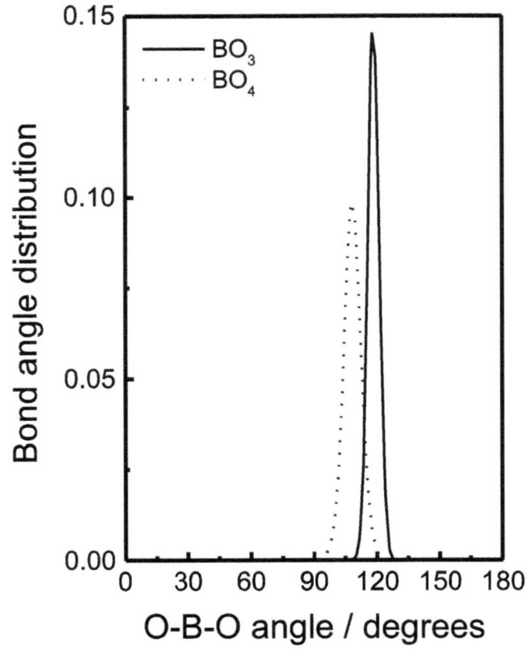

Figure 1. Distribution function of O–B–O angles for glass $0{\cdot}3Li_2O.0{\cdot}7B_2O_3$ *at* T=300 K

results are compared with experimental NMR and infrared findings. Then the microstructure of the metal ion hosting sites is investigated according to the type of structural borate unit with which they interact primarily. Moreover, we explore the possibility to distinguish different types of cations according to the particular environment in which they predominantly reside. This information is further elaborated to calculate the dynamic responses of ions in both the short and long time regime. Finally, we report also on the ionic conductivity mechanism based on the spatial distribution of cations in the glassy network and on their dynamical responses. The results are discussed in the framework of existing experimental or theoretical dynamic studies of ionic borate glasses.

Computational details

Molecular dynamics simulations were performed for $xM_2O-(1-x)B_2O_3$ glasses, with x=0·2–0·6 for M=Li and x=0·2–0·4 for M=Cs, and temperatures 300 K and 1250 K. Structures consisting of ~256 atoms (M=Li) and ~512 atoms (M=Cs) in a primitive cubic cell have been generated. For each composition, we used appropriate speciation of atoms whereas the lattice constant of the cubic cell was determined from experimental density data, as reported in detail in Ref. 50. The interaction potential was of the usual Born–Mayer–Huggins form augmented by an additional three-body harmonic interaction term for the O–B–O angles. We have employed the potential form and the values of the relevant parameters reported previously,[51] where a partial screening of the electrostatic charges for B and O atoms was found necessary for reproduction of experimental infrared and Raman spectra.

The simulations were carried out in the canonical ensemble with periodic boundary conditions, while long range Coulomb forces were handled by employing the Ewald summation method.[52] A fifth order Gear predictor–corrector integrator with a time step of 0·96 fs was used and the randomly generated melts at ~6000 K were quenched to the final temperature in five cooling cycles. Each cycle consisted of a fast cooling step of about 2 ps and an equilibration period of about 10 ps. The equilibration period in the final cooling cycle was 30 ps, whereas positions and velocities of each particle were accumulated for a final period of more than 180 ps in order to get reliable statistics for the diffusive ionic motion. Further details of the computational method can be found in Refs. 23, 50, 53.

Results and discussion

Short range order structure of Li- and Cs-containing borate glasses

The detailed mapping of the short range order structure of the glasses was determined by examining the B–O radial distribution functions, $g_{B-O}(R)$, at each glass composition and temperature investigated. Integration of these functions provides directly the average boron–oxygen coordination numbers, <B(O)>, from which it is possible to determine the molar fractions of the borate polyhedra. In doing so, we first examined the O–B–O angle distribution functions, a typical example of which is reported in Figure 1, for the $0{\cdot}3Li_2O{-}B_2O_3$ composition at T=300 K. As shown, these functions exhibit two sharp peaks at ~109·5° and 120° which are characteristic for the presence of tetrahedral and triangular borate units. It is also underlined that the profiles of these functions remain unaffected for each composition and temperature studied suggesting that borate tetrahedral and triangles constitute the only structural entities encountered in the present simulated glasses.

In order to quantify the SRO network structure, the molar fractions of tetrahedral borate units, X_4, were determined from the integration of the $g_{B-O}(R)$ functions by the expression X_4=<B(O)>−3. The results are depicted in Figure 2 for T=300 and 1250 K and compared with the theoretical curve $x/(1-x)$ and existing room temperature NMR data for Cs-[16,24] and Li-borate glasses.[16] As seen, simulated X_4 values are systematically lower than the experimental ones. This effect is attributed to the higher quenching rates of simulated glasses which result in the enhancement of the molar fractions of triangular units with nonbridging oxygen (NBO) atoms and, consequently, in lower X_4 values.[51,54,55] Inspection of Figure 2 shows that for the x=0·2 composition, room temperature simulated X_4 values for Li- and Cs-borates are practically the same and present the smallest deviation from both NMR data and the theoretical $x/(1-x)$ value. This finding indicates that the effect of the cooling rate on the SRO structure becomes less pronounced whenever the fraction of NBO-containing units is small. This is in close agreement with a recent temperature Raman study of K-borate glasses which demonstrated that the Raman profiles of the x=0·3 glass/melt are much more sensitive to temperature changes than those of the x=0·2 glass/melt.[56]

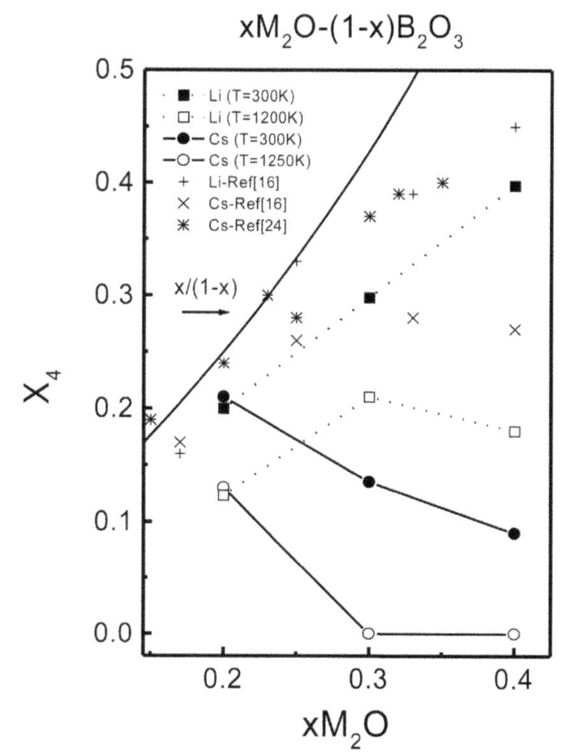

Figure 2. Molar fraction of $BØ_4^-$ tetrahedral units, X_4, in simulated Cs- and Li-borate glasses as a function of composition and temperature. Experimental room temperature NMR data reported for Li-[16] and Cs-borate glasses[16,24] are also included. The x/(1−x) curve gives the X_4 value if $BØ_4^-$ tetrahedra are the only charged units existing in borate glasses

For compositions with x=0·3, simulated X_4 values for Cs-borates are significantly lower than those for Li-containing glasses with the same alkali oxide content. This trend is consistent with the NMR data of Ref. 16 and is in contrast with those of Ref. 24. It is noted that differences in NMR data through the years have been attributed recently to variations in the quenching rates employed in different studies.[25] Within this context, it is underlined that the present simulations for Li and Cs-borate glasses were carried out under the same cooling procedure and, thus, whatever similarities (i.e. for x=0·2) or differences (for x≥0·3) were observed in the network structure between the two simulated glass systems cannot be

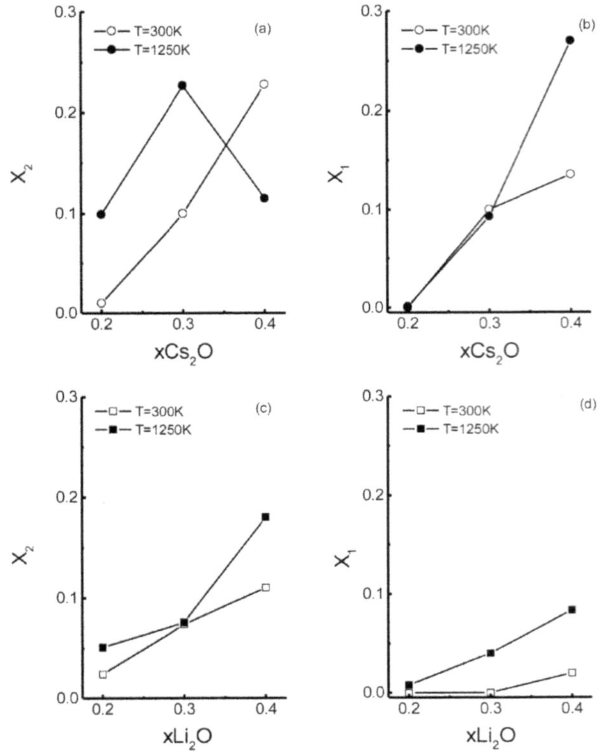

Figure 3. *Molar fractions of $B\varnothing_2O^-$, X_2, and $B\varnothing O_2^{2-}$, X_1, units in simulated Cs- and Li-borate glasses as a function of alkali content and temperature*

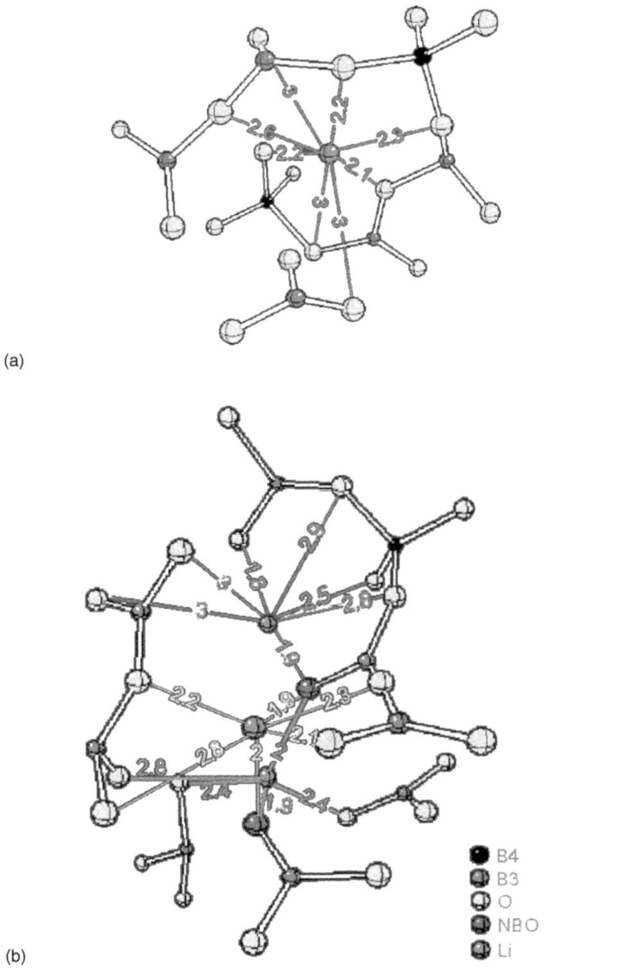

Figure 4. Simulated structures for Li hosting sites at T=300 K *for the* $0{\cdot}3Li_2O.0{\cdot}7B_2O_3$ *glass: (a) b-type site, and (b) nb-type site. Distances between Li ions and oxygen atoms in the first coordination shell within a radius of 3 Å are included.* B_4 *and* B_3 *denote fourfold and threefold coordinated boron atoms while BO and NBO denote bridging and nonbridging oxygen atoms, respectively*

attributed to cooling rate variations.

Deviations of X_4 values from the theoretical value $x/(1-x)$ indicate the presence of charged triangular borate units with NBO atoms. In order to determine their molar fraction, an oxygen atom was identified as nonbridging if no other boron atom was found in its vicinity within the first minimum of the B–O radial distribution curve. This minimum at ~2 Å was found to be independent of temperature and composition. This procedure provides the total number of NBO atoms, while the number of NBO atoms per charged triangular entity can be determined by their average charge evaluated from the integration of the first peak of the corresponding NBO–NBO correlation function, which gives the average number of NBOs seen by an NBO atom on the same unit.[50] The calculated molar fractions X_2 and X_1 for charged triangular borate units with one, $B\varnothing_2O^-$, or two NBO atoms, $B\varnothing O_2^{2-}$, respectively, are reported in Figure 3 for Li- and Cs-borate glasses at T=300 and 1250 K. As expected, for x=0·2 both systems have similar values while for higher compositions X_2 and X_1 are systematically enhanced in Cs-containing glasses.

The results reported in Figures 2 and 3 clearly show that the SRO structure depends on the temperature as well. In particular, for x=0·2, the temperature increase results in the transformation of $B\varnothing_4^-$ tetrahedral into charged $B\varnothing_2O^-$ triangular entities. Such a transformation can be fully described by invoking the following isomerisation reaction

$$B\varnothing_4^- \Leftrightarrow B\varnothing_2O^- \qquad (1)$$

which shifts to the right with increasing temperature. This finding is fully consistent with the results of statistical mechanical calculations[57] and of spectroscopic NMR,[58,59] infrared,[41] x-ray diffraction[60] and Raman[56] investigations which showed that the NBO-containing triangular units of modified borate glasses are favoured at high temperatures at the expense of $B\varnothing_4^-$ tetrahedra.

For compositions higher than x=0·2, increasing temperature results in the conversion of $B\varnothing_4^-$ tetrahedra into $B\varnothing_2O^-$ and $B\varnothing O_2^{2-}$ units. The transformation into $B\varnothing_2O^-$ units is described with Equation (1) while the creation of $B\varnothing O_2^{2-}$ units may result from the fol-

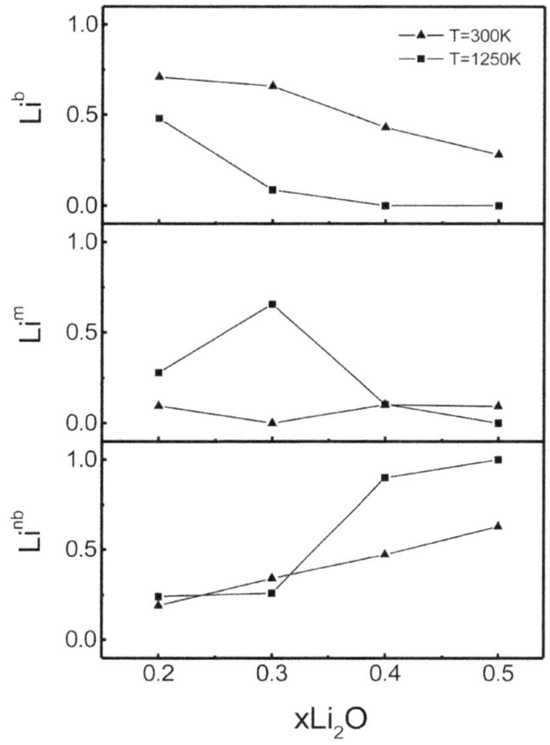

Figure 5. Molar fractions of Li metal ions in b-type, Li^b, nb-type, Li^{nb}, and mixed type, Li^m, environments. For details see text

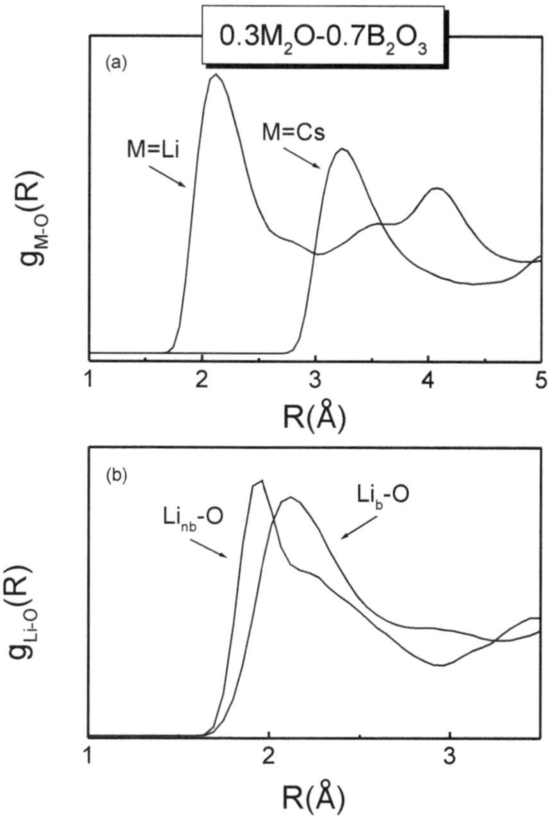

Figure 6. Comparison of the: (a) total M–O pair radial distribution functions, and (b) Li^{nb}–O and Li^b–O partial radial distribution functions, for simulated glasses with composition 0·3M$_2$O.0·7B$_2$O$_3$ at T=300 K (M=Li, Cs)

lowing disproportionation reaction

$$2BØ_4^- \Leftrightarrow BØO_2^{2-} + BØ_3 \qquad (2)$$

which was invoked in previous infrared investigations of magnesium borate glasses.[61]

Microstructure of metal ion hosting sites

Cation hosting sites are formed by oxygen atoms provided by the borate structural units. In order to maintain charge neutrality, metal ions are expected to be found in the vicinity of negatively charged units. As previously described, in the simulated glasses under investigation the charged structural moieties can be either borate triangles with NBO atoms or tetrahedra where all oxygen atoms are bridging. This classification provides a suitable criterion to distinguish metal ions sites according to the nature of oxygen atoms participating in their formation. It is noted that besides charged borate units, neutral borate triangles will be involved in the formation of cation hosting sites, as well. Analysis of the results showed that for all simulated glasses it is possible to distinguish long lived and well defined distinct sites in the borate network. In particular, there are sites formed exclusively by bridging oxygen atoms of $BØ_4^-$ and $BØ_3$ units, designated as bridging type sites (b-type), and sites consisting of NBO atoms of $BØ_2O^-$ or $BØO_2^{2-}$ units and bridging oxygen atoms of neutral $BØ_3$ entities, denoted as nonbridging type sites (nb-type).[23,50,53] Typical snapshots of b-type and nb-type hosting sites are depicted in Figure 4 in the case of the 0·3Li2O–0·7B$_2$O$_3$ glass at room temperature. The existence of different cation hosting sites implies the occurrence of structural heterogeneities at the nanoscale length in borate glasses.

Within this distinction, metal ions can be distinguished as M^b or M^{nb} if they predominantly reside (for more than 75% of the total simulation time) into b-type or nb-type sites, respectively. Ions that do not fulfil this criterion are classified as mixed type ions, M^m. Representative molar fractions of Li^b, Li^{nb} and Li^m ions are presented in Figure 5 for *T*=300 and 1250 K.[50] It is clear that the population of Li^{nb} increases with increasing the Li$_2$O content and the temperature for *x*=0·4 and 0·5. These changes are fully consistent with the increased number of NBO atoms with composition and temperature. Similar effects were also found in the case of simulated Cs-containing borate glasses.[23] It was also found that the rate of increase of M^{nb} ions as a function of composition is systematically higher than the one of the total number of NBO atoms, suggesting that each NBO atom can be coordinated to more than one metal ions. A similar effect was observed in molecular dynamics simulations of sodium silicate glasses.[62,63] The implications of this finding will be discussed in the forthcoming sections.

The classification of metal ions into M^b and M^{nb} allows further evaluation of differences in the metal ion–oxygen bonding characteristics in b- and nb-type of sites. This is effected by the calculation of the relevant pair radial distribution functions $Li^{nb}(O)$ and $Li^b(O)$ which are depicted in Figure 6 for the $0{\cdot}3Li_2O$–$0{\cdot}7B_2O_3$ glass at room temperature along with the total alkali metal–oxygen pair radial distribution functions $g_{M-O}(R)$ for both types of ions under same composition and temperature. As shown, the total $g_{M-O}(R)$ functions reach their maximum at 3·25 Å for M=Cs and at 2·12 Å for M=Li suggesting that the average metal ion–oxygen coordination is higher for Cs ions, in agreement with the size difference of Cs and Li ions. On the other hand, the Li^{nb}–O function peaks at ~1·92 Å and is better defined than the partial L^{ib}–O function which peaks at 2·12 Å. Integration of these partial distribution functions provides the coordination numbers of oxygen around Li^{nb} and Li^b ions, which were found to be about 5·5 and 8, respectively. This difference shows that nb-type sites are more organised and better defined compared to b-type sites.[50,53] Similar trends were found to be valid also in Cs-containing glasses, with Cs^{nb} and Cs^b coordination numbers being equal to ~9·2 and 11·6, respectively. In this case the partial Cs^{nb}–O and Cs^b–O radial distribution functions, although differing in their profiles, show their first peak at approximately the same position as the total Cs–O RDF.[23]

Short- and long-time dynamics of metal ions

The classification of metal ions into b- and nb-types, based on different structural characteristics encountered in their hosting sites, is expected to be reflected in distinct dynamic responses in both the short- and long-time regime. To this aim, short-time dynamics were further explored by considering the vibrational properties of metal ions as derived by the velocity autocorrelation function, Φ(t)

$$\Phi(t) = \left\langle \frac{1}{N} \sum_{j=1}^{N} \vec{\upsilon}_j(t) \cdot \vec{\upsilon}_j(0) \right\rangle \tag{3}$$

where $\vec{\upsilon}_j(t)$ is the velocity of metal ion j at time t and N is the total number of metal ions. Previous molecular dynamics studies showed that the vibrational density of states (VDOS) can be obtained through the Fourier transform of $\Phi(t)$.[21,64] Following this procedure, we calculated separately the $\Phi(\omega)$ power spectra for M^{nb} and M^b metal ions. The results for the power spectra in the $0{\cdot}3Li_2O$–$0{\cdot}7B_2O_3$ and $0{\cdot}4Cs_2O$–$0{\cdot}6B_2O_3$ glasses are reported in Figures 7 and 8, respectively, and compared with the corresponding far infrared experimental profiles of $0{\cdot}3Li_2O$–$0{\cdot}7B_2O_3$ and $0{\cdot}33Cs_2O$–$0{\cdot}67B_2O_3$ glasses.

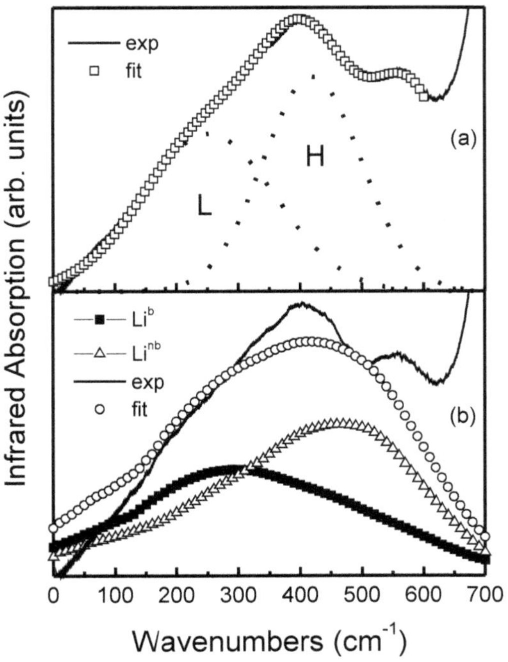

Figure 7. (a) Experimental far infrared spectrum of the $0{\cdot}3Li_2O.0{\cdot}7B_2O_3$ glass deconvoluted into Gaussian components (denoted by L and H). (b) Calculated power spectra of Li ions residing in bridging-type (Li^b) and nonbridging-type (Li^{nb}) sites in the simulated $0{\cdot}3Li_2O.0{\cdot}7B_2O_3$ glass at T=300K. A linear combination of the power spectra is compared with the experimental far infrared spectrum of the same composition

As seen, the responses associated with M^{nb} and M^b ions lie in the far infrared range and are quite broad. The important result is that power spectra of M^{nb} and M^b cations are clearly separated, with the spectral weight of M^{nb} ions being always transferred at higher frequency values. Moreover, there is a close correspondence between the simulated power spectra and the H and L Gaussian component bands into which the experimental far infrared absorption coefficient profiles were deconvoluted.[37–40] Thus, the molecular dynamics results strongly support the "two-site" model and provide for the first time the microscopic origin of the H and L component bands. Specifically, the H component band originates mostly from vibrations of ions in nb-type sites, whereas vibrations of cations in b-type sites contribute mostly to the intensity of the L component. It is noted that these results are also consistent with earlier propositions which attributed the origin of band H to M–O vibrations in sites of smaller coordination number and larger charge density compared to the sites contributing to the L component band.[38,39]

In the long-time regime, the dynamic properties of ions were investigated at T=1250 K. This is due to the fact that at room temperature hopping events

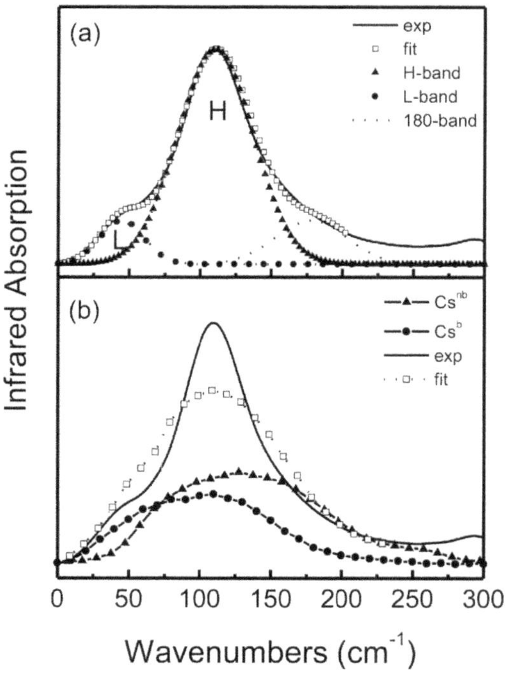

Figure 8. (a) Experimental far infrared profile of the $0{\cdot}33Cs_2O.0{\cdot}67B_2O_3$ glass deconvoluted into Gaussian components (denoted by L and H) originating from Cs–O vibrations in different local environments. The third component band at ~180 cm^{-1} was attributed[38] to vibration modes of disconnected segments of the borate network. (b) Calculated power spectra of Cs ions in bridging-type (Cs^b) and nonbridging-type (Cs^{nb}) sites in the simulated $0{\cdot}4Cs_2O.0{\cdot}6B_2O_3$ glass at T=300 K. A linear combination of the power spectra is compared with the experimental far infrared spectrum of the $0{\cdot}33Cs_2O.0{\cdot}67B_2O_3$ glass

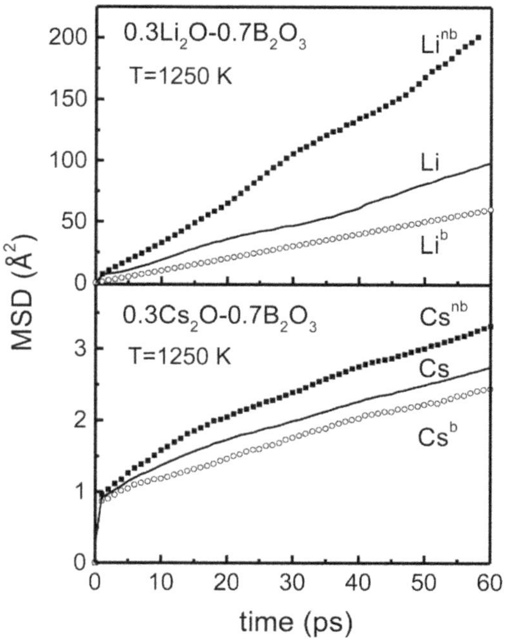

Figure 9. Mean square displacements of Li and Cs ions at T=1250 K in $0{\cdot}3Li_2O.0{\cdot}7B_2O_3$ and $0{\cdot}Cs_2O.0{\cdot}7B_2O_3$ glasses. Li and Cs denote the average behaviour of all corresponding ions while M^b and M^{nb} indicate MSD curves of alkali ions in b- and nb-type sites, respectively

are rare, and extremely long computational times would be required in order to get reliable results. The relevant quantity that represents the diffusive properties of metal ions is the mean square displacement (MSD) function, $\langle r^2(t)\rangle$. Typical MSD curves for compositions $0{\cdot}3Li_2O–0{\cdot}7B_2O_3$ and $0{\cdot}3Cs_2O–0{\cdot}7B_2O_3$ are plotted in Figure 9 and include the responses of all metal ions as well as of b-type and nb-type ions. It is noted that the corresponding curves for boron and oxygen atoms at 1250 K have no detectable slopes, suggesting that only metal ions contribute to the diffusive properties of glasses. Though the plots in Figure 9 are on a linear scale for reasons of clarity, it was found[50] that, in the log–log representation, the MSD curves exhibit the three typical time regimes described for instance by Funke,[65] i.e. the short, dispersive and long time regimes. As shown in Figure 9, nb-type cations are significantly more mobile than b-type ions suggesting that the diffusion process is mostly NBO-assisted. Such a statement appears to contradict physical intuition since the interaction of metal ions with NBO atoms is stronger than that with tetrahedral $BØ_4^-$ units (where the negative charged is delocalised) and one should therefore expect their diffusivity to be lower. This paradox can be resolved by invoking two different arguments: one is the greater free volume available around NBO-containing units with respect to the more compact space around borate tetrahedra, and the second is the tendency of NBO atoms to aggregate. This latter argument, which will be considered explicitly below, originates from the fact that the <NBO–NBO> coordination numbers were found to be higher than one for all compositions studied.[23,50] This tendency towards NBO aggregation results in the overall lowering of the intervening energy barrier as seen in previous model calculations.[66] The finding that M^{nb} ions are more mobile is consistent with experimental optical basicity[67,68] and ion exchange studies[69] which showed that nb-type of sites facilitate metal ion transport more effectively than the b-type sites.

Ionic conductivity mechanism in ionic borates

In order to address the mechanism of ionic transport in ionic borate glasses we first calculated the self-van Hove (SVH) correlation functions $G_s(r,t)$ given by the expression

$$G_s(\vec{r},t)=\frac{1}{N}\left\langle \delta\left[\vec{r}-\left|\vec{r}_i(0)-\vec{r}_i(t)\right|\right]\right\rangle \tag{4}$$

where the average is extended over all N particles and time origins. The SVH functions represent the

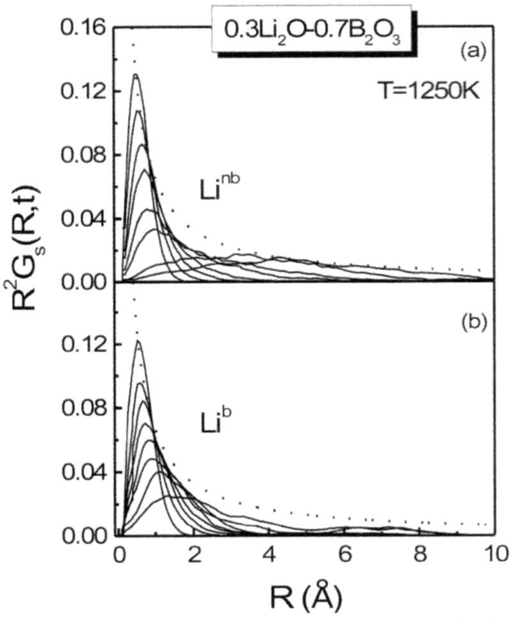

Figure 10. Self van Hove correlation functions for (a) Li^{nb} and (b) Li^{b} ions in $0{\cdot}3Li_2O.0{\cdot}7B_2O_3$ glass at T=1250 K and times 0·048, 0·096, 0·29, 0·82, 3·87, 7·74, 19·35 and 30·96 ps, in ascending order from the up left to the bottom right curve. The loci of maxima for random walk diffusion are represented by dotted curves

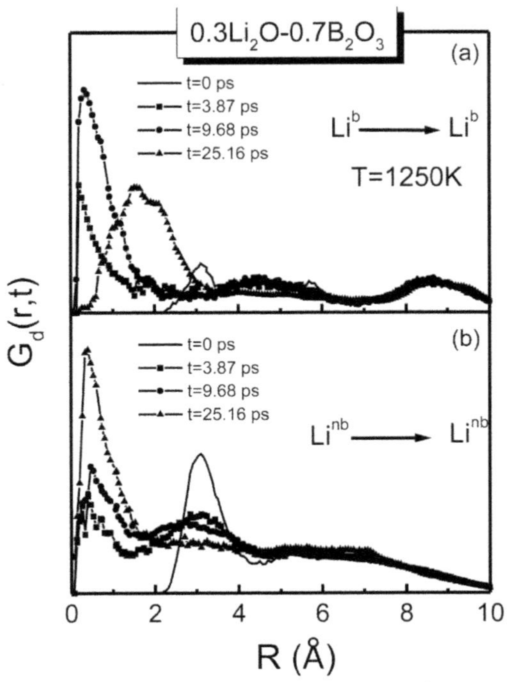

Figure 11. Distinct van Hove correlation functions for hopping processes into similar sites in $0{\cdot}3Li_2O.0{\cdot}7B_2O_3$ glass at T=1250 K and selected times in ps: (a) $Li^{b}\rightarrow Li^{b}$, and (b) $Li^{nb}\rightarrow Li^{nb}$

probability of finding a single particle at a position *r* at time *t*, given that it was at the position *r*=0 at time *t*=0. These functions were calculated separately for M^b and M^{nb} ions and typical examples are plotted in Figure 10 for the $0{\cdot}3Li_2O–0{\cdot}7B_2O_3$ composition at *T*=1250 K. In the same figure, the loci of maxima of Gaussian SVH curves corresponding to an ideal random walk motion have been also included.[70] Inspection of Figure 10 shows that for both types of ions and for most times the position of the maxima of their SVH correlation curves lags behind the corresponding Gaussian SVH functions suggesting that the diffusion mechanism differs from the random walk motion. Instead, this time lag signifies the occurrence of structural arrests between successive hopping of ions which is consistent with a jump diffusion mechanism.

It is also interesting to investigate the paths followed by individual cations in their diffusive motion through the glassy network. This information can be obtained from the distinct van-Hove (DVH) correlation functions $G_d(r,t)$, expressed as

$$G_{\mathrm{d}}^{\alpha\beta}(\vec{r},t)=\frac{1}{N_\alpha}\sum_{i=1}^{N\alpha}\sum_{j=1}^{N_\beta}\left\langle\delta\left[r-r_i^{\alpha}(0)+r_j^{\beta}(t)\right]\right\rangle \qquad (5)$$

where α and β denote either M^{nb} or M^b cations. The DVH functions is the probability to find an ion at time t at a distance r from the position occupied by another ion at *t*=0. The terms for $\alpha=\beta$ or $\alpha\neq\beta$ correspond to hopping of cations into similar or dissimilar sites, respectively. Typical plots of DVH curves are depicted in Figure 11 for jumps between similar sites for Li^b and Li^{nb} ions for the $0{\cdot}3Li_2O–0{\cdot}7B_2O_3$ composition at *T*=1250 K. It is evident that in both cases the initial profile at *t*=0 relaxes very fast and the intensity is progressively building up near the origin. This finding shows that cations diffuse by hopping to empty sites occupied previously by ions of the same type suggesting that sites retain their identity for a certain time, as pointed out in terms of a unified theory for ion migration in glasses.[45] The probability to have hopping events between dissimilar sites is depicted in Figure 12 with the DVH curves for $\alpha\neq\beta$ derived for the $0{\cdot}3Li_2O–0{\cdot}7B_2O_3$ composition at *T*=1250 K. Inspection of Figure 12 shows that there is a substantial probability for Li^b ions to hop into the nb-type of sites, whereas the inverse process has negligible probability. Such a difference can be traced to both topological and energetic arguments. On one hand, the tendency of NBO atoms to aggregate facilitates the migration of M^{nb} ions to similar rather than to dissimilar sites while on the other, aggregation results in the lower potential energy of metal ions in these sites.[71] Thus, the hopping process from a higher energetically state (b-type site) to a state of lower energy (nb-type site) is more probable than the inverse process.

Due to the important role of NBO atoms in the ion migration process, we further considered their spatial distribution in the glassy network. To this aim, we

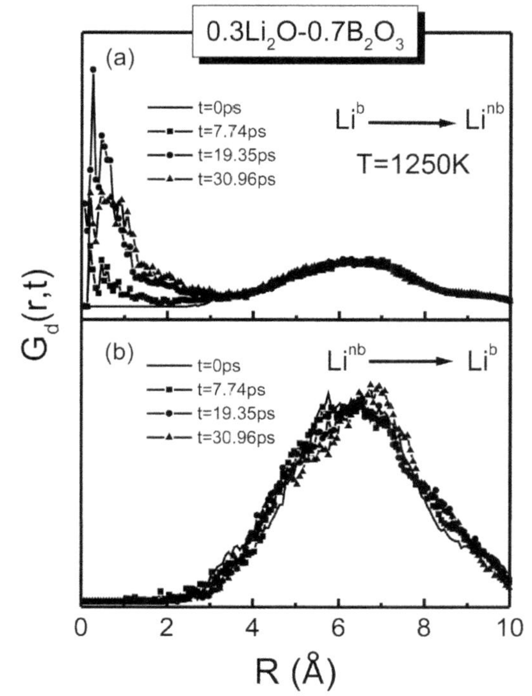

Figure 12. Distinct van Hove correlation functions for hopping processes into dissimilar sites in 0·3Li_2O.0·7B_2O_3 glass at T=1250 K and selected times in ps: (a) $Li^b \rightarrow Li^{nb}$, and (b) $Li^{nb} \rightarrow Li^b$

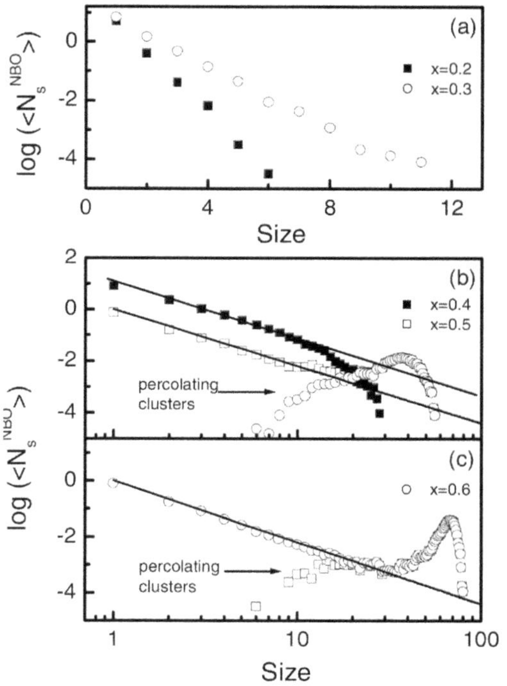

Figure 13. Time averaged distributions curves of clusters of NBO atoms with size s, $<N_s^{NBO}>$, in Li-borates as a function of cluster size (s) for (a) x=0·2, 0·3 in a semi-logarithmic scale, (b) x=0·4, 0·5, and (c) x=0·6 in a log–log plot. The corresponding curves for the percolating clusters are also included

counted the clusters formed either by NBO atoms or metal ions of size *s*, i.e. the number of particles in the particular cluster.[72] The calculated time averaged cluster size distribution curves are plotted in Figure 13 for Li-containing glasses.[72] Curves for low concentrations (x=0·2 and 0·3) are in a semi-logarithmic plot whereas for x>0·3 curves are depicted in a double logarithmic plot. Similar results were obtained for metal ions in both glassy systems investigated. These results provide a clear evidence for the existence of structural inhomogeneities at the nanoscale size and, since metal ions are found in the vicinity of NBO atoms, of NBO/metal rich regions in the glassy network. This structural picture is in full agreement with the modified random network model developed by Greaves.[29] Furthermore, the results reported in Figure 13 indicate that the cluster size distribution curves follow to a very good approximation an exponential law for x=0·2 and x=0·3 while they obey a power law at higher concentrations. These distinct behaviours behaviors are explained in the framework of standard percolation theory,[73] which predicts that the cluster size distribution curves exhibit different behaviors according to whether the system is below or above the percolation threshold. On these grounds, the present results suggest that the ionic transport mechanism in alkali borate glasses is of percolative nature. The existence of percolating clusters above the percolation threshold is also inferred from the very pronounced peak in the large size region which characterises the cluster size distribution curves. These peaks coincide perfectly with the distributions of percolating NBO clusters that have been independently calculated and superimposed on the same figure. Further analysis of the results showed that the percolation threshold can be placed between x=0·3 and 0·4 for Li-borate glasses and between x=0·2 and 0·3 for Cs-borates.[74] This difference is due to the increased population of NBO atoms in Cs-containing glasses, as earlier discussed. It is noted also that the estimation of the percolation threshold in Li-borates is in close agreement with the interpretation of the experimental results of time domain reflectometry in the same family of glasses.[75]

Conclusions

Structural and dynamic properties of xM_2O–$(1-x)B_2O_3$ glasses, with x=0·2–0·6 for M=Li and x=0·2–0·4 for M=Cs and at temperatures 300 K and 1250 K, were revisited in view of recent molecular dynamics simulations. The short range order structure of simulated glasses was found to consist of $BØ_4^-$ tetrahedral and a variety of triangular borate entities, neutral $BØ_3$, and charged, $BØ_2O^-$ and $BØO_2^{2-}$. The molar fractions of these units were found to depend on the alkali content, the temperature, as well as the nature of the alkali ion. In particular, the increase of alkali content causes the depolymerisation of the borate network with the formation of NBO-containing triangular units. The temperature increase results in the progressive transformation of $BØ_4^-$ tetrahedral moieties into the less thermodynamically stable $BØ_2O^-$ and $BØO_2^{2-}$ units. This effect was explained by invoking chemical equilibration reactions operating between the network building units. The nature of the alkali ion was found to affect the network structure and, in particular, NBO-containing units are favoured at the expense of borate tetrahedra with increasing the alkali size, i.e. from Li to Cs.

Investigation of the microstructure of cation hosting sites revealed that long lived distinct sites exist in ionic borate glasses. These can be formed either by only bridging oxygen atoms of the network, b-type sites, or by both nonbridging and bridging oxygen atoms of triangular borate units, nb-type sites. Accordingly, metal ions were classified as M^b or M^{nb} if they predominantly reside in sites of b- or nb-type, respectively. These findings point to the existence of structural inhomogeneities at the nanoscale size in ionic borates.

These structural inhomogeneities were found to be reflected in distinct dynamic responses of metal ions, as well. In the short time regime, the vibrational responses of M^b and M^{nb} alkalis are well separated, with those corresponding to M^{nb} ions being situated at higher frequency values. These findings are in close agreement with earlier far infrared spectroscopic results and provide the microscopic explanation of the 'two site' model. In the long-time regime, it was also seen that the diffusive properties of the two types of ions are distinct. In particular, ions embedded in nb-type of sites are characterised by a higher mobility than those found in b-type sites.

The investigation of the ion transport mechanism showed that ions perform successive hopping events that obey a jump diffusion mechanism. Moreover, metal ions hop preferentially into neighbouring sites of the same nature. Closer inspection of the spatial distribution of NBO atoms in the glassy network revealed the tendency of NBO atoms to aggregate and form clusters. The calculated cluster size distribution curves follow the predictions of percolation theory suggesting that ion conduction in borate glasses is of percolative nature. These results provide a quantitative description of the modified random network model in ionic borate glasses.

Acknowledgements

Partial support of this work through the EU (project HPMD-CT-2000-00033) and the project "Excellence in the Research Institutes Supervised by the General Secretariat for Research and Technology / Ministry of Development" in Greece (ΟΠΣ 64769) is gratefully acknowledged.

References

1. For a review article on early studies of borate glasses see: Griscom, D. L. *Borate Glasses: Structure, Properties and Applications*. 1978. Edited by L. D. Pye, V. D. Frechette and N. J. Kreidl. Plenum Press, New York. pp 11-149.
2. Ingram, M. D. *Curr. Opin. Solid State Mater. Sci.*, 1997, **2**, 339.
3. Krogh-Moe, J. *Phys. Chem. Glasses*, 1965, **6**, 46.
4. Silver, A. H. & Bray, P. J. *J. Chem. Phys.*, 1958, **29**, 984.
5. Bray, P. J. & O'Keefe, J. G. *Phys. Chem. Glasses*, 1961, **4**, 37.
6. Jellison, G. E., Feller, S. A. & Bray, P. J. *Phys. Chem. Glasses*, 1978, **19**, 52.
7. Konijnendijk, W. L. & Stevels, J. M. *J. Non-Cryst. Solids*, 1975, **18**, 307.
8. Li, P. C., Ghose, A. C. & Su, G. J. *J. Am. Ceram. Soc.*, 1962, **45**, 83; ibid. **45**, 89.
9. McSwain, B. D., Borelli, N. F. & Su, G. J. *Phys. Chem. Glasses*, 1963, **4**, 1.
10. Easteal, A. J. & Udy, D. J. *Phys. Chem. Glasses*, 1973, **14**, 107.
11. Duffy, J. A. & Ingram, M. D. *J. Non-Cryst. Solids*, 1976, **21**, 273.
12. Karsch, K. K. *Glastech. Ber.*, 1962, **35**, 234.
13. Shelby, J. E. *J. Am. Ceram. Soc.*, 1983, **66**, 225.
14. Uhlmann, D. R., Kolbeck, A. G. & DeWitte, D. L. *J. Non-Cryst. Solids*, 1971, **5**, 426.
15. Kodama, M. & Kojima, S. in *Proceedings of the Second Int. Conference on Borate Glasses, Crystals and Melts*, 1997. Edited by A. C. Wright, S. A. Feller and A. C. Hannon. Society of Glass Technology, Sheffield, UK, pp. 181–188.
16. Zhong, J. & Bray, P. J. *J. Non-Cryst. Solids*, 1989, **111**, 67.
17. Chryssikos, G. D., Kamitsos, E. I. & Karakassides, M. A. *Phys. Chem. Glasses*, 1990, **31**, 109.
18. Kamitsos, E. I., Chryssikos, G. D. & Karakassides, M. A. *Phys. Chem. Glasses*, 1988, **29**, 121.
19. Youngman, R. E. & Zwanziger, J. W. *J. Am. Chem. Soc.*, 1995, **117**, 1397.
20. Majerus, O., Cormier, L., Calas, G. & Beneu, B. *Phys. Rev. B*, 2003, **67**, 024210.
21. Verhoef, A. H. & den Hartog, H. W. *J. Non-Cryst. Solids*, 1995, **182**, 235.
22. Cormack, A. N. & Park, B. *Phys. Chem. Glasses*, 2000, **41**, 272.
23. Vegiri, A., Varsamis, C. P. E. & Kamitsos, E. I. *J. Chem. Phys.*, 2005, **123**, 014508.
24. Clarida, W., Berryman, J. R., Affatigato, M., Feller, S. A., Kroeker, S., Ash, J., Zwanziger, J. W., Meyer, B., Borsa, F. & Martin, S. W. *Phys. Chem. Glasses*, 2003, **44**, 215.
25. Ratai, E.-M., Janssen, M., Epping, J. D., Chan, J. C. C. & Eckert, H. *Phys. Chem. Glasses*, 2003, **44**, 45.
26. Zachariasen, W. H., *J. Am. Chem. Soc.*, 1932, **54**, 3841.
27. Warren, B. E & Biscoe, J. *J. Am. Ceram. Soc.*, 1938, **21**, 259.
28. Warren B. E. *J. Am. Ceram. Soc.*, 1941, **24**, 256.
29. Greaves, G. N. *J. Non-Cryst. Solids*, 1985, **71**, 203.
30. Krogh-Moe, J. *Ark. Kemi*, 1959, **14**, 1.
31. Krogh-Moe, J. *Phys. Chem. Glasses*, 1962, **3**, 208.
32. Block, S. & Piermarini, G. J. *Phys. Chem. Glasses*, 1962, **3**, 208.
33. Kamiya, K., Sakka, S., Mizuno, T. & Matusita, K. *Phys. Chem. Glasses*, 1981, **22**, 1.
34. Exarhos, G. J. & Risen, M. W. *Solid State Commun.*, 1972, **11**, 255.
35. Kamitsos, E. I. & Risen, M. W. *J. Non-Cryst. Solids*, 1984, **65**, 333.
36. Kamitsos, E. I. *J. Phys. Chem.*, 1989, **93**, 1604.
37. Kamitsos, E. I., Patsis, A. P., Karakassides, M. A & Chryssikos, G. D. *J. Non-Cryst. Solids*, 1990, **126**, 52.
38. Kamitsos, E. I., Patsis, A. P. & Chryssikos, G. D. *J. Non-Cryst. Solids*, 1993, **152**, 246.
39. Kamitsos, E. I. & Chryssikos, G. D. *Solid State Ionics*, 1998, **105**, 75.
40. Chryssikos, G. D., Liu, L., Varsamis, C. P & Kamitsos, E. I. *J. Non-Cryst. Solids*, 1998, **235–237**, 761.
41. Varsamis, C. P., Kamitsos, E. I. & Chryssikos, G. D. *Phys. Rev. B*, 1999, **60**, 3885.
42. Uchino, T. & Yoko, T. *Proc. Second Int. Conf. on Borate glasses, crystals and melts*, 1997. Edited by A. C. Wright, S. A. Feller & A. C. Hannon, Society of Glass Technology, Sheffield, UK, pp. 417–24.

43. Wright, A. C., Vedishcheva, N. M. & Shakhmatin, B. A. *J. Non-Cryst. Solids*, 1995, **192&193**, 92.
44. Ingram, M. D., Mackenzie, M. A., Muller, W. & Torge, M. *Solid State Ionics*, 1988, **28–30**, 677.
45. Maas, P., Bunde, A. & Ingram, M. D. *Phys. Rev. Lett.*, 1992, **68**, 3064.
46. Elliott, R. J., Perondi, L. & Barrio, R. A. *J. Non-Cryst. Solids*, 1994, **168**, 167.
47. Ingram, M. D. *J. Non-Cryst. Solids*, 1997, **222**, 42.
48. Montani, R. A. *J. Non-Cryst. Solids*, 1997, **215**, 307.
49. Schulz, B. H., Dubiel, M. & Schulz, M. *J. Non-Cryst. Solids*, 1998, **241**, 149.
50. Varsamis, C. P. E., Vegiri, A. & Kamitsos, E. I. *Phys. Rev. B*, 2002, **65**, 104203.
51. Verhoef, A. H. & den Hartog, H. W. *J. Non-Cryst. Solids*, 1995, **182**, 235.
52. Allen, M. P. & Tildesley, D. J. *Computer Simulations of Liquids*, Clarendon, Oxford, 1986, p. 156.
53. Varsamis, C. P. E., Vegiri, A. & Kamitsos, E. I. *Cond. Matter Phys.*, 2001, **4**, 119.
54. Soules, T. F. *J. Non-Cryst. Solids*, 1982, **49**, 29.
55. Cormack, A. N. & Park, B. *Phys. Chem. Glasses*, 2000, **41**, 272.
56. Akagi, R., Ohtori, N. & Umesaki, N. *J. Non-Cryst. Solids*, 2001, **293–295**, 471.
57. Araujo, R. J. *J. Non-Cryst. Solids*, 1983, **58**, 201.
58. Stebbins, J. F. & Ellsworth, S. E. *J. Am. Ceram. Soc.*, 1996, **79**, 2247.
59. Sen, S., Xu, Z. & Stebbins, J. F. *J. Non-Cryst. Solids*, 1998, **226**, 29.
60. Herms, G. & Sakowski, J. J. *Phys. Chem. Glasses*, 2000, **41**, 309.
61. Yiannopoulos, Y. D., Chryssikos, G. D. & Kamitsos, E. I. *Phys. Chem. Glasses*, 2001, **42**, 164.
62. Smith, W., Greaves, G. N. & Gillan, M. J. *J. Chem. Phys.*, 1995, **103**, 3091.
63. Oviedo, J. & Sanz, J. F. *Phys. Rev. B*, 1998, **58**, 9047.
64. Karthikeyan, A. & Rao, K. J. *J. Phys. Chem. B*, 1997, **101**, 3105.
65. Funke, K. *Prog. Solid State Chem.*, 1993, **22**, 111.
66. Knodler, D., Pendzig, P. & Dietrich, W. *Solid State Ionics*, 1996, **86–88**, 29.
67. Kamitsos, E. I., Chryssikos, G. D., Patsis, A. P. & Duffy, J. A. *J. Non-Cryst. Solids*, 1996, **196**, 249.
68. Duffy, J. A., Harris, E., Kamitsos, E. I., Chryssikos, G. D. & Yiannopoulos, Y. D. *J. Phys. Chem. B*, 1997, **101**, 4188.
69. Yano, T., Nagano, T., Lee, J., Shibata, S. & Yamane, M. *J. Non-Cryst. Solids*, 2000, **270**, 163.
70. Varsamis, C. P. E., Vegiri, A. & Kamitsos, E. I. *J. Non-Cryst. Solids*, 2002, **307–310**, 956.
71. Chemla, M. *Electrochim. Acta*, 1990, **35**, 1761.
72. Vegiri, A. & Varsamis, C. P. E. *J. Chem. Phys*, 2004, **120**, 7689.
73. Stauffer, D. *Phys. Rep.*, 1979, **54**, 1.
74. Varsamis, C. P. E., Vegiri, A. & Kamitsos, E. I. *Phys. Chem. Glasses*, 2005, **46**, 72.
75. Verhoef, A. H. & den Hartog, H. W. *Solid State Ionics*, 1994, **68**, 305.

Proc. Fifth Int. Conf. on Borate Glasses, Crystals and Melts *Phys. Chem. Glasses: Eur. J. Glass Sci. Technol. B*, August 2006, **47** (4), 430–434

In situ evolution of the structure of alkali borate glasses and melts by infrared reflectance and Raman spectroscopies

Laurent Cormier,[1] Domingos De Sousa Meneses,[2] Daniel R. Neuville[3] & Patrick Echegut[2]*

[1] *Institut de Minéralogie et de Physique des Milieux Condensés, CNRS UMR 7590, Université Paris 6, Université Paris 7 et IPGP, 140 rue de Lourmel, 75015 Paris, France*

[2] *Centre de Recherche sur les Matériaux à Haute Température, CNRS-CRMHT, 1D avenue de la Recherche Scientifique, 45071 Orléans Cedex 2, France*

[3] *Physique des Minéraux et des Magmas, CNRS UMR 7047, IPGP, 4 place Jussieu, 75005 Paris, France*

The structural modifications induced by temperature are investigated for alkali borate glasses by infrared reflectance and Raman spectroscopies. The borate network is significantly changed above the glass transition temperature with a decrease of the proportion of BO_4 tetrahedra. The analysis of the infrared reflectance spectra indicates that alkali environment is also affected. In the liquid state, nonbridging oxygens are formed and their abundance around alkali increases, favouring ionic mobility.

Binary alkali borate glasses have long attracted considerable interest due to the important variations with compositions of both their structure and physical properties. With the addition of alkali oxide up to ~30 mol%, trigonal BO_3 units are converted into BO_4 tetrahedral units, which are charge compensated by alkali. These positively charged ions have a high mobility, while the boron network is essentially stationary. The motion of these ions is the origin of the fast ionic conductivity that can be used in vitreous electrolytes in solid state batteries, chemical sensors and fuel cells. Detailed understanding on the local alkali environment and its spatial distribution are central issues to understand ionic transport and in optimising glass compositions. Furthermore, as alkali migration is appreciable already far below the glass transition temperature, T_g,[1] the alkali borate structure must be investigated over a large temperature range, including the glassy and the supercooled liquid state, to assess ion transport properties.

The structure of alkali borate glasses has been studied only recently by various *in situ* experimental techniques. Neutron and x-ray diffraction were used to investigate pure B_2O_3 melt[2,3] and these studies have shown that temperature induces a modification of ring statistics, with the gradual opening of the boroxol rings above the glass transition temperature. These findings were also confirmed by Raman spectroscopy.[4] In alkali borate glasses, ^{11}B NMR,[5] Raman spectroscopy,[6,7] x-ray and neutron diffraction[8,9] have evidenced the partial conversion of BO_3 to BO_4 above T_g. All these studies have focused their attention on the borate network modifications upon temperature. Using neutron diffraction coupled with Li isotopic substitution, we specifically determined the temperature induced structural changes for the Li environment in Li diborate glasses and melts.[10,11] These investigations have shown that the $BO_4 \rightarrow BO_3$ conversion is accompanied by a modification of the structural role of Li from charge compensator in the glass to modifier in the liquid.[12] However, the structural variations of the alkali environments with temperature are still poorly known despite their importance for conductivity. The purpose of this paper is to better assess the structural reorganisation of alkali borate glasses with increasing temperature. We present new infra red reflectance data from room temperature up to T_g. Raman spectra extend our information up to the liquid state.

Experimental

Sample preparation

Glasses of composition $M_2O–2B_2O_3$ and $M_2O–4B_2O_3$ with M=Li, Na or K were prepared from dried reagent grade powders of boron oxide and alkali carbonate (Li_2CO_3, Na_2CO_3, K_2CO_3). The starting powders were melted for 1 h at 900°C, quenched by immersion of the bottom of the Pt crucible in water, then crushed and remelted for 15 min at 1050°C and quenched again, to ensure glass homogeneity. Analysis of the chemical compositions

[1] Corresponding author. Email: cormier@impmc.jussieu.fr

Table 1. Composition of the glasses investigated (in mol%), determined by atomic absorption (Na, Li) and spectrophotometrically (B). The densities and glass transition temperature are from Ref. 13

Sample	B_2O_3	M_2O (M=Li, Na)	d ($g\,cm^{-3}$)	T_g (K)
NB4	80·82±0·01	19·18±0·01	2·19	702
LB4	81·44±0·01	18·56±0·01	2·10	708
LB2	66·60±0·01	33·40±0·01	2·27	773

Uncertainties are standard deviations

are reported in Table 1, with the glass transition temperatures.[13]

Infrared reflectance

The infrared reflectance spectra have been acquired with a Bruker IFS 113v interferometer, with an experimental resolution of 4 cm^{-1}. Samples were cut with a disc shape of 5 mm of radius and 1 mm thick and optically polished. Glasses were again carefully polished just before measurements to avoid hydrolysis. Measurements at high temperature were obtained with a home made furnace. The reflectance spectra are only reported for temperatures below T_g. At higher temperatures, the softening of the samples led systematically to the deformation of their surface which precludes quantitative measurements. Above room temperature, the thermal radiation emitted by the sample is no longer negligible. Hence, its contribution to the sample interferogram must be explicitly removed for a correct evaluation of the reflectivity spectrum. This parasitic contribution was easily obtained experimentally by acquiring a supplementary interferogram with the same sample configuration and the same temperature, but without the incident source.

Raman spectroscopy

Raman spectra were recorded using a T64000 Jobin-Yvon confocal micro-Raman spectrometer equipped with a charge coupled device (CCD) detector for the Li-diborate sample. The emission line at 514·532 nm of a Coherent 70-C5 Ar^+ laser was used for sample excitation, with an incident power of about 2·8 W and an integration time of 30 s. All reported spectra are unpolarised. High temperature measurements were obtained with a home made heating device and the spectra were corrected as previously described.[12] The corrected Raman intensities were normalised to the data point of the greatest absolute intensity. Spectra were recorded between room temperature and up to the liquid state at 1350 K.

Results and discussion

IR spectra are reported in Figure 1 over a large frequency range (20–5000 cm^{-1}) allowing the study of

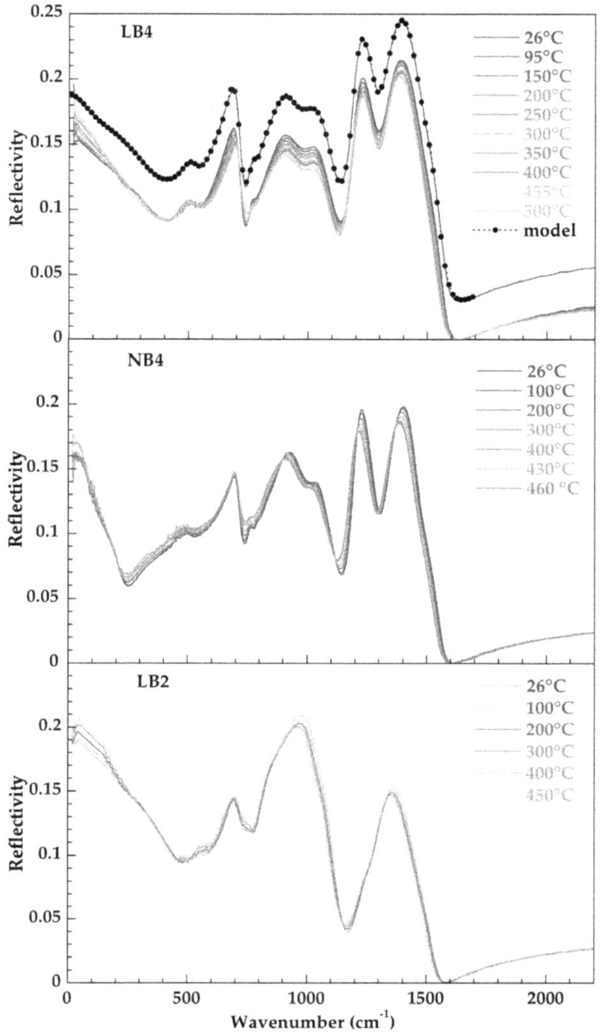

Figure 1. Infrared reflectance spectra for Li diborate (LB2), Na and Li tetraborate (NB4 and LB4) as a function of temperature. For the LB4 glass, the spectrum at 26°C is also plotted and shifted vertically for comparison with the deconvolution model

both the borate network in the mid-frequency region (500–1600 cm^{-1}) and the dynamics of alkali ions in the far-IR (20–500 cm^{-1}). The ambient spectra are similar to those already reported for Li and Na.[14,15] The spectra present a large number of overlapping bands whose determination and assignment are difficult. The IR reflectance spectra were directly fitted with the dielectric function model proposed by Meneses *et al*[16] using the curve fitting software FOCUS and minimising the number of Causal Voigt profiles to obtain a good agreement with the experimental spectra. The accuracy of the fitted parameters are related to that of the approximation used to evaluated the Voigt function and discussed by Meneses *et al.*[16] Figure 2 shows a typical example of the imaginary part of the complex dielectric function calculated from the band deconvolution for the lithium tetraborate glass at room temperature. The number of bands used and

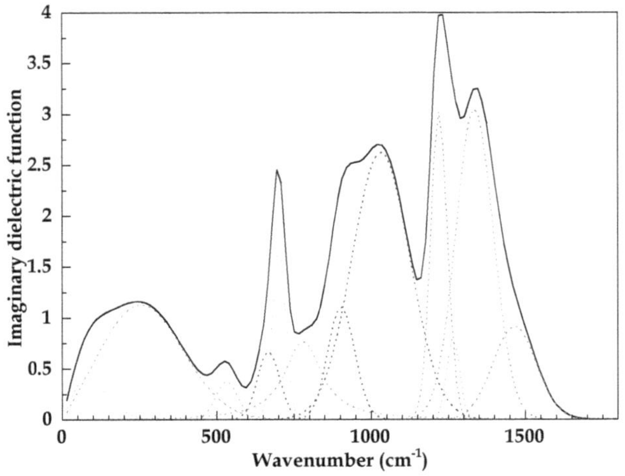

Figure 2. Imaginary part of the dielectric function at room temperature used to fit the reflectance spectrum for the lithium tetraborate glass ($Li_2O.4B_2O_3$). The band deconvolution using Gaussian functions is shown

their position agree well with a previous fit of the infrared absorption profiles of lithium borate glasses.[14] However some discrepancies appear because we are not fitting the absorption coefficient obtained after Kramers–Krönig transformation as in Ref. 14, but we are instead directly fitting the reflectance spectra. The absorption coefficients of the calculated dielectric functions are shown in Figure 3 for the three glasses at room temperature and close to T_g. The ambient spectra are similar to the previously published experimental data.[14,15] Major changes are observed with the alkali content while the nature of the alkali affect less the IR spectra.

The ambient temperature reflectance spectra were widely investigated by Kamitsos & Chryssikos *et al* and, though the band assignment is based on comparison with crystalline references, some general views emerge.[15] The region 1200–1500 cm^{-1} is assigned to B–O bond stretching of trigonal BO_3 units. The band at 1400 cm^{-1} may be due to a ring stretching vibrations while the band at 1250 cm^{-1} may correspond to B–O stretching vibrations involving mainly the oxygen connecting different groups. The region 850–1200 cm^{-1} originates from B–O bond stretching of tetrahedral BO_4 units associated with various borate groups (tetraborate for bands at 1050 and 900 cm^{-1} and diborate for a band at 1000 cm^{-1}). The bands between 600–800 cm^{-1} are assigned to bond bending vibrations of B–O–B linkages. The small band at 760 cm^{-1} has been assigned to [3]B–O–[4]B bending of bridges. With increasing the alkali content, important changes in the IR spectra are observed but variations with the nature of the alkali are less striking for a given composition.

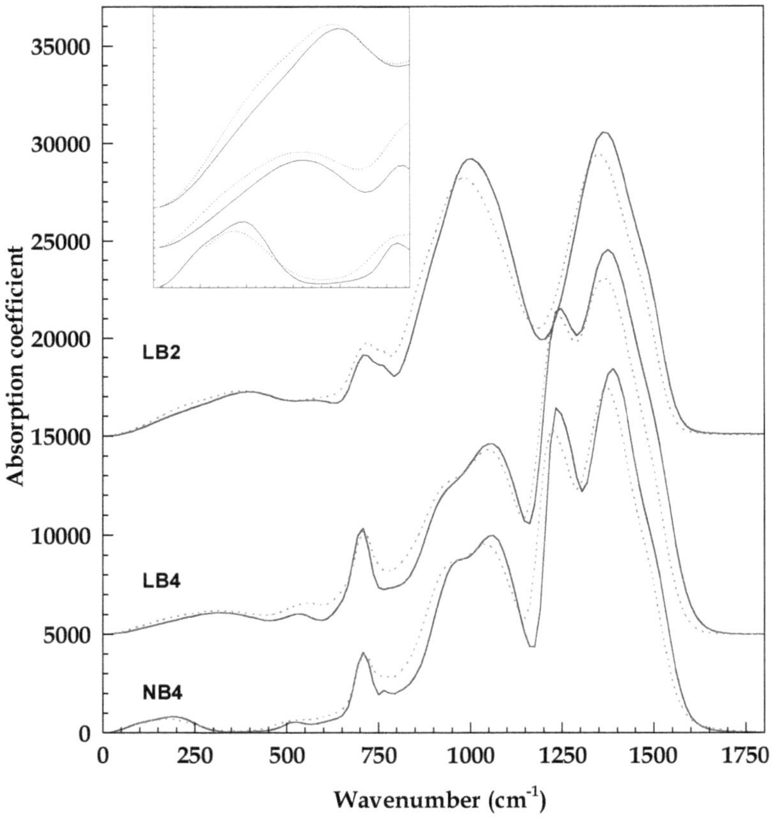

Figure 3. Absorption coefficients calculated from the fitted dielectric function at room temperature (plain curve) and at the highest measured temperature (dashed curve) for Li diborate (LB2), Na and Li tetraborate (NB4 and LB4). The insert show a zoom on the far-IR region of the alkali rattling modes

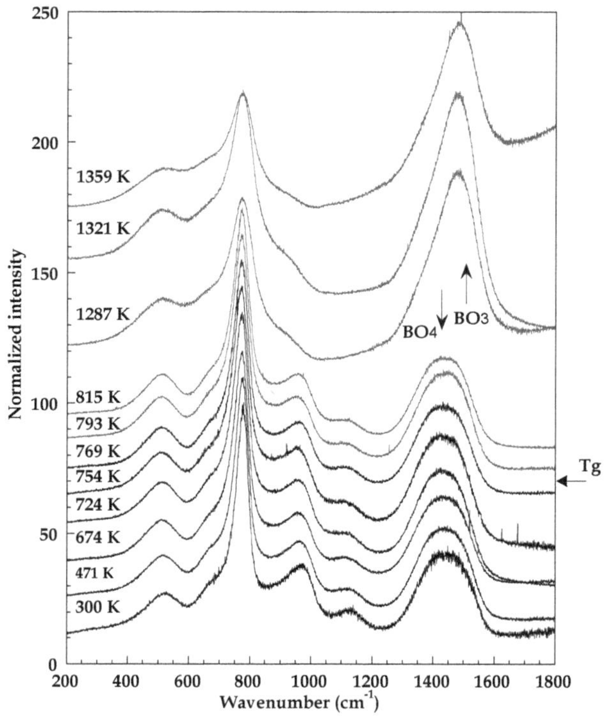

Figure 4. Normalised Raman spectra for Li diborate (LB2) as a function of temperature

With increasing temperature, the high frequency bands are shifted towards lower frequencies except those at 700 and 760 cm^{-1}, whose position is not affected. This shift can be due to the thermal expansion of the B–O bonds. We do not observe a significant change in the width of the band but the intensity decreases except for the bands at 700 and 760 cm^{-1}.

On the far-IR part, two bands are clearly observed (see insert in Figure 3), and assigned to vibrations of alkali cations in two distinct environments within the borate network.[17] The low and high frequency contributions have been associated to a charge compensating role around the BO_4 units and to a modifying role near NBOs, respectively, which was confirm by molecular dynamics (MD) simulation.[18]

Raman spectra of Li diborate glass and melt are shown in Figure 4. No spectra are presented between 1280 and 1359 K because crystallisation occurs in this temperature range. In the glass, a strong peak at 770 cm^{-1} is assigned to the localised breathing vibration of six-membered rings containing one or two BO_4 tetrahedra,[19] though this band alone cannot be clearly attributed to a particular borate group. Diborate groups are likely to be present due to a band at 1130 cm^{-1}, and tetraborate groups could have their signature in the bands at 520, 670 and 975 cm^{-1}.[19] These groups have also been evidenced as the dominant ones for diborate compositions by NMR studies.[20] The high frequency range of the Raman spectra is characterised by two main bands at 1410 and 1500 cm^{-1}. They are assigned to B–O^- (O^-= nonbridging oxygen, NBO)

Table 2. Parameters used for modeling the infrared reflectance spectrum for LB4 glass at room temperature. In this model, ω_{Oj}, γ_{Gj} and γ_{Lj} are the transverse optical wave number, the Gaussian and Lorentzian full width at half maximum of the jth band. A_j is the maximum amplitude of the imaginary part of the band. The fitted high frequency dielectric constant is ε_∞=2·375

Band	A_j (±0·001)	ω_{Oj} (±0·1)	γ_{Gj} (±0·1)	γ_{Lj} (±0·1)
1	0·245	32·4	187·8	1
2	1·132	253·9	319·4	1
3	0·372	532·0	83·1	1
4	1·686	700·7	48·9	1
5	0·769	779·1	1·0	140·8
6	0·673	667·3	83·2	1
7	1·109	903·4	102·3	1
8	2·627	1032·1	219·7	1
9	0·803	1265·3	48·8	1
10	3·056	1335·6	143·9	1
11	0·917	1464·7	163·7	1
12	3·024	1220·2	62·9	1

For uncertainties see text

bond stretching vibrations,[21] with the lower one related to $BØ_2O^-$ triangles (Ø=bridging oxygen) attached to $BØ_4^-$ units and the higher one to $BØ_2O^-$ triangles connected to other trigonal BO_3 units.[7] Using Gaussian fitting of this region, it is possible to determine the relative peak areas (A4 and A3 for the low and high frequency band) and thus to have a good indication on the proportion of BO_4 units, as shown in previous studies.[6,7,12]

With increasing temperature, the bands present a systematic shift towards lower frequencies due to the thermal expansion. The band intensities between 500–1100 cm^{-1} decreases noticeably above T_g, and we observe the disappearance of the diborate contribution (1100 cm^{-1}). However, the band at 770 cm^{-1} remains in the liquid state. Since this band is related to rings involving BO_4 tetrahedra, its persistence in the liquid state suggests that BO_4 units in rings groups could be stabilised. Drastic modifications appear above T_g in the range 1200–1800 cm^{-1}, with an important increase in intensity of the high frequency band. These findings indicate the conversion from BO_4 units to BO_3 units.

The IR spectra indicate the lack of coordination change which would be apparent in the relative intensities of the bands in the two regions 800–1150 cm^{-1} (associated with BO_4) and 1200–1500 cm^{-1} (associated with BO_3). Indeed, IR reflectance spectra have shown to be sensitive to variations in the proportion of BO_4 due to cooling rate gradient from the surface to the core.[22] The absence of significant changes in the boron coordination below T_g is in agreement with Raman spectra in Figure 4 and with previous neutron data.[10,12] Changes in Raman spectra appear above T_g and become important in the liquid state. The increase in intensity of the mode at 1500 cm^{-1} clearly indicate the conversion of BO_4 units to BO_3 units. This coordination change around boron implies the simultaneous formation of NBOs in the liquid.

Below T_g, in the absence of boron coordination change, the environment around alkali may be affected (Figure 3, insert) as shown for the far-IR bands due to the alkali rattling modes. A shift towards lower frequencies for the high frequency band (band H) seems to occur when approaching T_g. Modifications in the far-IR region are important as they are intimately related to the mobility of ions, since alkali diffusion is a thermally activated process. An increase in the number of mobile ions in borate glasses has been detected from glassy to liquid state by a ^{23}Na spin lattice relaxation study.[23] Since a strong link between alkali and NBOs seems responsible for the band H,[17] changes with temperature could be related to alkali motion towards environment where the number of NBOs is optimised to favour jumps. Indeed, MD simulations indicate that alkali mobility increases when NBOs are present in local alkali site of various oxide glasses.[19,24,25]

At higher temperature (liquid state), we have shown that the environment around alkalis, specifically that of Li ions, is significantly modified.[11,12] Alkali ions are increasingly found in modifying positions in borate melts, that is associated with NBOs. This connection between the enhanced mobility of Li ions in the melt and NBOs could arise from the existence of regions enriched in alkalis and NBOs, in which alkali cations could move by hopping. Such preferential diffusive pathways (either static or dynamic) have been proposed[26,27] and could still be persistent in the liquid state.[28] MD simulations of borate glasses and melts also indicate the possibility of spatial segregation of alkali and NBOs.[18] Understanding of the melt properties and ion diffusion requires thus a detailed characterisation of the alkali environments at high temperatures and structural modifications between glasses and melts.[28] Evidently, more systematic studies at high temperature by IR reflectance are required and over larger temperature range.

Conclusions

In this study, we report the variation with temperature of infrared reflectance and Raman spectra obtained on alkali borate glasses. Both techniques agree on a lack of boron coordination change below the glass transition temperature. The alkali environment may be affected at such temperatures but future work is needed to relate the observed variation to ionic mobility. In the liquid state, the Raman spectra clearly indicate a partial boron conversion from a tetrahedral to a trigonal environment. This coordination change is associated with the formation of NBOs and a modification of the structural role of alkali. Alkali ions are increasingly found in environments with NBOs that should favour their diffusion.

Reference

1. de Marchi, G., Mazzoldi, P. & Miotello, A. *J. Non-Cryst. Solids*, 1990, **125**, 302.
2. Misawa, M. *J. Non-Cryst. Solids*, 1990, **122**, 33.
3. Sakowski, J. & Herms, G. *J. Non-Cryst. Solids*, 2001, **293–295**, 304.
4. Hassan, A. K., Torell, L. M. & Börjesson, L. *Phys. Rev. B*, 1992, **45**, 12797.
5. Sen, S., Xu, Z. & Stebbins, J. F. *J. Non-Cryst. Solids*, 1998, **226**, 29.
6. Akagi, R., Ohtori, N. & Umesaki, N. *J. Non-Cryst. Solids*, 2001, **293–295**, 471.
7. Yano, T., Kunimine, N., Shibata, S. & Yamane, M. *J. Non-Cryst. Solids*, 2003, **321**, 147.
8. Herms, G. & Sakowski, J. *Phys. Chem. Glasses*, 2000, **41**, 309.
9. Majérus, O., Cormier, L., Calas, G. & Beuneu, B. *Phys. Rev. B*, 2003, **67**, 024210.
10. Majérus, O., Cormier, L., Calas, G. & Beuneu, B. *J. Phys. Chem. B*, 2003, **107**, 13044.
11. Majérus, O., Cormier, L., Calas, G. & Soper, A. K. *Physica B*, 2004, **350**, 258.
12. Cormier, L., Majérus, O., Neuville, D. R. & Calas, G. *J. Am. Ceram. Soc.*, 2006, **89**, 13.
13. Feller, S. A., Kottke, J., Welter, J. , Nijhawan, S., Boekenhauer, R., Zhang, H., Feil, D., Parameswar, C., Budhwani, K., Affatigato, M., Bhatnagar, A., Bhasin, G., Bhowmik, S., Mackenzie, J., Royle, M., Kambeyanda, S., Pandikuthira, P., Sharma, M. In *Borate glasses, crystals and melts. 1997*. Edited by A. C. Wright, S. A. Feller, A. C. Hannon, Society of Glass Technology, Sheffield. Pp 246–253.
14. Kamitsos, E. I., Patsis, A. P., Karakassides, M. A. & Chryssikos, G. D. *J. Non-Cryst. Solids*, 1990, **126**, 52.
15. Kamitsos, E. I., Karakassides, M. A. & Chryssikos, G. D. *J. Phys. Chem.*, 1987, **91**, 1073.
16. de Sousa Meneses, D., Gruener, G., Malki, M. & Echegut, P. *J. Non-Cryst. Solids*, 2005, **351**, 124.
17. Kamitsos, E. I., Karakassides, M. A., Chryssikos, G. D. *J. Phys. Chem.*, 1987, **91**, 5807.
18. Varsamis, C.-P. E. Vegiri, A. & Kamitsos, E. I. *Phys. Rev. B*, 2002, **65**, 104203.
19. Meera, B. N. & Ramakrishna, J. *J. Non-Cryst. Solids*, 1993, **159**, 1.
20. Bray, P. J. *J. Non-Cryst. Solids*, 1987, **95 & 96**, 45.
21. Chryssikos, G. D., Kamitsos, E. J., Patsis, A. P., Bitsis, M. S., Karakassides, M. A. *J. Non-Cryst. Solids*, 1990, **126**, 42.
22. Machowski, P. M. Varsamis, C. P. E. & Kamitsos, E. I. *J. Non-Cryst. Solids*, 2004, **345 & 346**, 213.
23. Sen, S. & Stebbins, J. F. *Phys. Rev. B*, 1997, **55**, 3512.
24. Oviedo, J. & Sanz, J. F. *Phys. Rev. B*, 1998, **58**, 9047.
25. Karthikeyan, A., Vinatier, P., Levasseur, A. & Rao, K. J. *J. Phys. Chem. B*, 1999, **103**, 6185.
26. Greaves, G. N. & Ngai, K. L. *Phys. Rev. B*, 1995, **52**, 6358.
27. Jund, P., Kob, W. & Julien, R. *Phys. Rev. B*, 2001, **64**, 134303.
28. Majérus, O., Cormier, L., Calas, G. & Beuneu, B. *Chem. Geol.*, 2004, **213**, 89.

Glass formation and microheterogeneous structure in the B_2O_3–MoO_3–Me_nO_m (Me=W, Mn, Cu) systems

Y. Dimitriev,[1] *E. Kashchieva, D. Ilieva*

University of Chemical Technology and Metallurgy, 1756 Sofia, Bulgaria

R. Iordanova

Institute of General and Inorganic Chemistry, Bulgarian Academy of Sciences, 1113 Sofia, Bulgaria

The glass forming tendency in the B_2O_3–MoO_3–Me_nO_m (Me=W, Mn, Cu) systems was investigated with different cooling rates (10^2–10^5 K/s) of the melts. The glasses were characterised by x-ray diffraction (XRD), infrared spectroscopy (IR), transmission electron microscopy (TEM), scanning electron microscopy (SEM), electron microprobe analysis (EMRA) and x-ray photoelectron spectroscopy (XPS). Liquid phase separation regions and samples with a microheterogeneous structure were observed in the compositions containing more than 20 mol% B_2O_3. Droplet like and more complex microstructures were formed during the cooling. MoO_3, WO_3, $CuO.B_2O_3$ and $MnO.B_2O_3$ crystal phases were detected near the boundaries of the glass formation regions. The reason for the formation of microheterogeneous structures is that transition metal ions could not convert $B_3O_6(BO_3)$ to BO_4 units in the amorphous matrix. The existence of BO_3, MoO_4, and CuO_4 structural units forming the glass network were proved.

Introduction

In a previous study[1] of the B_2O_3–V_2O_5–MoO_3 system it was found a wide region of liquid macrophase separation accompanied with different forms of microheterogenities dispersed in the amorphous or polycrystalline matrix. It was suggested that vanadium and molybdenum do not stimulate the formation of superstructural borate units with simultaneous participation of BO_3 and BO_4 groups, and the mixed bridging bonds B–O–V and B–O–Mo are not dominant in the network.[1] These results initiated a long-term research program in order to solve the following problems:

(i) to obtain more exact information about the influence of transition metal ions on the structure of the low melting borate glasses;

(ii) to accumulate more fundamental knowledge for the formation of the microphases in amorphous borate glasses which are suitable as model systems;

(iii) to improve and to create new methods for the generation of microheterogeneous structures with controlled particle size distribution that is important for the synthesis of nanostructured materials.

The aim of the present investigation is to study the glass formation and immiscibility regions in B_2O_3–MoO_3–Me_nO_m (Me=W, Mn, Cu) systems in order to verify the competitive role of transition metal oxides MenOm in the formation of microheterogeneous structures. The eutectic B_2O_3–MoO_3 system was selected as a basic one possessing a wide region of metastable phase separation situated under the liquids.

Experimental

All batches were prepared using MoO_3, WO_3, MnO_2, CuO, H_3BO_3 (reagent grade) as starting materials. The homogenised batches were melted for 20 min in air at 900–1300°C in platinum crucibles. The samples are presented according to their nominal compositions. Manganese appeared to be in two valence oxidising states. Two different cooling rates of the melts were applied: fast quenching by the roller technique (10^4–10^5 K/s) and free annealing of the crucible after removal from the furnace (10^2 K/s). The amorphous state of the samples was verified by x-ray diffraction (Philips APT-15 diffractometer, Cu K_α radiation). The glasses obtained were investigated by IR spectroscopy using the KBr technique (Nicolet-320, FTIR spectrometer). Transmission electron microscopy (EM-400, Philips) experiments were made on fresh fractured and chemically treated (2 vol% HF acid for 10 s) surfaces of the bulk samples. Scanning electron microscopy (JEOL, JEM-200, CX) and electron microprobe analysis EMRA, (JEOL, Superprobe 733) were performed on polish samples. The error in

[1] Corresponding author. Email yanko@uctm.edu

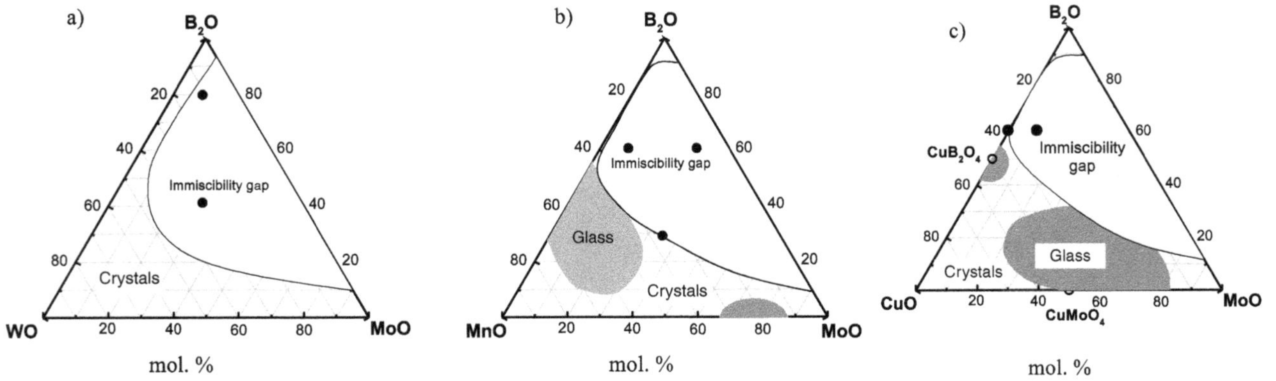

Figure 1. Immiscibilty gap and glass formation regions in the systems (a) B_2O_3–WO_3–MoO_3, (b) B_2O_3–MnO–MoO_3, (c) B_2O_3–CuO–MoO_3. • samples investigated by SEM (a) and TEM (b), (c)

these measurements was estimated at ~2%. The x-ray photoelectron spectroscopy measurements were carried out by electron spectrometer (ESCALAB-MKII, VG Scientific).

Results

The glass formation and the liquid phase separation regions are presented in Figure 1. In the B_2O_3–MoO_3 system transparent glasses are not obtained. In the three component system B_2O_3–MoO_3–WO_3 processes of phase separation begin below 90 mol% B_2O_3 and it is difficult to obtain homogeneous samples. Below 20 mol% B_2O_3 different crystalline phases of solid solutions $W_{1-x}Mo_xO_3$ are formed. The immiscibility gaps of all investigated systems are compatible. Actually the boundary that divides the regions of homogeneous from the heterogeneous samples, is conditional and strongly dependent on the cooling rates. The regions near the B_2O_3 corner should be subjected to special experiments due to the hygroscopity of the glasses and the separation of H_3BO_3 on the surface. The increase in MoO_3, WO_3 and CuO contents stimulates the crystallisation while B_2O_3 provokes the appearance of stable or metastable liquid immiscibility. Glasses are obtained near the MnO–B_2O_3 site at slow cooling of the melts in the B_2O_3–MoO_3–MnO system. The glass formation region with the participation of copper oxide is smaller. At high cooling rates the glass formation region is situated near the MoO_3–CuO site. Typical XRD patterns of amorphous samples in the B_2O_3–MoO_3–CuO system are shown in Figure 2. By TEM observations, several types of separate structures were established (Figures 3, 4). In the most cases formation of several different microphases was detected. They may be considered as a result of primary and multiple separation processes. For some samples the formation of shells around droplets is observed. SEM investigations on the system B_2O_3–MoO_3–WO_3 show several phases penetrating one another (Figures 5, 6). The EMRA data on the chemical composition of the separate phases are summarised in Table 1. The inner fine crystalline layer of the sample, $80B_2O_3.10MoO_3.10WO_3$,

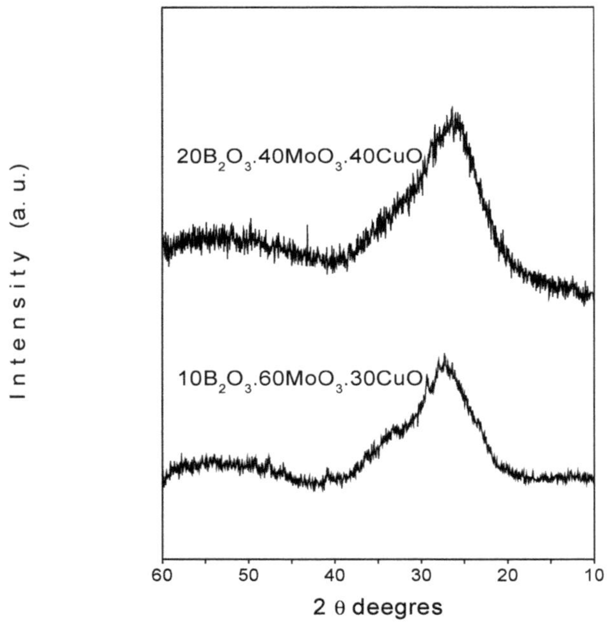

Figure 2. XRD pattern of glasses in the B_2O_3–MoO_3–CuO system

Table 1. Electron microprobe analysis

Sample Nominal composition mol%	Electron microprobe analysis B_2O_3 wt%	mol%	MoO_3 wt%	mol%	WO_3 wt%	mol%
$80B_2O_3.10MoO_3.10WO_3$						
P 1	59	35	21	25	20	40
P 2	62	38	19	24	19	38
P 3	13	5	11	8	76	88
P 4	58	34	21	25	21	41
P 5	72	48	13	18	15	34
P 6	27	12	33	29	40	59
P 7	65	41	16	21	18	38
P 8	76	53	11	17	13	30
$40B_2O_3.30MoO_3.30WO_3$						
P 1	23	9	28	24	49	67
P 2	40	19	23	22	37	59
P 3	58	32	15	17	27	51
P 4	22	9	30	26	48	65

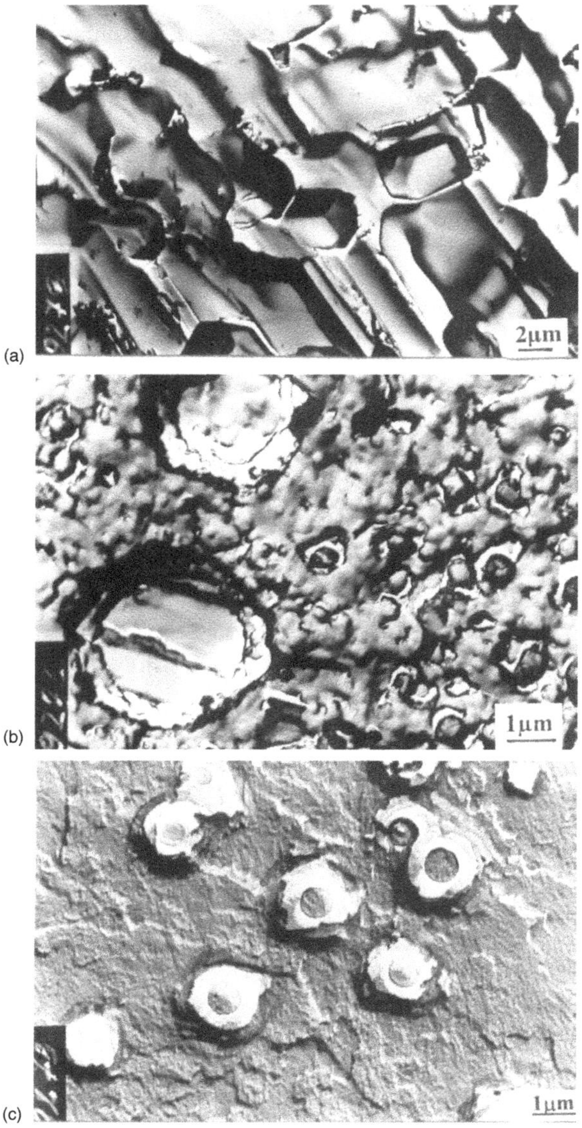

Figure 3. TEM micrographs of slowly cooled samples (a) $30B_2O_3.35MnO.35MoO_3$; (b) $60B_2O_3.10MnO.30MoO_3$; (c) $60B_2O3.30MnO.10MoO_3$

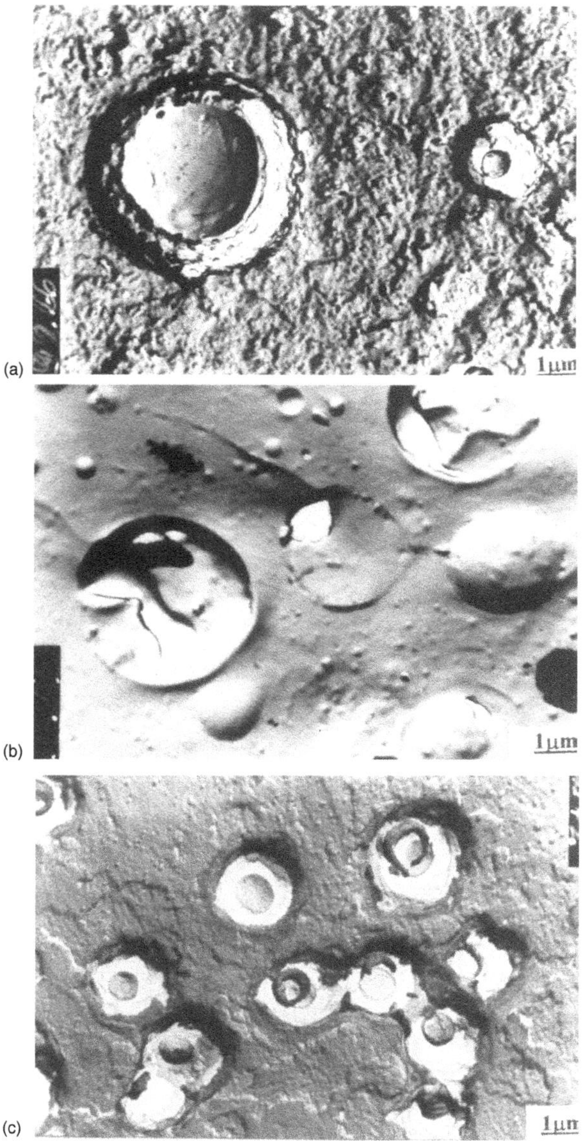

Figure 4. TEM micrographs of slowly cooled samples: (a) $60B_2O.30CuO.10MoO_3$ (upper layer); (b) $60B_2O_3.30CuO.10MoO_3$ (bottom layer); (c) $60B_2O_3.40CuO$

(Figure 5) contains about 35 mol% B_2O_3 (p. 1, 2, 4). It is covered by another one in which crystallised dendrites containing mainly WO_3 (about 88 mol%, p.3, 6) are imbedded in the amorphous borate matrix (48–53 mol% B_2O_3, p. 5, 7, 8). In the central part of the sample, $40B_2O3.30MoO_3.30WO_3$, (Figure 6) and near the surface WO_3 is the dominant (65–67 mol% p.1, 4). Between these layers there is a spinodal like microheterogeneous structure in which the content of B_2O_3 is in the range 19–32 mol% (p.2, 3).

The infrared spectra of amorphous B_2O_3–MoO_3–CuO samples are selected as more representative in order to understand the structural transformation with the composition (Figure 7). The compositions are situated outside the immiscibility regions with low B_2O_3 content. The absorption bands may be divided into three regions: 1400–1100 cm^{-1}, 1000–650 cm^{-1}, and 650–500 cm^{-1}. The XPS spectrum of the glass sample $10B_2O_3.30MoO_3.60CuO$ is presented in Figure 8.

Discussion

The results presented on the glass formation regions and the detected microheterogeneity of the samples give rise to questions about the reason for the liquid separation and glass formation ability. The problem will be discussed from thermodynamic, kinetic and structural points of view.

The phase separation phenomenon in glasses was an important topic many years ago (1960–1970). The most complete summary of historical information, experimental data and theoretical treatment can be found in several monographs and review papers.[2–6]

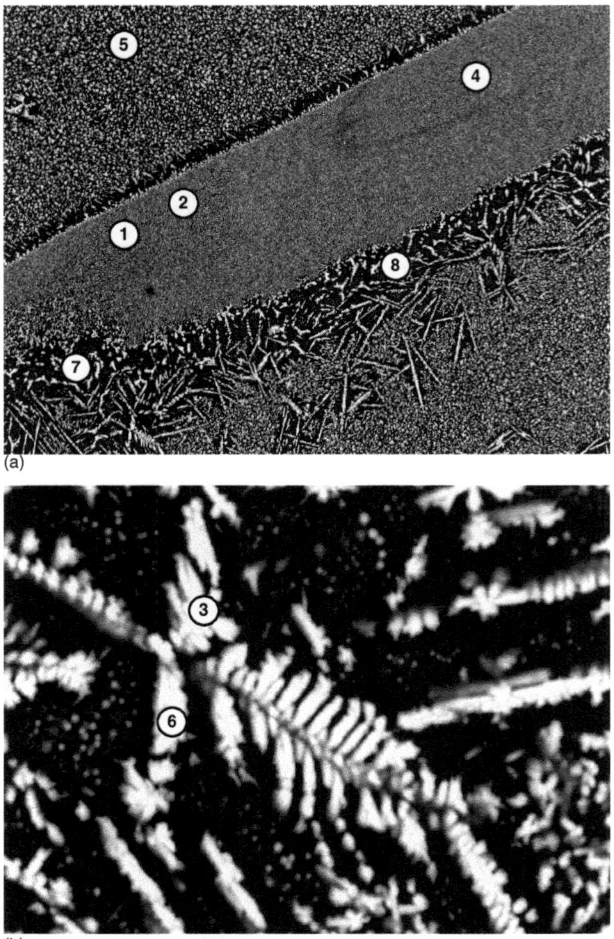

Figure 5. SEM micrographs of slowly cooled $80B_2O_3.10MoO_3.10WO_3$ sample (a) ×360; (b) ×3600. The microprobe analysis of point 1×8 are presented in Table 1

Some detailed considerations of the special aspect concerning borate systems containing nontraditional network forming oxides (TeO_2, V_2O_5, MoO_3) have been made previously.[7–11] Using the phenomenological theory of classical thermodynamics, the compositions of both stable coexisting liquid phases may be presented as a function of the temperature by binodal curves.[12] It is well known that the obtained immiscibility gap may be situated above the liquids (macro liquid phase separation) or below liquids (metastable immiscibility). In more complicated systems the immiscibility phases may be combined together in more complex diagrams. In our case the $MnO–B_2O_3$[13] and $CuO–B_2O_3$[14] systems are representative for a stable monotectic type of separation. In the other systems, $B_2O_3–MoO_3$[15] and $B_2O_3–WO_3$,[16] the metastable immiscibility occurs below the liquids. Our experimental results show that stable and metastable regions of separation are interconnected in one immiscibility gap ($B_2O_3–CuO–MoO_3$, and $B_2O_3–MoO_3–MnO$). The situation is the same for the system $B_2O_3–MoO_3–WO_3$ where two metastable regions are unified. To our opinion this is the thermodynamical explanation for the observed complex macro- and microheterogeneous structure (Figures 3–6). The other reason is the consequence of the nonequilibrium conditions of cooling (kinetic factors) which leads to formation of the structure with several steps of separation. For the first time Vogel[2] observed and explained the multiple phase separation of six microphases with different compositions. The observed shell around droplets (Figures 3, 4) may be considered as a result of incomplete transport processes during cooling.[2] In our previous studies it was shown that the complicated microstructures were typical for tellurium borate glasses.[7,8]

The main idea of our experiments is that due to the competitive action of the network forming components it is not possible to establish in advance if mixed chains are formed or if there are micro domains richer in one of the components.[17] As was pointed out by Vogel in his excellent book,[2] "usually, a melt containing two kinds of structural elements with different special requirements will tend to phases separate". The obtained wide regions of immiscibility confirm this suggestion, i.e. that the selected compounds (B_2O_3, CuO, MnO, MoO_3, WO_3) do not stimulate the formation of homogeneous phases. The reason is the incompatibility of the building polyhedra. That is to say it is necessary to present some information about the structural units forming the amorphous network. It was discussed elsewhere by modelling[11]

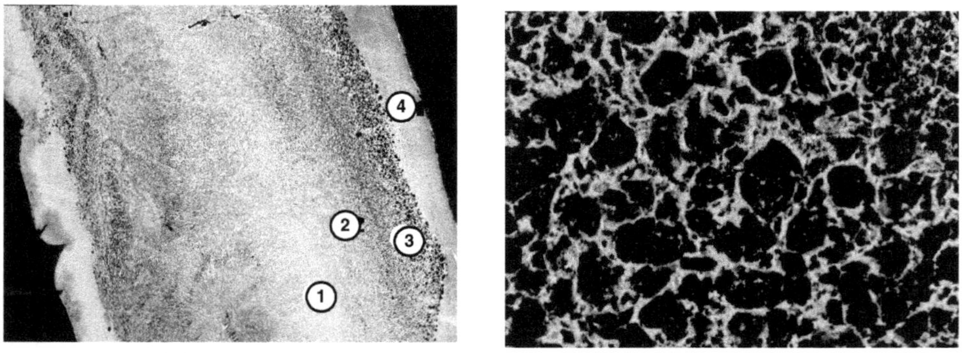

Figure 6. SEM micrographs of slowly cooled $40B_2O_3.30MoO_3.30WO_3$ sample (a) ×60; (b) ×1500. The microprobe analysis of point 1×4 is presented in Table 1

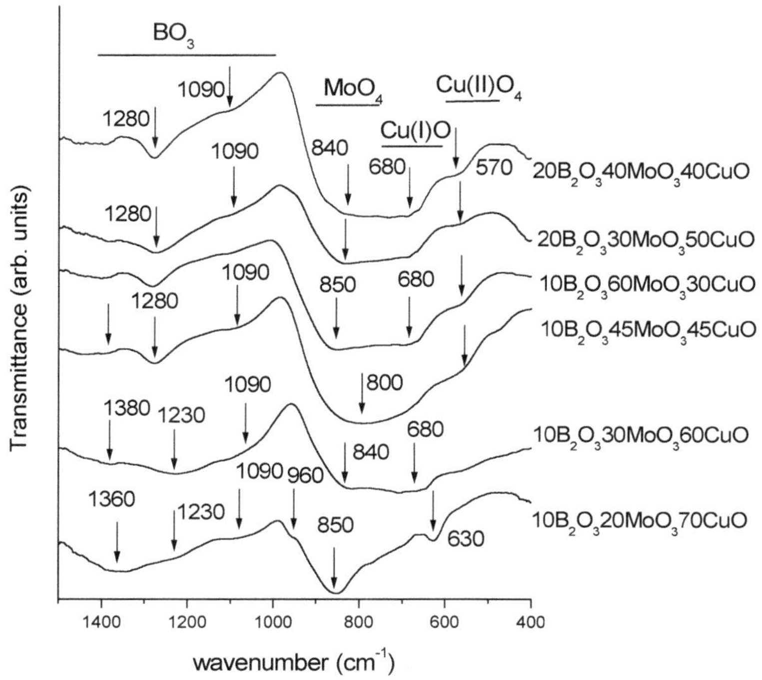

Figure 7. IR spectra of glasses in the B_2O_3–MoO_3–CuO system

and TEM[7,8] observation that the tendency of cluster formation is typical for the borate network containing boroxol rings (pure B_2O_3, or compositions with small amounts of additives). According to presented IR spectra (Figure 7), MoO_3 and CuO do not stimulate the formation of BO_4 units, because the absorption bands in the 1400–1100 cm^{-1} region are connected with the vibration of BO_3 units.[18,19] The change in relative intensity of the bands at 1280–1230 cm^{-1} and 1380–1360 cm^{-1} is attributed to the rearrangement of the triangular borate units. The increase in band intensity near 1280 cm^{-1} with the decrease in CuO content may be connected with the formation of the isolated BO_3 units, and that at 1380 cm^{-1} with the vibration of associated BO_3 units. The bands at about 850–780 cm^{-1} may be assigned to the vibration of strongly distorted MoO_4 groups.[20] The band at 570 cm^{-1} is due to vibration of CuO_4 "square planar CuO_4 complexes" with D_{4h} point symmetry and that at 630 cm^{-1} to vibration of Cu(I)–O bonds.[21,22] On the basis of these preliminary structural results, it is supposed that in a wide concentration range transition metal

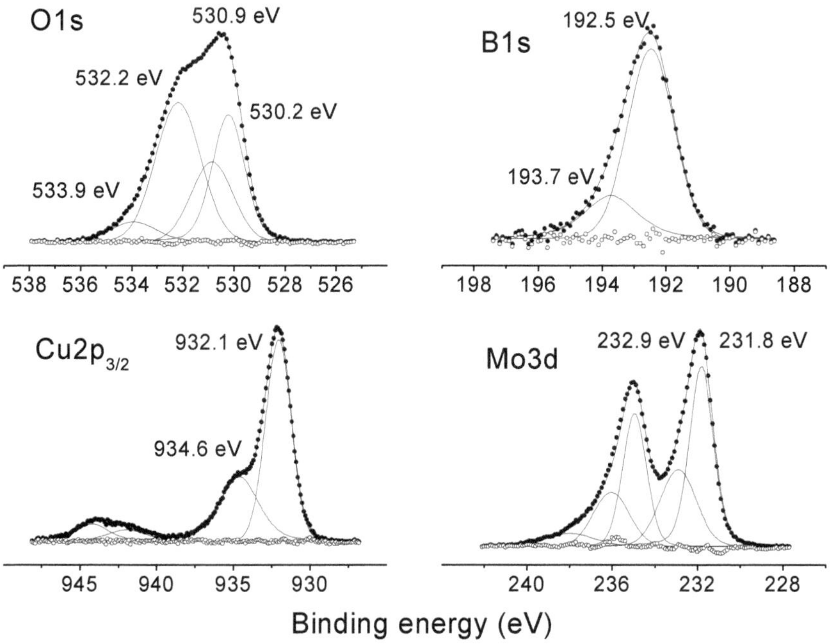

Figure 8. Curve fitting of O1s, B1s, Cu2p3/2 and Mo3d for $10B_2O_3.60CuO.30MoO_3$ amorphous sample

ions are not able to attack the boroxol units. B_3O_6 are preserved, which leads to cluster formation and phase separation. When the content of B_2O_3 in the melts is lower (below 20 mol%) isolated or associated BO_3 units are formed. They are distributed randomly in the network and homogeneity of the amorphous sample increases. In contradiction, the crystallisation tendency of these compositions rapidly increases because in the network mainly isolated MeOn units (CuO_4, MoO_4, BO_3) will participate. That is to say for the vitrification is necessary to apply fast quenching methods.

The behaviour of manganese will be discussed separately because it is found to stimulate the formation of superstructural units containing BO_4. Probably this is the reason for a better glass formation tendency in B_2O_3–MoO_3–MnO (Figure 1).

Additional information about the building of the amorphous network and bridging bonds may be obtained from XPS analysis. The O 1s spectrum of one representative amorphous sample will be discussed as the most informative (Figure 8). Both low binding energy components at 530·2 eV and 530·9 eV can be assigned to Me–O–Me (Me=Mo, Cu) bridging bonds.[23, 24] The higher binding energy at 532·2 eV should be attributed to the B–O–Mo and B–O–Cu mixed bridging bonds.[1,25] The shoulder at 533·9 eV is most probably due to hydroxide contamination.[1,26] The peak positions at 932·1 eV and 934·6 eV are due to Cu(I) and Cu(II) respectively.[26,27] The formation of mixed B–O–Me bridging bonds explains the obtaining of amorphous samples without liquid immiscibility.

Conclusion

Single phase glasses are obtained only in a narrow concentration range after fast quenching of the melts in the systems. It has been found that the liquid phase separation covers a wide region of compositions. The reasons for the complex character of the immiscibility phenomena are: (i) the feature of phase diagrams of the binary systems (stable and metastable regions); (ii) incomplete transport processes during the cooling of the melts; (iii) the incompatibility of the building units. Based on the IR spectra the existence of MoO_4, BO_3, and CuO_4 has been proved in the network. MoO_3, WO_3 and CuO do not stimulate the formation of superstructural units containing BO_3, BO_4 groups.

Acknowledgement

This work has been carried out with the financial support of the National Foundation of Sciences, Bulgaria, Contract 806, (2005).

References

1. Dimitriev, Y., Kashchieva, E., Iordanova, R. & Tyuliev, G. *Phys. Chem. Glasses*, 2003, **44**, 155.
2. Vogel, W. *Glass Chemistry*, 1992, Springer-Verlag.
3. *Phase separation in glass*, 1984, edited by Mazurin, O. & Porai-Koshits, E., North-Holland, Amsterdam.
4. Milyukov, E. & Kasymova, S. *Immiscible melts and glasses* (in Russian), 1981. Fan, Tashkent.
5. Varshal, B. *Two phase glasses :structure, properties and application*, (in Russian), 1991.
6. Polyakova, I. *Phys. Chem. Glasses*, 2000, **41**, 247–258. In: *Borate glasses, crystals and melts*, 2001. Edited by Y. Dimitriev & A. Wright, Society of Glass Technology, Sheffield, p. 37.
7. Kashchieva, E., Hinkov, P., Dimitriev, Y. & Miloshev, S. *J. Mater. Sci. Lett.*, 1994, **13**, 1760.
8. Kashchieva, E. & Dimitriev, Y. *J. Mater. Sci. Lett.*, 1995, **14**, 732.
9. Kashchieva, E., Pankova, M. & Dimitriev, Y. *Ceram. Silikaty*, 2001, **3**, 111.
10. Iordanova, R., Dimitriev, Y., Kashchieva, E. & Klissurski, D. *Ceram. Silikaty*, 2001, **3**, 115.
11. Kashchieva, E., Shivachev, B. & Dimitriev, Y. *J. Non-Cryst. Solids*, 2005, **351**, 1158.
12. West, D. *Ternary equilibrium diagrams*, Chapman & Hall, 1982.
13. Levin, E., Robbins, C. & McMurdie, H. *Phase Diagram for Ceramics*, American Ceramic Society, 1964, No. 13.
14. Toropov, N. *Phase diagrams in silicate systems*, 1969, Nauka.
15. Malzev, V., Chobanan, N. & Volkov, V. *Z. Neorg. Chem.*, 1974, **XIX**, 1617.
16. Malzev, V., Chobanan, N. & Volkov, V. *Z. Neorg. Chem.*, 1973, **XIX**, 2010.
17. Zarzycki, J. *Glasses and vitreous state*, Cambridge Solid State Science Series, Cambridge Univerity Press, New York, 1991, p. 229.
18. Varsamis, C., Kamitsos, E. & Chryssikos, G. *Solid State Ionics*, 2000, **136–137**, 1031.
19. Kamitsos, E. *Phys. Chem. Glasses*, 2003, **44**, 79.
20. Iordanova, R., Dimitriev, Y., Dimitrov, V. & Klissurski, D. *J. Non-Cryst. Solids*, 1994, **167**, 74.
21. Nakamoto, K. in *Infrared and Raman spectra of Inorganic and Coordination Compounds*, Wiley, New York, 1978.
22. Kliche, G. & Popovic, Z. *Phys. Rev. B*, 1990, **42**, 10060.
23. Sawatzky G. & Post, D. *Phys. Rev.*, 1979, **B 20**, 1546.
24. Barr, T. *Modern ESCA, The Principles and Practice of X-Ray Photoelectron Spectroscopy*, CRS Press, Boca Raton, Fl, 1994.
25. Honma, T., Benino, Y., Komatsu, T., Sato, R. & Dimitrov, V. *Phys. Chem. Glasses*, 2002, **43**, 32.
26. Salim, M., Khattak, G., Tabert, N. & Wenger, L. *J. Electron Spectrosc. Related Phenomena*, 2003, **128**, 75.
27. Losev, A., Kostov, K. & Tyuliev, G. *Surf. Sci.*, 1989, **213**, 564.

Ab initio calculations on borate systems

G. Ferlat,[1] L. Cormier, F. Mauri, E. Balan, G. Calas

Institut de Mineralogie et de Physique des Milieux Condenses, UMR CNRS-IPGP-Universites Paris 6 et 7 n° 7590, case 115 4 place Jussieu, 75252 Paris cedex 05, France

T. Charpentier

Laboratoire Claude Frejacques, URA CNRS-CEA n° 331 CEA Saclay, 91191 Gif-sur-Yvette cedex, France

E. Anglada

Departamento de Fisica de la Materia Condensada, Universidad Autonoma de Madrid E-28049 Madrid, Spain

This paper is part of an ongoing effort aimed at modelling the structure of liquid and vitreous B_2O_3 within a first principles framework. Using density-functional theory, pseudoatomic orbitals basis sets and pseudopotentials, we carried out a series of calculations on known borate systems, namely boric acid $B(OH)_3$ and the two polymorphic B_2O_3 crystals. The results obtained show that rather small basis sizes can compete with the much more expensive planewave ones. This paves the way for an efficient description of the disordered phases using ab-initio *molecular dynamics simulations.*

1. Introduction

Borate glasses, crystals and melts are systems of considerable interest from both fundamental and technological point of views. Boron trioxide B_2O_3, as a key component of such systems, has attracted much attention. Regarding its structural description, it is however striking that even in such a simple case, there remain much debated issues such as the existence of phase transitions between amorphous,[1,2] and even liquid phases[3] or the importance of medium range superstructural units, i.e. the so-called boroxol (B_3O_6) rings, below the glass transition.[4,5] Regarding the latter issue, the theoretical description of B_2O_3 glasses, initiated more than two decades ago by molecular dynamics (MD) simulations,[6–8] appeared to be quite challenging. Indeed, while most of the experimental information tends to establish that a high proportion (~80%) of the boron atoms are involved in boroxol units, the numerical investigations predicted so far a much lower figure (in between 0 and 30%). This failure has been attributed to the use of empirical force fields and possibly to the use of unrealistic quenching rates in the simulated samples.

In fact, simulations using potentials restricted to pair interactions did not succeed at all in reproducing boroxos-containing structures and it appeared that the inclusion of higher order (three- and even four-body) interactions are required in order to get some of the BO_3 triangles to connect into boroxol rings.[9–11] It has been suggested that the potential should include forces necessary to account for the ring stabilisation energy which arises from delocalised π-bonding.[4] There has been several *ab-initio* total energy calculations undertaken on finite-size clusters[12–14] or crystals[15–18] and such types of calculations have been used to derive empirical potentials for MD simulations.[10,18,19] However, the obtained fraction of boroxols was in any case below 30%. There has not been yet any full *ab-initio* MD simulations of bulk liquid and glassy B_2O_3 although it is clear from the above considerations that a first-principles description is highly desirable. This lack stems from the highly time demanding needs of such simulations which until recently have limited the size of the simulated system to several tens of atoms. In addition, the accessible time-scale over which the simulations are carried out are typically of the order of few picoseconds. These limitations in size and time severely hinder the simulation of a realistic glass.

However, significant progress toward more realistic system sizes (and simulation times) have been made recently, among which the development of the SIESTA code.[20] This code, based on the the density functional theory (DFT), allows for a very efficient resolution of the Kohn–Sham equations thanks to the use of pseudo-atomic orbitals basis; it is thus very adapted to simulations of disordered systems for which large unit cells are required. In addition, the solution to the eigenvalue problem can be performed either by diagonalisation or with the linear scaling solver of Kim *et al.*[21] This allows for simulation times scaling essentially like the number of particles instead of the usual cubic behaviour (see for instance Fernández-Serra *et al*[22] for a successful

[1] Corresponding author

application to water).

These considerations motivated us to undertake an extensive description of B_2O_3 disordered phases with the SIESTA code. The purpose of the present work is to assess the validity of this approach by undertaking calculations on a series of known crystalline borate systems. To this end, results obtained with different numerical atomic orbitals (NAO) basis sets are confronted with calculations obtained using a plane-wave (PW) basis as well with experimental data. This allowed us to obtain efficient NAO basis sets to be used in forthcoming *ab-initio* molecular dynamics simulations of liquid and glassy B_2O_3.

2. Calculations

2.1. Technical details

First principles calculations were performed within the density functional theory (DFT) framework using the generalised gradient approximation (GGA) PBE exchange correlation functional.[23] Core electrons are replaced by norm-conserving Troullier–Martins pseudopotentials,[24] in the Kleinman–Bylander[25] form. Core radii of 1·4 and 1·7 a.u. were used for oxygen and boron, respectively. For calculations on $B(OH)_3$ and B_2O_3 crystals, a 2×2×2 Monkhorst–Pack points[26] grid was used to sample the Brillouin zone.

PW calculations on $B(OH)_3$ were carried out with the PWSCF code[27] using a 50 Ry energy cutoff for the plane-wave expansion of the electronic wavefunctions. For SIESTA[20] calculations, the real-space integrations were carried out on a grid whose fineness is defined by a 200 Ry energy cutoff. The basis sets used in SIESTA are made of localised atomic orbitals: we used DZP (double-ζ polarised) basis sets which represent a good compromise between efficiency and accuracy. They amount to two 2s orbitals, two 2p shells and one 3d shell for both oxygen and boron atoms.

2.2. Generation of the NAO basis sets

The performance of several DZP basis sets were tested and we present here the results obtained for two of them. These two basis differ in the cutoff radii of the support regions of their wave functions.

The first basis, hereafter referred as DZP_{auto} is obtained using the now standard procedure among SIESTA users: all radii are defined by one single parameter, the energy shift, i.e. the energy penalty payed by the orbital when confined.[28] This parameter was set to 150 meV. The second basis, hereafter referred as DZP_{opt} is obtained using the variational basis optimisation scheme proposed in Anglada *et al*:[29] using a downhill-simplex minimisation, parameters defining the basis are varied so as to minimise the enthalpy (E+PV) of a given reference system. The fictitious pressure, used to compress the orbitals radii and thus to reduce the computational efforts, was set to 0·2 GPa. The parameters that were allowed to vary were the orbitals cutoff radii and fractional charges for each atom type. The latter parameters influence the shape of an orbital: in cations, orbitals tend to shrink while they tend to swell for anions. Therefore, introducing a δQ parameter in the basis generating free atom calculations gives orbitals better adapted to ionic situations in the condensed system.[30]

The reference system is a molecule or solid in which the atoms considered have a prominent role and which is small enough to allow many self consistent calculations with different basis parameters. We choose the boroxol molecule with hydrogens to cap the nonbridging oxygens as reference system ($B_3O_6H_3$). The hydrogens pseudopotentials and basis parameters were taken from a recent study of liquid water.[22,31]

The parameters used to define each basis set are summarised in Table 1. It is observed that the range of extension of the orbitals radii is about the same for both basis (in between 5 and 6 a.u. for the occupied ones) but while DZP_{auto} provides cutoff radii longer for boron than for oxygen, DZP_{opt} does the opposite. As will be shown in the next sections, the latter basis provides better structural results. In addition, the computational cost using DZP_{opt} is reduced by about 50% as compared to calculations carried out with DZP_{auto}.

As for the boroxol ring molecule, both basis allowed to obtain interatomic distances which match the experimental values (obtained in the glass) and Hartree–Fock values[12] (obtained with a STO-3G basis) within 1%. The highest discrepancy (1·2%) is found for the B–O intra-ring distance (1·386 *versus* 1·37 Å).

We shall now test the transferability and the performances of the obtained DZP basis by carrying out calculations on boric acid and boron trioxide crystals.

2.3. Boric acid $B(OH)_3$

Boric acid $B(OH)_3$, also known as sassolite mineral, is a molecular crystal having a layered structure with hydrogen bonding within the layers. It was determined by x-ray diffraction[32] that the crystals are triclinic with four molecules in the unit cell and $P\bar{1}$ space group symmetry. Periodic Hartree–Fock

Table 1. Parameters describing the DZP basis sets

	r_c^B (a. u.)			r_c^O (a. u.)			δQ_B	δQ_O
	s	p	d	s	p	d	|e|	|e|
DZP_{auto}	5·16	6·30	6·30	3·56	4·24	4·24	0	0
DZP_{opt}	4·52	5·54	4·33	5·43	6·30	3·84	1·60	−0·16

calculations have been carried out previously.[33] Using the experimental structure,[32] we performed relaxations of the coordinates using a PW basis and the two DZP basis derived in the previous subsection. The results are compared in Table 2.

We note that DZP results (for either basis) compare very well with the PW ones. This means that the errors due to the use of localised basis sets is very small. It is indeed smaller that the errors introduced by the use of pseudopotentials and/or the use of the PBE exchange-correlation functional (as seen from the comparison of the PW results with the experimental data). Errors arising from whichever calculation on the B–O distance are about 2% (errors involving hydrogen atoms are much bigger due to the use of a pseudopotential for this atom but since our aim is the study of B_2O_3 we are not concerned by these errors).

2.4. B_2O_3 crystals

We now turn to the description of B_2O_3 crystals. There are two known polymorphs, hereafter referred as B_2O_3-I and B_2O_3-II, in which the boron atoms have different coordination numbers[34,35] (see for instance Figures 1 and 2 in Takada *et al*[15]). B_2O_3-I is stable at ambient pressure while B_2O_3-II is a high pressure phase. The former has hexagonal symmetries while the latter is orthorhombic.

A former *ab-initio* (plane wave pseudopotential) study carried out within the local density approximation (LDA) did not succeed in reproducing the experimental energy ordering and phase transition pressure between the two phases.[15] It was later shown that this failure was due to the use of the LDA: a proper account could be obtained[17] using the Perdew & Wang's GGA.[36] We shall use the more recent PBE GGA functionality and show that our localised basis sets can well reproduce the transition features as well.

The equilibrium curves for each crystal were obtained by allowing the atomic coordinates to relax from the experimental ones for different values of the volume cell. As in previous studies,[15,17] we considered only isotropic volume changes.

The data obtained with the DZP_{opt} basis are shown Figure 1 (very similar curves were obtained with DZP_{auto}). The values for the equilibrium volume (relative

Table 2. Interatomic distances (in Å) obtained in $B(OH)_3$ crystal using different basis sets

	DZP_{auto}	DZP_{opt}	PW (50 Ry)	Expt, Ref. 32
B–O	1·381	1·376	1·387	1·36
O_1–H_1	1·01	1·00	1·01	0·96
O_2–H_2	1·01	1·00	1·01	0·83
O_3–H_3	1·01	1·00	1·01	0·91
B–$B_{intralayer}$	4·08	4·08	4·08	4·09
B–$B_{interlayer}$	7·64	7·65	7·64	7·66

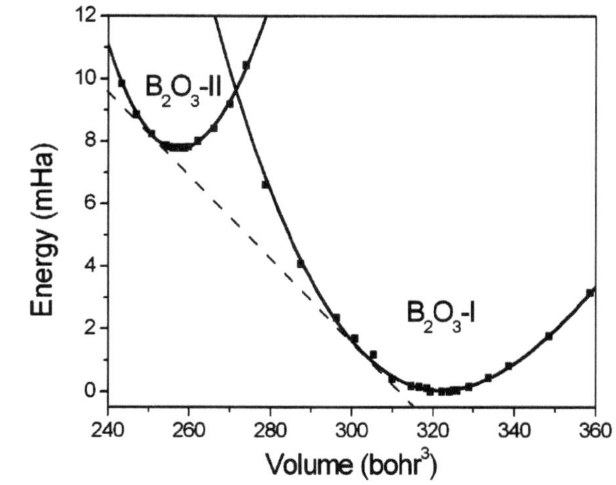

Figure 1. Energy (per B_2O_3 unit) as a function of the volume cell for B_2O_3-I and B_2O_3-II, crystals. The continuous lines are polynomial fits to the data and provided as guide for the eye. The values have been shifted to make the minimum of the B_2O_3-I curve equal to zero. The dashed straight line indicates the tangent line to both equilibrium curves (Gibbs construction): its slope is used to determine the transition pressure (4·1 GPa)

to the experimental ones) are reported in Table 3 as well as the energy and volume difference between the two polymorphs.

A slight expansion of the equilibrium volume cell is obtained (of the order of 5 and 2% for B_2O_3-I and B_2O_3-II, respectively) which results from the tendency of GGA functionals to overestimate the interatomic bond lengths. However, the B_2O_3-I crystal appears more stable than the B_2O_3-II, one, in agreement with experimental observations. Moreover, the values for the pressure transition between the two polymorphs (obtained from the Gibbs construction, see Figure 1) agree fairly well with the somehow uncertain experimental determination of 2 GPa.[37,38] We note however a dependency on the obtained value with the basis used (see Table 3).

It is worthwhile to note that by reducing the B_2O_3-I volume cell to 73% of its original value, a transition to a new structure is observed in which part of the boron atoms became fourfold coordinated (and part of the oxygens became threefold coordinated). This

Table 3. Equilibrium volumes (V_{eq}) and energies (E_{eq}) obtained in B_2O_3 crystals using different basis sets. V_{exp} stand for the experimental known equilibrium volumes.[34,35] The transition pressures are obtained from the Gibbs construction (slope of the tangential line to the equilibrium curves)

	DZP_{auto}	DZP_{opt}	PW (Engberg[17])
B_2O_3-I: $(V_{eq}-V_{exp})/V_{exp}$ (%)	4·7	5·5	5·4
B_2O_3-II: $(V_{eq}-V_{exp})/V_{exp}$ (%)	1·5	2·7	2·0
$V_{eq}(B_2O_3\text{-II})-V_{eq}(B_2O_3\text{-I})$ (bohr)	61	65	54·8
$E_{eq}(B_2O_3\text{-II})-E_{eq}(B_2O_3\text{-I})$ (mHa)	2·65	7·8	5·8
$P_{transition}$ (GPa)	1·4	4·1	3·1

is a pressure induced transition although the final structure is not exactly the same as B_2O_3-II.

Both phases are found to be insulators: electronic gap of 7·1 and 9·1 eV are obtained for B_2O_3-I and B_2O_3-II: respectively (a previous study,[16] carried out within the LDA, gave values of 6·2 and 8.85 eV, respectively).

3. Conclusions

Ab-initio calculations using the SIESTA code have been carried out on a series of borate systems: the boroxol ring molecule: the boric acid crystal and the two known B_2O_3 crystals. Special efforts were devoted to obtaining and testing different atomic orbitals basis sets by comparing the obtained results to plane-wave calculations and experimental data. In all cases: it is shown that results of comparable accuracy to PW ones can be obtained with a rather small basis (DZP like). Indeed: the errors introduced by the use of localised orbitals appeared to be smaller than other sources of errors in the calculations (such as the use of the PBE exchange-correlation functional and pseudopotentials). The magnitude of these errors were found to be within the typical error-bar of DFT-GGA calculations (i. e. ~2% on interatomic distances: ~5% on volume cell).

The derived basis sets have been used for molecular dynamics simulations of B_2O_3 liquid and the results will be reported elsewhere. Our aim is an extensive exploration of the liquid at high pressure and high temperature as well as the description of the amorphous phases.

We note that the characteristics of the high pressure transition between B_2O_3 polymorph are well reproduced by our calculations: a value of about 4 GPa is obtained for the transition pressure and the change of boron from three- to fourfold coordination can be reproduced by the compression of B_2O_3-I. This is particularely encouraging for the on-going high pressure studies of the liquid and amorphous phases.

Acknowledgements

Technical support from A. P. Seitsonen is acknowledged. Part of the calculations have been performed at the French supercomputing center IDRIS. G.F. would like to thank M. Fernandez-Serra for interesting discussions.

References

1. Nicholas, J., Sinogeikin, S., Kieffer, J. & Bass, J. *Phys. Rev. Lett.*, 2004, **92**, 215701.
2. Lee, S. K., Mibe, K., Fei, Y., Cody, G. D. & Mysen, B. O. *Phys. Rev. Lett.*, 2005, **94**, 165507.
3. Brazhkin, V., Katayama, Y., Inamura, Y., Kondrin, M., Lyapin, A., Popova, S. & Voloshin, R., *JETP Lett.*, 2003, **78**, 393. See also erratum in *JETP Lett.*, 2004, **79**, 308.
4. Hannon, A. C., Grimley, D. I., Hulme, R. A., Wright, A. C. & Sinclair, R. N. *J. NonCryst. Solids*, 1994, **177**, 299.
5. Swenson, J. & Borjesson, L. *Phys. Rev. B*, 1997, **55**, 11138.
6. Soules, T. F. *J. Chem. Phys.*, 1979, **71**, 4570.
7. *J. Chem. Phys.*, 1980, **73**, 4032.
8. Soules, T. F. & Varshneya, A. K., *J. Am. Ceram. Soc.*, 1981, **64**, 145.
9. Inoue, H., Aoki, N. & Yasui, I., *J. Am. Ceram. Soc.*, 1987, **70**, 622.
10. Takada, A., Catlow, C. R. A. & Price, G. D. *J. Phys., Condens. Matter*, 1995, **7**, 8693.
11. Maranas, J. K., Chen, Y., Stillinger, D. K. & Stillinger, F. H. *J. Chem. Phys.*, 2001, **115**, 6578.
12. Zhuang, H.-Z., Zou, X.-W., Jin, Z.-Z. & Tian, D.-C. *Phys. Rev. B*, 1995, **52**, 829.
13. Uchino, T. & Yoko, T. *J. Chem. Phys.*, 1996, **105** (10), 4140.
14. Zwanziger, J. W. *Solid State Nucl. Magn. Reson.*, 2005, **27**, 5.
15. Takada, A., Catlow, C. R. A., Lin, J. S., Price, G. D., Lee, M. H., Milman, V. & Payne, M. C. *Phys. Rev. B*, 1995, **51**, 1447.
16. Li, D. & Ching, W. Y. *Phys. Rev. B*, 1996, **54**, 13616.
17. Engberg, U. *Phys. Rev. B*, 1997, **55**, 2824.
18. Teter, M. *Ab-initio* based studies of borate glasses. *Proc Second Int. Conf on Borates Glasses, Crystals and Melts*. Soc. Glass Tech., Sheffield, 1997, 407–16.
19. Cormack, A. N. & Park, B. *Phys. Chem. Glasses*, 2000, **41** (5), 272–7.
20. Soler, J. M., Artacho, E., Gale, J. D., Garcia, A., Junquera, J., Ordejon, P. & Sanchez Portal, D. *J. Phys., Condens. Matter*, 2002, **14**, 2745.
21. Kim, J., Mauri, F. & Galli, G. *Phys. Rev. B*, 1995, **52**, 1640.
22. Fernández-Serra, M. V. & Artacho, E. *J. Chem. Phys.*, 2004, **121**, 11136.
23. Perdew, J. P., Burke, K. & Ernzerhof, M. *Phys. Rev. Lett.*, 1996, **77**, 3865.
24. Troullier, N. & Martins, J. L. *Phys. Rev. B*, 1991, **43**, 1993.
25. Kleinman, L. & Bylander, D. M. *Phys. Rev. Lett.*, 1982, **48**, 1425.
26. Monkhorst, H. J. & Pack, J. D. *Phys. Rev. B*, 1976, **13**, 5188.
27. Baroni, S. PWSCF. Http://www.pwscf.org.
28. Artacho, E., Sanchez-Portal, D., Ordejon, P. & Garcia, A., *Phys. Status Solidi B*, 1999, **215**, 809.
29. Anglada, E., Soler, J. M., Junquera, J. & Artacho, E. *Phys. Rev. B*, 2002, 66, 205101.
30. Junquera, J., Paz, O., Sanchez-Portal, D. & Artacho, E. *Phys. Rev. B*, 2001, 64, 235111.
31. Fernández-Serra, M. V., Ferlat, G. & Artacho, E., *Mol. Simul.*, 2005, **31**, 361.
32. Zachariasen, W. H. *Acta Crystallogr.*, 1954, **7**, 305.
33. Zapol, P., Curtiss, L. A. & Erdemir, A. *J. Chem. Phys.*, 2000, **113**, 3338.
34. Gurr, G. E., Montgomery, P. W., Knutson, C. D. & Gorres, B. T. *Acta Crystallogr.*, 1970, **B26**, 906.
35. Prewitt, C. T. & Shannon, R. D., *Acta Crystallogr*, 1968, **B24**, 869.
36. Perdew, J. P., Burke, K. & Wang, Y. JOURNAL TITLE? 1991.
37. Dachille, F. & Roy, R. *J. Am. Ceram. Soc.*, 1959, **42**, 78.
38. Mackenzie, J. D. & Claussen, W. F. *J. Am. Ceram. Soc.*, 1961, **44**, 79.

Proc. Fifth Int. Conf. on Borate Glasses, Crystals and Melts Phys. Chem. Glasses: Eur. J. Glass Sci. Technol. B, August 2006, 47 (4), 445–448

XAS study of barium borate glasses and crystals

Katsumi Handa[1]

Synchrotron Radiation Center, Ritsumeikan University, 1-1-1 Nojihigashi Kusatsu, Shiga 525-8577, Japan

Junko Ide, Yoko Nishiyama, Kazuhiko Ozutsumi

Department of Applied Chemistry, Faculty of Science and Engineering, Ritsumeikan, University, 1-1-1 Nojihigashi, Kusatsu, Shiga 525-8577, Japan

Giuseppe Dalba

INFM and Dipartimento di Fisica dell'Universita di Trento, Via Sommarive 14 – 38050 Povo (Trento), Italy

Norikazu Ohtori

Graduate School of Science and Technology, Niigata University, Niigata 950-2181, Japan

Norimasa Umesaki

Japan Synchrotron Radiation Research Institute, Spring-8, Mikazuki, Sayo 679-5198, Japan

Barium borate $xBaO$–$(1-x)B_2O_3$ and lithium borate xLi_2O–$(1-x)B_2O_3$ glasses were investigated by x-ray absorption spectroscopy at the boron K-edge and barium L_{III}-edge. This is the first XAS work at the K edge of boron in lithium borate and barium borate glasses. A resonant peak at the B K-edge was observed at around 194·6 eV; its intensity and area decrease with increasing the amount of four-coordinated B changed by increasing the modifier oxide content and reach their minima at 45 mol% BaO and 40 mol% Li_2O. Ba $N_{II,III}$-edges were also observed. With increasing BaO content in barium borate glasses, the local structure around Ba changes from a single shell structure of six-coordinated O atoms at a Ba–O interatomic distance of 2·77 Å to a two-shell structure with distances 2·67 and 2·90 Å and coordination numbers four and 3–4, respectively.

1. Introduction

The physical properties of barium borate (BBO) glasses, like those of many glass forming families, depend strongly on the addition of metal ions. The structure of the glasses is also expected to change with changing modifier. Barium borate xBaO–$(1-x)B_2O_3$ is well known as one of the binary alkaline earth borate systems with wide glass forming range going from x=16·7–66·7 mol%.[1] The crystalline phases α- and β-BBO (BaB_2O_4) have been also extensively studied in order to elucidate its structural and optical properties.[2–6]

We have prepared barium borate glasses xBaO–$(1-x)B_2O_3$ with x ranging from 10·0–60·0 mol% by the usual quenching method, without resorting to the twin-roller rapid quenching method. Lithium borate (LBO) glasses were made with Li_2O content ranging from 5–65 mol%.

X-ray absorption spectroscopy is a powerful tool for the investigating the electronic structure and the local geometric structure around x-ray absorbing atom.[7–10] We measured and examined the XAS spectra of B K-edge and Ba L_{III}-edge of BaO–B_2O_3 glasses, and compared B K-edge spectra of BBO glasses with those of LBO glasses.

2. Experimental procedures

Thirteen barium borate and thirteen lithium borate glasses were prepared from extra-reagent grade barium carbonate, lithium carbonate and boric acid by usual quenching method. The samples were heated in platinum crucibles in an electric furnace at temperatures in the range 1073–1623 K for periods of 30 min. All samples were checked by Raman spectroscopy.[11–12]

B K-edge and Ba L_{III}-edge XAS measurements were carried out at the beamlines BL-2 and BL-4 of the synchrotron facility in Ritsumeikan University.[13,14] The stored electron beam energy was 575 MeV, the radius of the beam 0·5 m, and the initial beam current is 300 mA. BL-2 is a VUV beamline with a 1200 grooves/mm concave curved grating, and works in the photon energy range 50–700 eV. XAS spectra

[1] Corresponding author

of B were measured by total electron yield mode (TEY) at BL-2. Powder samples for the B K-edge measurements were spread on a carbon double sided tape put on the first dynode of an electron multiplier (Hamamatsu, R595). Ba L_{III}-edge XAS spectra were measured by transmission mode at BL-4, which is a XAFS beamline with a Si(220) double crystals monochromator (Toyama). Ionisation chambers filled with N_2 gas were employed as detectors for measuring the incident and the transmitted x-ray beam intensities. The Ba L_{III}-edge XAS data were analysed using standard software packages, Athena and Artemis,[15] REX2000 (Rigaku), and EXAFS (Technos), and they were also checked by FEFF8.2 code.[16] B_2O_3, BaO powder, α-, β-BBO (Casix) and LBO (Chichibu Fuji) single crystals and powders were used as model compounds.

3. B K-edge XAS

B K-edge x-ray absorption near edge structure spectra (XANES) of $BaO–B_2O_3$ glasses and some crystalline reference compounds are shown in Figure 1(a); Figure 1(b) shows B K-edge XANES spectra of $Li_2O–B_2O_3$ glasses together with the spectrum of a reference crystalline compound. The XANES spectra were normalised at 210·0 eV. In the case of B_2O_3 glasses and their reference crystalline compounds, two peaks were observed: one sharp peak located at 194·6 eV, which is assigned to the transition from 1s to $p\pi^*$ state and a broader peak at 203·6 eV. The sharp peak slightly shifts to lower energies with increasing the modifier content x and its intensity reaches a minimum at 40 mol% Li_2O and 45 mol% BaO concentrations for $Li_2O–B_2O_3$ systems and $BaO–B_2O_3$ systems, respectively. Moreover, the FWHM value of the sharp peak increases with increasing the modifier content. The area of the first peak reaches a minimum at 35 mol% Li_2O and 40 mol% BaO concentrations, respectively. As for the second broader peak, its position is strongly influenced by the concentration of the modifier. However, since the shape of this peak can be influenced by both three- and four-coordinated B–O transitions induced by the modifier content,[7–10] a deconvolution analysis is not simple and suitable.

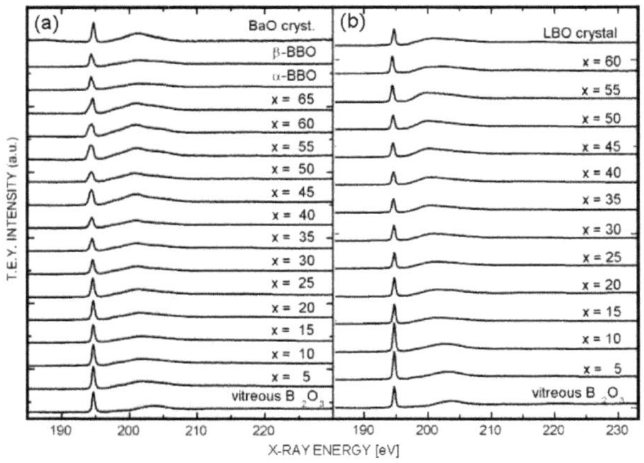

Figure 1. B K-edge XANES spectra of barium borate glasses $xBaO–(1-x)B_2O_3$ and lithium borate $xLi_2O–(1-x)B_2O_3$ systems. A sharp resonance peak related to 1s-pπ^* transitions of three coordinated boron and a subsequent broader peak are evident

Total amounts of three-coordinated B are decreased with increasing four-coordinated B. Therefore, corresponding to the procedure of four-coordinated B in lithium borate glasses, the area and the intensity of the first peak are influenced strongly as shown in Figure 1(b). However, it is difference from the case of barium borate glasses. As shown in Figure 1(a), Ba $N_{II,III}$-edge of BaO is located at 192 and 178·6 eV, respectively. Therefore, XAS spectra as shown in Figure 1(a), are composed of B K-edge of B–O and Ba $N_{II,III}$-edge of Ba–O in barium borate glasses.

B XANES spectrum of BPO_4 crystal as a reference sample of four-coordinated B shows a sharp peak at 199·2eV and a broad peak at 203·8 eV, however it cannot be ignored the chemical shift because of the phosphorous atoms and also the first peak is observed as a bit broader than threee-coordinated B peak at 194·6 eV. Therefore, the details of the first resonance peak of three-coordinated B were examined. However, the sharp peak cannot be deconvoluted with both Gaussian and Lorentzian curves, precicely. This indicates that the first resonance peak includes the unit structures of three-coordinated B such as a BO_3 unit, a BO_3 unit with NBO, and a boroxol ring, etc. On the other hand, the position, the intensity and the values of FWHM of the first resonance peak contain information of not only three-coordinated B but also four-coordinated B.

4. Ba L_{III}-edge XAFS

Figure 2 shows the radial distribution functions obtained from Ba L_{III}-edge XAFS spectra; the structural parameters are listed in Table 1. The local structure around Ba of systems with low BaO content is similar to the structure of crystalline BaO where the Ba-O inter-atomic distance is 2·77 Å with six oxygen atoms. With increasing the BaO content in $BaO–B_2O_3$ glasses, two Ba–O distances are observed one at 2·67 Å with four oxygens and 2·90 Å with 3–4 oxygens. The structure around Ba changes becoming similar to the structure of BBO crystals at high BaO content.

5. Conclusion

B K-edge XANES spectra of barium borate glasses show changes of the local structure around B atoms similar to those already observed in alkali borosilicate

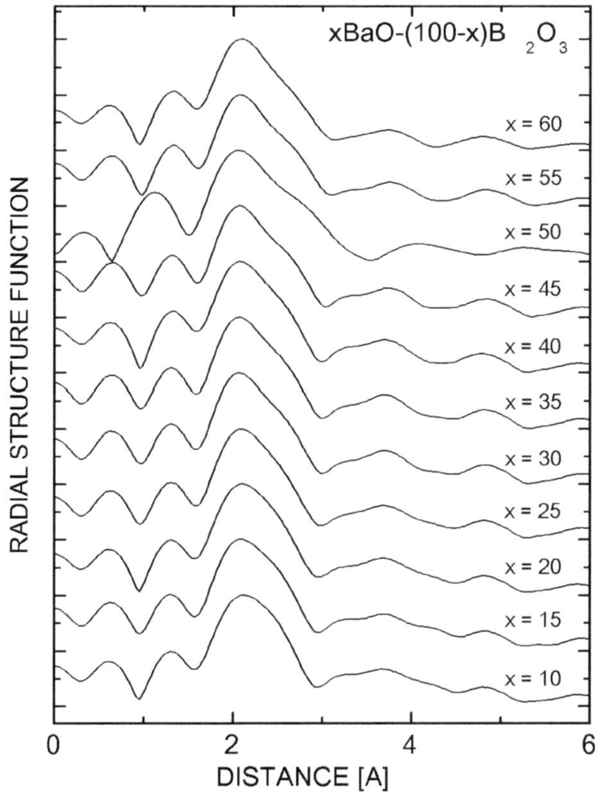

Figure 2. Radial structure function of Ba–O in barium borate glasses, which is obtained Ba L_{III}-edge XAS and corrected phase shift of EXAFS

glasses.[9] The resonance peak at 194·6 eV of B K-edge XAS shifts slightly to lower energy with addition of modifier oxide and the peak intensity reaches a minimum at 40 mol% Li_2O concentrations in xLi_2O–$(1-x)B_2O_3$ and 45 mol% BaO in xBaO–$(1-x)B_2O_3$. The peak position of the broader peak is strongly influenced by the concentration of the modifier oxide. However, it is composed of complex transitions of both three- and four-coordinated B. With increasing four-coordinated B content, the intensity and the area of the first resonance peak of B K-edge decrease, and it reach their minima at 40 mol% Li_2O and 45 mol% BaO in litium borate and barium borate glasses respectively. Unfortunately, Ba $N_{II,III}$-edges are located in the same region and the XAS spectra in Figure 1(a)

Table 1. Local structure parameters around Ba obtained from Ba L_{III}-edge XAS

mol%	$N_{B\text{-}O}$ (atoms)	$R_{Ba\text{-}O}$ (Å)
10	6·1±0·2	2·77±0·02
15	5·7±0·2	2·76±0·03
20	5·9±0·3	2·76±0·02
25	5·6±0·3	2·74±0·03
30	5·7±0·2	2·74±0·02
35	3·4±0·2	2·68±0·02
	2·8	2·89
40	3·6±0·2	2·68±0·02
	3·0	2·89
45	3·4±0·2	2·67±0·02
	3·2	2·88
50	3·7±0·2	2·68±0·02
	3·1	2·90
55	3·6±0·2	2·67±0·02
	3·2	2·90
60	3·8±0·2	2·68±0·02
	3·1	2·90
65	3·9±0·2	2·69±0·02
	3·4	2·92

were convoluted by both Ba $N_{II,III}$-edge and B K-edge. For this reason, the relationship with the peak and four-coordinated B are still unknown. Ba L_{III}-edge EXAFS show the local structural change around Ba atoms. The Ba–O inter-atomic distances changed from 2·77 Å with six oxygen atoms to 2·67 Å with four oxygen atoms and 2·90 Å with 3–4 oxygen atoms with increase of the BaO ratio in BaO–B_2O_3 glasses.

References

1. Kapoor, S., George, H. B., Betzen, A., Affatigaton, M. & Feller, S. *J. Non-Cryst. Solids*, 2000, **270**, 215–22.
2. Xue, D. F. & Zhang, S. Y. *Acta Crystallogr.*, 1998, **B54**, 652–6.
3. Mighell, A. D., Perloff, A. & Block, S. *Acta Crystallogr.*, 1966, **20**, 819–23.
4. Tsvetkov, E. G., Khranenko, G. G. & Solntsev, V. P. *J. Cryst. Growth*, 2005, **275**, 2123–8.
5. Meshalkin, A. B. & Kaplum, A. B. *J. Cryst. Growth*, 2005, **275,** (2005) 301-305.
6. Honma, T., Benino, Y., Fujiwara, T., Sato, R. & Komatsu, T. *J. Phys. Chem. Solids*, 2004, **65**, 1705–10.
7. Dalba, G., Fornasini, P., Kuzmin, A., Monti, F., Sanson, A., Sipr, O. & Rocca, F. *Phys. Scr.*, 2005, **T115**, 149–51.
8. Li, D. & Ching, W. Y. *Phys. Rev.*, 1996, **B54,** 13616–22.
9. Fleet, M. E. & Muthupari, S. *J. Non-Cryst. Solids*, 1999, **255**, 233–41.
10. Fleet, M. E. & Muthupari, S. *Am. Miner.*, 2000, **85**, 1009–21.
11. Konijnendijk, W. L. & Stevels, J. M. *J. Non-Cryst. Solids*, 1975, **18,** 307.
12. Kamitsos, E. I. *J. Non-Cryst. Solids*, 1990, **126**, 52.
13. Ozutsumi, K. & Handa, K. *Rev. Sci. Instrum.*, 2004, **75**, 111–18.
14. Handa, K., Ozutsumi, K. & Kojima, K. *Phys. Scr.*, 2005, **T115**, 992–4.
15. Ravel, B. & M. Newville *J. Synchrotron Rad.* **12** (2005), , 537-541.
16. Rehr, J. J. & Albers, R. C. *Rev. Mod. Phys.*, 2000, **72**, 621–54.

Proc. Fifth Int. Conf. on Borate Glasses, Crystals and Melts   Phys. Chem. Glasses: Eur. J. Glass Sci. Technol. B, August 2006, 47 (4), 449–454

A parameterisation for the composition dependence of N_4 in borate glasses

*Alex C. Hannon**

ISIS Facility, Rutherford Appleton Laboratory, Chilton, Didcot, Oxon, OX11 0QX, UK

Diane Holland

Department of Physics, University of Warwick, Coventry, CV4 7AL, UK

N_4 is defined to be the fraction of borons in a borate which are four-coordinated, and this can be determined reliably by NMR measurements. Although the composition dependence of N_4 has been measured for numerous glass forming systems, little work has been done to understand its behaviour. We have used the most extensive set of N_4 data in the literature [Jellison Jr., G. E., Feller, S. A. & Bray, P. J. Phys. Chem. Glasses, *1978,* **19**, *52], for lithium borate glasses, to develop a parameterisation for the composition dependence of N_4. This involves a conversion factor, α, which has the maximum value of two for low modifier content, but decays exponentially above a limiting composition, which occurs when there is a network of alternating BO_3 and BO_4 units. Above this composition the creation of BO_4 units is not favoured because this brings centres of negative charge into close proximity. Our parameterisation is empirical, but has potential for use in the characterisation and comparison of N_4 data from a wide variety of borate glass systems.*

Pure vitreous boron oxide, B_2O_3, has a random network structure formed by corner sharing between planar BO_3 units.[1] All of the boron atoms are three-coordinated and all of the oxygen atoms are bridging (i.e. they are bonded to two borons). The addition of modifier oxide, M_aO_b, to the glass introduces surplus oxygen to the network which may be accommodated in two different ways, as shown in Figure 1. The additional oxygen may either be incorporated into the network as nonbridging oxygens (NBOs) by breaking oxygen bridges (Figure 1(b)), or it may be incorporated by the conversion of borons from three-coordination to four-coordination (Figure 1(c)).

Conventionally the coordination in a borate glass is described in terms of N_4, the fraction of borons which are four-coordinated. As discussed below, several different experimental methods have been used to measure N_4 as a function of modifier content for a variety of different modifiers. The ubiquitous result is that, as modifier is added to the glass, N_4 increases initially, until a maximum is reached and then decreases. Biscoe & Warren[2] first observed this behaviour by using x-ray diffraction to measure the B–O coordination number in several sodium borate glasses, and they concluded that the maximum in N_4 provides an explanation of the maxima and minima which are observed in the thermophysical properties of the glasses – the so-called 'borate anomaly'. A particularly clear example of this is given by the glass transition temperature data[3] and density data[4] for lithium borate glasses, which are compared with N_4 data[5] in Figure 2.

Values for the composition-dependence of N_4 have been determined experimentally by a number of techniques, including neutron diffraction,[6–10] Raman scattering,[11,12] infrared,[13] ultrasonic velocity[14] and XANES.[15] However, the most reliable method for its determination is NMR. The first use of NMR to measure N_4 was reported by Silver & Bray,[16] since when there have been numerous reports of NMR measurements of N_4 values for a wide variety of borate glasses. However, despite the fact that NMR has been used for several decades to measure the composition dependence of N_4 in borates, there have been very few attempts either to produce structural models which explain the observed behaviour, or to investigate trends across the large range of systems that has been studied.

The most promising structural model for the composition dependence of N_4 is that due to Wright *et al*,[7] the key assumption of which is that negatively charged BO_4 tetrahedra cannot exist in close proximity to other centres of negative charge. According to this model, the number of BO_4 tetrahedra is the maximum possible subject to the two constraints that there are no bridges between BO_4 tetrahedra and that there are no BO_4 tetrahedra with nonbridging oxygens. The thin line in Figure 2(c) shows the prediction for N_4 in lithium borate glasses for Wright *et al*'s model.[7] Although this model shows considerable potential,

[1] Corresponding author. Email: a.c.hannon@rl.ac.uk

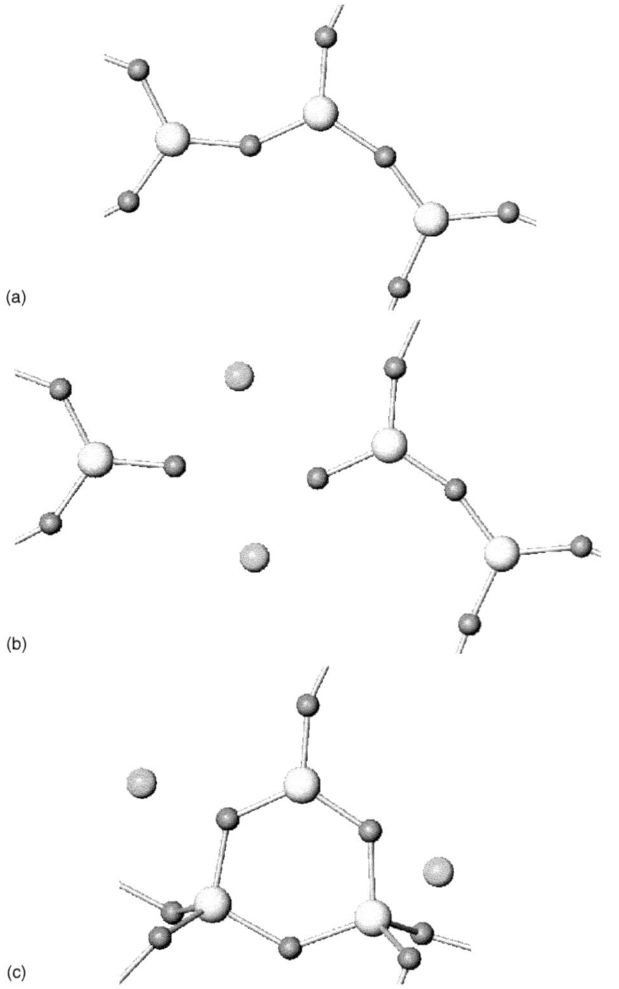

Figure 1. (a) A fragment of borate network – boron atoms are shown larger than oxygen atoms. (b) The network fragment in Figure 1(a) is shown after the incorporation of one unit of M_2O modifier by the conversion of one bridging oxygen to two nonbridging oxygens – modifier cations, M, are shown without bonds. (c) The network fragment in Figure 1(a) is shown after the incorporation of one unit of M_2O modifier by the conversion of two BO_3 units to BO_4. (Note that this arrangement of atoms has been drawn to illustrate simply the way in which four-coordinated boron arises; It should not be taken to indicate whether or not BO_4 units exist in close proximity to each other, or to imply the presence of a particular type of superstructural unit)

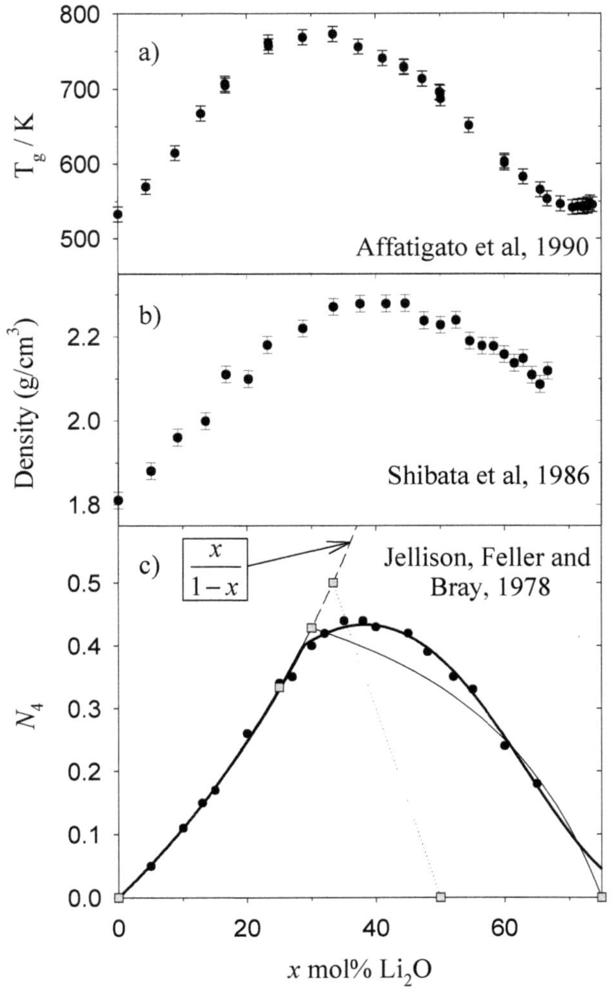

Figure 2. The glass transition temperature, T_g, data[3] (a), density data[4] (b) and N_4 data[5] (c) for lithium borate glasses, all shown as solid circular points. The thin line shows the prediction of the composition dependence of N_4 according to Wright et al*'s model,[7] whilst the thick line shows the result derived in this paper. The square points (connected by a dotted line) show the N_4 data for lithium borate crystals. The dashed line shows the curve which corresponds to maximum creation of BO_4 units in alkali borates*

it predicts the position of the maximum in N_4 at too low a lithia content, and above this composition its prediction for N_4 is mostly too low.

We have two ultimate aims in the study of N_4 data; firstly to gain understanding of the mechanism which drives the growth and subsequent decline in N_4, and secondly to be able to predict the value of N_4 for any particular borate glass composition (i.e. for any value of x in $(1-x)B_2O_3.xM_aO_b$, and for any modifier cation M). The data shown in Figure 2 demonstrate that this has direct relevance for understanding of the variation in physical properties. In this paper we give a preliminary report of an empirical parameterisation of the composition-dependence of N_4, which has been developed using NMR data for lithium borate glasses.[5] We have chosen this system because these experimental N_4 data for lithium borate glasses cover a wider range of composition, with more data points, than for any other system.

Table 1. N4 data for lithium borate crystal structures

Crystal compound		*Reference*	*x (mol% Li_2O)*	N_4
B_2O_3	(19)	0	0	
LiB_3O_5	(20)	25	0·3333	
$Li_3B_7O_{12}$	(21)	30	0·4286	
$Li_2B_4O_7$	(22)	33·33	0·5	
$LiBO_2$	(23)	50	0	
Li_3BO_3	(24)	75	0	

Lithium borate crystal structures

The study of related crystal structures can often shed light on the structure of amorphous materials, and hence the values of N_4 for the known crystals from the Li_2O–B_2O_3 system are given in Table 1. (The structure of the high pressure form of pure B_2O_3,[17] in which the borons are all 4-coordinated, has not been included because it is of doubtful relevance to the present discussion.) The coordination numbers of the boron atoms in each crystal were determined using the XTAL[18] program for calculating interatomic distances and coordination numbers. The N_4 data for the crystal structures are shown in Figure 2(c) by square markers. To a limited extent the crystal data show the same behaviour as for the glasses, in that N_4 rises and then falls as the lithia content increases. As for the glasses, the crystal data at low x follow the curve which is predicted (see below) for maximum conversion of BO_3 units to BO_4 units, with no NBOs. However, beyond 33·33 mol% Li_2O the crystals have no BO_4 units, i.e. N_4 falls discontinuously to zero, unlike the situation for the glasses where N_4 falls away from the curve and then decreases in a gradual and continuous fashion. It is thus concluded that in this case crystal structures are not a good guide to the structural trend in the glasses. It is probable that crystal structures are constrained by the requirements of symmetry, and hence cannot show the variability of N_4 that is shown by glasses. This is because in a glass there is a distribution of atomic environments, whereas in a crystal there is only a small number of different atomic environments.

Parameterisation of the composition dependence of N_4

In order to parameterise the composition dependence of N_4, we will reformulate the problem in terms of two new variables, the conversion factor, α, and the fractional oxygen excess, f_O. For these calculations we will only consider binary borate systems, with composition $(1-x)B_2O_3.xM_aO_b$ ($2b/a$ is the valence of the modifier cation, M).

We define the conversion factor, α, to be the number of BO_3 units converted to BO_4 per added oxygen. (This is not the same as the 'conversion rate' defined by Bray,[25] which is the derivative of our factor α.) We will assume here that all added oxygens are involved in bonding to boron, i.e. we assume there are no oxygens which bridge solely between modifier cations, M. Note that the value of α must lie in the range from zero to two, because one added oxygen can create at most two BO_4 units, as illustrated by Figures 1(a) and (c).

Consider a sample containing N formula units of $(1-x)B_2O_3.xM_aO_b$. The number of added oxygens in this sample is Nxb, and hence the number of 4-coordinated borons is αNxb, according to the definition of α. The total number of borons in this sample is $2N(1-x)$, and hence the fraction of borons which are 4-coordinated is

$$N_4 = \frac{\alpha Nxb}{2N(1-x)} = \frac{\alpha xb}{2(1-x)} \tag{1}$$

The boron–oxygen coordination number is then given by

$$n_{BO} = 3 + N_4 = 3 + \frac{\alpha bx}{2(1-x)} \tag{2}$$

Since it is the addition of oxygen which drives the conversion of BO_3 units to BO_4 units, due to the need to accommodate the added oxygens, we choose to define the fractional oxygen surplus, f_O, by

$$f_O = \frac{bx}{3(1-x)}, \quad x = \frac{3f_O}{b+3f_O} \tag{3}$$

f_O is the ratio of the amount of modifier oxygen (i.e. oxygen in M_aO_b) to the amount of borate oxygen (i.e. oxygen in B_2O_3). f_O is an alternative composition variable, and the composition of the glass can be written as $3f_OM_{a/b}O.B_2O_3$. (Bray[25] and co-workers have preferred to express the composition of binary borate glasses in the form $RM_aO_b.B_2O_3$, where is the ratio of alkali to boron for alkali borate glasses.) Combining Equations (2) and (3) gives

$$N_4 = \frac{3\alpha f_O}{2} \tag{4}$$

and

$$\alpha = \frac{2N_4(1-x)}{bx} \tag{5}$$

Equation (4) makes the important point that the maximum in the plot of N_4 versus x arises simply from a competition between two effects: Firstly the amount of added oxygen rises steadily as x increases; that is to say f_O increases steadily as x increases, according to Equation (3). Secondly the conversion factor α decreases steadily as x increases, as we will show. Thus the maximum N_4 arises when the effect of the decrease in α becomes greater than the effect of the increase in f_O.

At any particular composition, the maximum possible value of N_4 occurs if α=2, in which case $N_4=3f_O=bx/(1-x)$. For low x the data for lithium borates (for which b=1) follow this equation, as shown by the dashed line in Figure 2(c), indicating that the addition of oxygen results solely in the conversion of BO_3 units to BO_4 units, with no significant creation of NBOs.

Figure 3(a) shows the data from Figure 2c plotted in the form of α versus f_O. The conversion factor, α, changes little for low values of f_O, and is close to the maximum conversion limit of two. However, for higher values of f_O there is a steady decline of α. The plot of $\log_e(\alpha)$ versus f_O appears to be linear in this

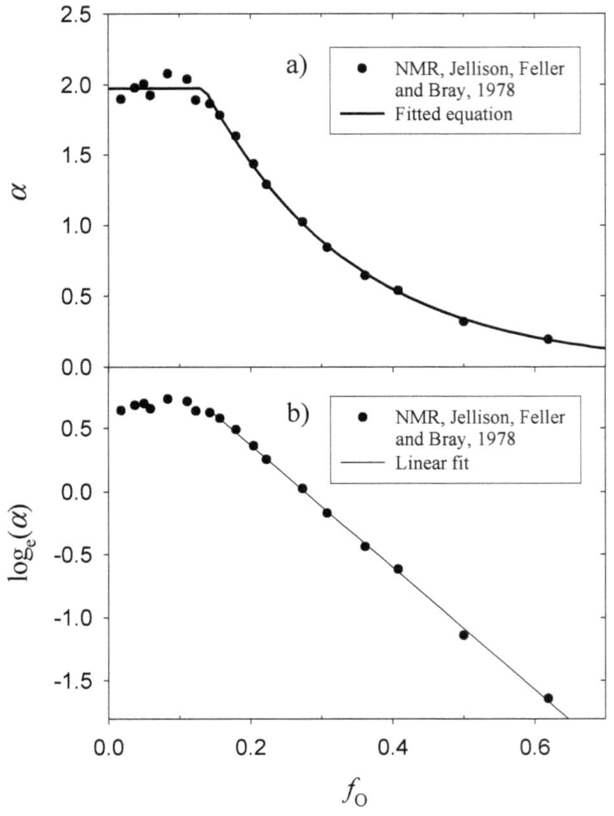

Figure 3. The conversion factor (a) for lithium borate glasses,[5] *α, and its natural logarithm (b), $\log_e(\alpha)$, plotted as a function of fractional oxygen excess, f_O. The thin line indicates a linear fit to the $\log_e(\alpha)$ values for higher f_O, whilst the thick line indicates the fit of the function given in Equation (6)*

region, as shown by the straight line fit in Figure 3(b). This behaviour may thus be expressed as follows

$$\alpha(f_O) = \alpha_{max} \quad \text{if} \quad f_O < f_c$$
$$\alpha(f_O) = \alpha_{max} \exp\left(-\left(\frac{f_O - f_c}{f_{max}}\right)\right) \quad \text{if} \quad f_O > f_c \qquad (6)$$

where f_c represents a critical value of f_O, below which α is constant. f_{max} is the rate of decay of α in the region where f_O is greater than f_c, and is also the value of f_O for which N_4 is at a maximum. That is to say N_4 is at its maximum when $f_O=f_{max}$, or $x=x_{max}=3f_{max}/(b+3f_{max})$. Equation (6) represents our parameterisation for the composition dependence of N_4.

Discussion

The expression given by Equation (6) was fitted to the $\alpha(f_O)$ data shown in Figure 3(a), using a Simplex algorithm to achieve convergence.[26,27] The thick line in both Figure 2(c) and Figure 3(a) shows the fitted function, whilst the fitted parameters are given in Table 2.

Table 2. Parameters from fitting Equation (6) to the $\alpha(f_O)$ data[5] *for lithium borate glasses*

α_{max}	fc	f_{max}
1·975±0·017	0·1354±0·0034	0·2064±0·0064

The fitted value of α_{max} is very close to two, consistent with the observation that for low x the addition of oxygen results solely in the conversion of BO_3 units to BO_4 units, with no significant creation of NBOs. We have found that fixing the value of α_{max} at exactly two makes almost no difference to the other fitted parameters.

It is important to note that according to our parameterisation in Equation (6), the position, x_{max}, of the maximum in N_4 arises from, f_{max}, the exponential decay rate of α, and at this composition the creation of BO_4 units starts to become less favoured. This is to be contrasted with Wright *et al*'s model,[7] which predicts that x_{max} arises when the creation of BO_4 units becomes impossible. Using the fitted value of f_{max} in Equation (3) yields a value of 38·2 mol% Li_2O for x_{max}, which is consistent with the N_4 data shown in Figure 2(c).

Of greatest interest from a structural point of view, is the value for the limiting composition, f_c, above which α decays with composition. The fitted value of f_c corresponds to a composition of 28·9 mol% Li_2O. This value is very close to the composition of 30 mol% Li_2O (or f_O=1/7≈0·143), at which Wright *et al*[7] predict that N_4 starts to deviate from the $x/(1-x)$ curve for maximum creation of BO_4 units. According to Wright *et al*'s model[7] this limit arises when there is a network of alternating BO_3 units and BO_4 units, at which point the value of N_4 is 3/7≈0·429. However, Wright *et al*'s model[7] underestimates both the value and the position for the maximum of N_4. The reason for this is that the model assumes that there are no bridges between BO_4 tetrahedra and that there are no BO_4 tetrahedra with NBOs. The fact that N_4 continues to increase beyond the predicted limit of the Wright *et al* model,[7] but more slowly than at lower x, is evidence that one or both of these configurations can occur, but that they are not favoured. Thus we conclude that the decline in α for values of f_O beyond f_c arises because these configurations are not favoured in this composition range.

Figure 4 shows the diborate group which is found in the crystal structure of lithium diborate.[22] The diborate group involves a pairing of BO_4 units, and its presence in this crystal structure is clear evidence that bonds can occur between two BO_4 units, contrary to one of the simple assumptions of Wright *et al*'s model.[7] The two relevant assumptions of Wright *et al*'s model[7] are firstly that there are no bridges between two BO_4 units, and secondly that there are no BO_4 units with NBOs, and the fact that the diborate group is found in the lithium diborate

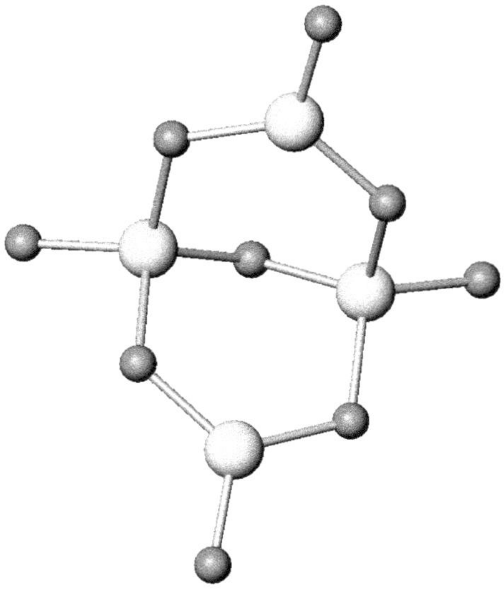

Figure 4. The diborate group in crystalline lithium diborate,[22] *$Li_2O.B_2O_3$*

crystal structure[22] suggests that it is the first of these assumptions which is not fully obeyed. That is to say it is the formation of a few bridges between two BO_4 units, probably in diborate groups, which allows N_4 to increase for compositions beyond f_c, due to the allowed but not favoured formation of some additional BO_4 units. The relatively low value of N_4 for lithium borate glass, compared to the diborate crystal (which has N_4=0·5), indicates that it is likely that the diborate group is less common in the glass than in the crystal. There is qualitative evidence from Raman measurements that diborate groups may exist in lithium borate glasses,[28] but the identification of the signature of diborate groups in the Raman data is inconclusive,[29] and certainly there is no quantitative evidence as to their importance in the structure. Most of the few previous considerations of the composition dependence of N_4 (for example, see Bray *et al*[30]) have ascribed great significance to superstructural units, such as the boroxol group[1] and the diborate group. However, whilst it is clear from Raman scattering results[28,29] that superstructural units do occur in borate glasses, it is not clear whether N_4 is sensitive to their occurrence, and they play no role in our parameterisation.

It should be noted that our parameterisation for the composition dependence of N_4 is empirical, and we have not derived it from a more fundamental consideration. Nevertheless, the exponential form suggests a Boltzmann form with an activation energy origin, similar to that described by Kerner.[31]

The NMR data for borate glasses actually show evidence for two types of BO_3 site,[32,33] and furthermore some recent studies show evidence for two types of BO_4 site.[34] It would be most worthwhile to perform a re-investigation of the lithium borate NMR data, over the same wide composition range as originally employed by Jellison *et al*,[5] to determine the role that these different sites play in the evolution of N_4.

We have already begun a survey of the literature N_4 data from NMR measurements for as wide a range of modifiers as possible,[35] and it appears that there may be chemical effects on N_4, not included in topological models such as that due to Wright *et al*.[7] Our investigations indicate that modifier cations with high valence are likely to show the greatest variation in N_4 behaviour, due to their larger cation potential.[36] The parameterisation described here provides a useful means of characterising this behaviour.

Although we have only considered binary borate glasses here, our parameterisation can also be applied to multicomponent borate glasses. In evaluating the fractional oxygen excess, f_O, only modifiers should be taken into account because the evidence is that, as should be expected, the addition of glass formers to borate glasses does not affect the formation of BO_4 units.[37] However, the theory will require greater modification if the borate glass contains any conditional glass formers.

Finally we note that the N_4 data for lithium borate glasses[5] used here have now been made available on the ISIS Disordered Materials Database.[38]

Conclusions

The most extensive set of data for N_4 in the literature is that reported by Jellison *et al*[5] for lithium borate glasses, and we have used this dataset to develop an empirical equation for the composition dependence of N_4. This involves a conversion factor, α, representing the number of BO_3 units converted to BO_4 per added oxygen. For low modifier content, α has the maximum possible value of two, but above a limiting composition α decays exponentially as the fractional oxygen excess increases. The limiting composition arises when there is a network of alternating BO_3 and BO_4 units, and above this composition the creation of further BO_4 units is possible but not favoured because this brings centres of negative charge (either units or NBOs) into close proximity. Our parameterisation for N_4 shows potential for the characterisation and comparison of the N_4 data for a wide variety of borate glass systems.

Acknowledgements

We are indebted to Professor Steve Feller of Coe College for discussions and provision of experimental data.[5] Professor Adrian Wright of Reading University is acknowledged for useful and stimulating discussions. We also gratefully acknowledge the receipt of a Warwick University Research and Teaching Development grant, number RTDF-SCO10009.

References

1. Hannon, A. C., Grimley, D. I., Hulme, R. A., Wright, A. C. & Sinclair, R. N. *J. Non-Cryst. Solids*, 1994, **177**, 299.
2. Biscoe, J. & Warren, B. E. *J. Am. Ceram. Soc.*, 1938, **21**, 287.
3. Affatigato, M., Feller, S. A., Khaw, E. J., Feil, D., Teoh, B. & Mathews, O. *Phys. Chem. Glasses*, 1990, **31**, 19.
4. Shibata, M., Sanchez, C., Patel, H., Feller, S., Stark, J., Sumcad, G. & Kasper, J. *J. Non-Cryst. Solids*, 1986, **85**, 29.
5. Jellison Jr., G. E., Feller, S. A. & Bray, P. J. *Phys. Chem. Glasses*, 1978, **19**, 52.
6. Vedishcheva, N. M., Shakhmatkin, B. A., Wright, A. C., Grimley, D. I., Etherington, G. & Sinclair, R. N. In: *Fundamentals of Glass Science and Technology* (ESG, Venice), 1993, p. 459–462.
7. Wright, A. C., Vedishcheva, N. M. & Shakhmatkin, B. A. *Mater. Res. Soc. Symp. Proc.*, 1997, **455**, 381.
8. Stone, C. E., Wright, A. C., Sinclair, R. N., Feller, S. A., Affatigato, M., Hogan, D. L., Nelson, N. D., Vira, C., Dimitriev, Y. B., Gattef, E. M. & Ehrt, D. *Phys. Chem. Glasses*, 2000, **41**, 409.
9. Stone, C. E., PhD Thesis, 2001, University of Reading, UK.
10. Ohtori, N., Takase, K., Akiyama, I., Suzuki, Y., Handa, K., Sakai, I., Iwadate, Y., Fukunaga, T. & Umesaki, N. *J. Non-Cryst. Solids*, 2001, **293**, 136.
11. Tatsumisago, M., Takahashi, M., Minami, T., Tanaka, M., Umesaki, N. & Iwamoto, N. *J. Ceram. Soc. Jpn*, 1986, **94**, 464.
12. Yiannopoulos, Y. D., Chryssikos, G. D. & Kamitsos, E. I. *Phys. Chem. Glasses*, 2001, **42**, 164.
13. Kamitsos, E. I. Phys. Chem. Glasses, 2003, 44, 79.
14. Kodama, M., Hirashima, T. & Matsushita, T. *Phys. Chem. Glasses*, 1993, **34**, 129.
15. Handa, K., Ide, J., Nishiyama, Y., Ozutsumi, K., Dalba, G., Ohtori, N. & Umesaki, N. *Phys. Chem. Glasses : Eur. J. Glass Sci. Technol. B*, 2006, **47**, 445–448.
16. Silver, A. H. & Bray, P. J. *J. Chem. Phys.*, 1958, **29**, 984.
17. Prewitt, C. T. & Shannon, R. D. *Acta Cryst. B*, 1968, **24**, 869.
18. Hannon, A. C. In: Rutherford Appleton Laboratory Report RAL-93-063, 1993.
19. Gurr, G. E., Montgomery, P. W., Knutson, C. D. & Gorres, B.T. *Acta Cryst. B*, 1970, **26**, 906.
20. Ihara, M., Yuge, M. & Krogh-Moe, J. *J. Ceram. Soc. Jpn*, 1980, **88**, 179.
21. Jiang, A.-D., Lei, S.-R., Huang, Q.-Z., Chen, T.-B. & Ke, D. *Acta Cryst. C*, 1990, **46**, 1999.
22. Natarajan-Iyer, M., Faggiani, R. & Brown, I. D. *Cryst. Struct. Commun.*, 1979, 8, 367.
23. Kirfel, A., Will, G. & Stewart, R. F. *Acta Cryst. B*, 1983, **39**, 175.
24. Stewner, F. *Acta Cryst. B*, 1971, **27**, 904.
25. Bray, P. J. In: *Tenth International Congress on Glass*, vol. 13, The Ceramic Society of Japan, Kyoto, 1974, p. 1–20.
26. Press, W. H., Teukolsky, S. A., Vetterling, W. T. & Flannery, B. P. In: *Numerical Methods in Fortran 77*, Cambridge University Press, Cambridge, 1992, p. 402–406.
27. Sivia, D. S. Private communication, 2002.
28. Kamitsos, E. I., Karakassides, M. A. & Chryssikos, G. D. *Phys. Chem. Glasses*, 1987, **28**, 203.
29. Irion, M., Couzi, M., Levasseur, A., Reau, J. M. & Brethous, J. C. *J. Solid State Chem.*, 1980, **31**, 285.
30. Bray, P. J., Feller, S. A., Jellison Jr., G. E. & Yun, Y. H. *J. Non-Cryst. Solids*, 1980, **38–39**, 93.
31. Kerner, R. *J. Non-Cryst. Solids*, 1991, **135**, 155.
32. Ratai, E.-M., Janssen, M., Epping, J. D., Chan, J. C. C. & Eckert, H. *Phys. Chem. Glasses*, 2003, **44**, 45–53.
33. Holland, D, Smith, M. E., Howes, A. P., Davies, T. & Barrett, L. *Phys. Chem. Glasses*, 2003, 44, 59.
34. Kroeker, S., Feller, S., Affatigato, M., O'Brien, C., Clarida, W. & Kodama, M. *Phys. Chem. Glasses*, 2003, 44, 54.
35. Holland, D., Hannon, A. C., Smith, M. E., Johnson, C. E., Thomas, M. F. & Beesley, A. M. *Solid State Nucl. Mag.*, 2004, **26**, 172.
36. Warren, B. E. & Pincus, A. G. *J. Am. Ceram. Soc.*, 1940, **23**, 301.
37. Baugher, J. F. & Bray, P. J. *Phys. Chem. Glasses*, 1972, **13**, 63.
38. ISIS Disordered Materials Database. http://www.isis.rl.ac.uk/disordered/Database

The structure of borates between glassy and liquid state – results of diffraction methods

Gerhard Herms & Jan Sakowski*

Rostock University, Institute of Physics, D -18051 Rostock, Germany

Dedicated to Professor Dr Werner Vogel

Undercooled melts of boron oxide and sodium borate with 29 mol% Na_2O were studied by high temperature diffraction experiments, applying synchrotron radiation and image plates as position sensitive detectors. The main object was to determine the functions $\sigma_{struct}(T)$ and $\sigma_{shape}(T)$, which describe the intensity of structural changes per temperature step ΔT. The maxima of these functions were interpreted on the basis of existing knowledge.

Introduction

In the last decade, high temperature diffraction experiments have shown that glass and melt have different structures (see e.g. Refs 1–7). The question of how the differences develop during cooling down and heating respectively is not easy to answer. It requires studying the whole range of undercooling, and is connected with several difficulties. The equilibrium state "melt" must be compared with the non-equilibrium state "glass", which is fortunately long lasting (the half-life period being much longer than a man's life). In the range of undercooling, the half-life of a non-equilibrium state varies by many orders of magnitude between glass transformation temperature T_g and liquidus temperature T_l. A structure analysis of such a state is of no interest and, besides, is mostly not feasible. Only the metastable states are of importance to describe the structure and how it changes with temperature. Their lifetimes are given by the time-lag of nucleation, which has its minimum around the middle of the interval (T_g, T_l), and increases very strongly towards both ends of the temperature interval (see e.g. Ref. 8). The main requirement for accurately measuring the metastable state is a measuring time smaller than the time-lag. Only in favourable circumstances, the time lag in the middle of the temperature interval exceeds more than a few minutes. Therefore the use of modern experimental means (such as synchrotron radiation, fast working position sensitive detectors) is inevitable. If the measuring time is too long, the beginning of crystallisation is recorded. On the other hand, just above T_g another difficulty arises, since the experimenter cannot wait till the metastable state is really reached.

To follow the structure as a function of temperature, many diffraction patterns at gradually increased temperatures must be recorded. Small temperature steps, however, can cause only small changes of the structure. Therefore the measurement must have a very small statistical error. But even if this is fulfilled, it is difficult concluding something from the family of only slightly changing scattering curves (and radial distribution curves respectively). On that account a special method for ranking must be applied.

Experimental technique

The x-ray diffraction experiments were performed at the high energy beam line BW 5 of HASYLAB at DESY. The image plate system of MarResearch served as a two-dimensional, fast working detector with about 6 millions of applicable pixels. The measuring time was, as a rule, 100 s. The containerless high temperature device[9] allowed nearly inertia free adjustment of sample temperature (rate about 200 K/s). Before recording a scattering curve, the melt was always held for 30 s clearly above the liquidus temperature T_l, and then rapidly cooled down to the desired temperature. Closely above T_g, several scattering curves were recorded immediately one after the other, and those of them which show no more relaxation were used.

Materials under investigation were boron oxide and sodium borate with 29 mol% Na_2O. The preparation of the glasses was performed as described elsewhere.[1,3]

Method

To be able to draw quantitative conclusions from families of scattering curves or radial distribution functions, we have developed a reliable method, which starts from the difference of two curves suc-

[1] Corresponding author. Email gerhard.herms@gmx.de

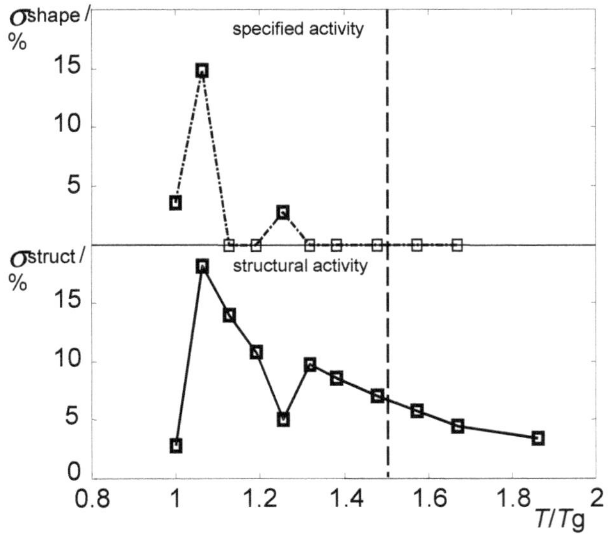

Figure 1. Overall structural activity σ_{struct} (solid line) and specified activity σ_{shape} (dash-dotted line) of molten B_2O_3 as a function of T/T$_g$. The straight broken line marks the liquidus temperature. The values for σ_{shape} are mostly not calculable since the structural differences are drowned by the statistical noise. The functions are calculated over the Q-range (5 nm^{-1}, 65 nm^{-1})

ceeding one another.[5] By means of the standard deviation σ, the so called "overall structural activity", σ_{struct}, was defined, which describes the magnitude of the structural change per temperature step. In this connection all kinds of changes are considered, not least the broadening and flattening of the diffraction peaks due to increasing disorder with rising temperature. Another quantity, the so called "specified structural activity", σ_{shape}, can be defined, which ignores this effect, and so is a measure of the reorganisation of structure (i.e. for qualitative changes because of an otherwise connecting of structural units or / and conversion of structural units).[10]

Results and discussion

Structural activity

The study of vitreous boron oxide provides the results shown in Figure 1. The overall activity has a sharp peak slightly above T_g, and a somewhat lower and much broader maximum between $1{\cdot}25T_g$ and $1{\cdot}5T_g$. The available measurements do not meet the requirements for calculating the whole curve of the specified activity σ_{shape}, because the statistical noise is still too high. Nevertheless an essential conclusion can be drawn. From the fact that the curve is calculable only in the vicinity of T_g, but not above $1{\cdot}3T_g$, it can be recognised that the shape of the σ_{shape} curve is undoubtedly similar to that of GeO_2 (see Figure 2). This means that qualitative changes worth mentioning only occur closely above T_g. In case of GeO_2, a comparison of σ_{struct} with σ_{shape} (Figure 2) leads to the

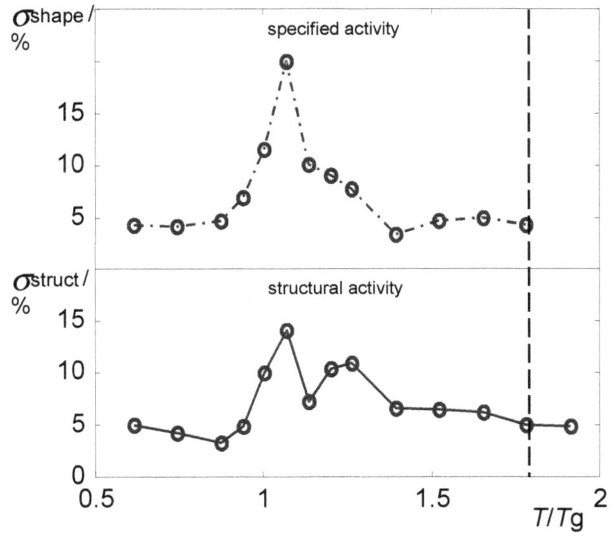

Figure 2. The same functions as in Figure 1, but for GeO_2 (calculated over the same Q-range)

conclusion that, during cooling down, the degree of order noticeably increases between $1{\cdot}4T_g$ and $1{\cdot}2T_g$, whereas any qualitative changes are moderate in this interval. It can be assumed also in the case of molten B_2O_3 that an ordering effect is the main reason for the second maximum in σ_{struct}.

The overall structural activity, σ_{struct}, of sodium borate with 29 mol% Na_2O is presented in Figure 3. Also, in this case, the largest changes occur slightly above T_g as indicated by the sharp peak in the temperature interval $1{\cdot}05T_g$ and $1{\cdot}15T_g$. At higher temperatures, between $1{\cdot}23T_g$ and $1{\cdot}35T_g$, the structural activity has a second maximum. This seems to correspond to the second maximum of B_2O_3 (Figure 1). However, in case of sodium borate the maximum is noticeably more pronounced, perhaps indicating that not only quantitative changes occur. Of course, this supposition could actually be corroborated or refuted easily

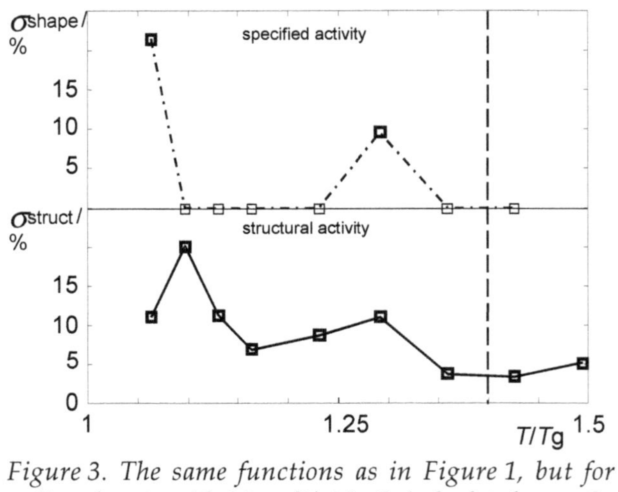

Figure 3. The same functions as in Figure 1, but for sodium borate with 29 mol% Na_2O (calculated over the same Q-range)

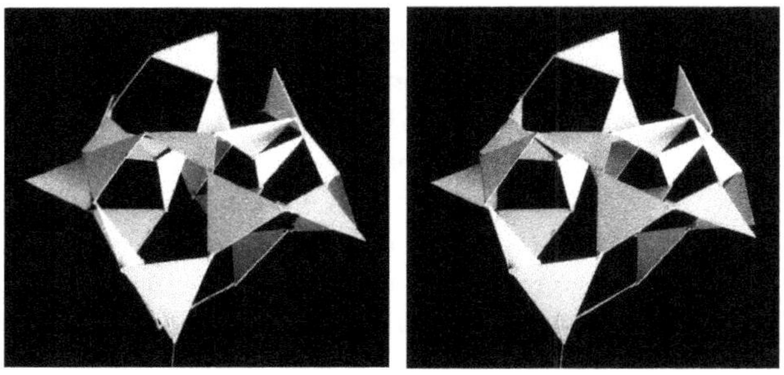

Figure 4. Stereoscopic view of the three-dimensional network made of corner linked triangles

by calculating the specified structural activity σ_{shape}, but, up to now, suitable measurements are not available. At least, the fact that at $1\cdot3T_g$ a measuring point is calculable indicates stronger qualitative changes of the structure occurring in this temperature range.

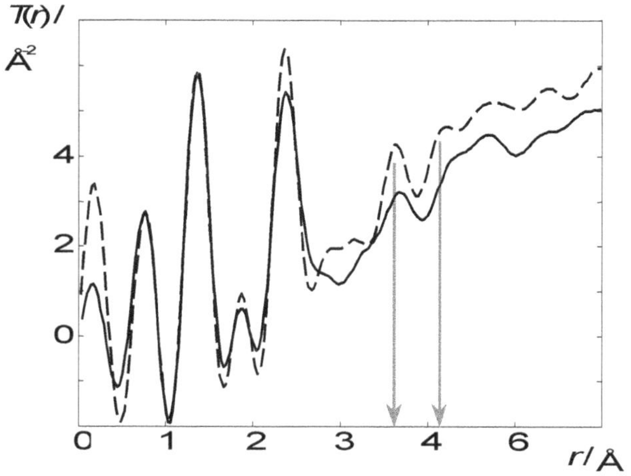

Figure 5. Total distance correlation function of B_2O_3 in the vitreous state (broken line) and in the molten state (solid line)[4]

Interpretation of structural changes

Vitreous boron oxide

In case of B_2O_3, we have to deal with a two-dimensional structure unit. This tempts us to assume two-dimensional networks. In its most glaring example, in the structural model of Bell & Carnevale,[11] there were even well defined positional relations between neighbouring, well ordered layers. A more realistic and apparently widespread view considers the structure, so to speak, as a ball of creased paper, in which the sheets represent the two-dimensional networks. However, there is an argument against two-dimensional networks. The peak at 4·8 Å reflecting the r_{OO} distance in second coordination (twice the side of the BO_3 triangle) should be much more pronounced than is actually found in the experimental radial distribution curve.[1] A three-dimensional network consisting only of "independent" BO_3-groups, fastened together with their corners, is shown in Figure 4. However, diffraction studies[12,13] as well as NMR and Raman studies[14–17] have shown that, in actual fact, the network of vitreous B_2O_3 is essentially built up by boroxol groups, apart from a small fraction of independent BO_3 groups. Meanwhile it seems to be clear that denying the existence of boroxol groups[18] or advocating a very small fraction f of such groups[19] is out of place, and that in the glass 60%–80% of the boron atoms are constituents of boroxol groups.[20] The most accurate value should be the NMR result by Zwanziger *et al*,[21] namely f=0·70±0·08.

Molten B_2O_3 was studied by Misawa[22] in 1989 by means of neutron scattering. In spite of the fact that the scattering was measured over a much wider Q-range (Q=36 Å^{-1}) and consequently well resolved radial distribution functions should be attained, these show no remarkable differences between glass and melt. However, Sakowski & Herms[4] had found distinct differences. Two peaks, one at 3·6 Å and the other at 4·1 Å, are distinctly decreased in the melt (Figure 5). They correspond to certain distances (r_{BO} and r_{OO}, respectively) which are characteristic for the boroxol group. This indicates that the number of boroxol groups is reduced in the melt. Misawa concluded the opening of the boroxol rings above T_g only from the changed position of the first peak of the structure factor and from peculiarities of the fluctuation of the second peak in the radial distribution function. In the same paper,[4] by adopting the distance correlation function of a well founded structure model to the experimental correlation function, evidence could be produced that a quasi-planar network with closed rings of independent BO_3 groups is an inappropriate model for the structure of molten B_2O_3. Instead of this, it could be shown that a network which contains "broken up" boroxol rings is in good agreement with the experimental correlation function. By breaking up the boroxol ring, the distance between the O atoms 1 and 5 and the distance between B atom 6 and O atom 4

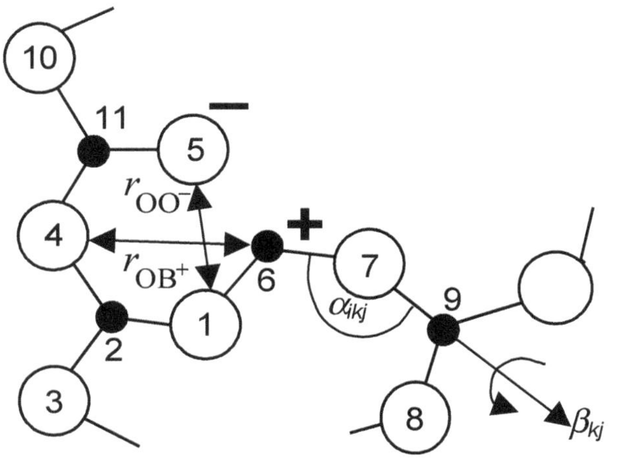

Figure 6. A broken-up boroxol group (left hand) linked with a BO_3 group (right hand)[4]

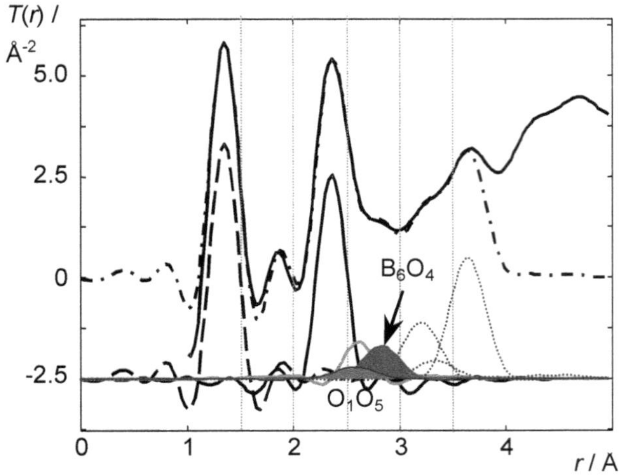

Figure 7. Experimental distance correlation function (thick solid line) and model correlation function (thick dash-dotted line). In addition the partial model correlation functions are given (for the sake of clarity shifted downwards by 2·5 Å^{-2}). The shaded areas emphasise the two partial correlation functions which are caused by the opening of the boroxol ring[4]

(see Figure 6) increase, which is the cause for two new partial model correlations represented by the shaded areas in Figure 7. They are responsible for the fact that the second real peak in the distribution curve of the melt is distinctly shifted to larger distances.

In spite of the fragmentary character of the σ_{shape} curve it can be said that there are no remarkable maxima apart from the high, sharp peak closely above T_g. This means that the process of conversion from the glass structure to the melt structure (or vice versa) is completed for the most part in a relatively small temperature interval slightly above T_g. Cooling down molten B_2O_3, the glass transition begins with a structural reorganisation, in the course of which network defects (see atoms 5 and 6 in Figure 6) heal and the open boroxol rings close.

Sodium borate

Nearly 70 years ago, the classical diffraction study by Warren *et al*[23] proved that in binary borate glasses the mean coordination number of the boron atom increases with the content of modifier oxide. This knowledge, later on repeatedly corroborated by diffraction,[24] but also by NMR,[25] is generally accepted. The question, however, whether the coordination changes with temperature, was for a long time a subject of controversy. Already in 1942, Dietzel advocated the idea that in borate glasses the concentration of 4-coordinated boron atoms decreases to zero at higher temperatures.[26] This was supported by conclusions from different methods (e.g. solubility of silver halide,[27] isothermal viscosities[28]). Whereas earlier spectroscopic high temperature measurements[16,29] seemed to disprove the temperature induced coordination change, newer ones[30–31] have corroborated such a change. Also the diffraction studies present a similar appearance. Japanese authors[32] who had established neutron experiments found no difference between glass and melt concerning the composition dependent coordination change. In contrast with these studies the x-ray study of Herms *et al*[3] produced evidence of the temperature induced coordination change. In order to increase the detection efficiency, not only the mean number of oxygen atoms around the boron atom, NBO, but also the mean number of oxygen atoms around the oxygen atom, N_{OO}, was determined, since a coordination change has a higher effect on N_{OO} than on N_{BO}. Admittedly, contributions of other atomic pairs to the OO peak must be taken into consideration, which was done by modelling the structure at short distances (r<4 Å). The study of lithium borate, sodium borate and potassium borate of roughly diborate composition provided nearly identical results. The same borates but with somewhat higher modifier content were investigated in the neutron experiment by Majerus *et al.*[6] The very careful evaluation with detailed error considerations produced convincing proof of the BO_4 to BO_3 conversion in the melt. The fraction of 4-coordinated boron atoms were assessed in two independent ways, on the one hand by means of the coordination numbers, on the other hand from the shift of the BO peak. The results were in good agreement with one another, but also with Bray's NMR results.[14]

All things considered, there is no doubt that a process of $BO_4 \leftrightarrow BO_3$ conversion takes place in the undercooled melt of binary borates. It can be assumed that the sharp peak of the curve of structural activity (between $1{\cdot}05T_g$ and $1{\cdot}15T_g$ in Figure 3) is caused by processes which are comparable with the structural reorganisation in the under cooled melt of B_2O_3. If the temperature rises above T_g, the superstructural units (diborate groups and tetraborate groups) begin to break up and a looser structure with an increasing number of network defects develops. This is mainly a matter of medium range order. The $BO_4 \leftrightarrow BO_3$ conversion, however, is a phenomenon of the short range order. Presumably, the second peak of the activity curve (Figure 3) is mostly caused by this conversion. In a former paper,[5] the conclusion was drawn from the overall structural activity that the convolution has its highest activity at about 700°C. However, this can only be a rough estimate for two reasons: Firstly, this determination should be made by means of the specified structural activity (up to now not available), and secondly, the spacing between the measuring points should be narrower than in Figure 3. But a narrower spacing would have the consequence that the structural differences between two succeeding measuring points would become still smaller, so that the difficulties concerning the statistics would considerably increase. Since under no circumstances can the exposure time be prolonged, there is nothing for it but to use a much more intensive radiation source.

The σ-formalism is also applicable to the families of radial distribution functions. In this way one can discriminate between changes of the short range order and the medium range order, i.e. one can determine a function σ_{short} and a function σ_{med}. In case of sodium borate, the former should have a distinct maximum near $1{\cdot}3T_g$ in contrast to boron oxide.

In the past, the distribution functions had usually been determined for two arbitrarily chosen temperatures. However, it is more effective to pick out such temperatures which correspond to the beginning and end of a certain reorganisation process, which can be discerned in the so-called difference quotient curves of the scattering. Detailed studies of this kind will be a future task.

Conclusions

The concept of "structural activity" enables us to assess the development within families of scattering curves. This makes it possible to follow structural changes in their dependence on temperature, which is very important for studying vitreous melts. It is established that B_2O_3 and alkali borates show their largest changes slightly above T_g. Including previous results, we show that with rising temperature superstructural units break up and a looser structure with an increasing number of network defects develops. In the case of alkali borates, a second maximum of the structural activity at higher temperatures (about $1{\cdot}3T_g$) reflects the coordination change of the boron atom.

Acknowledgements

Financial support of the DFG is gratefully appreciated (Grant HE 3023/1-3).

We wish to thank Prof. Dr I. Gutzow and Dr J. Schmelzer for valuable discussions, Prof. Dr H. J. Pentinghaus and DP U. Precht for preparing the samples and Dr J. Neuefeind and Dr P. Jovari (DESY, HASYLAB) for supporting the measurements.

References

1. Herms, G., Sakowski, J., Gerike, W. & Stachel, D. *Z. Naturforsch.*, 1998, **A 53**, 874.
2. Herms, G., Sakowski, J., Gerike, W, Hoppe U. & Stachel, D. *J. Non-Cryst. Solids*, 1998, **232–234**, 427.
3. Herms, G. & Sakowski, J. *Phys. Chem. Glasses*, 2000, **41**, 309.
4. Sakowski, J. & Herms, G. *J. Non-Cryst. Solids*, 2001, **293–295**, 304.
5. Sakowski, J. & Herms, G. *Phys. Chem. Glasses*, 2002, **43C**, 13.
6. Majerus, O., Cormier, L., Calas, G. & Beuneu, B. *Phys. Rev. B*, 2003, **67**, 024210.
7. Beaufils, S., Cormier, L., Bionducci, M., Ecolivet, C., Calas, G., Le Sauze A. & Marchand, R. *Phys. Rev. B*, 2003, **67**, 104201.
8. Gutzov, I. & Schmelzer, J. *The Vitreous State*, 1995. Springer-Verlag, Berlin Heidelberg . P 269.
9. Herms, G., Sakowski, J., Gerike, W. & Weyer, S. *Meas. Sci. Technol.*, 1996, **7**, 627.
10. Herms, G. & Sakowski, J. *J. Non-Cryst. Solids*, 2004, **345–346**, 551.
11. Bell, R. J. & Carnevale, A. *Phil. Mag. B*, 1981, **43**, 389.
12. Mozzi, R. L. & Warren, B. E. *J. Appl. Crystall.* 1970, **3**, 251.
13. Sinclair, R. N., Desa, J. A. E., Etherington, G., Johnson, P. A. V. & Wright, A. C. *J. Non-Cryst. Solids*, 1980, **42**, 107.
14. Jellison, G. E., Panek, L.W., Bray, P. J. & Rouse, G. B. *J. Chem. Phys.*, 1977, **66**, 802.
15. Feller, S. A., Dell, W. J. & Bray, P. J. *J. Non-Cryst. Solids*, 1982, **51**, 21.
16. Furukawa, T. & White, W. B. *Phys. Chem. Glasses*, 1980, **21**, 85.
17. Walrafen, G. E., Samanta, S. R. & Krishnan, P. N. *J. Chem. Phys.*, 1980, **72**, 113.
18. Williams, S. J. & Elliot, S. R. *Proc. R. Soc. London A*, 1982, **380**, 427.
19. Swenson, J. & Börjesson, L. *Phys. Rev. B*, 1996, **55**, 11138.
20. Hannon, A. C., Grimley, D. I., Hulme, R. A., Wright, A. C. & Sinclair, R. N. *J. Non-Cryst. Solids*, 1994, **177**, 299.
21. Joo, C., Werner-Zwanziger, U. & Zwanziger, J. W. *J. Non-Cryst. Solids*, 2000, **271**, 265.
22. Misawa, M. *J. Non-Cryst. Solids*, 1990, **122**, 33.
23. Bicoe, J. & Warren, B. E. *J. Am. Ceram. Soc.*, 1938, **21**, 287.
24. Becherer, G., Brümmer, O. & Herms, G. *Z. Naturforsch.*, 1961, **16a**, 1253.
25. Jellison, G. E., Feller, S. A. & Bray, P. J. *Phys. Chem. Glasses*, 1978, **19**, 52.
26. Dietzel, A. *Z. Elektrochemi*, 1942, **48**, 9.
27. Araujo, R. J. *Phys. Chem. Glasses*, 1979, **20**, 115.
28. Visser, T. J. M. & Stevels, J. M. *J. Non-Cryst. Solids*, 1972, **7**, 376.
29. Soppe, W., Ph. D. Thesis (Rijksuniversiteit Groningen), 1989, P. 62.
30. Sen, S., Xu, Z. & Stebbins, J. *J. Non-Cryst. Solids*, 1998, **226**, 29.
31. Akagi, R., Ohtori, N. & Umesaki, N. *J. Non-Cryst. Solids*, 2001, **293–295**, 471.
32. Umesaki, N., Kita,Y., Kirihara, T., Iida, T., Fukunaga, T. & Misawa, M. *J. Non-Cryst. Solids*, 1994, 177, 200.

Binary alkaline earth borates: phase diagram correction and low thermal expansion of crystallised stoichiometric glass compositions

Rafael M. Hovhannisyan[1]

Institute of Material Science, 17 Charents st., 375025 Yerevan, Armenia

The phase diagrams of the $BaO–B_2O_3$, $SrO–B_2O_3$ and $CaO–B_2O_3$ systems have been redetermined. The new binary eutectic compositions 27·5CaO.72·5B_2O_3 (975°C) and 36·5CaO.63·5B_2O_3 (980°C) in the $CaO–B_2O_3$ system, 68·5SrO.31·5B_2O_3 (1165°C) in the $SrO–B_2O_3$ system, 31·5BaO.68·5B_2O_3 (890°C) and 68·5BaO.31·5B_2O_3 (1125°C) in the $BaO–B_2O_3$ system have been revealed. Melting points of the congruent melted Sr_2B (1185°C), Sr_3B (1285°C) and Ba_2B (1150°C) have been determined. The low, ultralow and negative TEC values (20–300°C) for $CaO.B_2O_3$ (25×10^{-7}, K^{-1}), $SrO.B_2O_3$ (12×10^{-7}, K^{-1}), $BaO.B_2O_3$ (-11×10^{-7}, K^{-1}) and BaB_2 (-13×10^{-7}, K^{-1}) have been observed. The possible mechanism of low and ultralow thermal expansion of the studied crystallised alkaline earth borate glasses is discussed.

1. Introduction

Binary alkaline earth borates continue to be interesting to scientists. Objects of investigation are structural features of crystalline borates,[1] coordination changes of boron atoms in glasses,[2] and processes of phase separation[3] are interesting to others.

Phase diagrams of binary RO(Ba,Sr,Ca)–B_2O_3 systems are well investigated.[4] Seven compounds are known in the $BaO–B_2O_3$ system. Four congruent melting binary compounds were found by Levin & McMurdie:[5] $3BaO.B_2O_3$ (Ba_3B), $BaO.B_2O_3$ (BaB), $BaO.2B_2O_3$ (BaB_2), $BaO.4B_2O_3$ (BaB_4) with melting points (m.p.) 1383, 1105, 910, 889°C, respectively.[5] Green & Wahlers[6] studied $BaO–B_2O_3–Al_2O_3$ system and found new $2BaO.5B_2O_3$ (Ba_2B_5) compound. Hubner[7] confirmed the existence of the congruent melted Ba_2B_5 compound with m.p. 890°C, and revealed two new compounds $Ba_4B_2O_6$ (Ba_4B), $Ba_2B_2O_5$ (Ba_2B). The BaB_2, Ba_2B_5 and BaB_4 compounds formed stable glasses by traditional methods of glass formation.[8] The Ba_2B, BaB compositions can be obtained in glassy state through twin roller quenching method.[9] From the listed compounds, the greatest interest is in β-BaB_2O_4, which has strongly pronounced nonlinear optical properties for which single crystals have been discovered and studied recently.[10] Authors[11] have obtained and studied the transparent glass ceramics with nonlinear optical properties, in which crystalline phase was β-BaB_2O_4. Phase transition of BaB from low temperature β-forms to the high temperature form occurs at 925°C.[7] However, we have used the melting diagram of the $BaO–B_2O_3$ system created by Levin & McMurdie[5] up to now, without the indication in it the specific areas of existence of new compounds.[4]

Witzmann with co-workers[12,13] constructed the phase diagram of the $SrO–B_2O_3$ system and four binary congruent melted compounds indicated in it: $3SrO.B_2O_3$ (Sr_3B), $2SrO.B_2O_3$ (Sr_2B), $SrO.B_2O_3$ (SrB), $SrO.2B_2O_3$ (SrB_2). Chenot also studied this system, partly constructed its phase diagram and synthesised incongruent melted at 890°C the $SrO.3B_2O_3$ (SrB_3) compound.[14] However, the m.p. of binary compounds either are not determined (Sr_3B), or essentially differ by various authors data 1115 (1177), 1100 (1155), 930 (994)°C accordingly to Sr_2B, SrB, SrB_2.[12,13,15]

In the $CaO–B_2O_3$ system four binary compounds have been known: $3CaO.B_2O_3$ (Ca_3B), $2CaO.B_2O_3$ (Ca_2B), $CaO.B_2O_3$ (CaB), $CaO.2B_2O_3$ (CaB_2) with m.p. 1479, 1298, 1154 and 986°C, respectively.[4,16] However, a liquidus curve between two Ca_2B and CaB compounds is necessary for specifying an eutectic point which in our opinion should be there.

The glass forming areas for $SrO–B_2O_3$ system from 24 up to 43 mol% SrO and for $CaO–B_2O_3$ system changes from 27 up to 41 mol% CaO are obtained by traditional methods.[8]

Many authors had investigated various physical and chemical properties of binary alkaline-earth borate glasses. Data are summarised in Mazurin *et al.*[8] Kluev & Pevzner have studied thermal expansion of crystallised barium diborate glasses and have revealed that its crystallised glass powder samples, pressed and heated at 700–800°C, have negative TEC

[1] Email address: hovhannisyan@netsys.am

data by cooling from 700–800°C.[17] Properties of the binary alkaline earth compounds and crystallised glasses on their basis are not studied sufficiently and require additional research. For the best understanding of the received data we think it is also necessary to redetermine some fragments of the binary alkaline-earth borates systems phase diagrams.

2. Experimental

All compositions were prepared from 'chemically pure' grade $BaCO_3$, $SrCO_3$, $CaCO_3$ and H_3BO_3. Most compositions at first are obtained as glass by traditional cooling methods: casting on metallic plate, or granulation in water with the subsequent crystallisation. Some compositions with high crystallisation ability and lying outside the glass formation field are obtained as glass only through the twin roller quenching method. The glasses (50–70 g batch, 50 ml Pt crucible, electric furnace 'Supertherm 17/08', air atmosphere, 1 h) are melted in an interval of temperature 1200–1400°C. The batch components losses are controlled by traditional chemical analysis methods, and are corrected subsequently. In parallel the solid state sintering was carried out for binary alkaline earth borates synthesised for comparison their x-rays and DTA data. Mixes (15–20 g) were carefully grained in an agate mortar, were pressed as tablets, were located on platinum plates and passed for the first thermal treatment in Naber electric muffles. After regrinding the powders were tested by DTA and x-ray methods. Tablets of the pressed samples passed repeated thermal treatment on the basis of obtained analyses data at 800–1200°C within 24 h.

The phase diagram redetermination in the binary $RO–B_2O_3$ system has been made from DTA and x-ray diffraction data. The DTA analysis (platinum crucible, powder weight 400–600 mg, heating rate 15 K/min) were carried out on Q-1500 type derivatograph. Reproducibility of temperatures of the effects on DTA curves from melting to melting were ±10 K. The accuracy of temperature measurement is ±5 K.

X-ray patterns were obtained on a DRON-3 type diffractometer (powder method, Cu K_α radiation, Ni-filter). Samples for glass crystallisation were prepared with glass powder pressed in the form of cane or tablets. Crystallisation was done in the Naber electrical muffles by a single stage heat treatment. This was done within from 6–12 h around a temperature at which the maximum exothermal effects on glasses by DTA were observed.

TEC (at intervals 20–300°C) measurements were made with a heating rate of 3 K/min on a DKV-4A type vertical quartz dilatometer graduated by the quartz glass and sapphire standards. The accuracy of TEC measurement is $\pm3\times10^{-7}$, K^{-1}.

Crystalline phases of binary compounds formed both at glasses crystallisation and at solid state sintering were identified by using JCPDS-ICDD database.

3. Results

First of all we have carried out correction of some parts of phase diagrams of the binary systems.

The areas of compositions containing 80–60 and 40–25 mol% B_2O_3 have been tested in the $BaO–B_2O_3$ system and its revised version is submitted in Figure 1. Congruent melting of Ba_2B_5 compound at 900°C has been confirmed and the new eutectic $31{\cdot}5BaO.68{\cdot}5B_2O_3$ composition with m.p. 890°C between Ba_2B_5 and BaB_2 compounds has been revealed. In the area of BaO high content we have determined the m.p. of Ba_2B compound, which melted congruently at 1150°C. The new eutectic $68{\cdot}5BaO.31{\cdot}5B_2O_3$ composition with m.p. 1125°C between Ba_2B and Ba_3B compounds has also been revealed.

Areas of compositions containing 80–65 and 25–40 mol% B_2O_3 have been revised in the $SrO–B_2O_3$ system and its revised version is submitted in Figure 2. Similarly to the data,[15] the SrB_3 compound melts incongruently at 890°C and has liquidus temperature 980°C. In the high SrO content area we have determined the m.p. of the Sr_2B and Sr_3B compounds, which melted congruently at 1185 and 1285°C accordingly. The new eutectic $68{\cdot}5SrO.31{\cdot}5B_2O_3$ composition with m.p. 1165°C between S_3B and S_2B compounds has been revealed.

We had studied a multi-borate part of the system phase diagram of the $CaO–B_2O_3$ from the end of the sta-

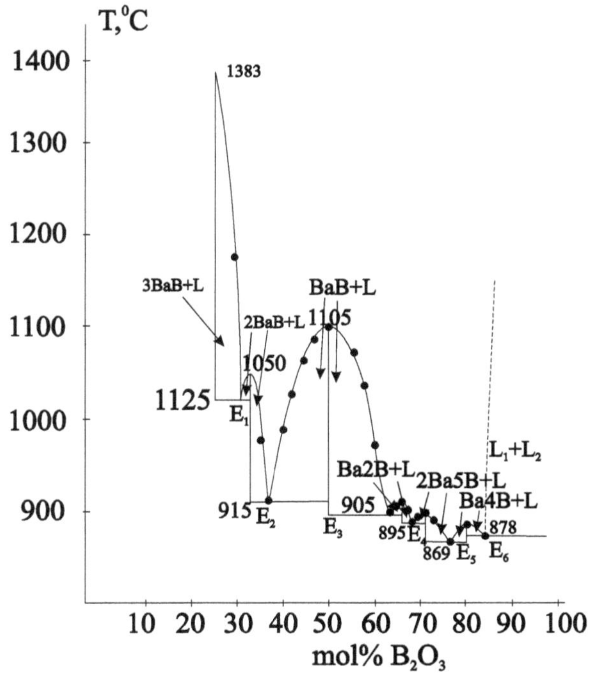

Figure 1. Corrected version of the $BaO–B_2O_3$ system phase diagram

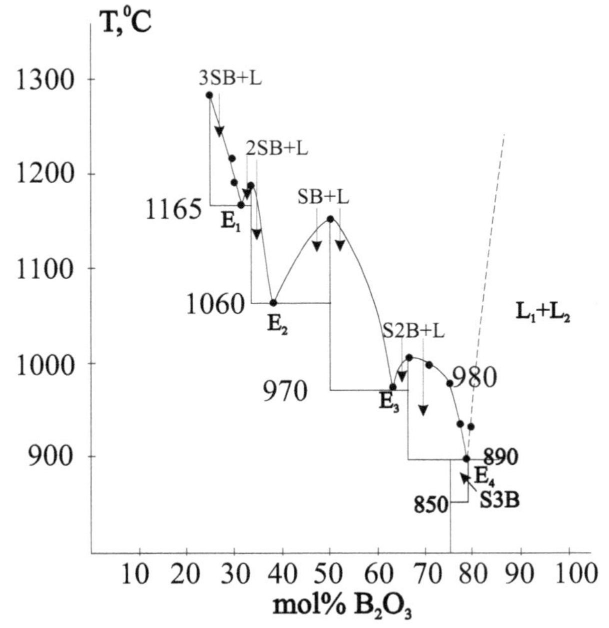

Figure 2. Corrected version of the SrO–B_2O_3 system phase diagram

ble phase separation area containing 27·5CaO.72·5B_2O_3 up to calcium metaborate composition. The new eutectics compositions E_1 27·5CaO.72.5B_2O_3 with m.p. 975°C and E_2 36·5CaO.63·5B_2O_3 with m.p. 980°C has been determined on the liquidus curve (Figure 3). The maximum at 1005 °C corresponding to congruent melting of the Ca_2B compound is clearly observed on the liquidus curve between this new eutectics E_1 and E_2 (Figure 3).

The diborate and high borate content compositions have good glass forming ability and formed glasses at various cooling rates. The metaborate compositions are obtained in the glassy state only using fast cooling. On the DTA curves of the synthesised glasses we observe one strong exothermic effect of crystallisation and one strong endothermic effect of melting. It testifies to crystallisation and melting of one crystalline phase. Only on DTA curve of SB_3 compositions do we observe one exothermic effect of glass crystallisation at 765°C and two endothermic effects at 890 and 980°C. These endothermic effects associated with crystalline SB_3 incongruent melting at 890°C with formation of SB_2 and melt. The dissolution of SB_2 in a melt leads to the appearance on a DTA curve the second endothermic effect at 980°C. Only in a narrow area of compositions (21–25 mol% SrO) and narrow interval of crystallisation (850–890°C) can we obtain a pure SB_3 compound. The products of crystallisation of other glasses of stoichiometric compositions at temperatures of maximums of exothermic effects on their DTA curves were corresponding to their crystalline binary alkaline earth borates, identified by the x-ray analysis.

The compositions are divided into three groups: metaborates, diborates and others. The TEC curves of crystallised glasses presented on Figures 4 and 5. Crystallised glasses of SB and CB compositions have low TEC value (20–300°C) 12×10^{-7}, K^{-1} and 25×10^{-7}, K^{-1}

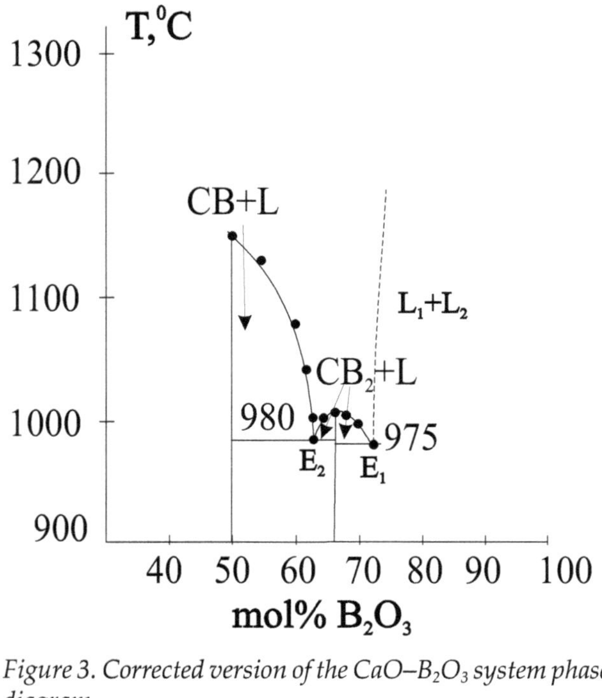

Figure 3. Corrected version of the CaO–B_2O_3 system phase diagram

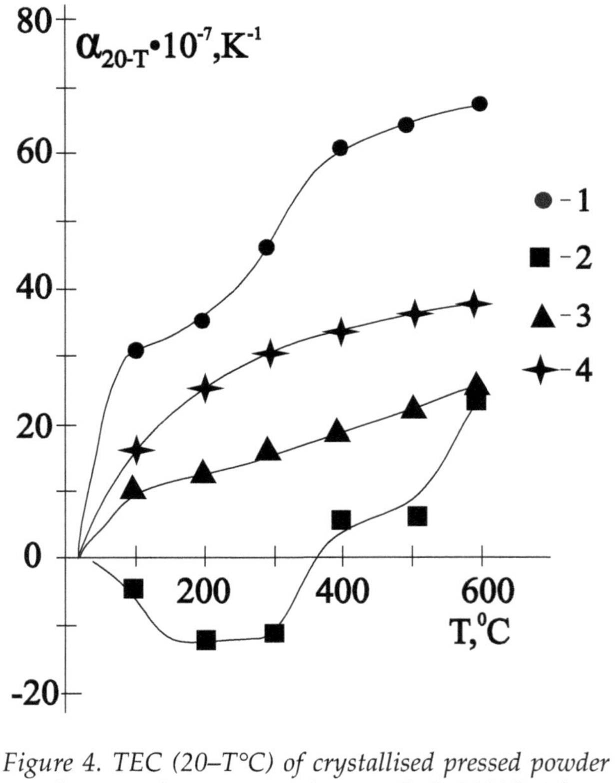

Figure 4. TEC (20–T°C) of crystallised pressed powder metaborates glass compositions: 1. BaB 900°C 6 h, 2. BaB 900°C 6 h+950°C 24 h, 3. SB 1100°C 12 h, 4. CB 900°C 6 h

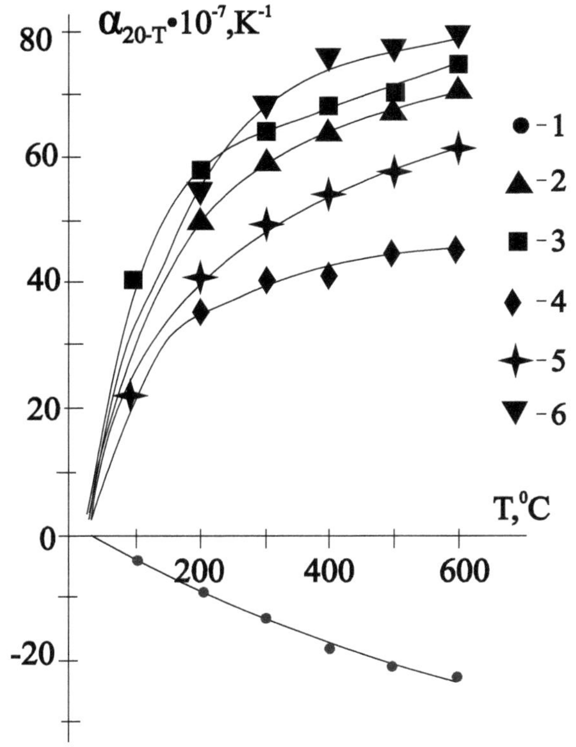

Figure 5. TEC (20–T°C) of crystallised pressed powder stoichiometric glass compositions: 1. BaB_2 850°C 6 h, 2. CB_2 700°C 6 h, 3. SB_2 730°C 6 h, 4. BaB_4 750°C 6 h, 5. Ba_2B_5 700°C 12 h, 6. SB_3 785°C 12 h

accordingly (Figure 4, curves 3, 4). The TEC changes of barium metaborate crystallised glasses have other characteristics: low temperature form of BaB, which formed up to 925°C have positive TEC value (Figure 4, curve 1), which is higher than at calcium and strontium metaborates. Crystallisation of BaB glasses above temperature of phase transition results in sharp reduction of TEC crystallised glasses, which get negative values at an interval 20–300°C (Figure 4, curve 4).

From diborates group first of all the crystallised (850°C, 6h) BaB_2 glass composition has strongly negative TCE data, which decrease up to −23 at an interval 20–600°C (Figure 5, curve 1). Our data differ from data in Kluev & Pevzner,[17] because we observe negative thermal expansion of crystallised BaB_2 sample by heating from room temperature. Crystallised from glasses strontium (730°C, 6h) and calcium (700°C, 6 h) diborates have higher TEC values, but very close to each other (Figure 5, curves 2,3). The Ba_2B_5 and BaB_4 crystallised glasses have lower TEC values, than at calcium and strontium diborates (Figure 5, curves 5,4). The highest TEC values from all crystallised binary alkaline earth borates stoichiometric glasses has SB_3, which TEC size reaches up to 80×10^{-7}, K^{-1}, at an interval 20–600°C (Figure 5, curve 6).

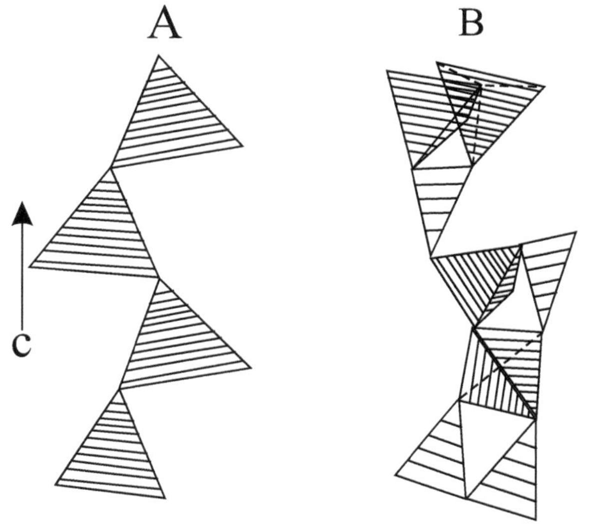

Figure 6. Arrangements of BO_4-tetrahedrons and BO_3-triangles in borate groups of A: Ca(Sr)O.B_2O_3, B: BaO.2B_2O_3

4. Discussion

Revision of phase diagrams of binary alkaline-earth borate systems has allowed us to specify them more precisely. Revision from our point of view is important not only for binary systems, but also important for ternary systems, since sometimes authors did not fix fields of crystallisation of the some alkaline earth binary borates.[4] Study of thermal expansion of crystallised alkaline earth binary borates has shown the timeliness of the lead correction on an example of BaB_2. Exact location of the revealed narrow area of the BaB_2 crystallisation important in connection with the big difference of TEC of the crystallised BaB_2 glass composition (-13×10^{-7} K^{-1}) and its nearest neighbour Ba_2B_5 (48×10^{-7} K^{-1}).

Comparing the received data on thermal expansion of the crystallised glasses with structure of crystal binary borates it is impossible to come to a general conclusion. However, some analogies can be carried out. Trying to explain the nature of TEC low sizes of calcium and strontium metaborates we prefer Bayer's model which he suggested for barium titanium (zirconium, tin) borates in which boron atoms are in triangular coordination.[19] All boron atoms in the structure of CB and SB[16,18] have triangular coordination. Atoms of calcium and strontium have coordination numbers equal to eight. BO_3-triangles, connected by two tops form $[BO_2]_\infty$ chains extended along 'C' axis (Figure 6A). According to Bayer's model low thermal expansion can be connected with abnormal compression on one of axes of a crystal cell and simultaneous rotation of BO_3-triangles at heating.[19]

The basis of monoclinic BaB_2 structure is made

with three-dimensional network of BO_3-triangles (50%) and BO_4-tetrahedrons (50%) in which it is possible to indicate unary and double rings.[20] The single ring will consist of two tetrahedrons and one triangle, the double ring will consist of two tetrahedrons and three triangles (Figure 6B). The barium atoms have coordination numbers equal to nine and ten. This radical is repeated for barium diborate structure.[20] In our opinion for this structure it is also possible to accept Bayer's model, but with the amendment, that at heating there is a partial (unary or double rings) or complete rotation of all structural radicals.

The fullest review of existing glasses and glass ceramics having low and negative TEC is carried out by Roy with co-authors.[21] Our earlier data on glass ceramics on the alkaline earth aluminum borate basis,[22–24] which was not reviewed by authors and data published later by MacDowell[25] was not reviewed either.

For an explanation of the nature of low and negative thermal expansion crystallised alkaline earth aluminoborate glasses we also offered Bayer's[26] model, though Prof. Ruessel offered his model of low thermal expansion for crystallised barium aluminoborate glasses.[27]

The received data on low and negative thermal expansion of some crystallised binary alkaline earth stoichiometric glass compositions demand additional consideration. However, it is possible to ascertain, that the list of before known materials with low and negative TEC is supplemented by calcium, strontium and barium metaborates, and also barium diborate having low and negative TEC sizes.

5. Conclusion

The phase diagrams in the $BaO–B_2O_3$, $SrO–B_2O_3$ and $CaO–B_2O_3$ systems have been determined. The new binary eutectic compositions $27{\cdot}5CaO.72{\cdot}5B_2O_3$ (975°C) and $36{\cdot}5CaO.63{\cdot}5B_2O_3$ (980°C) in the $CaO–B_2O_3$ system, $68{\cdot}5SrO.31{\cdot}5B_2O_3$ (1165°C) in the $SrO–B_2O_3$ system, $31{\cdot}5BaO.68{\cdot}5B_2O_3$ (890°C) and $68{\cdot}5BaO.31{\cdot}5B_2O_3$ (1125°C) in the $BaO–B_2O_3$ system have been revealed. Melting points of the congruent melted Sr_2B (1185°C), Sr_3B (1285°C) and Ba_2B (1150°C) have been determined.

The TEC data of crystallised glasses of alkaline-earth borate stoichiometric compositions have been studied. The low, ultra low and negative TEC (20–300°C) values for CB (25×10^{-7} K^{-1}), SB (12×10^{-7} K^{-1}), BaB (-11×10^{-7} K^{-1}) and BaB_2 (-13×10^{-7} K^{-1}) have been revealed.

Bayer's model of low and ultralow thermal expansion of the studied crystallised alkaline earth borate glasses is discussed.

Acknowledgements

This work was supported by the International Science and Technology Centre (Projects # A-288 and A-952).

References

1. Leonyuk, N. I. & Leonyuk, L. I. *Crystallochemistry of Anhydrous Borates*. MSU Press, Moscow, 1983.
2. Kamitsos, E. I. Infrared Studies of Borate Glasses. *Borate Glasses, Crystals & Melts*, Edited by M. Affatigato, S. A. Feller, N. M. Vedishcheva & A. C. Wright, Society of Glass Technology, Sheffield, 2004, 79–87.
3. Crichton, S. N. & Tomozawa, M. *J. Non-Cryst. Solids*, 1997, **215**, 244.
4. Phase Equilibria Diagrams, CD-ROM Database, Version 3.0, A CerS-NITS, 2004.
5. Levin, E. M. & McMurdie, H. F. *J. Res. Nat. Bur.Standards*, 1949, **42**, 131.
6. Green, C. H. & Wahler, R. L. Crystallisation of barium aluminium borate glasses-rhodium as a specific catalist. *Kinetics of Reactions in Ionic Systems. Proc. Intern. Symp*. 18-23 IV 1967, New York, 1969 pp.545–53.
7. Hubner, K.-H. *Neues Jahrb. Mineral., Monatsch*, 1969, 335.
8. Mazurin, O. V., Streltzina, M. V. & Shvaiko-Shvaykovskaya, T. P. Properties of Glass and Glass Forming Melts. *Unicomponents and Binary Non Silicate Oxides Systems*, Leningrad, Nauka Press, 1975, **Vol. II**.
9. Kapoor, Sh., George, H. B., Betzen, A., Affatigato, M. & Feller, S. *J. Non-Cryst. Solids*, 2000, **270**, 215.
10. Chen, C. & Liu, G. *Ann. Rev. Mater. Sci.*, 1986, **16**, 203
11. Kao, Y.-H., Hu, Y., Zheng, H., Mackenzie, J. D., Perry, K., Bourhill, G. & Perry, J. W., *J. Non-Cryst. Solids*, 1994, **167**, 247.
12. Witzmann, H. & Beulich, W. *Z. Phys. Chem.*, 1964, **225** (5–6), 336.
13. Witzmann, H. & Herzog, G. *Z. Phys. Chem.*, 1964, **225** (3–4), 197.
14. Chenot, C. F. *J. Am.Ceram. Soc.*, 1967, **50** (2), 117.
15. Toropov, N. A. *et al. State Diagrams of Silicate Systems. Handbook*. Binary Systems, L. Nauka Press, Russia, 1972.
16. Marezio, M., Piettinger, H. A. & Zachariasen, W. H. *Acta Crystallogr.*, 1963, **16**, 390.
17. Kluev, V. P. & Pevzner, B. Z. *Phys. Chem. Glasses.*, 2004, **45** (2), 146–8.
18. Block, S., Perloff, A. & Weir, C. R. *Acta Crystallogr.*, 1964, **17**, 314.
19. Bayer G., *Z. Kristallogr.*, 1971, **133**, 85.
20. Block, S., Perloff, A. & Weir, C. R. *Acta Crystallogr.*, 1965, **19**, 297.
21. Roy, R., Agraval, D. K. & McKinstry, H. A. *Ann. Rev. Mater. Sci.*, 1989, **19**, 59.
22. Oganesyan (Hovhannisyan), R. M., Toroyan, V. P. & Knyazian, N. B. The USSR copyright certificate №1144332. Nov 8, 1984.
23. Oganesyan (Hovhannisyan), R. M. & Geodakyan, J. A. The USSR copyright certificate №1351013. July 8, 1987.
24. Oganesyan (Hovhannisyan), R. M. The USSR copyright certificate №1512022. June 1, 1989.
25. MacDowell, J. F. *J. Am. Ceram. Soc.*, 1990, **73**, 2287.
26. Hovhannisyan, R. M. Melt Chemistry, Relaxation, and Solidification Kinetics of Glasses. *Proc. 106 Annual Meeting of the Am. Ceram. Soc.*, Indianapolis, Indiana, USA (2004), *Ceram. Trans.*, 2004, **170**, 201.
27. Ruessel, C., Tauch, D., Garkova, R., Woltz, S. & Volksch, G. Proc. Fifth Borate Conference, Trento, Italy, 10–14 July 2005, (in Press).

Proc. Fifth Int. Conf. on Borate Glasses, Crystals and Melts Phys. Chem. Glasses: Eur. J. Glass Sci. Technol. B, August 2006, 47 (4), 465–470

Borate anomaly and anharmonicity in sodium borate glasses

M. Kodama,[1] A. Ono
Department of Applied Chemistry, Sojo University, Ikeda 4-22-1, Kumamoto 860-0082, Japan
S. Kojima
Institute of Materials Science, University of Tsukuba, Tsukuba, Ibaraki 305-8573, Japan
S. A. Feller & M. Affatigato
Physics Department, Coe College, Cedar Rapids, IA 52402, USA

The thermal expansion coefficient in each binary system of alkali borate glasses shows a minimum termed the borate anomaly. The borate anomaly relates essentially to anharmonic vibrational properties. In order to study the vibrational anharmonicity in sodium borate glasses $xNa_2O-(1-x)B_2O_3$ ($0\leq x\leq 0{\cdot}35$), the following physical quantities have been measured: the density, the longitudinal and transverse ultrasonic velocities, the linear expansion coefficient, and the temperature coefficients of travel times for longitudinal and transverse ultrasonic waves. From these physical quantities, the four different Anderson–Grüneisen constants representing vibrational anharmonicity are calculated. Each of these constants first decreases to a distinct minimum at $x\approx 0{\cdot}09$ with increasing x, and then increases gradually in the range $0{\cdot}09\leq x\leq 0{\cdot}17$ and steeply in the range $0{\cdot}17\leq x\leq 0{\cdot}35$ with further increasing x. The presence of the minimum at $x\approx 0{\cdot}09$ is a new type of borate anomaly relating to vibrational anharmonicity.

Introduction

In each binary system of alkali borate glasses, the thermal expansion coefficient against composition exhibits a minimum termed the borate anomaly.(1) The borate anomaly relates to anharmonic vibrational properties since thermal expansion is caused by the anharmonic terms in the potential energy. For the shape of the thermal expansion minimum, Uhlmann & Shaw(2) reported a broad and shallow minimum while Shelby(1) reported a broad but distinct minimum at a composition of about 20 mol% alkali oxide. A great number of early studies on the linear expansion coefficient and the borate anomaly have been reviewed in these papers. In order to characterise more clearly the borate anomaly, it may be important to investigate other kinds of anharmonic vibrational properties.

If the potential energy of a solid is expanded in powers of the interatomic displacements and all terms past those which are quadratic in the interatomic displacements are neglected, we have what is known as the harmonic approximation to the potential energy. Some consequences in the harmonic theory of lattice vibration(3) are (1) there is no thermal expansion and (2) the elastic constants are independent of pressure and temperature. In real solids, none of these consequences is satisfied accurately owing to the presence of anharmonic (higher than quadratic) terms in the interatomic displacements. In order to represent the anharmonicity, the Anderson–Grüneisen constant(4–6) is often used. This constant, which is originally defined in terms of the bulk modulus, is extended in the present paper to become applicable to other elastic constants. These extended Anderson–Grüneisen constants are used to evaluate the anharmonicity.

In the present paper, the following physical quantities have been measured for sodium borate glasses as a function of composition: density, both velocities of longitudinal and transverse ultrasonic waves, linear expansion coefficient, and temperature coefficients of travel times for longitudinal and transverse ultrasonic waves. From these physical quantities, the Anderson–Grüneisen constants are determined for sodium borate glasses.

Anderson–Grüneisen constant

Let T, p, and V be thermodynamic temperature, pressure, and volume, respectively. If anharmonic terms of atomic vibration are not present, the cubic expansion coefficient defined as $\alpha_V=(1/V)(\partial V/\partial T)_p$ vanishes and the bulk modulus defined as $K=-V(\partial p/\partial V)_T$ is independent of temperature and pressure.(3)

Grüneisen(4) defined a constant representing

[1] Corresponding author. Email kodama@chem.sojo-u.ac.jp

anharmonicity (Equation 29 of Grüneisen's original paper[4]) by

$$\delta_K = -\frac{1}{\alpha_V}\left(\frac{\partial \ln K}{\partial T}\right)_p \quad (1)$$

The constant δ_K is often called the Anderson–Grüneisen constant.[5,6] We see from Equation (1) that the Anderson–Grüneisen constant can be extended more generally to other types of elastic constants. Let E be an elastic constant. We can then define the Anderson–Grüneisen constant more generally by

$$\delta_E = -\frac{1}{\alpha_V}\left(\frac{\partial \ln E}{\partial T}\right)_p \quad (2)$$

Suzuki *et al*[6] extended the Anderson–Grüneisen constant to the shear modulus G, that is, to E=G in Equation (2). We may use Equation (2) for other types of elastic constants such as the longitudinal modulus and the Young's modulus. In this paper, we evaluate the anharmonicity of sodium borate glasses more generally on the basis of Equation (2).

By measuring the travel times, τ_l and τ_t, for one round trip of longitudinal and transverse ultrasonic waves in a specimen whose length is denoted by l, we obtain the velocities, c_l and c_t, of longitudinal and transverse ultrasonic waves from $c_l=2l/\tau_l$ and $c_t=2l/\tau_t$, respectively. With the mass density ρ and c_l and c_t, we can calculate elastic constants such as longitudinal modulus L, shear modulus G, bulk modulus K, and Young's modulus Y by the equations[7]

$$L=\rho c_l^2 \quad (3)$$

$$G=\rho c_t^2 \quad (4)$$

$$K=L-(4/3)G \quad (5)$$

$$Y=G(3L-4G)/(L-G) \quad (6)$$

In order to determine the Anderson–Grüneisen constants given by Equation (2), we measured the linear expansion coefficient $\alpha_l=(1/l)(dl/dT)$ and the temperature coefficients of travel times, $(1/\tau_l)(d\tau_l/dT)$ and $(1/\tau_t)(d\tau_t/dT)$, for longitudinal and transverse ultrasonic waves, respectively. Differentiating $c_l=2l/\tau_l$ and $c_t=2l/\tau_t$ with respect to temperature, we obtain the temperature coefficients of the velocities for longitudinal and transverse ultrasonic waves

$$k_l=(1/c_l)(dc_l/dT)=-(1/\tau_l)(d\tau_l/dT)+\alpha_l \quad (7)$$

and

$$k_t=(1/c_t)(dc_t/dT)=-(1/\tau_t)(d\tau_t/dT)+\alpha_l \quad (8)$$

We can now calculate the Anderson–Grüneisen constants for the four elastic constants given by Equations (3)–(6). Differentiating the natural logarithm of Equation (3) with respect to temperature and noticing the relation that the cubic expansion coefficient is given by $\alpha_V=-(1/\rho)(d\rho/dT)=3\alpha_l$, we have the Anderson–Grüneisen constant, δ_L, for the longitudinal modulus L

$$\delta_L \equiv -\frac{1}{\alpha_V}\left(\frac{\partial \ln L}{\partial T}\right)_p = 1 - 2k_l/\alpha_V \quad (9)$$

Similarly, we obtain the Anderson–Grüneisen constants, δ_G, δ_K and δ_Y, for shear modulus, bulk modulus and Young's modulus, respectively from

$$\delta_G \equiv -\frac{1}{\alpha_V}\left(\frac{\partial \ln G}{\partial T}\right)_p = 1 - 2k_t/\alpha_V \quad (10)$$

$$\delta_K \equiv -\frac{1}{\alpha_V}\left(\frac{\partial \ln K}{\partial T}\right)_p = \frac{1}{K}\left(L\,\delta_L - \frac{4}{3}G\,\delta_G\right) \quad (11)$$

$$\delta_Y \equiv -\frac{1}{\alpha_V}\left(\frac{\partial \ln Y}{\partial T}\right)_p \quad (12)$$

$$= \delta_G + \frac{1}{3L-4G}\left(3L\,\delta_L - 4G\,\delta_G\right) - \frac{1}{L-G}\left(L\,\delta_L - G\,\delta_G\right)$$

Experimental

A series of sodium borate glasses denoted by the composition formula xNa$_2$O–(1−x)B$_2$O$_3$ was prepared in the bulk-glass formation range $0 \le x \le 0{\cdot}35$. In order to obtain good transmission of ultrasonic waves, it is necessary to prepare glasses with high homogeneity and without strains or bubbles. With the aim of preparing glasses with high homogeneity, the starting materials were initially made to react in an aqueous solution. First, 50 wt% solution of NaOH was prepared by adding distilled water to analytical reagent grade sodium hydroxide (more than 97·0% purity, the main residual being sodium carbonate). Then the solution was diluted, for example, to a concentration of about 1·0 mol/dm^3 by adding distilled water and this diluted solution was used as the starting material for Na$_2$O. After the concentration of the diluted solution had been determined with a potentiometric titration method, the required volume was taken from a burette. As the starting material for B$_2$O$_3$, analytical reagent grade boric acid (more than 99·7% purity) was used. Amounts of the starting materials calculated to give 30 g in the melts were perfectly dissolved by adding distilled water in a beaker made of polytetrafluoroethylene. Then the mixed solution was transferred to a dry box and after the complete evaporation of the water a chemically synthesised powder was obtained.

The powder was fused in a 20 cm^3 platinum crucible at temperatures from 900 to 1300°C for about 4 h by heating in an SiC resistance electric furnace. The melt was then poured into a cylindrical graphite

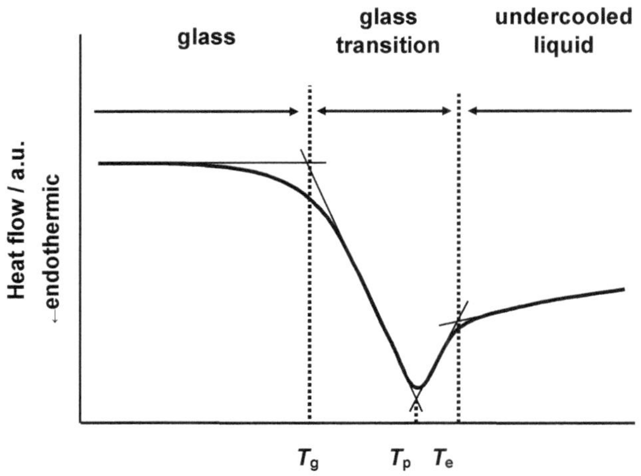

Figure 1. Schematic DSC heating curve of glass transition. The extrapolated onset temperature, the extrapolated peak temperature, and the extrapolated end temperature of glass transition are denoted by T_g, T_p, *and* T_e, *respectively*

mould, 15 mm in diameter and 30 mm deep, which had been preheated at 300°C in an electric muffle furnace. Subsequently, the glass cast in the mould was cooled at a rate of 1°C/min to room temperature. The residual melt was poured onto an aluminium plate and later used for chemical analysis and for differential scanning calorimetry (DSC). The chemical compositions of all the prepared glasses were analysed with respect to the contents of Na_2O and B_2O_3 with a potentiometric titration method described in the previous paper,[8] from which the analysed composition x in the composition formula $xNa_2O–(1-x)B_2O_3$ was determined and used throughout the present paper.

The glass transition temperature of sodium borate glasses as a function of x was measured by a DSC apparatus (Seiko Instruments, DSC 6200). Figure 1 shows, schematically on an enlarged scale, a part of the DSC heating curve in the vicinity of the glass transition, which also defines the onset temperature T_g, the peak temperature T_p, and the end temperature T_e of glass transition in a manner consistent with a phenomenological theory of glass transition;[9] T_g is often referred to as the glass transition temperature. Approximately 30 mg glass fragment of each composition was fused in an approximately 120 mg DSC pan made of platinum at 1000°C for 5 min by heating in an electric furnace. The DSC heating curve of glass transition was recorded at a heating rate of 10°C/min and subsequently the DSC cooling curve was recorded at the cooling rate of 10°C/min. In the first run of DSC heating curve, T_e could not be determined exactly due to the presence of a residual strain in the glass specimen. Thus, the second run of DSC heating curve was recorded and, from this second DSC heating curve, the three characteristic temperatures of glass transition shown in Figure 1 were determined.

Each cast glass should be carefully annealed for the following purposes: (1) all thermal history of the cast glass should be erased, (2) all residual strain in the cast glass should be removed, and (3) the glass should be stabilised for measurement of the physical properties. In order to erase the thermal history and at the same time remove the residual strain, each cast glass should first be heated near to T_e in view of the fact that undercooled liquid is in a state of thermodynamic equilibrium. On cooling the liquid, the physical properties depend upon the cooling rate through the entire temperature range of glass transition so that a slow cooling rate of 1°C /min through the entire temperature range of glass transition is recommended for the standard cooling rate and therefore for the standard stabilisation condition.[10] Thus, every cast glass was annealed in an electric furnace in a high purity nitrogen gas atmosphere according to the following annealing schedule: (1) the temperature was raised from room temperature to T_e at a constant rate of 1°C/min, (2) the temperature was kept constant at T_e for 2 h, and (3) the temperature was lowered from T_e to room temperature at a constant rate of 1°C/min. Every annealed glass was transparent and showed no opalescence. Density, linear expansion coefficient, ultrasonic velocity, and temperature coefficient of ultrasonic travel time were measured for the bulk glass annealed in this manner. As a result, the measured values should be interpreted in the strict sense as representing the physical properties inherent in the standard stabilisation condition.

The density of each annealed glass was measured at 25°C by a hydrostatic weighing method. In brief, a silicon single crystal was used as a density standard[11] for calibrating the density of the immersion liquid (cyclohexane) so that the density of each glass could be measured within an accuracy of $1{\cdot}0\times10^{-3}$ g/cm^3. In order to measure the ultrasonic velocity, each annealed glass was ground and polished to give a pair of end faces that were flat and parallel. Inspection with a strain viewer showed that all specimens were free from strains or bubbles. The ultrasonic travel time was measured at a frequency of 10 MHz and at a temperature of 25°C by means of the pulse-echo overlap method.[12] X-cut and Y-cut quartz transducers resonating at a fundamental frequency of 10 MHz were used for the longitudinal wave and for the transverse wave, respectively. The transducer was bonded to the specimen on one of the two parallel faces with phenyl benzoate and this single transducer was used for both generation and detection of ultrasound. Adjacent two echoes were overlapped according to the McSkimin criterion[13] in order that the ultrasonic travel time could be measured with high accuracy: a travel time measurement within an error of 0·02% is attainable when the round trip travel time in the

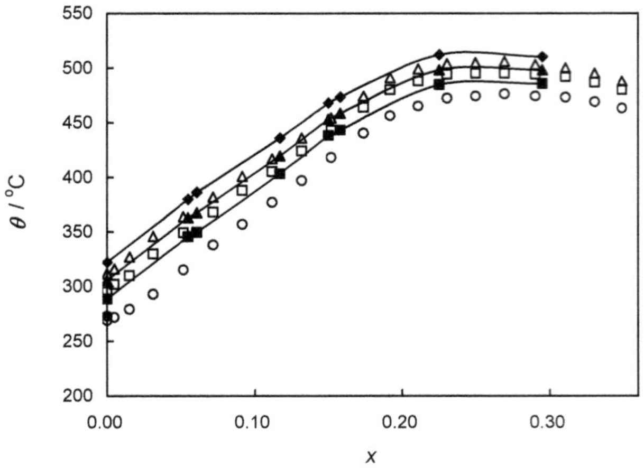

Figure 2. Characteristic temperatures, T_g*,* T_p*, and* T_e*, of glass transition determined by the present DSC measurement and the temperatures*[15] *at which* $\log(\eta/\mathrm{Pa\,s})=12$*, 11, 10, 9*

○ T_g □ T_p △ T_e

● $\log(\eta/\mathrm{Pa\,s})=12$ ■ $\log(\eta/\mathrm{Pa\,s})=11$ ▲ $\log(\eta/\mathrm{Pa\,s})=10$

◆ $\log(\eta/\mathrm{Pa\,s})=9$

All of the temperatures, θ, are denoted by degrees Celsius. Each smooth curve is shown as a guide for the eye

specimen is greater than 5 μs. The velocities, c_l and c_t, of longitudinal and transverse ultrasonic waves were determined at 25°C from τ_l and τ_t measured in this way and the specimen length measured with a micrometre reading to 1 μm. Subsequently, τ_l and τ_t were measured as a function of temperature in the temperature range 0 to 60°C. Both τ_l and τ_t changed linearly with temperature so that $(1/\tau_l)(d\tau_l/dT)$ and $(1/\tau_t)(d\tau_t/dT)$ were evaluated from their respective slopes at 25°C. After the ultrasonic properties had been measured for each glass, the linear expansion coefficient was determined in the temperature range 50 to 150°C by a standard horizontal vitreous silica dilatometer; the change in specimen length with temperature was measured by a differential transformer. The dilatometer was calibrated beforehand against the thermal expansivity of platinum[14] by making use of a platinum rod.

Results and discussion

Figure 2 shows the characteristic temperatures, T_g, T_p, and T_e, of glass transition for sodium borate glasses determined by the present DSC measurement. Visser & Stevel[15] prepared, in a vacuum furnace, a series of sodium borate melts essentially free from water and measured the change in viscosity, η, with temperature since viscosity and hence glass transition temperature are sensitive to the water contamination in the specimen. The temperatures at which $\log(\eta/\mathrm{Pa\,s})=12, 11, 10, 9$ are also shown in Figure 2. Although the amount of water which may be present in the prepared sodium borate glasses is not determined, we see that there is

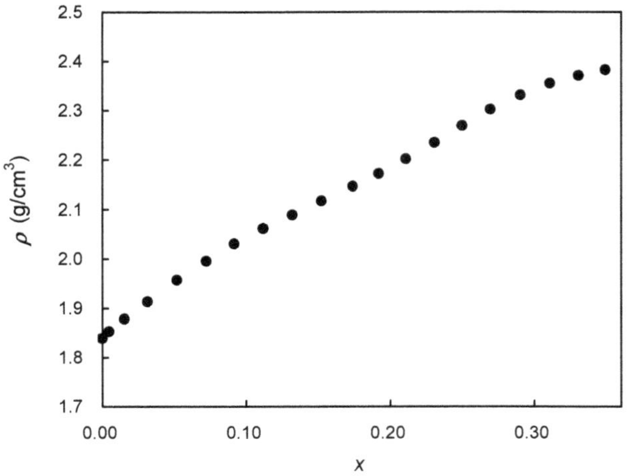

Figure 3. Density, ρ, of sodium borate glasses against composition, x

a close correlation between the one set of T_g, T_p, and T_e and the other set of temperatures at which the viscosity takes the given values; namely, $\log(\eta/\mathrm{Pa\,s})\approx12$ at T_g, $\log(\eta/\mathrm{Pa\,s})\approx10{\cdot}5$ at T_p, and $\log(\eta/\mathrm{Pa\,s})\approx9{\cdot}8$ at T_e. These estimated values of viscosity at T_g and T_e are in close agreement with the viscosity values of $\log(\eta/\mathrm{Pa\,s})\approx11{\cdot}3$ at T_g and $\log(\eta/\mathrm{Pa\,s})\approx9{\cdot}2$ at T_e determined by Moynihan[9] for many different kinds of glasses. We may conclude from this fact that a small amount of water may be present in the prepared sodium borate glasses but it does not affect the physical properties to a great extent. Figure 3 shows the density, ρ, as a function of composition. The density increases with an increase in x in the whole composition range; the density also shows a slight depression in the range $0{\cdot}09\leq x\leq0{\cdot}27$. Figure 4 shows the longitudinal and transverse velocities against composition. Each of the velocities first increases with increasing x, passes through a maximum at $x\approx0{\cdot}32$, and then decreases with further increasing x. Each velocity also shows a slight depression in the composition range $0{\cdot}09\leq x\leq0{\cdot}27$. Figure 5 shows four elastic constants,

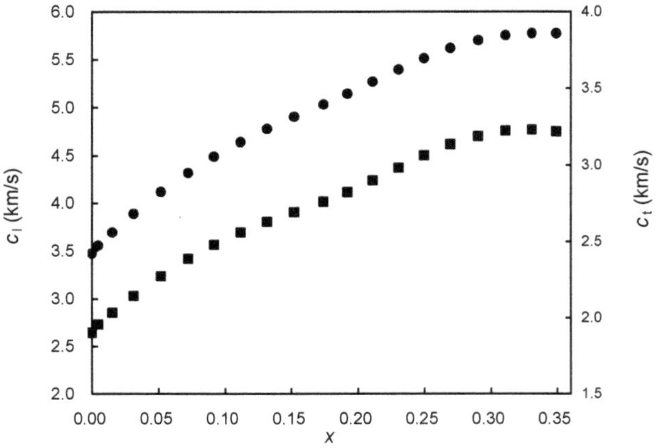

Figure 4. Longitudinal velocity, c_l*, and transverse velocity,* c_t*, in sodium borate glasses against composition* x

● c_l ■ c_t

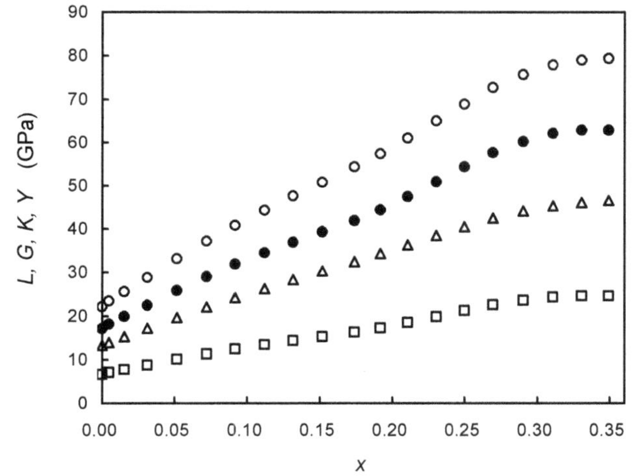

Figure 5. Four elastic constants of sodium borate glasses against composition x

○ L □ G Δ K • Y

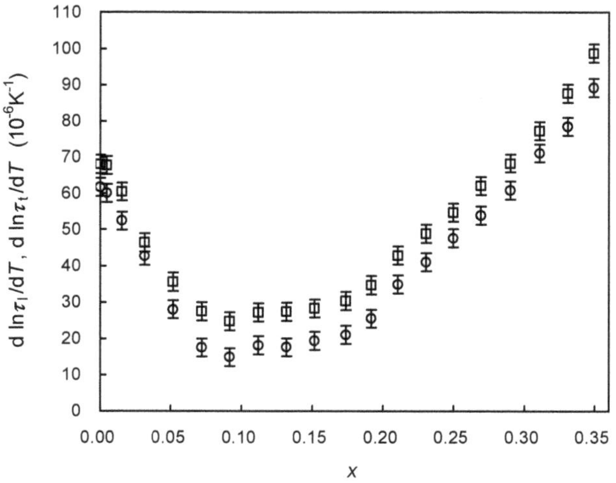

Figure 7. Temperature coefficient of travel time, $(1/\tau_l)(d\tau_l/dT)$, for longitudinal ultrasonic wave and temperature coefficient of travel time, $(1/\tau_t)(d\tau_t/dT)$, for transverse ultrasonic wave in sodium borate glasses against composition x

○ $(1/\tau_l)(d\tau_l/dT)$ □ $(1/\tau_t)(d\tau_t/dT)$

L, G, K and Y, as a function of composition. Each of these elastic constants has the same characteristic as that of the velocity of sound. The behaviour of each elastic constant is similar to that of the fraction, N_4, of boron atoms in tetrahedral coordination.[16] The correlation between the elastic constant and N_4 can be easily understood since the presence of four coordinated boron atoms strengthens the glass network and therefore increases the elasticity. Figure 6 shows the linear expansion coefficient against composition. The linear expansion coefficient shows a broad minimum at around x=0·17. This behaviour of linear expansion coefficient is in close agreement with that measured by Shelby[1] for sodium borate glasses. Figure 7 shows the temperature coefficients of travel times for longitudinal and transverse ultrasonic waves. Each of the temperature coefficients shows a minimum at x=0·09 and a distinct change in slope at x=0·17. Figure 8 shows the Anderson–Grüneisen constants, δ_L, δ_G, δ_K and δ_Y, of sodium borate glasses as a function of composition. Each of these Anderson–Grüneisen constants shows a minimum at x=0·09 and a distinct change in slope at x=0·17. The composition of x=0·17 at which the change in slope occurs coincides with the composition of x=0·17 at which the thermal expansion coefficient exhibits the broad minimum. An explanation for the cause of thermal expansion minimum has already been given by Shelby.[1] The occurrence of minima in the Anderson–Grüneisen constants at x=0·09 is a new type of borate anomaly relating to anharmonicity. It should be noted that the Anderson–Grüneisen constants of lithium borate glasses studied in the previous paper[17] have the same characteristics as those of sodium borate glasses shown in Figure 8.

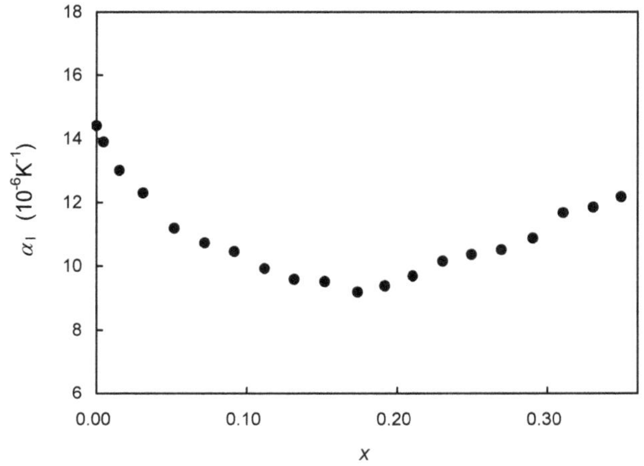

Figure 6. Linear expansion coefficient, α_l, of sodium borate glasses against composition x

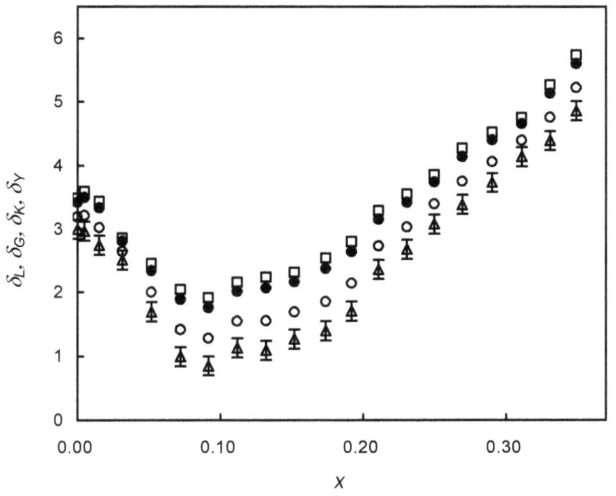

Figure 8. Anderson–Grüneisen constants of sodium borate glasses against composition x

Δ δ_L □ δ_G ○ δ_K • δ_Y

Error bars are attached only to δ_K. The uncertainty for δ_L, δ_G, and δ_Y is the same as that for δ_K

An analysis[18] of the crosslinking density in sodium borate glasses gives a reason why the minima are present at $x \approx 0.09$. Sodium borate glasses are formed by two structural units, $BØ_3$ and $Na^+BØ_4^-$, in the composition range $0 \leq x \leq 0.28$, where Ø represents a bridging oxygen. In the composition range $0 \leq x \leq 0.09$, sodium ions are initially incorporated in holes outside the boroxol rings. This incorporation converts non-ring $BØ_3$ units located outside the boroxol rings into $Na^+BØ_4^-$ units, which results in an increase in crosslink between boroxol rings. Above $x \approx 0.09$, $BØ_4^-$ tetrahedra are preferentially included in six-membered rings by transforming boroxol rings into pentaborate rings, tetraboratre rings and so forth, which results in a decrease in the crosslinking density between the rings. As a result, the crosslinking density of sodium borate glasses reaches a maximum at $x \approx 0.09$. If we assume that anharmonicity decreases as crosslinking density increases, the presence of minima at $x \approx 0.09$ is ascribable to the maximum crosslinking density at this composition. Although the relationship between the glass structure and the physical properties are beyond the scope of the present paper, it should be noted that the composition of $x=0.09$ coincides closely with the enneaborate composition $Na_2O.9B_2O_3$, or $x=0.10$, and the composition of $x=0.17$ coincides with the pentaborate composition $Na_2O.5B_2O_3$, or $x=0.166$.

Conclusion

Borate anomaly in sodium borate glasses $xNa_2O-(1-x)B_2O_3$ ($0 \leq x \leq 0.35$) has been studied with special reference to the anharmonic vibrational property. The vibrational anharmonicity has been determined from the four different Anderson–Grüneisen constants expressed in terms of longitudinal modulus, shear modulus, bulk modulus and Young's modulus. The Anderson–Grüneisen constants have been calculated from the measured quantities of the density, the linear expansion coefficient, the longitudinal and transverse ultrasonic velocities, and the temperature coefficients of travel times for longitudinal and transverse ultrasonic waves. Each of these Anderson–Grüneisen constants shows a minimum at $x=0.09$ and a change in slope at $x=0.17$.

References

1. Shelby, J. E. *J. Am. Ceram. Soc.*, 1983, **66**, 225.
2. Uhlmann, D. R. & Shaw, R. R. *J. Non-Cryst. Solids*, 1969, **1**, 347.
3. Leibfried, G & Ludwig, W. *Solid state physics. Vol. 12*. 1961. Edited by F. Seitz & D. Turnbull. Academic Press, New York. Pp 275–444.
4. Grüneisen, E. *Ann. Phys. (Germany)*, 1912, **39**, 257.
5. Anderson, O. L., Schreiber, E., Liebermann, R. C. & Soga, N. *Rev. Geophys.*, 1968, **6**, 491.
6. Suzuki, I., Ohno, I. & Anderson, O. L. *Am. Miner.*, 2000, **85**, 304.
7. Mason, W. P. *Physical acoustics and the properties of solids*. 1958. D. Van Nostrand, Princeton. Pp 12–16.
8. Kodama, M., Iizuka, K., Miyashita, M., Nagai, N., Clarida, W., Feller, S. A. & Affatigato, M. *Glass Technol.*, 2003, **44**, 50.
9. Moynihan, C. T. *J. Am. Ceram. Soc.*, 1993, **76**, 1081.
10. Vogel, W. *Chemistry of Glass*. 1985. American Ceramic Society, Columbus. Pp 16–24.
11. Bowman, H. A., Schoonover, R. M. & Carroll, C. L. *Metrologia*, 1974, **10**, 117.
12. Papadakis, E. P. *Physical acoustics. Vol. 12*. 1976. Edited by W. P. Mason & R. N. Thurston. Academic Press, New York. Chap. 5, Pp 277–374.
13. McSkimin, H. J. *J. Acoust. Soc. Am.*, 1961, **33**, 12.
14. Oldfield, L. F. *Glass Technol.*, 1964, **5**, 41.
15. Visser, T. J. M. & Stevels, J. M. *J. Non-Cryst. Solids*, 1972, **7**, 376.
16. Bray, P. J. & O'Keefe, J. G. *Phys. Chem. Glasses*, 1963, **4**, 37.
17. Kodama, M., Kojima, S, Feller, S. & Affatigato, M. *Phys. Chem. Glasses*, 2005, **46**, 190.
18. Kodama, M., Feller, S. & Affatigato, M. *J. Therm. Anal. Cal.*, 1999, **57**, 787.

Medium range order in lead borophosphate glasses

L. Koudelka,[1] *P. Mošner, J. Jirák*

Faculty of Chemical Technology, University of Pardubice, 532 10 Pardubice, Czech Republic

M. Zeyer-Düsterer

University Hamburg, Institute for Applied Physics, D-20355 Hamburg, Germany

C. Jäger

Federal Institute for Materials Research and Testing, Project Group I.3903, D-12489 Berlin, Germany

The structure of 50PbO–xB_2O_3–(50−x)P_2O_5 glasses with x=0–25 mol% B_2O_3 was refined by Raman and ^{11}B MQMAS NMR. The analysis of Raman spectra, using the concentration step of 2 mol% B_2O_3 within the range of 0–12 mol% B_2O_3, showed that the evolution of changes in the type of oxygen bridges can be observed from the Raman bands within the range of 600–800 cm^{-1}. ^{11}B MQMAS NMR spectra of glasses with x=15, 20 and 25 mol% B_2O_3 revealed the presence of two BO_4 sites in these glasses. The first site B^1O_4, characterised by the chemical shift of δ=−3·8 — (−3·4) ppm, was assigned to tetrahedral $B(OP)_4$ structural units. Within the concentration range x=15–25 mol% B_2O_3, the site B^1O_4 is gradually replaced by the site B^2O_4 with the chemical shift of δ=−1·8 — (−1·1) ppm. This second site was assigned to the presence of $B(OP)_3(OB)$ units in which the central boron atom is linked to three PO_4 groups and one BO_3 group.

Introduction

Borophosphate glasses reveal better chemical durability than phosphate glasses, because the addition of B_2O_3 into phosphate glasses improves their water resistance through crosslinking phosphate chains. Practical applications for borophosphate glasses are for example vitreous electrolytes in solid state batteries(1) and low temperature hermetic seals.(2) Mainly NMR spectroscopy, Raman spectroscopy and XPS spectroscopy have been applied to studying their structures.

NMR spectroscopy has evolved during the last two decades into one of the most powerful techniques applied for the study of the glass structure. Borophosphate glasses offer the presence of two sensitive NMR active nuclei for these studies – ^{31}P and ^{11}B, and many studies of ^{11}B and ^{31}P NMR spectra have been reported.(2–9)

Koudelka *et al*(7) studied ^{31}P and ^{11}B MAS NMR and Raman spectra of two series of lead borophosphate glasses of the composition 50PbO–xB_2O_3–(50−x)P_2O_5 and 55PbO–yB_2O_3–(45−y)P_2O_5. The study showed that within the composition range of x=0–15 mol% B_2O_3, boron forms exclusively BO_4 groups, whereas above 15 mol% B_2O_3 NMR spectra show that the amount of BO_3 groups starts to increase.

Recently, Zeyer-Düsterer *et al*(8) investigated the medium range order in sodium borophosphate glasses using combined ^{17}O NMR and ^{11}B–^{31}P double resonance NMR. With these techniques, the authors(8) were able to identify and distinguish five possible oxygen sites P–O–P, Na···O–P, P–O–B, B–O–B and Na···O–B in these glasses. Moreover, two different BO_4 sites were identified with different number of links to PO_4 tetrahedra. From ^{11}B–^{31}P HETCOR NMR, the authors(8) concluded that the assignment of a ^{31}P signal at about −20 ppm to Q_2 units as in sodium phosphate glass is wrong in some of the glasses and that the PO_4 unit belonging to this resonance must be connected to borate groups.

Elbers *et al*(9) investigated intermediate range order in silver borophosphate glasses of the composition 50Ag_2O–xB_2O_3–(50−x)P_2O_5 with ^{11}B–^{31}P and ^{31}P–^{11}B REDOR NMR and came to the conclusion that BO_3 groups in these glasses are exclusively bonded only to BO_4 groups and do not form B–O–P bonds to PO_4 groups. They also claim that no $B(OP)_4$ units are formed in these glasses and suggest the formation of B–O–B links between $B(OP)_3$ units.

These new findings initiated the presented study of crystallisation, and the Raman spectra and MQMAS NMR spectra of lead borophosphate glasses in the 50PbO–xB_2O_3–(50−x)P_2O_5 glasses with the aim to further elucidate the intermediate range order in these glasses.

[1] Corresponding author. Email ladislav.koudelka@upce.cz

Experimental

Samples of the composition $50PbO–xB_2O_3–(50–x)P_2O_5$ glasses with 0, 2, 4, 6, 8, 10, 12, 15, 20 and 25 mol% B_2O_3 were prepared by the reaction of PbO with H_3BO_3 and H_3PO_4, and heating slowly the reaction mixture up to 1000–1200°C in a Pt crucible. Clear homogeneous glasses were obtained by slow cooling in air.

Thermal behaviour of the glasses was studied with the Perkin-Elmer DTA-DSC equipment, model 1700, working in the DSC mode in the temperature range of 20–800°C at a scan rate of 10°C min^{-1}. Crystalline samples obtained by annealing above the crystallisation temperature were investigated by x-ray diffractometer Bruker D8 Advance. Glass transition temperature, T_g, was determined from the dilatation curves measured on the equipment of TMA CX04R (RMI Pardubice) at the heating rate of 5°C min^{-1}.

Raman spectra were measured on a FT IR spectrometer, Bruker model IFS 55, with the Raman attachment, FRA 106, under excitation with a Nd:YAG laser radiation using a slit width of 4 cm^{-1} with the power of 300 mW at the sample surface. The Raman spectra were measured at room temperature on bulk samples using 400 scans.

^{11}B MQMAS NMR spectra were acquired at 9·4 T using a BRUKER AMX400 spectrometer with a 4 mm standard Bruker probe head and with a z-filter pulse program. The spinning speed was 12·5 kHz. The length of the first pulse was set 6 μs, the length of the second and third pulse was 2 μs. The application of a "hard" third pulse was used instead of a so-called "soft" pulse which is to be preferred if second order quadrupolar broadening determines the spectrum (e.g. for BO_3 units). Rotorsynchronised t1 incrementation (80 μs) was used. Typically 128 scans were acquired.

Results and discussion

The addition of B_2O_3 by replacing P_2O_5 in the $50PbO–xB_2O_3–(50–x)P_2O_5$ glasses significantly changes the thermal behaviour of these glasses (see Figure 1). All glasses crystallise on heating, as can be observed from the presence of one or two exothermic crystallisation peaks on the DSC curves. At low B_2O_3 content the crystallisation peak shifts very rapidly to higher temperatures from 330°C for the glassy $Pb(PO_3)_2$ to 480°C for the glass with 4 mol% B_2O_3. Such a steep increase in the crystallisation temperature T_c reveals a significant increase in the thermal stability of these glasses. Simultaneously with T_c, the glass transition temperature increases from 257 to 347°C, but this increase is not as steep as the increase in T_c.

The incorporation of boron oxide into lead borophosphate glasses at low B_2O_3 content is realised by an incorporation of BO_4 units into metaphosphate chains. This conclusion results mainly from NMR and Raman spectra.[7] Nevertheless, indirect information on the structural units present in the glasses can also be obtained from the compositional analysis of crystallised glass samples. Therefore we have annealed all glass samples at 600°C and subsequently analysed the crystalline products with x-ray diffraction and tried to identify the obtained crystalline phases. The results of this analysis are summarised in Table 1.

Table 1. The compounds identified in crystallised samples of the $50PbO–xB_2O_3–(50–x)P_2O_5$ glasses

B_2O_3 content (mol%)	*Identified phases*
0	$Pb(PO_3)_2$
4	$Pb(PO_3)_2$, $Pb_3P_4O_{13}$,
8	$Pb_3P_4O_{13}$, $Pb(PO_3)_2$, BPO_4
12	$Pb_3P_4O_{13}$, $Pb_2P_2O_7$, BPO_4
15	$Pb_2P_2O_7$, BPO_4, $Pb_3P_4O_{13}$
20	$Pb_2P_2O_7$, BPO_4, $Pb_3P_4O_{13}$, Pb_3O_4

The x-ray diffraction analysis showed that the starting metaphosphate glass crystallised to the monoclinic lead metaphosphate $Pb(PO_3)_2$. In the samples with $x \geq 5$ mol% B_2O_3 diffraction lines of BPO_4 started to appear in the diffraction pattern. Simultaneously, the decreasing amount of phosphate anions resulted in the shift of the composition of crystalline phosphate phases towards the polyphos-

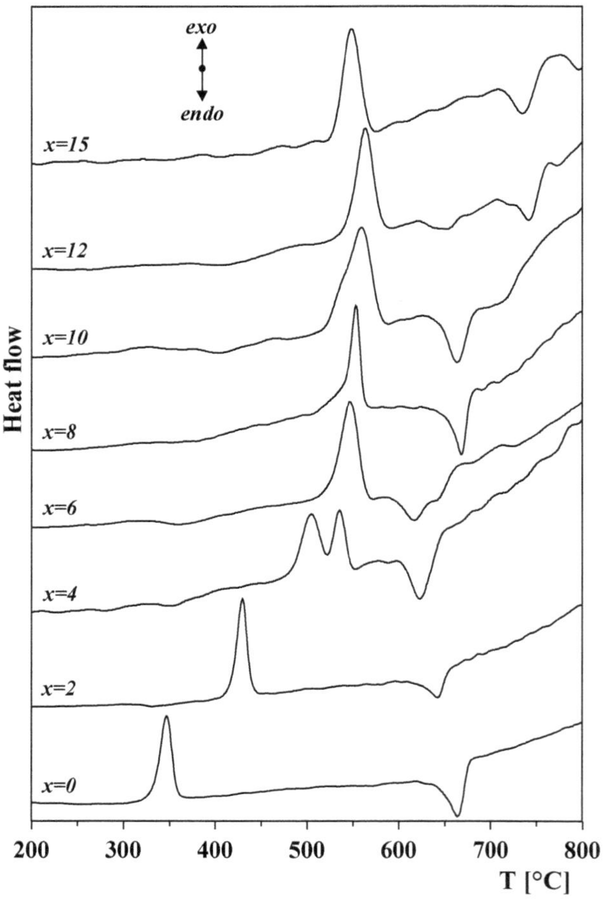

Figure 1. DSC curves of $50PbO–xB_2O_3–(50–x)P_2O_5$ glasses on heating in the range of 200–800°C

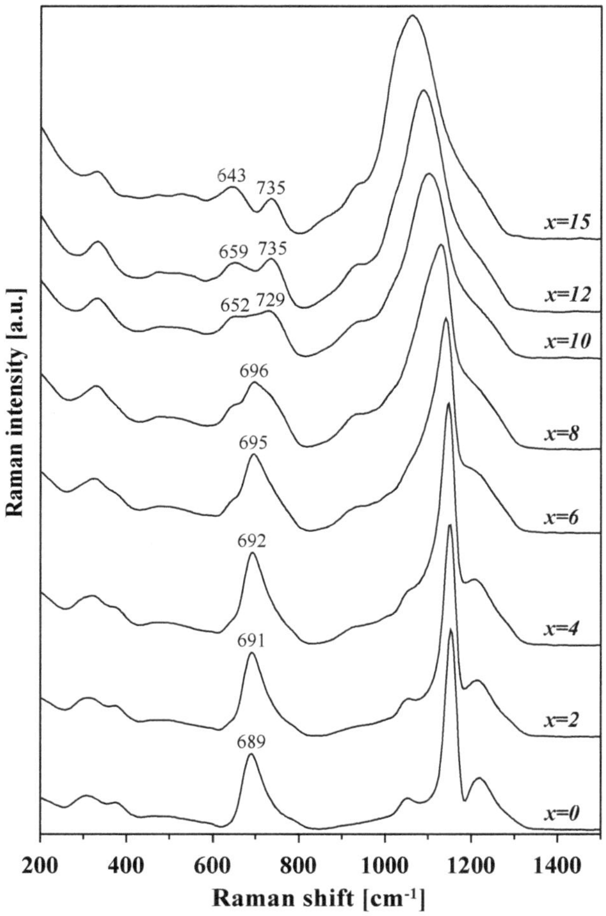

Figure 2. Raman spectra of 50PbO–xB_2O_3–(50–x)P_2O_5 glasses

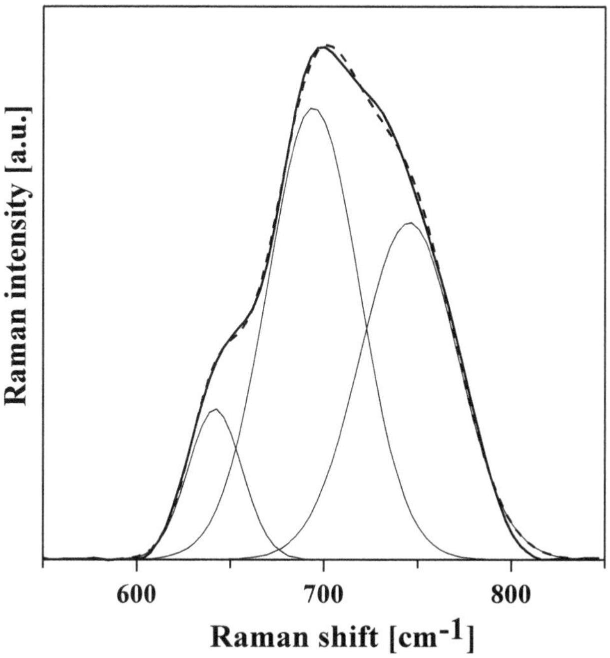

Figure 3. Example of the decomposition of Raman spectra of the 50PbO.8B_2O_3.42P_2O_5 glass within the spectral region of 600–800 cm^{-1}

phate region, i.e. in the diffraction patterns the lines of $Pb_3P_4O_{13}$=3PbO.2P_2O_5 with 60 mol% of PbO started to increase their strength. The lines of this compound dominated the diffraction patterns within the compositional region with x=8–12 mol% B_2O_3. In the crystallised products of glasses with x≥15 mol% B_2O_3, the diffraction lines of the compound $Pb_2P_2O_7$ with 66 mol% PbO started to dominate due to an increasing amount of phosphate anions present as BPO_4. The x-ray pattern of the crystallised sample with 25 mol% B_2O_3 contained many lines and their proper assignment could not be done.

Raman spectra of 50PbO–$x$$B_2O_3$–(50–$x$)$P_2O_5$ in the region with a low B_2O_3 content are shown in Figure 2. The Raman spectra reflect only the changes in the vibrations of phosphate skeleton of the glass structure, because Raman scattering cross section of borate ions is generally much weaker than that of phosphate ions. The incorporation of BO_4 units into the phosphate chain structure results in the shortening of phosphate chains and thus in a decrease in the number of Q_2 units and their conversion into Q_1 units. This transfer can bee seen also in the ^{31}P MAS NMR spectra, where distinct two signals can be seen at about −25 ppm (Q_2) and −10 ppm (Q_1) at the glasses with x=0–10 mol% B_2O_3, whereas for the glasses with 15–25 mol% B_2O_3 ^{31}P MAS NMR spectra reveal only one broad signal. On the other side, Raman spectra of the 50PbO–$x$$B_2O_3$–(50–$x$)$P_2O_5$ glasses in the compositional region of x=0–15 mol% B_2O_3 reveal distinct changes in the spectral region of 600–800 cm^{-1} characteristic of the vibrations of bridging oxygen atoms in P–O–P bridges. In the starting glassy $Pb(PO_3)_2$, its maximum lies at 689 cm^{-1}. Nelson & Exarhos[11] took this band as constant, but we have found a relatively broad span of frequencies from 672 cm^{-1} for g-$LiPO_3$, 679 cm^{-1} for g-$NaPO_3$, 687 cm^{-1} for g-KPO_3, 679 cm^{-1} for g-$Pb(PO_3)_2$ up to 703 cm^{-1} for g-$Zn(PO_3)_2$. The position of this band in lead polyphosphate glasses $(PbO)_x(P_2O_5)_{1-x}$ also depends on the ratio of PbO/P_2O_5;[12] with increasing x, it shifts towards higher wavenumbers and for x=0·66 amounts 731 cm^{-1}.

The transformation of Q_2-type P–O–P phosphate bridges into Q_1-type P–O–P bridges can be seen in the Raman spectra of 50PbO–$x$$B_2O_3$–(50–$x$)$P_2O_5$ glasses within the compositional region of x=6–12 mol% B_2O_3. The strength of the vibrational band at 695–696 cm^{-1}, characteristic of Q_2-type bridges, decreases with increasing B_2O_3 content, whereas the strength of another two Raman bands at 729–735 cm^{-1} and 651–652 cm^{-1} increases. The position of the first band is characteristic of Q_1-type units and the second band was previously[7] ascribed to the vibrations of oxygen atoms in P–O–B bridges. The decomposition of the Raman spectrum into these three components can be

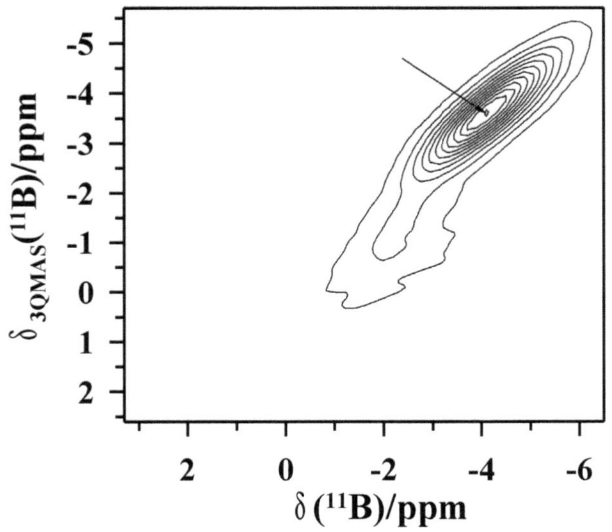

Figure 4. ^{11}B MQMAS NMR spectrum of the $50PbO.15B_2O_3.35P_2O_5$ glass

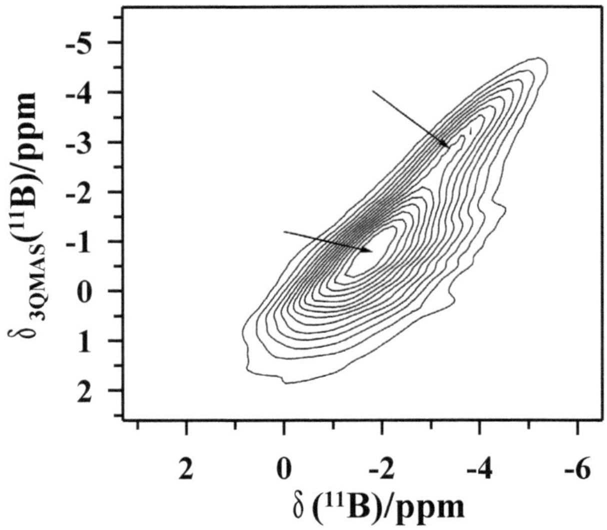

Figure 5. ^{11}B MQMAS NMR spectra of the $50PbO.25B_2O_3.25P_2O_5$ glass

seen in Figure 3 for the glass with 8 mol% B_2O_3. On the Raman spectra of the glass with 15 mol% B_2O_3, we do not see the band of 695–696 cm^{-1} which results in the conclusion that there is no Q_2–Q_2 connection in this glass. With a further increase in B_2O_3 content, the strength of the vibrational bands of P–O–B and Q_1–Q_1 bonds decreases, which we assume to be due to an increased rigidity of the three-dimensional network and the subsequent damping of vibrations involving bridging oxygen atoms.

One-dimensional ^{11}B NMR spectra of the 50PbO–xB_2O_3–$(50-x)P_2O_5$ glasses[7] show a broader asymmetrical signal in the region of BO_4 resonances, probably composed of two resonances for the glasses with x=15–25 mol% B_2O_3. Therefore, ^{11}B MQMAS NMR was carried out in this compositional region. In Figures 4–5 ^{11}B MQMAS NMR spectra of the glasses with 15 and 25 mol% B_2O_3 are shown. They reveal two different BO_4 units denoted as B^1O_4 and B^2O_4. The first site B^1O_4 with the chemical shift in the single-quantum dimension within the range from −3·8 to −3·4 ppm dominates at the glass containing 15 mol% B_2O_3 (see Figure 4), whereas the second site B^2O_4 with the chemical shift in the single-quantum dimension within the range from −1·8 to −1·1 ppm prevails at the glass with 25 mol% B_2O_3 (see Figure 5). The spectrum of the glass with x=20 mol% B_2O_3 reveals the presence of both BO_4 sites with nearly equal amount.

It is clear that the presence of these two sites in the glasses under study is associated with different coordination around BO_4 units, i.e. with a changing medium range order in the glasses with x=15–25 mol% B_2O_3. Our previous study[7] showed that BO_4 sites are the only structural units formed by boron atoms in the glasses within the composition range x=0–15 mol% B_2O_3. In the borophosphate glasses with low B_2O_3 content, boron atoms form most probably $B(OP)_4$ units[8] with the central boron atoms coordinated by four phosphorus atoms. The formation of these units is associated with the replacement of Q_2-type P–O–P bridges by Q_1-type P–O–P bridges which is reflected in the Raman spectra (Figure 2). Because B_2O_3 brings only three oxygen atoms per one boron atom, we assume that boron is bonded partly to non-bridging oxygen atoms of Q_2 units and partly enters also metaphosphate chains and forms P–O–B bonds with bridging oxygen atoms. Therefore, we assume that B^1O_4 resonances represent $B(OP)_4$ structural units. The presence of such units in the glasses with a low B_2O_3 content is supported also by the formation of BPO_4 in the crystallised 50PbO–xB_2O_3–$(50-x)P_2O_5$ glasses (see Table 1). It does not mean that the glass structure starts to be similar to that of BPO_4, because phosphorus atoms do not have $P(OB)_4$ coordination, so we do not expect such units to appear in the glass structure. The reason for this statement is a low value of the chemical shift of the ^{31}P signal (between −20 and −10 ppm) for the 50PbO–xB_2O_3–$(50-x)P_2O_5$ glasses containing 10–20 mol% B_2O_3[7] as compared with the value of −31·2 ppm observed by Ducel *et al*[5] for crystalline BPO_4, where all phosphorus atoms have $P(OB)_4$ coordination. Nevertheless, due to the reorganisation of chemical bonds during the crystallisation process $P(OB)_4$ units are present as crystalline BPO_4 in some recrystallised glasses.

The assignment of the B^2O_4 sites is not an easy task. These units have less than four (OP) groups bonded to central boron atoms, i.e. $B(OP)_3(OB)$. The presence of such units has been reported also for sodium borophosphate glasses by Zeyer *et al.*[8] Another explanation is related to the experimental observation that the number of B^2O_4 sites increases with increasing B_2O_3 content within the range of x=15–25 mol% B_2O_3 (see Figures 4 and 5). The ^{11}B MAS NMR spectra of $50PbO–xB_2O_3–(50–x)P_2O_5$ glasses(7) reveal in the glasses with 20 and 25 mol% B_2O_3 a broad resonance within the chemical shift range of 0–17 ppm, characteristic of the presence of BO_3 trigonal units. Thus, boron atoms start to form BO_3 units in the glasses with x>15 mol% B_2O_3 and their number steadily increases with increasing B_2O_3 content. Therefore, we assume that BO_3 units are associated with the formation of B^2O_4 units and propose the formation of mixed units in which boron atom is bonded via oxygen bridges to three phosphorus atoms and one BO_3 unit, i.e. $B(OP)_3(OB)$ units. As mentioned earlier, this assignment is in agreement with NMR studies published recently,(8) where the presence of $B(OP)_3(BO)$ units with different numbers of connected phosphate groups was determined. Furthermore, it does not contradict the results of Elbers *et al*,(9) who proposed that BO_3 units are exclusively bonded only to BO_4 units and do not form B–O–P connections. The formation of two BO_4 sites is thus similar to sodium borophosphate glasses(8) and also to zinc borophosphate glasses,(2) where a similar splitting of BO_4 resonance in the ^{11}B NMR spectra was observed.

Conclusions

The results presented here show that medium range order in the structure of lead borophosphate glasses can be monitored by a suitable combination of various techniques. Raman spectra sensitively revealed the changes in oxygen bridges between various types of structural units. The presence of two different BO_4 sites in the $50PbO–xB_2O_3–(50–x)P_2O_5$ glasses within the compositional region of x=15–25 mol% B_2O_3 was confirmed by the measurement of ^{11}B MQMAS NMR spectra. These spectra are a valuable tool for distinguishing different BO_4 sites and, in combination with ^{11}B MAS NMR spectra, their population in lead borophosphate glasses.

Acknowledgements

The authors L.K. and P.M. are grateful for the financial support to the Grant Agency of Czech Republic (Grant No. 104/04/0711) and to the research project No. 0021627501 of the Ministry of Education of Czech Republic.

References

1. Brow, R. K. & Tallant, D. R. *J. Non-Cryst. Solids*, 1997, **222**, 396.
2. Magistris, A., Chiodelli, G. & Villa, H. *J. Power Sources*, 1985, **14**, 87.
3. Beekenkamp, P. & Hardeman, G. E. G. *Verres Refr.*, 1966, **20**, 419.
4. Yun, Y. H. & Bray, P. J. *J. Non-Cryst. Solids*, 1978, **30**, 45.
5. Ducel, J. F., Videau, J. J., Suh, K. S. & Senegas, J. *Phys. Chem. Glasses*, 1994, **35**, 10.
6. Villa, M., Scagliotti, M. & Chiodelli, G. *J. Non-Cryst. Solids*, 1987, **94**, 101.
7. Koudelka, L., Mošner, P., Zeyer, M. & Jäger, C. *Phys. Chem. Glasses*, 2002, **43C**, 102.
8. Zeyer-Düsterer, M., Montagne, L., Palavit, G. & Jäger, C. *Solid State Nucl. Magn. Resonance*, 2005, **27**, 50.
9. Elbers, S., Strojek, W., Koudelka, L. & Eckert, H. *Solid State Nucl. Magn. Resonance*, 2005, **27**, 65.
10. Schulze, G. E. R. *Z. Phys. Chem.*, 1934, **24**, 215.
11. Nelson, B. N. & Exarhos, G. J. *J. Chem. Phys.*, 1979, **71**, 2739.
12. Le Saout, G., Simon, P., Fazon, F., Blin, A. & Vaills, Y. *J. Raman Spectrosc.*, 2002, **33**, 740.

Glass property calculations and prediction of new compounds on the basis of raman spectroscopy of borate glasses

Tatiana S. Markova,[1] *Oleg V. Yanush*

Saint-Petersburg State Technological University of Plant Polymers, 4 Ivan Chernykh St., St-Petersburg, 198095, Russia

Irina G. Polyakova & Boris Z. Pevzner

Institute of Silicate Chemistry, Russian Academy of Sciences, 29 Odoevskogo St., St-Petersburg, 199155, Russia

Structural studies of B_2O_3–BaO–Al_2O_3 glasses and crystals have been carried out by Raman spectroscopy to investigate interaction of initial oxides resulted in so-called constant stoichiometry groupings (CSG) formation. The Raman spectra of glasses are interpreted as the superposition of relatively small numbers of unchangeable spectral forms belonging to CSG. The composition dependences of CSG mole fractions afford an opportunity to explain, calculate and predict dependences of refractive index, density, structural thermal expansion coefficient, the fraction of four coordinated borons and other properties of barium aluminoborate, Na_2O–B_2O_3, and Sb_2O_5–B_2O_3 glasses in agreement with experimental data. The Raman spectra analysis developed enabled us to predict the existence of unknown compounds $Na_2O.8B_2O_3$, $Na_2O.4SiO_2$ and others, which have been discovered now. Based on the measured Raman spectra we have proposed an explanation, which is able to predict the Boson peak frequency and its intensity correlations with the glass transition temperature. This prediction gives excellent agreement with the experimentally obtained data in B_2O_3–BaO, Na_2O–B_2O_3, and PbO–GeO_2 glasses.

Introduction

Borate based compositions are of interest for application because of their piezoelectric, electro-optic and fast ionic properties, ability to the second harmonic generation of glass samples poled, etc.[1] Recently, remarkable interest has been devoted to glass ceramic materials containing polar phases characterised by nonlinear optical activity. These materials combine the advantages of glass technology and special properties of borate crystals. The nonlinear optical crystals of β-BaB_2O_4 exhibit the second harmonic generation of six times higher intensity than that for potassium dihydric phosphate crystal.[2] However optimisation of borate glasses microphases properties is the most complicated technological problem.[3,4]

Vibrational spectroscopy is one of the most useful experimental techniques available for structural studies of glasses.[5] This method is sensitive to the smallest variation of chemical bonds determining properties of glasses. The majority of the investigators discuss assignments of vibrational spectra of borate glasses focusing attention on the effect of conversion of the triangular BO_3 structural units to BO_4 tetrahedra and the formation of nonbridging oxygen atoms on characteristic bands. It is conventional to explain the 'boron anomaly' of some physical properties of borate glasses by this model. Alternatively, as early as 1962 Krogh-Moe[6] postulated that the structure of borate glasses can be described as a random network of large borate groups (such as diborate, triborate, pentaborate, etc). This 'molecular' approach is presently accepted widely in chemistry.

In the progress of this idea the approach has been proposed[7–10] to interpret vibrational spectra (mainly Raman, but also IR) of glasses as superposition of relatively small numbers of unchangeable spectral forms (principal spectral components, PSC[10]), which being additive are summarised from the measured spectrum. These PSCs are interpreted as belonging to the stable products of the interaction of initial oxides resulted in formation of so-called constant stoichiometry groupings (CSG). The analysis of the intensities of the PSCs, which are extracted from Raman spectra by the technique described in Ref. 10, makes it possible to calculate the composition dependences of CSGs mole fractions ("diagram" of CSGs). These diagrams of CSG content afford an opportunity to explain, calculate and predict many properties of glasses in agreement with experimental data. It should be noted that the concept of CSG is

[1] Corresponding author

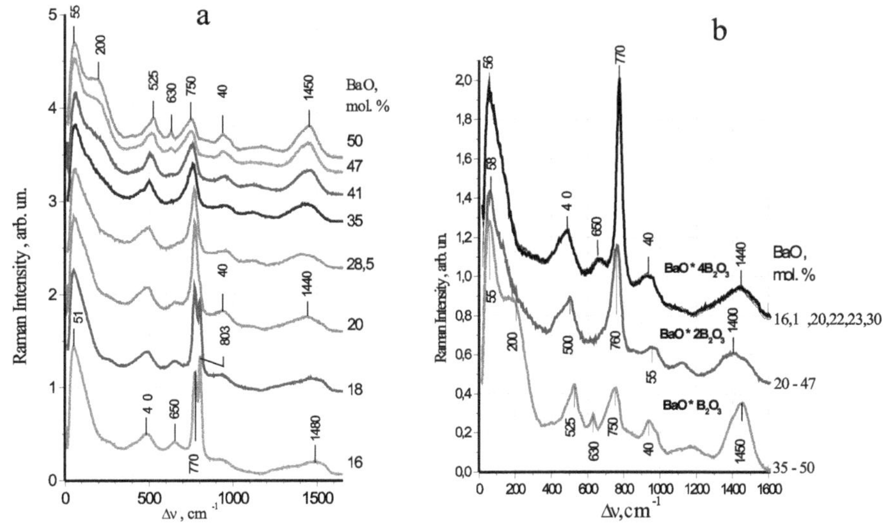

Figure 1. Raman spectra (a) and demonstration of constancy of PSCs belonging to CSGs (b) in xBaO–(100–x) B_2O_3 glasses (VV polarisation). Numerals are compositional ranges where the PSCs were resulted

based on the experimentally measured vibrational spectra exclusively without using any supplementary data (for example, vibrational spectra of crystals,(11) phase diagram, thermodynamic properties of crystals, constitutional relations of mineral phases, etc) in contrast to techniques developed respectively in Refs. 12–15 and others.

The aim of the present paper is to develop our concept of CSG for B_2O_3–BaO–Al_2O_3 glassforming system. It will be shown that correlations between glass transition temperature (T_g) and the Boson peak features are the same for some borate and germinate glasses.

Experimental

The compositions of glass and crystal samples used in the present Raman spectroscopy study were $(100-x)B_2O_3-x$BaO with $16\leq x\leq 65$ mol%, and also glass series xBaO–$(100-x)B_2O_3+yAl_2O_3$ where x=16, 18, 20, 33, 36, 50, doped with $0\leq y\leq 25$ mol% Al_2O_3.

The glasses were synthesised in a platinum crucible under 1200–1250°C; the initial ingredients were high purity $BaCO_3$, $Al(OH)_3$ and H_3BO_3. Bulk glasses and powders were crystallised in electrical furnaces at proper temperature–time conditions. Phase composition of the crystallised samples was established by x-ray powder diffractometry. The preparation and characterization of the Na_2O–B_2O_3 and Sb_2O_3–B_2O_3 glasses are described in Refs. 16–17.

Raman scattering was excited by argon laser (λ=0·488 μm, 0·5 W). Polarised Raman spectra were measured with 3 cm^{-1} step at room temperature by a two beam spectrometer using photomultiplayer in photon counting mode. The spectral width of the slit was no greater than 3 cm^{-1}. In the calculations of spectra even derivatives, smoothing interval was chosen in such a way that for several spectra of glasses with close compositions the same maxima could be clearly observed.

Relative intensities of Raman spectra of barium borate glasses were measured at frequency 51 and 770 cm^{-1}.(18) The sample made of fused silica was used as a reference and the Raman scattering (VV), intensity of which at frequency 440 cm^{-1} was taken as 1 (I^{SiO_2} (440 cm^{-1}) =1). Details of the measurement procedure were described in Ref. 4.

Results

The Raman spectra (VV) of binary barium borate glasses are illustrated in Figure 1(a). Considering that glass molar volume decreases and refractive index increases with the addition of barium oxide,(19) so that Vn^2 (4) does not vary significantly with BaO content, all experimental spectra were not multiplied by Vn^2 and their intensities correspond to fused silica standard (I^{SiO_2} (440 cm^{-1})=1).

The most striking feature in the Raman spectra of BaO–B_2O_3 glasses is the rapid decrease of a strong peak at about 803 cm^{-1} and the concurrent increase of a peak at about 770 cm^{-1} for increasing barium oxide content (Figure 1(a)). At 25 mol% BaO the 803 cm^{-1} peak has completely disappeared. Krogh-Moe(20) and Bril(21) attributed the 806 cm^{-1} peak to the symmetrical breathing vibration of the boroxol ring of vitreous boron oxide. In Ref. 16 it was assumed that these vibrations can shift to the frequency 803 cm^{-1} when boron oxide takes part in solvation of tetraborate groupings. The CSG which is manifested by the band developing at 803 cm^{-1} will be represented by "$B_2O_3^{solv}$", in the following.

Raman spectra analysis by Wallace–Katz technique(22) shows three composition ranges (16–18,

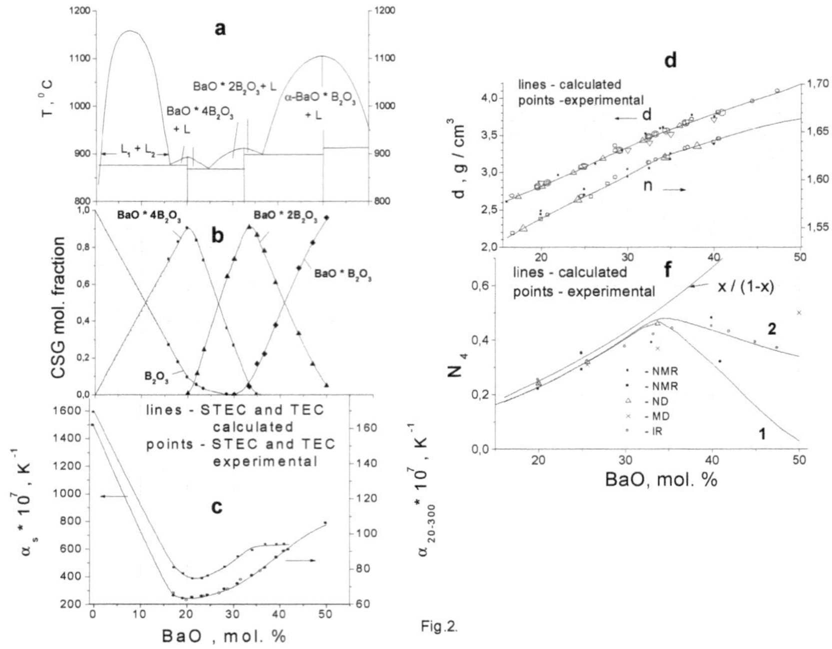

Figure 2. Demonstration of Phase Diagram (a – (23)), the composition dependences of CSG molar fractions (b, points – present experimental values, lines – calculations using Equation (1)), TEC α20-300 and STEC αs (c, lines – calculations, points – experimental data from (19) and (24)), density, refractive indexes (d, curves "d' and "n", lines – calculations, points – experimental data of different authors (19)) and the fraction of four coordinated borons N4 (f, calculations on condition to z=0 (curve 1) and 0·33 (curve 2) in the Equation (4)), points – experimental data obtained from NMR (26,27), infrared spectroscopy (28,29), neutron diffraction and molecular dynamics simulations (30)) of xBaO–(100−x)B_2O_3 glasses

23–33 and 33–50 mol% BaO) where experimental spectra can be interpreted as the sum of two PSCs only. Demonstration of constancy of the PSCs obtained from Raman spectra and belonging to the CSGs (barium tetra-BaO.4B2O3, di-BaO.$2B_2O_3$ and metaborate BaO.B_2O_3) is shown in Figure 1(b). The analysis of the composition dependences of PSC intensities let us determine the stoichiometry of every CSGs which are in coincidence to relevant phase diagram[23] (Figure 2(a)).

Finally the composition dependences of Raman intensities of every PSC were converted to diagram of CSGs of barium borate glassforming system (Figure 2(b)) using equations like this

$$\frac{I^{\text{solv}}_{B_2O_3}}{\sigma_{B_2O_3}}+\frac{4}{5}*\frac{I_{BaO.4B_2O_3}}{\sigma_{BaO.4B_2O_3}}+ \frac{2}{3}*\frac{I_{BaO.2B_2O_3}}{\sigma_{BaO.2B_2O_3}}+\frac{1}{2}*\frac{I_{BaO.B_2O_3}}{\sigma_{BaO.B_2O_3}}=x_{B_2O_3} \quad (1)$$

where I_{CSG} is the relative intensity of PSC, σ_{CSG} is a specific scattering cross section for Raman spectra (or molar extinction for absorption). The coefficients in Equation (1) fit the expression of the glass composition in molar fractions of initial oxides implying that 1 mole of glass contains the Avogadro's number of molecules of initial oxides. If the glass, for example, of barium tetraborate is comprised of the CSG BaO.$4B_2O_3$ only, dissociation of which we ignore, 1 mole of glass will contain 1/5 mole of BaO, 4/5 mole of B_2O_3 and correspondingly 1 mole of the CSG BaO.$4B_2O_3$.

Discussion

In Refs. 8–10 it was shown (see Table 1) that glass molar volume is linearly dependent on CSG content

$$V=\sum \upsilon_i * x_i^{CSG} \quad (2)$$

where V is the glass molar volume (cm^3/mole), υ_i is the molar volume of the ith CSG ("partial" molar volume) and x_i^{CSG} is the molar fraction of the CSG (being expressed as molar fractions of initial oxides). The density data were used to find the best fit values

Table 1.

Mode of expression of glass composition	*Values related to* Refractive index	*Density*	*Molar volume*	*Raman intensity of PSC*	*Sound velocity*	*Thermal expansion*
	n	d	V	I	c	α
Volume fractions	*n*	*d*	*1/V*	*I*	*1/c*	*α*
Mole fractions	*nV*	*dV=M*	*V*	*I*V*	*V/c*	*αV*
Mass fractions	*n/d*	*1/d*	*1/dV*	*I/d*	*1/dc*	*α/d*

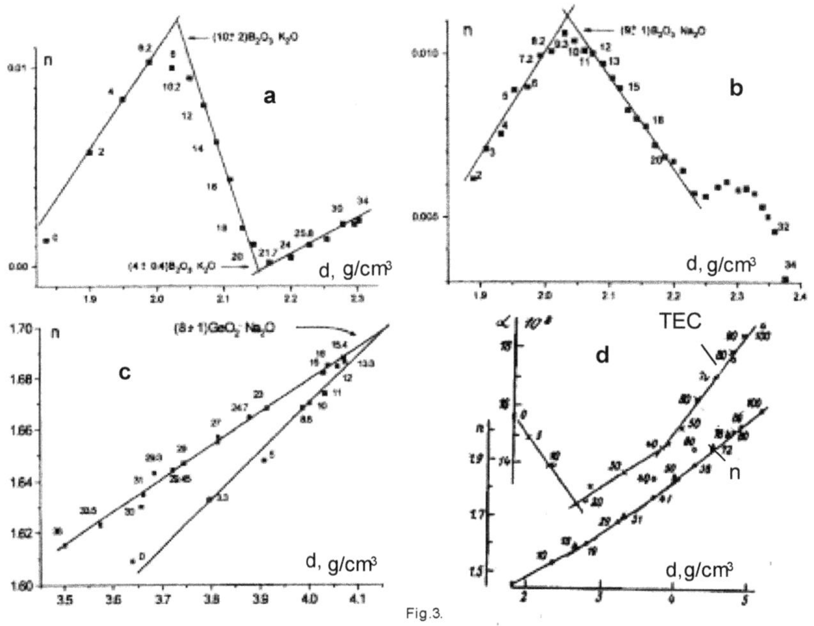

Figure 3. Estimation of partial properties (ni, di and αi) and the stoichiometry of CSG $K_2O.10B_2O_3$ in potassium borate glasses (a), $Na_2O.8B_2O_3$ in sodium borate glasses (b), $Na_2O.8GeO_2$ in sodium germanate glasses (c), $Sb_2O_3.4B_2O_3$ and $2Sb_2O_3.3B_2O_3$ in Sb borate glasses (d). Points – experimental data from different authors in Ref. 19. Numerals – M_2O where M is alkali metal, mol%. The tendency resulted is illustrated by rotating the plots. Lines – guides to the eye

for the "partial" molar volume of CSGs using a compilation of results from the literature (Ref. 19, Figure 2, curve "d").

Refractive index (as well as density) is not linearly dependent on the CSG content[8] but the product Vn is already linearly dependent on CSG content being expressed as molar fractions of initial oxides[8] that allows calculation of Vn and n of glasses

$$n=\sum_{i}\upsilon_i n_i x_i^{CSG} \Big/ \sum_{i}\upsilon_i x_i^{CSG} \qquad (3)$$

where n_i is the refractive index of the ith CSG ("partial" refractive index).

Thermal expansion coefficient (TEC) α_g and structural thermal expansion coefficient (STEC) α_s [24] as well as refractive index are additive in the volume fractions of CSGs and can be calculated by using the equation similar to Ref. 3. The CSG "partial" properties obtained are assigned to properties of model glasses contained the only CSG and listed in Table 2. The experimental and calculated TEC and STEC, densities and refractive indexes for $BaO–B_2O_3$ glass-forming system are presented in Figure 2(c,d) curves "d" and "n".

It should be emphasised that when glass composition is expressed in mole fractions, only the molar properties of glass are linearly dependent on this glass composition. Similarly, the specific magnitudes are to be related to mass fractions and in much the same way as volume fractions (Table 1). In any one line of Table 1 all the magnitudes are linearly interdepended. Consequently, in some composition range, where the glass consists of only two CSGs, the straight line has to be presented on the appropriate diagrams "property–property". To obtain this diagram any pair of glass properties magnitudes from one line in Table 1 can be used as coordinates, for example, n and d, α/d and $1/d$, etc. Disregard of these rules (using "incorrect" coordinates) will result in nonlinear dependences even for binary (or pseudobinary) ideal systems.

The potentialities of above ideas are demonstrated in Figure 3(a-d). On these diagrams, points of intersection of every two lines indicate the "partial" properties for certain CSG formed. Evident advantage of the method proposed[8] is the possibility to determine the "partial" properties for every CSG using "property–property" diagrams when glass content is not known with certainty. Examples of efficiency of this

Table 2. "Partial" properties of CSGs in barium borate glasses

CSG	"Partial" properties of CSGs Density d_i, g/cm³	Molar volume υ_i, cm³/mole	TEC α_{20-300} ×10⁷, K⁻¹	STEC α_s ×10⁷, K⁻¹	Refractive index n_i
B_2O_3	1·83*	37·52*	161·6*	1600	1·467*
$B_2O_3^{solv}$	1·947	35·75	77	1000	1·467
$BaO.4B_2O_3$	2.858	30·22	61	310	1·570
$BaO.2B_2O_3$	3·533	27·60	74	620	1·622
$BaO.B_2O_3$	4·181	26·66	106·5	**	1·666

* data from Ref. 19

** "partial" $\alpha^s_{BaO.B2O3}$ was incompatible with experimental α_s[24] and the CSG diagram

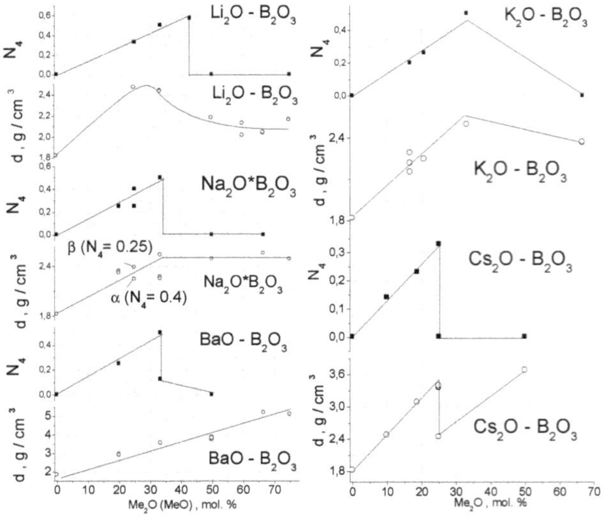

Figure 4. Comparison of compositional dependences of the fraction of four coordinated borons N_4 and densities (d) trends for alkali (Li, Na, K and Cs) and barium borate crystals. Points – experimental data from ICDD. Lines – guides to the eye

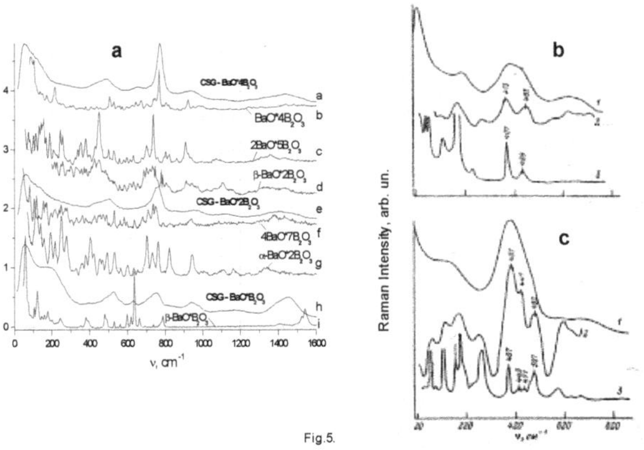

Figure 5. Raman spectra of polycrystals of $BaO–B_2O_3$ (a) system (b $BaO.4B_2O_3$, c-$2BaO.5B_2O_3$, d – β-$BaO.2B_2O_3$, f - $4BaO.7B_2O_3$, g – α-$BaO.2B_2O_3$ and i – β-$BaO.B_2O_3$) in comparison to PSCs (a – $BaO.4B_2O_3$, e - $BaO.2B_2O_3$ and h - $BaO.B_2O_3$) extracted from Raman spectra of glasses, and correlation of Raman spectra (1) and their second derivatives (2) in $Sb_2O_3.4B_2O_3$ (b) and $2Sb_2O_3.3B_2O_3$ (c) CSGs and corresponding crystals (3)

technique have been demonstrated for $BaO–B_2O_3$,[25] $Na_2O–B_2O_3$,[16] $B_2O_3–Sb_2O_3$,[17] $K_2O–B_2O_3$, $Na_2O–GeO_2$, and others glassforming systems.[8,10]

Traditionally, the fraction of four-coordinated borons (N_4) trends are used to explain properties of borate glasses. Although NMR is sensitive enough for determination of N_4, there is a significant disagreement between various NMR techniques for barium borate glasses[26,27] (Figure 2, curves N_4). At the same time the composition dependences of N_4 values obtained from NMR, infrared spectroscopy,[28,29] neutron diffraction and molecular dynamics simulation[30] correlate well. The CSG diagram resulting from the Raman spectra analysis (Figure 2(b)) makes possible another method of determination of the composition dependences of N_4 values. Assuming that N_4 values for CSGs are equal those for corresponding crystals (N_4=¼ for $BaO.4B_2O_3$;[31] ½ for $BaO.2B_2O_3$[32] and 0 for α- and β-$BaO.B_2O_3$[33,34]) N_4 for CSGs can be calculated from the Equation

$$N_4 = \frac{\left(\frac{1}{5}x^{BaO.4B_2O_3} + \frac{1}{3}x^{BaO.2B_2O_3} + z * \frac{1}{2}x^{BaO.B_2O_3}\right)}{\left(x^{B_2O_3} + \frac{4}{5}x^{BaO.4B_2O_3} + \frac{2}{3}x^{BaO.2B_2O_3} + \frac{1}{2}x^{BaO.B_2O_3}\right)} \quad (4)$$

where z is N_4 value (for the CSG $BaO.B_2O_3$) taken equal to 0 or 0·33. Composition dependences of N_4 values calculated (curves 1, 2 in Figure 2(f)) show a good agreement with the available NMR results[26,27] and correlate well with the theoretical curve $x/(1-x)$ in the composition range 16>x>33 mol% BaO.

According to Ref. 28 for alkaline earth borate glasses of the same MO content, N_4 was found to decrease from Ba to Mg, i.e. upon increasing cation field strength, a trend opposite to that exhibited by alkali borate glasses. However in contrast to alkali borates, for alkaline earth borate glasses of the same MO content T_g increases, i.e. from Ba to Ca and this occurs despite the fact that N_4 decreases in the same order.[28,35]

It is worthwhile to compare compositional dependences of N_4 and densities for alkali (Li, Na, K and Cs) and barium borate crystals (Figure 4). Despite the extreme dependences of N_4 observed for all above crystals, dependences of densities of potassium and cesium borate crystals only agree with N_4 trends. But

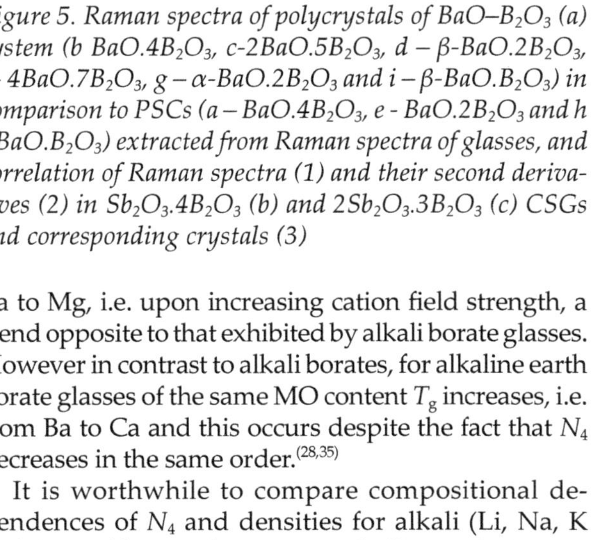

Figure 6. The Boson peak frequency and its intensity correlations with the glass transition temperature in $BaO–B_2O_3$ (a), $Na_2O–B_2O_3$ (b) and $PbO–GeO_2$ (c) glass forming systems. Points – experimental data from different authors in Refs 19, 24, 35. Lines – guides to the eye

for Li, Na and Ba borate crystals clear distinctions between N_4 and densities trends are evidenced. So N_4 is not the only cause of physicochemical glass properties behavior in glasses and even in crystals.

On Raman spectra treatment we hold to the hypothesis that CSGs are "the smallest particles" which compose the glass defining it properties.[10,16,36,37] Vibrational spectroscopy cannot deliver the information concerning the size of CSGs. But in binary borate glasses with heavy modifier cations (Cs, Ba) the cation clustering (with Ba–Ba distance of 6·9 Å) was detected by x-ray diffraction.[38] It means that CSGs are of nanoscale range and their real chemical formulas, for example, are not $BaO.4B_2O_3$ but $n(BaO.4B_2O_3)$ where $n \geq 2$. At least, this note is important for the Mass Action Law uses.

Comparison of Raman spectra of CSG and crystals

The idea is well known that glass structure shows similarities with the structure of congruently melting crystals, which are close to the glass in composition. According to this idea, one can expect that the structure of barium borate glasses includes some fragments of barium meta-, di- and tetra-borates. In reality, Raman spectra do not entirely confirm this suggestion (Figure 5(a)).

All of the glasses of barium borate system were crystallised in powder under subsolidus temperatures. Depending on initial glass composition, x-ray investigation showed $BaO.4B_2O_3$, $BaO.2B_2O_3$ and $BaO.B_2O_3$ compounds and corresponding mixtures of this compounds. Figure 5(a) demonstrates essential distinctions in the spectra of the PSCs and the congruently melting crystals ($BaO.2B_2O_3$ and $BaO.B_2O_3$) with the exception of the tetraborate.

Careful investigation of the system shows that a few low temperature compounds exist. $2BaO.5B_2O_3$ has been established by Hübner.[39] The new low temperature form of the diborate, β-$BaO.2B_2O_3$, was synthesised by the authors of Ref. 40. It was shown that the known before modification of barium diborate has some structural similarity with the surface of glass with the stoichiometric composition, but new β-diborate has resemblance to the structure of internal glass volume. The same is valid for another new compound established by us in the system under discussion, $4BaO.7B_2O_3$. This compound is close to the eutectic in composition and realises low temperature eutectic interaction in glass. Similar to β-$BaO.2B_2O_3$, it may be produced by bulk glass crystallisation only; therefore it is possible to expect again a similarity between the crystal structure and a structure of inside volume of the glass. Raman investigation suggests this assumption and spectra of the low temperature crystals (Figure 5(a) curves c, d and f) have a better correspondence with the glass ones (Figure 5(a) curve e) that it takes place for the congruently melting compounds. Seemingly in barium borate system β-meta- and α-diborates (Figure 5(a) curves i and g) have structures, which are not correspond to the structure of the internal volume of glass, but to its surface layers.

Sb_2O_3–B_2O_3 glass forming system may serve as another example of glass and crystal structure similarity. It was shown that two types of CSGs ($Sb_2O_3.4B_2O_3$ and $2Sb_2O_3.3B_2O_3$) are formed in the composition range 0–90 mol% Sb_2O_3. Based on the Raman spectra it was also established the existence of two types of crystals in the same range of Sb_2O_3 concentration. Figure 5(b) demonstrates very good correspondence between Raman spectra of these crystals and even derivatives of the PSCs belonging to the CSGs determined. That provides evidence for the reality of CSGs occurrence.

The Raman spectra analysis developed enables one to predict even the existence of unknown compounds. The CSG $Na_2O.8B_2O_3$ discovered by authors of Ref. 16 using Raman spectroscopy data exclusively has been verified recently by XRPD.[1] Two high boric sodium compounds, $Na_2O.5B_2O_3$ and $Na_2O.9B_2O_3$, were established by Ref. 41, with three polymorphic modifications for each of them. However the authors did not present any XRPD data to confirm the fact of $Na_2O.5B_2O_3$ existence. The x-ray pattern for β-$Na_2O.5B_2O_3$ corresponds to crystal hydrate $Na_2O.5B_2O_3.5H_2O$, and synthesised by Milman & Bouaziz α-$Na_2O.5B_2O_3$ is a well known phase α-$Na_2O.4B_2O_3$. On the other hand, our own investigation has shown $Na_2O.5B_2O_3$ and $Na_2O.8B_2O_3$ volume crystallisation in the glasses close in composition to the stoichiometries. The both of established compounds have individual x-ray patterns, which may be indicated in monoclinic syngony. The comparison with Milman & Bouaziz data shows that their β-$Na_2O.9B_2O_3$ contains some peaks of our $Na_2O.5B_2O_3$ in mixture with an unknown phase. The both compounds, $Na_2O.5B_2O_3$ and $Na_2O.8B_2O_3$, are metastable and under heating up to 600°C they decompose into α-$Na_2O.4B_2O_3$ and liquid.

The CSG $Na_2O.4SiO_2$ was found on the base of vibrational spectroscopy by authors of Ref. 37. Though $Na_2O.4SiO_2$ crystal was mentioned by McCulloch in 1952, the fact of $Na_2O.4SiO_2$ existence was not confirmed in the later study[42] of the sodium silicate phase diagram.

Boson peak features and T_g correlations

The nature of the Boson peak in borate and other glasses is still much debated. In the context of this

[1] Private presentation of I. G. Polyakova

paper, two hypotheses describing the Boson peak are of most interest. The first, based on the medium range order model, describes the Raman density of states as belonging to vibrations of the regular structural elements (see CSGs) within some correlation length l of disordered glass network.[43–46] The second one analyses the features of the Boson peak in the framework of the approach of intermolecular interactions.[43] In the first case, the low wavenumber vibrations could be perceived as being localised inside CSGs, while in the second one the Boson peak vibrations would belong to weak CSG–CSG bonds.

Based on the carefully measured Boson peak intensities and frequencies in the wide composition range of a glass forming system we have proposed an explanation, which is able to predict the Boson peak frequency and its intensity correlations with the glass transition temperature. In Refs 47 and 48 it was shown that the Boson peak frequency in Raman spectra of vitreous SiO_2 linearly increases in the range from 49 to 76 cm^{-1} when the "fictive" temperature T_f increases from 1100°C to 2700°C. Since T_g is known to correlate with T_f one can expect that the composition dependence of the Boson peak frequency will correspond to T_g behaviour. This correlation between T_g and υ_{Bos} is really observed in Figure 6 for $BaO–B_2O_3$ (a), $Na_2O–B_2O_3$ (b), and $PbO–GeO_2$ (c) glasses. Because of strong conformity of the low frequency Raman scattering intensity with the Bose–Einstein thermal population factor for the Stokes spectrum, low frequency shift of Boson peak density of states as T_g increases will result in a rise of the Boson peak intensity. This prediction gives simple explanation of the experimentally obtained data in $B_2O_3–BaO$, $Na_2O–B_2O_3$, and $PbO–GeO_2$ glasses (Figure 6).

The concept of CSGs has been also applied to ternary borate glass forming systems $B_2O_3–BaO–P_2O_5$[49] and $B_2O_3–BaO–Al_2O_3$. In Raman spectra of $BaO–B_2O_3–xAl_2O_3$ glass series four types of CSGs $2BaO.4B_2O_3.Al_2O_3$, $BaO.2B_2O_3.Al_2O_3$, $5BaO.6B_2O_3.2Al_2O_3$ and $BaO.B_2O_3.Al_2O_3$ have been revealed. Since Al is known to form structures like $MeO/(M_2O).Al_2O_3.nR_xO_y$ (Me and M are alkaline earth or alkali metal and R is glass forming cation) we proposed that in $BaO–B_2O_3–xAl_2O_3$ glass series Al will compete with B for oxygen atoms supplied by BaO. Consequently, on addition of Al_2O_3 to $xBaO–(1–x)B_2O_3$ glass series, $B_2O_3^{solv}$ will release on condition that $((1–x)/x)>n$. This prediction appeared to be in a good agreement with experimental Raman spectroscopy data.

Conclusions

Based on the vibrational spectra, the concept of constant stoichiometry groupings (CSG) is demonstrated for $B_2O_3–BaO–Al_2O_3$, $Na_2O–B_2O_3$ and $Sb_2O_3–B_2O_3$ glass forming systems. Calculated glass properties (refractive index, density, structural thermal expansion coefficient, the fraction of four coordinated borons) are in a good agreement with experimental data. The Raman spectra analysis developed here enabled us to predict the existence of unknown compounds $Na_2O.8B_2O_3$ and others that have been discovered now. Raman spectra treatment allowed us to predict the Boson peak frequency and intensity correlations with the glass transition temperature. This prediction agrees with the experimentally obtained data in $B_2O_3–BaO$, $Na_2O–B_2O_3$, and $PbO–GeO_2$ glasses.

Acknowledgement

This work was partially supported by grants from INTAS (Project No. 03-51-5360) and RFFI (Project No. 02-03-32499).

References

1. *Borate glasses, crystals & melts 2002. Proc. 3rd Int. Conf. on Borate Glasses, Crystals and Melts*, Ed. Ingram, M. D. 2004. Society of Glass Technology, Sheffield. P. 282.
2. Fedorov, P. P., Kokh, A. E. & Kononova, N. G. *Uspehi Himii*, 2002, **71** (8), 741. (In Russian.)
3. Gode, G. K. *Dr. Sc. Thesis*, 1974. Riga. P.600. (In Russian.)
4. Lipovskii, A. A., Tagantsev, D. K., Vetrov, A. A. & Yanush, O. V. *Opt. Mater.*, 2003, **21** (4), 749.
5. Wong, J. & Angell, C. A. *Glass structure by spectroscopy*. 1976. Marcel Dekker Inc., New York. Chapter 7. P.409.
6. Krogh-Moe, J. *Phys. Chem. Glasses*, 1962, **3**, 101.
7. Gukasyan, S. B., Kostanyan, K. A., Mukhitdiniva, I. A. & Yanush, O. V. Zh. *Prikladnoi Spektroskopii*, 1982, **36** (3), 487. (In Russian.)
8. Kabanov, V. O. & Yanush, O. V. *Fiz. Khim. Stekla*, 1987, **13** (4), 524. (In Russian.)
9. Karapetyan, G. O., Maksimov, L. V. & Yanush, O. V. *Fiz. Khim. Stekla*, 1992, **18** (6), 10. (In Russian.)
10. Yanush, O. V. *Spectroscopy Studies of Glass structure. Proc. 18th Int. Congr. on Glass*, 1998. P.75
11. Krogh-Moe, J. *Phys. Chem. Glasses*, 1962, **3**, 1.
12. Vedishcheva, N. M., Shakhmatkin, B. A., Shultz, M. M. & Wright, A. C. In: *Borate Glasses, Crystals and Melts*, Eds. Wright, A. C., Feller, S. A. & Hannon A. C. 1997. Society of Glass Technology, Sheffield. P.215.
13. Vedishcheva, N. M., Shakhmatkin, B. A. & Wright, A. C. *J. Non-Cryst. Solids*, 2001, **293–295**, 312.
14. Priven', A. I. *Fiz. Khim. Stekla*, 2003, **29** (1), 93. (In Russian.)
15. Conradt, R. *J. Non-Cryst. Solids*, 2004, **345–346**, 16.
16. Yanush, O. V., Kabanov, V. O. & Mukhitdiniva, I. A. *Fiz. Khim. Stekla*, 1988, **14** (3), 330. (In Russian.)
17. Zubkova, L. V., Pivovarov, M. M. & Yanush, O. V. *Fiz. Khim. Stekla*, 1994, **20** (1), 36. (In Russian.)
18. Markova, T. S., Yanush, O. V., Polyakova, I. G., Pevzner, B. Z. & Klyuev, V. P. *Fiz. Khim. Stekla*, 2005, **31** (5), in press. (In Russian.)
19. Mazurin, O. V., Strel'tsina, M. V. & Shvaiko-Shvaikovskaya, T. P. (Eds.) *Properties of Glasses and Glass-forming Melts. Handbook. Vol. 2*. 1975. Nauka, Leningrad. (In Russian.)
20. Krogh-Moe, J. *Phys. Chem. Glasses*, 1965, **6**, 46.
21. Bril, T. W. *Philips Res. Repts. Suppl.*, 1976, 2.
22. Wallace, R. M. & Katz, S. M. *J. Phys. Chem.*, 1964, **68** (12), 3890–2.
23. Toropov, N. A., Barzakovskii, V. P., Lapin, V. V. & Kurtseva, N. N. (Eds.) *Phase diagrams of silicate systems. Handbook. Vol. 1*. 1969. Nauka, Leningrad. P.822. (In Russian.)
24. Klyuev, V. P. & Pevzner, B. Z. *Fiz. Khim. Stekla*, 1998, **24** (4), 532. (In Russian.)
25. Kabanov, V. O., Sedmale, G. P., Yanush, O. V. *Fiz. Khim. Stekla*, 1990, **16** (2), 174. (In Russian.)
26. Dell, W. J. & Bray, P. J. *Phys. Chem. Glasses*, 1982, **23** (4), 126.
27. Stallworth, P. E., Bray, P. J. In: Uhlmann, D. R., Kreidl, N. J. (Eds.). *Glass: Science and Technology. Vol. 4B*. Academic press, Boston. MA.

1990. P.77.
28. Yiannopoulos, Y. D., Chryssikos, G. D. & Kamitsos, E. I. *Phys. Chem. Glasses*, 2001, **42** (3), 164.
29. Kamitsos, E. I. *Phys. Chem. Glasses*, 2003, **44** (2), 79
30. Ohtory, N., Takase, K., Akiama, I., Suzuki, Y., Handa, K., Sakai, I., Iwadate, Y., Fukunaga, T. & Umesaki, N. *J. Non-Cryst. Solids*, 2001, **293–295**, 136.
31. Krogh-Moe, J. *Acta Cryst. B*, 1969, **25** (10), 2153.
32. Block, S. & Perloff, A. *Acta Cryst.*, 1965, **19** (3), 297.
33. Mighell, A. D., Perloff, A. & Block, S. *Acta Cryst.*, 1966, **20** (6), 819.
34. Frohlich, R. *Z. Kristal.*, 1984, **168**, 109.
35. Kapoor, S., George. H. B., Betzen, A., Affatigato, M. & Feller, S. A. *J. Non-Cryst. Solids*, 2000, **270**, 215.
36. Karapetyan, G. O., Maksimov, L. V. & Yanush, O. V. *J. Non-Cryst. Solids*, 1990, **126**, 93.
37. Mukhitdinova, I. A. & Yanush, O. V. *Fiz. Khim. Stekla*, 1989, **15** (1), 34. (In Russian.)
38. Krogh-Moe, J. *Phys. Chem. Glasses*, 1962, **3** (6), 208.
39. Hübner, K.-H. N. *Jb. Miner. Abh.*, B, 1970, **112** (2), 150.
40. Polyakova, I. G. & Pevzner, B. Z. *Phys. Chem. Glasses*, 2005, **46**, (2), 138.
41. Milman, T. & Bouaziz, R. *Ann. Chem.*, 1968, **3**, 311.
42. Williamson, J. & Glasser, F. P. *Phys. Chem. Glasses*, 1966, **7**, 127.
43. Baranov, A. V., Perova, T. S., Petrov, V. I., Vij, J. K. & Nielsen, O. F. *J. Raman Spectrosc.*, 2000, **31**, 819.
44. Martin, J. & Brenig, W. *Phys. Status Solidi B*, 1974, **64**, 163.
45. Shuker, R. & Gammon, R.W. *Phys. Rev. Lett.*, 1970, **25**, 222.
46. Malinovsky, V. K. & Sokolov, A. P. *Solid State Commun.*, 1986, **57**, 767.
47. Pevnitskii, I. V. *Fiz. Khim. Stekla*, 1990, **16** (4), 667. (In Russian.)
48. Konstantinov, A. V., Maksimov, L. V., Silin, A. R. & Yanush, O. V. *J. Non-Cryst. Solids*, 1990, **123**, 286.
49. Kabanov, V. O., Sedmale, G. P. & Yanush, O. V. *Fiz. Khim. Stekla*, 1992, **18** (4), 44. (In Russian.)

A NMR, conductivity and molecular dynamics study of AgI–Ag_2O–B_2O_3 glasses

Piercarlo Mustarelli,[a,*] *Corrado Tomasi,*[b] *María del Pilar Infante García*[a] *& Aldo Magistris*[a]

[a] *Department of Physical Chemistry "M. Rolla" and INSTM, University of Pavia, Via Taramelli 16, I-27100 Pavia, Italy*

[b] *IENI-CNR, Section of Pavia, Via Taramelli 16, I-27100 Pavia, Italy*

Fast ion conducting (FIC) AgI-based borate glasses $xAgI:(1-x)[Ag_2O.nB_2O_3]$ (n=1, 2) *were studied by means of* ^{109}Ag *solid-state NMR, impedance spectroscopy (IS) and molecular dynamics (MD) simulations. Diffusion coefficients in the range* 10^{-8}–10^{-10} cm^2/s *were obtained from conductivity data by means of the Nernst–Einstein equation. The activation energies for NMR line narrowing and d.c. conductivity differ by a factor of approximately ~3. Standard (ST) and non-equilibrium (NE) MD investigations suggest that this difference is due to the effect of cooperative processes.*

1. Introduction

Fast ion conducting (FIC) glasses belonging to the family AgI–Ag_2O–B_2O_3 display ambient conductivity up to 10^{-2} ohm^{-1} cm^{-1} [(1)] and are interesting both for their possible applications in electrochemical devices and as model systems to investigate cation dynamics, and its relationships with the short/medium range structure of glass.[(2)]

While the local coordination of silver cations in borate glasses has been addressed by means of experimental techniques[(3)] and modelling,[(4)] the medium range order and connectivity are still under debate. We have recently shown that the unique transport properties of AgI-based FICs are due to cation mobility, and that silver diffusion takes place along nanochannels, with fractal structure, of oxygen/iodine coordinated sites.[(5)]

However, several details of cation dynamics are still obscure. For example, it is well known that activation energies obtained from NMR linewidth narrowing (which is related to the spin–spin relaxation time, T_2) may be systematically lower than those from the d.c. conductivity.[(6,7)] While it was pointed out that these discrepancies can be the consequence of the non-exponential nature of the relaxation,[(8)] it was also shown that correlation effects, or the presence of a complex energy landscape characterised by a distribution of shallow traps, must be considered when discussing migration mechanisms.[(9)] A new correlated ion hopping model for the scaling properties of the ionic conductivity was recently proposed for AgI-based phosphate glasses.[(10)]

A relevant help to understanding the details of ion dynamics can be given by simulation. In particular, MD seems to be the best choice for investigating transport phenomena, because it retains the time arrow and allows calculation of instantaneous and time averaged properties on a given ensemble. However, to date, only a couple of studies on AgI-based borate glasses have appeared in literature.[(11,12)]

In this paper, we discuss dc conductivity and ^{109}Ag NMR results obtained on fresh samples of $xAgI:(1-x)[Ag_2O.nB_2O_3]$ glasses (n=1, 2). A comparison with microscopic observables obtained by MD simulations is performed, chiefly with the aim of elucidating the discrepancies between the activation energy values extracted from the above techniques. Because of the heaviness of the methods (standard and non-equilibrium), only two compositions of AgI:$Ag_2O.2B_2O_3$ system were simulated.

2. Experimental section

Sample preparation

In order to avoid the composition uncertainties caused by the use of hygroscopic B_2O_3 or H_3BO_4,[(13)] boron was introduced as $(NH_4)B_4O_7.4H_2O$, whose stoichiometry was checked by means of thermogravimetric analysis (TGA). Such a reagent was intimately mixed with suitable amounts of $AgNO_3$ (Metalli Preziosi, bicrystallised) and AgI (Aldrich 99+%). All mixtures were heated up to 900°C and the melt was kept for 15 min at this temperature before quenching to room temperature. For samples with n=2, glassy $Ag_2B_4O_7$ was prepared as a precursor and subsequently mixed and melted with proper amounts of AgI. In all the examined compositions the stoichiometric deviation was less than 4%.[(13,14)]

* Corresponding author. Email mustarelli@matsci.unipv.it

Measurements

Differential scanning calorimetry (DSC) scans were performed at 5°C/min, under a N_2 purge, by means of a 2910 DSC, and controlled by a Thermal Analyst 5000 (TA Instruments). The instrument was previously calibrated both in temperature and in enthalpy.

Impedance spectroscopy measurements were carried out with a 1255HF frequency analyser, connected with a SI 1287 electrochemical interface by Solartron (Schlumberger) between 1 and 10^6 Hz, with a voltage amplitude of 100 mV. Bulky pellet samples (13 mm of diameter and 2 mm thick) were covered with graphite on both faces to optimise electrical contact, and put into a cell with stainless steel plates. Low temperature data were collected by introducing the cell into a home made cryostat fed with liquid nitrogen.

^{109}Ag solid state NMR spectra were acquired on a AMX400WB spectrometer (Bruker, Germany), at the Larmor frequency of 18·61 MHz, with a wideline probe (Bruker, Germany). The spectra were referenced to a 10 M aqueous solution of $AgNO_3$. A 90° pulse of 16 μs, a repetition time varying from 100 to 500 s, and a bandwidth of 200 kHz were chosen. Sixty four accumulations were generally enough to obtain acceptable S/N ratios. Both the glassy samples and the reference solution were kept in standard thin wall, 10 mm glassy tubes (Bruker). The glassy samples were kept at the centre of the NMR coil (19 mm length, 9·5 mm inner diameter). The spectra were obtained both by single pulse and Hahn-echo sequences. Care was taken to avoid signal saturation and to minimise the baseline distortions. Where not differently specified, the experiments were performed at 297±1 K.

Simulation

MD simulations were performed with the DL_POLY_2 package (Smith & Forester; The DL_POLY_2 User Manual v2.14,2003). Configurations containing 1344 (n=2, x=0·33) and 1440 (n=2, x=0·5) particles underwent the following thermal treatment under NPT ensemble: heating to 1200 K, quenching to 300 K and heating plus three thermalisation runs every 50 K, for a total of ~2×10^5 timesteps. Additional production runs (1×10^4 timesteps of 1 fs) were performed at each T value, by averaging every 50 timesteps. More details are reported elsewhere.(15) A mixed BHM-like/Vashishta–Rahman pair potential was employed.(15) Despite of the quite good predictions obtained with this potential for pure and doped $Ag_2B_4O_7$, some results suggest a need of further optimisation of the I^- interaction treatment.(16)

Diffusion coefficients and activation energies were computed from Ag^+ mean square displacement (MSD) curves within both standard (ST) and non-equilibrium (NE) frameworks. In the former, equilibrium is supposed and Einstein relation holds

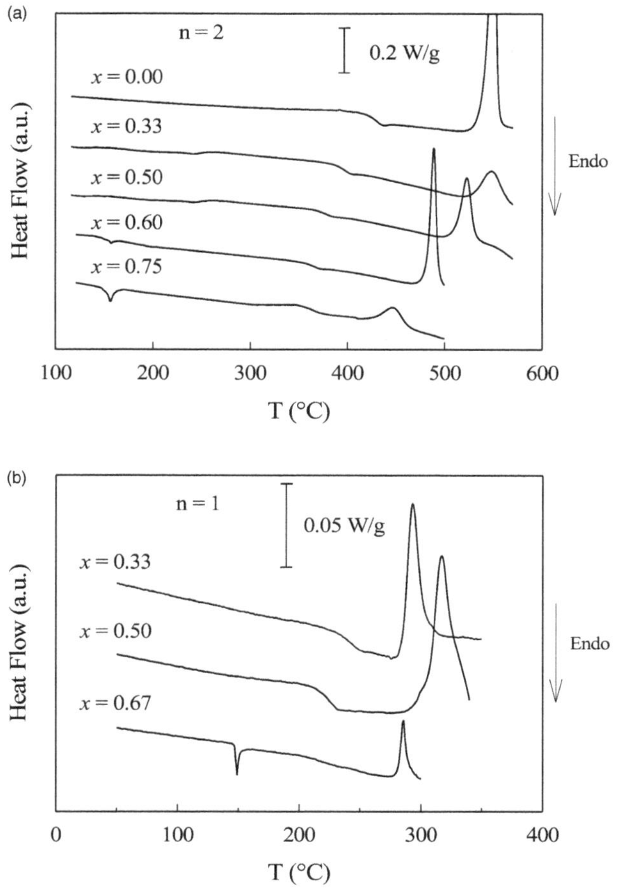

Figure 1(a). DSC thermograms of the glasses and glass-ceramics with n=2. *(b) DSC thermograms of the glasses and glass-ceramics with* n=1

$$D = \lim_{t\to\infty} \frac{1}{6t} MSD(t)$$

Instead, in the NE approach, a steady state is reached after the application of an electric field.(17) Linear response theory allows one to extract self diffusion coefficients from the extrapolation of fluxes obtained under different stimuli. Production runs for NE were repeated by varying the stimulus within 6 orders of magnitude and data collected every 500 timesteps along 25 ps runs. The computational cost of the method limited its application to n=2, x=0·5 sample in a small temperature range.

3. Results and Discussion

Thermal properties

The DSC thermograms of n=2 and n=1 as prepared glassy and glass-ceramic samples are reported in Figure 1(a) and 1(b), respectively.

All samples with n=2 exhibit a more or less evident glass transition temperature, T_g, followed by an exothermic crystallisation peak. Both T_g and the

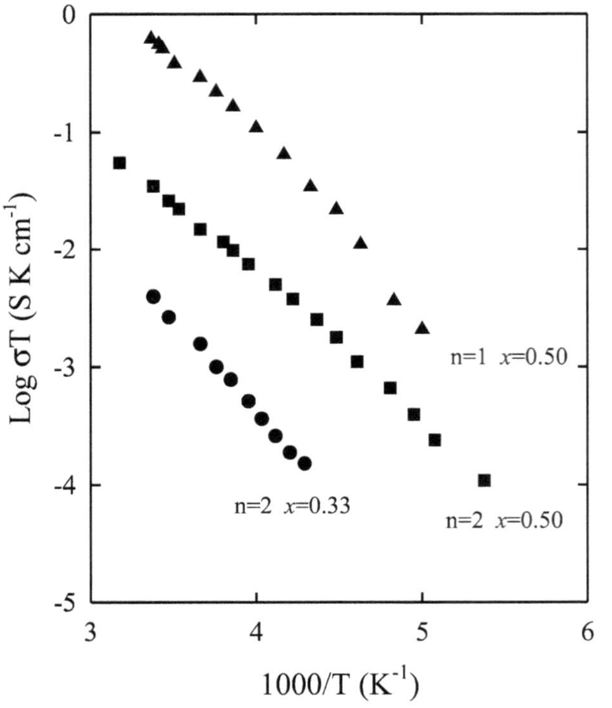

Figure 2. Arrhenius plot of the ionic conductivity for some selected samples

crystallisation temperature, T_x, decrease as the AgI content increases. The T_g values, reported in Table 1, are systematically higher than those reported in literature.[14] The discrepancy likely stems from the different preparation route followed in the present work. For the x=0·60 sample a tiny endotherm is observed around 150°C, more evident in case of x=0·75. This feature, related to the β→α AgI solid–solid transition, along with the invariancy of T_g allows to establish the upper limit of the glass formation region at x≅0·60.

For n=1, samples with x=0·33 and x=0·50 exhibit very similar DSC traces where T_gs (at ~230°C and ~215°C, respectively) are followed by crystallisation peaks at about 280°C and 300°C, respectively. For both compositions, the T_g values are lower than those reported in Ref. 14. In case of sample with x=0·67, an endothermic peak at ~150°C, ascribed to the β→α AgI transition is followed by a shallow deflection and a small exotherm related to T_g and cold crystallisation, respectively. These evidences clearly show that the latter composition falls out of the glass forming region.

Ionic conductivity

Figure 2 shows the Arrhenius plots of the ionic conductivity of samples x=0·33 and x=0·50 for n=2, and x=0·50 for n=1. For a given value of x the conductivity increases as n decreases, and, at a given n, the conductivity increases with the AgI content. This is consistent with our recent conclusion that the conductivity of AgI-based glasses is simply related to the iodine/oxygen ratio in the glass.[5] Room temperature conductivity values are in good agreement with those reported in Ref. 14. All samples exhibit a more or less Arrhenian behaviour. Mean activation energies are reported in Table 1, together with the values reported in Refs. 7 and 14 for glasses of the same nominal composition. Our values are 15–20% lower than those of Ref. 14. The diffusion coefficients, D_σ, can be estimated by means of the Nernst–Einstein equation

$$D_\sigma = \frac{\sigma k T}{q^2 n}$$

where σ is the d.c. conductivity, q and n are charge and number of carriers, respectively, while the other symbols have the usual meaning. By using for n the values obtained from ^{109}Ag NMR,[2] we get the room temperature values $D_{\sigma,rt}$ shown in Table 1.

NMR

Figure 3 shows, for several glasses, the behaviour of the NMR linewidth versus the reciprocal tem-

Table 1. $E_{a,\sigma}$=activation energy of the d.c. conductivity. $E_{a,NMR}$=activation energy of the NMR line narrowing. $E_{a,MD}$=activation energy of the diffusion coefficients extracted by MD simulations (see text). $D_{\sigma,rt}$=room temperature diffusion coefficients estimated with the Nernst–Einstein equation. $D_{MD,rt}$=room temperature diffusion coefficients by MD simulations

Sample	T_g *(K)*	$E_{a,\sigma}$ *(eV)*	$E_{a,NMR}$ (eV)	$E_{a,MD}$ *(eV)*	$D_{\sigma,rt}$ *(cm²/s)*	$D_{MD,rt}$ *(cm²/s)*
n=2, x=0·33	661± 2	0·31	0·11	0·16	5·7×10^{-10}	6·5×10^{-8}
n=2, x=0·50	642±2	0·24$_5$	0·07	0·08	4·4×10^{-9}	2·4×10^{-8}
	618$^+$	0·31$^+$	0·10*	0·27§		1·2×10^{-8} §
n=2, x=0·60	628±2	-	0·08		2·4×10^{-8}	
		0·27*	0·09*			
	606$^+$	0·27$^+$				
n=1, x=0·33	522±2	0·30$^+$	0·24		2·0×10^{-9}	
n=1, x=0·50	488±2	0·29	0·08$_5$		6·1×10^{-8}	
	543$^+$	0·35$^+$				
		0·30*	0·12*			

* From ref. 7. The NMR data are taken at the Larmor frequency of 16.8 MHz in the temperature range ~150K to ~ 450K.
+ From ref. 14.
§ Non-equilibrium MD

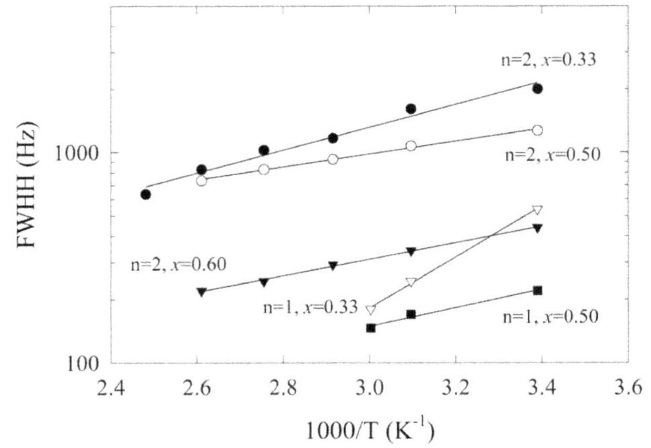

Figure 3. Full width at half height (FWHH) of the ^{109}Ag *NMR line versus reciprocal temperature.* ●: n=2, x=0·33; O: n=2, x=0·50; ▼: n=2, x=0·60; ∇: n=1, x=0·33; ■: n=1, x=0·50

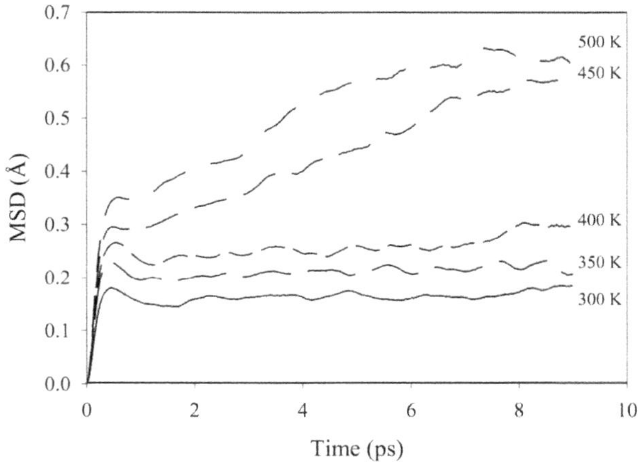

Figure 4. Ag^+ *MSD curves for* n=2, x=0·33 *glass from Standard MD procedure. Solid line* T=300 K, *long dash* T=350 K, *short dash* T=400 K, *dash-dot* T=450 K, *dash-dot-dot* T=500 K

perature. It is well accepted that the narrowing of the NMR line with temperature is due to ion diffusion, which averages out both the isotropic and the anisotropic parts of the chemical shift.(7) Our data are taken in a temperature region where complete line narrowing has not yet been attained, and the values of activation energy, $E_{a,NMR}$, for the narrowing process are reported in Table 1. They are in good agreement with those of Ref. 7, obtained on a wider temperature range extending nearly down to the rigid lattice region. We observe that the $E_{a,NMR}$ values are lower than the $E_{a,\sigma}$ (but for the glass n=1, x=0·33) by a factor of about ~0·3, i.e. roughly the value of β of the stretched exponential (KWW) correlation function expected for these glasses(7)

$$g(t)=\exp\left[-\left(\frac{t}{\tau}\right)^{\beta}\right]$$

While low β values of the phenomenological KWW equation are simply reflecting the existence of a wide distribution of correlation times, we can suppose that the observed E_a discrepancies are related to the different timescales probed by NMR line narrowing and d.c. conductivity. In fact, the ambient temperature correlation time for ionic conductivity of these glasses is in the range 10^{-6}–10^{-8} s,(18) whereas the KWW fits of the NMR lines narrowing reported by Chung *et al*(7) give ambient temperature correlation times of the order of 10^{-10} s for the chemical shift average.

Molecular dynamics

Figure 4 reports the Ag^+ MSD curves for the x=0·33 glass computed with standard MD in the temperature range 300–500 K. At low temperatures, MSD behaviour is typical of a solid whose characteristic cage frequency lies well within the simulation timescale (~ps). In contrast, for T>400 K, a more liquid like behaviour is found. The same feature of Ag^+ motion is not shared by the glass matrix components (B and O, not shown), which seem to retain a more solid like character. Although further analysis is still in progress, this evidence suggests that our simulation indicates a T_g above the reported temperature range, in agreement with the experimental thermal behaviour (see Figure 1).

The simulated diffusion coefficients at 300 K $D_{MD,rt}$ and the related activation energies $E_{a,MD}$ extracted by both the ST and the NE methods are reported in Table 1. The direct comparison of $D_{MD,rt}$ with the experimental $D_{\sigma,rt}$ is limited by the non-optimised I^- effect in the simulation, with a consequent different weight on each glass composition. In contrast, the activation energy values can be considered more reliable since they are extracted from a consistent set of data.

Comparison among ST and NE methods on n=2, x=0·5 glass provides a significant result. The former method gives a higher $D_{MD,rt}$ while the latter predicts a diffusion coefficient of the order of the experimental $D_{\sigma,rt}$. Besides, the NE method seems to predict a more activated Ag^+ motion than the ST one, giving a activation energy value ~3 times higher. These results are consistent with the experimental dc conductivity and NMR, respectively.

The interpretation of the above data can help one to understand the experimental results. The nature and timescale of the ST procedure make it sensitive to the single ion self diffusion, at least for the current box size and time scales. Instead, the NE method is based on a perturbation–response approach that is able to highlight more cooperative effects on time scales as short as 10 ps, and to split the frequency dependence response.(17)

The parallels between MD and experimental results suggest that the discrepancies between $E_{a,NMR}$ and $E_{a,\sigma}$ derive from the same phenomenological differences to which the two MD methods are sensitive. Hence, such discrepancies can be explained in terms of cooperative processes which are probed to a different extent by NMR and d.c. conductivity, probably because of the different characteristic correlation times.

4. Conclusions

A combined NMR, IS and MD study on AgI-based silver borate glasses is reported. We show that MD is a useful technique to investigate the transport properties of FIC glasses. In particular, our ST and NE molecular dynamics simulations provide a key for interpreting the E_a discrepancies arising from the non-exponential nature of the relaxation.

Acknowledgements

This research is funded by Italian MIUR (PRIN project 2004, M. Cutroni Leader).

References

1. For a review, see Magistris A. Ionic conduction in glasses, *NATO ASI Series, Series E: Applied Sciences*, 1993, **250**, 213, and references therein.
2. Mustarelli, P., Infante Garcia, M. P., Tomasi, C., Magistris, A. & Linati, L. *Phys. Chem. Glasses*, 2003, **44**, 159.
3. See Mustarelli, P., Linati, L., Tartara, V., Tomasi, C. & Magistris, A. *Solid State Nucl. Magn. Resonance*, 2005, **27**, 112, and references therein.
4. Swenson, J. & Adams, S. *Phys. Rev. B*, 2001, **64**, 024204.
5. Mustarelli, P., Tomasi, C. & Magistris, A. *J. Phys. Chem. B*, 2005, **109**, 17417.
6. Martin, S., Bishof, H. J., Mali, M., Roos, J. & Brinkmann, D. *Solid State Ionics*, 1986, **18&19**, 421.
7. Chung, S. H., Jeffrey, K. R., Stevens, J. R. & Borjesson, L. *Phys. Rev. B*, 1990, **41**, 6154.
8. Martin, S. W. *Mater. Chem. Phys.*, 1989, **23**, 225.
9. Tatsumisago, M., Angell, C. A. & Martin, S. W. *J. Chem. Phys.*, 1992, **97**, 6968.
10. Funke, K., Banhatti, R. D. & Cramer, C. *Phys. Chem. Chem. Phys.*, 2005, **7**, 157.
11. Abramo, C., Pizzimenti, G. & Consolo, A. *Phil. Mag. B*, 1991, **64**, 495.
12. Abramo, M. C., Caccamo, C. & Pizzimenti, G. *J. Phys. Cond. Matter*, 1993, **5**, 397.
13. Tomasi,C. *PhD Thesis*, 2002.
14. Chiodelli, G., Campari Viganò, G., Flor, G., Magistris, A. & Villa, M. *Solid State Ionics*, 1983, **8**, 311.
15. Infante García, M. P., Tomasi, C., Mustarelli, P., Romano, S. & Magistris, A. *in preparation*.
16. Infante García, M. P. *PhD Thesis*, 2004.
17. Boon, J. P. & Yip, S. in *Molecular Hydrodynamics*.1980. Mcgraw-hill, New York.
18. Villa, M., Chiodelli, G., Magistris, A. & Licheri, G. *J. Chem. Phys.*, 1986, **85**, 2392.

Proc. Fifth Int. Conf. on Borate Glasses, Crystals and Melts Phys. Chem. Glasses: Eur. J. Glass Sci. Technol. B, August 2006, **47** (4), 489–492

Magnetic resonance studies on amorphous and crystalline lanthanum aluminoborates

S. Simon

Babes-Bolyai University, Faculty of Physics, M. Kogalniceanu Str. 1, 400084 Cluj-Napoca, Romania

The development of an $LaAl_{2\cdot03}B_4O_{10\cdot5}$ (124) crystalline phase from amorphous xerogels and glasses doped with iron and gadolinium has been characterised by magic angle spinning nuclear magnetic resonance (MAS NMR) and electron paramagnetic resonance (EPR) spectroscopies. The crystallisation of amorphous precursors takes place between 760 and 860°C in the case of xerogels and around 1000°C for glasses. The magnetic resonance results indicate that the local structure dramatically changes when crystalline phases are developed in amorphous precursors. The major changes are evidenced in the boron environment. In amorphous precursors boron atoms are predominantly three-coordinated, while in the crystalline samples almost only tetra-coordinated species are detected. As regards aluminium, it occurs in amorphous samples as hexa-, penta- and tetra-coordinated, while in the crystalline samples it appears preponderantly penta-coordinated. Three different sites for Fe^{3+} ions were identified in crystalline samples by EPR measurements with only one preponderant site for Gd^{3+} ions.

Introduction

Amorphous and crystalline aluminates are intensively studied for properties that make them suitable for applications such as support materials in catalytic reactions.[1,2] It was found that the presence of rare earths such as lanthanum preserves their porosity and boron stabilises the penta-coordinated aluminium species, both effects being important for catalytical uses.[3]

Up to now, only one stable crystalline phase is known in the ternary lanthanum aluminium boron oxide system.[4,5] The composition of this crystalline phase is $LaAl_{2\cdot03}B_4O_{10\cdot54}$ (124), and in its structure the La, Al and B atoms are surrounded by O atoms in trigonal prismatic, pyramidal and tetrahedral arrangements, respectively. It is an interesting oxide compound in which all cations exhibit less usual coordinations: lanthanum is hexa-coordinated, aluminium penta-coordinated and boron tetra-coordinated.

The local arrangements of the structurally disordered borates have been investigated by different techniques such as nuclear magnetic[6–9] and electron paramagnetic resonance,[10–12] Raman[13,14] and infrared spectroscopies.[15–17] Due to the lack of structural long range order in amorphous borates, an unambiguous assignment of the structural groups is not possible without comparing their spectra with those of corresponding crystalline counterparts.

The structurally disordered form of the oxide compound with a composition corresponding to $LaAl_{2\cdot03}B_4O_{10\cdot54}$ crystalline phase can be obtained by quickly undercooling the mixture melted at 1450°C, or by a sol-gel procedure.[18] The latter method is preferred for following the local structural changes during the transformation from disordered system to the well defined crystalline phase and for preparing the porous aluminates for catalytic applications.

The aim of this study is to characterise the local structural evolution in iron and gadolinium doped amorphous lanthanum aluminoborates of composition $LaAl_2B_4O_{10\cdot5}$ in view of the heat treatment applied up to 1100°C using MAS NMR and EPR spectroscopic data.

Experimental

The systems investigated, $0{\cdot}5M_2O_3.99{\cdot}5[LaAl_2B_4O_{10\cdot5}]$ (mol %), where M=Fe or Gd) were prepared by the sol-gel method from aqueous solutions of $La(NO_3)_3.6H_2O$, $Al(NO_3)_3.9H_2O$, H_3BO_3 and $Fe(NO_3)_3.6H_2O$ or $Gd(NO_3)_3.6H_2O$ of analytical purity grade by thermal decomposition of starting nitrates and boric acid accompanied by simultaneous oxidation of glycerol. The reagent mixture de solved in desalinated water appears as a transparent solution at room temperature. A viscous gel is obtained after 5 h heating at 95°C. Further heating at this temperature leads to a spongeous solid sample, initially amorphous and later crystalline. Samples were obtained after heat treatment for 30 min at different temperatures, T_t, between 150 and 1100°C.

Glass samples were prepared by melting the corresponding mixtures of the same starting reagents at

[1] Corresponding author

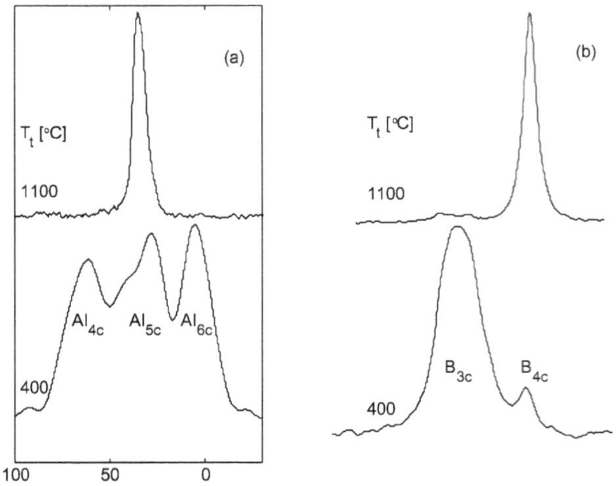

Figure 1. ^{27}Al (a) and ^{11}B (b) MAS NMR spectra of vitreous and crystalline samples

1450°C for 30 min followed by rapid cooling to room temperature. The crystalline samples were been obtained by heat treatment applied at 1100°C for 2 h.

The ^{27}Al and ^{11}B MAS-NMR measurements were carried out at room temperature on Chemagnetics Infinity 600 MHz spectrometer equipped with a solid state accessory, with MAS frequency of 15 kHz, using as references aqueous solutions of boric acid and aluminium nitrate, respectively. The EPR measurements were carried out in X band (9·4 GHz) on powder samples at room temperature using a Bruker spectrometer.

Results and discussion

Thermal analysis results obtained for the xerogel sample heat treated for 30 min at 150°C show that up to 800°C the weight of the sample is decreasing as a result of losing water, nitrogen oxides and CO_2 caused by nitrate decomposition and glycerol pyrolysis, and later as a result of dehydroxylation. The differential thermal curve indicates that structural changes related to the development of crystalline phases takes place above 750°C. The crystallisation process starts at 760°C and different changes appear up to 950°C as result of successive phase transformations to the preponderant final crystalline phase of composition $LaAl_{2\cdot 03}B_4O_{10\cdot 5}$ as was confirmed by x-ray diffraction studies.

The xerogels after heat treatments applied at temperatures up to 750°C are amorphous from x-ray diffraction point of view. The 124 phase is prevalent in the samples after heat treatment at temperatures above 1000°C. Some other phases like $Al_{18}B_4O_{33}$ and $La(Al)BO_3$ are also developed in small amounts (less than 10%). No crystalline phase with significant iron or gadolinium content has been indicated by x-ray diffraction, showing that the added small amounts these oxides are well dispersed in crystalline lanthanum aluminoborates.

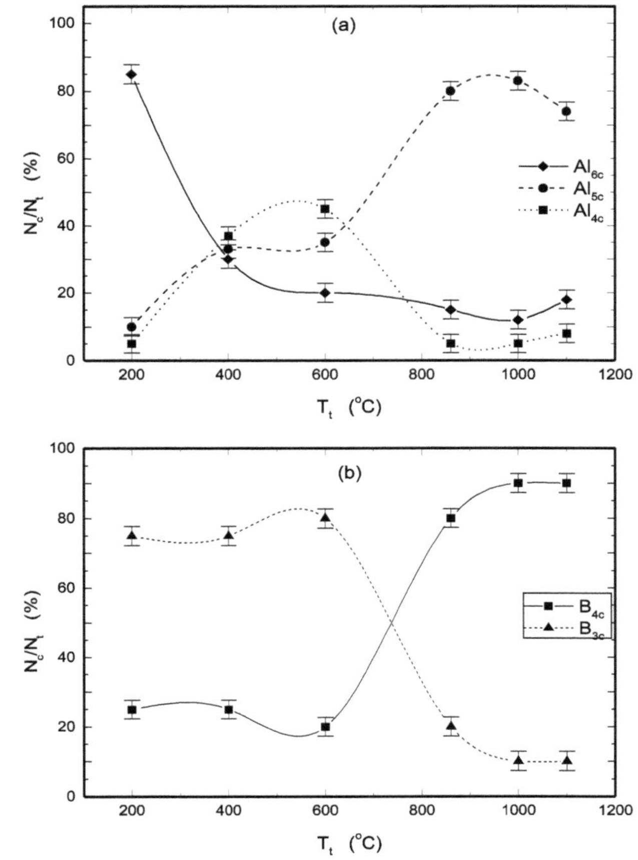

Figure 2. The coordination change of aluminium (a) and boron (b) versus heat treatment temperature in samples prepared by sol-gel method. (N_t is the total number of aluminum, respectively boron atoms and N_c is the number of atoms with the specified coordination)

According to the x-ray diffraction patterns the glass samples do not contain any crystalline phase. In this case, almost only the 124 phase is obtained after 2 h heat treatment at 1100°C.

The ^{27}Al and ^{11}B MAS-NMR spectra of the amorphous sample prepared by melt undercooling and heat treated at 400°C and of the crystalline sample obtained after 2 h heat treatment at 1100°C are shown in Figure 1. One remarks that in the amorphous sample aluminium occurs unpreferentially distributed in tetra-, penta- and hexa-coordinated sites, and boron is mainly three-coordinated, while the crystalline sample contain only penta-coordinated aluminium and mainly tetra-coordinated boron.

The evolution of aluminium and boron coordination in the samples prepared by sol-gel method, as was obtained by deconvolution of the corresponding NMR spectra, is presented in Figure 2. In amorphous xerogels, like in the glass sample, tetra-, penta- and hexa-coordinated aluminium, and three- and tetra-coordinated boron are indicated. In the crystalline samples, there are almost only penta-coordinated aluminium and tetra-coordinated boron species.

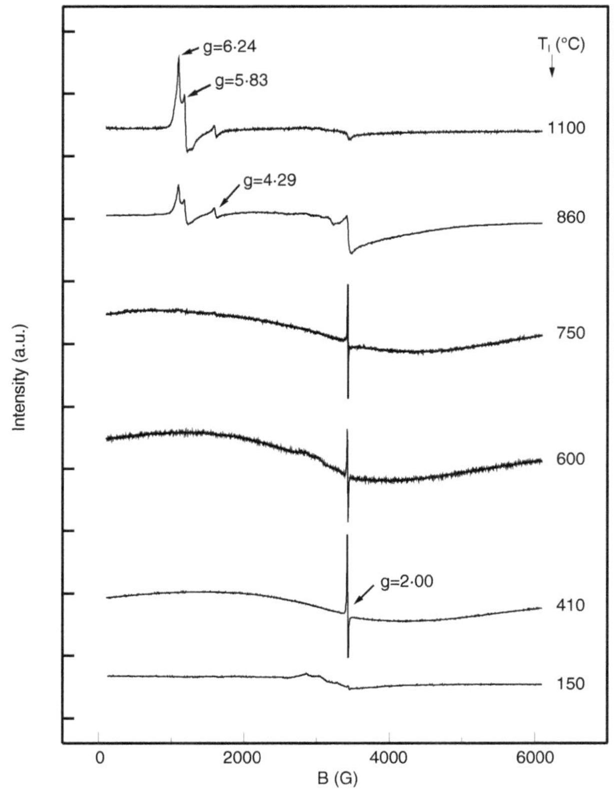

Figure 3. EPR spectra of $LaAl_2B_4O_{10\cdot5}$ xerogel samples heat treated at different temperatures

The EPR spectra (Figure 3) of amorphous samples consist of a very broad line arising from rich iron precipitates and from some organic radicals resulted during glycerol pyrolysis that are responsible for the very narrow line,[19] both centred at $g\approx2\cdot0$. The structural changes induced by the heat treatment applied at higher temperatures make possible the diffusion of iron ions in the crystalline samples where three types of sites with different local surrounding having g factor values between 2·0 and 6·24 have been identified. The peaks at $g\approx6\cdot24$ and 5·83 may be assigned to Fe^{3+} ions disposed in sites of rhombic symmetry[20] with relatively strong crystalline field. Due to their similar ionic radius, it is expected that the majority of the Fe^{3+} ions occupy the sites of aluminium in 124 phase and give rise to the above mentioned EPR lines. The line with $g\approx4\cdot3$ can be assigned to Fe^{3+} ions situated in the relatively distorted sites in a still noncrystalline phase. A small part of Fe^{3+} ions are located in the sites with weak crystalline fields of octahedral symmetry and give rise to the line located at $g\approx2\cdot0$.

The EPR spectra of iron doped glasses is the 50 years known spectra of Fe^{3+} ions in glasses[21] characterised by an asymmetric line with $g\approx4\cdot3$ (Figure 4). After crystallisation, besides the two peaks assigned to Fe^{3+} ions disposed in sites of rhombic symmetry from the 124 phase, the line with $g\approx4\cdot3$ is much narrower and more symmetric. This reflects a much relaxed and uniform environment for a small part

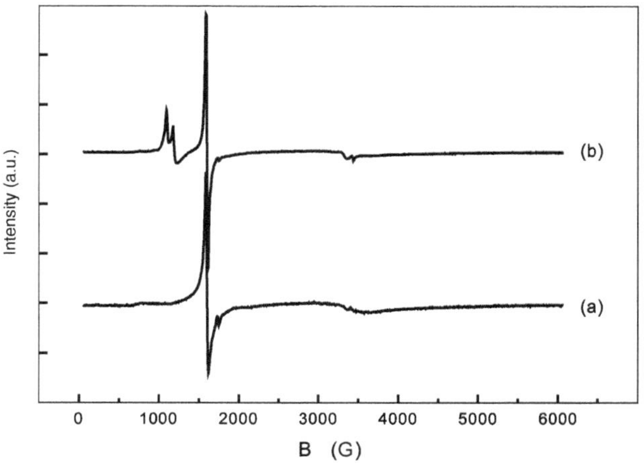

Figure 4. EPR spectra of iron doped $LaAl_2B_4O_{10\cdot5}$ glass (a) and crystalline (b)samples

of Fe^{3+} ions not incorporated in to the 124 phase. In the case of gadolinium doped xerogels the EPR spectra (Figure 5) are similar to the well-known U spectrum[22–27] specific for noncrystalline matrices containing Gd^{3+} ions.

After crystallisation the EPR spectra mainly consist of a relatively large line with $g\approx2\cdot0$ suggesting that the surrounding of Gd^{3+} ions are experiencing weak crystal fields as a result of structural relaxation. The line is not narrow enough to conclude that the rare earth environments in the 124 phase are well defined, despite the fact that according to x-ray data for the124 phase[4] the vicinity of lanthanum is built by six oxygens situated at equal distances from the rare earth site.

Conclusions

The results obtained from MAS NMR investigations indicate, in all amorphous lanthanum aluminoborates synthesised by sol-gel method or by rapid cool-

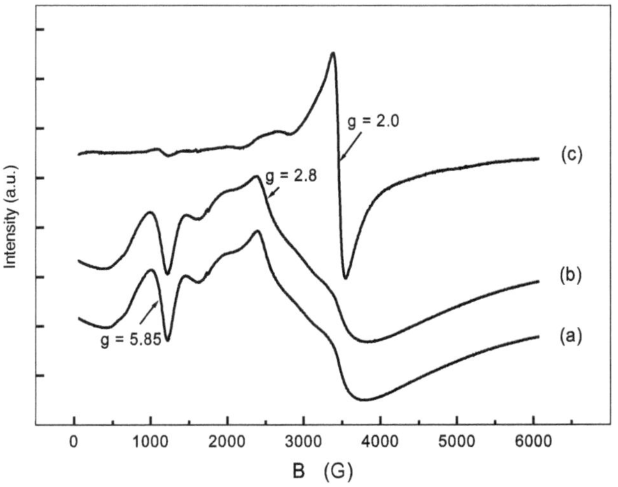

Figure 5. EPR spectra of $LaAl_2B_4O_{10\cdot5}$ xerogel samples heat treated at different temperatures: 400°C (a), 750°C (b) and 1100°C (c)

ing of the oxide melts, the occurrence of boron in BO_3 triangles and BO_4 tetrahedra and of aluminium polyhedra with aluminium tetra-, penta- and hexa-coordinated by oxygens. While boron appears mainly as three-coordinated, there is no special preference of aluminium species for a certain coordination. After crystallisation of the glass sample only penta-coordinated aluminium and mainly tetra-coordinated boron are indicated.

The local order is strong affected on passing from amorphous xerogels to crystalline lanthanum aluminoborate phases. The most pronounced transformations are indicated for the heat treatments applied between 750 and 860°C when, by crystallisation, the environment of boron atoms changes to tetra-coordinated and of aluminium atoms to predominantly penta-coordinated.

These structural changes induced by heat treatment are also reflected in EPR spectra that show that iron ions have diffused through the crystalline samples, and mainly occupy sites of rhombic symmetry, while gadolinium ions, which are disposed on sites of the lanthanum ions, show in the crystalline phase a structurally relaxed environment, which is but not as expected for lanthanum in a well defined 124 crystalline phase.

Acknowledgements

The author would like to thank Professors E. de Boer and A. P. M. Kentgens from the University of Nijmegen for providing access to facilities of their laboratories.

References

1. Smith, M. E. *Appl. Magn. Reson.*, 1993, **4**, 1.
2. Sato, R. K., McMillan P. F., Dennison P. & Dupree R. *J. Phys. Chem.*, 1991, **95**, 4483.
3. Iuga, D., Simon, S., Kentgens, A. P. M. & de Boer, E. *J. Phys. Chem. B*, 1999, **103** (36), 7591.
4. Yang, P., Yu, W., Wang J. Y., Wei, J. Q. & Liu, Y. G. *Acta Cryst.*, 1998, **C54**, 11.
5. Simon, S. *J. Optoelectr. Adv. Mat.*, 2003, **5** (1), 147–153.
6. Stallworth, P. E. & Bray, P. J. in *Glass Science and Technology*, Eds. Uhlmann, D. R. & Kreidl, N. J. 1990. Acad. Press Inc., Boston. P. 77–100.
7. Brow, R. K., Tallant, D. R. & Turner, G. L. *J. Am. Ceram. Soc.*, 1996, **79** (9), 1410.
8. Simon, S., van Moorsel, G. J. M. P., Kentgens, A. P. M. & de Boer, E. *Solid State NMR*, 1995, **5**, 163.
9. Züchner, L., Chan, J. C. C., Müller Warmuth W. & Eckert, H. *J. Phys. Chem. B*, 1998, **102**, 4495.
10. Griscom, D. L. *J. Non-Cryst. Solids*, 1980, **40**, 211.
11. Simon, S., van der Pol, A., Reijerse, E. J., Kentgens, A. M. P., van Morsel, G. J. M. P. & de Boer, E. *J. Chem. Soc. Faraday Trans.*, 1995, **91** (10), 1519.
12. Simon, S., Pop, R., Simon, V. & Coldea, M. *J. Non-Cryst. Solids*, 2003, **331** (13), 1.
13. Konijnendijk, W. L.& Stevels, J. M. *J. Non-Cryst. Solids*, 1975, **18**, 307.
14. Meera, B. N. & Ramakrishna, J. *J. Non-Cryst. Solids*, 1993, **159**, 1.
15. Krogh-Moe, J. *Phys. Chem. Glasses*, 1965, **6** (2), 46.
16. Kamitsos, E. I., Patsis, A. P., Karakassides, M. A.& Chryssikos, G. D. *J. Non-Cryst. Solids*, 1990, **126**, 52.
17. Moustafa, Y. M., Doweidar, H. & El-Damrawi, G. *Phys. Chem. Glasses*, 1994, **35** (2), 104.
18. Abbattista, F., Delmarstro, A., Gozzelino, G., Mazza, D., Vallino, M., Busca, G. & Lorentzelli, V. *J. Chem. Soc. Faraday Trans.*, 1990, **86**, 3653.
19. Simon, S., van der Pol, A., Reijerse, E. J., Kentgens, A. P. M., van Moorsel, G. J. M. P. & de Boer, E. *J. Chem. Soc. Faraday Trans.*, 1994, **90**, 2663.
20. Cordischi, D., Valigi, M., Gazzoli, D. & Indovina V. *J. Solid State Chem.*, 1975, **15**, 82.
21. Sands, R. H. *Phys. Rev.*, 1955, **99**, 1222.
22. Nicklin, R. C., Johnstone, J. K., Barnes, R. G. & Wilder, D. R. *J. Chem. Phys.*, 1973, **59**, 1652.
23. Cugunov L.& Kliava, J. *J. Phys. C*, 1982, **15**, L993.
24. Iton, L. E., Brodbeck, C. M., Suib S. L.& Stucky, G. D. *J. Chem. Phys.*, 1983, **79**, 1185.
25. Brodbeck G. M. & Iton, L. E. *J. Chem. Phys.*, 1985, **83**, 4285.
26. Legein, C., Buzaré, J. Y., Silly G. & Jacoboni, C. *J. Phys.: Cond. Matter*, 1996, **8**, 4339.
27. Simon, S. *Mod. Phys. Lett. B*, 2001, **15**, 375.

Modelling of B_2O_3 glass structure by coupled MD/MC simulation

A. Takada[1*†]

Research Center, Asahi Glass Co. Ltd, 1150 Hazawa-cho, Yokohama 221-8755, Japan
*Department of Earth Sciences, University College London, Gower Street, London W1E 6BT, UK
†Department of Chemistry, University College London, Gower Street, London W1E 6BT, UK

The difficulties in theoretical modelling of vitreous B_2O_3 result from the complexity of B–O bonding and the resulting slow process of glass network formation. The previously developed interatomic potentials are applied to model the complex B–O bonding. Moreover, new methodology that couples molecular dynamics and Monte Carlo-like techniques has been developed to search effectively for a promising structure of vitreous B_2O_3 among the abundant possible structures. The role of Monte Carlo-like techniques is to generate randomly modified structures varying a B–O ring size distribution successively and update to the more stable structure. The function of molecular dynamics techniques is to relax each structure and calculate its minimum energy. A key feature of this study is the estimation of the fraction of boron atoms in boroxol rings (highly planar three-membered ring) as 74%.

1. Introduction

The structure of vitreous B_2O_3 is still the subject of controversy. Two types of structural model have been proposed. One is the structure that contains a high fraction of boroxol ring (highly planar three-membered ring).[1] The other is a structure that mainly comprises independent BO_3 triangles.[2]

Most experimental investigations using diffraction techniques,[1,3] Raman scattering[4] and NMR[5–7] have supported a high fraction (70–80%) of boron atoms in boroxol rings. On the contrary, most simulation studies failed to reproduce the existence of a high fraction of boroxol rings. Only a few of them showed 30–50% fractions of boron atoms in boroxol rings.[8–10] However, there is still a large difference in the estimates of the fraction of boron atoms in boroxol rings between experiments and simulation studies.

The difficulty in modelling vitreous B_2O_3 originates from the complexity of B–O bonding, which is not only partially covalent, but also allows the boron coordination number to change depending on its environment.[11–13] At present it is thought to be necessary to include the three-body[8,10] or the polarisation[9] term for precise modelling. Moreover, the relaxation of pressure compacted vitreous B_2O_3 at ambient pressure over a period of 1–2 days[14] has suggested the slow process of glass network formation and its recovery. The corresponding simulation study showed that it would require much more simulation time to get fully stabilised vitreous structures than in other glass systems.[15]

On the other hand, the previously proposed methodology named 'computer synthesis',[11] where new structures were modified from the existing crystal structures, has succeeded in generating such new structures that comprise a high fraction of boroxol rings (75%) and reproduce the experimental glass density at the same time.[16] Although the calculated structure is thought to be a promising structural model of vitreous B_2O_3, it would be more appropriate to refer the structure to as a 'pseudo-crystal' than as an amorphous structure, because it lacks the 'randomness' of intermediate-range and long-range orders.

In this study, new simulation techniques that couples molecular dynamics and Monte Carlo-like techniques has been developed to search for promising structures of vitreous B_2O_3 among an abundant possible structures effectively. Characteristics of calculated structure employing these new techniques are discussed in detail.

2. Computational model

In previous studies, I had already developed a interatomic potential model for crystalline and vitreous B_2O_3.[8,11,15,16] This potential model is employed in this study. First, one random configuration where all boron and all oxygen atoms have three-fold and two-fold coordination respectively is prepared as an initial structure. Here, one pseudo-crystal structure[16] generated by 'computer synthesis' is used, because this structure satisfies the above condition. This initial structure has originally been modified from the real B_2O_3–I crystal[17] and finally comprised 50% fractions of boron atoms in boroxol rings. Of course, it is also possible to use one of other pseudo-crystal structures as an initial structure. Secondly, a new coupled MD/MC simulation is performed. The simulation procedure has two parts. The role of the first simulation part is to modify the

[1] Corresponding author. Email address: akira-takada@agc.co.jp

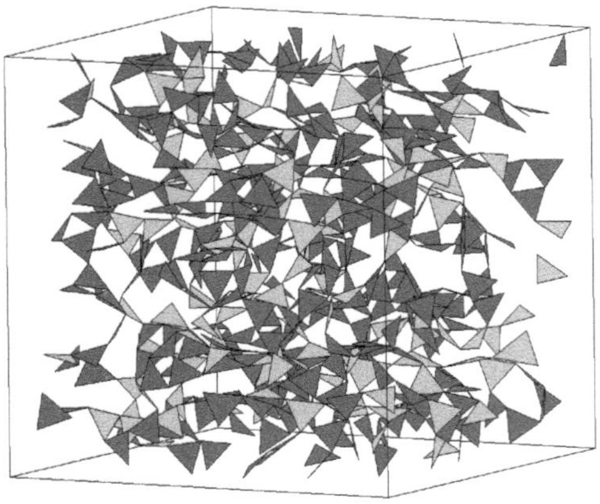

Figure 1. Atomic configuration of calculated structure of vitreous B_2O_3

latest structure and generate two different structures (structure *a* and *b*). Any one of the BO_3 units is picked up randomly in the latest structure. If an independent BO_3 unit is picked up, it is replaced by a B_3O_6 unit. On the other hand, if a BO_3 inside a B_3O_6 unit is picked up, another random selection occurs, whether the B_3O_6 is replaced with a new independent BO_3 unit or the BO_3 is replaced with a B_3O_6 unit. The structure so obtained is named structure *a*. Further, the above procedure with another random number is applied, and the obtained structure is then named structure *b*. It is important to note that even such a simple procedure can increase or decrease ring sizes successively and generate any size of B–O rings after a long sampling. The role of second simulation part is to geometrically optimise structures *a* and *b* by constrained MD techniques. Two structures are temperature controlled at 500 K over 1000 steps (1 ps) and then annealed to 0·1 K gradually over 1000 step (1 ps). It is important to avoid bond breakage, therefore, all the B–O bond lengths are constrained during the heating period at 500 K and then are released at the annealing stage down to 0·1 K. If a bond breakage occurs during annealing, the structure is rejected.

After the second simulation part, two lattice energies for annealed structures *a* and *b* are compared. The transition probability for going from structure to structure may be replaced with the criterion based on the Metropolis method.[18] However, it is important to note that this MC-like procedure in this study does not follow the usual Boltzmann sampling, because the sampling is performed only within the restricted configuration space through the replacement procedure of BO_3 and B_3O_6 units. Hence, two factors are neglected in this study. The first is that such a restricted configuration space would neglect part of entropy effect. The second is that the energy barrier between two structures is not considered. In this study the structure with lower energy is adopted after each sampling and updated to the latest structure because of simplicity and efficiency in simulation. Therefore, this procedure is expected to search for such structure that has the lower energy, hopefully, close to the global energy minimum. The total number of atoms increases or decreases during each MC step. A constant density method, that is where the simulation box is varied so as to keep its experimental glass density, is used for each cycle. It is very important to update the latest configuration so that all boron and all oxygen atoms have three-fold and two-fold coordination respectively, from beginning to end. This restraint enables a substantial decrease in the number of searching steps compared with the simple MD or simple MC simulation.

Over 500 coupled MD/MC cycles, the latest structures almost converge and fluctuate around an averaged configuration.

3. Results

One converged structure is shown in Figure 1. Total number of atoms is 1540. Light grey and dark grey triangles show BO_3 units outside and inside B_3O_6 unit, respectively. It comprises 74% fractions of boron atoms in boroxol rings. The initial structure was generated from the chain-like structure of real B_2O_3–I crystal, however, the final structure leads to the three-dimensional networks comprised of independent BO_3

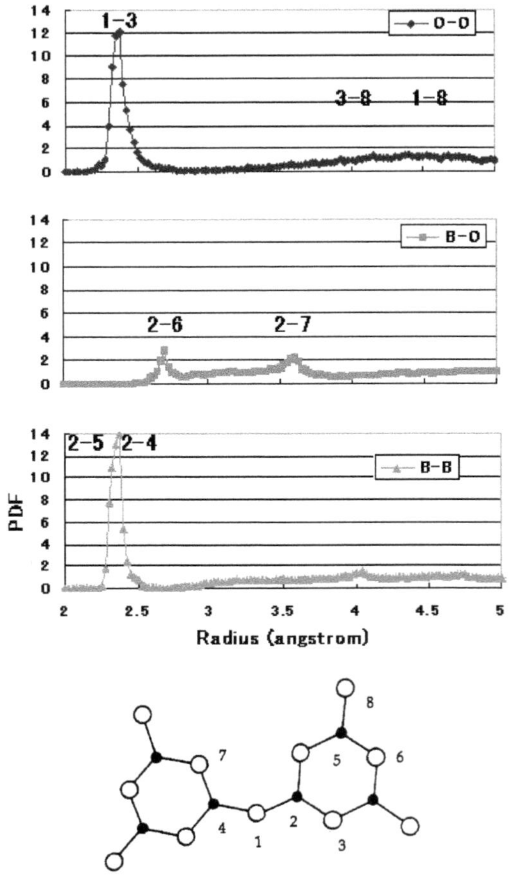

Figure 2. Pair distribution functions of calculated structure of vitreous B_2O_3

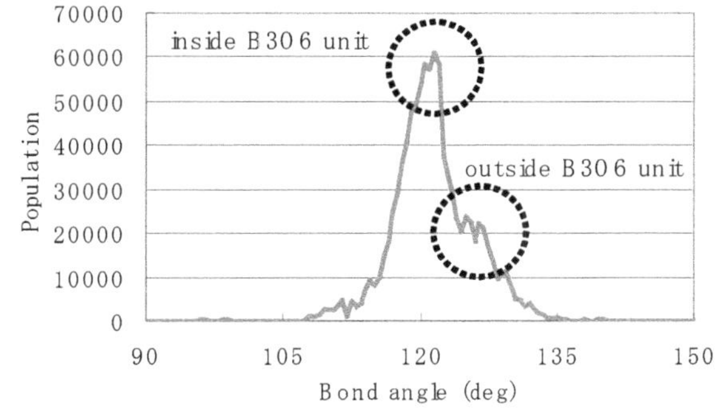

Figure 3. Bond angle distribution of B–O–B calculated structure of vitreous B_2O_3

and B_3O_6 units.

The pair distribution functions and bond angle distribution at 0·1 K are shown in Figures 2 and 3. Peaks peculiar to boroxol rings are obscure in pair distribution functions compared with the computer-synthesised structure.[16] However, two characteristic peaks are distinguishable in bond angle distributions. The torsional angle distribution at 0·1 K is also shown in Figure 4. Smaller and larger angles originate from those inside and outside B_3O_6 units, respectively.

The B–O ring statistics are shown in Figure 5. The dominant ring size is three, that is the boroxol ring is the dominant structure unit. Other ring sizes are observed but their populations are small. Typical shapes of three-, four-, five- and six-membered rings are shown in Figure 6. The larger its ring size, the more deviated from planarity its ring shape is.

Next, statistics of numbers of B_3O_6 units connecting to B_3O_6 and BO_3 units are counted respectively and shown in Figure 7. Most connections are such that one or two B_3O_6 units, that is two BO_3 or one BO_3 unit, are connected either to B_3O_6 or BO_3 units.

4. Discussion

All the calculated results support the structure with high fraction of boroxol rings proposed by experimental studies[1,3–7] and by several previous simulation studies.[16] The feature of the structure

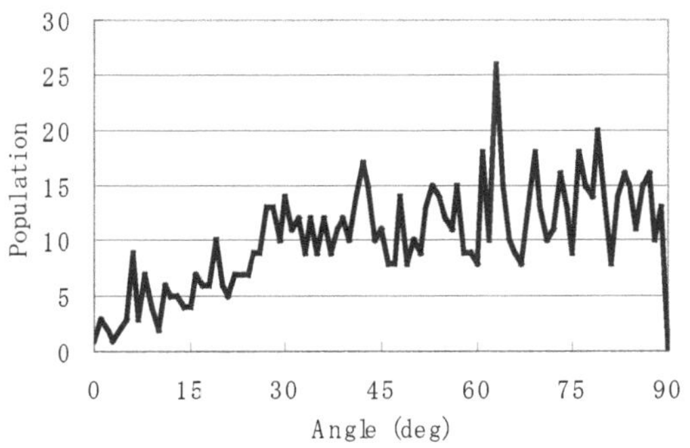

Figure 4. Torsional angle distribution of calculated structure of vitreous B_2O_3. Torsional angle is calculated in terms of intersecting angle of two BO_3 planes

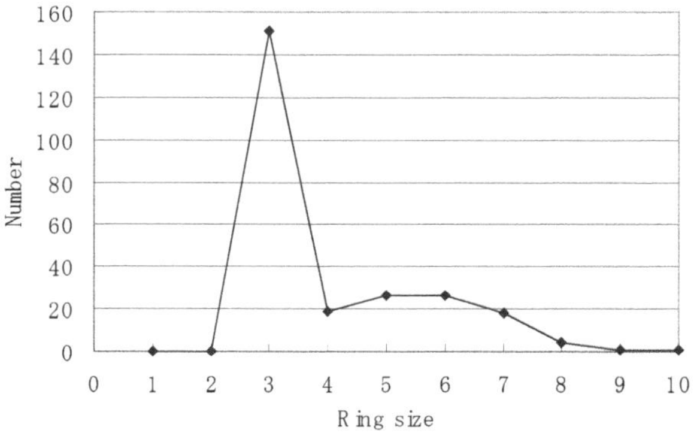

Figure 5. Distribution of B–O ring sizes. Total number of boron atoms is 616. The number of three-membered rings is 151 and the number of boron atoms in boroxol rings 453

with such high fraction of boroxol rings is discussed as follows.

The snapshot of Figure 1 shows that the structure with a high fraction of boroxol rings does not have any planar sub-structure. This feature also agrees with the previous simulation study.[16] Regarding the torsional angle distribution shown in Figure 4, it is interesting to note that the broad torsional angle distribution enables the structure to accommodate the compact packing needs of BO_3 and B_3O_6 units. Otherwise, a structure with a high fraction of boroxol rings expands and shows much lower density compared with the experimental density.[8,16]

Next, the ring statistics shown in Figure 5 are compared with other structures. When ring statistics are analysed for B_2O_3–I crystal and the pseudo-crystal structure model[16] slightly modified from Cs_2O–$9B_2O_3$ crystal, the formal structure does not have smaller rings than eight-membered, and the latter has only three-membered rings. Small B–O rings other than

Figure 6. Typical shapes of B–O rings. Three-, four-, five- and six-membered rings are shown. Triangle shows BO_3 unit

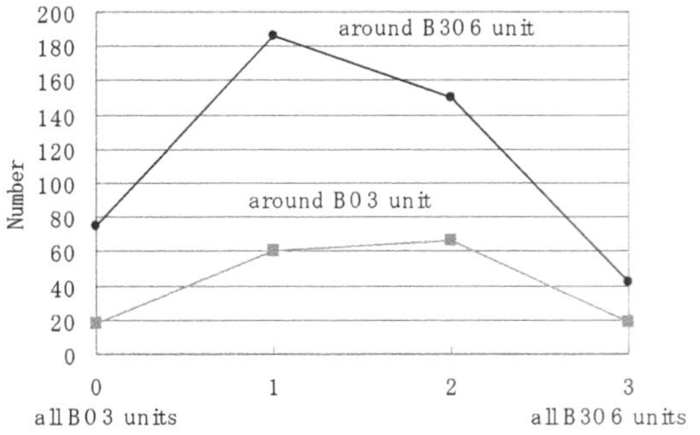

Figure 7. Statistics of connection between independent BO_3 and B_3O_6 units. The numbers of B_3O_6 units connecting to B_3O_6 and BO_3 units are counted

three-membered seem undesirable for the naturally existing structures. The one-dimensional chain-like structure of B_2O_3–I crystal or less interfering interpenetrating network of Cs_2O–$9B_2O_3$ crystal seems to reduce the number of bridging in network. As for vitreous structure, the calculated results show geometrical distortions of small rings. This distortion is closely related to the broad torsional angle distribution. Consequently, the compatibility between a high fraction of boroxol rings and the experimental density can be satisfied only under such conditions that the less interfering network and the distorted ring topology. The less interfering network means that small rings other than three-membered are a few. The distorted ring topology is the same with the large torsional angle distribution. These characteristics are expected to be a key factor for intermediate range order of vitreous B_2O_3.

Finally, the statistics of numbers of B_3O_6 units connecting to B_3O_6 and BO_3 units shown in Figure 7 are compared with other structures. As for the pseudo-crystal structure model[16] slightly modified from Cs_2O–$9B_2O_3$ crystal, where a fraction of boron atoms in boroxol rings is 75%, two B_3O_6 units are always connected to both BO_3 and B_3O_6 units, respectively. On the other hand, for the crystal structure of B_2S_3, where a fraction of boron atoms in three-membered rings is 75%, two B_3S_6 units are always connected to both B_2S_4 and B_3S_6 units, respectively. It is interesting to note that two crystal structures have the common connection statistics. It is thought that the intrinsic randomness of vitreous structure would allow more flexible connection compared with two crystal structures.

5. Conclusions

The combination of the previously developed interatomic potential model and the coupled MD/MC simulation techniques demonstrates a promising structure model of vitreous B_2O_3. The three-body potential model contributes to model the short range order and intermediate range order, that is a high fraction of boroxol rings. On the other hand, the coupled MD/MC techniques contribute efficiently to the search for structures with minimum energy.

The calculated results suggest that a fraction of boron atoms in boroxol rings is 74%. This value almost agrees with those suggested by experiment.[1,3,5–7] The vitreous structure model has several characteristics peculiar to vitreous B_2O_3. The first point is that the structure has a few smaller B–O rings other than three-membered. It is strongly related to such three-dimensional structure that interfere less in network. The second is that the calculated results show geometrical distortions of small rings and the broad torsional angle distribution. These two points are expected to be key factors for intermediate range order of vitreous B_2O_3, because these enable the structure to show compatibility between a high fraction of boroxol rings and the experimental density. Finally, the selectivity of connections between B_3O_6 and BO_3 units are analysed. The vitreous model showed more randomness in the selectivity of connections.

More precise evaluation of energy and entropy is a further aspect of the physics of the glass structure that needs to be clarified.

Acknowledgement

Thanks are due to the referee who pointed out that the explanation on MC-like sampling procedure was insufficient in the first draft, and also let the author recognise two important papers on recent NMR studies.[6,7]

References

1. Johnson, P. A. V., Wright, A. C. & Sinclair, R. N. *J. Non-Cryst. Solids*, 1982, **50**, 281.
2. Swenson, J. & Borjesson, L. *Phys. Rev. B*, 1997, **55**, 11138.
3. Hannon, A. C., Grimley, D. I., Hulme, R. A., Wright, A. C. & Sinclair, R. N. *J. Non-Cryst. Solids*, 1994, **177**, 299.
4. Bronswijk, J. P. & Strijks, E. *J. Non-Cryst. Solids*, 1977, **24**, 145.
5. Jellison, Jnr., G. E., Panek, L. W., Bray, P. J. & Rouse, G. B. *J. Chem. Phys.*, 1977, **66**, 802.
6. Youngman, R. E. & Zwanziger, J. W. *J. Non-Cryst. Solids*, 1994, **168**, 293.
7. Joo, C., Zwanziger, U. W. & Zwanziger, J. W. *Solid State NMR*, 2000, **16**, 77.
8. Takada, A., Catlow, C. R. A. & Price G. D. *J. Phys., Condens. Matter*, 1995, **7**, 8693.
9. Maranas, J. K., Chen, Y, Stillinger, D. K. & Stillinger, F. H. *J. Chem. Phys.*, 2001, **115**, 6578.
10. Kashchieva, E., Shivachev, B. & Dimitriev, Y. *J. Non-Cryst. Solids*, 2005, **351**, 1158.
11. Takada, A., Catlow, C. R. A. & Price, G. D. *J. Phys., Condens. Matter*, 1995, **7**, 8659.
12. Takada, A., Catlow, C. R. A., Lin, J. S., Price, G. D., Lee, M. H., Milman, V. & Payne, M. C. *Phys. Rev. B*, 1995, **51**, 1447.
13. Takada, A., Catlow, C. R. A., Price, G. D. & Hayward, C. L. *J. Phys. Chem. Miner.*, 1997, **24**, 423.
14. Wright, A. C., Stone, C. E., Sinclair, R. N., Umesaki, N., Kitamura, N., Ura, K., Ohtori, N. & Hannon, A. C. *Phys. Chem. Glasses*, 2000, **41** (5), 296–9.
15. Takada, A. *Phys. Chem. Glasses*, 2004, **45** (2), 156–9.
16. Takada, A., Catlow, C. R. A. & Price, G. D. *Phys. Chem. Glasses*, 2003, **44** (2), 147–9.
17. Gurr, G. E., Mongomery, P. W., Knutson, C. D. & Gorres, B. T. *Acta Crystallogr. B*, 1970, **26**, 906.
18. Frenkel, D., & Smit, B. *Understanding Molecular Simulation*. Academic Press, San Diego, 1996, p 26.
19. Mozzi, R. L. & Warren, R. L. *J. Appl. Cryst.*, 1970, **3**, 251.

Proc. Fifth Int. Conf. on Borate Glasses, Crystals and Melts Phys. Chem. Glasses: Eur. J. Glass Sci. Technol. B, August 2006, 47 (4), 497–506

Thermodynamic modelling of the real-space correlation function for four sodium borosilicate glasses

*Adrian C. Wright**

J.J. Thomson Physical Laboratory, University of Reading, Whiteknights, Reading, RG6 6AF, UK

Natalia M. Vedishcheva & Boris A. Shakhmatkin

Institute of Silicate Chemistry of the Russian Academy of Sciences, Ul. Odoevskogo, 24, Korp. 2, Sankt Petersburg, 199155, Russia

The interpretation of even the first peak in the total correlation function, T(r), for a multicomponent oxide glass is extremely complex, especially when the glass contains more than one glassforming oxide and hence a number of different structural unit species. In the case of alkali borosilicate glasses, there are five possible borate and five possible silicate structural unit species, all of which have slightly different bond lengths and internal angles. In addition, these bond lengths and angles vary in a systemmatic way when the basic borate structural units are incorporated into superstructural units. In the present paper, the model of associated solutions is used to predict the detailed shape of the first two peaks in T(r) for four sodium borosilicate glasses with nominal Na_2O:B_2O_3:SiO_2 ratios of 2:2:1, 1:1:1, 1:1:6 and 1:1:8. It is concluded that the agreement between the thermodynamic model and experiment is within their combined uncertainty. The main limitation at present is lack of knowledge concerning the structure of relevant crystalline phases, especially their high temperature polymorphs.

1. Introduction

In a previous paper,[1] thermodynamic modelling, using the model of associated solutions,[2] was employed to explain apparently anomalous bond lengths extracted from fits to the neutron total correlation function, *T(r)*, for two sodium borosilicate glasses, of composition $2Na_2O.2B_2O_3.SiO_2$ and $Na_2O.B_2O_3.8SiO_2$. For both glasses, the apparently anomalous peak parameters were explained by comparing the neutron data and peak fit with histograms representing the distributions of B–O and Si–O bond lengths predicted by the thermodynamic modelling. The purpose of the present paper is to extend this approach to predict the detailed shape of the first two peaks in *T(r)*, by including estimates of static and thermal disorder broadening, together with that arising from the experimental resolution as represented by the real-space peak function, *P(r)*.

Borate glass systems may contain up to five possible borate structural units, usually designated $B^{(n)}$, *n* being the number of bridging oxygen atoms. These comprise $BØ_4^-$, $BØ_3$, $BØ_2O^-$, $BØO_2^{2-}$ and BO_3^{3-} (corresponding to *n*=4, 3, 2, 1 and 0, respectively), where Ø represents a bridging and O a nonbridging oxygen atom. In addition, the situation is complicated by the fact that borate glass structures may also include superstructural units, which comprise well defined arrangements of the basic borate structural units, with no internal degrees of freedom in the form of variable bond or torsion angles. The structural units present in each superstructural unit are summarised in Table 1 and the superstructural units are illustrated schematically in Figure 1 of Ref. 1. Not only do each of the five basic structural units have unique bond lengths and angles (cf. Table 2 of Ref. 1) but, as indicated by Krogh-Moe,[3] the B^{III}–O and B^{IV}–O bond lengths and intrastructural unit bond angles vary systemmatically when they are incorporated into superstructural units, and there is also some evidence that external B–Ø bond lengths may be affected by the type of boron atom (3-fold co-ordinated, B^{III}, or 4-fold co-ordinated, B^{IV}) to which they bridge.

In silicate glasses, there are also five possible silicate structural units, denoted $Si^{(n)}$ ($SiØ_4$, $SiØ_3O^-$, $SiØ_2O_2^{2-}$, $SiØO_3^{3-}$ and SiO_4^{4-}, again corresponding to *n*= 4, 3, 2, 1 and 0), all of which are found in crystalline alkali silicate polymorphs. Each $Si^{(n)}$ species has a maximum of two discrete bond lengths, Si–Ø ($r_Ø$) and Si–O (r_O), and two independent internal bond angles Ø–Si–Ø ($\beta_{ØØ}$) and O–Si–O (β_{OO}). {Note that these four parameters fix the Ø–Si–O bond angle ($\beta_{ØO}$).} The average values of $r_Ø$, r_O, $\beta_{ØØ}$ and β_{OO}, extracted from

[1] Corresponding author. Email A.C.Wright@reading.ac.uk

Table 1. Borate (super)structural units and anions

Unit/Anion	Formula	x_M*	$B\varnothing_4^-$	$B\varnothing_3$	$B\varnothing_2O^-$	$B\varnothing O_2^{2-}$	BO_3^{3-}
-	$B\varnothing_3$	0·000	-	1	-	-	-
Boroxol	$B_3\varnothing_6$	0·000	-	3	-	-	-
Pentaborate	$B_5\varnothing_{10}^-$	0·167	1	4	-	-	-
Triborate	$B_3\varnothing_7^-$	0·250	1	2	-	-	-
Di-pentaborate	$B_5\varnothing_{11}^{2-}$	0·286	2	3	-	-	-
Diborate	$B_4\varnothing_9^{2-}$	0·333	2	2	-	-	-
Tri-pentaborate	$B_5\varnothing_{12}^{3-}$	0·375	3	2	-	-	-
Di-triborate	$B_3\varnothing_8^{2-}$	0·400	2	1	-	-	-
Triborate + NBO	$B_3\varnothing_6O^{2-}$	0·400	1	1	1	-	-
-	$B\varnothing_4^-$	0·500	1	-	-	-	-
-	$B\varnothing_2O^-$	0·500	-	-	1	-	-
Metaborate	$B_3\varnothing_9^{3-}$	0·500	3	-	-	-	-
Cyclic metaborate	$B_3\varnothing_3O_3^{3-}$	0·500	-	-	3	-	-
-	$B\varnothing O_2^{2-}$	0·667	-	-	-	1	-
Pyroborate	$B_2\varnothing O_4^{4-}$	0·667	-	-	-	2	-
-	BO_3^{3-}	0·750	-	-	-	-	1

* Mole fraction of network modifying oxide for a structure containing only the unit/anion

accurate crystallographic studies of ambient pressure alkali (or silver) silicate polymorphs performed at ambient temperature, are compared in Table 2 of Ref. 1, from which optimum values have been estimated for each $Si^{(n)}$ species.

To model the correlation function for sodium borosilicate glasses, therefore, it is not only necessary to obtain structural parameters for all of the independent borate and silicate structural units, but also for all of the possible borate superstructural units that occur in the crystalline phases in the Na_2O–B_2O_3–SiO_2 phase diagram. Of the 22 known binary sodium borate crystalline phases,[4] however, crystal structures have only been published for seven, for which the (super)structural unit contents are given in Table 2, the entries indicating the fractions of the structural units in the crystal that are incorporated into each (super)structural unit species. Each superstructural unit has a specific number of independent internal parameters, comprising bond lengths and angles. These are outlined in Ref. 1, which also discusses the derivation of the optimum value for each parameter, as obtained from a detailed survey of alkali borate crystal structures.

2. Modelling of equilibrium bond length distributions

The additional complication introduced by the presence of superstructural units in vitreous borate networks can be seen from Figure 1, which shows the interatomic distances from each unit. To use the structural parameters from Ref. 1 in the analysis of diffraction data, it is necessary to resort to a model which predicts the fraction of the Si and/or B atoms in a given glass that are involved in each (super)structural unit present. One such model is that of associated solutions[2] and this paper will illustrate its use for four sodium borosilicate glasses, namely $2Na_2O.2B_2O_3.SiO_2$, $Na_2O.B_2O_3.SiO_2$, $Na_2O.B_2O_3.6SiO_2$ and $Na_2O.B_2O_3.8SiO_2$, using neutron diffraction data from Ref. 5.

The model of associated solutions describes the structure of glasses in terms of chemical groupings, which are assumed to have the same stoichiometry as, and similar structures to, crystalline compounds formed in the same system, at least in terms of the basic structural units and co-ordination polyhedra. It is further assumed that the mix of (super)structural units and co-ordination polyhedra is the same as in the crystalline phase with the same stoichiometry. The calculation of the chemical structure involves no adjustable parameters and the accuracy of the final model, expressed in terms of chemical groupings, depends only on the uncertainty in the input thermodynamic parameters (standard Gibbs free energies of formation).

Each chemical grouping contains a fraction, x_S, of one or more (super)structural units and hence the correlation function per network forming cation for that grouping is[1]

$$T_G(r)=\sum_S x_S T_S(r) \qquad (1)$$

where $T_S(r)$ is the correlation function for

Table 2. Distribution of (super)structural units in crystalline Na_2O–B_2O_3 phases[1]

Polymorph	$B_5\varnothing_{10}^-$	$B_3\varnothing_7^-$	$B_5\varnothing_{11}^{2-}$	$B_4\varnothing_9^{2-}$	$B_3\varnothing_6O^{2-}$	$B\varnothing_4^-$	$B_3\varnothing_3O_3^{3-}$	$B\varnothing O_2^{2-}$	BO_3^{3-}
α-$Na_2O.4B_2O_3$	5/8	3/8	-	-	-	-	-	-	-
α-$Na_2O.3B_2O_3$	5/9	-	-	4/9	-	-	-	-	-
β-$Na_2O.3B_2O_3$	5/9	3/9	-	-	-	1/9	-	-	-
α-$Na_2O.2B_2O_3$	-	-	5/8	-	3/8	-	-	-	-
α-$Na_2O.B_2O_3$	-	-	-	-	-	-	1	-	-
$2Na_2O.B_2O_3$	-	-	-	-	-	-	-	1*	-
$3Na_2O.B_2O_3$	-	-	-	-	-	-	-	-	1

* Present in pairs as $B_2\varnothing O_4^{4-}$ pyroborate anions

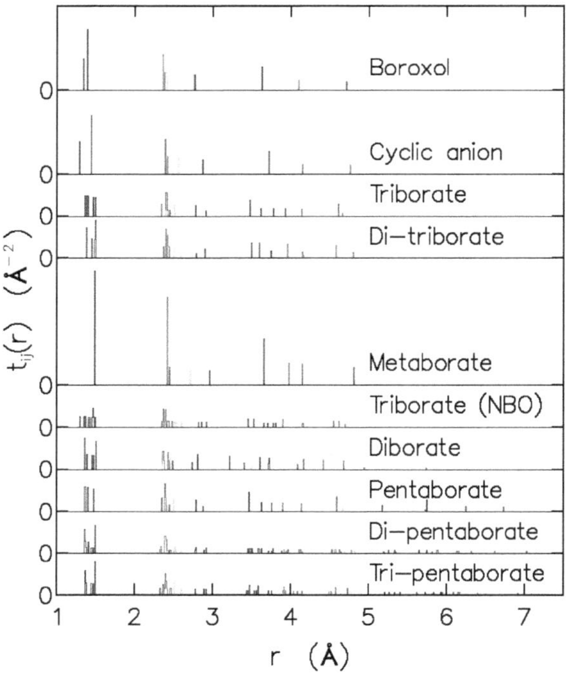

Figure 1. Contributions, $t_{ij}(r)$, to the real space correlation function from the various borate superstructural units, normalised to one structural unit. (In the electronic version of this paper B–O distances are in blue, O–O in red and B–B in cyan)

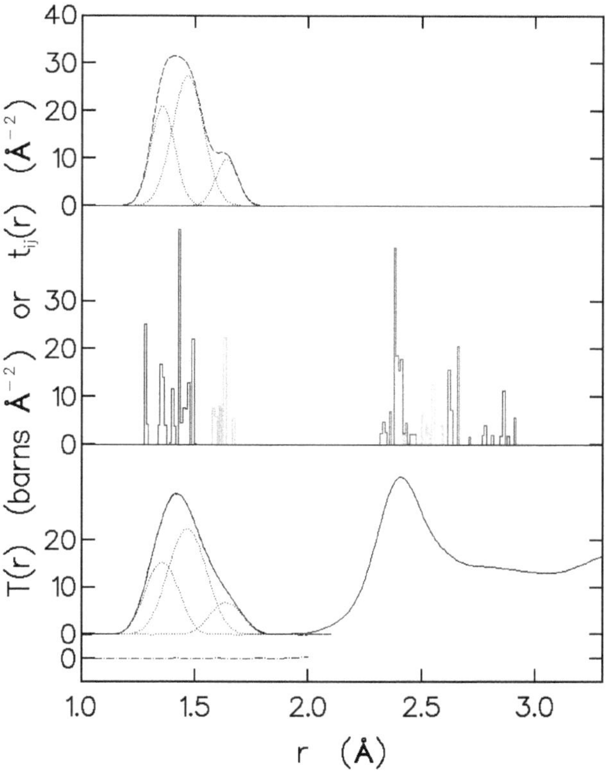

Figure 2. Peak fit for vitreous $2Na_2O.2B_2O_3.SiO_2$, together with the interatomic distances predicted from the chemical structure.[1] Top, unbroadened peaks from fit; centre, thermodynamic prediction; bottom, fit to experimental data. ———, experimental data; – – – –, fit; ········, individual peaks from fit; –·–·–·–, fit residual and histogram, interatomic distances from chemical structure. (The electronic version has the following colours: blue, B–O; green, Si–O; red, O–O and cyan, B–B)

(super)structural unit S, normalised to one structural unit. If the formula fraction of a grouping in the glass is x_G and n_G is the number of network forming cations in one formula unit of the grouping, the total contribution of the (super)structural units to $T(r)$ for the glass becomes

$$T(r)=\sum_G x_G n_G T_G(r)/\sum_G x_G n_G \qquad (2)$$

The procedure for generating the total contribution of the various (super)structural units is outlined in Ref. 1. The chemical structure for the four glasses is given in Table 3 and the (super)structural unit fractions in each glass are summarised in Table 4. In Ref. 1, these thermodynamic predictions were plotted as a histogram and used to explain why an attempt to fit the first peak in $T(r)$ with three components, corresponding to the B^{III}–O, B^{IV}–O and Si–O bonds, led to bond lengths at variance with the generally accepted values. This is illustrated by the results for vitreous $2Na_2O.2B_2O_3.SiO_2$ shown in Figure 2,[1] which shows that a number of different bond lengths contribute to the first peak. The fit has separated these into three arbitrary groups (dotted lines), two for the borate and one for the silicate component, but the groups do not contain a single bond species. In particular,

Table 3. Chemical structure (structural unit fractions)

Chemical grouping	$2Na_2O.2B_2O_3.SiO_2$	$Na_2O.B_2O_3.SiO_2$	$Na_2O.B_2O_3.6SiO_2$	$Na_2O.B_2O_3.8SiO_2$
B_2O_3	0·0006	0·0013	0·0084	0·0095
SiO_2	0·0136	0·0393	0·3526	0·4339
$Na_2O.4B_2O_3$	0·0003	0·0033	0·0081	0·0047
α-$Na_2O.3B_2O_3$	0·0051	0·0130	0·0067	0·0039
β-$Na_2O.3B_2O_3$	0·0045	0·0116	0·0060	0·0035
$Na_2O.2B_2O_3$	0·4378	0·4403	0·0479	0·0280
$Na_2O.B_2O_3$	0·3218	0·1276	0·0029	0·0017
$2Na_2O.B_2O_3$	0·0118	0·0008	-	-
$3Na_2O.8SiO_2$	-	0·0072	0·0027	0·0009
$Na_2O.2SiO_2$	0·1217	0·1888	0·0436	0·0305
$Na_2O.SiO_2$	0·0462	0·0228	0·0008	0·0005
$Na_2O.B_2O_3.2SiO_2$	0·0366	0·1440	0·1588	0·1116
$Na_2O.B_2O_3.6SiO_2$	-	-	0·3615	0·3713

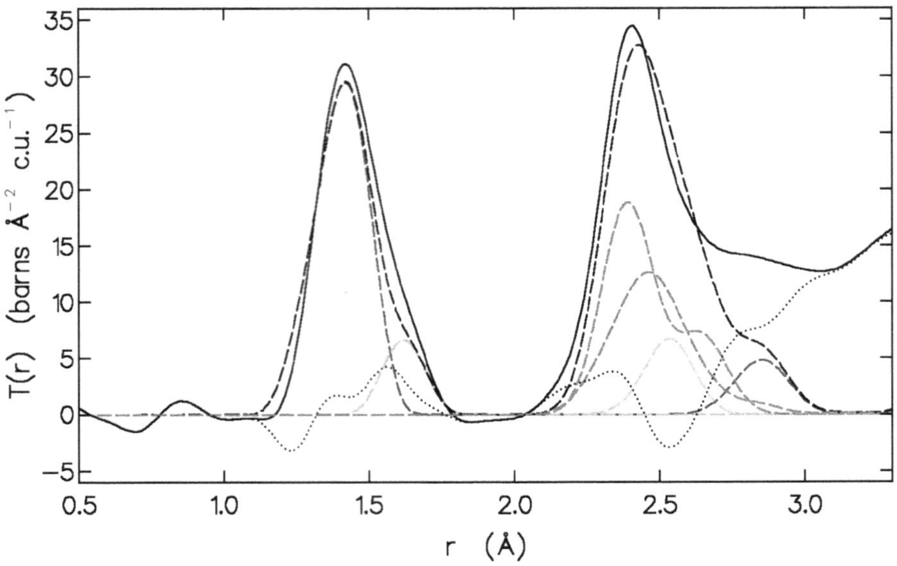

Figure 3. A comparison of the neutron correlation function for vitreous $2Na_2O.2B_2O_3.SiO_2$ (solid line) with the predictions of the thermodynamic model (dashed line), together with the residual (dotted line). In the electronic version, the individual components are shown in colour: Blue, B–O; green, Si–O; red, O–O; cyan, B–B and magenta, Na–O

the use of three peaks does not allow for the presence of both B^{III}–Ø and B^{III}–O^- bonds. Hence it was concluded in Ref. 1, that the examples considered clearly demonstrated that thermodynamic modelling, using the model of associated solutions, is an invaluable aid in interpreting diffraction data for multicomponent glasses.

3. Calculation of $T(r)$

The comparison between model and experiment outlined in the preceeding section was based on the predicted distribution of equilibrium bond lengths. However, to model the detailed shape of the first two peaks in $T(r)$, it is necessary to include both static and thermal disorder broadening, together with that arising from the experimental resolution as represented by the real-space peak function, $P(r)$. It should be noted that the structural parameters from Ref. 1 refer to an idealised equilibrium structure for each (super)structural unit that does not take account of any static strain introduced by it being incorporated into a 3-dimensional network.

The root mean square (rms) B–O and Si–O bond length variations of 0·045 Å were taken from experimental data for vitreous B_2O_3 and SiO_2, whilst the rms Na–O bond length variation was estimated to be 0·07 Å. In the simulation of the O–O contribution to the second peak, the oxygen atoms were assigned an isotropic rms vibrational amplitude of 0·085 Å. Obviously, these values are somewhat arbitrary, but reasonable variations do not significantly affect the final comparison between model and experiment and the overall conclusions. The real-space peak function, $P(r)$, was generated[6] using the appropriate maximum scattering vector magnitude, Q_{max} (30 or 40 Å^{-1}), and the model correlation functions are compared to experiment in Figures 3 to 6.

4. Discussion

Before discussing the results in detail, it should be emphasised that no adjustable parameters have been used in calculating the model correlation functions.

Table 4. (Super)structural unit fractions

(Super)structural unit	*$2Na_2O.2B_2O_3.SiO_2$*	*$Na_2O.B_2O_3.SiO_2$*	*$Na_2O.B_2O_3.6SiO_2$*	*$Na_2O.B_2O_3.8SiO_2$*
$SiØ_4$	0·032	0·113	0·704	0·768
$SiØ_3O^-$	0·122	0·194	0·046	0·031
$SiØ_2O_2^{2-}$	0·046	0·023	0·001	0·001
$BØ_4^-$	0·019	0·073	0·170	0·149
$BØ_3$	-	-	0·002	0·002
$B_3Ø_6$	-	0·001	0·006	0·007
$B_5Ø_{10}^-$	0·005	0·016	0·012	0·007
$B_3Ø_7^-$	0·002	0·005	0·005	0·003
$B_5Ø_{11}^{2-}$	0·274	0·275	0·030	0·018
$B_4Ø_9^{2-}$	0·002	0·006	0·003	0·002
$B_3Ø_6O^{2-}$	0·164	0·165	0·018	0·010
$B_3Ø_3O_3^{3-}$	0·322	0·128	0·003	0·002
$BØO_2^{2-}$	0·012	0·001	-	-

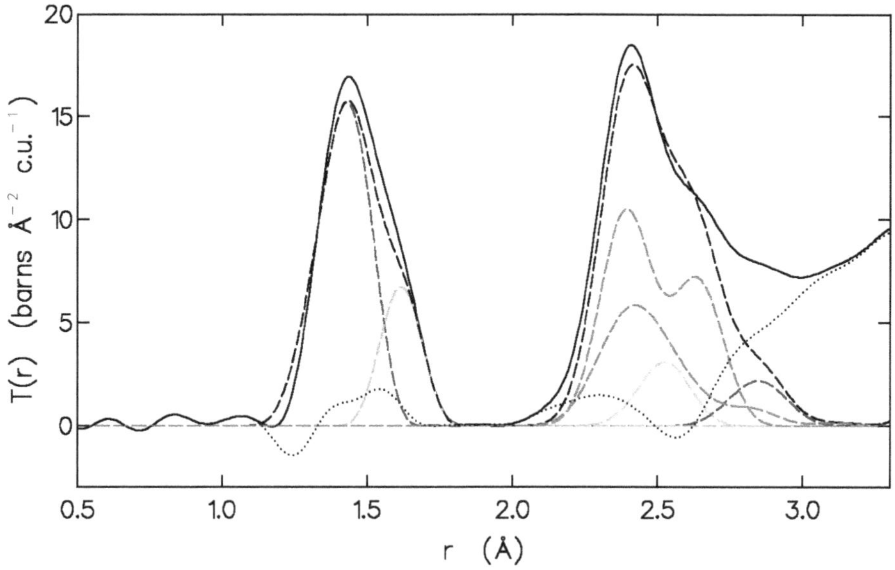

Figure 4. A comparison of the neutron correlation function for vitreous $Na_2O.B_2O_3.SiO_2$ with the predictions of the thermodynamic model (key as Figure 3)

The magnitude of the noise in the experimental correlation functions can be estimated from the region below the first peak and it can be seen that, for all four glasses, the difference between the model and experiment (dotted lines) is somewhat larger than the noise. The four glasses fall into two groups. For the borate rich glasses ($2Na_2O.2B_2O_3.SiO_2$ and $Na_2O.B_2O_3.SiO_2$), the difference curve exhibits a minimum at ~1·25 Å and a maximum at ~1·55 Å. This is consistent with the model predicting too many B^{III}–O^- bonds and not enough B^{IV}–Ø bonds. For the silicate rich glasses ($Na_2O.B_2O_3.6SiO_2$ and $Na_2O.B_2O_3.8SiO_2$), on the other hand, there is only a minimum at ~1·45 Å, which suggests too many B^{IV}–Ø bonds.

There are several possible reasons for the discrepancies between the model predictions and experiment. The first concerns the crystal structures used in calculating the chemical structure. As indicated in Section 1, of the 22 known crystalline phases in the Na_2O–B_2O_3 system, detailed crystal structures have been reported for only seven. The structure of $3Na_2O.8SiO_2$ is unknown, as is that of $Na_2O.B_2O_3.2SiO_2$, which makes a significant contribution to the chemical structure of vitreous $Na_2O.B_2O_3.SiO_2$, $Na_2O.B_2O_3.6SiO_2$ and $Na_2O.B_2O_3.8SiO_2$. In calculating $T(r)$ from the chemical structure, it was assumed that the structure of $3Na_2O.8SiO_2$ involves 75% of $Si^{(3)}$ units and 25% of $Si^{(4)}$ units and that the Na–O bond length distribution is the average of that for the other sodium silicates. The structure of $Na_2O.B_2O_3.2SiO_2$ was rep-

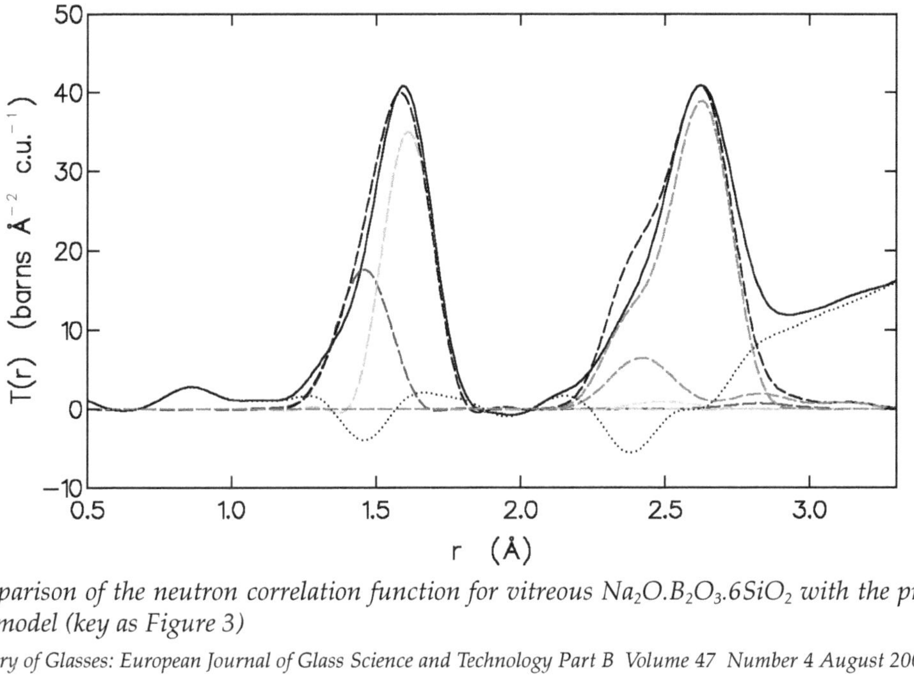

Figure 5. A comparison of the neutron correlation function for vitreous $Na_2O.B_2O_3.6SiO_2$ with the predictions of the thermodynamic model (key as Figure 3)

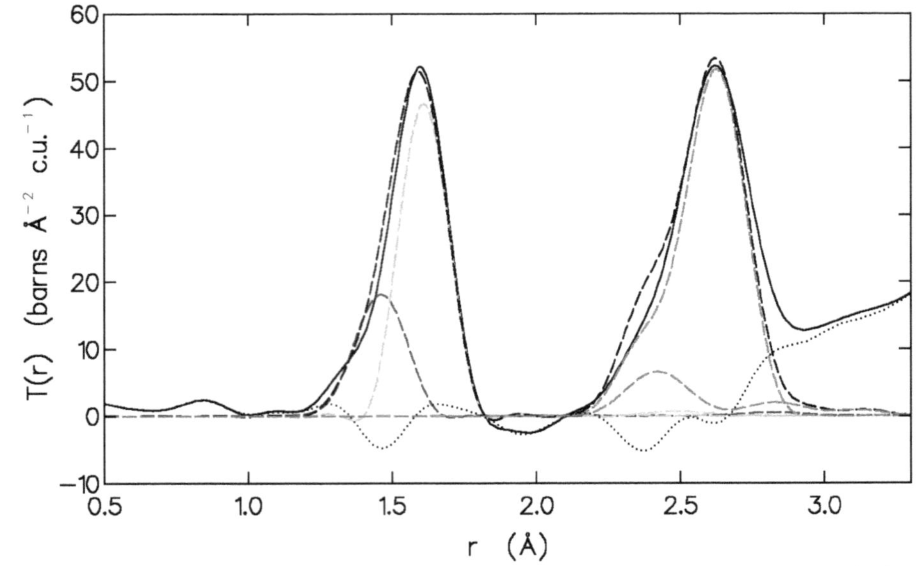

Figure 6. A comparison of the neutron correlation function for vitreous $Na_2O.B_2O_3.6SiO_2$ with the predictions of the thermodynamic model (key as Figure 3)

resented by equal numbers of $B^{(4)}$ and $Si^{(4)}$ units and the Na–O bond length distribution was taken to be the same as that for $Na_2O.B_2O_3.6SiO_2$.

Many crystalline borates and silicates exist as several polymorphs and frequently only the structures of the lower temperature polymorphs have been determined. This is important, since the thermodynamic modelling should be performed at the fictive temperature, T_f, which means that the crystal structure of the appropriate high temperature, ambient pressure polymorph should be used. Hence there is a problem if the polymorphic transitions are reconstructive, rather than displacive. It was assumed that unknown polymorphic transitions are displacive. However, it should be noted that this is not always correct, as may be seen by considering the case of $Na_2O.3B_2O_3$.[7,8] For this compound, the α–β transition is clearly reconstructive (cf. Table 1) and the structure of the highest temperature γ-polymorph is unknown. A further complication arises, if the high temperature structure cannot be quenched to a metastable state at ambient temperature, in that, to model the glass, structural parameters are required that relate to ambient temperature.

The accuracy of the thermodynamic modelling, i.e. of calculating the chemical structure, is determined by the uncertainties in the available thermodynamic data (standard Gibbs free energies of formation) of the various crystalline phases. It is estimated that these uncertainties would lead to a maximum error of 5% of the mole fraction value for each component of the chemical structure. As noted above, the uncertainties in the experimental correlation function, $T(r)$, can be seen from the error ripples below the first true peak. There is also an uncertainty in the sample composition, which affects the comparison between model and experiment in several ways. First, the data are normalised to the self scattering, which is calculated from the sample composition, and hence any inaccuracy will lead to an error in the absolute magnitude ot $T(r)$. Second, the modelling is performed for the assumed sample composition and, finally, an error in the B_2O_3:SiO_2 ratio will yield a similar error in the relative numbers of B–O and Si–O bonds. If it is assumed that any loss from the melt, during the sample preparation, was due to the evaporation of B_2O_3, weight loss measurements[9] suggest that the maximum uncertainty in the total borate structural unit content is small (≲1 mol%) for the borate rich compositions, but somewhat larger (≲4–5

Table 5. Structural unit fractions

Structural unit	*$2Na_2O.2B_2O_3.SiO_2$*	*$Na_2O.B_2O_3.SiO_2$*	*$Na_2O.B_2O_3.6SiO_2$*	*$Na_2O.B_2O_3.8SiO_2$*
$SiØ_4$	0·032	0·113	0·704	0·768
$SiØ_3O^-$	0·122	0·194	0·045	0·031
$SiØ_2O_2^{2-}$	0·046	0·023	0·001	0·001
$SiØO_3^{3-}$	-	-	-	-
SiO_4^{4-}	-	-	-	-
$BØ_4^-$	0·186	0·246	0·194	0·163
$BØ_3$	0·226	0·240	0·047	0·032
$BØ_2O^-$	0·376	0·183	0·009	0·005
$BØO_2^{2-}$	0·012	0·001	-	-
BO_3^{3-}	-	-	-	-

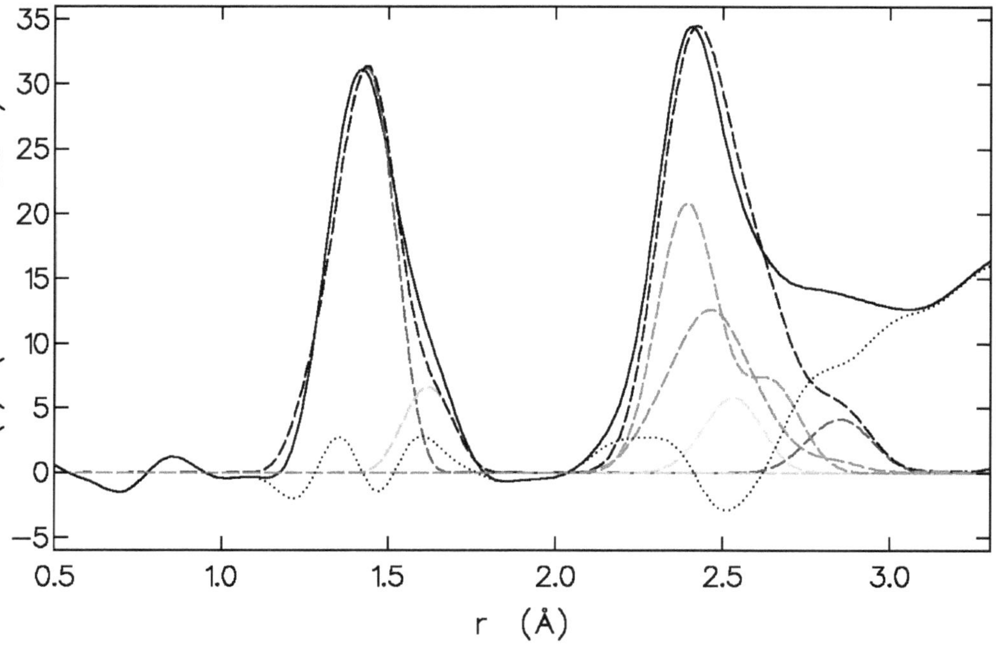

Figure 7. The effect of substituting $BØ_4^-$ tetrahedra for $BØ_2O^-$ triangles on the thermodynamic model for vitreous $2Na_2O.2B_2O_3.SiO_2$ (key as Figure 3)

mol%) for the silicate rich glasses, due to the higher melt temperature. It should also be noted that Na_2O is mainly lost from the melt in the form of sodium metaborate[10] and hence may lead to a reduction in the fraction of sodium metaborate chemical groupings, where these are present, and thus the number of $BØ_2O^-$ structural units.

Based on the various sources of error outlined above, an investigation was carried out to identify the possible source of the discrepancy between the thermodynamic model and experiment for each glass. This can best be understood by considering the fractions of the basic structural units present, as summarised in Table 5. For the two borate rich glasses, the B^{III}–O^- bonds are mainly due to $BØ_2O^-$ units, which are contributed by the α-$Na_2O.2B_2O_3$[11] chemical grouping, via the triborate group with a nonbridging oxygen atom ($B_3Ø_6O^{2-}$), and the α-$Na_2O.B_2O_3$[12] chemical grouping, which is based on the cyclic metaborate anion ($B_3Ø_3O_3^{3-}$). It is interesting to note that α-$Na_2O.2B_2O_3$ is the only known crystal structure to incorporate the $B_3Ø_6O^{2-}$ superstructural unit and that it has two higher temperature polymorphs of unknown structure, β-$Na_2O.2B_2O_3$ and

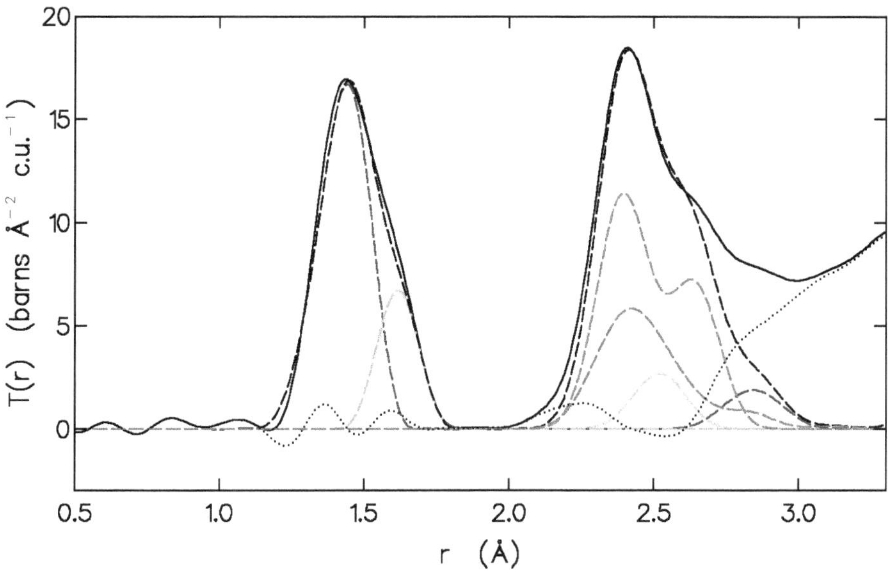

Figure 8. The effect of substituting $BØ_4^-$ tetrahedra for $BØ_2O^-$ triangles on the thermodynamic model for vitreous $Na_2O.B_2O_3.SiO_2$ (key as Figure 3)

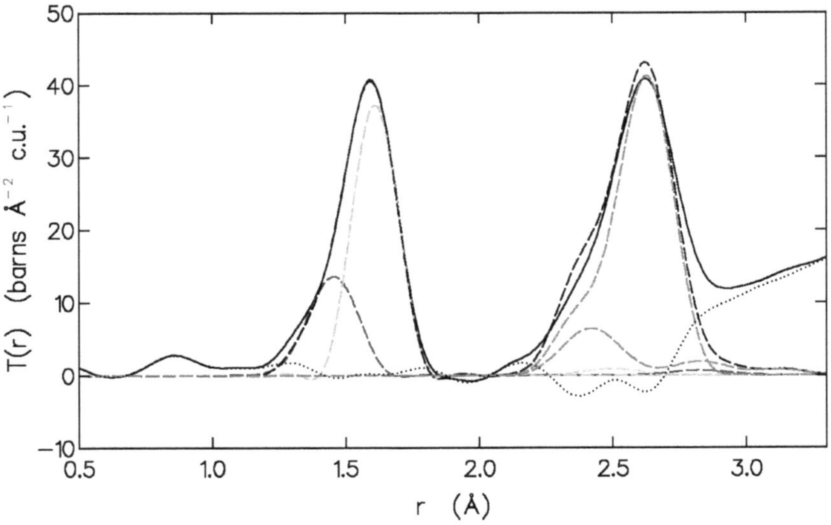

Figure 9. The effect of a reduced B_2O_3 content (i.e. of substituting $SiØ_3O^-$ tetrahedra for $BØ_4^-$) on the thermodynamic model for vitreous $Na_2O.B_2O_3.6SiO_2$ (key as Figure 3)

γ-$Na_2O.2B_2O_3$. Hence it is entirely possible that the the α–β and/or β–γ transitions are reconstructive and that the $B_3Ø_6O^{2-}$ group is not present in the γ-polymorph, thus substituting $BØ_4^-$ structural units for $BØ_2O^-$ groups; i.e. B^{IV}–Ø bonds for B^{III}–O^- bonds. α-$Na_2O.B_2O_3$ also has a high temperature polymorph, β-$Na_2O.B_2O_3$ although, in this case, it is unlikely that the α–β transition would lead to a reduction in the number of B^{III}–O^- bonds. Figures 7 and 8 show the effect of substituting $BØ_4^-$ units for $BØ_2O^-$. For vitreous $2Na_2O.2B_2O_3.SiO_2$, the fraction of $BØ_4^-$ units was increased by 0·1 and that of $BØ_2O^-$ decreased by the same amount while, for vitreous $Na_2O.B_2O_3.SiO_2$, the corresponding increase/decrease was 0·077. It can be seen that the discrepancy between the thermodynamic model and experiment is reduced, particularly in the case of vitreous $Na_2O.B_2O_3.SiO_2$, for which a larger percentage of the the B^{III}–O^- bonds originate from the $B_3Ø_6O^{2-}$ superstructural unit. The difference between the modified thermodynamic models and experiment, especially for vitreous $Na_2O.B_2O_3.SiO_2$, is not much larger than the magnitude of the experimental noise and is probably not significant, given the arbitrary nature of the bond substitution. The magnitude of the above substitutions is, however, larger than the uncertainties in the glass composition and/or thermodynamic modelling (see above).

For the two silicate rich glasses, it was assumed that they had a reduced B_2O_3 content, as a result of evaporation from the melt. Hence the $BØ_4^-$ structural unit fractions for vitreous $Na_2O.B_2O_3.6SiO_2$ and $Na_2O.B_2O_3.8SiO_2$ were reduced to 0·144 and 0·123, respectively, and the fractions of $SiØ_3O^-$ were increased to 0·096 and 0·071 (i.e. assuming no loss of Na_2O). These adjustments are entirely consistent with the maximum uncertainty in the B_2O_3 content derived from the weight loss measurements discussed above. As may be seen from Figures 9 and 10 this leads to an excellent fit to the first peak in $T(r)$ and a much smaller residual under the second peak, which is comparable to the experimental uncertainty. The fact that this residual is negative, and in the region of the Na–O bond length, may reflect a loss of Na_2O, in addition to B_2O_3.

A similar approach to that employed here could also be used in the interpretation of NMR data, given a knowledge of the chemical shifts for each atom in the various (super)structural units. One of the problems

Table 6. Structural unit fractions, x_n^{Si} and x_n^B, obtained by considering the borate and silicate subnetworks separately. x^{Na} is the fraction of the sodium ions associated with the borate subnetwork

x_n	$2Na_2O.2B_2O_3.SiO_2$	$Na_2O.B_2O_3.SiO_2$	$Na_2O.B_2O_3.6SiO_2$	$Na_2O.B_2O_3.8SiO_2$
x_4^{Si}	0·160	0·343	0·938	0·960
x_3^{Si}	0·609	0·588	0·061	0·039
x_2^{Si}	0·231	0·069	0·001	0·001
x_1^{Si}	-	-	-	-
x_0^{Si}	-	-	-	-
x_4^B	0·232	0·367	0·777	0·814
x_3^B	0·283	0·359	0·188	0·160
x_2^B	0·470	0·273	0·035	0·026
x_1^B	0·015	0·001	-	-
x_0^B	-	-	-	-
x^{Na}	0·732	0·642	0·811	0·839

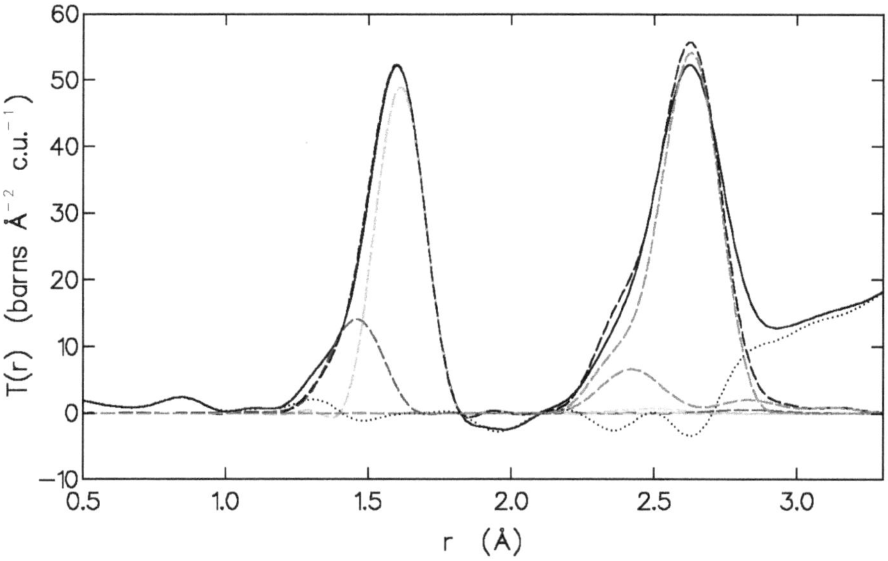

Figure 10. The effect of a reduced B_2O_3 content on the thermodynamic model for vitreous $Na_2O.B_2O_3.8SiO_2$ (key as Figure 3)

with most NMR studies of alkali borosilicate glasses to date is that the borate and silicate subnetworks have often been considered separately, which gives an incomplete and hence distorted picture of the glass structure.(13) For example, the glass composition that contains the maximum number of $BØ_4^-$ tetrahedra is *not* the same as that for which the fraction of the boron atoms that are 4-fold coordinated, x_4^B, is a maximum. This can be seen by comparing Table 5, which shows the combined distribution of the structural units in the present thermodynamic models, with Table 6, where the distributions of the borate and silicate structural unit species are calculated separately. This is extremely important, for example, when discussing the influence of the number of $BØ_4^-$ tetrahedra on the glass properties, such as density. Note that this comment refers to any system that comprises more than one glass forming oxide. Thus, an undeniable advantage of the thermodynamic approach is the possibility of quantitatively modelling the combined distribution of all of the basic structural units in such glasses, over extended composition and temperature regions.

Table 6 also shows the distribution of the Na^+ ions between the borate and silicate subnetworks and it can be seen that, for all four glasses, the majority of the Na^+ ions is associated with the borate subnetwork. This effect is seen in extremum in the Vycor region of the $Na_2O–B_2O_3–SiO_2$ phase diagram, where phase separation occurs into regions that are composed essentially of sodium borate and silica.

An important point raised by the present study is

Table 7. Crystalline compounds containing superstructural units. (Compounds of the form mM_2O.nB_2O_3 *are denoted by the abbreviation* mM.nB*, and Roman numerals are used for polymorphs formed at different pressures.)*

Unit	*Formula*	*Crystalline Compound(s)* *Single Unit*	*Multiple Units*
Structural units			
BO_3 Triangle	$BØ_3$	B-I	-
BO_4 Tetrahedron	$BØ_4^-$	γ-Li.B	-
BO_3 Triangle with NBO	$BØ_2O^-$	Li.B, Ca.B	-
Superstructural units			
Boroxol	$B_3Ø_6$	-	Cs.9B (+$B_3Ø_7^-$)
Pentaborate	$B_5Ø_{10}^-$	K.5B, Rb.5B, Cs.5B	-
Triborate	$B_3Ø_7^-$	Li.3B, Cs.3B	-
Di-pentaborate	$B_5Ø_{11}^{2-}$	-	Na.2B (+$B_3Ø_6O^{2-}$), Ba.2B (+$B_3Ø_8^{2-}$)
Diborate	$B_4Ø_9^{2-}$	Li.2B, Zn.2B, Mg.2B	-
Tri-pentaborate	$B_5Ø_{12}^{3-}$	-	2Ca.3B (+$BØ_4^-$)
Di-triborate	$B_3Ø_8^{2-}$	-	3Li.7B (+$B_3Ø_7^-$+$BØ_3$), K.2B & Rb.2B (+$B_4Ø_9^{2-}$+$BØ_3$), Ba.2B (+$B_5Ø_{11}^{2-}$)
Triborate with NBO	$B_3Ø_6O^{2-}$	-	Na.2B (+$B_5Ø_{11}^{2-}$)
Metaborate	$B_3Ø_9^{3-}$	Ca.B-IV, Sr.B-IV, Cu.B	-
Anions			
Cyclic Metaborate	$B_3Ø_3O_3^3$	Na.B, K.B, α-Rb.B, Cs.B, Ba.B	-
Pyroborate (2 × $BØO_2^{2-}$)	$B_2ØO_4^{4-}$	2Na.B, 2Mg.B, 2Ca.B, 2Sr.B	-
Orthoborate	BO_3^{3-}	α-3Li.B, 3Na.B, 3Ag.B-I, 3Tl.B, 3Zn.B, 3Mg.B, 3Ca.B, 3Sr.B	-

that, whilst there have been many crystal structure studies of complex synthetic borates and borate minerals, there is a surprising lack of structural data for the simple anhydrous binary borates required for the interpretation of experimental data for the corresponding glasses. The thermodynamic modelling requires accurate structural parameters (bond lengths and angles) for the various borate superstructural units and, similarly, the interpretation of both NMR and vibrational (IR, Raman and inelastic neutron scattering) spectra depends on a precise characterisation of the response of each unit. It is important to note that the crystalline borate networks necessary for this characterisation of each superstructural unit should ideally be formed by a single superstructural unit species, with no independent BO_3 triangles or BO_4 tetrahedra. Table 7 lists such crystalline compounds or, for superstructural units where no known crystalline compound exists, one or more compounds containing the given superstructural unit, together with a list of the other units present. As is clearly apparent, there is a great need for crystal structure studies of anhydrous binary borates, particularly of the high temperature polymorphs, to yield both the distribution of (super)structural units and accurate structural parameters.

5. Conclusions

It may thus be concluded that thermodynamic modelling, using the model of associated solutions, is an invaluable aid in interpreting diffraction data for complex multicomponent glasses. In the case of the sodium borosilicate glass system, the model predicts the distribution of both structural unit and superstructural unit species and hence experimentally measurable quantities, such as the real space correlation function, $T(r)$, and the fraction of 4-fold co-ordinated boron atoms, x_4^B. For the two silica-rich samples considered here, the agreement between the thermodynamic model and experiment is within their combined uncertainties. However, the discrepancy for the borate rich samples is somewhat larger, but this may be due to inadequate crystal structure data.. The main limitation at present is lack of knowledge concerning the structure of relevant crystalline polymorphs, especially their high-temperature phases. To understand the glass structure, it is first necessary to understand the phase relationships of the corresponding crystalline materials and hence there is a great need for accurate structural studies of a wide range of crystalline compounds relevant to glassforming systems, particularly at higher temperatures in the region of T_f.

References

1. Wright, A. C., Shaw, J. L., Sinclair, R. N., Vedishcheva, N. M., Shakhmatkin, B. A. & Scales, C. R. *J. Non-Cryst. Solids*, 2004, **345&346**, 24.
2. Shakhmatkin, B. A., Vedishcheva, N. M. & Wright, A. C. *J. Non-Cryst. Solids*, 2001, **293–295**, 220.
3. J. Krogh-Moe, *Acta Crystallogr.*, 1972, **B28**, 168.
4. Heller, G. *Gmelin Handbuch der Anorganischen Chemie*, Vol. 28, *Borverbindungen Part 7* (Springer-Verlag, Berlin, 1975).
5. Shaw, J. L., Wright, A. C., Sinclair, R. N., Kiczenski, T. J., Feller, H. A. Feller, S. A. & Scales, C. R. *Phys. Chem. Glasses*, 2002, **43C**, 59.
6. Wright, A. C. in: *Experimental Techniques of Glass Science*, Eds C. J. Simmons & O. H. El-Bayoumi (Amer. Ceram. Soc., Westerville, 1993), p. 205.
7. Krogh-Moe, J. *Acta Crystallogr.*, 1974, **B30**, 747.
8. Krogh-Moe, J. *Acta Crystallogr.*, 1972, **B28**, 1571.
9. Feller, S. A. Private communication (2005).
10. Vedishcheva, N. M., Shakhmatkin, B. A. & Shultz, M. M. *J. Thermal Anal.*, 1990, **36**, 2055.
11. Krogh-Moe, J. *Acta Crystallogr.*, 1974, **B30**, 578.
12. Marezio, M., Plettinger, H. A. & Zachariasen, W. H. *Acta Crystallogr.*, 1963, **16**, 594.
13. Vedishcheva, N. M., Shakhmatkin, B. A. & Wright, A. C. *Phys. Chem. Glasses*, 2003, **44**, 191.

EELS study of boron coordination in alkali borosilicate glasses under extensive electron irradiation

G. Yang, G. Möbus & R. J. Hand*

Immobilisation Science Laboratory, Dept Engineering Materials, University of Sheffield, Mappin Street, S1 3JD, Sheffield, UK

Analytical transmission electron microscopy, especially electron energy loss spectroscopy (EELS), has the potential to extract similar glass structure and glass chemistry information as XAS due to the complementary processes of inelastic electron scattering and x-ray absorption. The main advantage is higher spatial resolution, while the higher irradiation sensitivity to electrons as compared to photons is a drawback. We have studied a series of ternary alkali borosilicate (ABS) glasses with A=Li, Na, K, Cs and constant ratios between R=[A_2O]/[B_2O_3] and K=[SiO_2]/[B_2O_3], in order to follow the trend of tetrahedral versus planar borate fingerprints in the EELS fine structure of the boron-K-edge. We found that Cs glasses are stable under the electron beam, while all other alkali borosilicate glasses suffer from alkali removal which is however not instantaneously linked to BO_4 conversion.

Introduction

Alkali borosilicate glass (ABS) systems have been widely used for many years, such as laboratory glass wares and also in nuclear waste immobilisation.[1,2] Many techniques have been used to characterise glass structure, such as extended x-ray absorption fine structure (EXAFS),[3,4] nuclear magnetic resonance (NMR),[5] infrared spectroscopy (IR),[6] photoluminescence (PL)[7] and raman spectroscopy.[8] However, these techniques are all based on the analysis of an average microstructure. In the last decade electron energy loss spectroscopy (EELS)[9] has been developed and employed in the study of various oxide minerals.[10] However, little research on EELS analysis of glass structure has been conducted, and this may be due to the vulnerability of glass to the electron beam. Thus spectroscopy of glasses at nanometre resolution is still not established, despite the availability of appropriate instrumental capabilities.

So far, three dominant types of electron irradiation damage on glasses have been identified: oxygen bubble formation,[11] phase separation,[12] and crystallisation.[13] For ABS glasses further more specific damage types have to be accounted for, such as local alkali depletion and the conversion of tetrahedral boron (BO_4) to trigonal boron (BO_3) under electron beam irradiation.[14] The sensitivity of EELS to coordination of boron in minerals is very strong. Sauer *et al*[10] have obtained the energy loss near edge structure (ELNES) of reference materials with pure trigonal boron and tetrahedral boron, thereby providing "fingerprint"[9] spectra for boron coordination. Within a representative experimental spectrum (Figure 1) the following features are distinguished: According to molecular orbital (MO) theory, the intense peak at 193·6 eV is called 'π^* peak' which is characteristic of trigonal boron (BO_3), while the so-called 'σ^* peak' at 198·6 eV is the typical feature of tetrahedral boron (BO_4). A further peak at 204·1 eV is also mainly attributed to trigonal boron. From this point of view, observation of changes in the ELNES of boron K edge can be used as a criterion for the evidence of structural changes. For example, Jiang *et al*[15] found that after electron irradiation tetrahedral boron was depleted, and molecular oxygen formed. Sun *et al*[16] found that a boron rich phase was separated from a Si rich phase, and a mixture of tetrahedral with trigonal boron formed under electron irradiation.

The purpose of this study is to follow general trends in the boron K-edge fine structure for ABS glasses as a function of irradiation time and of alkali choice (Li, Na, K and Cs). We then attempt to (i) correlate alkali depletion and boron structure variation; (ii) establish the most robust conditions for TEM observations of ABS glasses.

Experimental

The alkali borosilicate glass composition studied here was A_2O:B_2O_3:SiO_2=16·6:20:63·4 (mol%), in which A=Li, Na, K and Cs and R=[A_2O]/[B_2O_3]=0·83 and K=[SiO_2]/[B_2O_3]=3·17. This choice is close to the maximum of tetrahedral boron fraction[17] while simultaneously representing a typical nuclear waste glass composition which seeks a compromise between high durability (high K) and low melting temperature (low K). For all the glass melts reagent grade alkali carbonates (Sigma Aldrich, 99%), boric acid (Fisher

* Corresponding author. Email g.moebus@sheffield.ac.uk

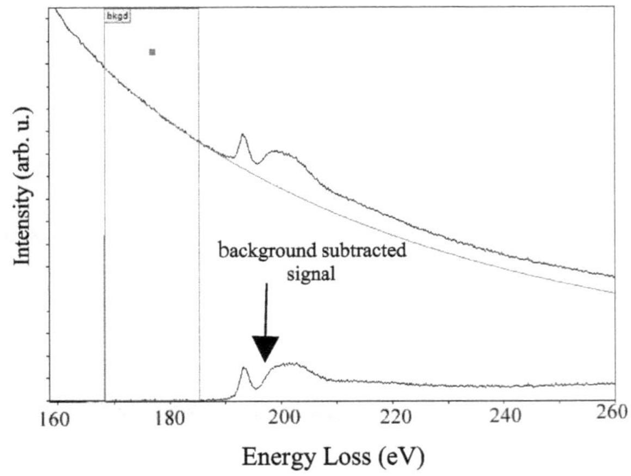

Figure 1. EELS of boron K-edge

Scientific, 99+%) and silica sand (Tilcon, 99·8%) were batched to an accuracy of ±0·005 g, to give a 300 g glass melt. They were placed in a Pt crucible and put into a 147 cm^3 (internal volume) electric furnace with SiC elements. The melting temperature was 1200°C and the total melting time was 5 h with 1 h batch free time and 4 h stirring. After melting all the glasses were poured into a preheated steel mould to form a rectangular glass block approximately 2×2×10 cm in size. This block was then transferred to an annealing furnace where it was held at 570°C for 1 h and then cooled to room temperature at the rate of 1°C/min to relieve the stresses in the glass. The bulk specimen was sliced into small pieces for sample preparation.

In order to minimise the damage of glass during sample preparation, the specimens were prepared by crushing bulk glass in agate mortar and pestle with acetone, and small amount of suspended particles were mounted on carbon film covered copper grids.

A Jeol 2010F Field Emission Gun TEM with an accelerating voltage of 200 kV was used for all the analysis. EELS were obtained using a Gatan Image Filter (GIF) using DigitalMicrograph software. The energy resolution of the EEL spectra is 1·6 eV at a dispersion of 0·3 eV/pixel. In order to quantify the fraction of tetrahedral boron (N_4) in these glasses, the experimental results were compared with different superposed reference spectra.

Figure 1 shows a typical boron K-edge EEL spectrum and the background subtraction spectrum using exponential fits to the pre-edge background.

Results

Figures 2(a)–(d) show the ELNES spectra of boron K-edge as a function of electron irradiation time in ABS glasses. The experimental results mainly show a broad peak at 193·6 eV rather than the very sharp peak seen in the reference spectra (Figure 3, labelled Peak "A"), which we attribute mainly to peak broadening by lower energy resolution. Bond length variation in glasses as opposed to crystalline minerals would be a second possible broadening effect, see Sipr *et al*[18] on

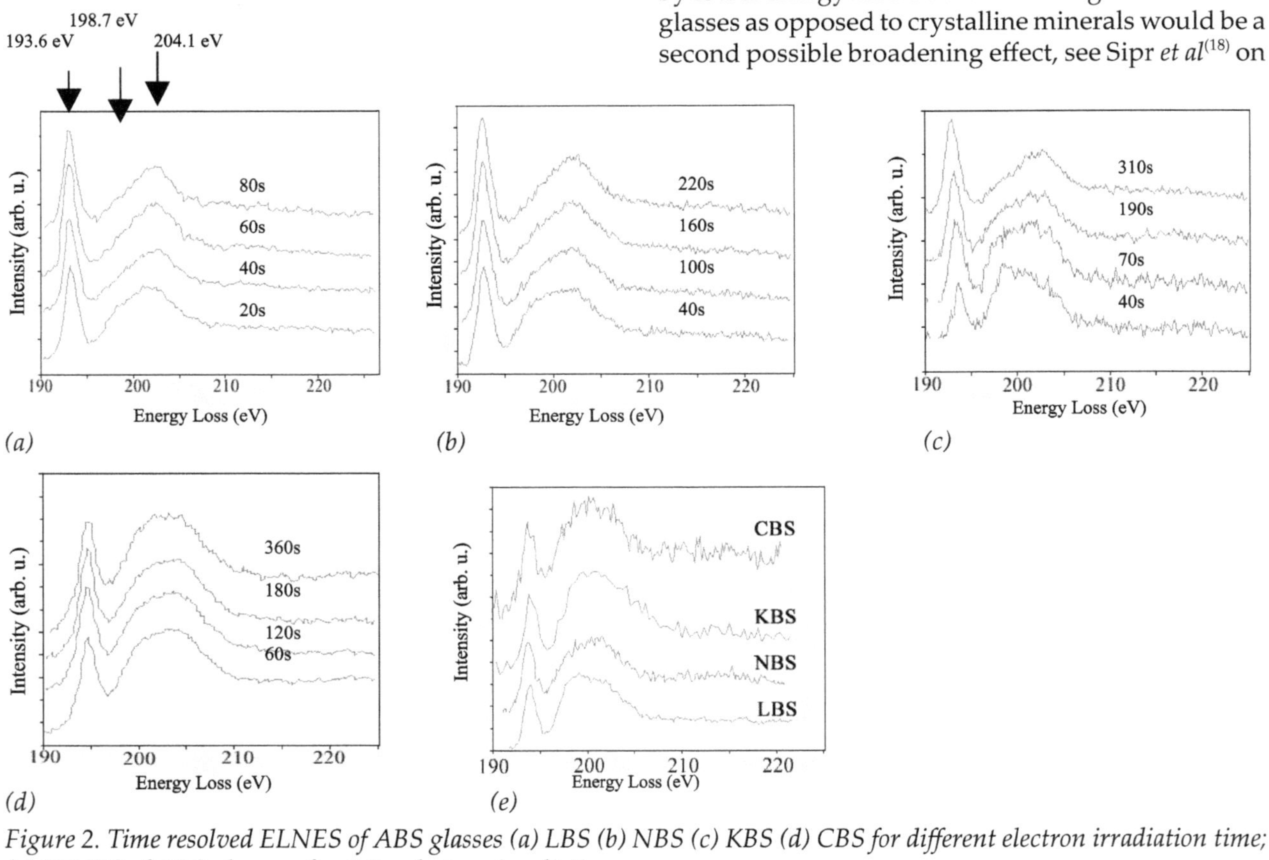

Figure 2. Time resolved ELNES of ABS glasses (a) LBS (b) NBS (c) KBS (d) CBS for different electron irradiation time; (e) ELNES of ABS glasses after 1·3 s electron irradiation

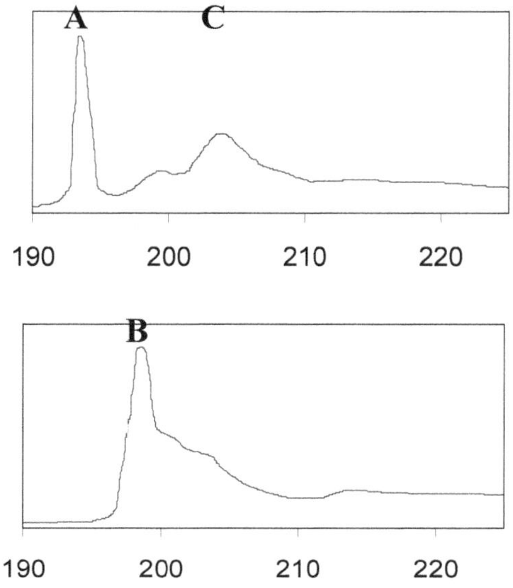

(a) BO_3

(b) BO_4

Figure 3. Boron K-edge ELNES fingerprints for BO_3 and BO_4 (after Sauer et al[10])

this topic. Figure 2(e) shows separately the ELNES spectra of boron K-edge of all ABS glasses acquired with the shortest exposure times (here 1·3 s) of electron irradiation manageable. The detailed acquisition sequence was as follows: For 2(a), four individual spectra, with 5 s exposure each, were integrated into a 20 s spectrum. The following exposure times are then achieved by accumulation of several 20 s groups of spectra. For 2(b)–(d), six individual spectra, with 5 s exposure each, were integrated into 30 s spectra. An offset of 10 s exposure before the first such spectrum was applied for (b) and (c) only. The exposure times longer than 30 s are accumulated as for 2(a).

In Figure 2(e) the spectra are single acquisitions only, which make the signal vulnerable to experimental condition fluctuations, such as electron beam fluctuations, speed of beam positioning, and acquisition delays. These effects all become pronounced for sub-second work, and made spectra of the kind of Figure 2(e) look different for repeated experiments at different glass positions.

Overall, from Figure 2 it is clear that there are significant changes of the boron K-edge under the electron beam since the peak at 198·7 eV (labelled Peak "B" in Figure 3) which indicates the presence of BO_4 rapidly diminishes, although this peak B is mostly merged with the peak at 204·1 eV (labelled Peak "C" in Fig. 3), which is supported by the presence of BO_3 (however has a finite intensity from BO_4 as well). This latter effect of peak merging B+C is not only attributed to low energy resolution, but was also observed as irradiated related for the crystalline boron mineral colemanite.[10]

Figure 3 reproduces the data of Sauer *et al*[10] for the purely coordinated crystalline minerals vonsenite (a) and rhodizite (b). Combining these two references with different ratios, boron K-edge ELNES superposition reference spectra can be obtained, such as in Figure 4. Figure 5 shows an example of the N_4 quantification method. The bottom spectrum is an experimental EEL spectrum of the boron K-edge in the potassium borosilicate glass after 40 s electron irradiation, and the top spectrum is the matched reference superposition which was generated by IDL software. All the spectra shown in Figure 2 were evaluated by this visual method, detailed below, and the quantification results are shown in

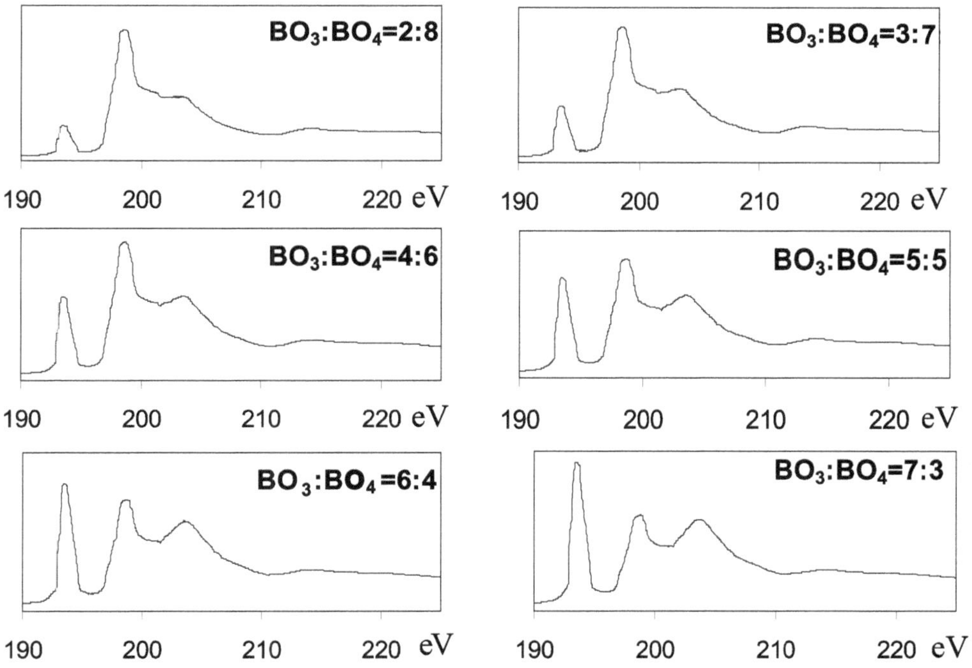

Figure 4. ELNES superposition spectra of different ratio of boron coordination; Intensity (arb. Units) versus energy loss (eV)

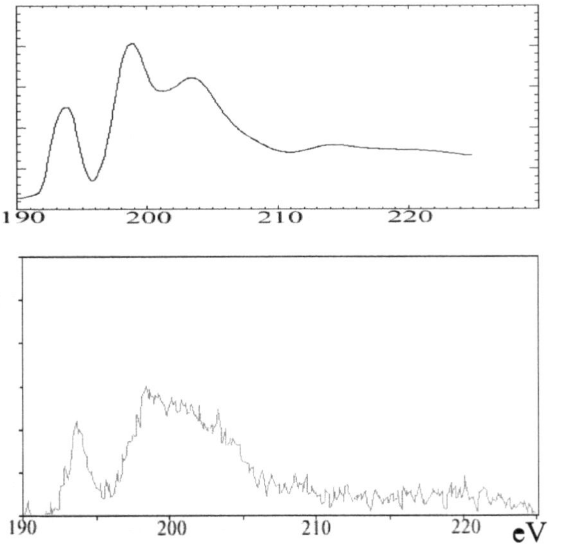

Figure 5. Matched reference superposition (top, N_4=0·5) and experimental glass spectrum (bottom, KBS glass after 40 s exposure to electron beam); Intensity (arb. units) versus energy loss (eV)

Figure 6. It is clear that tetrahedral boron changes to trigonal boron with electron beam irradiation. In borosilicate glasses, the borate network is interlinked with a silicate network. Both trigonal boron and tetrahedral boron exist in the network. The addition of alkali ions to borosilicate glasses will modify the two networks. When alkali ions are bonded with the silicate network, nonbridging oxygens (NBO) are formed. As long as they are associated with the borate network, tetrahedral boron is increased up to a limit (while nonbridging oxygen atoms (NBOs) may be generated above this limit[19]) so that the addition of alkali oxide to borosilicate glass changes trigonal boron into tetrahedral boron. Correspondingly a reduction of alkali atoms is expected to change the boron from tetrahedral to trigonal.

In order to proceed towards a quantitative estimation of N_4 for all four ABS glasses, we use a visual fit between experiments and superimposed references, which concentrates on two alternative criteria, leading to slightly different results:
Criterion I uses the peak-ratio A versus (B+C), where C has contributions from both boron coordinations. Criterion II uses the trace of peak B as it stands out from the background defined by A+C. An automated curve fit or cross correlation would not be able to distinguish sensibly between these two criteria and was avoided. N_4 values for zero irradiation were estimated from the data shown in Figure 6 by extrapolation to the y-axis, i.e. t=0. Second order polynomial extrapolation was used as the best fitting method for all glass spectra.

*Table 1. Comparison of N_4 values extrapolated from EELS with NMR Ref. 17 in which all K=3 (except *, which is from Refs 22, 23)*

Glass	LBS	NBS	KBS	CBS
(a) N_4	0·37	0·40	0·54	0·44
(b) N_4	0·43	0·46	0·63	0·40
N_4 from NMR	0·43	0·65/0·55*	0·51	-

(a) Estimates based on Peak Area Ratio A / (B+C)
(b) Estimates based on Peak Area Ratio B / (A+C)

(i) Li borosilicate glass

Figure 2 (a) shows the changes of boron coordination with irradiation time in lithium borosilicate glasses. The strong peak "B" at 198·7 eV becomes invisible after 80 s. However, further electron beam irradiation after 80 s did not lead to any further changes in the ELNES spectra, and the structure seems to have reached a stable state. No lithium K-edge was detectable within the electron illuminated spot by EELS, even in fresh, previously un-irradiated regions and using 0·64 s as acquisition time. This is not an instrumental problem. The possible reason may be that the lithium ions, which are at a low concentration (16·6 mol% Li_2O) in the glass anyway, were driven out of the observation area, due to their high mobility (e.g. as known from SIMS), in a time shorter than our minimum acquisition time. The rapid alkali depletion is attributed by Jiang *et al*[14] primarily to a charging induced migration process of alkali ions out of the central electron spot into the unirradiated neighbourhood, while sputtering of alkali ions out of the sample is a secondary effect. However, for most of our intermediate irradiation times (Figure 2(e)), there were still BO_4 groups in the glasses. At the same time, from the absence of any Li K-edge we conclude that Li was totally depleted inside the illuminated area. This means the tetrahedral boron is not fully dependent on the presence of alkali atoms. These intermediate glass states (see also our results in Ref. 20) do not follow the linear dependency of N_4 versus R, as postulated for R<0·5 from NMR results.[17] The glass structure obtained after medium times of irradiation with incomplete structural unit conversion but complete alkali depletion, seems to be inaccessible by manufacturing routes other than irradiation. The quantitative N_4 versus time estimation for criteria I and II is printed as Figure 6 (a) and (b). The main difference between the two criteria is the final asymptotic state, which approaches N_4≈0–0·1 for criterion II, while N_4≈0·2 for criterion I. The error bars are at least ±0·1.

(ii) Na borosilicate glass

Figure 2(b) is the time resolved ELNES spectrum of the boron K-edge in our sodium borosilicate glass. The change of boron coordination was similar but significantly slower than seen in LBS, and it took over 220 s for the spectra to become stable, that means the structure again reached a final state of structural

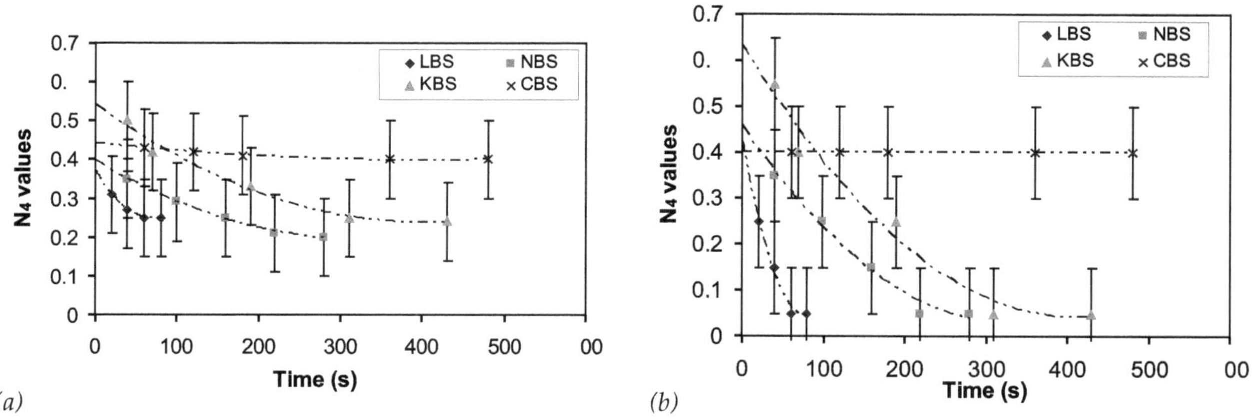

Figure 6. Estimated N_4 values of ABS glasses according to time resolved ELNES from superposition of reference spectra. (a) Estimates based on Peak Area Ratio A/(B+C), (b) Estimates based on Peak Area Ratio B/(A+C)

transformation. For the same reasons as discussed for LBS, we failed to detect any trace of the Na K-edge. However, we observed that the EDX signal for the same focused spot never drops to exact zero unlike EELS, probably due to fluorescent scattering of x-rays from neighbouring regions, which are un-illuminated and still contain enough Na (a typical EDX artefact). The quantitative N_4 estimates and the interpolated curve in Figure 6(a) and (b) again reveal the same curvature and also the same discrepancy for the final minimum N_4 value between criteria I and II, as already described with LBS. The obvious difference between LBS and NBS is the time scale involved in the BO_4 to BO_3 transformation, while initial N_4 values (Table 1) are found to be similar.

(iii) K borosilicate glass

Figure 2(c) is the time resolved EEL spectra in our potassium borosilicate glass. It is obvious that KBS is significantly more stable than either of our LBS and NBS analogue glasses under electron beam irradiation. The quantitative N_4 estimate reveals that the N_4 content after 300 s is clearly non-zero, valid for both criterion I and II unlike for LBS and NBS. Furthermore it was found that the presence of potassium within the illuminated spot can be directly confirmed using potassium L-edge EELS unlike for Li and Na. The change from detection of K-edges to L-edges is not considered to be significant in this aspect. This K L-edge was visible for any irradiation time, confirming that the lateral migration of K ions out of the beam or the sputtering out of the sample must be significantly reduced compared to Li and Na. The significantly larger values of KBS initial N_4 as compared to LBS and NBS still lacks explanation.

(iv) Cs borosilicate glass

Omitting rubidium in our series of alkali borosilicates we conclude with caesium borosilicate as the largest of alkali species. The irradiation time series of Figure 2(d) shows for the first time in our experiments that there is comparatively minor difference between ELNES spectra after 60 s and 360 s. Further confirmed by an easily visible direct Cs-signal by means of the Cs M-edge, we conclude that Cs is unable to migrate or sputter out of the sample at our standard 200 kV irradiation conditions. This finding corresponds to the theory for alkali leaching from glasses via activation energies.[21] Quantified, the fits show a moderate BO_4 to BO_3 transformation and N_4 remains between $N_4 \approx 0{\cdot}5$ and 0·4 depending on different criteria.

After all, one common feature to all four glasses is the effect that peaks B+C lose their separation, which they would have for all possible superposition ratios of the two references (see Figure 4).

Discussion

The main known effects of irradiated ABS glasses, alkali migration and boron unit conversion, have been confirmed and quantified. In addition, we emphasise on three unexpected but clear key results:

1. Within the time series of irradiation, no indication of any boron depletion in the glasses was found, as confirmed by the total edge area of the boron K-edges. This indicates that for 200 kV electrons, the stability of the boron–oxide network is not much inferior to the silicon–oxide network.
2. The boron K-edge fine structure, from which we derive boron coordination, is, before irradiation, similar for all alkali species (not necessarily transferable to the entire boron network structure). This finding possibly supports an entirely "functional" role by valence of the alkali atoms linked as ionic pairs to BO_4 anion units, rather than having a major role by their ionic radii in glass network configuration.
3. The question remains whether EELS is useful for glass chemistry studies "per se", which do not aim at examining irradiation damage. As mentioned in the results section, ultra-short exposure times, correspond-

ing to un-irradiated glass structures (Figure 2(e)) are difficult and of limited reproducibility. Therefore, two alternative road maps seem to be feasible:

Firstly, extend the interpolated series results of Figure 4 (N_4 versus time) for all accumulative spectra backwards before the first measurement point. Such a "zero irradiation" extrapolation seems to be feasible, wherever the curvature of the plots is regular. We are optimistic that the so obtained zero time results conform to the unirradiated glass structure.

Secondly, it seems that the replacement of Li or Na by Cs (or possibly Rb) as a surrogate would solve alone much of the irradiation damage problems, subject to the condition, that the targeted glass chemistry parameters do not depend on quantities like ion radius or diffusion coefficient itself. This concept might be applicable to other glasses beyond borosilicate.

4. The crucial test is finally then of how well our extrapolation data for zero irradiation times do match with NMR data.[17] It can be seen that reasonable agreement was obtained for LBS (criteria II) and KBS (criteria I) glasses. For NBS the N_4 values extrapolated from both criteria are lower than $N_4 \approx 0{\cdot}65$[17] or $N_4 \approx 0{\cdot}55$[22] and the N_4 value for CBS was not available. For XAS only one NBS value was found with similar glass composition,[23] which at $N_4 \approx 0{\cdot}55$ (interpolated published data) is not too far from our EELS results.

Comparing with other irradiation damage studies by EELS, we refer to the collection of boron minerals by Sauer *et al.*[10] The general transformation pattern of boron units, seems to be the same for crystals and for glass. Except, we are not sure whether for the glasses it would be possible to take the spectra of the remaining material after irradiation as the pure BO_3 reference spectrum as suggested for colemanite and howlite,[10] assuming that the depletion of tetrahedral boron is complete. In the EELS studies by Jiang & Silcox[14] for a R=0·5 and K=3·5 NBS glass, it seems that the agreement between their EELS data and the N_4 value obtained by NMR is better.

In another study by Sun *et al*,[16] the irradiation intensity rather than irradiation time was varied (at 200 kV), for which the peak B, characteristic for tetrahedral boron, increased rather than decreased with irradiation dose, an effect which we could not observe for our (different) compositions and lower total doses.

Conclusion

Electron beam irradiation effects on alkali borosilicate glasses were studied using EELS. From the evaluation of boron K-edge in time resolved ELNES spectra, the variation of boron coordination by tetrahedral to trigonal transformation can be clearly observed. The glasses are more stable to electron irradiation as the alkali ion size increases from Li over Na, K to Cs, and tetrahedral boron becomes stable for caesium borosilicate glass even under prolonged electron irradiation. Extrapolation of the N_4 values (coordination ratio) back to zero time is recommended as a method to estimate unirradiated results which are comparable with the reported NMR values for Li and K borosilicate glasses. Agreement for NBS glasses was less well for different NMR references, while no values were available for CBS glasses. The great potential and advantage for EELS/ELNES over XAS/XANES for future work, lies in the excellent spatial resolution on the nm-scale, which promises sensitivity to suspected variations of glass parameters, e.g. in thick/thin areas of a glass foil, or near versus far from precipitates or surfaces or for the examination of nanoscale phase separation phenomena.[24]

References

1. Plodinec, M. J. *Glass Tech.*, 2000, **41** (6), 186–192.
2. Duddrige, A., Holland, D., Dixon, S. M. & Scales, C. R. *Glass Tech.*, 2003, **44** (2), 85–59.
3. Dalba, G., Fornasini, P., Rocca, F. & Monti, F. *Phys. Chem. Glasses*, 2000, **41** (5), 290–295.
4. Dalba, G., Fornasini, P., Grisenti, R., a Beccara, S., Daldosso, N., Sanson, A., Rocca, F., Monti, F. & Kuzmin, A. *Phys. Chem. Glasses*, 2003, **44** (2), 75–78.
5. Ratai, E. M., Janssen, M., Epping, J. D., Chan, J. C. C. & Eckert, H. *Phys. Chem. Glasses*, 2004, **44** (2), 45–53.
6. Kamitsos, E. I. *Phys. Chem. Glasses*, 2004, **44** (2), 79–87.
7. Harami, T., Ogawa, A. & Handa, K. *Glass Tech.*, 2003, **44** (2), 90–93.
8. Kapoutsis, J. A., Kamitsos, E. I., Chryssikos, G. D., Feller, H. A., Lower, N., Affatigato, M. & Feller, S. A. *Phys. Chem. Glasses*, 2000, **41** (5), 321–324.
9. Egerton, R. F. *Electron Energy Loss Spectroscopy in the Electron Microscope*, Plenum Press, New York, 1996.
10. Sauer, H., Brydson, R., Rowley, P. N., Engel, W. & Thomas, J. M. *Ultramicroscopy*, 1993, **49**, 198–209.
11. Manara, A., Antonini, M., Camagni, P. & Gibson, P. N. *Nucl. Instr. Methods Phys. Res. B*, 1984, 1, 475–480.
12. Sun, K., Wang, L. M. & Ewing, R. C. *Microsc. Microanal.*, 2003, **9** (2), 128–129.
13. Jiang, N., Qiu, J., Ellison, A. & Silcox, J. *Phys. Rev. B.*, 2003, (68), 064207.
14. Jiang, N. & Silcox, J. *J. Non-Cryst. Solids*, 2004, **342**, 12–17.
15. Jiang, N. & Silcox, J. *J. Appl. Phys.*, 2002, **92** (5), 2310–2316.
16. Sun, K., Wang, L. M., Ewing, R. C. & Weber, W. J. *Nucl. Instr. Methods Phys. Res. B*, 2004, **218**, 368–374.
17. Stallworth, P. E. & Bray, P. J. in *Glass Science and Technology*, 4B edited by D. R. Uhlmann, N. J. Kreidl, Academic Press, 1990, 77–149.
18. Sipr, O. Simunek, A. Vackar, J. *Phys. Chem. Glasses*, 2006, **47**, 412–418.
19. Nanba, T. & Miura, Y. *Phys. Chem. Glasses*, 2003, **44** (3), 244–248.
20. Möbus, G., Short, R. & Hand, R. J. *Microsc. Microanal.*, 2004, **10** (Suppl 2), 796–798.
21. Ojovan, M. I. & Lee, W. E. J. *Nucl. Mater.*, 2004, **335**, 425–432.
22. Martens, R. & Mueller-Warmuth, W. *J. Non-Cryst. Solids*, 2000, **265**, 167–175.
23. Fleet, M. E. & Muthupari, S. *J. Non-Cryst. Solids*, 1999, **255**, 233–241.
24. Yang, G., Möbus, G. & Hand, R. J. *Micron*, 2006, **37**, 433–441.

Correlation of lead borate glass structure with photoelastic response

J. L. Shaw, U. Werner-Zwanziger & J. W. Zwanziger[1]

Department of Chemistry and Institute for Research in Materials, Dalhousie University, Halifax, NS B3H 4J3, Canada

Lead borate glasses were examined in order to find correlations with their photoelastic response. Both ^{11}B and ^{207}Pb NMR spectra were acquired on several compositions spanning the range from positive to negative stress-optic coefficient. The ^{11}B spectra showed the familiar dependence of N_4 on composition, as had been seen in earlier studies of lead borates; the ^{207}Pb spectra showed a distinct change in coordination number from high to low, with a compositional trend that agrees with that of the stress-optic coefficient better than that of either N_4 or the elastic modulus.

1. Introduction

The element lead stands out as giving the strongest compositional dependence to the stress optic response in glass,[1,2] where the stress optic coefficient is a measure of the optical path length difference in a sample when uniaxial stress is applied. Under ambient conditions this path length difference is zero, because glass is optically isotropic, but the application of stress breaks this symmetry and induces birefringence. The standard glass formers, including boron oxide, exhibit positive stress-optic coefficients, which means that the change in index of refraction due to stress is greater in the stress direction and less when orthogonal to it. Various modifiers reduce the response, but lead is notable in that significant addition of lead oxide reduce the stress-optic coefficient to zero, and further addition actually drives it to negative values.

The technological interest in this effect arises from the desire to fabricate glass pieces that are effectively insensitive to applied stress. This response is achieved for compositions where the stress-optic coefficient is zero or at least very small. For these compositions it is not the case that the glass does not respond to stress at all, but rather, its response in the stress direction and orthogonal to it are identical, yielding zero induced birefringence irrespective of stress direction. As noted, lead oxide is the best (and one of the only) additives known that confers this response, but lead is heavy and expensive, and difficult to recycle. The recycling issue has led the European Union to impose stringent requirements, fully implemented by 2006, on lead in products, intensifying the search for replacements.

Lead oxide is known for other unusual characteristics in glass science. For example, lead oxide forms glasses together with the standard glass formers (oxides of boron, silicon, and phosphorus) over impressively wide composition ranges. Perhaps more importantly in the present context, it is known to undergo a coordination number change from six to eight when it is present in small amounts, to three to four at high concentrations. In other words, it acts as an ionic modifier at low concentrations, and then switches in character to being covalently bound, and hence network former, at high concentrations. Note in both cases that the formal oxidation state remains +2. This switching behaviour is known, but also not understood, in structural inorganic chemistry, where it is known as the inert pair effect.[3] All the heavy main group elements, but most notably thallium, lead and bismuth, are stabilised in low oxidation states by the relativistic contraction of the low lying core electrons. The resulting lone pair ($6s^2$ in the case of lead) may or may not be stereochemically active, meaning that it may or may not act to direct the ligands around it. The former case gives rise to low coordination numbers and relatively covalent bonding; the latter case, to high coordination numbers and ionic bonding. It is an open question as to which factors determine one behaviour or the other in a given situation.

Various theories have been proposed to account for the compositional dependence of the stress-optic coefficient, including separation of the response into ion polarisation and lattice effects,[4] directional polarisation of the oxide ions in different environments and directional polarisation of the modifying ions themselves,[5] and several works that focus on the contributions of different bond types.[6–8] Of particular importance we believe is the work of Kastner, which showed the qualitative difference that results in systems with and without lone electron pairs on the

[1] Corresponding author. E-mail address: jzwanzig@dal.ca

cations[8] Our contention is that insufficient data at the atomic level exist to test these theories at a deeper level than simply correlating various bulk response measurements with composition. Because of the excellent sensitivity of NMR to local electronic and structural changes, we are developing experimental and *ab initio* tools to use NMR as a probe of stress effects in glass. The present contribution details our results on the structures in lead borate glasses, focusing primarily on ^{107}Pb NMR in compositions covering the full range of stress-optic coefficients seen in this system.

2. Experimental and computational methods

Glass samples were prepared by melting PbO and $B(OH)_3$ together in a platinum crucible at 1000°C. After 15 min the progress of reaction and composition were confirmed by weight loss measurements. Further melting for 10–30 min was then carried out and weight loss rechecked, after quenching onto a stainless steel plate. Compositions ranging from 25–75 mol% PbO were prepared and studied in this way.

^{11}B magic angle sample spinning NMR was used to quantify the fraction of four-fold coordinate boron, N_4. Experiments were performed on a Bruker Avance spectrometer operating at 16·44 Tesla (224·69 MHz ^{11}B resonance frequency), with sample spinning at 30 kHz. Spectra were acquired using single pulse excitation, with a 2/sec pulse length and 10 sec recycle time. The probe used gave a broad boron background, and this was subtracted from the spectra shown.

^{207}Pb spectra were acquired on a Bruker DSX spectrometer operating at 9·4 Tesla (83·69 MHz ^{207}Pb resonance frequency) using a wideline probe. The very broad lines were acquired using the Variable Offset Cumulative Spectrum method (VOCS, Fayon *et al*[9]). Spectra were acquired at 500 ppm shifts and then added to cover the full spectral range. This procedure typically required acquiring 10–16 spectra per sample, each with 720 scans and 20 sec recycle times, for a total acquisition of 48 h per sample.

Computational results were obtained through use of the ABINIT code, which is based on pseudopotentials and planewaves.[10] This is a density functional theory approach to finding the energies of periodic solids, in which the valence electron wavefunctions are expanded in terms of planewaves, and the effect of the core electrons are approximated through the use of pseudopotentials. Pseudopotentials of the Teter extended norm conserving type were used,[11] as downloaded from the ABINIT web site. The properties of each system were computed using the local density approximation (LDA) for the exchange and correlation functionals, and reciprocal space was sampled using a 4×4×4 Monkhorst–Pack grid, shifted by 0·5 units. The planewave cutoff energy was 30 Ha. For the models studied, structures were optimised (both unit cell and ion positions) to give residual forces less than 2×10^{-5} Ha/bohr, for a final residual pressure of 2×10^{-5} GPa. The dielectric tensors were computed from the optimised structures by calculating the response of the system to a homogeneous electric field. Both the optimisation and electric field response are standard features of ABINIT and were used without modification.

Two model compounds were considered with *ab initio* calculations. PbO, also known as litharge, is a tetragonal solid (space group P4/nmm[12]). It is a layered compound in which each lead atom is covalently bonded to four oxygen atoms at the base of a square pyramid. The oxygen atoms are coordinated to two lead atoms each; the resulting Pb–O–Pb layers are offset from each other so that the Pb lone pairs, which are stereo chemically active in this compound, avoid each other.

We also studied PbO_2, known as plattnerite, which is a Pb^{IV} compound in which each Pb atom is octahedrally coordinated by oxygen (*P*4(2)/nmn[13]). The octahedra are distorted to give two long axial bonds and four shorter equatorial bonds.

3. Results

Figure 1 shows the ^{11}B magic angle spinning NMR spectra of a series of lead borate glasses, as a function of composition. These spectra were acquired at a relatively high field of 16·44 Tesla (224·59 MHz resonance frequency) so that the second order quadrupolar broadening is almost fully suppressed, and the threefold and fourfold coordinate boron sites are completely resolved. The fractions of four-fold coordinate boron, N_4, extracted from these spectra are listed in Table 1.

Figure 2 shows the ^{207}Pb NMR spectra of the lead borate glasses, acquired with nonspinning samples and the VOCS method outlined above. Note the very wide chemical shift range necessary to capture the full spectra.

The relevant *ab initio* results on the model systems PbO and PbO_2 are the dielectric tensors. As both crystals are tetragonal, both are naturally birefringent. We found for PbO that it is negatively birefringent, with dielectric constant along the *c* axis of 6·43 and in the *a–b* plane of 6·94, while PbO_2 is positively birefringent, with dielectric constants of 8·77 along the *c* axis and 8·66 in the *a–b* plane.

4. Discussion

Our objective is to find if possible structural elements in the glass that, as a function of composition, correlate with the stress-optic response of this system.

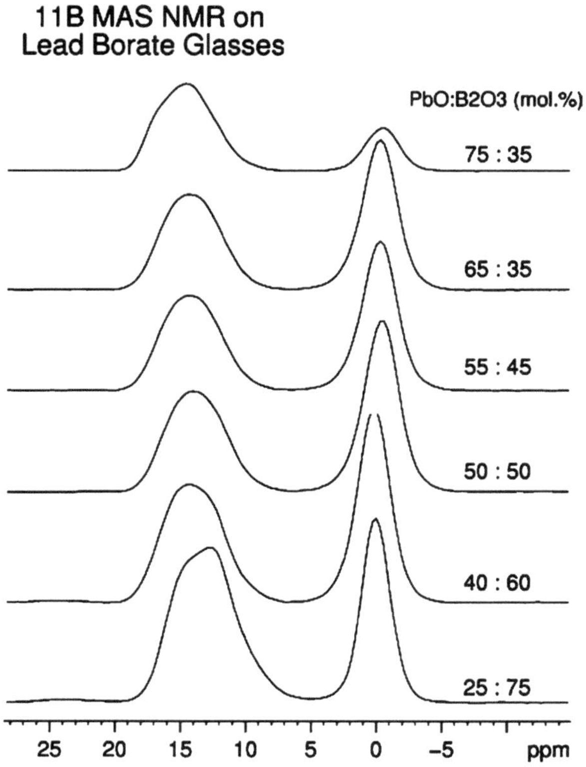

Figure 1. ^{11}*B magic angle spinning NMR spectra of lead borate glasses, taken at 16·44 Tesla (224·59 MHz resonance frequency), spinning at 30 kHz. Referenced to* BF_3 *etherate at 0 ppm. At this field the fourfold coordinate sites at about 0 ppm are completely resolved from the threefold coordinate sites at about 14 ppm*

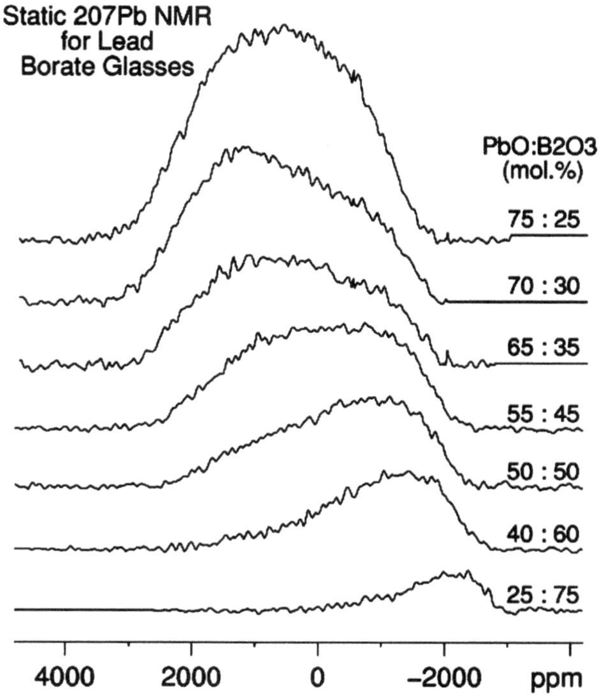

Figure 2. ^{207}*Pb static NMR spectra of lead borate glasses, taken at 9·4 Tesla (83·69 MHz resonance frequency). These spectra were acquired with the Variable Offset Cumulative Spectrum method. Referenced to* $Pb(CH_3)_2$ *at 0 ppm*

Our hope is that such a correlation would provide an important step towards an atomistic understanding of photoelastic response and the special role that lead and related ions might play.

The ^{11}B spectra shown in Figure 1 show the typical resolution into three and four-fold coordinate boron; at the high field used here the sites are completely non-overlapping, since the second order quadrupole interaction is almost fully suppressed. Integrating these sites allows the determination of the fraction of four-coordinate boron, the well-known quantity N_4. Values so obtained from our data are listed in Table 1, and plotted along with earlier results of Kim *et al.*[14] in Figure 3. As had been seen before,[14] N_4 peaks as a function of lead oxide content at about 50 mol%. This is also the composition where the stress-optic coefficient vanishes (Table 1 and Matusita[1]). However, after reaching its maximum, N_4 decreases again (as

Table 1. Compositional trends in lead borate glasses. Column one lists the mol% PbO content; column two gives the fraction of four-coordinate boron as determined in the present study, error estimate ±0·05; column three lists the stress-optic coefficient, in Brewsters, from Matusita et al;[1] *and column four gives the elastic modulus in GPa, from Osaka* et al[17]

PbO fraction	N_4 *(this work; ±0·05)*	*C (Brewster)*[1]	*E (GPa)*[17]
0·25	0·36	0·30	56·51
0·35			62·31
0·40	0·48		64·1
0·45		0·62	61·41
0·50	0·52	0·04	60·32
0·55	0·50		54·16
0·60		−1·28	49·55
0·65	0·41		44·17
0·70	0·34		
0·75	0·23		

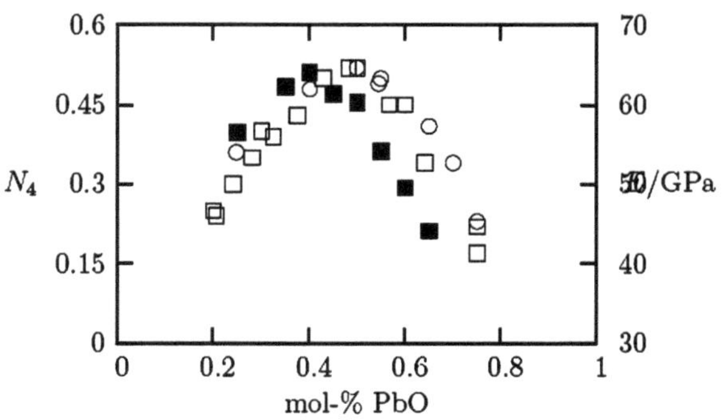

Figure 3. Plot of glass structure, as assessed simply through the fraction of fourcoordinate boron N_4*, and the elastic modulus E, as a function of lead oxide content. Left axis:* N_4*; open circles: data from present study; open squares: data of Kim* et al.[14] *Right axis: E in units of GPa; filled squares, data of Osaka* et al[17]

is always found in modified borates). Therefore, the stress-optic coefficient is not likely to result from the response of four-coordinate boron structures.

The ^{207}Pb NMR spectra (Figure 2) do not show multiple separated resonances, but do show a clear progression from lower shift to higher shift as well as the appearance at intermediate compositions of at least two different species. We found that the spectra could be decomposed satisfactorily into a sum of two gaussians, the parameters for which are listed in Table 2 for three compositions. There the shift progression as PbO is added is quite clear. The shift trend was also observed in earlier work by Bray *et al.*[15] In that study, a field swept spectrometer was used, with derivative mode observation, so it is not apparent from the spectra that multiple sites are present (and were not so assigned in Bray *et al*[15]). To assign the features observed in Figure 2 we use results for ^{207}Pb NMR published in Fayon *et al.*[16] First, the feature at −2226 ppm in the 25:75 composition, based on both its shift and width, is assigned to coordination number eight. This shift fits very well to the correlation found for ionic lead environments.[16] Next, the feature present in all three samples in the −842–−1432 range we assign to coordination number 4–6, where at the most negative shift the majority of sites are probably more ionic with higher coordination, while the less negative shifts would correspond to four coordinate sites with long Pb–O bonds.[16] Finally, the positively shifted sites are assigned to coordination number three, similar to the sites observed in $PbSiO_3$ and Pb_2SiO_4.[16] The down field side of these sites overlap with the shifts for linked PbO_4 units; the spectra are thus consistent with the assignment that at high lead content, there is not only covalent lead–oxygen bonding but also lead clustering, in the form of linked Pb–O–Pb units.

Concerning the correlation with the stress-optic coefficient, we recall that the boron oxide framework, as assessed simply through N_4, does not vary in the same way as the stress-optic response. Neither, it turns out, does the stress response itself, as measured by the elastic modulus E: in Figure 3 we plot in addition to N_4, data on E as reported in Osaka *et al.*[17] This graph shows that E also goes through a maximum as a function of lead oxide content. Had E continually increased, it may have been taken as the marker for the photoelastic response. Of the various features considered here, it is the lead coordination that undergoes a continuous change as a function of composition, in the same way that the stress-optic coefficient does. In this context it is important to note that PbO itself, which has a Pb coordination number of four and covalent Pb–O bonding, is a negatively birefringent crystal, while PbO_2, with a coordination number of six and ionic bonding, is positively birefringent. In PbO, as noted above, the dielectric tensor elements in the *a–b* plane are larger than in the *c* direction, and the crystal is thus more polarisable in that plane; it therefore seems likely that such structures would give larger changes when stressed in plane, and this could give rise to a negative stress response. PbO_2 by the same token could give a negative stress response along the *c* axis but positive in the *a–b* plane; thus this structure might be expected to give a weaker overall contribution to reduction in the stress-optic coefficient than the structure found in PbO. Indeed, these predictions agree with the experimental results, in that the glasses with low PbO content have highly coordinated lead, and show reduced but still positive stress-optic coefficients, while the glasses with high PbO content have lead in low coordination environments and negative stress-optic response.

Table 2. Decompositions of the ^{207}Pb NMR spectra (see also Figure 4). Each spectrum was fit by two gaussians; the table reports their shifts and widths in ppm, the spectral weight of each component, and the assignment to Pb–O coordination number.

PbO fraction	*Shift (ppm)*	*Width (ppm)*	*Spectral fraction*	*Coordination number*
0·25	−2224±4	875±15	0·53±0·03	7–8
	−1400±30	1550±70	0·47±0·03	4–6
0·50	−1220±15	1440±30	0·38±0·03	4–6
	210±60	2390±80	0·62±0·03	3
0·65	−842±15	1328±30	0·23±0·03	4–6
	962±60	963±80	0·77± 0·03	3

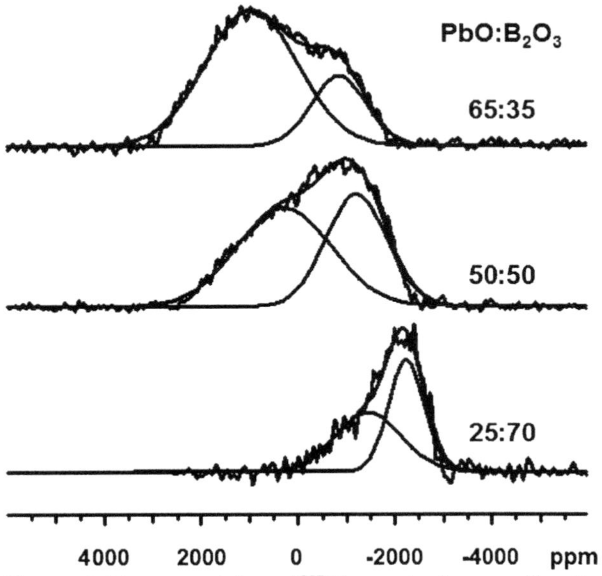

Figure 4. Decomposition of ^{207}Pb spectra by coordination number. Assignments of different shifts to coordination numbers follows the data from Fayon et al.[16] *Due to the overlapping shift ranges, no quantitative fits were attempted*

5. Conclusions

We showed that the N_4 fraction in lead borate glasses

follows the typical trend, and that neither it nor the elastic modulus correlates well with the stress-optic response. The lead coordination, on the other hand, switches smoothly from high coordination at low PbO content to low coordination at high content. Furthermore, we showed by ab initio calculations that PbO is a negatively birefringent material, while PbO_2, which certainly contains Pb^{4+} rather than Pb^{2+}, but does model the high coordination and ionic bonding in lead oxides, is positively birefringent. We argued that both structures could give to reduction in the stress optic response, but the PbO structure much more strongly because it could show negative response for a wider range of stress distortions (the full *a–b* plane, as opposed to just the *c* direction). We tentatively conclude, therefore, that it is not just the lead ion that gives rise to the striking compositional dependence of the stress-optic coefficient in PbO modified glasses, but the special polarisability of lead in a low coordination environment, with a stereo chemically active lone pair. We must caution, however, that this conjecture is dependent on details of the elastic tensor in both the crystal models and the glasses; this dependence will be considered in detail in a future publication.

Acknowledgments

Financial support for this research came from the the Canada Foundation for Innovation, the Canada Research Chairs program, and the Natural Sciences and Engineering Research Council of Canada.

References

1. Matusita, K., Yokota, R., Kimijima, T., Komatsu, T. & Ihara, C. Compositional trends in photoelastic constants of borate glasses. *J. Am. Ceram. Soc.*, 1984, **67**, 261–5.
2. Vedam, K. Elastic and photoelastic properties of some optical glasses. *Proc. Ind. Acad. Sci.*, 1950, **A31**, 450–8.
3. Shimoni-Livny, L., Glusker, J. P. & Bock, C. W. Lone pair functionality in divalent lead compounds. *Inorg. Chem.*, 1998, **37**, 1853–67.
4. Mueller, H. Theory of photoelasticity in amorphous solids. Physics. 1935, 6, 179–84.
5. Tashiro, M. The effects of the polarisation of the constituent ions on the photoelastic birefringence of the glass. *J. Soc. Glass Tech.*, 1956, **40**, 353–62T.
6. Lines, M. Photoelastic trends from halides to pnictides by a bond-orbital method. *J. Appl. Phys.*, 1986, **60**, 1472–8.
7. Weinstein, B. A., Zallen, R., Slade, M. L. & deLozanne, A. Photoelastic trends for amorphous and crystalline solids of differing network dimensionality. *Phys. Rev. B*, 1981, **24**, 4652–65.
8. Kastner, M. Pressure dependence of the refractive index of amorphous lone-pair semiconductors. *Phys. Rev. B*, 1972, **6**, 2273–9.
9. Fayon, F., Landron, C., Sakurai, K., Bessada, C. & Massiot. D. Pb^{2+} environment in lead silicate glasses probed by Pb-L_{III} edge XAFS and ^{207}Pb NMR. *J. Non-Cryst. Solids*, 1999, **243**, 39–44.
10. Gonze, X., Beuken, J.-M., Caracas, R., Detraux, F., Fuchs, M., Rignanese, G.-M., Sindic, L., Verstraete, M., Zerah, G., Jollet, F., Torrent, M., Roy, A., Mikami, M., Ghosez, P., Raty, J.-Y. & Allan. D. C. First-principles computation of material properties: the ABINIT software project. *Comp. Mater. Sci.*, 2002, **25**, 478–92.
11. Teter, M. Additional condition for transferability in pseudopotentials. *Phys. Rev. B*, 1993, **48**, 5031–41.
12. Wychoff, R. W. J. *Crystal Structure*. Interscience, New York, 1978, **Vol. 1**.
13. Waychunas. G. A. Oxide minerals: petralogic and magnetic significance. *Rev. Miner.*, 1991, **25**, 38–46.
14. Kim, K. S. & Bray, P. J. Nuclear magnetic resonance studies of the glasses in the system $PbO–B_2O_3–SiO_2$. *J. Chem. Phys.*, 1976, **64**, 4459–65.
15. Bray, P. J., Leventhal, M. & Hooper, H. O. Nuclear magnetic resonance investigations of the structure of lead borate glasses. *Phys. Chem. Glasses*, 1963, **4** (2), 47–66.
16. Fayon, F., Farnan, I., Bessada, C., Coutures, J., Massiot, D. & Coutures, J. P. Empirical correlations between ^{207}Pb NMR chemical shifts and structure in solids. *J. Am. Chem. Soc.*, 1997, **119**, 6837–43, 1997.
17. Osaka, A., Soga, N. & Kunugi, M. Elastic constants and Vickers hardness of lead borate glasses. *J. Soc. Mater. Sci. Jpn.*, 1974, **23**, 128–31, 1974.

XAFS study on the local order of Pb in PbO borate glasses

G. Dalba, P. Fornasini, R. Grisenti, F. Rocca,[1] N. Afify & M. Affatigato[2]

Dipartimento di Fisica dell'Università di Trento, Italy
[1] *I.F.N. Istituto di Fotonica e Nanotecnologie del Centro Nazionale delle Ricerche sezione CeFSA, Trento, Italy*
[2] *Coe College, Cedar Rapids (Iowa), USA*

An XAFS study on several samples of borate glasses with increasing PbO concentrations was performed in order to detect a possible clustering of Pb atoms or PbO structures. Samples with concentrations from 28·6 up to 75% of PbO were examined at the L_3 edge of Pb. Analysis of data was carried out by experimental comparison with low temperature crystalline references and with the results of theoretical evaluation of amplitude and phases of possible Pb–Pb and Pb–O structures. It was observed that, even at the highest concentration, metallic Pb segregation is incompatible with the obtained XAFS spectra. The most probable structure of Pb neighbours is a tetrahedral replica of the PbO structure (litharge or red PbO). This order weakly extends also to second neighbours, except for the least concentrated sample in which only a distorted first PbO shell was reproduced.

Introduction

The formation of metallic clusters in oxide glasses remains a very important question in the field of glass nano-science. Besides its applications in optoelectronics (quantum dots), clustering is a process that affects a large number of practical properties, from conductivity to thermal expansion and the glass transition temperature. The formation of these metallic groups is also relevant to computer simulations, where the threshold and extent of clustering heavily influences the model results. This is even more true in the case of lead silicate and lead borate glasses and crystals.

One of the authors[(1)] has already investigated the structure of lead containing glasses up to high lead contents using high resolution laser time of flight mass spectrometry (LITOF MS). This technique has provided valuable information about glass structure: several structural and superstructural network units have been identified. Moreover, evidence of a lead substructure beginning to form in the glass with increased lead content has been shown. In particular lead/lead oxide clusters of high masses (>1000 amu) have been observed in lead borate glasses but not in the corresponding crystals. The aim of this work is to use XAFS spectroscopy to study the local order of lead in order to determine what type of local structure is formed as a function of the glass modifier concentration. The open questions involve: (a) whether lead and lead oxide clusters are formed at relatively high (PbO>50 mol%) lead contents; (b) the likely size distribution of the clusters; and (c) if there is a threshold at which the clustering begins. The threshold represents the point at which the lead cations convert from being modifiers to form their own structural units.

Experiment

XAFS measurements were carried out in transmission mode at the L_3 edge of Pb (13·035 KeV) at the beam line BM08 at the European Synchrotron Radiation Facility in Grenoble (France). The line was equipped with a Si (311) monochromator. The energy was 6 GeV with a typical current of 180 mA. The energy step was varied during a single measurement from 2 eV in the pre-edge region, 0·5 eV in the edge region and 1 to 4 eV from 50 eV past the edge up to 14100 eV. Glasses were prepared by conventional rapid cooling, roller quenching techniques. The glasses and the crystals were then finely milled and deposited on Millipore™ filter membranes: these were finally stacked till to obtain a satisfactory x-ray absorption. The measurements on PbO (tetragonal and orthorhombic), Pb_3O_4 as well as Pb metallic foil were performed at 77 K to use as reference spectra. Four samples of lead borate glasses, JPbO.100–$x$$B_2O_3$+3$H_2O$ with J=0·5, 1·2, 2 and 3 (which correspond to 33·3%, 54·5%, 66·6% and 75% PbO concentrations), were measured at room temperature. Two–three spectra were recorded for each composition, in order to achieve enough S/N ratio and accuracy.

Results

An x-ray absorption near edge structure (XANES) analysis on all samples was carried out first. The

[1] Corresponding author

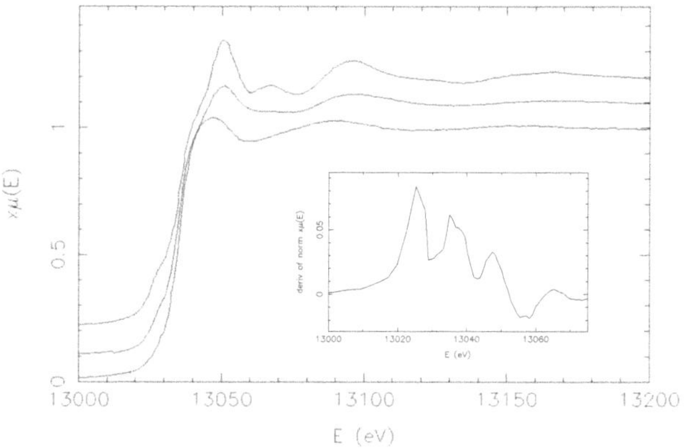

Figure 1. Normalised XANES of the least concentrated sample (33% molar of PbO) and of 66% molar sample with the reference PbO red spectrum (from bottom to top). The curves were upward shifted for sake of clarity. In the inset the derivative of the absorption signal of crystalline sample is reported

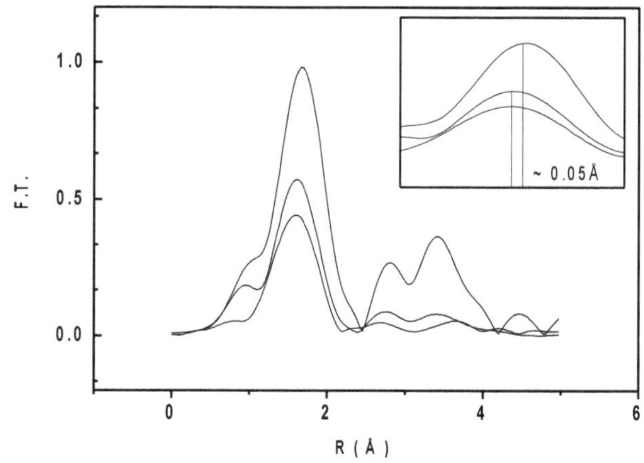

Figure 2. Modulus of the Fourier transform of χ(k) signal of the 33% and 66% molar PbO concentration plus the red PbO reference (from bottom to top). A weight of k^2 and a "Keiser–Bessel" window were applied to the signal before the transform

check on the edge position gave a relative error of 0·5 eV with respect to the Pb L_3 edge, independent of the PbO concentration of the glasses samples. This error is comparable with the step reproducibility of the monochromator. Successively, a range of about 100 eV around the same edge was examined by analysing the derivative of the spectra. It is well known that in orthorhombic (yellow) PbO, two peaks in the derivative of the absorption spectrum aside the L_3 main peak are present: one in the pre-edge region at −8·1 eV and the other one past the edge at 12·5 eV. These features are related to the 2p→6d and 2p→non-bonding d levels transitions respectively.[2,3] They are well reproduced in the XANES spectrum of the yellow PbO sample, which we measured as reference, and in the red PbO sample. In this latter case, the same features appear in the same positions with a minor difference of 1 eV only for the pre-edge peak (see Figure 1). In the tetragonal PbO (red) there are four oxygen atoms at a distance of 2·316 Å, while in the orthorhombic PbO (yellow) two atoms are at 2·21 Å and two at 2·49 Å. This is probably the origin of the small difference in the first peak position. However, in all the glass samples, the three peaks appeared clearly: they do not change the energy position, as does the main edge, with varying the PbO concentration. In the sample with 33·3% PbO molar concentration, the peak at 12·5 eV appeared very weakened. In figure 1, the XANES region for the reference PbO and the four glasses samples is presented. Note that, past the main peak, the detail of the oscillation vanishes as the PbO concentration gets lower.

The EXAFS signal was extracted and analysed using the IFEFFIT[4] software package. The spectra, when compared with the low temperature references, exhibited a marked similarity with the tetragonal PbO (from now on referenced simply as red PbO). This can be seen in Figure 2 where the modulus of the Fourier transform of the signal is reported. In the figure the FT of the $k^2\chi(k)$ of two glasses and the red PbO is shown. The two glasses have 33·3% and 66·6% molar concentration of PbO, respectively. The uppermost curve represents the crystal sample and the others the two glasses with diminishing PbO concentration from top to bottom. The feature past the highest peak originates from the outer shells in the crystal. It is also present for the 66·6% sample, being absent in the 33·3% PbO sample. All the peak magnitudes decrease with reducing quantity of PbO in the glass. This is time also for the intermediate samples, which are not reported in the figure for reasons of clarity. In all samples, the position of the first peak is shifted toward shorter distances with respect to the crystalline sample. In the figure inset, a detail of the displacement is shown together with a measure of its value. This means that the Pb–O distance in the glass is slightly contracted with respect to the crystal.

In order to validate and to quantify the degree of similarity with the red PbO, a further analysis based on FEFF algorithm was carried out later.

The results and the successive fit procedure indicated a good agreement with the previous qualitative analysis. A first shell analysis, obtained by filtering out the first peak in the Fourier transform, evidenced a tetrahedral structure with the Pb atom in the vertex of the pyramid. It was well reproduced by the first shell in the red PbO. An indication about the order behind the first shell was obtained extending the fitting procedure to larger distances. A fit with theoretical amplitude and phases of metallic lead was absolutely incompatible with the experimental signal. Other plausible structures, like that of Pb_2O_3 and massicot (orthorhombic, yellow) PbO, were ruled out because they led to very poor fits.

Conclusions

The present EXAFS measurements were performed on borate glasses obtained by varying the content of PbO evidenced some precise characteristics. The lead oxide that is added to the borate matrix tends to dispose itself in the tetragonal form and, more precisely, it tends to the red PbO structure. This is confirmed by the analysis of the derivative of XANES which reveals the two peaks in the pre- and post-edge region typical of a PbO structure. It is, at the same time, confirmed by EXAFS, both by the experimental comparison with the reference and by the signal fits of the several glass samples with the theoretical EXAFS reconstruction. This last analysis confirms that the first shell is entirely constituted by oxygens in the red PbO structure. Furthermore, the indication that also the second shell starts to follow the same structure of PbO is the second result. More precisely, the higher the concentration of PbO, the higher the number of oxygen atoms that assume the second shell PbO structure. As for the most diluted sample, the litharge PbO structure is confirmed for the nearest oxygen atoms, but no kind of longer range order was indicated by the XAFS technique. In conclusion, it can be stated that for the PbO concentrations examined in the present work, there was no evidence of Pb metallic clustering even at the higher concentrations of PbO, the Pb atoms preferring to retain the in original bonds to oxygen.

References

1. Stentz, D., Blair, S., Goater, C., Feller, S. A. & Affatigato, M. *Appl. Phys. Lett.*, 2000, **76** (1), 61–63.
2. Rao, K. J. & Wong, J. *J. Chem. Phys.*, 1984, **81**, 11.
3. Witkowska, A., Rybicki, J., Trzebiatowski, K., Di Cicco, A. & Minicucci, M. *J. Non-Cryst. Solids*, 2000, **276**, 19–26.
4. Newville, M., Ravel, B., Haskel, D., Rehr, J. J., Stern, E. A. & Yacoby, Y. *Physica B*, 1995, **208&209**, 154–155.

Proc. Fifth Int. Conf. on Borate Glasses, Crystals and Melts *Phys. Chem. Glasses: Eur. J. Glass Sci. Technol. B*, August 2006, **47** (4), 521–523

XAS study of barium aluminoborate glasses

Junko Ide, Kazuhiko Ozutsumi

Department of Applied Chemistry, Faculty of Science and Engineering, Ritsumeikan, University, 1-1-1 Nojihigashi, Kusatsu, Shiga 525-8577, Japan

Katsumi Handa

Synchrotron Radiation Center, Ritsumeikan University, 1-1-1 Nojihigashi Kusatsu, Shiga 525-8577, Japan

Giuseppe Dalba

INFM and Dipartimento di Fisica dell'Universita di Trento, Via Sommarive 14 – 38050 Povo (Trento), Italy

Norikazu Ohtori

Graduate School of Science and Technology, Niigata University, Niigata 950-2181, Japan

Norimasa Umesaki

Japan Synchrotron Radiation Research Institute, SPring-8, Mikazuki, Sayo 679-5198, Japan

Barium aluminoborate glasses with 10–60 mol% BaO content and 1–40 mol% Al_2O_3 content have been investigated by x-ray absorption spectroscopy (XAS) at the Ba L_{III}-edge, B K-edge and Al K-edge. With increasing BaO content, the Ba–O distance, R_{Ba-O}, and the coordination number, N_{Ba}, change from a single coordination shell of 6 oxygen atoms at 2·74 Å to a double shell of 3·5 oxygen atoms at 2·68 Å and 3·0 oxygen atoms at 2·91 Å. The local structure of B and Al atoms changes in the Al_2O_3 high content region of these glasses. With increase of Al_2O_3 content, both local structures around B and Al atoms change from BO_4 to BO_3 units and from AlO_4 to AlO_6 units, respectively.

1.Introduction

Barium borate is an attractive material because of its optical properties and the wide glass forming range.[1,2] It is well known that an addition of Al to glasses extends the glass forming range. However, the effects of Al addition to these glasses are not yet clearly understood. Therefore, we have prepared $y\text{Al}_2\text{O}_3$–(1·0–y/100)[xBaO–(100–x)B_2O_3] glasses by the usual quenching method, and have investigated the local structure of the Ba, B and Al atoms by the XAS method.

2.Experimental

The composition of barium borate (BBO) and barium aluminoborate (Al-BBO) glasses are expressed by the general formula: $y\text{Al}_2\text{O}_3$–(1·0–y/100)[xBaO–(100–x)B_2O_3] with x=10, 15, 20, 25, 30, 35, 40, 45, 50, 55, 60 mol% and y=0, 1 , 5 mol%, and x=50 mol% and y=1, 5, 10, 20, 30, 40 mol%. The glasses were prepared from extra-reagent grade powder of Al_2O_3, $BaCO_3$ and H_3BO_3. Raw materials were melted in platinum crucibles at 900–1650 K for 20 min and quenched by the usual methods.

Fine powders of the glasses were investigated by x-ray absorption experiment at Ba L_{III}-edge, Al K-edge and B K-edge using transmission method at Synchrotron Radiation Center of Ritsumeikan University. Ba L_{III}-edge and Al K-edge were measured by Si(220) and Quartz(1010) at BL-4, and B K-edge were measured at BL-2.[3]

EXAFS spectra were analysed using the standard curve fitting method.

3.Results and discussion

3.1. Ba L_{III}-edge XAS

Figure 1(a) and 1(b) shows the Fourier transform of the EXAFS signals at the Ba L_{III}-edge of Al_2O_3–0·99[xBaO–(100–x)B_2O_3], and $y\text{Al}_2\text{O}_3$–(1·0–y/100)(50BaO–50B_2O_3) glasses.

In the case of Al_2O_3–0·99[xBaO–(100–x)B_2O_3] glasses, the Ba–O peak at about 2·1 Å shifts to shorter distance with an increase of BaO content from 10 to 30 mol%. In the range 30–60 mol% BaO a convoluted shoulder peak appears at about 2·7 Å whose intensity grows up and whose position shifts to longer distance with increasing of BaO content.

On the other hand, in the case of $y\text{Al}_2\text{O}_3$–(1·0–y/100)(50BaO–50B_2O_3) glasses (Figure 1(b)), the

[1] Corresponding author

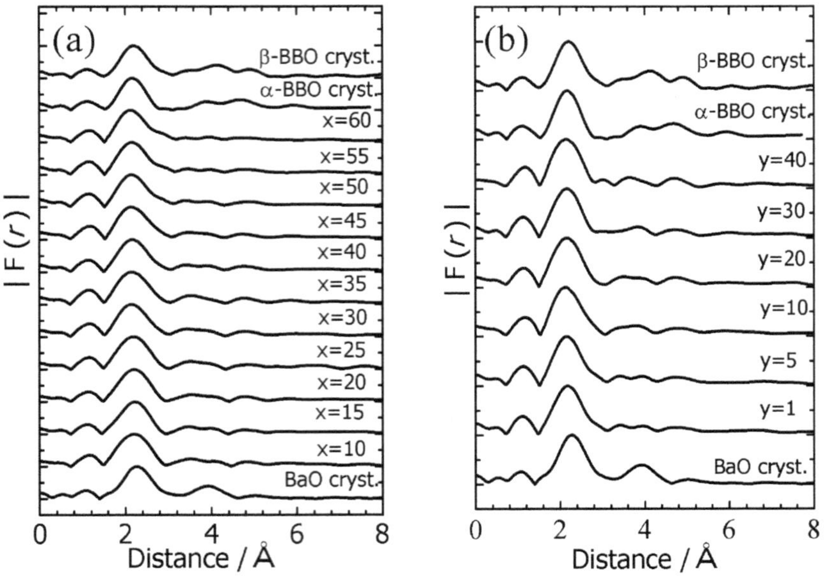

Figure 1. The Fourier transforms (F.T.) of the EXAFS signals: (a) Al_2O_3–0·99[xBaO–(100–x)B_2O_3] {x=10, 15, 20, 25, 30, 35, 40, 45, 50, 55, 60 mol%}; (b) yAl_2O_3–(1·0–y/100)(50BaO–50B_2O_3) {y=1, 5, 10, 20, 30, 40 mol%}

Ba–O peak located at about 2·1 Å. In the range 1–20 mol% Al_2O_3 a convoluted shoulder peak appears at about 2·7 Å.

The coordination number of oxygen atoms around Ba, N_{Ba-O}, and Ba–O interatomic distance, R_{Ba-O}, are summarised in Table 1. In Al_2O_3–0·99[xBaO–(100–x)B_2O_3] glasses and 5Al_2O_3–0·95[xBaO–(100–x)B_2O_3] with 10–30 mol% BaO concentration, N_{Ba-O} is 5·8–6·1 and R_{Ba-O} is 2·74–2·77 Å as obtained for xBaO–(100–x)B_2O_3 glasses. In the case of 35–60 mol% BaO, the first shell is located at 2·68 Å and N_{Ba-O} is 3·4–3·6. The second shell is located at 2·88–2·91 Å and N_{Ba-O} is 2·7–3·0. With increasing the BaO content, the coordination shell of oxygen atoms around Ba atoms separates from a single shell at 2·74–2·77 Å with 5·8–6·1 atoms to double shell at 2·68 Å with 3·4–3·6 atoms and 2·88–2·91 Å with 2·7–3·0 atoms. While in $y$$Al_2O_3$–(1·0–$y$/100)(50BaO–50$B_2O_3$) glasses, the coordination shell of oxygens around Ba atoms is double shell in the range of 1–20 mol% Al_2O_3, the first shell is located at 2·68–2·69 Å and N_{Ba-O} is 3·5–3·7 and the second shell is located at 2·89–2·91 Å and N_{Ba-O} is 2·8–2·9, and it is single shell in the range of 30–40 mol% Al_2O_3, N_{Ba-O} is 4·1–4·3 and R_{Ba-O} is 2·71–2·72 Å. The change around Ba atoms in BBO and Al-BBO glasses corresponds to the change from BaO-like structure to BBO single crystal like structure. Therefore, it is considered that oxygen atoms in the second shell surrounding Ba atoms are used for the structural changes of barium aluminoborate glasses.

3.2.B K-edge XAS

It is well known that B atoms are surrounded by three oxygen atoms in vitreous B_2O_3 glass and a sharp

Table 1. Structural parameters for barium alminoborate glasses obtained from EXAFS analysis

xBaO–(100–x)B_2O_3			Al_2O_3–0·99[xBaO–(100–x)B_2O_3]			5Al_2O_3–0·95[xBaO–(100–x)B_2O_3]			yAl_2O_3–(1–y/100)[50BaO–50B_2O_3]		
x (mol%)	N (atoms)	R (Å)	x (mol%)	N (atoms)	R (Å)	x (mol%)	N (atoms)	R (Å)	y (mol%)	N (atoms)	R (Å)
10	6·1	2·77	10	6·0	2·76	10	6·0	2·76	1	3·5	2·68
15	-	-	15	6·0	2·76	15	6·0	2·76		2·9	2·90
20	5·9	2·76	20	5·8	2·75	20	6·0	2·75	5	3·7	2·69
25	-	-	25	6·1	2·75	25	6·0	2·74		2·9	2·91
30	5·7	2·74	30	6·1	2·74	30	3·5	2·69	10	3·7	2·68
							2·7	2·88		2·9	2·90
35	3·4	2·68	35	3·5	2·68	35	3·5	2·68	20	3·7	2·68
	2·8	2·89		2·8	2·89		2·8	2·89		2·8	2·89
40	3·6	2·68	40	3·6	2·69	40	3·5	2·68	30	4·1	2·71
	3·0	2·89		2·7	2·89		2·8	2·88	40	4·3	2·72
45	3·4	2·67	45	3·4	2·68	45	3·4	2·68			
	3·2	2·88		2·9	2·90		2·9	2·90			
50	3·7	2·68	50	3·5	2·68	50	3·6	2·68			
	3·1	2·90		3·0	2·90		2·9	2·90			
55	3·6	2·67	55	3·4	2·68	55	3·6	2·68			
	3·2	2·90		2·9	2·90		3·0	2·91			
60	3·8	2·68	60	3·6	2·68	60	3·5	2·68			
	3·1	2·90		2·9	2·91		3·0	2·90			
65	3·9	2·69									
	3·4	2·92									

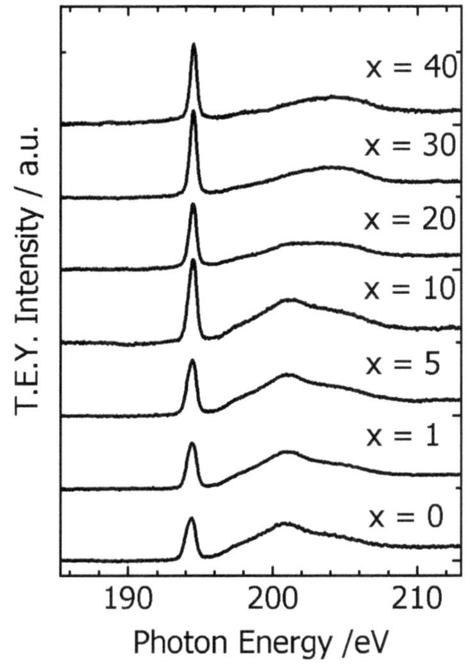

Figure 2. B K-edge XAS spectra of xAl_2O_3–(1·0–x/100)(50BaO–50B_2O_3) glasses

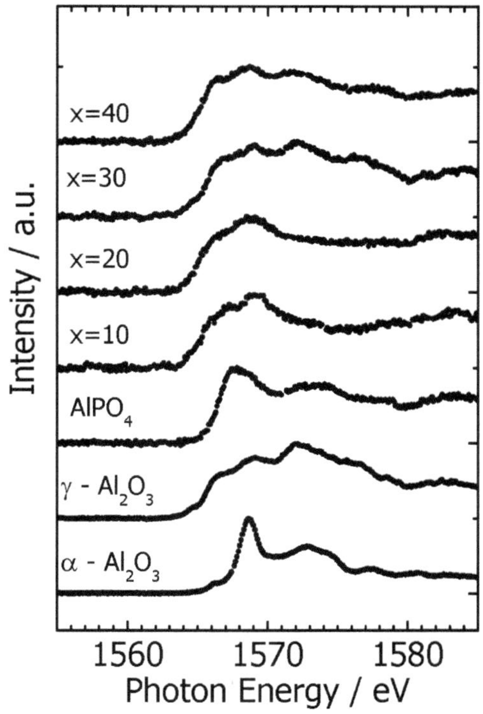

Figure 3. Al K-edge XANES spectra of xAl_2O_3–(1·0–x/100)(50BaO–50B_2O_3) glasses

resonance peak related to the transition from B 1s to $p\pi^*$, appears in B K-edge XANES spectra of pure B_2O_3 glass and crystal.[4] Figure 2 shows B K-edge XAS spectra of yAl_2O_3–(1·0–y/100)(50BaO–50B_2O_3) glasses. The main peak at 194 eV shifts to lower energy side and its shape becomes similar to one of vitreous B_2O_3 with increasing the Al_2O_3 content. Therefore, the addition of Al_2O_3 to BBO glasses modifies the local structure around B atom changing from 4-coordinated B to 3-coordinated B. The linkage of Al–O bonding is much stronger than B–O bonding, therefore it is supposed that Al atoms take oxygen atoms from BO_4 unit in order to change from AlO_4 to AlO_6 units. Unfortunately, Ba $N_{II,III}$-edges are located at 178·6 eV and 192 eV, respectively and so the relative XAS spectra overlap the B K-edge spectrum making difficult to analyse the Ba environment.

3.3.Al K-edge XAS

Figure 3 shows Al K-edge XANES spectra of yAl_2O_3–(1·0–y/100)(50BaO–50B_2O_3) glasses. The change in the shape of spectra indicates that the structure around Al atoms changes from AlO_4 unit to AlO_6 unit of γ-type alumina with an increase of Al_2O_3 content. This results agree with the results of B K-edge XANES study of Al-BBO glasses.

4.Conclusions

We conclude that in yAl_2O_3–(1·0–y/100)[xBaO–(100–x)B_2O_3] glasses with x=10, 15, 20, 25, 30, 35, 40, 45, 50, 55, 60 mol% and y=0, 1 , 5 mol%, and x=50 mol% and y=1, 5, 10, 20, 30, 40 mol%, the coordination numbers of the Ba atom, $N_{Ba\text{-}O}$, and the Ba–O distances, $R_{Ba\text{-}O}$, change with increase in BaO content, especially for glasses containing a high BaO content, where the coordination shell of oxygen atoms surrounding Ba atoms separate into a double shell. In x=10–30 mol% glasses, $N_{Ba\text{-}O}$ is 5·8–6·1 and $R_{Ba\text{-}O}$ is 2·74–2·77 Å, while in x=35–60 mol% glasses $N_{Ba\text{-}O}$ is 3·4–3·6 and $R_{Ba\text{-}O}$ is 2·68 Å in the inner shell and $N_{Ba\text{-}O}$ is 2·7–3·0 and $R_{Ba\text{-}O}$ is 2·88–2·91 Å in the outer shell. On the other hand, yAl_2O_3–(1·0–y/100)(50BaO–50B_2O_3) glasses with y=1, 5, 10, 20, 30, 40 mol%, $N_{Ba\text{-}O}$ changes from 3·7 and 2·8 to 4·1, and $R_{Ba\text{-}O}$ changes from 2·68 Å and 2·89 Å to 2·71 Å between y=20 and 30 mol%. The structural modifications are associated with the change of the coordination number of B and Al atoms from 4 to 3 and 4 to 6, oxygen atoms, respectively. Therefore, it is considered that oxygen atoms in the outer shell surrounding Ba atoms are involved in the structural changes occurring in barium aluminoborate glasses with changing composition.

5. References

1. da Rocha, M. S. F., Pontuschka, W. M. & Blak, A. R. *J. Non-Cryst. Solids*, 2003, **321**, 29–36.
2. Pernice, P., Esposito, S. *et al. J. Non-Cryst. Solids*, 1999, **258**, 1–10.
3. Ozutsumi, K. & Handa, K. *Rev. Sci. Instrum.*, 2004, **75**, 1.
4. Fleet, M. E. & Muthupari, S. *J. Non-Cryst. Solids*, 1999, **255**, 233–241.

Proc. Fifth Int. Conf. on Borate Glasses, Crystals and Melts Phys. Chem. Glasses: Eur. J. Glass Sci. Technol. B, August 2006, 47 (4), 524–529

Influence of the cation nature on some dilatometric glass properties in the systems $RO–Al_2O_3–B_2O_3$ (R=Mg, Ca, Sr, Ba)

V. P. Klyuev, B. Z. Pevzner** & I. G. Polyakova**

**Institute of Silicate Chemistry of the Russian Academy of Science, St.-Petersburg, Russia*
***Laboratory of Glass Properties, St.-Petersburg, Russia*

Generalisations are made from numerous data obtained by investigating systematically such dilatometric glass properties as the glass transition temperature T_g, *the thermal expansion coefficient TEC, and the structural thermal expansion coefficient STEC for the systems* $RO–B_2O_3$ *and* $RO–Al_2O_3–B_2O_3$ *(R=Mg, Ca, Sr, Ba) in wide concentration areas covering the glass formation regions of the systems under study. It has been shown that although concentration dependencies of the properties look complicated in these systems, for example, they have extrema and saddle points; they resemble one another in these systems with different alkaline earth cations. This similarity in the dependence of a considered property on composition indicates similar anion network changes taking place in the systems. In terms of the results of direct structural investigations it allows evaluation of some structural rearrangements occurring in them. Since glass anion networks are alike in these systems with different cations, the intensity of a given property is determined by the nature of a given cation.*

Introduction

It is well known that such "simple" properties as density, refractive index, molar volume, and thermal expansion in borate glasses exhibit distinctly non-additive dependencies on a given composition, the dependencies showing extremes in many cases. This is due to the changes in the coordination state of some of the boron atoms caused by introducing an appropriate oxide- modifier. Addition of aluminium to borate glasses leads to further complications of the composition-property dependences and consequent structural interpretations.

The paper represents the results of a comparative systematic study of dilatometric properties for more than 300 alkaline earth aluminoborate glasses, which makes it possible to draw some conclusions about how structure changes with composition. The data on the glass properties of the systems $BaO–Al_2O_3–B_2O_3$ and $CaO–Al_2O_3–B_2O_3$ investigated before[1–5] were complemented by new results for glasses of the systems $SrO–Al_2O_3–B_2O_3$ and $MgO–Al_2O_3–B_2O_3$.

Systems under study, methods and measured properties

The required glasses were melted in accordance with the conventional technique[1] in platinum crucibles. The regions of homogeneous glass formation in the systems explored are expanded by increasing the cation atomic order number, i.e. from Mg to Ba. They are restricted from one side, i.e. the triangle vertex B_2O_3, by liquid phase separation gaps, and from the other ones related to the vertices RO and Al_2O_3 correspondingly, by increasing melt crystallisation upon cooling.

The following properties were determined with a quartz dilatometer: the glass transition temperature T_g, the glass thermal expansion coefficient (TEC) in the region 20–300°C and denoted by α_{20-300}, and the structural TEC (STEC) denoted by α_s, the technique of determining the last property being described in Ref. 5. STEC is the difference between the TEC above the glass transition region and that below it (α_g, measured in the temperature interval $\Delta_{T<T_g}$). The standard errors of determination do not exceed 2% for TEC (α_{20-300}), 5% for STEC (α_s), and 3 K for T_g. Our data on TEC and T_g are in reasonable agreement with results of another investigators of $RO–B_2O_3$ glasses[6] and $RO–Al_2O_3–B_2O_3$ glasses.[7,8] All figures in this paper are based on the numerical data of the Tables 1 and 2 (SrO- and MgO-containing glasses) and of the tables in Refs 1, 2, 5 (BaO- and CaO-containing glasses).

Results and discussion

Structure–property relationship in appropriate glass binary systems $RO–B_2O_3$

Figure 1 represents the properties of the glass binary borate systems and the fraction of $BO_{4/2}$ (N_4) groupings in them depending on RO content, the data

[1] Corresponding author

Table 1. Compositions and properties of glasses in the systems $SrO–B_2O_3$ and $SrO–Al_2O_3–B_2O_3$

Al_2O_3, mol%	SrO, mol%	SrO/B_2O_3	$\alpha_{20\text{-}300}$ $\times10^7$, K^{-1}	$\Delta_{T>T_g}$, °C	$\alpha_g\times10^7$, K^{-1}	$\alpha_s\times10^7$, K^{-1}	T_g, °C
0	20	0·25	62·2	430–480	76	350	569
0	22·5	0·29	60·8	450–500	80	355	588
0	25	0·33	61·1	470–520	86	426	597
0	27·5	0·38	64·2	450–500	86	511	605
0	30	0·43	68·2	460–510	88	606	614
0	33·3	0·50	71·5	470–520	91	654	630
0	36	0·56	75·4	500–550	92	786	630
0	39	0·64	80·5	500–550	100	757	630
0	41	0·69	84·0	480–530	102	822	627
0	43	0·75	88·0	—	--	--	--
2·5	19·5	0·25	61·2	380–430	72	439	559
5	19	0·25	61·8	380–430	75	488	554
10	18	0·25	59·7	360–410	74	670	553
15	17	0·25	56·7	370–420	69	908	563
20	16	0·25	54·6	430–480	69	939	586
25	15	0·25	51·2	430–480	66	850	616
5	23·8	0·33	62·4	450–500	72	396	590
10	22·5	0·33	60·1	400–450	71	480	579
15	21·3	0·33	58·4	400–450	72	614	577
20	20	0·33	58·3	410–460	70	678	586
25	18·8	0·33	54	450–500	69	608	606
5	28·6	0·43	65·9	460–510	81	534	604
10	27·1	0·43	63·4	410–460	78	512	591
15	25·6	0·43	61·5	410–460	75	536	587
20	24·1	0·43	59·5	410–460	71	543	591
25	22·6	0·43	55·8	410–460	64	593	600
5	31·6	0·50	69·9	420–470	83	595	605
10	30	0·50	67·2	400–450	81	579	594
15	28·3	0·50	64·8	400–450	78	550	586
20	26·6	0·50	62·1	410–460	74	485	587
25	25	0·50	58	420–470	70	509	594
27·5	24·1	0·50	56·8	415–465	68	493	611
5	38	0·67	78·2	435–485	91·6	687	604
10	36	0·67	75·5	420–470 490–540	90	583	595
15	34	0·67	72·7	400–450	84·6	554	588
20	32	0·67	68·4	415–465	83	493	582
25	30	0·67	66·5	420–470	75·4	371	591
30	28	0·67	58·4	410–460	71·2	294	608
5	47·5	1·0	93·1	400–450	113	542	582
10	45	1·0	91·8	430–480	114	616	578
15	42·5	1·0	86·5	430–480	104	522	574
20	40	1·0	81·8	410–460	94	433	573
25	37·5	1·0	74·6	400–450	82	382	578

being taken from Refs 9, 8 and 11. The data of Ref. 10 were corrected on account of a small content SiO_2 in some barium borate glasses. As for strontium borate glasses, different authors represent of different data on N_4.[11,12] In Ref. 12 the values on N_4 for strontium borate glasses practically coincide with those for barium borate glasses. As regards appropriate magnesium borate glasses, whose glass formation region is very narrow, N_4 is estimated to lie near 0·2.[13]

Vitreous B_2O_3 is known to be made up of BO_3 triangles integrated in boroxol rings thus introducing two-dimensional character into the structure. The valence bond via the bridge oxygen $B_3–O–B_3$ is strong in such triangles and rings, as well as between all the rings inside the layer, but weak between the layers, the latter belonging to Van der Waals forces. It is this weak bond that is responsible for the low glass transition temperature as well as for the high TEC and STEC of vitreous boron oxide.

On introducing RO, an oxide of alkaline earth metal, the structure of borate glass undergoes changes, and can be represented as a row of structural chemical groupings (SCG)

$$\{2[BO_{3/2}]^0 + RO\} \rightarrow [BO^-_{4/2}]_2R^{2+} \rightarrow [BO_{2/2}O^-]^2R^{2+} \qquad (1)$$

Going in parallel to this development, changes in the properties under varying composition is due to the following two factors. The first is changing the anion network strength; the second is changing the ion bond between R^{2+}, an alkaline earth cation, and the network. The process of changing the properties includes a number of stages.[3,5] The composition region from boron oxide to 20–25 mol% RO exhibits formation of $BO_{4/2}$ tetraheda; such structures combine with $BO_{3/2}$ triangles, "sewing together" the weakly connected boron layers into the three-dimensional anion network, the linkages being formed by the bonds $B_3–O–B_4$ via the oxygen bridge atom. As the connectivity increases, T_g becomes higher, but TEC and STEC decrease. Increasing the content of RO

Table 2. Compositions and properties of glasses in the system $MgO–Al_2O_3–B_2O_3$

Al_2O_3, mol%	MgO, mol%	MgO/B_2O_3	α_{20-300} $\times10^7$, K^{-1}	$\Delta_{T>T_g}$, °C	$\alpha_g\times10^7$, K^{-1}	$\alpha_s\times10^7$, K^{-1}	T_g, °C
15·0	21·3	0·33	48	360–410	62	1189	601
17·5	20·6	0·33	47·2	440–490	60	1124	615
20·0	20·0	0·33	48·7	400–450	58	1152	620
22·5	19·4	0·33	47·6	450–500	61	1139	636
25·0	18·8	0·33	50·1	450–500	63	1001	647
5	31·7	0·5	47·2	450–500	70	850	626
7·5	30·8	0·5	52·7	460–510	64	831	633
10	30	0·5	49·4	460–510	69	832	634
12·5	29·2	0·5	51·3	465–515	66	823	632
15	28·3	0·5	47·2	465–515	62	833	633
17·5	27·5	0·5	50·1	470–520	65	838	640
20	26·7	0·5	50·3	470–520	64	833	644
22·5	25·8	0·5	50·7	470–520	60	817	647
25	25	0·5	50·8	475–525	61	769	656
27·5	24·2	0·5	46·6	480–530	61	726	667
2·5	39	0·67	56·1	470–520	67	798	643
5	38	0·67	59·0	480–530	70	762	638
10	36	0·67	55·6	480–530	69	734	638
15	34	0·67	52·0	480–530	68	699	637
20	32	0·67	50·9	500–550	67	688	644
25	30	0·67	50·1	510–560	67	638	658
27·5	29	0·67	49·8	510–555	62	630	662
28·6	28·6	0·67	44·7	520–570	64	550	670
7·5	46·2	1·0	61·1	450–500	74	621	632
10	45	1·0	59·6	460–510	75	589	631
12·5	43·8	1·0	59·2	475–525	74	571	634
15	42·5	1·0	57·4	480–530	74	580	637
17·5	41·3	1·0	56·4	490–540	69	540	640
20	40	1·0	55·3	490–540	66	542	647
22·5	38·8	1·0	54·3	490–540	69	504	652
25	37·5	1·0	53·6	490–540	67	489	656
27·5	36·2	1·0	52·3	--	--	--	--

further (up to 30–37 mol%), leads to an increase in the structure of such tetrahedra of $BO_{4/2}$ that are connected with one another by their vertices via the bridge oxygen; the link B_4–O–B_4 existing between the vertices is weaker than the link B_3–O–B_4.[14] The connectivity still increases, but the bonds are weaker. This leads to a slowing down in the rise of the glass transition temperature. Though the anion network connectivity keeps increasing, this composition region begins to exhibit an enhancing of TEC and STEC. This is due to the fact that if there are no weakly connected structure elements, i.e. those fastened together by Van der Waals forces, the structural expansion is determined by the thermal cation oscillations and continues as the total cation content increases, as the strength of bond between R^{2+} and the corresponding borate groupings decreases. At the same time the TEC is lower, the smaller the cation radius and, therefore, the higher the electric field strength of the cation and the stronger is its bond with the glass anion network. In case the content of RO exceeds, depending on a given system, the respective value from 30–37%, the tetrahedra content decreases which leads to forming $BO_{2/2}O^-$ triangles. Such a triangle has the nonbridging oxygen O^-; the bond O^-–R^{2+}, which has become the weakest by this time, is formed via this atom. The network connectivity decreases, and having passed the maximum, the value of T_g starts decreasing.

Thus, in case of sequence,[1] the glass exhibits weakening of the bond between R^{2+} and the borate grouping, the structure gets rid of its two-dimensional character, and instead of this feature three-dimensional connectivity, appears and disappears. The connectivity begins to be destroyed when the content of a given RO has exceeded the corresponding value from 30–37%. The manner of changing a property by varying the composition is due to transforming the anion fragment constituents that are important for this property. The value of each feature is determined by the nature of the given cation modifier. When moving along the row Ba–Sr–Ca, the cation radius and its coordination number with respect to oxygen decrease, but the cation field strength and the energy of its bond with oxygen increase. In the same order, one can observe T_g increasing and TEC decreasing.

The factors influencing the STEC are more difficult to determine. The both two- and three-dimensional structural character increases the STEC, but the influence of pressure of the two-dimensional character is several times as large in intensity. The explanation of the STEC structural sensitivity should be done in terms of a specific mechanism of the glass transition. In our papers, for example Ref. 2, the emphasis is placed on the principal factors responsible for increasing STEC, namely: the exhibiting of layer character of the structure and the transformation in the glass transition region of some four-fold boron atoms in to three-fold ones, the transformation being

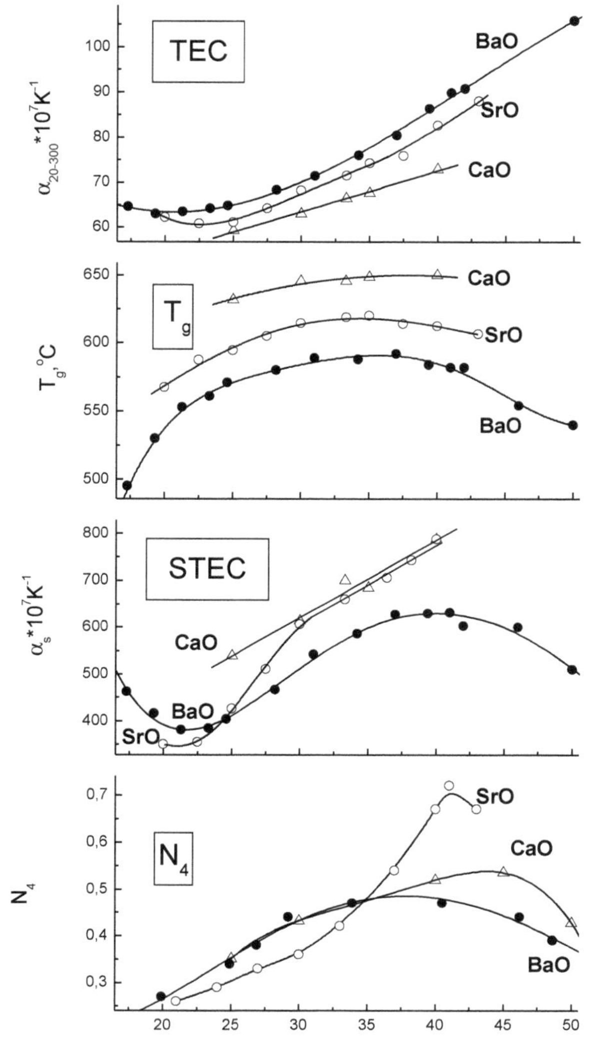

Figure 1. Properties obtained by dilatometry way and, proportion of N_4 for binary glasses in $RO–B_2O_3$ systems. For TEC and Tg error limits are within the symbols. The errors for the STEC do not exceed ±5%. N_4 is plotted on data;[(6–10)] the errors do not exceed ±5–10%. The lines are drawn as a guide for the eye

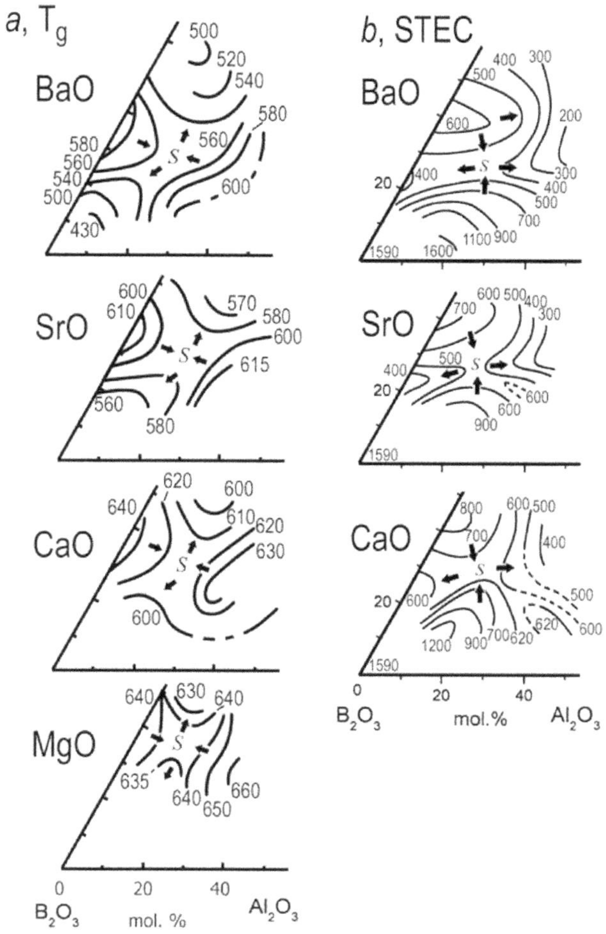

Figure 2. Isolines of T_g (a) and STEC (b) for glasses in systems $RO–Al_2O_3–B_2O_3$

accompanied by increasing molar volume (see, for example, Ref. 15). Thus, four-fold boron prevalence in the structure of glasses in the region of the boron anomaly provides many structural elements which are able to the described transformation at the T_g region, and assures a high intensity of the STEC value. The extent of these factors influencing STEC may be different for different cations and cannot yet be estimated reliably.

Patterns of changing the properties with varying the composition in ternary systems $RO–Al_2O_3–B_2O_3$

The concentration triangles of Figure 2(a) and (b), demonstrate T_g and STEC isolines for the systems studied. The STEC triangle for the MgO-containing system was not constructed because of too sharp a diminution of the property with MgO and Al_2O_3 content increases: in this case, even 32 investigated compositions are not enough for constructing a few isolines.

The shapes of T_g and STEC surfaces are rather complicated, but for different cations resemble one another for each of the properties. The common distinctive feature of the both T_g and STEC surfaces is the presence of a saddle point (denoted by S on Figure 2) in one and the same concentration region where the ratio Al_2O_3:RO is close to 0·6. The left sides of triangles correspond to the binary systems described above. The arrows across the isolines show the directions for reducing T_g and STEC. In the region of the maximal T_g for binary systems, adding Al_2O_3 is seen to diminish the glass transition temperature, which is evidence of destruction of the existing three-dimensional borate network. Aluminium, while competing with boron for the oxygen brought by an appropriate cation-modifier, forms its own structural-chemical groupings $[AlO^-_{4/2}]_2R^{2+}$. With all this going on, the amount of $BO_{4/2}$ is reduced as fast as

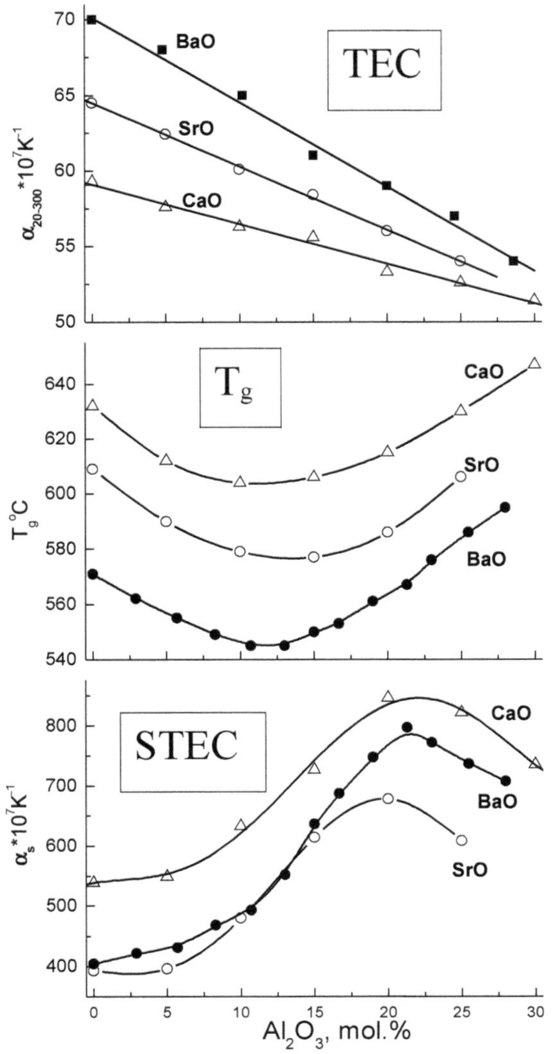

Figure 3. Influence of Al_2O_3 on the dilatometric properties of glasses for the systems $RO–Al_2O_3–B_2O_3$ with the ratio RO/B_2O_3 being equal to 1:3. For TEC and T_g *errors limits are within the symbols. The errors for the STEC do not exceed ±5%. The lines are drawn as a guide for the eye*

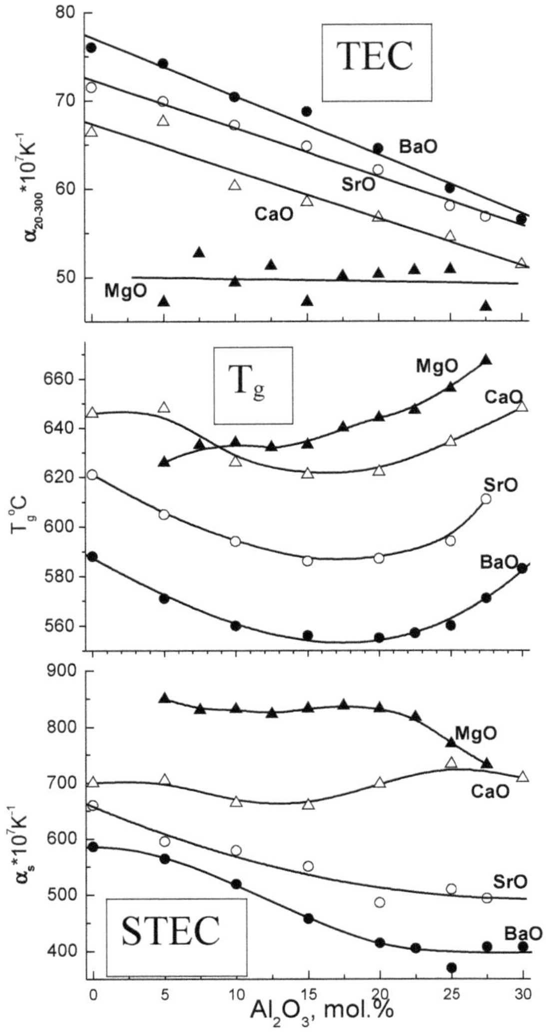

Figure 4. Influence of Al_2O_3 on the dilatometric properties of glasses from the systems $RO–Al_2O_3–B_2O_3$ the ratio RO/B_2O_3 being equal to 1:2. For TEC and T_g, *errors limits are within the symbols. The errors for the STEC do not exceed ±5%. The lines are drawn as a guide for the eye*

that of $BO_{3/2}$ increases. In the vicinity of S, the saddle point of surfaces $RO–Al_2O_3–B_2O_3$, T_g does not depend significantly on the composition, and if the content of Al_2O_3 keeps increasing, the parameter begins to increase. In this region, there develops a process of forming a new structure which seems to also have a three-dimensionally connected anion framework. Resemblances between such surfaces, as well as STEC and TEC ones, in several appropriate systems confirms the fact that their anion networks develop similarly under changing the composition.

The effect of the described structural changes on the properties of appropriate glasses, depends on both the ratio $RO:B_2O_3$ and the amount of added Al_2O_3. The influence of cation nature and the interpretation of network structure development, can be considered with the help of separate cuts indicated in Figure 2: 1:3 (the cut starts from the region of minimal TEC and STEC in appropriate binary borate systems) and 1:2 (it starts from that of maximal T_g in such systems).

The properties of glasses in the cuts with the ratio Al_2O_3:RO equal to 1:3 and 1:2

The curves are presented in Figures 3 and 4 as functions of Al_2O_3 content. Comparison of these cuts makes it possible to find out that in appropriate systems with different cations Al_2O_3 has a similar influence on TEC and T_g, but may influence the STEC differently. For all the discussed systems except the magnesium containing ones, TEC decreases smoothly by introducing Al_2O_3, and the function of T_g has the minimum if Al_2O_3:RO is close to 0·6, i.e. in the region of the saddle point S. If considering the cut with the ratio $RO:B_2O_3$ equal to 1:2, addition of Al_2O_3 to magnesium borate glasses changes TEC negligibly, whereas T_g increases smoothly. Among the cations under discussion, magnesium, which has the smallest radius and highest field strenght, is capable of forming $MgO_{4/2}$ tetrahedra and acts, to some extent, as a network former.[(13)]

The dependence of STEC on the content of Al_2O_3 shows a distinct maximum in the cut with the ratio 1:3, if Al_2O_3:RO is within the interval of 1·1–1·2. In the cut with the ratio 1:2, the dependencies look smoother. Increasing STEC in the cross section RO:B_2O_3=1:3 (Figure 3) is due to the fact that the content of B_2O_3 in the cross section is high enough to exhibit a two-dimensional character in the structure. There are no distinct extremes of STEC in the cross section 1:2 (Figure 4), because the B_2O_3 content is insufficient in the glass to exhibit two-dimensional character of the structure. Some other factors may be important because of higher sensitivity to the nature of an appropriate cation.

In case of the cross section RO:B_2O_3=1:1 (marked in Figure 2), introduction of aluminium oxide leads to smooth decreases in TEC, creating low minima in T_g, and exhibiting the tendency of STEC to decrease, all of this being evidence of the absence of any fragments of two-dimensional borate structures.

The difference in the influence of RO manifests itself in appropriate ternary systems in a sequence similar to that in binary ones: as the cation radius gets lower from Ba to Mg, TEC decreases and T_g increases. Since each of these metal oxides influences STEC in a particular way, it is still impossible to find any distinctive patterns of relationship. Nevertheless it is possible to say that in some cases the tendency is observed of STEC increasing when ionic radius, R^{2+}, is decreased (Figure 4).

Conclusion

The manner of changing glass properties in appropriate binary and ternary systems supports the fact that their anion networks develop in similar ways with changes in composition, while the intensity of changing a property is determined by the nature of a cation. The structural sensitivity of the properties enhances along the row TEC–T_g–STEC. As the radius of a cation modifier diminishes from Ba to Mg, the structural strength is enhanced, which is responsible for decreasing TEC and increasing T_g along this line, in both binary and ternary appropriate systems. No complete regularity has been found in the influence of the nature of an appropriate cation modifier on the value of STEC.

References

1. Klyuev, V. P. & Pevzner, B. Z. *Glass Phys. Chem.*,1998, **24**, 372.
2. Klyuev, V. P. & Pevzner, B. Z. *Phys. Chem. Glasses*, 2000, **41**, 380.
3. Klyuev, V. P. & Pevzner, B. Z. *Phys. Chem. Glasses*, 2001, **42**, 79.
4. Klyuev, V. P. & Pevzner, B. Z. *Glass Phys. Chem.*,2001, **27**, 54.
5. Klyuev, V. P. & Pevzner, B. Z. *Phys. Chem. Glasses*, 2003, **44**, 121.
6. Shermer., H. F. *J. Res. Nat. Bur. Stand.*, 1956, **56** (2), 73.
7. Naruse, A., Abe, Y. & Takami, A. *J. Ceram. Soc. Jpn.*, 1971, **79**, 225.
8. Brow, R. K. & Tallant, D. R. *J. Non-Cryst. Solids*,1997, **222**, 396
9. Bishop, S. G. & Bray, P. J. *Phys. Chem. Glasses*, 1966, **7**, 73.
10. Greenblatt, S. & Bray, P. J. *Phys. Chem. Glasses*, 1967, **8**, 190.
11. Park, M. J. & Bray, P. J. *Phys. Chem. Glasses*, 1972, **13**, 50.
12. Yiannopoulos, Y. D., Chryssikos, G. D. & Kamitsos, E. I. *Phys. Chem. Glasses*, 2001, **42**, 164.
13. Dell, W. J. & Bray, P. J. *Phys. Chem. Glasses*, 1982, **23**, 98.
14. Wright, A. C., Sinclair, R. N., Crimley, D. I., *et al. Glass Phys. Chem.*, 1996, **22**, 268.
15. Nagel, S. R. *Crystallization kinetics in binary borate glass melts*. University of Illinois at Urbana-Champagn. 1973. 253 p.

The B_2O_3/SrO ratio influence on the structural and magnetic properties of the $CuO–B_2O_3–SrO$ glasses

Raluca Ciceo Lucacel & I. Ardelean*

Faculty of Physics, Babes Bolyai University, Cluj Napoca, 400084, Romania

Glasses from the $xCuO.(100–x)[yB_2O_3.SrO]$ *system with* $0≤x≤30$ *mol% and* y=1, 2 *were prepared and studied by spectroscopic (FT-IR, Raman, EPR) and magnetic susceptibility measurements. Both, FT-IR and Raman spectroscopy have revealed changes in the glasses structure when the composition was modified. The FT-IR data revealed the presence of tri-, tetra-, pyro- and penta-borate groups in structures of both series of glasses. Supplementary, ortho-borate groups were detected for* y=1*. For the same x value, highest values of* BO_4/BO_3 *ratio were observed for samples with* y=2*. Raman data confirm the FTIR results and complete the information about these glasses, by detecting chain type metaborate groups. The EPR spectra show the tendency of the* Cu^{2+} *ions to insert in these two glass matrix networks in axially distorted octahedral environments. Both EPR and magnetic measurements point out the simultaneous presence, for* $x≤10$ *mol%, of* Cu^{2+} *and* Cu^{+} *species. The Curie behaviour of magnetic susceptibility indicates the presence of* Cu^{2+} *ions isolated and/or dipolar coupled.*

1. Introduction

Since B_2O_3 based glasses present interesting structural particularities, they have been widely studied over the years in order to clarify the role of the boron oxide component in determining the glasses properties.[1–3] The presence in these glasses of transition metal ions confers it different physical properties.[4] In the present work we study two series of glasses from the $xCuO.(100–x)[yB_2O_3.SrO]$ system. These complex glasses were studied in order to find out the influence of y on both BO_4/BO_3 ratio and copper ion behaviour at different doping levels (knowing the fact that copper ions can be found in the structural sites of vitreous network modifiers[5]).

Having different selection rules, FT-IR and Raman spectroscopies prove to be powerful complementary methods in elucidating glass structure. EPR investigations of Cu^{2+} ions that are present as a spin probe in oxide glasses have received considerable attention, the EPR parameter being sensitive to the local symmetry, the character of the chemical bonds, as well as to other factors.[6] Information about the valence states and magnetical interactions involving the copper ions in vitreous systems had been also obtained by means of magnetic measurements.[7]

2. Experimental

Two series of samples from $xCuO.(100–x)[yB_2O_3.SrO]$ system with y=1, 2 were prepared using pure reagent grade compounds, i.e. H_3BO_3, $SrCO_3$ and CuO in appropriate ratios. The preparation method is described in Ref. 8. The structure of the samples was tested by means of x-ray diffraction and no crystalline phase was detected up to 30 mol% CuO. The FT-IR absorption spectra were recorded with an Equinox 55 Bruker spectrometer in the manner described in Ref. 8. The Raman spectra have been recorded using a GDM 1000 double monochromator equipped with an ILA 120-1 argon ion laser in a 90° geometry at room temperature. We could obtain Raman spectra only for $x≤3$ mol% because they were recorded in the visible spectral range and the sample green colour became darker with increasing values of x. EPR measurements were performed using an Adani PS8400 spectrometer in the X frequency band at room temperature. Equal quantities of samples were studied. The magnetic measurements were performed using an Oxford Instruments, MagLab2000 balance in the 10–250 K temperature range.

3. Results and discussion

3.1 FT-IR and Raman data

The FT-IR spectra for samples, with x=0 and 30 mol% for both series, are presented in Figure 1. The absorption bands detected in the FT-IR spectra and their assignments are summarised in Table 1. The results are discussed on the basis of the method given by Tarte,[9] by comparing the experimental data of glasses with those of related crystalline compounds. The characteristic absorption lines for the vitreous B_2O_3[10–13] and crystalline CuO and SrO[10] were used as a reference point in the discussion of the results.

[1] Corresponding author. Email rluca@phys.ubbcluj.ro

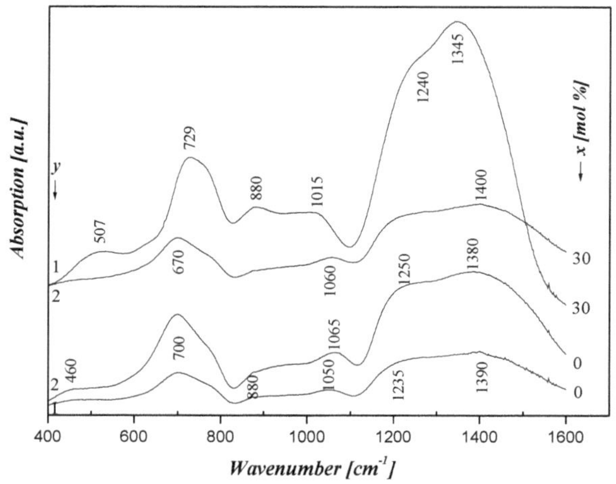

Figure 1. FT-IR spectra of glasses with x=0 *and* x=30 *mol%,* y=1 *and 2*

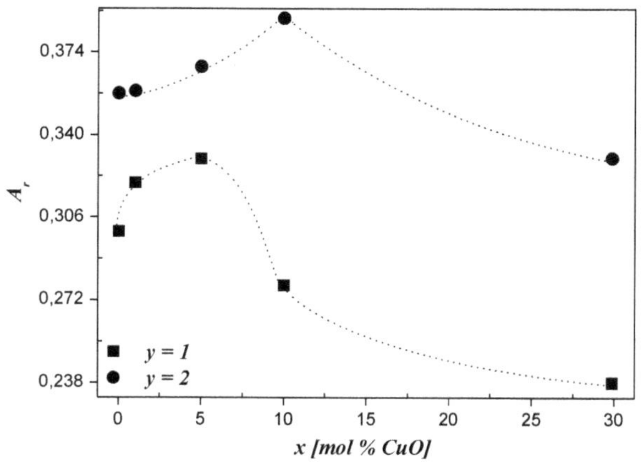

Figure 2. Relative integrated absorption $Ar=A_4/A_3$ *as an* x *function*

We propose a structure mainly built by tri- ($B_3O_5^-$), tetra-($B_8O_{13}^{2-}$), penta-($B_5O_8^-$), pyro-($B_2O_5^-$) and ortho-borate (BO_3^{3-}) groups for glasses with y=1 and tri-, tetra-, penta- and ortho-borate groups[(7–9)] for glasses with y=2 (Ø: oxygen atom bridging two boron atoms and O^-: nonbridging oxygen atom). We choose to discuss the structural changes involved by the CuO content addition on the basis of $Ar=A_4/A_3$ ratio (A_4 and A_3 were calculated as the integral of the absorption signal in the 800–1150 cm^{-1} (A_4) and 1150–1550 cm^{-1} (A_3) spectral ranges).[(13)] The A_4 and A_3 reflect the relative population of tetrahedral ($BØ_4^-$) and triangular ($BØ_3$ and $BØ_2O^-$) borate species, respectively.

The *Ar* ratio versus CuO content is given in Figure 2. For all the glasses investigated, the predominance of BO_3 units is revealed. The *Ar* values are higher in case of y=2 than for y=1. Moreover, the ascending trend in the *Ar* ratio in the first part of the compositional range stops at a lower value of x, for y=1. Until x=5 mol% for y=1 and x=10 mol% for y=2, the *Ar* ratio increased, indicating the progressive change in boron coordination number from three to four. Beyond these concentrations Ar values decrease, which means that the BO_4 units are gradually destroyed. The glasse structure contains a higher number of nonbridging oxygen atoms linked, very possible, with the Cu atoms. Copper ion additions (for $1 \le x \le 10$ mol%) in both glassy matrices, involved increasing intensity of all FT-IR absorption bands, revealing a numerical increasing of the detected structural units. The addition of strontium atoms to the B_2O_3 glass results in a higher degree of depolymerisation of the glass network. Introducing CuO in these glassy matrices seems to favour for compositions ≤10 mol% the polymerisation process of the network as a result of the presence of more oxygen atoms. In both series of glasses, for x>10 mol%, the tendency of absorption bands to increase intensity ceases. At high CuO contents, the glasses with y=1 present a structure with a higher degree of polymerisation than those with y=2 (Figure 1). Probably, at these values of CuO content, a greater oxygen atom deficit appears in samples with y=2, so that the boron atoms have not enough oxygen atom in those neighbourhoods to form some structural units. Compared with the results given in paper,[(13)] our studies indicate that the copper ions presence in yB_2O_3.SrO glass matrix favours the obtaining of homogeneous glasses, and the *Ar* ratio changes.

Table 1. FT-IR frequencies and their assignments

IR Wavenumber [cm^{-1}] *y=1*	*IR y=2*	*Raman y=1*	*Raman y=2*	*IR assignments*	*Raman assignments*
~507	~465	~450	~470	B–O–B bending vibration	B–O–B bending vibration
~700	~705	~720	~710	$O_3B–O–BO_3$ bending vibration	chain type metaborate groups
~883	~880	~770	~770	B–Ø bonds stretching vibration of $BØ_4^-$ tetrahedra from tri-, tetra- and pentaborate groups	symmetric breathing vibration of six member ring with a $BØ_3$ triangle replaced by a $BØ_4^-$ tetrahedra (tri-, tetra- and pentaborate groups)
~1050	~1066				
~1235		~980		asymmetric stretching vibration of B–Ø and / or B–O^- bonds in borate triungular units ($BØ_3$ and $BØ_2O^-$) from pyro- and orthoborate groups	orthoborate triangles
		~880	~880		pentaborate groups
	~1250		~1050	asymmetric stretching vibration of B–O bonds from orthoborate groups	diborate groups
~1390	~1380	~1340	~1350	B–O bonds stretching vibration of BO_3 units from various borate groups	stretching vibration of terminal B–O- bonds from various borate groups

* for details and references see text

The Raman spectra are noisy and seem to contain an envelope of several distinct signals (Figure 3). However, Raman spectroscopy partially confirms the glass structures suggested by the FT-IR results. Supplementary structural groups were detected in glasses from both series, e.g. chain type metaborate ($B_3O_6^{3-}$).[12] As an interesting fact, no boroxol rings were detected in the glass structures by Raman spectroscopy. A comparative view between Raman spectra of these two series of glasses has revealed that ortho-borate groups[8,12,13] were detected only for samples with y=1, while diborate groups ($B_4O_7^{2-}$)[12,13] were evident only in glasses with y=2. Also, in case of y=2, the band from ~770 cm^{-1} is a little better defined than in case of y=1, signifying a greater tendency of glasses with y=2 to include $BØ_4^-$ units in their structure. Raman spectra suggest that the addition of the copper ions in glasses of both series causes the break of the B–O–B links and an increasing number of nonbridging oxygen atoms. Therefore, progressive copper ion additions imply an increasing degree of vitreous network depolymerisation.

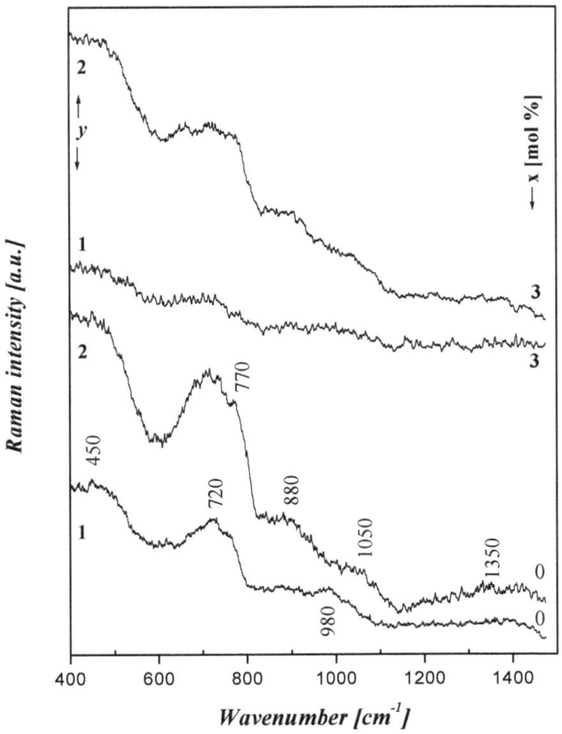

Figure 3. Raman spectra of glasses with x=0 *and* x=3 *mol%,* y=1 *and* 2

3.2. EPR and magnetic data

EPR spectra of Cu^{2+} ions in the glasses investigated are given in Figure 4, showing the evolution of the absorption line with the CuO content and the changes implied by the y ratio (Figure 4). The EPR spectra are asymmetric, as is characteristic for Cu^{2+} ions in axially distorted octahedral symmetric sites. The spectrum preserves its asymmetric character within the entire concentration range, suggesting a great structural stability of the vitreous matrix when accepting Cu^{2+} ions in units having more or less the same symmetry and configuration. Within the low concentration range (x≤5 mol% for y=1 and x≤10 mol% for y=2), the spectra show hyperfine structure (hfs) due to the interaction of the unpaired electron spin with the nuclear spin, I=3/2, of the Cu^{2+} ions. The hfs is well resolved in the parallel band of the spectra. For series with y=2 the compositional range in which the hfs is resolved is larger than in the case of y=1. So that it is obvious that an increasing number of boron atoms confers a higher grade of isolation for Cu^{2+} ions in the glass matrix structure. As the concentration of Cu^{2+} ions increases, the resolution diminishes due to the individual lines broadening. The hfs smears out, so for samples with high Cu^{2+} ion contents the line reduces to a broad envelope of all contributions to the EPR absorption. The estimated values of $g_{\parallel}$ and $g_{\perp}$ satisfy the relationship $g_{\parallel}>g_{\perp}>g_e$=2·0023 characteristic to Cu^{2+} ions coordinated with six ligand atoms in a distorted octahedron, elongated along one axis. The value of the g factors and hyperfine constants reveal the predominantly ionic character of bonding between Cu^{2+} and the ligand atoms.[4]

The intensity (I) of the absorption lines have the concentration dependence given in Figure 5. For x>5 mol%, I does not follow linearly the doping level of the sample, suggesting for both series of samples, the presence of Cu^+ ions, which do not contribute to the EPR absorption. These new ions balance the Cu^{2+} species so that, the addition of CuO changes the slope of the I=f(x) graph. The higher values obtained for I

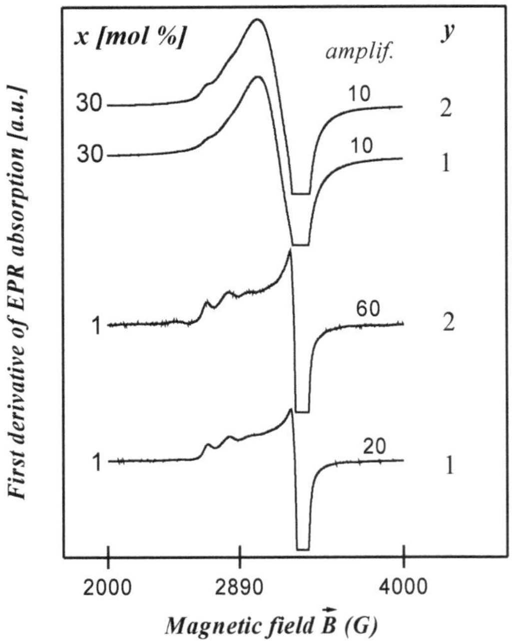

Figure 4. Evolution of hfs in parallel band for x=1 *mol% and* x=30 *mol%,* y=1 *and* 2

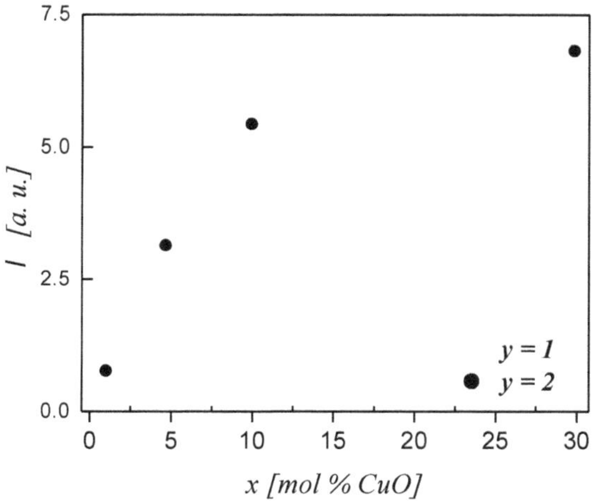

Figure 5. The EPR line intensity versus x for y=1 and y=2

parameter in case of series with y=2, indicates, that a greater quantity of boron atoms favours the existence of Cu^{2+} ions in the melting process.

Variation with the temperature of the reciprocal magnetic susceptibility for some glasses from system with y=2 is given in Figure 6. The Curie behaviour of χ indicates the presence in the glasses studied of isolated and/or dipolar coupled Cu^{2+} ions. Similar behaviour was detected for glasses with y=1. Values of the molar Curie constant (C_M) and effective magnetic moment (μ_{eff}) were determined (Table 2). The μeff values were calculated using the $\mu_{eff}=2{\cdot}827(C_M/x)^{1/2}$ relation, considering that all copper ions introduced in the batch contribute to these values. For x≥10 mol% the μ_{eff} values are lower than the magnetic moments of the Cu^{2+} in the free ion state ($\mu_{Cu^{2+}}$), indicating the presence of Cu+ ions in these glasses. Using the $\mu_{Cu^{2+}}=2{\cdot}827(C_M/w)^{1/2}$ relation, we estimate the molar fraction of the Cu^{2+} ions (w) that contribute to the magnetic moment (Table 2). The values obtained are higher in case of y=2, being in agreement with the EPR data (Figure 5).

4. Conclusions

The FT-IR data revealed the presence of BO_3 and BO_4 units in different borate groups, for all glasses investigated. Pyroborate groups were detected only for y=1. For all the glasses studied, the number of three coordinated boron atoms ($BØ_3$ and $BØ_2O^-$) is higher than that tetra coordinated ones. Moreover,

Table 2. Molar Curie constant, effective magnetic moment and molar fraction of $Cu^{2+}O$

CuO [mol%]	C_M [emu/mol] ×10^3 [±0·001]		μ_{eff} [μB]		w [mol%]	
	y=1	y=2	y=1	y=2	y=1	y=2
5	18·72	18·72	1·73	1·73	5	5
10	34·00	35·12	1·65	1·68	9·1	9·4
15	24·75	39·23	1·36	1·45	9·3	10·5
20	35·12	38·11	1·08	1·23	9·4	10·2
30	34·94	39·98	0·96	1·03	9·35	10·7

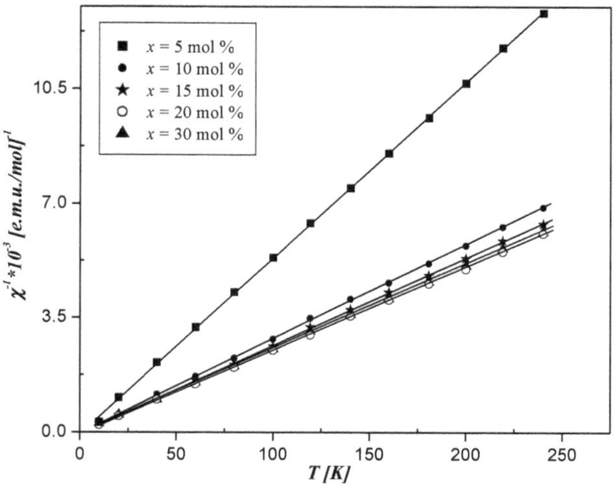

Figure 6. Temperature dependences of the reciprocal magnetic susceptibility for xCuO.(100–x)[2B_2O_3.SrO] glasses

the Ar values for y=2 are higher than those with y=1, showing the greater predisposition of these glasses to incorporate the boron atoms in BO_4 units. Raman spectroscopy detects new structural groups for both series of samples (chain type metaborate groups) and diborate groups only for series with y=2, this obtaining a better description of the glass structure.

The EPR spectra revealed the placement of Cu^{2+} ions in axially distorted octahedral sites. The compositional range, in which the hfs is resolved, increases with the modification of y from 1 to 2. This means that the structure of glasses with y=2 incorporates a higher number of isolated Cu^{2+} ions.

Both, EPR and magnetic data have indicated that the Cu^{2+} ions are either isolated or involved in dipolar interactions. For samples with x≥10 mol%, the simultaneous presence of Cu^{2+} and Cu^+ ions was indicated. The ratio of these ions depends on both x and y values. For glasses with same x, a higher content of Cu^{2+} ions was obtained for y=2.

References

1. Warren, B. E. *J. Am. Ceram. Soc.*, 1941, **24**, 256
2. Kreidl, N. J. *J. Non-Cryst. Solids*, 1990, **123**, 377
3. Yano, T., Kunimine, N., Shibata, S. & Yamane, M. *J. Non-Cryst. Solids*, 2003, **321**, 147
4. Griscom, D. L. *Mater. Sci. Res. Borate Glasses*, 1978, Eds. L. D. Pye, V. D. Frechette & N. J. Kreidl, Plenum, New York, P. 36.
5. Maniu, D., Ardelean, I. & Iliescu, T. Studia Univ. Babes-Bolyai, *Physica*, 1993, **2**, 13
6. Siegel, I. & Lorence, J. A. *J. Chem. Phys.*, 1966, **45**, 2315.
7. Ardelean, I., Cozar, O. & Ilonca, Gh. *Solid State Commun.*, 1984, **50**, 87
8. Ciceo Lucacel, R. & Ardelean, I. *Phys. Chem. Glasses*, 2004, **45** (4), 295–9
9. Tarte, P. *Physics of Non Crystalline Solids*, 1964, Ed. I. A. Prins, Elsevier, Amsterdam, P.549.
10. Bentley, F., Smithson, L. D. & Rozek, A. L. *Infrared Spectra and Characteristic Frequencies 700-300 cm^{-1}*, 1986, Interscience, New York, P. 103.
11. Varsamis, C. P., Kamitsos, E. I. & Chryssikos, G. D. *Solid State Ionics*, 2000, **136–137**, 1031.
12. Kamitsos, E. I., Karakassides, M. A. & Chryssikos,G. D. *J. Phys. Chem.*, 1987, **91**, 1073.
13. Yiannopoulos, Y. D., Chryssikos, G. D. & Kamitsos, E. I. *Phys. Chem. Glasses*, 2001, **42** (3), 164–72.

Peculiar properties of barium diborate polycrystals

B. Z. Pevzner, V. P. Klyuev, I. G. Polyakova

Institute of Silicate Chemistry of the Russian Academy of Science, St.-Petersburg, Russia

V. F. Borodzyulya

St.-Petersburg State Polytechnic University, St.-Petersburg, Russia

By crystallisation of stoichiometric composition glass two different crystal forms of $BaO.2B_2O_3$, the well known α-modification and a new β-one, can be obtained under the same temperature-time conditions depending on the pressure of dissolved gases and the state of the glass surface. The result depends on the presence or absence of dissolved gases in the glass and the state of its surface. Thermal expansion, density, and electric properties have been investigated in the range of 20–700°C for the glass and its crystal modifications. α-$BaO.2B_2O_3$, the known crystal modification, exhibits some peculiar properties. Thermal expansion anisotropy in the textured compact samples is so significant that the samples suffer contraction (the linear thermal expansion coefficient, TEC, is -49×10^{-7} K^{-1}) on heating, in a plane perpendicular to the crystal growth direction. The isotropic polycrystalline samples show dilatometric hysteresis, the range for the TEC lying from $+110\times10^{-7}$ to -210×10^{-7} K^{-1}. By proper heat treatment, compact isotropic samples of α-$BaO.2B_2O_3$ can be prepared with predetermined TEC, down to a uniform contraction under heating of -34×10^{-7} K^{-1}.

Introduction

Borate systems are characterised by a great variety of crystal phases and modifications. Different crystal compounds may be produced at the same glass depending on synthesis conditions and heat treatment.(1) Congruently melting stoichiometric compounds are of particular interest, because they melt without decomposition and the produced liquid and corresponding glass are similar in structure to the melted crystal. The data comparing the properties of stoichiometric-composition glasses with those of stoichiometric-composition crystals are rare, e.g. available and are not studied systematically. This type of information is necessary for developing glass-ceramic materials formed from appropriate glasses.

Crystallisation of barium borate melts has been studied before.(2–4) Reference 4 describes a few features of crystallisation of barium borates, their thermal expansion as well as electrical their resistance. The structure of known crystalline $BaO.2B_2O_3$ has been elucidated in Ref. 5, the substance being crystallised in the monoclinic system. Below it will be named as α-$BaO.2B_2O_3$ or the α-phase.

The melts were synthesised in a platinum crucible at 1200–1250°C; the initial substances were highly purified $BaCO_3$ and H_3BO_3. The samples were prepared for measurement by crystallising the balk glass and sintering the compacted glass powder. To obtain compact crystal samples the dissolved gases must be removed from the melt. The required procedure, which includes crystallising the glass and melting the crystals obtained, was performed twice for removing the gases. Reference 6 describes the conditions of obtaining various crystal modifications of barium diborate.

The phase transformations were investigated by differential thermal analysis (DTA). The compositions of the phases were determined by x-ray powder diffractometry (XRPD). Densities at room temperature were measured by hydrostatic weighing. The dilatometric properties were investigated with a quartz dilatometer; the rate of both heating or cooling was 3 K/min. The electric properties were explored with an immittance meter E7-14. Structures of the compact polycrystalline samples were examined by optical and electronic microscopy.

Results and discussion

Crystallisation

A bulk glass of the diborate composition crystallises from the surface: under heat treatment above the glass transition temperature the textured layers of α-phase grow inside the glass. If the initial glass have been carefully cleared of dissolved gases, and has a undisturbed surface to prevent α-$BaO.2B_2O_3$ nucleation, under the same treatment it crystallises by volume nucleation and the growth of a new phase, β-

[1] Corresponding author

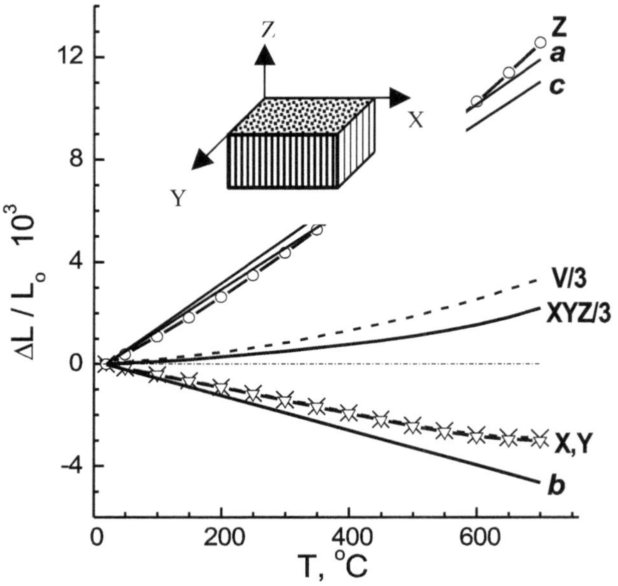

Figure 1. Dilatometric thermal expansion of the textured sample of α-BaO.2B$_2$O$_3$ in the direction of the crust growth (Z) and in a plane parallel to the outer surface of the crystallised bulk glass (X, Y); errors limits are within the symbols. On the dilatometric curves each tenth point is specified only, lines are received by computer spline. XYZ/3 *is the calculated effective value of relative linear thermal expansion. Thermal expansion of α-BaO·2B$_2$O$_3$ along the crystallographic axes (*a, b, c*) and the effective linear expansion (*V/3*) are calculated from the high temperature XRPD data of Ref. 7*

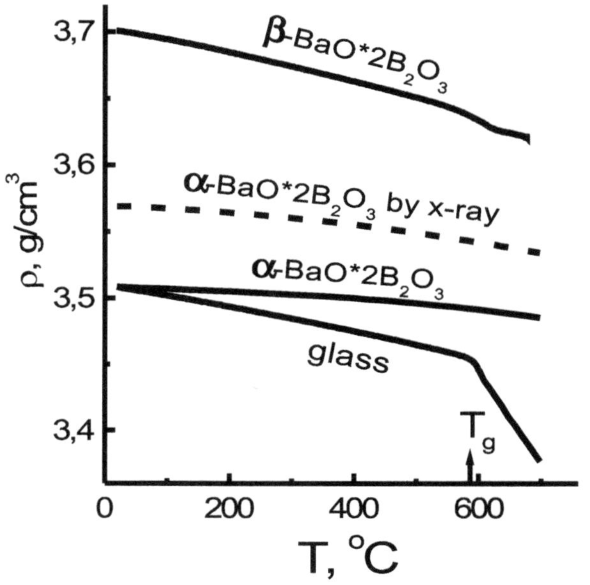

Figure 2. Thermal dependences of density for barium diborate glass, α-BaO.2B$_2$O$_3$, and β-BaO.2B$_2$O$_3$ obtained by dilatometry (the solid lines) and XRPD (the dash line). Error of measurement of density at 20°C is –0·005 g/cm^3

BaO.2B$_2$O$_3$. The x-ray data for β-phase are presented in Ref. 6, its melting temperature being 805·5°C. While heating, β-phase crystals with a disturbed surface, the phase partly transforms to the α-one, until the disturbed layer is exhausted. The transformation β→α is irreversible.[(6)]

Two kinds of compact samples of α-BaO.2B$_2$O$_3$ were made: textured and isotropic. While crystallising the bulk glass at 700°C for 2 h, the average size of individual crystallites in the obtained textured polycrystalline layer is 50–100 μm. Compact isotropic samples were produced by sintering and crystallising the pressed fine glass powder under the same heat treatment.

Thermal expansion and density

Figure 1 compares dilatometric thermal expansions of the textured α-BaO.2B$_2$O$_3$ with the x-ray data. Under heating the textured sample exhibits expansion (TEC, the thermal expansion coefficient, is positive) in the direction *Z*, while the directions *X* and *Y* are practically equivalent and show contraction (TEC is negative, -49×10^{-7}, K^{-1}). The effective value of relative thermal expansion (line *XYZ*/3 in Figure 1) is calculated as the arithmetic mean of the sum of relative thermal expansions along the axes. The results of high temperature XRPD for ground sample[(7)] are in correspond to the crystallographic axes for α-BaO.2B$_2$O$_3$. As indicated in the figure, the unit cell expands under heating along the axes *a* and *c*, but contracts along *b*. It follows that in the textured sample crystallites are oriented along the direction *Z* with the crystallographic axes *a* and *c*, and as a result the expansion occurs along *Z*. The axis b of the crystallites is oriented along directions *X* and *Y* and as a result the contraction occurs along these directions.

Figure 2 represents the thermal dependences of density for barium diborate glass and those for the served crystal modifications of this substance. The effective thermal expansion increases along the row "α-phase–glass–β-phase", while the density rises along "glass–α-phase–β-phase". Comparison of the x-ray density with the hydrostatic data has shown that samples of α-BaO.2B$_2$O$_3$ obtained by crystallizing an appropriate balk glass, have porosity of about 4%.

When investigating the thermal expansion of isotropic polycrystal barium diborate, the authors has observed dilatometric hysteresis. The magnitude and shape of a hysteresis loop under discussion depend on the thermal history that a sample has aquired during sintering, crystallisation, and in the course of a dilatometric experiment. The earliest information on this phenomenon was published by the authors in Ref. 8, where description of the specific features was given. An example will be given here (Figure 3), which demonstrates the TEC dependence, under

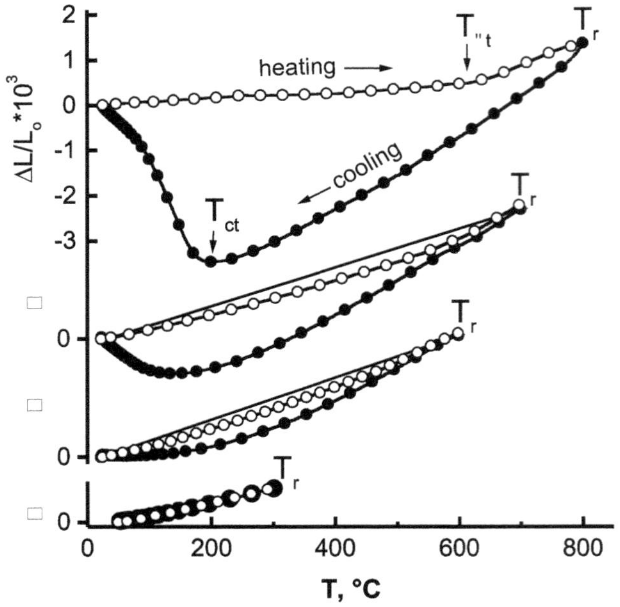

Figure 3. Effect of the temperature up to which the sample is heated in the dilatometric experiment on the hysteresis loop shape. Errors limits are within the symbols. On the dilatometric curves each tenth point is specified only, lines are received by computer spline. The temperature of sintering and crystallising the initial samples was 750°C. Sequence for the dilatometric cycles is from bottom-up

heating and cooling on the temperature up to which the sample is heated in the dilatometric experiment (T_r, reverse temperature). In all cases, XRPD has shown that one and only one crystal phase formed at room temperature is α-$BaO.2B_2O_3$.

What is the reason for the dilatometric hysteresis? One may suppose that there should be one more crystal phase of barium diborate besides α- and β-phases (let us name it the ε-phase, i.e. the enigmatic phase). The α-phase transforms into the β-one when heated above the heating transformation temperature, Tht (approximately 620°C); the higher T_r, the greater the extent of α→β transformation and the magnitude of hysteresis. The ε-phase should be considered to have the higher density and much higher TEC than the respective parameters for the α-phase. The opposite transition, while cooling, takes place at a certain lower temperature, T_{ct}, within the 250–200°C. To explain the hysteresis it should be suggested that at any given temperature below T_{ct} there is specific quantitative proportion of ε- and α-phases. Below the T_{ct}, the measured curve will result from two processes: the contraction determined by temperature coefficients of the phases, and the expansion caused by the transition of the denser phase into the lighter one. The second process is likely to make the more significant contribution, which is responsible for the observed expansion of the sample under cooling. We may obtain an isotropic polycrystalline sample by preliminary heat treatment at 800–860°C and

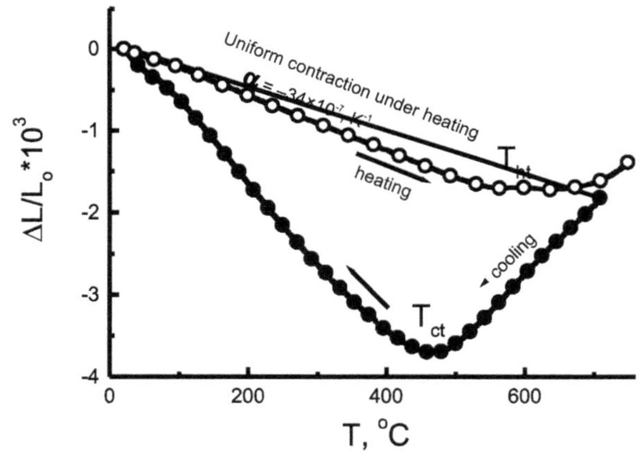

Figure 4. Dilatometric curves for a compact sample prepared by sintering of the preliminarily milled textured sample of α-$BaO.2B_2O_3$. Errors limits are within the symbols. On the dilatometric curves each tenth point is specified only; lines are received by computer spline

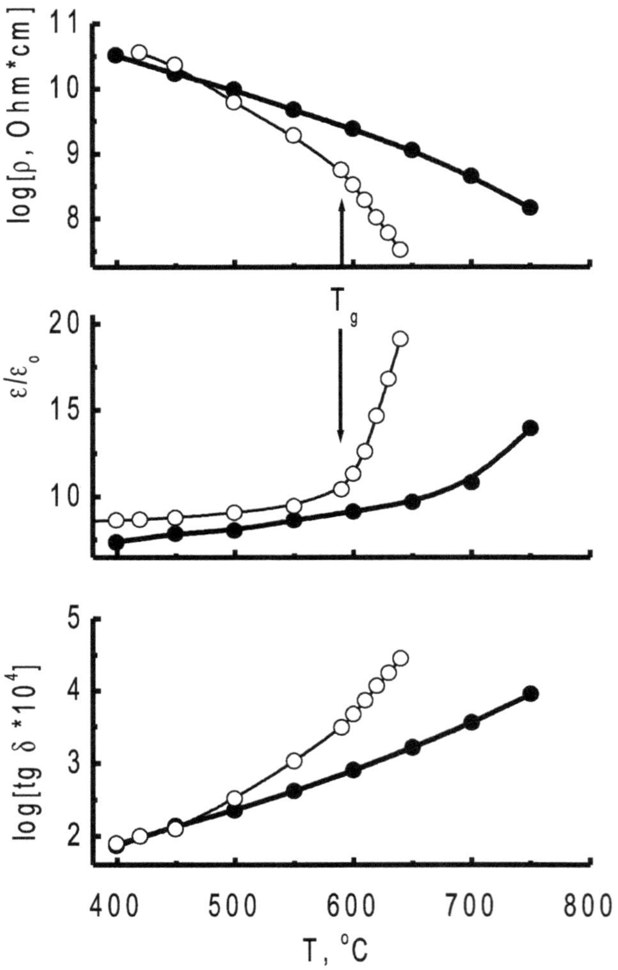

Figure 5. Comparison of thermal dependences of the electric are properties for glass of the diborate composition (the open circles) with those for polycrystalline α-$BaO.2B_2O_3$ (the solid circles). Frequency - 1 kHz. Errors limits are within the symbols; lines are received by computer spline

subsequent cooling to room temperature. The following heating from room temperature to T_{ht} leads to negative values of TEC(-34×10^{-7}, K^{-1}), i.e. to a uniform contraction (Figure 4).

The described dilatometric hystereses are absent for textured samples with relatively large crystallites of α-BaO.2B_2O_3. If an appropriate textured sample is dispersed and sintered, the obtained isotropic polycrystalline material acquires a pronounced dilatometric hysteresis. It may be concluded that the dilatometric hysteresis and mysterious ε-phase arise because of the dispersed state of the material. In terms of the dilatometric curves, calculations can be made of the effective (formal) TE coefficients for various temperature intervals. The range of such effective TE coefficients lies for powders under discussion from $+110\times10^{-7}$ to -210×10^{-7} K^{-1}.

Electrical properties

The electrical properties have been explored with polished glass samples having a thickness of 0·5–1 mm and structure of non-textured α-BaO.2B_2O_3. Figure 5 demonstrates that crystallisation enhances the dielectric properties, this improvement being most considerable at the temperatures higher than T_g.

Conclusion

Barium diborate exhibits some extraordinary properties, the nature of which needs further study.

1. Crystallising a degassed compact glass of barium diborate composition leads to formation of the more dense β-BaO·2B_2O_3, and crystallisation of an appropriate less dense glass gives the less dense α-BaO·2B_2O_3.
2. Large compact textured plates of α-BaO.2B_2O_3 can be synthesised with strongly pronounced anisotropy in TEC: under heating it will be expand along one direction, and will be diminished along two perpendicular directions.
3. Fine polycrystal samples of α-BaO.2B_2O_3 are isotropic. They have a thermal memory manifested by dilatometric hysteresis, and this seems to be caused by reversible transformation of α-phase into the new ε-modification. TEC of such a material can be controlled, and one can prepare appropriate samples exhibiting a uniform contraction under heating.

Acknowledgement

The work has been carried out with financial support of the Russian Foundation for Basic Research- RFBR, project 02-03-32499.

Reference

1. Polyakova, I. G. *Phys. Chem. Glasses*, 2000, **41**, 247.
2. Laird, J. A. & Bergeron, C. G. *J. Am. Ceram. Soc.*, 1970, **53**, 482.
3. Nagel, S. R. *Crystallization kinetics binary borate melts*. 1973. University of Illinois at Urbana-Champaing. P. 255.
4. Pevzner, B. Z. *Trudi VIII Vsesoyuzn. sov. po zharostoikim pokrytiyam*, 1979, 220. (In Russian.)
5. Block, S. & Perloff, A. *Acta. Cryst.*, 1965, **19**, 297.
6. Polyakova, I. G. & Pevzner, B. Z. *Glass. Phys. Chem.* (translated from Russian), 2005, **31**, 138.
7. Nikolaeva, N. V. *Tez. dokl. IV molodezhnoy nauchn. konfer. ISC RAS*, 2002, 67. (In Russian.)
8. Klyuev, V. P. & Pevzner, B. Z. *Phys. Chem. Glasses*, 2004, **45**, 146.

Proc. Fifth Int. Conf. on Borate Glasses, Crystals and Melts Phys. Chem. Glasses: Eur. J. Glass Sci. Technol. B, August 2006, **47** (4), 538–542

Structural interpretation of water content effect on the properties of one-alkali borate glasses

Olga Yu. Golubeva,[1] *Valerii F. Pavinich & Sergey V. Stolyar*

Institute of Silicate Chemistry of Russian Academy of Sciences, nab. Makarova, d.2, liter B, St.-Petersburg, 199395, Russia

The effect of water content on viscosity of binary sodium and potassium borate glasses is studied. Glasses with different water content were obtained under different melting conditions. The influence of the synthesis conditions on the spectral absorption in the wavelength range 2·5–4 μm, due to the presence of structurally bound water, is investigated. The synthesis under different conditions makes it possible to prepare glasses not only with different water contents but also with different ratios between the so-called free and bonded hydroxyl groups. It is demonstrated that the effect of water on viscosity depends on the ratio between free and bonded hydroxyl groups in glass structure.

1. Introduction

Any glass synthesised in air includes water in different forms. The amount of water formed in the glass in the course of synthesis can vary from hundredths of a percent to a few percents. The amount of water in glasses depends on the melting conditions and can be determined by such factors as purity of batch components, temperature and time of melting, and water vapour partial pressure. According to publications,[(1–7)] water in glasses can exist in molecular form and in the form of hydroxyl groups. The removal of water in any form from the glass is a rather complex problem.

The investigation into the structural features of glasses with residual water in different forms is especially topical for borate glasses, which are strongly hydroscopic, thus their properties are sensitive to the water content in glass.

As follows from our previous results[(8,9)] structure sensitive properties of borate glasses depend strongly on water content in the glass structure. It is worth noting that the effect produced on the viscosity by structural water is considerably more pronounced than that observed upon the introduction of an alkali oxide into a borate glass. For example, the data obtained in our study indicate that the change in the logarithm of viscosity by one order of magnitude corresponds to the introduction of approximately 2–3 mol% R_2O or 0·2–0·3 mol% H_2O into the structure of borate glasses. Thermal expansion coefficients of borate glasses are also substantially affected by water. The character of variations in the viscosity and thermal expansion with a change of the water content depends on the concentration of an alkali oxide entering into the glass composition.

Structural parameters (such as fragility parameters or the structural relaxation parameter), which depend directly on the content of alkali oxides entering into the glass compositions, remain unchanged under variations in the water content in the concentration range covered.

The paper presents some new results concerning the structure and properties of water containing alkali borate glasses.

2. Experimental

We have studied glasses of compositions $5{\cdot}5R_2O$–$94{\cdot}5B_2O_3$, $15R_2O$–$85B_2O_3$ and $25R_2O$–$75B_2O_3$. The series of binary alkali borate glasses containing 5·5, 15 and 25 mol% R_2O, where R=Na and K. Boric acid and carbonates of the corresponding alkali metals were used as initial components. The mixture of batch components was melted in a platinum crucible with a volume of 200 ml in the electric furnace in air.

The melting conditions of studied glasses are presented in Table. The glasses of different compositions were melted at temperatures of 950, 1200 and 1300°C. To obtain glasses with different residual water content, temperature and time of glass melting were varied. For the convenience of further discussion, identical synthesis conditions for glasses of different compositions in Table 1 are designated by identical letters (from A to F).

The chemical analysis of the glasses (performed at the laboratory of analytical chemistry, Institute of Silicate Chemistry) demonstrated that variations in the compositions of glasses synthesised under different conditions are insignificant and do not exceed 1–1·5 mol%.

1 Corresponding author. Email address: olga_isc@mail.ru

Table 1. Melting conditions of studied glasses

Synthesis denotation	Temperature (°C)	Time of heat treatment (h)
A	950	2
B	950	3
C	1200	1
D	1200	2·5
E	1200	4
F	1300	1 (after 3 h melting at 950°C)

The viscosity of the samples was measured with a viscometer by the beam bending method in the temperature range corresponding to a change in the viscosity (η) from 10^{10} to 10^{13} dPa s. The method consists of determining the bending rate of a rod horizontally mounted on two supports under a load applied to the cent of the sample.[10] The root-mean-square error of measurements using this method was equal to ±0·04 lg η.

The infrared absorption spectra of the glasses were obtained at room temperature with the use of samples in the form of 0·1–0·2 mm thick plates polished in kerosene. The experimental technique and preliminary processing were described in more detail in Startsev & Golubeva.[9]

3. Results and discussion

Figure 1 presents the results of viscosity measurements and measurements by infrared absorption method for the four series of studied glasses.

It can be seen from Figure 1 that, for the glasses of the same content of the main components (B_2O_3 and R_2O), the deviations in the logarithm of viscosity values can reach three orders of magnitude.

According to literature,[11–16] it is believed that the viscosity of glasses reduced on increasing the water in glass structure. The term 'water' usually means the water in the form of free hydroxyl groups. The decrease in viscosity is usually explained by the existence of broken bonds in glass structure due to formation of hydroxyl groups. Other forms of water content are not taken into consideration.

It was shown in our previous work[17] that glass melting conditions can affect the ratio between forms of water content in alkali borate glasses structure. These changes can influence the properties of glasses. Therefore a comparison was made of thermal dependencies of viscosity and infrared absorption spectra of glasses obtained under different glasses melting conditions.

Figure 1(e)–(h) depict the absorption spectra $K(\lambda)$ of studied glasses, where K is the absorption coefficient (cm^{-1}) and λ is the wavelength (μm). The absorption coefficients were calculated taking into account the losses by reflection from the sample surface.[9]

The wavelength range under investigation contains two bands associated with the stretching vibrations of hydroxyl groups in the glass structure with absorption maxima at 2·8–2·9 and 3·5–3·7 μm. On the basis of the experimental data for a number of oxide glasses, such as alkali silicate glasses and borate glasses containing alkali earth metal oxide, Scholze,[2] Franz[18] and Franz & Scholze[19] attributed the absorption band at 2·8–2·9 μm to nonassociated so called free (not linked by hydrogen bonds) hydroxyl groups, and assigned the band at 3·39–3·63 μm to so called bound hydroxyl groups linked by hydrogen hydrogen bonds in the glass structure. The structures of pure silica glass and vitreous boron oxide involve only free OH groups. The addition of an alkali metal oxide to silicate or borate glasses leads to the appearance of nonbridging oxygen atoms. These atoms and hydroxyl groups can form hydrogen bridging bonds. As a result, the second absorption band at a larger wavelength arises in the infrared spectra.

Different researches revealed other absorption bands associated with water. A number of authors[2,18,19] assign the absorption at 4·25 μm to vibrations of hydroxyl groups linked by very strong hydrogen bonds. Adams[3] rather convincingly justified the assumption that the absorption in this spectral range is attributed to CO_2 molecules of atmospheric air rather than to water in glass structure. Ernsberger[5] found the band at a wavelength of 6·1 μm in the spectra of glasses with very high water content. At present, this band is attributed to dissolved molecular water.

Furthermore, there are alternative interpretations of absorption spectra of glasses due to the presence of water. According to Efimov,[20] the infrared absorption spectra of inorganic glasses contain a considerably larger number of associated bands with water in the glass structure. The total number of these bands in the wavelength range from 2·7–6·5 μm can be as large as 15. Among these bands, at least four bands correspond to water in molecular form. Tomozawa[6] showed that, at water content up to 1 wt% in silicate glasses, water is predominantly represented by hydroxyl groups in the glass structure.

It can be seen from Figure 1(e)–(h) that increase in the temperature and time of synthesis (according to Table 1) leads to a decrease in the intensity of the first band (with absorption maximum at 2·8–2·9 μm) for all the glasses. These changes in the intensity indicate a decrease in the content of water in the form of free hydroxyl groups. However, the decrease of water in the form of free hydroxyl groups is not always accompanied by an increase of viscosity. In some cases (Figure 1(b) synthesis *F*, Figure 1(c) synthesis *F*, Figure 1(d) synthesis *F*) one can see a decreased viscosity on decreasing the intensity of fist band.

Such behaviour of viscosity can be explained by the changes in the ratio between free and bonded hydroxyl groups. As can be seen from Figure 1(e)–(h), simultaneously with the changes in the intensity of

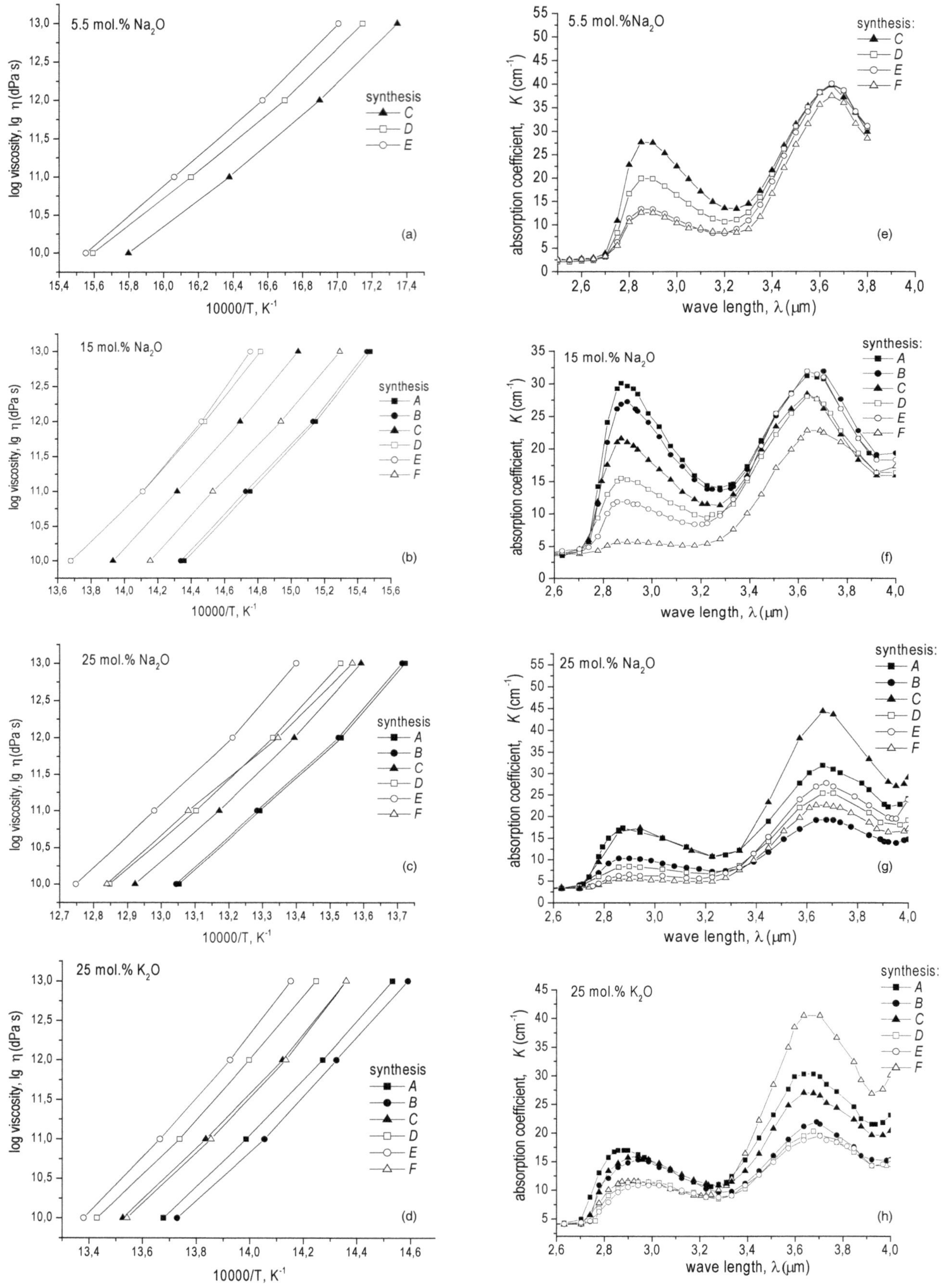

Figure 1. Temperature dependencies of viscosity and infrared absorption spectra of alkali borate glasses obtained under different glass melting conditions (according to Table 1)

the first band in absorption spectra, a variation in glass melting conditions leads to a change of the second band (with absorption maximum at 3·7 μm). However, this change in the intensity does not necessarily correlate with the change in the intensity if the first absorption band. Actually, in some cases, an increase in the intensity of the first band is accompanied by a decrease in the intensity of the second band. This tendency tendency is clearly seen in Figure 1(f) (synthesis conditions *D*, *F*) and Figure 1(g) (synthesis conditions *B*, *F*). It is especially pronounced in Figure 1(h), which shows the infrared absorption spectra of the glasses containing 25 mol% K_2O.

The spectrum of this glass, prepared at 1300°C (synthesis conditions *F*), is characterised by the lowest intensity of the first band (in addition to the glasses synthesised under conditions *D* and *E*) among the spectra of other glasses that have the same compositions but are synthesised under different conditions. This fact suggests a minimum water content in the sample under consideration, and consequently the highest values of viscosity in this series of glasses. However, as can we see from the Figure 1(d) the viscosity of the sample obtained under synthesis conditions *F* is lower than for glasses obtained under synthesis conditions *D* and *E*. As can be seen from the Figure 1(h) the intensity of the band with absorption maximum at 3·7 μm for this sample is highest compared to those for the other glasses of the series. Therefore, the inference that the water content in this sample is lowest cannot be made unambiguously. As a result, the viscosity values of this glass sample are rather higher than that should be expected form the consideration only the first absorption band.

The probability of the changes in ratio between free and bonded hydrogen groups in borate glass structure increases on the increasing alkali oxide content. This probability should correlate with the probability of forming hydrogen bonds in glass structure which depends on the number of the nonbridging oxygen atoms. It is well known that in borate glasses the introduction of alkali oxides into the network does not result in a drastic increase in the number of nonbridging oxygen atoms, because a number of boron atoms transform the three-fold coordinated state to the four-fold coordinated state.[21–23] Oxygen introduced into the glass with the alkali metal oxide is used for enhancing the network connectivity and is not accompanied by the appearance of nonbridging oxygen atoms. According to the data of different authors (for example Krogh-Moe[21]), this process only partially occurs, beginning with a content of 25–39 mol% R_2O. A part of oxygen atoms introduced with the alkali metal oxide then transforms bridging oxygen atoms into nonbridging oxygen atoms. However, at lower oxide contents, a few alkali metal atoms most likely fulfil a destructive function – lead to the breaking of the network and the appearance of nonbridging oxygen atoms.[24]

For glasses with small alkali oxide content (5 mol%) the number of nonbridging atoms is very low. As a result there is a direct connection between content of free hydroxyl groups in glass structure and viscosity. Further increases in alkali oxide content to 15–25 mol%, and increases in the number of nonbridging oxygen atoms in the structure of borate glasses, are accompanied by an increase of the bonded hydroxyl groups. These two types of hydroxyl groups have a double effect on the glass structure and properties. On the one hand, free hydroxyl groups decrease the connectivity of the network and decrease the viscosity, on the other hand, bonded hydroxyl groups by the formation the hydrogen bonds increase the connectivity of the network and increase the viscosity. When the content of bonded hydroxyl groups is high, the second effect becomes the main one. In the case of glasses with rather high alkali oxide contents (25 mol%), the effect of water content on properties is more complicated because of the more complicated glass structure. The network of these glasses is formed by different borate groups – boroxol groups, tri- and diborate groups. On increasing the water in glass structure, the double effect described above may be accompanied by the effects connected with the destruction or transformation of borate groups.

4. Conclusions

While studying the effect of water content on the borate glasses properties, it is necessary to take into account the existence of water in glass structure in the form of free and bonded hydroxyl groups. The correlation between free and hydrogen bonded groups on glass structure depends on the glass melting conditions. These two forms of water in glass structure exert a double effect on glass structure – free OH-groups decrease the connectivity of glass network while the hydrogen bonded OH-groups increase the connectivity of the network. As a result, changes in the ratio between free and bonded hydroxyl groups have an effect on structure sensitive properties of borate glasses.

Acknowledgments

The work was supported by the Russian Foundation for basic research (project no. 03-03-32269) and by Russian Science Support Foundation.

References

1. Sholze, H. *Glass. Nature, structure and properties*. New York, 1991.
2. Scholze, H. *Glastech. Ber.*, 1959, **32**, 81–88, 278–281, 314–320, 381–386.
3. Adams, R. V. *Phys. Chem. Glasses*, 1961, **2** (2), 39–49.
4. Tomlinson, J. W. *J. Soc. Glass. Tech.* 1956, **40**, 25–31T.

5. Ernsberger, F. M. *J. Am. Ceram. Soc.* 1977, **60** (1–2), 91–2.
6. Tomozawa, M. *J. Non-Cryst. Solids.* 1985, **73** (1–3), 197–204.
7. Shelby, J. E. *Handbook of gas diffusion in solids and melts.* ASM Int., 1992, 240 ps.
8. Golubeva, O. & Startsev, Y. *Phys. Chem. Glasses,* 2002, **43C**, 471–6.
9. Startsev, Y. & Golubeva, O. *Glass Phys. Chem.,* 2002, **28** (3), 159–69.
10. Kluev, V. P. & Totesh, A. S. *Methods and Instruments for Testing the Viscosity of Glasses,* Moscow, VNIESM, 1975 (in Russian).
11. Visser, Th. J. *Rheological Properties of Alkali Borate Glasses.* Thesis. Eindhoven, Netherlands. 1971. 88 ps.
12. Hetherington, G. & Jack, K. H. *Phys. Chem. Glasses,* 1962, **3** (4), 129–33.
13. Hetherington, G., Jack, K. H. & Kennedy, J. C. *Phys. Chem. Glasses.* 1964, **5** (5), 130–6.
14. Kimura, T. *Jpn. J. Appl. Phys.* 1969, **8 (12)**, 1397-1403.
15. Jewell, J. M. & Shelby, J. E. *J. Non-Cryst. Solids.* 1988, **102** (1–3), 24–9.
16. Jewell, J. M. & Shelby, J. E. *J. Am. Ceram. Soc.* 1992, **75** (4), 878–83.
17. Golubeva, O. Yu. & Pavinich, V. F. *CD Proc. XX Int. Congress of Glass.* 2004. Kyoto (Japan). P-07-062.
18. Franz, H. *J. Am. Ceram. Soc.* 1966, **49** (9), 473–7.
19. Franz, H. & Scholze, H. *Glastech. Ber.* 1963, **36** (9), 347–56.
20. Efimov, A. M. *J.Non-Cryst. Solids.* 1985, **43C**, 165–75.
21. Krogh-Moe, J. *Phys. Chem. Glasses.* 1962, **3** (1), 1-6.
22. Bray, P. J. & O'Keefe, J. G. *Phys. Chem. Glasses.* 1963, **4** (2), 37–46.
24. Zhong, J. & Bray, P. J. *J. Non-Cryst. Solids.* 1989, **111** (1), 67–76.
25. Balta T. & Radu, D. *Phys. Chem. Glasses,* 2000, **41** (5), 278–81.

Proc. Fifth Int. Conf. on Borate Glasses, Crystals and Melts *Phys. Chem. Glasses: Eur. J. Glass Sci. Technol. B*, August 2006, **47** (4), 543–547

Differential volume changes in trigonal and tetrahedral borate units

Gregory J. Ongie, Steven A. Feller, Mario Affatigato*

Physics Department, Coe College, Cedar Rapids, IA 52402, USA

Masao Kodama

Department of Applied Chemistry, Sojo University, 4-22-1 Ikeda Kumamoto, 860-0082 Japan

For the alkali borate systems, density trends have been used as an indicator for structural changes within a glass. With the use of high precision density data, we have been able to observe, indirectly, superstructural changes in the glass by, applying the differential volume technique used by Doweidar on various silicate and borate systems. This technique allows us to examine the differential changes of the volumes of the 3-coordinated trigonal units and the 4-coordinated tetrahedral units as a function of R *(up to* R<0·6*, where* R *is the mole ratio of alkali oxide to boron oxide). The relative volumes of the borate units are not constant in the low* R *region, but contain features that may correspond to short range units participating in large superstructural units. We applied this same analysis to the barium borates, and found that the trends of the unit volumes are nearly identical to those for the sodium borates, suggesting that the superstructural changes are also identical for both the alkali borate and alkaline earth borate systems. Finally, a comparison of our results with Kroeker's recent* ^{11}*B MAS NMR studies reveals that features in the unit volumes may correlate to changes in the abundance of the two observed tetrahedral units.*

Of recent interest in the study of alkali borates and related glass systems is the question of intermediate range structure. Currently, only a few methods have proven sensitive enough to detect the presence of such superstructural groups, such as NMR,[1,2] neutron scattering,[3] vibrational spectroscopy,[4] and laser desorption time-of-flight mass spectroscopy.[5] However, using the analysis employed in this study, it is possible to extract information that may be related to superstructural changes from a set of density data, provided it is of sufficient accuracy and resolution.

The aforementioned technique was originally devised by Doweidar,[6] which itself was an extension of the density model proposed by Shibata *et al.*[7] In Shibata's model, average values for the total volume of each of the structural units, including empty space and contributions from associated alkali, were calculated using a density data set and a model for the unit abundances. Doweidar, rather than calculating an average value for the volumes, calculated the volumes as a function of composition, which he was able to do because of the high precision of much of his density data.

The purpose of this study is to update and extend Doweidar's original work on the alkali borates using recent high precision density data, and to compare the results with recent NMR-based structural studies.

Procedure

Two sets of information are necessary for this analysis: high precision density data and the fractional abundances of the structural units as a function of composition. The density data used in this study were obtained from samples prepared for velocity of sound measurements using the solution method.[8–13] Each sample was cast and annealed in a cylindrical mould 15 mm in diameter and 30 mm deep, and then chemically analysed to determine its composition by means of the manninol method.[14] Measurements of density were made using a hydrostatic weighing method with a silicon single crystal as the density standard,[15] resulting in an estimated uncertainty of ±0·001 g/cc.

The model used for the fractional abundances of the structural units is alkali independent and was presented by Chong *et al*,[16] which is consistent with the results of many NMR studies,[17,18,4] The short range structural units relevant to this study are the f_1 unit, a trigonal BO_3 unit with all bridging oxygens, the f_2 unit, a tetrahedral BO_4 unit with all bridging oxygens and an associated alkali cation, and the f_3 unit, a trigonal BO_3 unit with one nonbridging oxygen and an associated alkali cation. In the alkali independent model, the fractional abundance of the f_1 unit in the region $0{\cdot}0 \le R \le 0{\cdot}7$ is given by $f_1 = 1-R$, where R is the mole ratio of alkali oxide to boron oxide. The fractional abundance of the f_2 unit in the region $0{\cdot}0 \le R \le 0{\cdot}4$ is given by $f_2 = R$, and for $0{\cdot}4 \le R \le 0{\cdot}7$,

[1] Corresponding author

$f_2=R/6+1/3$. Also, the fractional abundance of the f_3 unit in the region $0{\cdot}4\leq R\leq 0{\cdot}7$ is given by $f_3=5R/6-1/3$. However, the data sets used in this study primarily span the compositional region $0{\cdot}0\leq R\leq 0{\cdot}5$. Thus, the volumes of only the f_1 and f_2 units two units can be reliably analysed in this study, as they are the only two present in sufficient quantity in this region.

To begin the calculation, we define the glass density in terms of its dimensionless mass relative to pure borate glass (R=0), $m'(R)$, and its dimensionless volume relative to pure borate glass, $v'(R)$

$$\rho(R)=\frac{m(R)}{v(R)}=\frac{m'(R)}{v'(R)}\rho(0) \quad (1)$$

$$v'(R)=\frac{m'(R)}{\rho(R)}\rho(0) \quad (2)$$

where $\rho(0)$ is the density of vitreous B_2O_3. $m'(R)$ can simply be calculated by the ratio of the molecular weight of the glass at composition R to the molecular weight of the pure borate glass. Also, we can interpret $v'(R)$ as the sum of the fractional abundance of each unit multiplied by its volume relative to the f_1 unit (given by v_1 and v_2, respectively)

$$v'(R_1)=f_1(R_1)v_1+f_2(R_1)v_2 \quad (3a)$$

$$v'(R_2)=f_1(R_2)v_1+f_2(R_2)v_2 \quad (3b)$$

where R_1 and R_2 are the R-values of two adjacent points in the density data set. Replacing $v'(R_1)$ and $v'(R_2)$ with the expression derived in Equation (2) gives

$$f_1(R_1)v_1+f_2(R_1)v_2=\frac{m'(R_1)}{\rho(R_1)}\rho(0) \quad (4a)$$

$$f_1(R_2)v_1+f_2(R_2)v_2=\frac{m'(R_2)}{\rho(R_2)}\rho(0) \quad (4b)$$

All the variables except v_1 and v_2 in Equations (4a) and (4b) are known: $f_1(R_1), f_2(R_1), f_1(R_2), f_2(R_2)$ are defined by the fractional abundances model ($f_1=1-R$, $f_2=R$); $\rho(0)$, $\rho(R_1)$, and $\rho(R_2)$ are obtained from our density data set; and $m'(R_1)$ and $m'(R_2)$ can be calculated from the respective molecular weights. Therefore, since R_1 and R_2 are close in value, we can make the approximation that the solutions for v_1 and v_2 are defined at the composition $(R_1+R_2)/2$.

Additionally, some of the density data sets range up to approximately R=0·6. As noted earlier, in the region $0{\cdot}4\leq R\leq 0{\cdot}7$, trigonal units with one nonbridging oxygen (f_3 units) begin to appear. However, we found that extending this analysis to regions where three structural units exist presents some difficulties. The main problem is that the R values at which the unit volumes are defined would have to be obtained by averaging over three values, yielding results that would only be acceptable if the resolution of the density data was extraordinarily fine. However, a useful approximation is to hold one of the unit volumes constant (f_3), using the average volume derived from a least squares analysis over the entire data set, and proceed with the calculation letting the other two unit volumes vary (f_1 and f_2).

Also, we collected two sets of density data for the sodium borate system taken over the same composition range, one filling in the compositions between the other, resulting in data of twice the compositional resolution. To remove a very slight systematic error, we averaged adjacent points in the data set, yielding a set of the same resolution as the other alkali systems.

Finally, using the value of 1·838 g/cc for the density of vitreous B_2O_3,[13] the volume of the f_1 unit at R=0 can be calculated to be 31·45 Å^3 (referred to as V_0 in this paper), which can then be used to convert the relative volumes into absolute volumes.

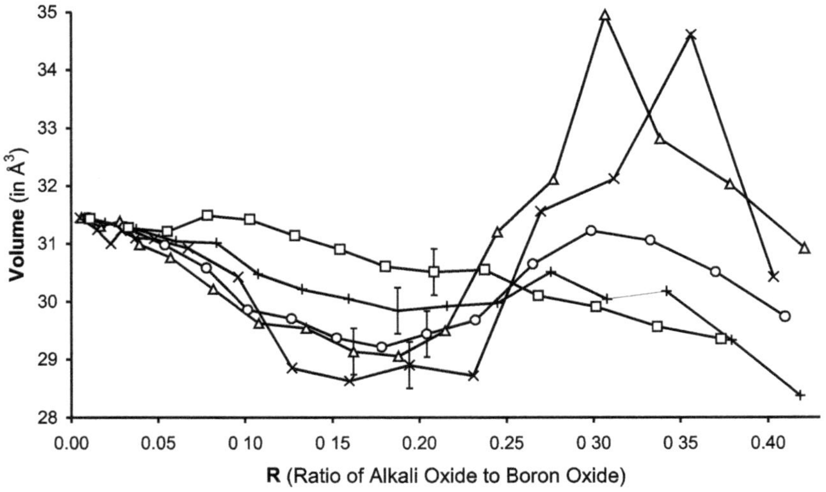

Figure 1. Volume of the f_1 unit as a function of R (ratio of alkali oxide to boron oxide) for the five alkali borate systems. The cesium system is represented by "△", rubidium by "×", potassium by "○", sodium by "+", and lithium by "□". Estimated uncertainty in the volume is ±0·4 Å^3, represented graphically by the error bars

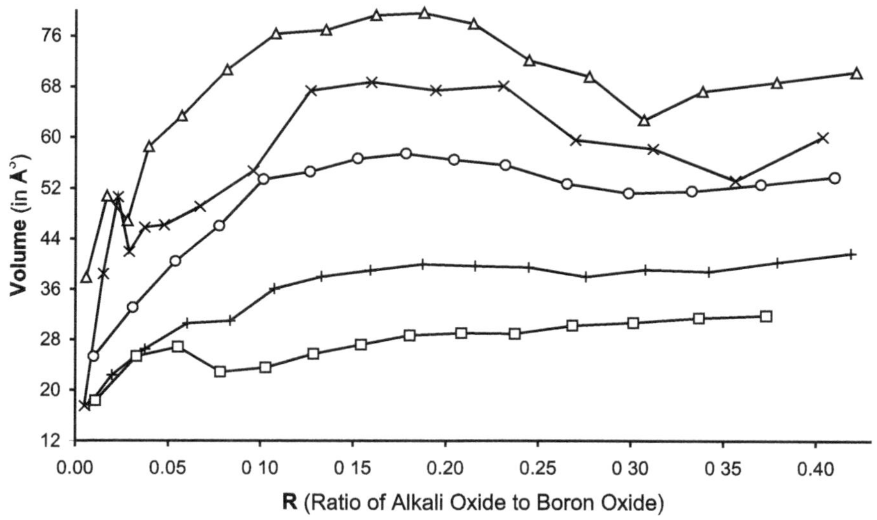

Figure 2. Volume of the f_2 *unit as a function of* R *(ratio of alkali oxide to boron oxide) for the five alkali borate systems. The cesium system is represented by "△", rubidium by "×", potassium by "○", sodium by "+", and lithium by "□". Estimated uncertainty in the volume is ±0·9 Å^3, approximately the size of the data markers*

Results

For the volume of the f_1 units in the region $0{\cdot}0 \leq R \leq 0{\cdot}4$, we find that the trends initially reach one minimum and then one maximum, with the exception of the lithium system, which appears to be monotonically decreasing, Figure 1. The minimum in the trends is at $R=0{\cdot}18\pm0{\cdot}05$, and the maximum is at $R=0{\cdot}30\pm0{\cdot}05$, where there is also a slight alkali dependence in the magnitude of the minimum and maximum greater for systems with larger alkali—and at most deviating approximately 10% from V_0, the volume of the f_1 unit at $R=0$. Nevertheless, the f_1 units for each system appear to have approximately the same average volume in this region of roughly $0{\cdot}97V_0$. This is consistent with the fact that the composition of the f_1 unit is alkali independent.

For the volume of the f_2 units in the region $0{\cdot}0 \leq R \leq 0{\cdot}4$, we find that trends initially reach a broad maximum and then a slight local minimum, again with the exception of lithium, which appears to be monotonically increasing (Figure 2). Additionally, there is a small sharp feature at low R in the trends (with the exception of potassium), which reaches a maximum at $R=0{\cdot}05\pm0{\cdot}01$ for the lithium and sodium systems, and at $R=0{\cdot}02\pm0{\cdot}01$ for the rubidium and cesium systems. The broad maximum in the compositional trends is at $R=0{\cdot}18\pm0{\cdot}10$, and the local minimum is at $R=0{\cdot}30\pm0{\cdot}05$, mirroring the changes seen in the f_1 unit.

Discussion

Unlike for the f_1 unit, there is a very clear alkali dependence on the volume of the f_2 unit. Such a dependency is explainable by the fact that alkali cations are only associated with f_2 units. To quantify this observation, we performed a least squares analysis to determine the best fit average value for v_2 over the entire range of available compositions. Plotting these values versus the volume of the alkali ion[19,20] yields a linear relationship given by

$$(\text{Best Fit } v_2 \text{ in Å}^3)=1{\cdot}43(\text{Volume of alkali ion in Å}^3)+26{\cdot}8 \quad (5)$$

with an R_2 value of 0·997 (Figure 3).

In addition, we applied the differential volume technique to data we obtained from the barium borates. Spectroscopic data indicate the fractional abundance of the f_1 and f_2 units in barium borates appear to follow the same $f_1=1-R$, $f_2=R$ dependence as in the alkali borates in the low R region.[21,17] Using this fact, we were then able to calculate the volumes

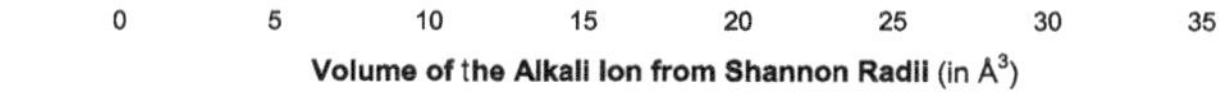

Figure 3. A comparison of the best fit average volume of the f_2 *unit over the entire range of available compositions versus the volume of the alkali ion calculated from Shannon radii for the five alkali borate systems. The cesium system is represented by "△", rubidium by "×", potassium by "○", sodium by "+", and lithium by "□". The equation of the linear fit is given by* $y=1{\cdot}43x+26{\cdot}8$*, with an* R^2 *value of 0·997*

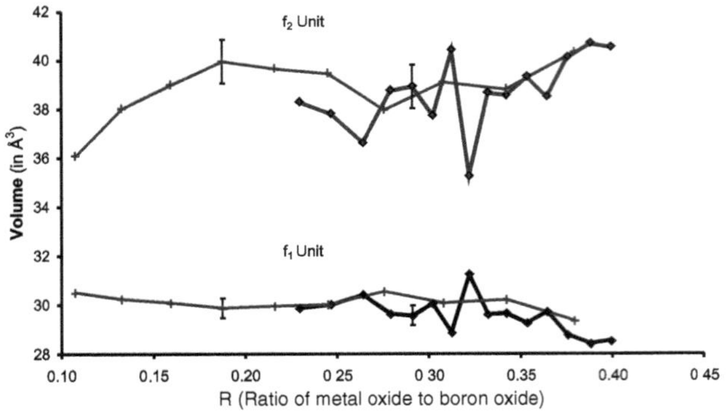

Figure 4. Volumes of the f_1 *and* f_2 *unit as a function of* R *(ratio of alkali oxide to boron oxide) of barium borates compared to sodium borates. The barium system is represented by "◆", and sodium by "+". Estimated uncertainty in the* f_1 *volumes is ±0·4 Å*^{3}*, and in the* f_2 *volumes, ±0·9 Å*^{3}*, represented graphically by the error bars*

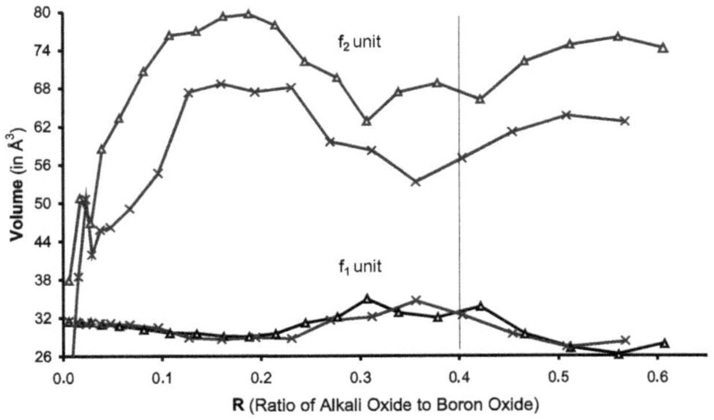

Figure 5. Volumes of the f_1 *and* f_2 *unit as a function of* R *(ratio of alkali oxide to boron oxide) of cesium borates and rubidium borates, extended to* R=0·6*. The cesium system is represented by "×", and rubidium by "Δ". All data points to the right of the vertical dotted line were calculated using the approximation of holding the* f_3 *unit volume constant. Estimated uncertainty in the* f_1 *volumes is ±0·4 Å*^{3}*, and in the* f_2 *volumes, ±0·9 Å*^{3}*, approximately equal to or less than the size of the data markers*

of its f_1 and f_2 units, and we proceeded to plot them with the results from the sodium borate system as a comparison (Figure 4). Sodium was chosen because the ratio of the cubes of the ionic radii of Ba to Na is roughly 2:1,[19] yet in sodium borates there are twice as many ions per mole than in barium borates, thus we would expect the volumes of the units to be comparable if the systems are isostructural. Due to the similarity observed between the unit volume trends in the sodium borates and the barium borates, we believe that the superstructural changes hypothesised for the alkali borates may extend to the alkaline earth borate systems, and could possibly be an intrinsic property of the borate network itself. Due to the phase separation in low R barium borates, we are limited in our comparison to the region $0{\cdot}2 \leq R \leq 0{\cdot}4$.

We also tried to extract information about the volume changes of the f_1 and f_2 unit in regions past R=0·4 by holding the f_3 unit volume constant. Of the data sets available, we found that rubidium and cesium were the only two which extended far enough past R=0·4 to warrant analysis. Thus, we performed a least squares analysis on these two data sets to find the best fit average value over the entire range of available compositions for the volume of the f_3 units. Comparing our values to those obtained by Lower *et al*,[22] we found that our values were within 10% of theirs. However, Lower's data cover the entire range of composition for which the f_3 unit is present ($0{\cdot}4 \leq R \leq 2{\cdot}1$), but our data are limited to only when the f_3 unit first appears ($0{\cdot}4 \leq R \leq 0{\cdot}6$). Since the difference between our values is small, though the difference between our ranges is large, we believe that holding the f_3 unit volume constant serves as a suitable approximation.

The above analysis reveals that the volumes of the f_1 unit in rubidium borates and cesium borates drop to another minimum at about R=0·53±0·05 of about $0{\cdot}85V_0$, and the f_2 unit volume trends reach a maximum in the same area of about $2{\cdot}4V_0$ for cesium and $2{\cdot}0V_0$ for rubidium (Figure 5).

Comparing our results to those in the original study by Doweidar,[6] we find that his results are remarkably similar to ours, albeit his are somewhat less accurate. For example, the maxima and minima in his unit volumes appear to be offset by roughly +0·5R from ours. Additionally, from our analysis we were able to observe fine peaks in the volume of the f_2 units at low R, but such features are absent from Doweidar's results.

Unlike when this analysis is performed on alkali silicates,[23] we find that in the alkali borates the volume of the units varies extensively as a function of composition. We hypothesise that this is due to the presence of certain superstructural units in which the f_1 and f_2 units are participating, affecting the net volume of these short range units.

Recent ^{11}B MAS NMR studies on cesium borates by Kroeker *et al*[1] and Ratai *et al*[4] indicate the existence of two distinct species of f_2 units present in the $0{\cdot}0 \leq R \leq 0{\cdot}6$ region, separated by roughly 1 ppm in the ^{11}B MAS NMR spectra, which are believed to correspond to f_2 units participating in two different superstructural environments. Quantitative analysis of Kroeker's data reveals that from $0{\cdot}0 \leq R \leq 0{\cdot}2$, the two species of f_2 are present in roughly equal quantities, from $0{\cdot}2 \leq R \leq 0{\cdot}35$ the species characterised by the lower chemical shift (the "lower frequency species") is created preferentially over species characterised by the higher chemical shift (the "higher frequency species") by approximately 3:2, from $0{\cdot}35 \leq R \leq 0{\cdot}45$ a crossover occurs in the trends, and from $0{\cdot}45 \leq R \leq 0{\cdot}6$ the higher frequency species is created preferentially over the lower frequency species by approximately 2:1.

The changes in abundances of the different f_2 species that occur from $0{\cdot}2 \leq R \leq 0{\cdot}35$ and $0{\cdot}45 \leq R \leq 0{\cdot}6$ appear to align with events in our analysis of the changes in the volumes. In the cesium system, we find that at approximately $R=0{\cdot}3$ the f_2 unit volume reaches a local minimum, and at approximately $R=0{\cdot}55$ reaches a local maximum. Such a correlation leads us to speculate that the lower frequency species may exhibit more efficient packing than the higher frequency species, resulting in the changes in net volume of the f_2 unit we observed in this study.

Conclusion

We performed an analysis of the differential volume changes of the f_1 and f_2 units of alkali borates glasses as a function of composition and found that the volumes of the units vary extensively over the composition $0{\cdot}0 \leq R \leq 0{\cdot}6$. The volume of the f_1 unit appears to be independent of alkali when compositionally averaged, while the f_2 unit average volume varies linearly with that of the associated alkali. The same analysis was then applied to barium borates which we found to exhibit trends similar to the alkali borates. Finally, we also compared our results to f_2 speciation observed in recent ^{11}B MAS NMR data, and we found that they contained compositional features which correlated to the minima and maxima of the unit volumes calculated in this study.

Acknowledgements

This work was supported by the National Science Foundation under grants DMR 0211718, DMR 0502051, NSFINT 0089510. Coe College is thanked for providing additional support.

References

1. Kroeker, S., Feller, S. A., Affatigato, M., O'Brien C. P. & Clarida, W. *Phys. Chem. Glasses*, 2003, **44** (2), 54–58.
2. Ratai, E.-M., Janssen, M., Epping, J. D., Chan, J. C. C. & Eckert, H. *Phys. Chem. Glasses*, 2003, **44** (2), 45–53.
3. Sinclair, R. N., Stone, C. E., Wright, A. E., Polyakova, I. G., Vedishcheva, N. M., Shakhmatkin, B. A., Feller, S. A., Johanson, B. C., Venhuizen, P., Williams, R. B., & Hannon, A. C. *Phys. Chem. Glasses*, 2000, **41** (5), 286–289.
4. Kamitsos, E. I. *Phys. Chem. Glasses*, 2003, **44** (2), 79–87.
5. Affatigato, M., Feller, S. A., Schue, A., Blair, S., Stentz, D., Smith, G., Liss, D., Kelley, M., Goater, C. & Leelesagar, R. *J. Phys.:Condens. Matter*, 2003, **15**, S2323–34.
6. Doweidar, H. J. *J. Mat. Sci.*, 1990, **25**, 253–258.
7. Shibata, M., Sanchez C., Patel, H., Feller, S. A., Stark, J., Sumcad, G. & Kasper, J. *J. Non-Cryst. Solids*, 1985, **85**, 29.
8. Kodama, M., Matsushita, T. & Kojima, S. *Jpn. J. Appl. Phys.*, 1995, **34** (5B), 2570–74.
9. Kodama, M. *J. Mater. Sci.*, 1991, **26** (15), 4048–53.
10. Kodama, M. *J. Non-Cryst. Solids*, 1991, **127** (1), 65–74.
11. Kodama, M. *J. Am. Ceram. Soc.*, 1991, **74** (10), 2603–09.
12. Kodama, M., Hirashima, T. & Matsushita, T. *Phys. Chem. Glasses*, 1993, **34** (4), 129–39.
13. Kodama, M., Ono, A., Kojima, S., Feller, S. A., & Affatigato, M. "Borate Anomaly and Anharmonicity in Sodium Borate Glasses." (submitted to *The Proceedings of the 5th International Conference on Borate Glasses, Crystals and Melts*)
14. Kodama, M., Iizuka, K., Miyashita, M., Nagai, N., Clarida, W., Feller, S. A. & Affatigato, M. *Glass Technol.*, 2003, **44** (2), 50–8.
15. Bowman, H. A., Schoonover, R. M. & Carroll, C. L. *Metrologia*, 1978, **10**, 117.
16. Chong, B. C. L., Choo, S. H., Feller, S. A., Teoh, B., Mathews, O., Khaw, E. J., Feil, D., Chong, K. H., Affatigato, M., Bain, D., Hazen, K. & Farooqui, K. *J. Non-Cryst. Solids*, 1989, **109**, 105–113.
17. Kroeker, S. C., Aguiar, P. M., Feller, S. A. & Affatigato M. "Exploring Medium-Range Order in Borate Glasses: New Experimental and Theoretical NMR Approaches." (submitted to *The Proceedings of the 5th International Conference on Borate Glasses, Crystals and Melts*).
18. Clarida, W. J., Berryman, J. R., Affatigato, M., Feller, S. A., Kroeker, S. C., Ash, J., Zwanziger, J. W., Meyer, B., Borsa, F. & Martin, S. W. *Phys. Chem. Glasses*, 2003, **44** (3), 215–217.
19. Shannon, R. D. *Acta. Cryst.*, 1976, **A32**, 751.
20. Giri, S., Gaebler, C., Helmus, J., Affatigato, M., Feller, S. A. & Kodama, M. *J. Non-Cryst. Solids*, 2004, **347**, 87–92.
21. Greenblatt, S. & Bray, P. J. *Phys. Chem. Glasses*, 1967, **8** (5), 190.
22. Lower, N. P., McRae, J. L., Feller, H. A., Betzen, A. R., Kapoor, S., Affatigato, M. & Feller, S. A. *J. Non-Cryst. Solids*, 2001, **293–295**, 669–675.
23. Doweidar, H. *J. Non-Cryst. Solids*, 1996, **194**, 155–162.

Proc. Fifth Int. Conf. on Borate Glasses, Crystals and Melts *Phys. Chem. Glasses: Eur. J. Glass Sci. Technol. B*, August 2006, **47** (4), 548–552

Luminescence and energy transfer properties of Sm^{3+}: borate glasses

S. Surendra Babu, C. K. Jayasankar, P. Babu*

Department of Physics, Sri Venkateswara University, Tirupati – 517 502, India

T. Tröster

University of Applied Sciences, Faculty of Automotive Systems and Production, D-50679 Cologne, Germany

W. Sievers & G. Wortmann

Universitat Paderborn, FB 6 Physik, 33095 Paderborn, Germany

The photoluminescent properties of borate based glasses that include: B_2O_3–PbO; B_2O_3–PbO–LiF; B_2O_3–ZnO; B_2O_3–ZnO–LiF; B_2O_3–ZnO–PbO and B_2O_3–ZnO–PbO–LiF with various Sm_2O_3 concentrations (0·01, 0·1, 1·0 and 2·5 mol%) have been investigated. The luminescence spectra and lifetimes of $^4G_{5/2}$ level of Sm^{3+} ions in these glasses have been measured using the 476·5 nm line of argon ion laser as an excitation source. In the binary glass systems studied, Sm^{3+}-doped B_2O_3–ZnO is having higher lifetimes for the $^4G_{5/2}$ level than B_2O_3–PbO system. When the oxyborate glasses are modified with LiF, a considerable increase in the lifetime has been noticed. The decay curves for these glasses are single exponential for lower (≤0·1 mol%) Sm^{3+} ion concentrations, but they become non-exponential with increasing concentration. The non-exponential decay curves have been well fitted to the Inokuti–Hirayama model and it is found that the interactions responsible for the energy transfer is of dipole–dipole type.

Introduction

Optical spectroscopic studies of glasses doped with trivalent lanthanide (Ln) ions have attracted great attention for the development of optical devices. Among the Ln ions Pr^{3+}, Nd^{3+}, Eu^{3+} and Er^{3+} are extensively studied ions as spectroscopically active centres for practical applications such as solid state lasers, high density storage devices, fibre amplifiers, probing the environment around the Ln ion, etc.[1] Relatively less studies are dedicated towards the optical spectroscopy of the trivalent samarium (Sm^{3+}) ions. Although the Sm^{3+} ion exhibits a strong luminescence in the orange region and finds application in high density optical storage, under sea water communication, colour displays, etc, it is one of the less investigated Ln ions. The complicated electronic structure of the $4f^5$ configuration of Sm^{3+} ion and the large number of energy levels lying close to each other makes the interpretation of the absorption spectra more cumbersome, which may be a reason to take up this ion for detailed analysis.[2–13] However, the fluorescence from the $^4G_{5/2}$ level of Sm^{3+} ion possess relatively high quantum efficiency and shows different emission quenching channels which make Sm^{3+} ion as an interesting case to analyse the energy transfer process.[4,10,12,14]

Among the glass hosts, borate glasses have attracted much attention due to their high transparency, low melting point, high thermal stability and good rare earth ion solubility, easy preparation in large scale, shaping and cost effective properties. However, interest in borate glasses is also limited due to their high phonon energy, and it is difficult to obtain high efficient infrared to visible upconversion emission in Pr^{3+}, Nd^{3+}, Tm^{3+}, Er^{3+} and Ho^{3+}-doped borate based systems. On the other hand, the high phonon energy in borate glass is not responsible for Sm^{3+} normal 4*f* emission transitions, as there exist a large energy gap of 7000 cm^{-1} between the fluorescent $^4G_{5/2}$ level and the next lower $^6F_{11/2}$ level and sometimes it can accelerate the relaxation processes, which is necessary and beneficial for visible emission.[11,15] More over, borate glasses make up one of the families that more clearly show the relationship between glass structure and physical properties. The aim of the present study is to investigate the influence of the chemical composition as well as the concentration of active ion on the luminescence intensities in the borate based glasses. In this paper, the studies of photoluminescence properties of Sm^{3+}-doped borate and fluoroborate glasses have been reported and the results are compared with those reported for other Sm^{3+} glasses.

* Corresponding author. Email ckjaya@yahoo.com

Experimental

Samarium doped glasses with varying compositions of B_2O_3–PbO; B_2O_3–PbO–LiF; B_2O_3–ZnO; B_2O_3–ZnO–LiF; B_2O_3–ZnO–PbO and B_2O_3–ZnO–PbO–LiF as well as various Sm_2O_3 concentrations (0·01, 0·1, 1·0 and 2·5 mol%) were prepared by melt quenching technique as explained in our earlier papers.[3,6] Table 1 shows the details of composition of the glasses along with their labels.

The glass samples were excited by the 476·5 nm line of Ar^+ laser and the resulting fluorescence was collected at right angles to the excitation source and recorded using a 0·85 m double grating monochromator equipped with a cooled photomultiplier. Decay curves were obtained using a mechanical chopper with a multichannel scalar interfaced to a personal computer that recorded and averaged the signal at room temperature.

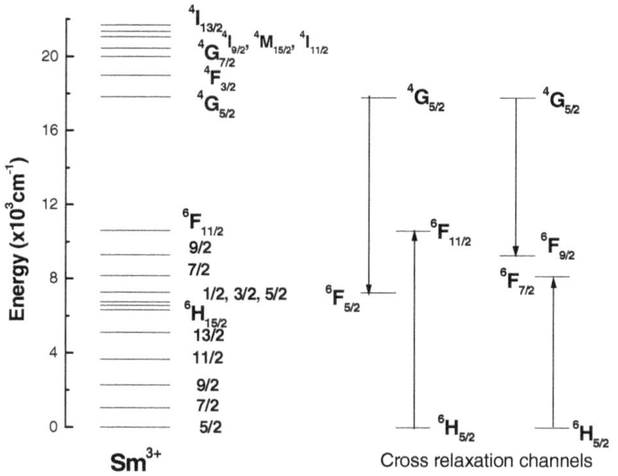

Figure 1. Energy level diagram and cross relaxation channels for Sm^{3+} ions in typical borate glasses

Results

The energy level diagram for 1·0 mol% of Sm^{3+} ions in lead borate glasses is presented in Figure 1, based on absorption and emission studies which are assigned with the help of free ion levels as well as earlier studies on Sm^{3+}-doped systems.[1–14] Emission spectra corresponding to an excitation at 476·5 nm of samples doped with 1·0 mol% of Sm^{3+} ions which include lead borate (BPS10), lead fluoroborate (BPLS10), lead zinc borate (BZPS10) and lead zinc fluoroborate (BZPLS10) glasses are presented in Figure 2. These emission spectra are normalised to have the same intensity in emission from fluorescent level. Figure 3 shows the luminescence decay obtained for the intense luminescent $^4G_{5/2}\rightarrow{}^6H_{7/2}$ transition after excitation at 476·5 nm for some of the title glasses with Sm^{3+} ion concentration equal to or greater than 1·0 mol% which includes BPS25, BZS10 and BZPLS10 systems. These profiles are non-exponential in nature and such non-exponential behaviour for high concentration of Sm^{3+} ions was also found in other reported Sm^{3+}: systems.[4,10,12–14] These decay profiles are well fitted to the Inokuti–Hirayama (I-H) model[16] for *S*=6. The I-H model fitting for all the three cases, i.e. for *S*=6, 8 and 10, is shown for BPS25 glass. The results derived from I-H model are presented in Table 1. The inset of Figure 3 shows the decay profiles of $^4G_{5/2}$ level in glasses with Sm^{3+} ion concentration equal to or lower than 0·1 mol% which includes BPS001, BZPS01, BZS01 and BZS001 systems to show the compositional dependence. The variation of lifetime with different glass compositions doped with 1·0 mol% of Sm^{3+} ions is shown in Figure 4.

Discussion

As seen from Figure 1, when any level above $^4G_{5/2}$ is optically excited, there is a fast non-radiative relaxation to this fluorescent level and consequently the radiative emission takes place from this level to the ground multiplets. Several emission bands

Table 1. Title of the glasses, compositions, experimental lifetimes (τ, μs) of the $^4G_{5/2}$ level, and I-H model fitting parameters (α, R_0 (Å)) and C_{DA} ($\times10^{-51}$ m^6s^{-1}) of Sm^{3+}-doped borate based glasses

Glass title	*Composition (mol%)* B_2O_3	*ZnO*	*PbO*	*LiF*	Sm_2O_3	τ (μs) ±10 μs	α	R_0 (Å)	C_{DA} ($\times10^{-51}$ m^6s^{-1})
BZS001	49·995	49·995	--	--	0·01	2427	--	--	--
BZS01	49·95	49·95	--	--	0·1	2133	--	--	--
BZS10	49·5	49·5	--	--	1·0	1181	1·487	522·5	8390
BZS25	48·75	48·75	--	--	2·5	624	2·150	508·1	7090
BPS001	49·995	--	49·995	--	0·01	1492	--	--	--
BPS01	49·95	--	49·95	--	0·1	1417	--	--	--
BPS10	49·5	--	49·5	--	1·0	908	0·693	461·1	6440
BPS25	48·75	--	48·75	--	2·5	485	1·667	458·9	6250
BZLS01	49·9	30	--	20	0·1	2429	--	--	--
BZLS10	49·5	29·5	--	20	1·0	1366	1·053	477·2	4860
BPLS01	49·9	--	30	20	0·1	1669	--	--	--
BPLS10	49·5	--	29·5	20	1·0	916	0·907	493·1	8620
BZPS01	49·9	25	25	--	0·1	1748	--	--	--
BZPS10	49·0	25	25	--	1·0	1007	0·821	462·5	5600
BZPLS01	49·9	15	15	20	0·1	2322	--	--	--
BZPLS10	49·9	15	15	20	1·0	1356	0·867	465·0	4350

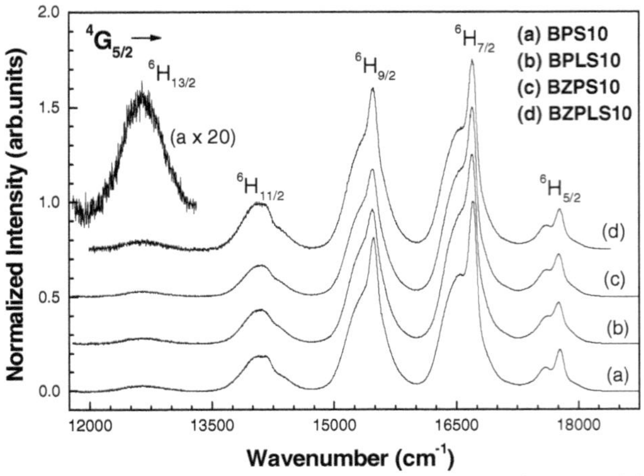

Figure 2. Normalised emission spectrum of 1·0 mol% Sm^{3+}-doped glasses. For details of the title glass compositions, see Table 1

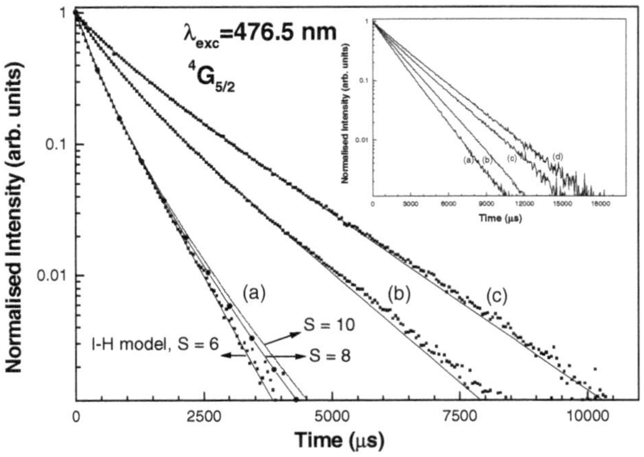

Figure 3. Decay curves for the $^4G_{5/2}\rightarrow{}^6H_{7/2}$ transition of Sm^{3+} for (a). BPS25, (b). BZS10 and (c). BZPLS10 glasses. The I-H model fits for S=6 (solid line), S=8 (line + circles) and S=10 (doted line) are shown for 2·5 mol% lead borate (BPS25) glass. Inset of the figure corresponds to the decay curves of (a) BPS001, (b) BZPS01, (c) BZS01 and (d) BZS001 glasses

due to the $^4G_{5/2}\rightarrow{}^6H_J$ (J=5/2, 7/2, 9/2, 11/2 and 13/2) transitions of Sm^{3+} ions have been observed and are shown in Figure 2. All the samples showed a strong reddish orange light emission which corresponds to $^4G_{5/2}\rightarrow{}^6H_{7/2}$ transition of Sm^{3+} ions in these glasses. The emission spectra obtained for all the labelled glasses are very much similar to each other and are comparable to those obtained for other reported Sm^{3+} glasses,[4–13] except for some small changes in intensity and shifts in emission band positions. Further, it is observed that there is a decrease in the intensity of luminescent bands with increase in the Sm^{3+} ion concentration which is attributed to concentration quenching of the luminescence.

The strong emission band with high branching ratio of 0·44 to 0·46 for all the glasses, corresponds to $^4G_{5/2}\rightarrow{}^6H_{7/2}$ transition at around 600 nm. The effective band width (in nm) of this transition in the labelled glasses with 1·0 mol% of Sm^{3+} ions follow the trend as BPLS10 (18·25±1)>BPS10 (17·73±1)>BZPS10 (17·68±1)>BZPLS10 (17·65±1)>BZLS10 (17·42±1)>BZS10 (17·17±1). These effective band widths are comparable with those values of 16·5 nm for $CdO–Al_2O_3–SiO_2–Sm_2O_3$,[8] 14·52 nm for $PbO–PbF_2–B_2O_3–Sm_2O_3$,[9] 17·0 nm for $Li_2O–BaO–La_2O_3–B_2O_3–Sm_2O_3$,[11] 6·67 nm for $H_3BO_3–Li_2CO_3–LiF–Sm_2O_3$[12] and 12·5 nm for $P_2O_3–Na_2O–Al_2O_3–B_2O_3–ZnF_2–Sm_2O_3$[13] systems. The branching ratio (β_R) and the peak stimulated emission cross section (σ_e), which depends on the effective band width are important in order to estimate the intensity of the probable lasing transitions for all the studied glass systems which will be reported in our future communication.[17]

From the inset of Figure 3, it is clear that the decay profiles of $^4G_{5/2}$ level with lower concentrations are perfectly single exponential, and indicates the absence of energy transfer processes among the Sm^{3+} ions at lower concentrations. For all the glass systems examined, the luminescence decay curves turns exhibit non-exponential behaviour with a quenching of lifetimes when the concentration is increased from 0·1 to 1·0 mol%. The quenching of lifetime and onset non-exponential behaviour of decay profiles at concentrations higher than 1·0 mol% is due to the non-radiative transfer of the excited energy.

A convenient way to explain the mechanism of non-radiative energy transfer processes is luminescence decay curve analysis. Since the interaction between the Sm^{3+} ions is negligible at lower concentration, the luminescence decay curves are single exponential and a determination of lifetime is straightforward. With Sm^{3+} ion concentration increasing, the experimental lifetimes of excited states significantly decrease, which indicates that activator–activator relaxation processes play an important role. Thus, one of the following two mechanisms of energy transfer processes resulting in luminescence quenching can be dominant.

The first one is attributed to the cross relaxation between pairs of Sm^{3+} ions. This mechanisms, which is indicated in Figure 1 is the cross relaxation with the following cross-relaxation channels

$$(^4G_{5/2}, {}^6H_{5/2}) \rightarrow (^6F_{5/2}, {}^6F_{11/2})$$

This cross relaxation is due to the energy transfer from the Sm^{3+} ions in an excited $^4G_{5/2}$ state, to a near by Sm^{3+} ion in the ground $^6H_{5/2}$ state. This transfer leaves the first ion in the intermediate level of $^6F_{5/2}$ at around 1375 nm and the second one in $^6F_{11/2}$ at around 942 nm, which occur in resonance with the $^4G_{5/2}\rightarrow{}^6F_{5/2}$ transition at around 947·6 nm in the case of BPS10 glass. Later both the ions quickly decay non-radiatively to the ground state. Figure 1 also demonstrates the other possible mechanism of cross relaxation process involving the channel

$(^4G_{5/2}, ^6H_{5/2}) \rightarrow (^6F_{9/2}, ^6F_{7/2})$

Similar arguments can be drawn in the case of other labelled glasses except for some changes in the energy values.

The second process is connected with the migration of the excitation energy, which can accelerate the decay by an energy transfer to the structural defects acting as energy sinks. In the case of high pressure dependent luminescence studies, Lavin *et al*[14] had given a detailed analysis of such energy migration processes, where the existence of pressure induced structural defect centres is a common phenomenon. Lavin *et al*[14] also concluded that the contribution due to energy migration process may be neglected.

In order to estimate the ion–ion interaction, the Inokuti–Hirayama model[16] was applied, which is applicable only for the analysis of energy transfer processes, where donor–acceptor transfer is faster than migration, for luminescence decay curves analysis. Considering the non-exponential character of the decay, the time evolution of the luminescence intensity was fitted to the following formula

$$I(t) = I_0 \exp\left[-(t/\tau_0) - \alpha(t/\tau_0)^{3/s}\right] \quad (1)$$

where I_0 is a constant, $I(t)$ is the luminescence intensity after pulse excitation, τ_0 is the intrinsic lifetime of donor in the absence of acceptor, S=6, 8 or 10 depending on whether the dominant mechanism of the interaction is dipole–dipole, dipole–quadrupole or quadrupole–quadrupole, respectively, and α is the energy transfer parameter and is given by the relation

$$\alpha = \frac{4\pi}{3}\Gamma(1-3/s)N_a R_0^3 \quad (2)$$

where Γ is the gamma function, N_a is the concentration of acceptor ions and R_0 is the critical transfer distance defined as a donor–acceptor separation for which the rate of energy transfer between a donor-acceptor is equal to the rate of intrinsic decay rate τ_0^{-1}.

Figure 3 shows the decay profiles fitted with I-H model and the associated parameters are presented in Table 1, where S=6 is the best fit case confirming the dominant interaction for energy transfer through cross relaxation of Sm^{3+} ions is of dipole–dipole type. The α, R_0 and C_{DA} parameters presented in Table 1 are in comparable with other reported Sm^{3+}:glass systems.[4,10,12–14,18] Similar results of non-exponential behaviour has been fitted to I-H model for S=6 confirming the dipole–dipole interaction for Sm^{3+} ions doped fluorozincate,[4] oxyfluoride[9] and tellurite[18] glasses.

Figure 4 shows the variation of lifetime with glass composition. It is evident from the Figure 4 that when lead borate glass is modified with LiF, the lifetime increases. From Table 1, it can be seen that except for lead borate system the C_{DA} are found to decrease with increasing lifetime. For example, the lifetime of BZS10 is 1181±10 μs and C_{DA} is 8390×10^{-51} m^{6}s^{-1} and when it is modified with LiF (BZLS10), the lifetime increases to 1366±10 μs with decreasing C_{DA} with a value of 4860×10^{-51} m^{6}s^{-1}. Similar trend of C_{DA} parameters are found for borate glasses doped with Dy^{3+} ions in our earlier work,[15] where lead borate glasses also showed the opposite trend. The decrease in lifetime with increasing C_{DA} also suggests that energy transfer by the dipole–dipole interaction between the Sm^{3+} ions is the major cause for the quenching of lifetime which leads to non-exponential behaviour of decay profiles with increasing Sm^{3+} ion concentration.

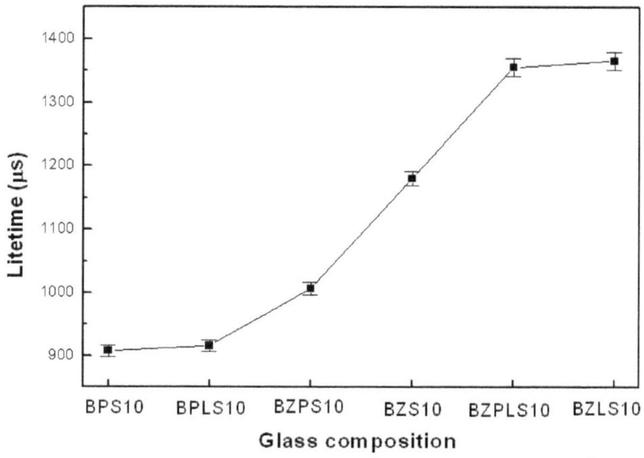

Figure 4. The variation of lifetimes for $^4G_{5/2}$ level of Sm^{3+} ion with glass composition of 1·0 mol% Sm^{3+}-doped glasses

As seen from Table 1, the experimental lifetime is much higher in the case of zinc borate glasses than in lead borate and lead zinc borate glasses. The lifetime of $^4G_{5/2}$ level in the glasses with particular concentration follows the trend as zinc borate>lead zinc borate>lead borate glasses. For example for 1·0 mol% Sm^{3+} doped system, zinc borate (BZLS10) glasses is having a lifetime of 1366±10 μs which is greater than the lifetime (1356±10 μs) of lead zinc borate (BZPLS10) glasses and lead borate (BPS10) glasses with lifetime of 908±10 μs. A clear knowledge of this trend can also be seen in the case of glasses with 0·1 mol% Sm^{3+} ions. Among the binary glasses with 0·1 mol% of Sm^{3+} ion concentration, ZnO based borate glasses are having very much longer lifetime with a value of 2427±10 μs than PbO based borate glasses with τ of 1496±10 μs. This trend of zinc borate glasses having longer lifetime than lead borate glasses persists for all the concentration of Sm^{3+} ions. This could be due to more ionic nature (strength) of Zn–O bond than Pb–O bond. When the pure zinc borate (~50 mol% of ZnO) is replaced with lead borate (~ with 25 mol% PbO) the strength is also expected to reduce from its initial strength and in turn resulting almost average value of lifetime, i.e. 1748±10 μs for 0·1 mol% of Sm^{3+} and 1007±10 μs for 1·0 mol% of Sm^{3+}.

By careful observation of the lifetimes in the labelled glasses, it is found that fluoride ions are having more significant influence on the lifetimes for the 0·1 mol% Sm^{3+}-doped glasses. For example, the difference in lifetimes between the BZS01 and BZLS01 glass is 296±10 μs whereas between BZS10 and BZLS10 glasses the difference is 185±10 μs. Similar differences are also noticed in the case of other compositions, except for BPS10 and BPLS10 where the LiF is not having any significant effect on the observed lifetimes. The main influence of fluoride ions is to reduce the phonon energy of the host and hence to increase the lifetime of the fluorescent level by decreasing the non-radiative decay. This is clearly seen in the case of 0·1 mol% glasses whereas in the case of 1·0 mol% glasses, this effect is masked by the energy transfer due to Sm^{3+}–Sm^{3+} ion interaction.

Conclusions

The luminescent properties of Sm^{3+} ion in borate based glasses are presented. The emission spectra of all the glasses are found to be similar to each other, and concentration quenching is found in luminescence intensity with increasing Sm^{3+} ion concentration. Further, the effective linewidth of $^4G_{5/2}\rightarrow{}^6H_{7/2}$ transition is found to be higher than for other reported Sm^{3+} glass systems. The lifetime of $^4G_{5/2}$ level has been found to depend strongly on Sm^{3+} ion concentration and composition of the glass. Among the binary glasses, zinc borate glasses exhibit the longer lifetimes. When these oxyborate glasses are modified with LiF, a considerable enhancement in the lifetime is noticed, except for lead borate glass with 1·0 mol% of Sm^{3+} ion concentration. The decay curves for these glasses are single exponential for glasses with Sm^{3+} ion concentrations lower than 0·1 mol% and turns to a non-exponential behavior with increasing concentrations. The non-exponential decay curves have been well fitted to the Inokuti–Hirayama model for S=6, indicating the type of cross relaxation is of dipole–dipole nature, leading to non-radiative energy transfer among Sm^{3+}–Sm^{3+} ions.

Acknowledgements

One of the authors (CKJ) is grateful to the University Grants Commission, New Delhi, India for the award of the Research Project (F.10-9/2002 (SR-1)) and to the Department Physik, Universität Paderborn, Germany, for the stay as a Guest Scientist.

References

1. Gorller-Walrand, C. & Binnemans, K. *Handbook on the Physics and Chemistry of Rare Earths. Vol. 25.* 1998. Elsevier B. V., North-Holland, Amsterdam. P 101.
2. May, P. S., Metcalf, D. H., Richardson, F. S., Carter, R. C., Miller, C. E. & Palmer, R. A. *J. Lumin.*, 1992, **51**, 249.
3. Jayasankar, C. K. & Rukmini, E. Opt. Mater., 1997, **8**, 193.
4. Rodriguez, V. D., Martin, I. R., Alcala, R. & Cases, R. *J. Lumin.*, 1992, **54**, 231.
5. Van Deun, R., Binnemans, K., Gorller-Walrand, C. & Adam, J. L. *SPIE. Proc.*, 1999, **3622**, 175.
6. Jayasankar, C. K. & Babu, P. *J. Alloys Compounds*, 2000, **307**, 82.
7. Binnemans, K., Van Deun, R., Gorller-Walrand, C. & Adam, J. L. *J. Non-Cryst. Solids*, 1998, **238**, 11.
8. Lin, H., Pun, E. Y. B., Huang, L.H. & Lin, X. R. *Appl. Phys. Lett.*, 2002, **80**, 2642.
9. Souza Filho, A. G., Mendes Filho, L., Melo, F. E. A., Custodio, M. C. C., Lebullenger, R. & Hernandes, A. C. J. *Phys. Chem. Solids*, 2000, **61**, 1535.
10. Nachimuthu, P., Jagannathan, R., Nirmal Kumar, V. & Narayana Rao, D. *J. Non-Cryst. Solids*, 1997, **217**, 215.
11. Lin, H., Yang, D., Liu, G., Ma, T., Zhai, B., An, Q., Yu, J., Wang, X., Liu, X. & Pun, E. Y. B. *J. Lumin.*, 2005, **113**, 121.
12. Mahato, K. K., Rai, D. K. & Rai, S. B. *Solid State Commun.*, 1998, **108**, 671.
13. Sharma, Y. K., Surana, S. S. L., Dubedi, R. P. & Joshi, V. *Mater. Sci. Eng. B*, 2005, **119**, 131.
14. Lavin, V., Martin, I. R., Jayasankar, C. K. & Tröster, T. *Phys. Rev. B*, 2002, **66**, 064207.
15. Jayasankar, C. K., Venkartamu, V., Surendra Babu, S. & Babu, P. *J. Alloys Compounds*, 2004, **374**, 22.
16. Inokuti, M. & Hirayama, F. *J. Chem. Phys.*, 1965, **43**, 1978.
17. Surendra Babau, S., Babu, P., Jayasankar, C. K., Wortmann, G., Tröster, T. & Sievers, W. (to be communicated).
18. Kumar, A., Rai, D. K. & Rai, S. B. *Spectrochim. Acta Part A*, 2003, **59**, 917.

Holmium doped borate glasses with high lead oxide content by spectroscopic studies

J. Pisarska,[1,*] G. Dominiak-Dzik,[2] W. Ryba-Romanowski,[2] T. Goryczka[3] & W. A. Pisarski[3]

[1] Silesian University of Technology, Department of Materials Science, Katowice, Poland
[2] Institute of Low Temperature and Structure Research, PAS Wrocław, Poland
[3] University of Silesia, Institute of Materials Science, Katowice, Poland

Holmium doped borate glasses with high lead oxide content have been investigated using spectroscopic methods. The radiative parameters were determined based on both absorption and emission measurements and Judd–Ofelt analysis. Green luminescence at 553 nm, corresponding to the $(^5S_2, ^5F_4) \rightarrow {}^5I_8$ transition of Ho^{3+} ions, was observed. Despite the presence of a heavy metal element, the quantum efficiency of emission is low. Strong luminescence quenching from the thermally coupled $^5S_2 + {}^5F_4$ excited level of Ho^{3+} ions is observed. It is due to the dominant role of a multiphonon relaxation process and presence of OH^- groups.

1. Introduction

Borate glasses belong to the group of the most extensively studied materials in the fields of physics, chemistry, materials science and engineering, because of their high potential for technological applications as a solid state battery, electrochromics display, phosphors and nonlinear optical (NLO) materials. Also, they are important materials for insulation (glass wool), textile (continuous filament) fibre glass, as well as being widely used in laser medicine, optical communication, signal processing and other fields of integrated optics.

The structure of borate glasses has been studied using infrared and Raman spectroscopy.[(1)] Structural differences can be caused by existence of various borate groups, correlated to chemical composition, kind of modifiers and conditions during glass manufacturing. Moreover, borate glasses create an excellent host to incorporate optically active ions. Thus, rare earth doped borate glasses present interesting spectroscopic properties.[(2)] The addition of a heavy metal oxide such as PbO to a borate glass significantly increases its thermal stability, IR transparency, refractive index, optical nonlinearity and radiative parameters. Especially, $PbO–B_2O_3$ based glasses, where differences between masses of the lead and the boron atom are significant, reveal improvement of spectroscopic properties. The effect of lead oxide on optical properties of Dy^{3+},[(3,4)] Sm^{3+}[(5)] and Pr^{3+}[(6)] ions in some borate based glasses has been stated.

Recently, the spectroscopic studies were focused on Pr^{3+} and Tm^{3+} ions in lead borate based glasses.[(7)] In the present work, it has been extended to Ho^{3+} doped lead borate glass in $PbO–B_2O_3–Al_2O_3–WO_3$ system. Spectroscopic analysis has been applied for Ho^{3+} ions in mixed alkali borate glasses.[(8,9)] Various optical glasses activated by trivalent holmium have been investigated as a potential light source in the visible or mid-IR spectral region. The spectroscopic parameters including measured and calculated oscillator strengths, Judd–Ofelt intensity parameters, radiative transition rates, branching ratios, measured and calculated radiative lifetimes, quantum efficiencies of excited states, spectral line widths as well as emission cross sections have been determined from absorption and emission measurements. Spectroscopic properties of holmium ions in borate glasses with high PbO content have been analyzed in the framework of the Judd–Ofelt theory.

2. Experimental

The nominal chemical composition of lead borate glass was as follows (in wt%): $72{\cdot}5PbO.18{\cdot}5B_2O_3.5Al_2O_3.3WO_3.1Ho_2O_3$. Appropriate amounts of lead oxide, boron oxide, aluminum oxide, tungsten oxide and rare earth oxide of high purity (99·99%, Aldrich Chemical Co.) were mixed homogeneously together in an agate ball mill for 2 h. The lead borate glasses containing rare earth ions (PBAW:Ho) were melted at 900°C, following by quenching. In order to eliminate an internal stresses, samples were annealed at a temperature below T_g.

Optical absorption spectra were recorded using a Varian 2300 UV-VIS-NIR spectrophotometer. Luminescence spectra were recorded using a Continuum Model Surelite I optical parametric oscillator pumped by a third harmonic of a Nd:YAG laser. Luminescence

1 Corresponding author. Email: Joanna.Pisarska@polsl.pl

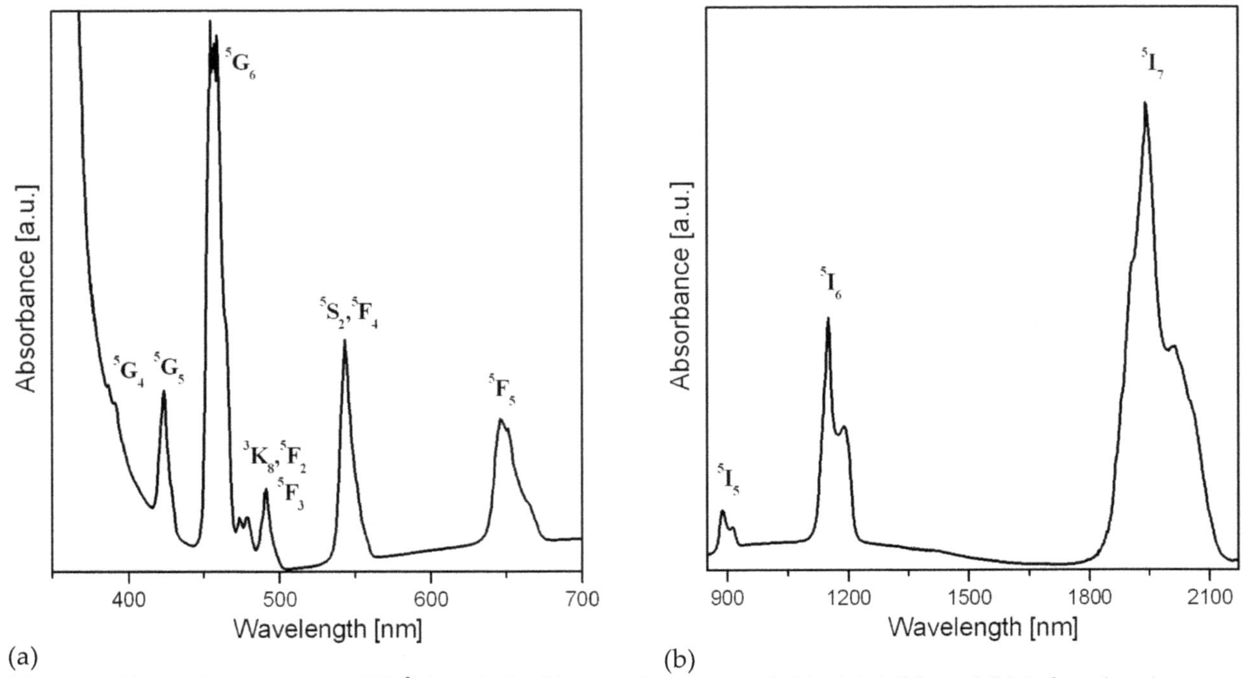

Figure 1. Absorption spectrum of Ho^{3+} ions in lead borate glasses recorded in (a)visible and (b) infrared region

was dispersed by a 1 m double grating monochromator and detected with a photomultiplier with S-20 spectral response. Luminescence spectra were collected using a Stanford SRS 250 boxcar integrator controlled by a computer. Luminescence decay curves were recorded and stored by a Tektronix TDS 3052 oscilloscope. All measurements were carried out at room temperature. The refractive index of the glass matrix at several wavelengths in the visible was measured using a prism method. The IR transmission spectra in the frequency region of 400–4000 cm^{-1} were taken on BIORAD spectrometer using the KBr pellet disc technique.

3. Results

Figure 1 shows absorption spectrum of Ho^{3+} doped lead borate glass (PBAW:Ho) registered at room temperature. Areas under absorption bands were numerically determined and the values of oscillator strengths have been derived. Next, the theoretical oscillator strengths for each transition of Ho^{3+} ions were calculated basing on the Judd–Ofelt theory[10,11] using the relation

$$P_{\text{calc}} = \frac{8\pi^2 mc(n^2+2)^2}{3h\lambda(2J+1)9n} \sum_{t=2,4,6} \Omega_t (< 4f^N J \| U^t \| 4f^N J' >)^2 \qquad (1)$$

where m is the mass of the electron, c is the velocity of light, h is the Planck constant and λ is the mean wavelength of the transition. In the analysis, a constant value of 1·92 was used for n, the refractive index of the medium, $\| U_t \|^2$ represents the square of the matrix elements of the unit tensor operator U_t connecting the initial and final states. The squared reduced matrix elements $\| U_t \|^2$ taken from Rukmini & Jayasankar[12] have been used for calculations. Equation (1) accounts for electric dipole line strengths only. Measured and calculated data are shown in Table 1. The three Ω_t intensity parameters were evaluated from the least square fit of measured P_{meas} and calculated P_{calc} oscillator strengths. The phenomenological Judd–Ofelt parameters for Ho^{3+} ions in lead borate based glasses are found to be; Ω_2=2·90, Ω_4=2·10, Ω_6=1·41 in 10^{-20} cm^2 units. The quality of the fit can be expressed by the magnitude of the root mean square (rms) deviation, defined by $(\Sigma(P_{\text{meas}}-P_{\text{calc}})^2/N)^{1/2}$, where N denotes number of levels used. The rms deviation of the fitted values equals $0·35\times10^{-6}$. The Ω_t values were used to calculate the radiative transition probabilities, branching ratios and radiative lifetimes. The radiative transition probability A_J for excited levels of rare earth

Table 1. Measured and calculated oscillator strengths for Ho^{3+} ions in lead borate glasses. Transitions are from the $^5I_8$ ground state to the indicated levels

Ω_2=2·90, Ω_4=2·10, Ω_6=1·41 (10^{-20} cm^2 units); rms=$0·35\times10^{-6}$

Levels	Average energy ν [cm^{-1}]	Oscillator strengths P_{meas} ($\times10^{-6}$)	P_{calc} ($\times10^{-6}$)
$^5I_7$	5050	1·520	1·512
$^5I_6$	8550	0·830	1·107
$^5I_5$	11100	0·180	0·203
$^5F_5$	15550	3·400	3·136
$^5S_2$, $^5F_4$	18600	4·220	4·012
$^5F_3$, ($^5F_2$, $^3K_8$), $^5G_6$	21400	18·770	18·767
$^5G_5$	23900	2·740	3·145

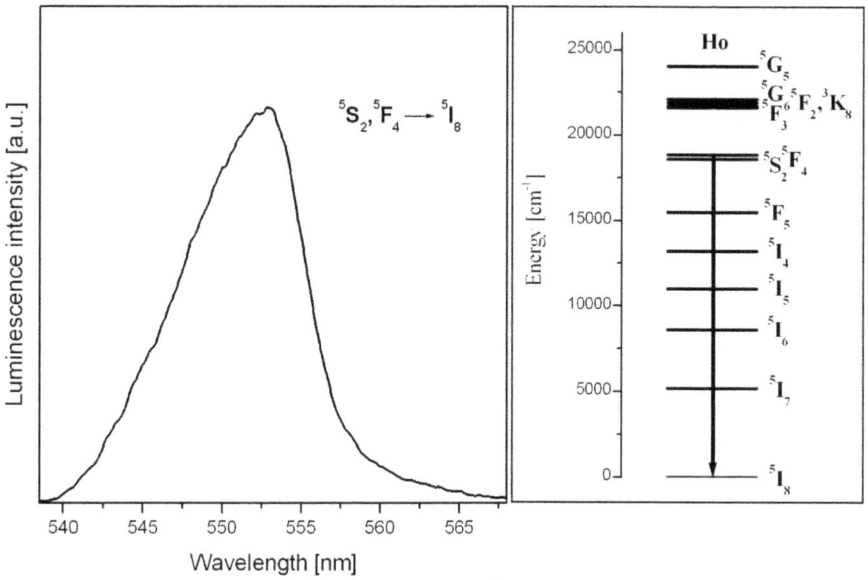

Figure 2. Luminescence spectrum and energy level scheme of Ho^{3+} ions in lead borate glass

ions from an initial state J to a final ground state J' can be calculated from formula

$$A_J = \frac{64\pi^4 e^2}{3h(2J+1)\lambda^3}\frac{n(n^2+2)^2}{9}\sum_{t=2,4,6}\Omega_t(<4f^N J\|U^t\|4f^N J'>)^2 \qquad (2)$$

The total radiative emission probability AT involving all the intermediate terms is given by the sum of the A_J terms calculated over all terminal states. Thus, radiative lifetime τ_{rad} of an excited level is given by the inverse of the total radiative emission probability

$$\tau_{rad} = \frac{1}{\sum_i A_{Ji}} = \frac{1}{A_T} \qquad (3)$$

The luminescence branching ratio β is defined as

$$\beta = \frac{A_J}{\sum_i A_{Ji}} \qquad (4)$$

Table 2. Calculated radiative transition rates A_J, luminescence branching ratios β and corresponding radiative lifetimes τ_{rad} for Ho^{3+} in lead borate glasses

Transition	*Energy ν [cm⁻¹]*	A_J [s⁻¹]	A_T [s⁻¹]	τ_{rad} [μs]	β
$(^5F_4, ^5S_2) \to {}^5F_5$	3050	13	6801	147	<0·01
$^5I_5$	7502	226			0·03
$^5I_6$	10050	451			0·08
$^5I_7$	13550	646			0·09
$^5I_8$	18083	5465			0·80
$^5F_5 \to {}^5I_5$	4450	13	3844	260	<0·01
$^5I_6$	6998	158			0·04
$^5I_7$	10504	730			0·19
$^5I_8$	15550	2943			0·77
$^5I_5 \to {}^5I_6$	2550	10	242	4132	0·04
$^5I_7$	6050	135			0·56
$^5I_8$	11100	97			0·40
$^5I_6 \to {}^5I_7$	3500	28	293	3413	0·10
$^5I_8$	8550	265			0·90
$^5I^7 \to {}^5I_8$	5050	110	110	9091	1·00

which indicates relative intensities of transitions from excited level to all terminal levels.

The calculated A_J and β values together with radiative lifetimes τ_{rad} for excited states of Ho^{3+} ions in the lead borate based glasses are given in Table 2.

Figure 2 shows luminescence spectrum of PBAW: Ho glass recorded at room temperature. Luminescence band located at 553 nm corresponds to the $(^5S_2, {}^5F_4) \to {}^5I_8$ transition of Ho^{3+} ions in lead borate based glass. The radiative parameters such as the effective linewidth $\Delta\lambda$ and the peak stimulated emission cross-section σ_{em} were determined.

The σ_{em} value of the $(^5S_2, {}^5F_4) \to {}^5I_8$ transition has been obtained from the calculated emission probability using the following equation

$$\sigma_{em} = \frac{\lambda_p^4}{8\pi c n^2 \Delta\lambda} A_J \qquad (5)$$

where λ_p is the peak emission wavelength, c is the velocity of light, n is the refractive index, and $\Delta\lambda$ is the effective linewidth defined as full width in half maximum (FWHM). Results are given in Table 3.

Figure 3 shows infrared spectrum of lead borate based glass containing Ho^{3+} ions. Characteristic bands at about 1600 cm^{-1} and 3400 cm^{-1} are connected with vibrations of the OH^- groups, whereas several infra-

Table 3. Radiative parameters for $(^5S_2, {}^5F_4) \to {}^5I_8$ transition of Ho^{3+} ions in lead borate glass

$(^5S_2, {}^5F_4) \to {}^5I_8$ transition (green)		
Peak emission wavelength	λ_p [nm]	553
Emission linewidth	$\Delta\lambda$ [nm]	9
Radiative transition rate	A_J [s⁻¹]	5465
Total radiative transition rate	A_T [s⁻¹]	6801
Luminescence branching ratio	β [%]	80
Measured lifetime	τ_{meas} [μs]	15
Quantum efficiency	η [%]	10
Emission cross-section	$\sigma_{em} \times 10^{20}$ [cm²]	2·04

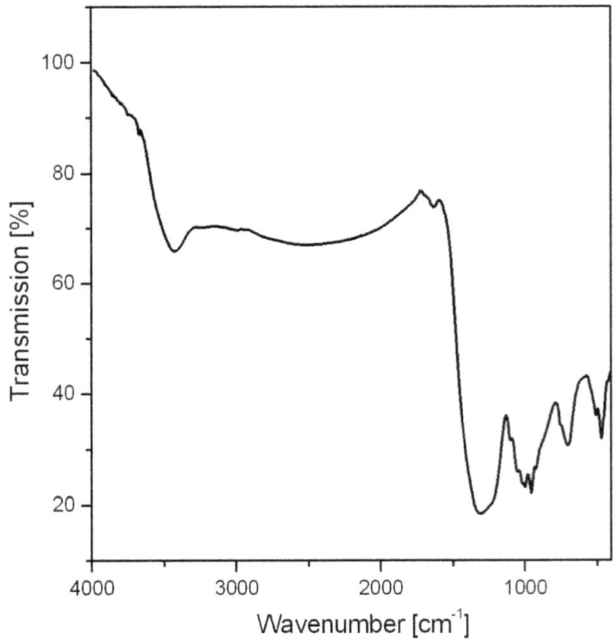

Figure 3. Infrared spectrum of lead borate glass containing Ho^{3+} ions

red bands due to the B–O vibrations are observed in the 400–1500 cm^{-1} spectral region.

4. Discussion

The absorption spectrum of Ho^{3+} ions in lead borate based glass (PBAW:Ho) is presented in Figure 1. The spectrum consists of inhomogeneously broadened absorption lines characteristic for $4f^{10}$–$4f^{10}$ transitions of trivalent holmium. Lines are located in the visible (a) and the infrared (b) ranges, respectively. They correspond to the transitions from the $^5I_8$ ground state to the $^5I_7$, $^5I_6$, $^5I_5$, $^5F_5$, ($^5F_4$, $^5S_2$), $^5F_3$, ($^5F_2$, $^3K_8$), $^5G_6$ and $^5G_5$ excited states. From absorption measurements together with Judd–Ofelt analysis, some spectroscopic parameters have been determined (Table 1 and 2).

In contrast to the borate glass with low PbO concentration,[13] the absorption edge is not located in UV region but near 400 nm. Thus, the absorption bands associated with transitions from the $^5I_8$ ground state to the $^3K_7$ and other high lying excited states of Ho^{3+} ions are completely masked by the absorption edge of the host, while band corresponding to the $^5I_8 \rightarrow {}^5G_4$ transition is located on the tail of the absorption edge. Increasing of PbO content in the borate glass shifts transparency in the direction of longer wavelengths.

On the other hand, radiative transition rates increase, which is useful for optical application, with increasing of heavy metal element (PbO) concentration. Thus, the total radiative transition rate from $^5S_2$ + $^5F_4$ excited levels of Ho^{3+} increases from 2909 s^{-1} [13] to 6801 s^{-1} value (Table 2), when the PbO:B_2O_3 ratio considerably increases in glass composition. Moreover, the $^5S_2$ and $^5F_4$ states are closely placed to each other, so in estimating the radiative transition rates of the $^5S_2$ + $^5F_4$ excited levels a thermal equilibrium effect must be taken into consideration. Thus, the level degeneracy and the energy difference between the $^5S_2$ and $^5F_4$ emitting levels should be known in order to calculate the effective spontaneous emission rate for thermally coupled $^5S_2$ + $^5F_4$ level. Unfortunately, accurate estimation of the energy gap between the $^5S_2$ and $^5F_4$ levels is difficult because of overlapping of the absorption bands, which corresponds to the $^5I_8 \rightarrow {}^5S_2$ and $^5I_8 \rightarrow {}^5F_4$ transitions (Figure 1). Thus, Free-ion calculation may be helpful.[12] However, this effect on the relaxation rate is too small and can be omitted.

Luminescence spectrum of PBAW:Ho glass is presented in Figure 2. Quite strong green luminescence at 553 nm corresponding to the ($^5S_2$, $^5F_4$) $\rightarrow$ $^5I_8$ transition of Ho^{3+} ions has been observed in borate glass with high lead oxide content, while other intensities of emission lines are negligible. It is consistent with Judd–Ofelt calculations, where line intensity ratio corresponds to luminescence branching ratios, presented in Table 2 (β=80% for ($^5S_2$, $^5F_4$) $\rightarrow$ $^5I_8$ transition). Since luminescence decay measurements were carried out at room temperature, the luminescence lifetime of the thermally coupled $^5S_2$ + $^5F_4$ level has been estimated. The measured and calculated lifetimes of the $^5S_2$ + $^5F_4$ excited level of holmium are close to 15 μs and 147 μs, respectively. This indicates that the quantum efficiency of emission is low (η=10%) in borate glass host even in the presence of heavy metal element. The low quantum efficiency can be associated with energy transfer process, multiphonon relaxation or occurrence of OH^- groups, which are effective quenching centers. The first one is rather not probable, because activator concentration is very low (1 wt% Ho) and Ho–Ho interaction is negligible. The shape of the luminescence decay curve was examined in detail. Single exponential shape confirms the absence of energy transfer process. The second one is more probable. For borate glasses, the multiphonon relaxation rates are usually 10^2 times larger than the radiative transition rates.[14] Also, two characteristic bands at about 1600 cm^{-1} and 3400 cm^{-1} associated with OH^- stretching vibrations have been observed in the infrared spectrum (Figure 3). From this point of view, multiphonon relaxation process together with OH^- groups occurrence play the dominant role in strong luminescence quenching from the $^5S_2$ + $^5F_4$ excited level of Ho^{3+} ions. It can be an explanation for shortening of the measured $^5S_2$ + $^5F_4$ lifetime of Ho^{3+} in lead borate glasses.

The high β value obtained for ($^5S_2$, $^5F_4$) $\rightarrow$ $^5I_8$ (green) transition characterises the lasing power of the potential laser transition. The radiative parameters for Ho^{3+} ions in borate glass with high PbO content are given in Table 3. The peak stimulated emission cross section σ_{em} and the effective linewidth $\Delta\lambda$ are presented. They belong to the most important parameters used to compute various spectroscopic properties, necessary for optical devices characterisation. The high σ_{em} and β values obtained for the ($^5S_2$, $^5F_4$) $\rightarrow$ $^5I_8$ transition of Ho^{3+} ions in lead borate based glass suggest that the host is promising as a green emitting material.

5. Summary

The spectroscopic properties of Ho^{3+} ions in borate glasses with high lead oxide content are reported. The radiative parameters are determined from absorption and emission measurements together with Judd–Ofelt analysis. Green luminescence at 553 nm corresponding to the ($^5S_2$, $^5F_4$) $\rightarrow$ $^5I_8$ transition of Ho^{3+} ions has been observed in lead borate based glass. Even in the presence of heavy metal element, the quantum efficiency of emission is low in borate glass host because of multiphonon relaxation process together with occurrence of OH^- groups. It plays the dominant role in strong luminescence quenching from the thermally coupled $^5S_2$ + $^5F_4$ excited level of Ho^{3+} ions. Increasing of PbO concentration shifts transparency in the direction of longer wavelengths. The proper choice of PbO concentration in borate glass composition is a compromise between light transparency in the UV-visible region and radiative parameters of the optically active ions.

Acknowledgments

This work was supported by the Polish Committee for Scientific Research (KBN) under grant No. 4 T08D 017 25.

References

1. Yiannopoulos, Y. D., Chryssikos, G. D. & Kamitsos, E. I. *Phys. Chem. Glasses*, 2001, **42**, 164.
2. Thyagarajan, K., Sooraj Hussain, N., Annapurna, K., Aruna, V., Rajamohan Reddy, K. & Buddhudu, S. *Phys. Chem. Glasses*, 1997, **38**, 223.
3. Srivastava, P., Rai, S. B. & Rai, D. K. *Spectrochim. Acta A*, 2003, **59**, 3303.
4. Jayasankar, C. K., Venkatramu, V., Surendra Babu, S. & Babu, P. *J. Alloys Compounds*, 2004, **374**, 22.
5. Srivastava, P., Rai, S. B. & Rai, D. K. *Spectrochim. Acta A*, 2004, **60**, 637.
6. Srivastava, P., Rai, S. B. & Rai, D. K. *J. Alloys Compounds*, 2004, **368**, 1.
7. Pisarski, W. A., Pisarska, J., Dominiak-Dzik, G. & Ryba-Romanowski, W. *J. Phys.: Condens. Matter*, 2004, **16**, 6171.
8. Ratnakaram, Y. C., Thirupathi Naidu, D., Vijaya Kumar, A. & Rao J. L. *J. Phys.: Condens. Matter*, 2004, **16**, 3779.
9. Ratnakaram, Y. C., Sreekanth Chakradhar, R. P., Ramesh, K. P., Rao, J. L. & Ramakrishna, J. *J. Mater. Sci.*, 2003, **38**, 833.
10. Judd, B. R. *Phys. Rev.*, 1962, **127**, 750.
11. Ofelt, G. S. *J. Chem. Phys.*, 1962, **37**, 511.
12. Rukmini, E. & Jayasankar, C. K. *Opt. Mater.*, 1995, **4**, 529.
13. Reddy, M. R., Raju, S. B. & Veeraiah, N. *J. Phys. Chem. Solids*, 2000, **61**, 1567.
14. Shojiya, M., Kawamoto, Y. & Kadono, K. *J. Appl. Phys.*, 2001, **89**, 4944.

Proc. Fifth Int. Conf. on Borate Glasses, Crystals and Melts *Phys. Chem. Glasses: Eur. J. Glass Sci. Technol. B*, August 2006, **47** (4), 558–562

An XAFS study of the local structure of europium ion in crystallised borosilicate glass

Toshihiro Okajima[1]

Kyushu Synchrotron Light Research Center, 8-7 Yayoigaoka, Tosu, Saga 841-0005, Japan

Norimasa Umesaki

Japan Synchrotron Radiation Research Institute, 1-1-1 Kohto, Mikazuki, Sayo, Hyogo 679-5198, Japan

Akio Konishi

New Glass Research Center, Nihon Yamamura Glass Co. Ltd., 2-1-18 Naruohama, Nishinomiya, Hyogo 663-8142, Japan/Advanced Industrial Science and Technology, Ikeda, Osaka 570-8501, Japan

Tetsuo Jin

Advanced Industrial Science and Technology, Ikeda, Osaka 570-8501, Japan

Isao Tanaka

Department of Materials Science and Engineering, Kyoto University, Sakyo, Kyoto 606-8501, Japan

Tetsuo Yazawa

Graduate School of Engineering, University of Hyogo, 2167 Shosha, Himeji, Hyogo 671-2201, Japan

The local atomic structure of Eu ions in the crystallised borosilicate glass, $62{\cdot}5SiO_2$–$28{\cdot}25B_2O_3$–$9{\cdot}25Na_2O$–$5Y_2O_3$, doped with 0·35 wt% Eu_2O_3 has been investigated through Eu L_{III}-edge x-ray absorption fine structure spectroscopy, using synchrotron radiation. The oxidation state of Eu ion has been proved to be +3 from the analysis of x-ray absorption near edge structure spectra. The bond length of Eu-O to the nearest neighbour in the borosilicate glass crystallised at 973 K is measured to be the same as that of Y–O in YBO_3 crystal by the analysis of extended x-ray absorption structure spectra. We discuss the environment of Eu ion in the borosilicate glass, taking the results of x-ray diffraction measurement into consideration. It has been clarified that the $EuBO_3$ crystal, substituted by Eu for Y, can be formed when the crystallisation temperature is relatively high.

1. Introduction

It is very important to develop a microsized next generation display device which drives with little power consumption, and shows a vivid colour. In order to develop the display device, we need a fluorescent material with long lived and high fluorescent strength. In our latest studies, we found that fluorescent strength from crystallised borosilicate glasses activated by europium (Eu) ions can be improved.[(1)] It was observed that the fluorescent characteristic is influenced by the heat treatment process for the crystallisation in the Eu ion doped borosilicate glass. In order to optimise the performance of the Eu ion doped borosilicate glass, detailed structural information in particular relating to the local structure of the optically active ion, that is Eu^{3+} ion, is required.

Because of its sensitivity to the short range order and atomic species surrounding to an absorbing site, x-ray absorption fine structure (XAFS) spectroscopy, including x-ray absorption near edge structure (XANES) and extended x-ray absorption fine structure (EXAFS), has been recognised as a powerful tool for determining the local structure around the absorbing site in the glasses. Furthermore, one can quantitatively obtain the structural parameters of one selected element in a complex material by XAFS. Therefore, XAFS is well suited for studying the local structure around the luminescent centres, which are always the low concentration dopants in the host.

In this study, we studied the change of local atomic structure of Eu ions in borosilicate glass as a function of crystallised temperature by XAFS analysis using synchrotron radiation.

[1] Corresponding author. Email address: okajima@saga-ls.jp

2. Experimental

62·5SiO_2–28·25B_2O_3–9·25Na_2O–5Y_2O_3 borosilicate glass doped with 0·35 wt% Eu_2O_3 was prepared. The glass was annealed at the temparatures between 773 and 1023 K for 24 h in air to obtain crystallised borosilicate glass. Eu_2O_3 powder was also used for a reference sample.

X-ray diffraction measurements were performed at room tempatrature, using Rigaku RAD-rC system. A conventional θ–2θ scanning technique in the symmetrical reflection mode was used, with Cu K_α radiation.

Eu L_{III}-edge XAFS spectra were obtained at BL01B1 of SPring-8 with a bending magnet system with two rhodium coated mirrors and a fixed-exit double-crystal monochromator using Si (311) planes.[2] The XAFS spectrum of Eu_2O_3 was obtained in the transmission mode. The incident and output beam intensities were monitored and recorded using a nitrogen gas and a 50% argon doped nitrogen flowing ionisation chamber. The XAFS spectra of Eu in doped borosilicate glass were obtained in the fluorescence yield mode. An ion chamber filled with N_2 gas was inserted into the optical path, and the photon flux was monitored. The emitted x-ray fluorescences from the sample were detected a multi-element solid state Ge detector. All measurements were done in air at room temperature. The spectra were scanned in the range of 6650–7560 eV, which covers the L_{III} absorption edge of Eu atoms. Samples were powdered before measurements. Borosilicate glass and a reference sample were diluted by high purity hexagonal BN powder.

Athena and Artemis interface to the Iffefit software were used for EXAFS data analysis.[3,4] Atomic absorption was removed, and the EXAFS spectra were extracted by using the Autbk routine.[5] The EXAFS spectra were apodised with a Kaizer-Bessel window and Fourier-transformed (FT) to R-space with k^2-weighting. Selected FT peaks were fitted by using phases and amplitude functions calculated using the FEFF code.[6]

3. Results and discussions

Figure 1 shows the XRD profiles of crystallised borosilicate glass with Eu ions annealed at the temperatures of 873, 923, 973 and 1023 K. All the profiles showed a broad peak at around θ=22°. The broad peak is attributed to the disordered structure of borosilicate glass. In the case of crystallised borosilicate glass annealed at 973 and 1023 K, an intense peak at θ=22° and some weak peaks were observed in addition to the broad peak. Peaks marked closed circle, which are containing the intense peak, are assigned to the peaks of cristobalite crystal given in JCPDS (No. 39–1425). The weak peaks marked closed triangle are also matched with YBO_3 standard values given in JCPDS (No. 16-0227). Cristobalite crystals are formed in the borosilicate glass by annealing at the temperature of 973 K and more. YBO_3 crystals are also formed in the borosilicate glass by annealing at the temperature of 1023 K. In contrast, the formations of these crystals weren't observed in the borosilicate glass at the annealing temperature of 923 K or less.

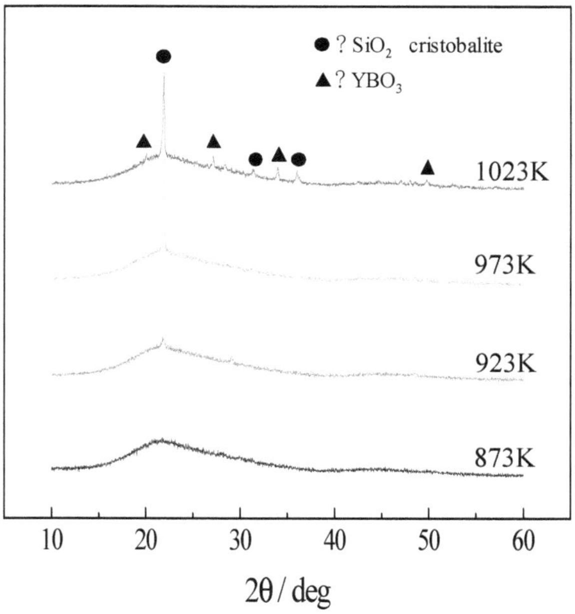

Figure 1. XRD profiles of crystallised Eu ions doped in borosilicate glass

The Eu L_{III}-edge XANES spectra for crystallised and noncrystallised borosilicate glass doped Eu ion with that for a reference sample are shown in Figure 2. The crystallised temparatures were 773 and 973 K The spectra obtained from borosilicate glass all showed an intense peak at 6980 eV and a weak peak at 6973 eV. The spectrum obtained from Eu_2O_3 showed an intense peak at 6981 eV. It is well known that the peak around L_{III} absorption edge of rare earths arise from the electron transition from Eu $2p_{2/3}$ to the Eu 4_f5_d final state, and the peak positions and features are sensitive to the chemical state and the valence state of the absorption atom.[7] The intense peak at 6980 eV and the weak peak at 6973 eV of borosilicate glass are attributed to Eu^{3+} ion, to Eu^{2+} ion, respectively.[8] This shows that most Eu ions in borosilicate glass are Eu^{3+}, with small amounts of Eu^{2+}. According to the previous study, it was reported that Eu^{3+} ion in glass changed to Eu^{2+} ion by x-ray irradiation.[9] We can consider that Eu^{2+} ion were produced by x-ray irradiation during measuring pre-edge region. Therefore, it would be concluded that the chemical state of most Eu ion in the borosilicate glass is the same as the chemical state of Eu_2O_3, and the valence state of Eu ion in the borosilicate glass is +3.

Figure 3 shows the k^2-weighted EXAFS spectra for

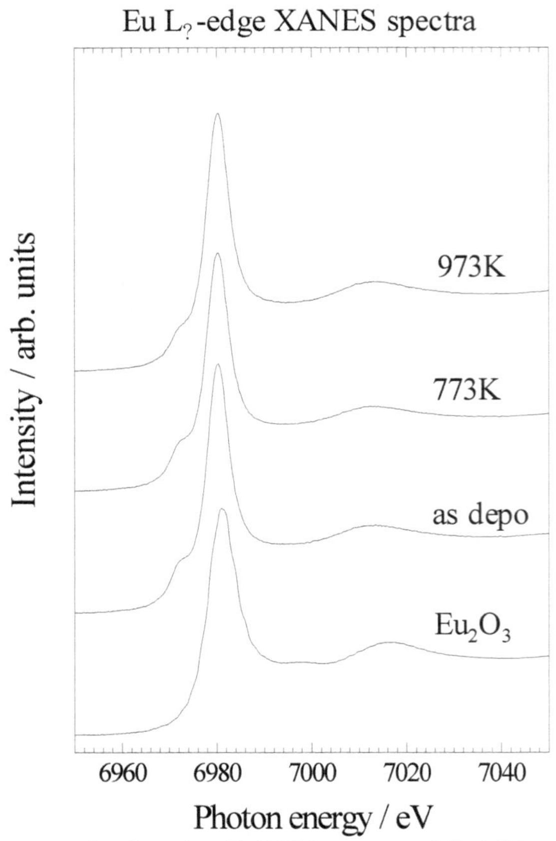

Figure 2. Eu L_{III}-edge XANES spectra of Eu2O3 and noncrystallised and crystallised Eu ion doped in borosilicate glass

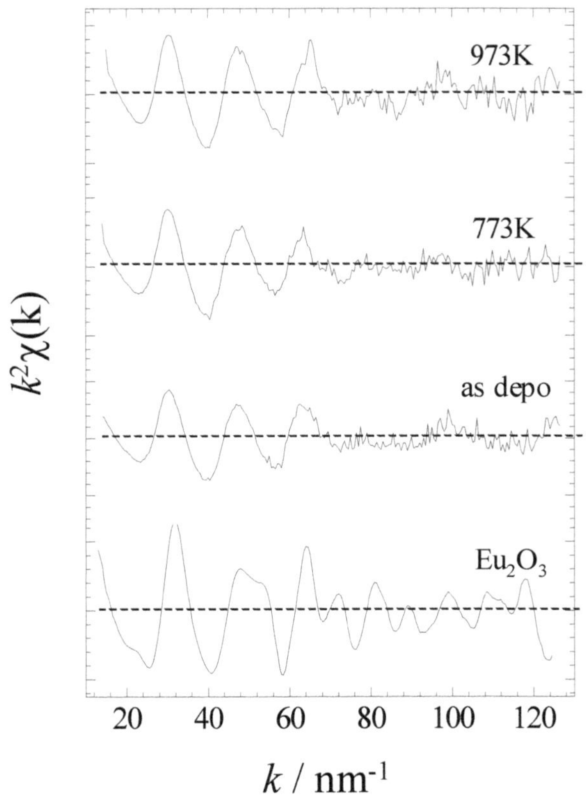

Figure 3. Eu L_{III}-edge EXAFS spectra of Eu_2O_3 and non-crystallised and crystallised Eu ion doped in borosilicate glass

crystallised borosilicate glass with that for reference sample. EXAFS oscilation obtained from reference sample was clearly observed in the range of 100 nm^{-1} and more. In contrast, the oscillations obtained from borosilicate glass were not clear in the range of 70 nm^{-1} and more. The EXAFS oscillations for borosilicate glass displayed also only single wave frequency, indicating the presence of a single ordered coordination sphere. In contrast, the oscillations for reference sample had distinct frequencies, indicating the presence of higher atomic shells. The radial distribution functions (RDF) were obtained from Fourier transform of $k^2\chi(k)$ and are shown in Figure 4. The Fourier transforms displayed a first peak around 0·2 nm, which could be related to contributions from oxygen atoms of nearest coordination sphere. High amplitude contributions at the range of 0·25 nm and more could be observed for Eu_2O_3 and originate mainly from next nearest Eu shells. In contrast, the FT peaks at the range of 0·25 nm and more were not observed for borosilicate glass. These phenomena resulted from the relatively disorder in the borosilicate glass.

The quantitative analyses of the first shell for all specimens are listed in Table 1. The fitting results for crystallised borosilicate glass also show in Figure 5. In the case of Eu_2O_3, the coordination number of first oxygen shell fixed to six from the structure of Eu_2O_3. The resulting bond length of Eu–O is 0·2345 nm was corresponding to the early study.[10] The amplitude reduction factor was set to the value obtained from the analysis of Eu_2O_3 to correctly reproduce the number of neighbouring atoms for the borosilicate glass. The Eu–O bond length was 0·24 nm for non crystallised borosilicate glass and for crystallised borosilicate glass at the temperature of 773 K, and was 0·237 nm for crystallised borosilicate glass at the temperature of 973 K. These bond lengths were longer than that of Eu_2O_3. In addition, the bond length of the sample crystallised at the temperature of 973 K was the shortest. On the other hand, the coordination number of first oxygen shell was about six, and increases according to the rise of crystallised temperature.

According to the analysis of XRD profiles, the YBO_3 crystal was expected to be generated in borosilicate glass at the crystallised temperature of 1023 K. The YBO_3 crystals gave the structure with two distinct Y–O, one for the six oxygen atoms located at the corner of a bicapped trigonal prism and the other for the two oxygen atoms located above the triangular faces.[11] The bond lengths of Y–O are 0·2386 and 0·2323 nm, respectively. That is to say,

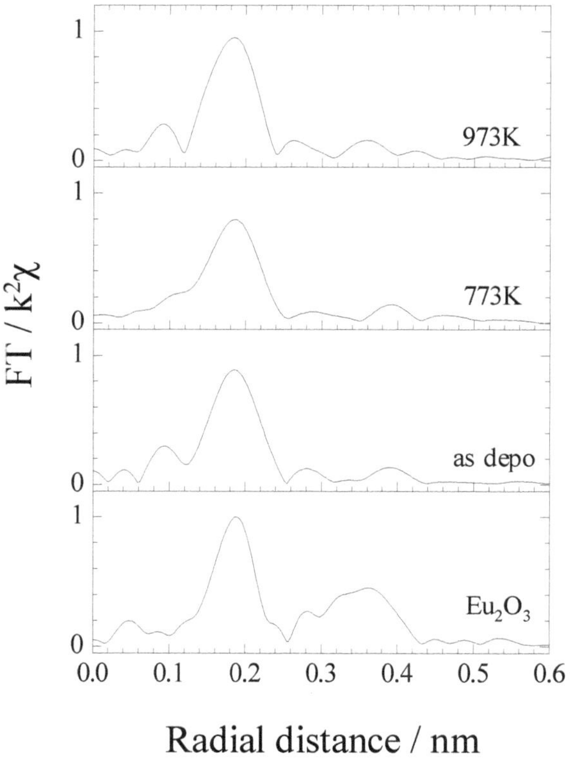

Figure 4. Eu L_{III}-edge Rasical distribution functions of Eu_2O_3 and noncrystallised and crystallised Eu ion doped in borosilicate glass

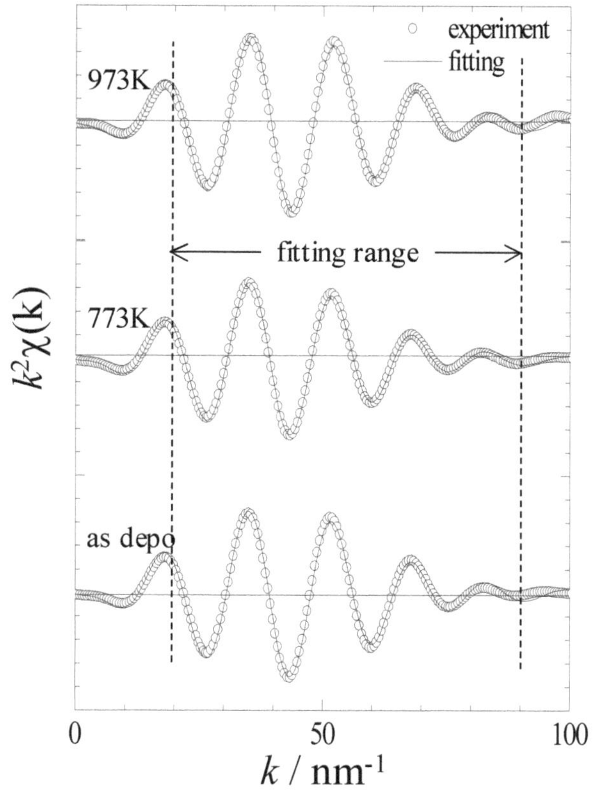

Figure 5. Fitted patterns of noncrystallised and crystallised Eu ion doped in borosilicate glass

we can consider the coordination number of first oxygen shell is eight in this system, and the averaged bond length is 0·2370 nm. The averaged bond length is corresponding to the bond length Eu–O in the borosilicate glass crystallised at the temperature of 973 K. From this result, it can be expected that the Eu ions substitute Y ions in YBO_3 crystals. However, the coordination number of the first oxygen shell is less than those of Y atom in YBO_3.

The result of XRD profile measurements showed that all the profiles had a broad peak. The broad peak is attributed to the disordered structure of borosilicate glass. In the disordered structure, it can be expected that the coordination number of first oxygen shell around Eu ion in borosilicate glass are less than that of YBO_3 crystal. On the other hand, YBO_3 crystals were formed in the borosilicate glass at crystallised temperature of 1023K. Because we can obtain averaged coordination number from EXAFS analysis, the coordination number of first oxygen shell around Eu ion in borosilicate glass is expected less than eight obtained YBO_3 structure. These results indicate that the most of Eu ions in borosilicate glass are in disordered structure of borosilicate glass, and YBO_3 crystals substituted by Eu ion for Y ion are formed with an increase of crystallised temperature.

Table 1. EXAFS Results for noncrystallised and crystallised borosilicate glasses and reference sample

Sample/ crystallisation temperature	R_{Eu-O} (nm)	N_0	σ^2 (nm^2)	R-factor
Eu_2O_3	0·2345 (18)	6·0 (fix)	0·00009 (8)	0·001
1/as depo	0·240 (4)	5·8 (2)	0·00010 (7)	0·007
2/773 K	0·240 (3)	5·9 (2)	0·00013 (3)	0·002
3/973 K	0·237 (4)	6·1 (3)	0·00010 (8)	0·008

4. Summary

Europium (Eu) doped borosilicate glass with 0·35 wt% Eu_2O_3 have been prepared, and the glass crystallised by annealing in air. X-ray absorption spectroscopic study was formed for noncrystallised and crystallised Eu doped in borosilicate glass annealed at the temperatures of 773 and 973 K. X-ray absorption near edge structure (XANES) spectra of the borosilicate glass showed that the chemical state of Eu ion in the borosilicate glass was +3 from the comparison of reference sample. XRD profiles were also obtained from borosilicate glasses. The profile from a borosilicate glass crystallised at the temperature 1023 K showed the pattern of YBO_3 crystals. The quantitative analysis of the first Eu–O path was performed for the extended x-ray absorption fine structure (EXAFS) spectra obtained from borosilicate glass annealed at the temperatures of 773 and 973 K. The bond length of Eu–O was almost the same as that of Y–O in the YBO_3 crystal at the temperatures of 973 K. These results are indicating that YBO_3 crystals substituted by Eu ion for Y ion are formed with an increase of crystallised temperature.

Acknowledgment

This work was supported in part by Grants in Hyogo Prefecture Corraboration of Regional Entities for the Advanced of Technical Excellence. This work has been performed under the approval of SPring-8 Advisory Committee (proposal Nos. 2003A0849-RI-np, 2003A0477-NX-np).

References

1. Private communications.
2. Uruga, T., Tanida, H., Yoneda, Y., Takeshita, K., Emura, S., Takahashi, M., Harada, M., Nishihata, Y., Yamamoto, T., Maeda, H., Kamishima, O., Takabayashi, Y., Nakata, Y., Goto, H. & Ishikawa, T. *J. Synchrotron Rad.*, 1999, **6**, 143.
3. Newville, M. *J. Synchrotron Rad.*, 2001, **8**, 322.
4. Ravel, B. & Newville, M. *J. Synchrotron Rad.*, 2005, **12**, 537.
5. Newville, M., Livins, P., Yacoby, Y., Rehr, J. J. & Stern, E. A. *Phys. Rev. B*, 1993, **47**, 14126.
6. Rehr, J. J. & Albers, R. C. *Rev. Mod. Phys., 2000,* **72,** 621.
7. *X-ray Absorption: Principles, Applications, Techniques of EXAFS, SEXAFS, and XANES in Chemical Analysis.* Edited by D. C. Koningsberger & R. Prins, John Wiley & Sons, 1988, **Vol. 92**.
8. Herbst, J. F. & Wilkins, J. W. *Phys. Rev. B*, 1982, **26**, 1689.
9. Shimizugawa, Y., Umesaki, N., Hanada, K., Sakai, I., & Qiu, J. *J. Synchrotron Rad.*, 2001, **8**, 797.
10. Schlegel, M. L., Pointeau, I., Coreau, N., & Reiller, P. *Environ. Sci. Technol.*, 2004, **38**, 4423.
11. Chadeyron, G., El-Ghozzi, M., Mahiou, R., Arbus, A., & Cousseins, J. C. *J. Solid State Chem.*, 1997, **128**, 261.

Proc. Fifth Int. Conf. on Borate Glasses, Crystals and Melts *Phys. Chem. Glasses: Eur. J. Glass Sci. Technol. B*, February 2008, **49** (1), 19–25

The vibrational excitations of glasses in the boson-peak region: application to borates

R. Vacher,[1] *B. Rufflé, B. Hehlen, G. Guimbrètiere, G. Simon & E. Courtens*

Laboratoire des Colloides, Verres et Nanomateriaux, UMR 5587 CNRS, Universite Montpellier II, F-34095 Montpellier Cedex 5, France

There exists a special interest in investigating acoustical and optical vibrations of glasses in the one THz frequency region. The main experimental tools that are used to this effect are briefly discussed. The current understanding of the boson-peak vibrations and of their resonance with acoustic branches are explained. Specific new results on the origin of the boson peak and on the end of acoustic branches are presented for borates.

1. Introduction

A main goal of the present paper is to explain in simple terms a question which is currently very much debated in the physics of glasses. The issue is that of the nature and properties of vibrational excitations at frequencies $\nu \equiv \Omega/2\pi$ of the order of 1 THz (or 30 cm^{-1}, or also $\hbar\Omega$=4 meV). As shown below, these excitations are mainly of two types: *optic* ones, which in a molecular picture can be thought of as the more or less local vibrations of structural units, and *acoustic* ones which are approximately collective rigid translations of the same units. It is legitimate to ask why is this question of such interest. The profound reason derives from a series of observations from which a sequence of conclusions can be derived. It is worth following these steps one by one.

(i) First it is known that there are *excess vibrational modes* in most glasses, which occur at relatively "low" frequencies.[1] The existence of these quasi-harmonic modes is best seen in the low temperature (T) specific heat C_p which shows a large "hump" when plotted as C_p/T^3 versus T.[2] In the Debye model of thermal properties, the Debye value $C_p{}^D$ is proportional to T^3 and it only depends on the sound velocities.[3] One speaks of an excess because around the hump C_p is much larger than $C_p{}^D$. The hump typically peaks at a T_{max} between ~3 and ~10 K. The corresponding dominant thermal phonons are given by $h\nu \sim 5k_B T$, with k_B the Boltzmann constant. This leads to a value of ν typically of the order of 1 THz, which is what was meant by "low" frequencies above. Indeed, these are low compared to the usual optic modes in crystals. One remarks that the Debye density of acoustic modes $Z^D(\omega)$ is simply proportional to ω^2, which is the reason for the T^3 law above. Note that we use ω for spectral frequencies and Ω for the frequency of specific modes. As opposed to $Z^D(\omega)$, the real density of vibrational states $Z(\omega)$ shows a peak when plotted as $Z(\omega)/\omega^2$ versus ω. It is this particular feature which is called the *boson peak* (*BP*). This somewhat unfortunate name has been given because in spectroscopies of this spectral region, the T dependence of the strength of this feature follows the Bose–Einstein statistics characteristic of harmonic modes,[4] as opposed to structural relaxations which do not. The angular frequency at the BP maximum will be written Ω_{BP}.

(ii) The second point is that *the energy in these excess vibrations does not propagate* like plane-wave acoustic modes would allow. This is known from the thermal conductivity $\kappa(T)$, which normally grows with some power of T at low T, but which starts to saturate, exhibiting an anomalous "plateau" in a logarithmic presentation of $\kappa(T)$ versus T.[1] This plateau is observed at temperatures around the same T_{max}. Hence, not only are there more modes than just the normal acoustic ones, but also the acoustic modes have lost their propagating plane-wave character at Ω_{BP}. One speaks of an *Ioffe–Regel* (IR)[5] crossover for acoustic modes, occurring at Ω_{IR}. This crossover can be defined as the frequency at which the *energy* mean-free-path l becomes equal to only half the acoustic wavelength, $l=\lambda/2$.[6] This is really a limit for plane-wave like propagation. As shown explicitly below, one finds experimentally that $\Omega_{IR} \approx \Omega_{BP}$. This is now understood as due to the hybridisation of the acoustic modes resonating with the optic ones.[7,8] Hence, in the glasses for which the BP is sufficiently strong (which is the case for all glasses of technical interest), there should be no plane-wave-like acoustic modes left above Ω_{IR}. One can say that the acoustic branch *terminates* at Ω_{IR}: what was "low" frequency for optic modes has become a "very high" frequency for acoustic modes.

[1] Corresponding author. Email rene.vacher@univ-montp2.fr

(iii) The special interest in this spectral region arises when one considers the *typical scale* at which this occurs. Extrapolating the wavelength of acoustic modes to Ω_{IR}, one typically finds values $\lambda\approx5$ nm. Of course, the BP vibrations need not be exactly of the same extent. They could be of smaller extent and nevertheless modify the acoustic modes owing to the strong resonance with them. In any case, the BP vibrations and their interaction with acoustic waves are likely to give information about the structure of glasses at medium to long range scales, about which so little is known otherwise. The most promising approach will presumably be to check the validity of simulated structures by calculating from them the observed BP spectra. This is expected to become an exacting test for structures. To our knowledge, such large scale simulations of the BP-region have not yet been achieved for the full collection of spectroscopies which have now been applied to this problem. To summarise, it is expected that detailed studies of the BP will be central to the understanding of the disorder particular to glasses.

In the case of borates, one has an excellent example in which vibrations have given decisive structural information which was difficult to obtain with classical structural tools.[9] This is the question of the existence and the number of boroxol rings in vitreous boron oxide, v-B_2O_3. The breathing motion of the rings produces a narrow Raman scattering feature at 808 cm^{-1} [10] whose relative strength has recently been used in relation with simulations[11] to definitively establish the fraction of boroxol rings in v-B_2O_3. In the case of the BP, it can be considered as having its origin in the lowest frequency optic-like vibrations of structural units. These generally are the rigid librations, which by a cog-wheel effect couple with translations. In the borates, the structural units are quite varied. They can be the BO_3 triangles, the B_3O_3 boroxols, and the BO_4 tetrahedra, among others. Hence, the study of the BP should be of special interest for borate glasses.

The plan of the paper will be the following. Section 2 gives a brief summary of the situation in borates and in the course of the development it presents the various experimental approaches that are traditionally used to gain information about the BP and the end of acoustic branches. In particular, it emphasises that each spectroscopy has its specific selection rules, explaining that apparently "different" BPs might be observed in diverse experiments. Section 3 describes a recent progress achieved in the observation of low lying optical modes of v-B_2O_3 using an alternate spectroscopy, namely hyper-Raman scattering. It also describes advances in the observation of acoustic excitations in lithium diborate that became possible by pushing the limits of Brillouin scattering spectroscopies, whether using x-ray or visible light excitation. Finally, Section 4 summarises the present understanding of the BP and of its hybridisation with the acoustic modes.

2. The boson peak and high frequency sound in borate

An unbiased observation of the density of states (DOS) $Z(\omega)$ is provided by measurements of the specific heat Cp at low temperatures T. At sub-helium T, one observes a contribution to C_p which is linear in T, $C_p=c_1T$. It is explained by two-level systems.[12] As T is raised, one expects to find in addition the Debye contribution which can be written $C_D=c_3^DT^3$. The value of c_3^D should be given by the sound velocities. However, it has been known for a long time that an adjustment of the low-T specific heat of glasses to $C_P=c_1T+c_3T^3$ leads to values of c_3 significantly larger than the Debye c_3^D.[12] This apparent discrepancy has recently been very well explained.[13] It is due to the low frequency tail of the BP-modes, whose density varies approximately as ω^4, leading to an extra term c_5T^5 in the specific heat. It has been shown that for v-B_2O_3, as well as for a series of alkali borates, adjusting C_p to $c_1T+c_3T^3+c_5T^5$ leads to $c_3\equiv c_3^D$. This confirms the validity of the Debye model for the acoustic mode contribution. It also shows that the low-ω onset of the BP DOS is well approximated by $Z(\omega)\propto\omega^4$. As T is further raised, a plot of C_p/T^3 goes then through a maximum at T_{max} which evidently ought to be related to the maximum in $Z(\omega)/\omega^2$, which we have called Ω_{BP}. It can indeed be shown that $\hbar\Omega_{BP}=ak_BT_{max}$, where a is a constant that depends slightly on the exact shape of the reduced distribution $Z(\omega)/\omega^2$. The latter difficulty arises because the contribution to C_p of a given mode of frequency Ω does not set in sharply in function of T, but rather gradually, smoothed by the Bose population factor $n(\Omega)$.[3] The value of a is typically around 5 or 6. Since k_BT at 1 K corresponds to a frequency of 20·86 GHz, this value of a gives $\Omega_{BP}/2\pi\approx21$ cm^{-1} for v-B_2O_3 in which $T_{max}\approx5{\cdot}2$ K,[14] which is a reasonable result.[15,16] Although the measurement of C_p gives an unbiased access to $Z(\omega)$, it is not a direct one. For this reason, and also because quality low-T specific heat measurements are demanding experiments, one finds in the literature many more measurements of BPs by spectroscopic techniques.

The vast majority of BP observations have in fact been made using Raman scattering (RS). The difficulty here is that one does not know for sure what is the Raman activity of the modes involved in producing the BP. As explained in Ref. 17, the Stokes intensity can be written as a sum over bands

$$I(\omega)=\sum_b C_b Z_b(\omega)\left[1+n(\omega)\right]/\omega \qquad (1)$$

where b is a band index, Z_b is the DOS of the particular band, and C_b is a light-vibration coupling coefficient. For simple Raman-active optic modes, C_b should be a constant, only depending on the band, but in more complicated cases it might depend on the modes within a given band. This is now well understood for vitreous silica, v-SiO_2. The BP is there related to rigid librations of SiO_4 tetrahedra, as known from inelastic neutron scattering,[18] hyper-Raman scattering,[19] or also simulations.[20] *Undistorted* SiO_4 tetrahedra are optically isotropic so that their pure librations are *inactive* in RS. Hence, the observed RS BP must be related either to small tetrahedra distortions or to the coupling of librations with Raman-active translations. In either case, it is obvious that the coupling to light depends then on the environment of the librating tetrahedra, and thus on the particular mode frequency. It became customary to use in this case an *effective* frequency dependent coupling coefficient $C(\omega)$, as reported in Ref. 21. For the borates, there are great many reports of BP observations with Raman scattering. For v-B_2O_3, the BP depends significantly on the water content and on the fictive temperature of the glass.[22,23] In series of alkali borates, one observes that Ω_{BP} evolves rapidly with the alkali concentration.[15,24–27] Mixing with alkali produces BO_4 tetrahedra at the expense of B_3O_3 rings. The trend on Ω_{BP} is easily understood as tetrahedra must have higher librational frequencies. They are also much less Raman active, so that the BP intensity drops rapidly with increasing concentration.[25,27] It is clear that in situations where there are several different structural units, such as BO_3 triangles, B_3O_3 boroxols, and BO_4 tetrahedra, Equation (1) must be used with all due care, and that a single $C(\omega)$ might not be physically significant.

The DOS entering in Equation (1) is most often obtained from inelastic neutron scattering (INS) experiments. The first important point in INS is that for borates the experiments must be performed with samples made of isotopically pure ^{11}B. Indeed, the boron isotope ^{10}B has an enormous thermal neutron absorption coefficient, incompatible with INS experiments. The second point is that one must consider carefully the meaning of the INS structure factor $S(Q,\omega)$, where Q is the scattering vector. It is incorrect to assume that $S(Q,\omega)$ is simply proportional to the density of states, $S(Q,\omega)\propto Z(\omega)[1+n(\omega)]/\omega$, as it would be in incoherent scattering. Indeed, $^{11}B_2O_3$ is a coherent scatterer, and librations imply correlated motions of nearby units. This is clearly a second difficulty, besides the one mentioned above, in using Equation (1) for determining a single physically meaningful light-vibration coupling coefficient $C(\omega)$. $S(Q,\omega)$ has been analysed in detail for v-B_2O_3, with the result that one can extract from the INS signal an incoherent part, also named *out-of-phase* in Ref. 28. This incoherent part matches the RS signal, implying that really $C(\omega)\propto\omega^0$ in this case. The coherent, or *in-phase* part, shows a BP at a lower frequency than the RS BP. It appears that no other INS results on borate glasses are available to date. On the other hand, neutron scattering was used extensively to investigate the structure of these glasses, e.g. in Refs 9, 29, 30, but this is a different subject.

The observation of hypersonic acoustic modes and of their broadening is most conveniently achieved with Brillouin scattering (BS) of electromagnetic waves. In these experiments, the momentum transfer $\mathbf{Q}$ is fixed by the experimental geometry, $\mathbf{Q}=\mathbf{k}_i-\mathbf{k}_f$, where $\mathbf{k}_i$ and $\mathbf{k}_f$ are the initial and final momentum of the photons, respectively. Energy conservation implies that one observes excitations at a frequency $\omega=\omega_i-\omega_f$ where $\omega_{i,f}$ are the photon frequencies. With this choice of sign, negative ω means that an excitation is annihilated in the material, corresponding to the anti-Stokes side of the spectrum. One sould note that the velocity of sound v is much smaller than the velocity of light c, and that both play an essential role in the experiment since $\omega_{f,i}=ck_{f,i}$ while for a sound wave $\Omega=vQ$. Since $c>>v$, it follows that $k_f\approx k_i$. For incident radiation of a given frequency, backscattering gives the highest achievable Q-value, which is $\approx 2k_i$. With visible light excitation, this limits Q to typically ~0·04 nm^{-1}, and thus the highest observable sound frequency Ω, which depends on v, typically ranges from ~20 to ~40 GHz for glasses. To observe higher frequency acoustic modes, it is necessary to use a higher frequency excitation. This has been attempted in silica using the second-harmonic of visible lasers and also synchrotron produced UV excitation,[31,32] but this is rapidly thwarted by the sample absorption that limits the definition of $\mathbf{Q}$. This approach allows at most an increase of Q by a factor of 3 to 4. Much higher frequencies can be achieved with x-rays, which then require the very high energy and angular resolutions that are only possible at advanced synchrotron sources. One speaks then of inelastic x-ray scattering (IXS).[33] If it is of interest to measure the linewidth of the scattered spectrum, Γ, it is then the resolution of the monochromator and analysers that is the limiting factor in IXS. The highest practical resolution to date has been achieved with 21·75 keV radiation (corresponding to ~$5{\cdot}3\times10^{18}$ Hz), and it is 1·5 meV full width (or ~0·36 THz).[34] This radiation penetrates sufficiently into samples of low atomic number to allow for many significant BS experiments. At such a high incident energy, k_i is *much larger* than any Q of interest in the glass. One is then in the opposite limit where one wants to work close to forward scattering to achieve at best Q~1 nm^{-1}. This is in the range of the IR-crossover wavevector, q_{IR}, of most glasses. Unfortunately, there exists no BS technique to access the potentially interesting region between ~0·1 and 1 nm^{-1}. In performing IXS experiments, it is therefore of prime interest to select materials for which q_{IR} is expected to be as large as possible, in order to be able to explore the region between 1 nm^{-1} and q_{IR}.

Up to now, among borates, only v-B_2O_3,[35] Li_2O–$4B_2O_3$,[36] and Li_2O–$2B_2O_3$,[36,37] have been investigated with IXS. In all three cases, a strong saturation of the linear dependence of Ω on Q was observed. It was interpreted as the manifestation of the IR-crossover. In that saturation regime, the full width Γ of a damped harmonic oscillator (DHO) lineshape adjusted to the spectrum is almost equal to the frequency Ω (both in angular frequency units). In fact, the definition of the crossover given above, $l=\lambda/2$, implies that $\Gamma=\Omega/\pi$ when one uses $\Gamma=v/l$ and $f=\lambda v$. The condition $\Gamma=\Omega/\pi$ occurs quite a bit before the saturation of $\Omega(Q)$, indicating that the saturation regime is far beyond the IR-crossover. We return to this in Section 3.2.

The above summarises the main experimental techniques that are presently available to investigate the frequency region of interest. As shown, there exists so far relatively little information on the boson peak region of borate glasses, except maybe for the always abundant Raman scattering literature.

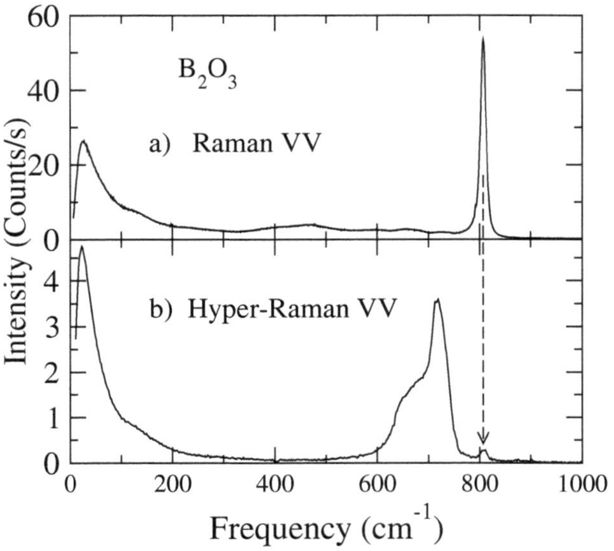

Figure l. Raman and hyper-Raman intensities observed in 90° scattering and in VV polarisation on v-B_2O_3 at room temperature. The vertical line shows the position of the 808 cm^{-1} breathing mode of the B_3O_3 rings which is a small peak in the HRS spectrum on this scale. The frequency of the BP is slightly lower in HRS than in RS

3. Recent progress

3.1. Hyper-Raman scattering from boron oxide

Hyper-Raman scattering (HRS) is a nonlinear optical spectroscopy in which two incident photons simultaneously create or destroy one quantised excitation in the material to produce one scattered photon.[38,39] The scattered photon is thus at an energy equal to twice that of the incident photon minus or plus the energy of the excitation. HRS obeys selection rules that are different from those of RS.[40] Therefore, the two techniques can advantageously be combined to identify which motions of the structural units produce the vibrational spectra, and in particular which motions are responsible for the BP. This was already described for the BP of v-SiO_2.[19] In the case of v-B_2O_3, the spectra can be indexed in terms of vibrations of structural units consisting principally of B_3O_3 rings and BO_3 triangles, in about 1:1 ratio.[11]

We recently showed that the BP observed both in RS and HRS (Figure 1) in v-B_2O_3 is produced by out-of-plane rigid librational motions of the structural units.[41] With flat structural units like these, out-of-plane librations are active both in RS and HRS, so that one would at first expect that the two BPs have somehow similar intensities. As found from an analysis of Figure 1, this turns out not to be the case. To compare the RS and HRS intensities, it is necessary to make use of a reference. Such a reference is provided by the 808 cm^{-1} narrow peak in both spectra. As already indicated above, this peak corresponds to the breathing motion of the B_3O_3 rings.[10] This is a local mode that should scatter equally well in both spectroscopies. In other words, it sets the scale to compare Figure 1(b) to Figure 1(a). It becomes then obvious that the BP scatters much more strongly in HRS than in RS, by nearly two orders of magnitude. As explained elsewhere, this enhancement of the HRS BP is produced by the coherence of the librational motions of many connected units.[41] For reasons of symmetry of the polarisability and hyper-polarisability tensors, the coherence affects differently the RS and HRS BPs. In fact, the RS BP signal appears to be dominated by incoherent scattering, as its shape matches well the out-of-phase component extracted from INS.[28] On the other hand, the HRS BP is slightly shifted to lower frequencies owing to the coherent motions. It should not be confused with the INS in-phase signal[28] which is produced by Umklapp scattering of the acoustic modes and which for that reason mimics the BP although it does not really reflect the BP density of states.

The reader will also have noticed the group of modes located between 600 and 750 cm^{-1} in Figure 1(b). These other strong modes are due to scattering by cooperative infrared-active LO–TO motions of the structural units. The particular motions leading to this group of modes are opposite displacements of B and O atoms perpendicular to the plane of the structural units.[41] The sharpest peak in this group can be assigned to the boroxols. It is interesting to remark that the BP and these other collective modes have similar activities.

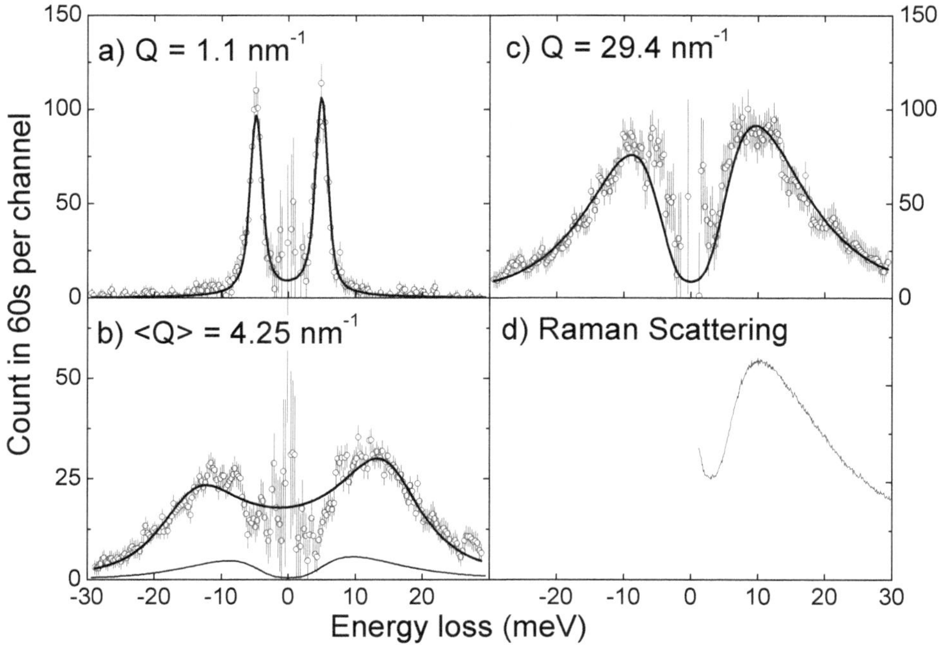

Figure 2. (a)-(c) The inelastic part of IXS spectra observed on Li_2O–$2B_2O_3$ at 573 K for three values of Q. In (a), the two peaks are produced by usual sound waves, while (b) is for Q well above q_{IR} so that the sound excitations become diffusive. In panel (c), the signal is dominated by the optic modes forming the boson peak. The solid lines in (a)–(c) are explained in the text. The thin line in (b) is an estimate of the optic-mode contribution to that spectrum. Panel (d) shows the Stokes side of the boson peak measured in RS. Its shape is very similar to the IXS spectrum in panel (c)

We hope that the observations of the relative strength of these BPs will stimulate large scale simulations aimed at obtaining quantitative estimates of the size of the coherently vibrating regions. This will presumably require a realistic model of the structure. This of course would be a major contribution towards clarifying the medium to long range order of these glasses.

3.2. The acoustic modes and IR-crossover of lithium diborate

We have recently investigated in detail, using IXS, the very high frequency acoustic modes of lithium diborate glass, Li_2O–$2B_2O_3$.[42] A few typical spectra are illustrated in Figure 2. The left-hand side of Figure 2 shows Brillouin spectra of acoustic modes in two extreme situations, well below crossover in (a) and well above it in (b). Figure 2(a) is at the lowest usable scattering vector, Q=1·1 nm^{-1}, where one observes a narrow Brillouin doublet corresponding to sound waves of frequency $\nu \approx 1$ THz. The apparent width in Figure 2(a) is essentially set by the instrumental resolution. As the scattering angle is increased, one first observes the usual linear increase of Ω with Q, accompanied by a rather unusual very fast increase of the linewidth Γ. The latter is obtained by adjusting the Brillouin spectra to the DHO lineshape convoluted with the instrumental function. Such DHO adjustments are shown by the solid lines in Figures 2(a)–(b). The initial increase of Γ with Q is approximately proportional to Q^4. This is observed up to $Q \approx 1{\cdot}8$ nm^{-1}, beyond which the fast increase in Γ slowly saturates. As explained above, the Q value at which $\Gamma/\pi = \Omega$ is used as definition of the IR-crossover, q_{IR}. We find for the crossover frequency $\Omega_{IR}/2\pi$=2·1 THz, at a wavevector q_{IR}=2·1 nm^{-1}. Beyond this crossover, the spectrum continues to broaden, while Ω starts to saturate. This is the other extreme case illustrated in Figure 2(b). The signal of Figure 2(b) is still mostly of acoustic origin, as explained below. However, being so far above crossover, it is not produced at all by plane waves, but rather by diffusive excitations. We recall that the DHO is the lineshape resulting from the scattering of radiation by exponentially damped plane waves. We have noticed that the DHO adjustment becomes progressively poorer as Q increases beyond q_{IR}. This can be seen in Figure 2(b), particularly in the central region of the spectrum, from about −12 to +12 meV. The origin of this departure from the DHO is now well understood.[43–45] It reflects the fact that this line shape is essentially *inhomogeneous,* being the sum of many individual excitations with ill defined wave vectors. The central dip in the observed spectrum of Figure 2(b) occurs because at small Ω the wave vector of the sound waves is well defined and small. Thus excitations of small frequency just cannot contribute to the scattering at this large experimental Q value.

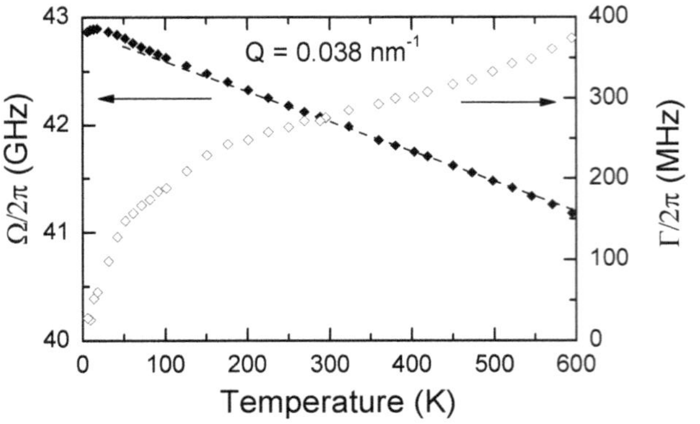

Figure 3. The temperature dependence of the frequency and full width of the longitudinal acoustic mode measured in Brillouin light scattering on Li_2O–$2B_2O_3$ at 0·038 nm^{-1}. The error bars are smaller than the symbols

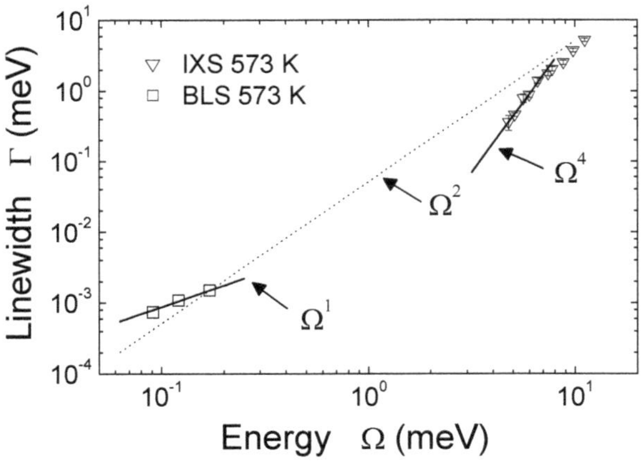

Figure 4. The frequency dependence of the linewidths measured on Li_2O–$2B_2O_3$ with BLS and INS at 573 K. The solid lines are guides to the eye emphasising the different regimes observed in the two spectroscopies. The dotted line is just there to show that an Ω^2 dependence would be a very poor approximation

As Q is further increased, the signal becomes progressively dominated by optic-like modes. This is illustrated in Figure 2(c) which shows essentially the BP whose maximum is located at $\Omega_{BP}/2\pi$=2·2 THz. This signal can be compared to the RS BP observed on the same glass, as shown in Figure 2(d). The fitted line in Figure 2(c) is a log-normal, a function which has been advocated for the BP.[46] The intensity of the optic-like scattering is proportional to Q^2 when due account is taken of the x-ray atomic form factors.[42] Extrapolating the optic-like signal to the Q-value of Figure 2(b) leads to the thin line drawn there. This confirms that the scattering at 4 nm^{-1} is still dominated by acoustic-like modes. To summarise, the approach of (q_{IR}, Ω_{IR}) from below is characterised by a very fast increase of $\Gamma \propto \Omega^4$. This is the region where specific-heat measurements reveal a contribution in $Z(\omega) \propto \omega^4$, as explained above. The two must obviously be related. Further, Ω_{IR} nearly coincides with the maximum of the BP, $\Omega_{IR} \approx \Omega_{BP}$, emphasising a common origin.

It is also of interest to examine what happens to Ω and Γ at lower Qs. Values around 0·04 nm^{-1} and below can be investigated by Brillouin scattering of light (BLS) in the visible region of the spectrum. The results of such an investigation, using the backscattering geometry and the green argon laser line as excitation, are shown in Figure 3.[47] These have been obtained with a very high resolution Fabry–Perot spectrometer of the type described in Ref. 48. This instrument includes a reference signal used for frequency calibration, a multiple pass planar interferometer acting as a filter, and in series a spherical interferometer that provides the very high resolution. With such an instrument the error bars in Figure 3 are smaller than the size of the data points. We observe that the sound velocity Ω/Q decreases almost linearly with increasing temperature T. This is typical of the anharmonic coupling of sound with thermal vibrations, an effect which is seen in crystals as well as in glasses.[49] The relative slope, dv/vdT, allows one to estimate the anharmonic contribution to the Brillouin linewidth provided the mean relaxation time of the thermal vibrations is known.[50] Making reasonable estimates, one finds that, although anharmonicity dominates the behaviour of Ω(T), its contribution to Γ is negligible in this frequency region compared to the observed broadening.[47]

The linewidth was measured with BLS at several scattering angles, i.e. at several values of Q and thus of Ω. As shown in Figure 4, we find around room temperature that $\Gamma \propto \Omega$. A line in Ω^1 is traced through these points. This behaviour is characteristic of damping produced by thermally activated relaxation (TAR) of "structural defects" at frequencies in the region of the maximum damping.[51] This is due to the fact that TAR gives $\Gamma \propto \Omega^2\tau/(1+\Omega^2\tau^2)$, where τ is a characteristic relaxation time of the defects. In general there is a broad distribution of τ which widens the region where $\Gamma \propto \Omega$.

The measured IXS linewidths are also shown in Figure 4. The onset of crossover in Ω^4 applies to the first five data points. It is followed by the progressive saturation that has been described above. One concludes that the behaviour of Γ(Ω) is complex. It cannot be approximated by a single power law $\Gamma \propto \Omega^2$, as illustrated by the dotted line. Beyond Ω_{IR}, the linewidth cannot be strictly defined since DHO fits become progressively meaningless. However, other workers have observed that enforcing fits to DHO, the linewidth varies then approximately in Ω^2. All this emphasises that different mechanisms producing the broadening might dominate in different frequency regions. The variation of Γ with Ω is fastest in the very rapid increase below the IR crossover.

4. Summary and outlook

The main facts emerging from the above presentation are the following:

(i) There exists an IR crossover beyond which acoustic modes cease to propagate as plane waves. This has now been clearly demonstrated in two cases, permanently densified silica glass d-SiO_2[45] and $Li_2O–2B_2O_3$. However, we have shown recently that such a crossover occurs in nearly all glasses on which we have the necessary experimental information.[42]

(ii) There also exists a boson peak which is produced by excess modes of low frequency. These excess modes appear to be of optic origin and involve collective motions of structural units, as found for v-SiO_2[18] and B_2O_3.[41]

(iii) With the definition given here for Ω_{IR}, the IR crossover nearly coincides with Ω_{BP}. This is shown above for $Li_2O–2B_2O_3$, but it is also found for all sufficiently strong glasses when Ω_{IR} is known.[42] These "strong" glasses are these where the BP is well developed.[52]

(iv) The approach of the crossover from below is characterised by a very rapid increase of the acoustic damping with the acoustic frequency. This was shown in the needed details for d-SiO_2[45] and $Li_2O–2B_2O_3$.[42] It seems related to the rapid increase of $Z(\omega)\propto\omega^4$ at similar frequencies. This points to a resonant interaction between the acoustic modes and the excess optic ones. It strongly supports the concept that the position and shape of the boson peak itself, and in particular the fact that $\Omega_{BP}\approx\Omega_{IR}$, result from the hybridisation of acoustic and optic modes.[7]

The existence of the excess vibrational excitations at low frequencies, and their coincidence with the end of acoustic branches, are not specific to borates. However, as shown above, borates have offered some very nice test cases to explore these facts. This will be worth pursuing, particularly in view of the structural modifications that can be produced in a controlled fashion in borates, for example by mixing boron oxide with alkali oxides. This will allow in the near future to perform what could be called some "creative" spectroscopy in which specific structural changes are performed to follow their influence on the spectral properties.

References

1. Zeller, R. C. & Pohl, R. O. *Phys. Rev. B*, 1971, **4**, 2029.
2. *Amorphous Solids: Low Temperature Properties*, edited by W. A. Phillips, Springer-Verlag, Berlin, 1981.
3. *Introduction to Solid State Physics*, C. Kittel, Wiley, New York, Fifth Ed., 1976.
4. Stolen, R. H. *Phys. Chem. Glasses*, 1970, **11**, 83.
5. Ioffe, A. F. & Regel, A. R. *Prog. Semicond.*, 1960, **4**, 237.
6. This definition of the crossover is identical to that used, e.g. in Taraskin, S. N. & Elliott, S. R. *J. Phys.: Condens. Matter*, 1999, **11**, A219, as these authors write $l_{IR}=\lambda_{IR}$, where l_{IR} is defined as the *amplitude* mean free path, which is twice the energy mean free path.
7. Gurevich, V. L., Parshin, D. A. & Schober, H. R. *Phys. Rev. B*, 2003, **67**, 094203.
8. Klinger, M. I. & Kosevich, A. M. *Phys. Lett. A*, 2002, **295**, 311.
9. Hannon, A. C., Grimley, D. I., Hulme, R. A., Wright, A. C. & Sinclair, R. N. *J. Non-Cryst. Solids*, 1994, **177**, 99.
10. Galeener, F. L., Lucovsky, G. & Mikkelsen, Jr. J. C. *Phys. Rev. B*, 1980, **22**, 3983.
11. Umari, P. & Pasquarello, A. *Phys. Rev. Lett.*, 2005, **34**, 137401.
12. Pohl, R. O. in Ref. 2., p 27.
13. Ramos, M. A. *Phil. Mag.*, 2004, **84**, 1313.
14. Perez-Enciso, E., Ramos, M. A. & Vieira, S. *Phys. Rev. B*, 1997, **56**, 32.
15. Lorosch, J., Couzi, M., Pelous, J., Vacher, R. & Levasseur, A. *J. Non-Cryst. Solids*, 1984, **69**, 1.
16. Brodin, A., Borjesson, L., Engberg, D., Torell, L. M. & Sokolov, A. P. *Phys. Rev. B*, 1996, **46**, 11511.
17. Shuker, R. & Gammon, R. W. *Phys. Rev. Lett.*, 1970, **24**, 222.
18. Buchenau, U., Prager, M., Nucker, N., Dianoux, A. J., Ahmad, N. & Phillips, W. A. *Phys. Rev. B*, 1986, **34**, 5665.
19. Hehlen, B., Courtens, E., Vacher, R., Yamanaka, A., Kataoka, M. & Inoue, K. *Phys. Rev. Lett.*, 2000, **84**, 5355.
20. Taraskin, S. N. & Elliott, S. R. *Phys. Rev. B*, 1997, **56**, 8605.
21. Fontana, A., Dell'Anna, R., Montagna, M., Rossi, F., Viliani, G., Ruocco, G., Sampoli, M., Buchenau, U. & Wischnewski, A. *Europhys. Lett.*, 1999, **47**, 56.
22. Surotsev, N. V., Wiedersich, J., Batalov, A. E., Novikov, N. V., Ramos, M. A. & Rossler, E. *J. Chem. Phys.*, 2000, **113**, 5891.
23. Surotsev, N. V., Shebanin, A. P. & Ramos, M. A. *Phys. Rev. B*, 2003, **67**, 024203.
24. Kojima, S. & Kodama, M. *Physica B*, 1999, **263–264**, 336.
25. Baranov, A. V., Perova, T. S., Petrov, V. I., Vij, J. K. & Nielsen, O. F. *J. Raman Spectrosc.*, 2000, **31**, 819.
26. Kojima, S., Novikov, V. N. & Kodama, M. *J. Chem. Phys.*, 2000, **113**, 6344.
27. Denisov, Yu. V. & Zubovich, A. A. *Glass Phys. Chem.*, 2003, **29**, 345.
28. Engberg, D., Wischnewski, A., Buchenau, U., Borjesson, L., Dianoux, A. J., Sokolov, A. P. & Torell, L. M. *Phys. Rev. B*, 1998, **58**, 9087.
29. Johnson, P. A. V., Wright, A. C. & Sinclair, R. N. *J. Non-Cryst. Solids*, 1982, **50**, 281.
30. Majerus, O., Cormier, L., Calas, G. & Soper, A. K. *Physica B*, 2004, **350**, 58.
31. Masciovecchio, C. Gessini, A., Di Fonzo, S., Comez, L., Santicci, S. C. & Fioretto, D. *Phys. Rev. Lett.*, 2004, **92**, 247401.
32. Benassi, P., Caponi, S., Eramo, R., Fontana, A., Giugni, A., Nardone, M., Sampoli, M. & Viliani, G. *Phys. Rev. B*, 2005, **71**, 172201.
33. Sette, F., Krisch, M. H., Masciovecchio, C., Ruocco, G. & Monaco, G. *Science*, 1998, **257**, 1550.
34. Verbeni, R., Sette, F., Krisch, M. H., Bergmann, U., Gorges, B., Halcoussis, C., Martel, K., Masciovecchio, C., Ribois, J. F., Ruocco, G. & Sinn, H. *J. Synchrotron Rad.*, 1996, **3**, 62.
35. Matic, A., Borjesson, L., Ruocco, G., Masciovecchio, C., Mermet, A., Sette, F. & Verbeni, R. *Europhys. Lett.*, 2001, **54**, 77.
36. Matic, A., Engberg, D., Masciovecchio, C. & Borjesson, L. *Phys. Rev. Lett.*, 2001, **86**, 3803.
37. Borjesson, L., Matic, A., Engberg, D. & Masciovecchio, C. ESRF experimental report HS-1147, 23/2/2000.
38. Terhune, R. W., Maker, P. D. & Savage, C. M. *Phys. Rev. Lett.*, 1965, **14**, 681.
39. Denisov, V. N., Marvin, B. N. & Podopedov, V. B. *Phys. Rep.*, 1987, **151**, 1.
40. Cyvin, S. J., Rauch, J. E. & Decius, J. C. *J. Chem. Phys.*, 1965, **43**, 4083.
41. Simon, G., Hehlen, B., Courtens, E., Longueteau, E. & Vacher, R. *Phys. Rev. Lett.*, 2006, **96**, 105502.
42. Ruffle, B., Guimbretiere, G., Courtens, E., Vacher, R. & Monaco, G. *Phys. Rev. Lett.*, 2006, **96**, 045502.
43. Vacher, R., Courtens, E. & Foret, M. *Phil. Mag. B*, 1999, **79**, 1763.
44. Rat, E., Foret, M., Courtens, E. Vacher, R. & Arai, M. *Phys. Rev. Lett.*, 1999, **83**, 1355.
45. Ruffle, B., Foret, M., Courtens, E. Vacher, R. & Monaco, G. *Phys. Rev. Lett.*, 2003, **90**, 095502.
46. Malinovsky, V. K., Novikov, V. N., Parshin, P. P. & Sokolov, A. P. *Europhys. Lett.*, 1990, **11**, 43.
47. Guimbretiere, G. Doctoral Thesis, University of Montpellier 2, France, Dec. 2005.
48. Vacher, R., Sussner, H. & v. Schickfus, M. *Rev. Sci. Instrum.*, 1980, **51**, 88.
49. Claytor, T. N. & Sladek, R. J. *Phys. Rev. B*, 1978, **18**, 5842.
50. Vacher, R., Courtens, E. & Foret, M. *Phys. Rev. B*, 2005, **72**, 214205.
51. Hunklinger, S. & Arnold, W. in *Physical Acoustics*, Vol. NII, W. P. Mason & R. N. Thurston, Eds., Academic Press, NY, 1976, p. 155.
52. Sokolov, A. P., Rossler, E., Kisliuk, A. & Quitmann, D. *Phys. Rev. Lett.*, 1993, **71**, 2068.

Proc. Fifth Int. Conf. on Borate Glasses, Crystals and Melts Glass Tech.: Eur. J. Glass Sci. Technol. A, October 2006, **47** (5), 127–132

Development of functional borate glass surfaces to inhibit bacterial growth

Desirae Leipply,[1] Bradley A. Melhus,[2] Michael R. Leonardo,[2] Steve A. Feller[1] & Mario Affatigato[1,]*

[1] *Physics Department and* [2] *Biology Department, Coe College, 1220 First Av. NE, Cedar Rapids, Iowa, USA*

*A variety of borate glass compositions are shown to be effective in inhibiting bacterial growth on their surfaces. Both Gram-negative (*Escherichia coli *and* Salmonella typhimurium*) and Gram-positive bacteria (*Stapphylococcus aureus*) were effectively killed by lithium borate and lithium potassium borate glasses of high and low alkali contents. The surface environment was shown to become acidic for low metal contents, halting growth, and alkaline for higher metal contents. The alkaline surface layer proved to be more efficient in eliminating bacteria. The more alkaline sample also appeared to act especially rapidly in preventing growth. Micro-Raman measurements were taken to characterise the surface layer responsible for the bacterial inhibition.*

1. Introduction

We report on our work developing borate glass systems with the ability to inhibit or eliminate bacterial growth on their surfaces. The mechanism needs to be fatal to bacteria but safe for human use. The practical application for such glasses is for use on surfaces, such as cutting boards and countertops, susceptible to bacterial contamination. This particular application joins a host of other biological uses[(1)] of borate/borosilicate glasses, which include drug and radioisotope delivery.[(2,3)] Though previous work has been done in developing bactericidal properties on glasses, these have primarily focused on silver-containing compositions.[(4–9)]

The primary focus of this study was to determine the bactericidal efficacy of a range of lithium and lithium-potassium borates. Alkali oxide was added to the borate glass to alter the dissolution rate of the material, and to strike a balance between the bactericidal efficiency and the practical survival of the material when exposed to atmospheric humidity. As alkali ions are added to the borate network (increasing R—the ratio of alkali oxide to bora), the three-coordinated boron units (f_1 units) are converted[(10)] into the more stable four-coordinated boron units (f_2 units) affecting the dissolution behaviour in these glasses. The idea was to search for an optimum composition where the alkali glass was soluble enough to create an acidic or basic layer to kill bacteria, but not so soluble that the glass would dissolve with prolonged exposure to water.

In this work we report on the effect of these glasses on *Escherichia coli* and *Salmonella typhimurium*, both Gram negative, and on the Gram-positive bacterium *Stapphylococcus aureus*. Aside from the measurements of the survival rate of the bacteria, we also performed surface characterisation by micro-Raman spectroscopy.

2. Experimental procedure

Glass making procedure

Glasses were made using boric acid (Aldrich Chemical Co., 99·5%) and alkali carbonate (Aldrich Chemical Co., 99+%). The appropriate stoichiometric amounts of the ingredients for each sample were weighed out into a platinum crucible according to the equations

$$RLi_2CO_3 + 2H_3BO_3 \rightarrow RLi_2O.B_2O_3 + RCO_2 + 3H_2O \quad (1)$$

for the lithium borates and

$$R[0{\cdot}5Li_2CO_3 + 0{\cdot}5K_2CO_3] + 2H_3BO_3 \rightarrow R[0{\cdot}5Li_2O.0{\cdot}5K_2O].B_2O_3 + RCO_2 + 3H_2O \quad (2)$$

for the lithium potassium borates. Table 1 shows all the compositions studied.

After the addition of the chemicals, the batch was mixed with a spatula for five minutes. The samples were then heated in a furnace at 1000°C for 12 min. Experimental weight losses were taken to verify com-

Table 1. Summary of the compositions used in the lithium and lithium potassium borate families. Not all compositions were used for each bacterium

Lithium borate R *values*	*Lithium potassium borates* R *values*
0	0
0·05	0·1
0·1	0·15
0·15	0·2
0·17	0·3
0·2	0·4
0·3	0·5
0·4	0·6
0·52	0·7
0·6	
0·7	

* Corresponding author. Email address maffatig@coe.edu

position, which was within 2% of the batch composition. The glass melt was then carefully poured into an aluminium mould to obtain glass beads for test tube work. Due to the rapid cooling of the glass, the melt quickly become too viscous to pour and had to be reheated for five minutes at a time to obtain more beads. Once cool, the beads were stored in a glass vial in a desiccator. The beads were clear, mushroom shaped, and roughly half a centimetre in diameter.

Bacterial assay procedure

E. coli, *Salmonella typhimurium*, and *S. aureus* were our principal test organisms. Sterile technique was used at all times, including use of 95% ethanol and a bunsen burner on all equipment. All test tubes and media were autoclaved before use. Tryptic soy broth (TSB, Difco) was the media chosen to culture the bacteria. Tryptic soy agar (TSA, Difco) was used for the bacteria platings. Bacterial source plates were replaced every 10–14 days.

The initial bacterial stock was prepared by using a sterilised inoculating loop to gather and spread an aliquot of frozen bacteria on a TSA plate. This source plate was incubated at 37°C for 24 h. In the initial experiments, single colonies were retrieved with an inoculating loop and placed in a 5 ml portion of TSB in a test tube as a starter culture. In later experiments, the colonies were retrieved from the source plate and placed in an Erlenmeyer flask containing 25 ml of TSB and allowed to grow in a 37°C incubator overnight. The beads were sterilised in 95% ethanol and flamed before insertion in the starter culture. The control was a small borosilicate bead (Fisher Scientific) that acted as an inoffensive substrate for bacterial attachment. After sterilisation, the beads were allowed to sit in the starter culture for one minute. The broth was then decanted into bleach, and the beads placed onto filter paper and allowed to dry for five minutes. Each bead was then transferred individually into 5 ml of fresh, sterile TSB medium in a test tube. After an overnight incubation at 37°C in an incubator, the optical densities of the resultant cultures were analysed with a Perkin Elmer UV/Vis spectrophotometer.

The spectrophotometer was calibrated before each use with the shutter for 0% T and a fresh TSB media set as the zero absorbance. The correlation between optical density and bacterial cell concentrations is well known, with each bacteria having their own correlation: The relationship for *E. coli* and *Salmonella typhimurium*: 1·0 OD=1×10^9 cells/ml is linear (in a logarithmic plot) between absorbance readings of 0·1 and 1·1, while for *S. aureus* the relationship is 1·0 OD=5×10^8 cells/ml. Thus, an undiluted absorbance reading of 0·523 corresponded to an *E. coli* cell concentration of $5{\cdot}09\times10^4$ cells/ml. Complete kills were readily apparent—the solutions that come out visually clear were assigned 0 cells/ml bacterial growth, and this was checked by plating the solutions in agar. The controls consistently showed the optimum growth possible, reaching the instrumental limit of 10^{10} cells/ml. Given the overnight (19–20 h) incubation, even a single bacterium would be able to grow to a detectable amount (division occurs every 20–40 min). If the control failed to grow, the trial was voided.

The final bacterial protocol followed the initial procedure up until the exposed, dried beads being immersed in their fresh TSB media. After 15 min, 100 μl of solution was drawn from near the bead's surface via micropipette and transferred to a new TSB. This withdrawing was then repeated after another 3 h. Thus, an idea of how long it takes the mechanisms to act was obtained.

Solubility played a pivotal role in the effectiveness of our glasses. Hence, rates of dissolution were monitored throughout the project in multiple ways. In several trials the masses of test beads were recorded before and after immersion to assess the practicality of the various compositions. A more thorough dissolution study was performed to obtain a more quantitative understanding of the dissolution rates. In this case, the mass of each bead was measured initially and then again after every half hour of soaking in deionized water for a total of 6 hours. The excess water was then removed from the beads using lint-free paper, and the mass of the beads measured. We do note that this experimental procedure did not shake the glass beads, and hence the reacted layers were not sloughed off until the measurements of mass were made. This may well have resulted in a slower reported dissolution rate. The water solution was changed every two hours to avoid the effect of the changing pH, which can alter the rate of dissolution.[11] We repeated the experiment three times for the R=0·1 lithium borate composition, and thus estimated an overall (batch-to-batch) error of ±0·05.

Raman spectroscopy

The Raman measurements were made on a JASCO NRS 3000 commercial micro-Raman instrument with confocal capabilities and CCD detection. Spectra were taken using a 532 nm Nd:YAG laser source, with a repeatable resolution better than 1 cm^{-1}. Depth mapping was also carried out on some samples using a confocal (0·05 mm diameter aperture) configuration that used a 100× objective lens.

3. Results and discussion

We started out with the belief that alkali borate glasses, or even pure boron trioxide glass, would be bactericidal because upon exposure to atmospheric water (humidity), boron oxide is well known to revert

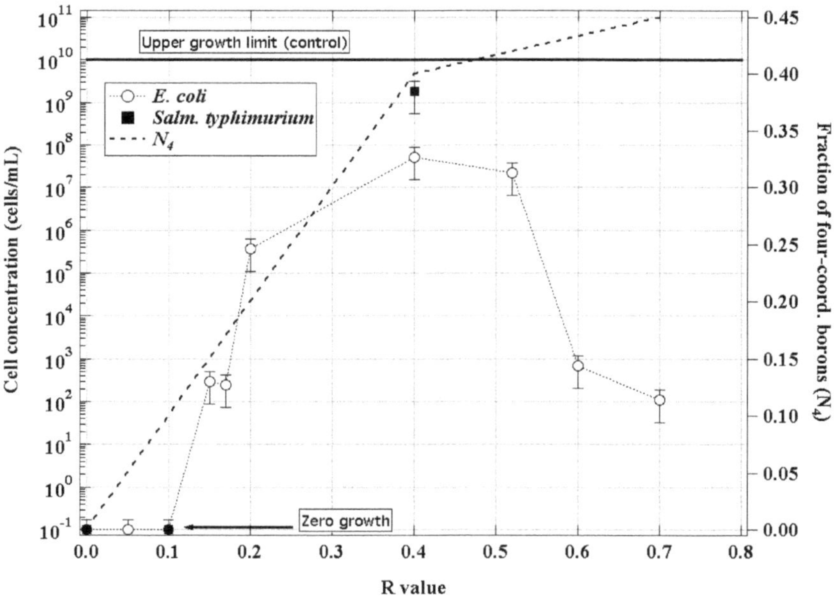

Figure 1. The survival rate of E. coli *and* S. typhimurium *as a function of composition for the lithium borate glass family. The right hand axis displays the fraction of four-coordinated borons,* N_4. *The upper bar shows the results from all the control beads, which reached the instrumental limit (10^{10} cells/ml). The error bars represent an overall variability in the survival rate estimated from all samples*

to boric acid. The reversible stoichiometry is

$$2H_3BO_3 \Leftrightarrow 2B_2O_3 + 3H_2O \qquad (3)$$

We expected that a thin layer would then form upon the glass surface, and this moist layer would have a low, acidic pH that would kill bacteria. Furthermore, wiping the surface would simply expose fresh glass to the atmospheric water, recreating the layer and its bactericidal properties for as long as glass remained.

Lithium borate glasses

Figure 1 shows the results as a function of composition. Two bacteria are shown (*E. coli* and a few *S. typhimurium* points) for the lithium borate (LiB) glass family. We note that the survival rate peaks at R=0·4, while it is zero early on (R<0·1). All lithium borate beads are capable of inhibiting growth, and the least effective (R=0·4) still manages to diminish the final number of cells/ml by a factor of at least 1000 for E. coli. The few *S. typhimurium* show a sturdier bacterium with a higher survival rate, though the beads with R<0·1 still hinder growth completely. These final cell concentrations are comparable to the silver doped AgBG glass developed by the Hench group.[4,7] As alkali oxide is added, the boron atoms go from three-coordinated to four-coordinated, making the glass more stable and less likely to dissolve and create the acidic layer, explaining the low bactericidal efficiency in this region. On the right hand axis we show the fraction of four-coordinated borons (N_4, denoted by the thin blue line). As R keeps increasing, however, the survival rate of the *E. coli* did not follow N_4. Rather, the glass evidently began to kill the bacteria more efficiently after the peak in R at 0·4.

To better understand this behaviour, we began a dissolution study. We immersed LiB beads in nanopure water and weighed them at regular time intervals, as described in the Experimental procedure section. Figure 2 shows the normalised results. The analysis of the graph allows us to obtain a slope for each of the roughly linear trends. These slopes are mass dissolution rates (mass per unit time), and they are shown in Table 2. The dissolution rates reached a minimum around R=0·4, and then began to rise again, in line with the inhibitory behaviour seen in Figure 1. This confirmed that the stability expected from the N_4 formation was not sufficient to explain the changes in dissolution or inhibiting efficiency.

We also monitored changes in the pH of the TSB broth as a function of time. A probe measured the average pH of the broth (not the region immediately surrounding the bead, which might be expected to have a different and more accentuated pH). The results are shown in Table 3. At low R values, the pH of the TSB broth does in fact decrease, albeit not

Table 2. Summary of the dissolution rates of several lithium borate glass beads. The second column displays the data as a percent function of the initial mass. The overall (batch-to-batch) error is estimated to be ±0·05

R value	*Mass dissolution rate [g/h]*	*% mass loss/bead/h*
0	0·0507	18·2
0·1	0·0331	9·27
0·2	0·00925	2·56
0·4	0·00355	0·835
0·7	0·0192	6·58

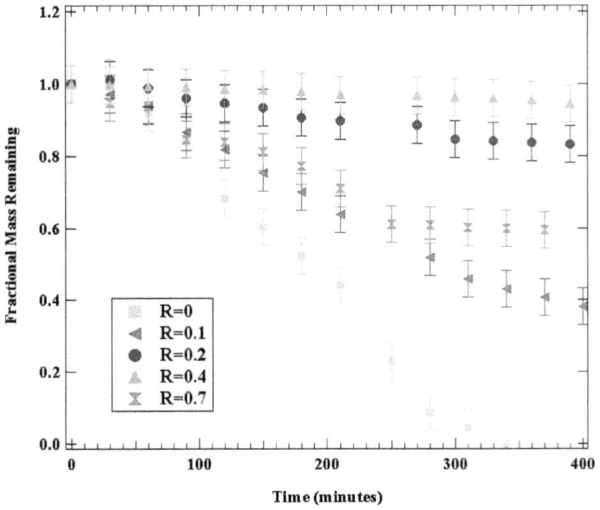

Figure 2. Dissolution results for lithium borate beads. The slope of the remaining mass constitutes the dissolution rate for that particular sample. The overall (batch-to-batch) error is estimated to be ±0·05

by much (~0·4). Indeed, it is surprising that *E. coli*, known to be a relatively sturdy bacterium, could not survive such a change. It is likely that this is because the local pH around the beads was much lower than that of the surrounding (and diluting) solution. Much more important was the obvious and significant turn to alkalinity of the broth at the higher *R* values, which we attributed to the formation of lithium hydroxide species (possibly arising from congruent dissolution(10)). The high pH broth then became a very inhospitable place for *E. coli*, resulting in a survival rate of 0%. We note that *Salmonella typhimurium* can only grow(12) in a pH higher than 4·7 at 25°C.

Significant changes in pH are fatal to most living organisms, bacteria included. Organisms have evolved their physical structure to maximise their success in a given chemical environment. The sudden introduction of highly acidic (positive) or highly basic (negative) species in the immediate living environment of a microorganism will disrupt the integrity of the membranes and proteins that constitute its physical structure. Cell membranes consist of a fluid phospholipid bilayer. The presence of highly charged species (imposition of an external electric field) diminishes the strength of electrical attraction holding the polar phospholipids of the phospholipid bilayer together, possibly lysing (popping) the cell. These charged species can also denature the integral (located within the membrane) transport proteins that help regulate the cell's internal environment, poisoning or suffocating the cell. The exact mode of death will vary with organism. Generally, the stronger the acid or base involved (the more dissociated charged species), the more significant the effect.

Table 3. Changes in the pH of the TSB broth for several glass compositions. The measurements were taken on four beads after an 18 h TSB soak

R value	*Change in pH*
0	7·37 to 6·90; Δ=−0·47
0·05, Li-B	7·26 to 6·95; Δ=−0·31
0·4, Li-B	7·28 to 8·09; Δ=+0·81
0·7, Li-K-B	7·27 to 9·83; Δ=+2·56

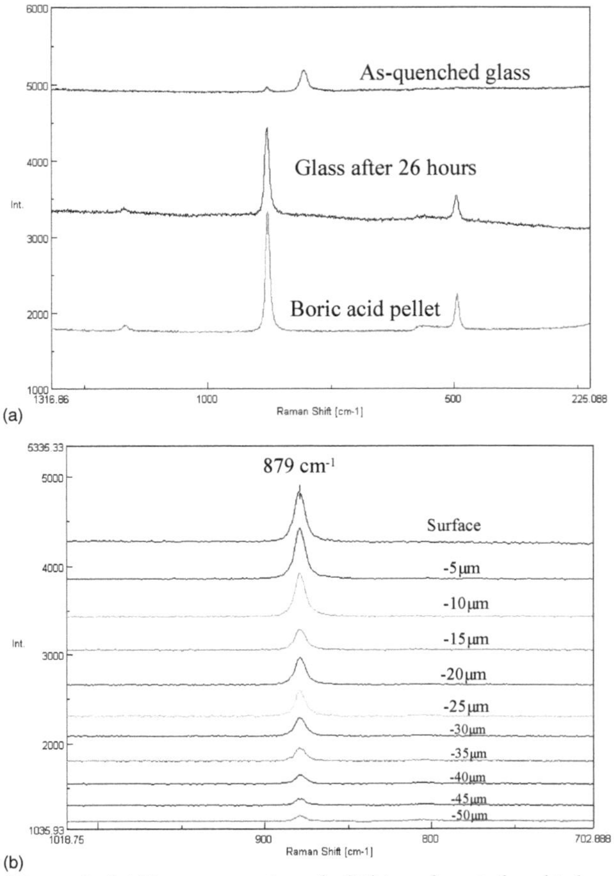

Figure 3. (a) Raman spectra of a lithium borate bead taken initially and after 26 h soaking in TSB, as well as the spectrum of a boric acid pellet. (b) Depth mapping of the presence of boric acid in a lithium borate R=0·1 glass bead. The latter sample had been previously exposed to water by a 5 min immersion followed by a 50 min in air

Raman results

Raman spectroscopy is a technique quite sensitive to surface analysis (and, equally important, mostly insensitive to water). Figure 3(a) compares the spectra of a pellet of boric acid, the surface of a lithium borate bead of low *R* value (*R*= 0·05) as soon as it was made, and the same pellet after 26 h in the TSB broth. The figure clearly shows that the glass surface of the bead approaches the boric acid spectrum, and the boroxol ring peak is eliminated. The peak at 880 cm^{-1} is typically(13) associated with the ν_{12} BO_3 breathing mode, while the 500 cm^{-1} mode is explained by the ν_{13} BO_3 bending mode. The bead began to react with the atmospheric water minutes after it was made (note

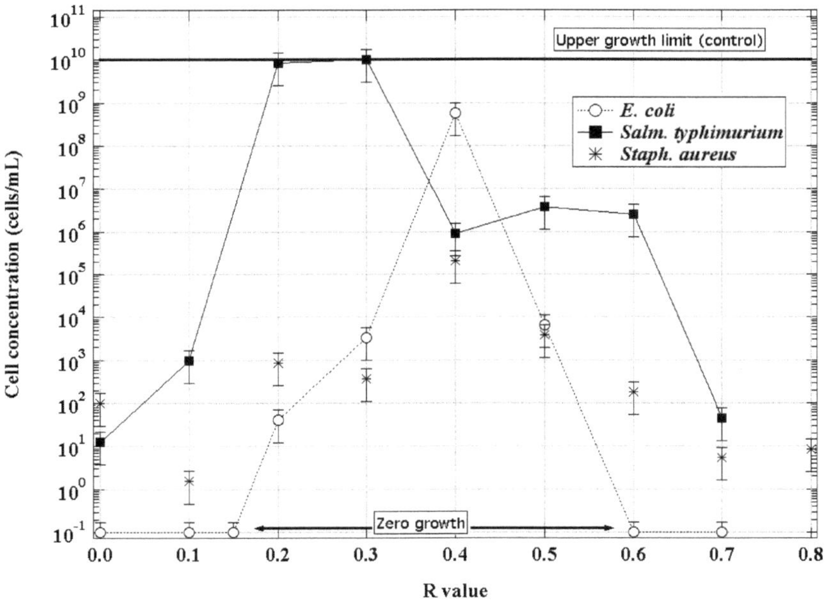

Figure 4. The survival rates of E. coli, S. typhimurium, *and* Staph. aureus *are plotted as a function of composition for the lithium potassium borate glass family. The upper bar shows the results from all the control beads, which reached the instrumental limit (10^{10} cells/ml). The error bars represent an overall variability in the survival rate estimated from all samples*

the small peak at 880 cm^{-1}). This confirmed the rapid formation of the acid layer.

We also carried out a measurement of the depth of the boric acid layer in a lithium borate sample of composition R=0·1. The glass bead was fully immersed in water for five minutes, and exposed to air for another 50 min. Using the confocal capability of the Raman microscope, we obtained spectra from different depths ranging from the surface to 50 μm below the surface, in steps of 5 μm. Figure 3(b) illustrates the steady decay of the boric acid peak at 880 cm^{-1}, indicating that the layer at this composition is approximately 50 μm deep. Other measurements at other compositions also pointed to this being a typical depth for many of the samples.

The lithium borate glass beads suffered from the obvious practical shortcoming: they dissolved too quickly for any kind of application. We therefore began to study a ternary glass system, the lithium potassium borates.

Lithium potassium borate glasses

The results were immediately promising. Figure 4 shows the low survival rates caused by these glass beads. New to this figure is the inclusion of *S. aureus,*

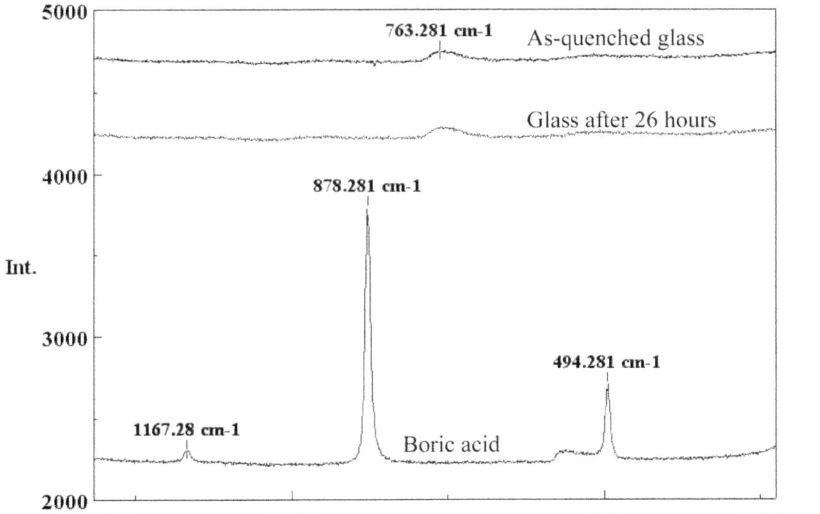

Figure 5. Raman spectra taken from a lithium potassium R=0·7 *sample. The initial sample is compared to the sample after 26 h immersion in TSB and a boric acid pellet*

a Gram-positive bacterium, which is known[14] to survive only above a pH of 4·15. This last bacterial species, with its tougher outer membrane, is indeed able to attain slightly higher final cell counts, though still below those of *S. typhimurium*. For *S. aureus* and *E. coli*, the beads inhibited growth by at least a factor of 15. Generally, with the exception of the R=0·4 composition, all final cell concentrations for *S. aureus* and *E. coli* were below those of the lithium-borate beads shown in Figure 1. In the case of the *S. typhimurium*, metal contents around the R=0·2–0·3 compositions were ineffective, yielding little growth inhibition compared to the control beads. Some of the tests were repeated for the new glasses and beads to ensure repeatability.

Again we noted that the higher R values resulted in efficient bactericidal action, with zero/small growth for R>0·6. We carried out a Raman study, and we compared (Figure 5) the boric acid pellet to a Li–K–B R=0·7 glass bead after it was initially made and then after 26 h soaking in TSB. Figure 5 compares well with Figure 3(a). The glass surface does not approach the boric acid spectrum, indicating that basic dissolution appears to fragment the entire glass network, resulting in no chemical changes on the surface, which points to congruent dissolution. This is typical of glasses, including silicates, exposed to alkaline environments, and has been known for many years.[15]

Finally, we obtained useful results from the time scale protocol. The trial was performed in duplicate, using R=0·7 and R=0·1 lithium potassium borate beads as well as a borosilicate control. The control media showed solid growth/solid growth after 15 min and 3 h, as expected. The R=0·1 media also showed growth/growth at 15 min and even after 3 h, in both trials. The R=0·7 media, on the other hand, showed no growth or very limited growth after 15 min in the trials. After 3 h, both R=0·7 media showed no growth. Apparently, the R=0·7 alkaline bactericidal mechanism is faster acting.

4. Conclusions

Both lithium and lithium potassium borate glasses proved to be bactericidal at low and high alkali concentrations. At intermediate alkali values the glass presence resulted in growth inhibition ranging, at a minimum estimation, from 100–1000. The effect was explained, at low R values, by the formation of a boric acid layer on the surface that proved inhospitable for bacterial growth. At higher alkali contents, hydroxide groups create a surface environment too basic for bacteria to survive. In both cases, the same compositions that formed an aqueous surface layer sufficiently different in neutral pH were also likely cause the glass to dissolve quickly in the presence of water. The bactericidal effect of the glasses affected both Gram-negative bacteria (*E. coli* and *Salmonella typhimurium*) and Gram-positive bacteria (*Stapphylococcus aureus*).

Raman measurements helped in the identification of the surface species, and also allowed us to determine that the acidic layer had a depth of approximately 50 μm. The killing action appeared to be relatively fast, obtaining bacterial eradication in as little as 15 min, for the lithium potassium borate glasses (R=0·7) that produced high alkalinity in the broth.

5. Acknowledgements

The authors acknowledge the funding of the McElroy Trust of Iowa, under a program administered by the Iowa College Foundation, and of the Murray Grant of Coe College. The College is also thanked for providing student housing, supplies, and other support. Ms Amy Marquardt and Mr James Borden are also thanked for their help with Raman measurements and sample preparation.

References

1. Osaka, A., Hayakawa, S., Tsuru, K. & Ohtsuki, C. Bioactivity of Alkali and Alkaline Earth Borosilicate Glasses, in *Borate Glasses, Crystals & Melts*, A. C. Wright, S. Feller and A. C. Hannon, eds. Society of Glass Technology, Sheffield, 1997, 490–497.
2. Day, D. E., White, J. E., Brown, R. F. & McMenamin, K. D. Transformation of borate glasses into biologically useful materials, *Glass Technol.*, 2003, **44** (2), 75–81.
3. Conzone, S. D., Hemrick, J. G. & Day, D. E. Glass Formation and Chemical Durability of Dysprosium Lithium Borate Glasses Intended for in vivo Radiation Synovectomy, *Glastech. Ber. Glass Sci. Technol.*, 2001, **74** (2), 39–45.
4. Bellantone, M., Coleman, N. J. & Hench, L. L. Bacteriostatic Action of a Novel Four-Component Bioactive Glass, *J. Biomed. Mater. Res.*, 2000, **51** (3), 484–490.
5. Bellantone, M., Coleman, N. J. & Hench, L. L. A Novel Sol-Gel Derived Bioactive Glass Featuring Antibacterial Properties, *Key Eng. Mater.*, 2001, **192–195**, 597–600.
6. Bellantone, M. & Hench, L. L. Bioactive Behavior of Sol-Gel Derived Antibacterial Bioactive Glass, *Key Eng. Mater.*, 2001, **192–195**, 617–620.
7. Bellantone, M., Williams, H. D. & Hench, L. L. Broad-Spectrum Bactericidal Activity of Ag_2O-Doped Bioactive Glass, *Antimicrobial Agents Chemotherapy*, 2002, **46** (6), 1940–45.
8. Verne, E., Nunzio, S. D., Bosetti, M., Appendino, P., Brovarone, C. V., Maina, G. & Cannas, M. Surface Characterization of Silver-Doped Bioactive Glass, *Biomaterials*, 2005, **26**, 5111–5119.
9. Kawashita, M., Tsuneyama, S., Miyaji, F., Kokubo, T., Kozuka, H. & Yamamoto, K. Antibacterial Silver-Containing Silica Glass Prepared by Sol-Gel Method, *Biomaterials*, 2000, **21**, 393–398.
10. Shelby, J. E. *Introduction to Glass Science and Technology*, second edition, RSC Paperbacks, Royal Society of Chemistry, Cambridge, 2005.
11. Doremus, R. H. *Glass Science*, John Wiley and Sons, New York, 1994.
12. Koutsoumanis, K. P., Kendall, P. A. & Sofos, J. N. Modeling the Boundaries of Growth of Salmonella Typhimurium in Broth as a Function of Temperature, Water Activity, and pH, *J. Food Protection*, 2004, **67** (1), 53–9.
13. Pereira, A. S., Perottoni, C. A. & da Jornada, J. A. H. Raman Spectroscopy as a Probe for In Situ Studies of Pressure-Induced Amorphization: Some Illustrative Examples, *J. Raman Spectrosc.*, 2003, 34 578-86.
14. Lanciotti, R., Sinigaglia, M., Gardini, F., Vannini, L. & Guerzoni, M. E. Growth/No Growth Interfaces of Bacillus cereus, Staphylococcus aureus and Salmonella enteridis in Model Systems Based on Water activity, pH, Temperature and Ethanol Concentration, *Food Microbiology*, 2001, **18**, 659–68.
15. Holland, L. *The Properties of Glass Surfaces*, John Wiley, New York, 1964.

Proc. Fifth Int. Conf. on Borate Glasses, Crystals and Melts Glass Tech.: Eur. J. Glass Sci. Technol. A, October 2006, 47 (5), 133–137

Thermal history of a low alkali borosilicate glass probed by infrared and Raman spectroscopy

Doris Möncke, Doris Ehrt*

Friedrich-Schiller-Universität, Otto-Schott-Institut für Glaschemie, Fraunhoferstr. 6, D-07743 Jena, Germany

Christos-Platon E. Varsamis, Efstratios I. Kamitsos

Theoretical and Physical Chemistry Institute, National Hellenic Research Foundation, GR-11635 Athens, Greece

Angelos G. Kalampounias

Foundation for Research and Technology Hellas, Institute of Chemical Engineering and High Temperature Chemical Processes, GR-26500 Patras, Greece

Infrared and Raman spectroscopies were employed to investigate the influence of the thermal history on the structure of a low alkali borosilicate glass (NBS2) with the composition $0{\cdot}74SiO_2.0{\cdot}21B_2O_3.0{\cdot}04Na_2O.0{\cdot}01Al_2O_3$. This glass is a model for technical glasses like Duran® and Borofloat® with similar low thermal expansion coefficients and high resistance to thermal shock. Previous studies including ^{11}B-NMR spectroscopy established that the fraction of $BØ_3$ units in ring formations is smaller in rapidly quenched samples as compared to annealed samples; whereas there is no change in the fraction of $BØ_4^-$ tetrahedra (Ø denotes a bridging oxygen atom). Samples were prepared for the present work at different melting temperatures (1650 to 1750°C) and different melt cooling temperatures (T_{mc}: 250 to 700°C). Differences in the thermal history of NBS2 samples result in glass structures with the same short range order (SRO) structural units while the medium range order (MRO) structure of the borosilicate network was found to be sensitive to changes in melt cooling temperature. With increasing melt cooling temperature the MRO of the silicate and borate structures increases, while the number of mixed Si–O–B linkages decreases. The better packing of the overall structure is reflected in a parallel increase in the density of the glasses annealed at higher T_{mc}.

1. Introduction

Borosilicate glasses are materials of great technical importance. Their applications include shock resistant household and laboratory glasses with high chemical and mechanical stability, as well as high performance optical glasses.(1–3) The boroxol ring is the most important structural unit in vitreous B_2O_3 where about 75% of boron atoms are found in those planar rings formed by three $BØ_3$ triangular groups (Ø denotes a bridging oxygen atom), and only one "loose" $BØ_3$ triangle is found for every boroxol ring.(4–6) Addition of alkali oxide to B_2O_3 initially increases the strength of the network as triangular $BØ_3$ groups are transformed into $BØ_4^-$ tetrahedra, but further increase of the alkali content results in the progressive depolymerisation of the network by creation of nonbridging oxygen (NBO) atoms on triangular borate units, $BØ_2O^-$ (O^-=NBO). Such structural changes are reflected by the appearance of extrema in property–composition curves, a phenomenon known as the "boron anomaly" effect. This peculiarity has been associated with a maximum connectivity of the modified borate network, and it is found usually in borate and borosilicate glasses with an alkali oxide to boron oxide ratio of R=0·2.(1–3)

The borosilicate glass with composition $0{\cdot}74SiO_2.0{\cdot}21B_2O_3.0{\cdot}04Na_2O.0{\cdot}01Al_2O_3$ (NBS2) is positioned along the line of the "boron anomaly" since Na_2O/B_2O_3=0·2, and can be seen as a model for technical glasses such as Duran® or borofloat®. All these glasses are characterised by an extremely low thermal coefficient of expansion and a high resistance to thermal shock. In addition, due to the absence of the easily polarisable nonbridging oxygen atoms in NBS2 glasses, their band gap extends far into the UV, and this makes them attractive candidates for high performance optical glasses. The small Al_2O_3 content in NBS2 glasses facilitates glass formation by suppressing the strong tendency towards phase separation.

^{11}B-NMR spectroscopy constitutes a powerful tool for the analysis of multicomponent systems and enables the determination of the fraction of $BØ_4^-$ units and the differentiation between $BØ_3$ groups in different environments.(5,6) For example, a recent NMR investigation(7) of NBS2 glasses concluded that the borate part of the network consists of tetrahedrally

[1] Corresponding author. Email dmoencke@email.de

coordinated $BØ_4^-$ units (18%), and neutral $BØ_3$ triangles in boroxol rings (40%) and in "loose" arrangements (42%). The substantial shift of the NMR signal of "loose" $BØ_3$ groups in NBS2 glass, as compared to $BØ_3$ entities in B_2O_3, suggested that "loose" $BØ_3$ units in the borosilicate structure are most likely linked to at least one $SiØ_4$ group. In addition, the NMR study showed no evidence for NBO-containing borate units.(7)

Quenching the high viscosity borosilicate glass NBS2 doped with structure indicator Co^{2+} or Ni^{2+} ions results in a "freezing in" of the high temperature structures with a tetrahedral or dodecahedral coordination of these ions (Co^{2+} grey-blue; Ni^{2+} orange-brown), as compared to their octahedral coordination (Co^{2+} lavender; Ni^{2+} green) favoured at room temperature.(7,8) Besides affecting the coordination of probe ions, high temperatures could favour also a smaller coordination number for the glass forming boron atoms according to the reaction(9)

$$BØ_4^- \Leftrightarrow BØ_2O^- \qquad (1)$$

However, the coordination change during the cooling or quenching process is normally so swift that no evidence of the coordination history of the glass constituents has thus far been observed in the glass NBS2.(7,8)

A ^{11}B-NMR study of NBS2 glasses either splat quenched or carefully cooled from 500°C showed that, while the fraction of $BØ_4^-$ tetrahedra remains fixed, every fourth $BØ_3$ group present in boroxol rings in annealed samples converts into "loose" $BØ_3$ triangle in splat quenched samples.(7,8) This "looser" packing of the structure is reflected in a decreased density of NBS2 glasses with higher fictive temperatures, indicating a higher degree of structural ordering in better annealed samples.(7,8)

The current work extends the investigation into the influence of thermal history of the structure of technically important NBS2 glasses. Infrared and Raman spectroscopic techniques were employed here as sensitive probes of structure, because of their proven effectiveness in revealing structural characteristics of borate,(10–21) silicate(22–26) and borosilicate glasses.(27,28) Band overlapping and coupling effects of vibrational modes attributed to borate and silicate structures often lead to conflicting assignments in vibrational studies of ternary borosilicate glasses. To facilitate this study, the spectra of SiO_2, B_2O_3 and $0{\cdot}2Na_2O.0{\cdot}8B_2O_3$ glasses were also measured and are compared with those of the NBS2 glasses investigated.

2. Experimental procedures

NBS2 glassy samples were prepared from high purity optical grade materials SiO_2, Na_2CO_3, H_3BO_3 and $Al(OH)_3$.(7,8) Reagent grade powders of SiO_2,

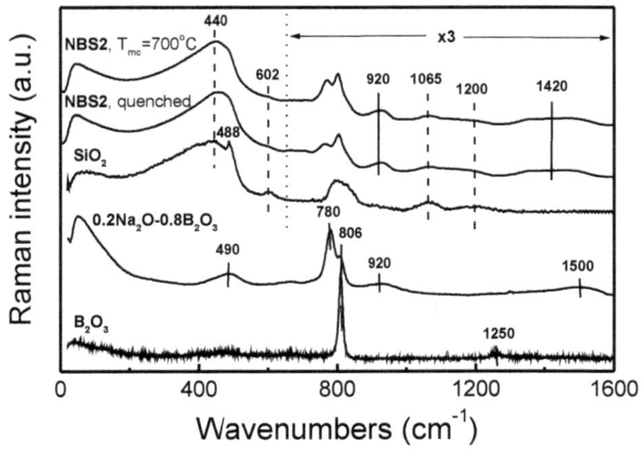

Figure 1. Raman spectra of melt quenched SiO_2, B_2O_3 and $0{\cdot}2Na_2O.0{\cdot}8B_2O_3$ NBS2 glasses and of NBS2 glasses prepared by melt cooling at T_{mc}=700°C and by quenching to room temperature. For clarity, the spectral intensity in the range 650–1600 cm^{-1} has been multiplied by a factor of three for SiO_2 and NBS2 glasses

Na_2CO_3 and B_2O_3, were used to prepare SiO_2, B_2O_3 and $0{\cdot}2Na_2O.0{\cdot}8B_2O_3$ glass by the conventional melt-quenching technique. 100 g to 250 g batches of NBS2 were melted in Pt or Pt–Rh20 crucibles at 1600°C in a resistance heated electric furnace. The melts were quenched in water and the dried frits were remelted to ensure a sufficient homogeneity. The melting temperature was finally increased to either 1650, 1700 or 1750°C for 30 min before the melt was poured onto brass plates or into graphite moulds preheated to different temperatures (melt cooling temperatures, T_{mc}, varied from 250 to 700°C), and they were subsequently cooled down to room temperature with an initial cooling rate of about 30 K/h. In addition, a quenched glass sample was prepared by pouring the melt onto a brass block and rapidly stamping it with a brass plate, with no further annealing. It is noted that high melting temperatures are needed for handling the melt, even though the T_g of the NBS2 glass is relatively low (440°C). Because of the high temperature gradient between the melt containing crucible and room temperature the melt cools down very rapidly and the high viscosity of the melt limits significantly the time available for handling the melt. Despite the high melting temperatures wet chemical analysis indicated that the glass compositions were within the analytical error of the batch compositions.

The density of NBS2 glasses was determined by the Archimedes method at 25°C in distilled water. Density values were found to depend on the thermal history of samples and varied between 2·174 and 2·160 g/cm^3 (±0·2%) for the glass with T_{mc}=700°C and the quenched glass, respectively. The refractive index, n_e=1·47±0·0006, and Abbe number, ν_e=65±2, were measured with a pulfrich refractometer. The glass transition temperature was determined with a Shimadzu DTA instrument (error: ±2 K). T_g values were determined by dilatometry (error: ±5 K), together with the thermal expansion coefficient, $\alpha=35\times10^{-7}$ K^{-1}.

Because of their low coefficient of thermal expansion, even the quenched samples could be mechanically handled easily. Polished glass plates with dimensions of 20×10×2 mm were used for infrared measurements on a vacuum Fourier transform spectrometer. Spectra were collected in the specular reflectance mode in the range 30 to 5000 cm^{-1} with a resolution of 2 cm^{-1}. Reflectance spectra were analysed further by the Kramers–Kronig transformation to yield the absorption coefficient spectra, $\alpha(\nu)$. Raman spectra were recorded in 90° scattering geometry with a T-64000 micro-Raman system operating in double subtractive mode, and using the 514·5 nm line of an Argon ion laser for excitation at a power of 400 mW.

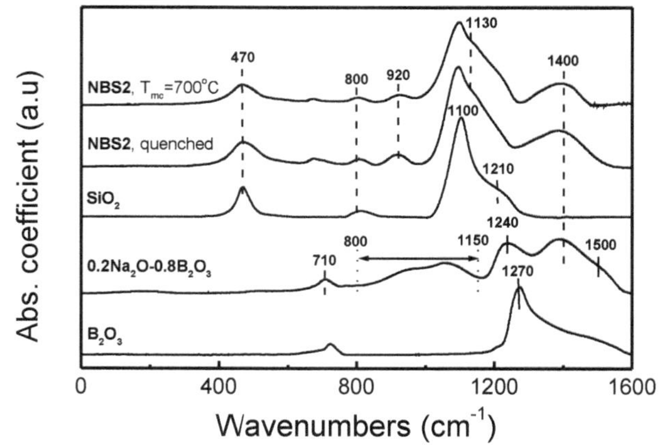

Figure 2. Absorption coefficient spectra, $\alpha(\nu)$, of melt-quenched SiO_2, B_2O_3 and $0{\cdot}2Na_2O.0{\cdot}8B_2O_3$ glasses and of NBS2 glasses prepared by melt cooling at T_{mc}=700°C and by quenching to room temperature

3. Results

3.1 Raman spectra

The Raman spectra of NBS2 glasses prepared by melt cooling at T_{mc}=700°C and by quenching to room temperature are shown in Figure 1, where they are compared with the spectra of melt quenched SiO_2, B_2O_3 and $0{\cdot}2Na_2O.0{\cdot}8B_2O_3$ glasses. For clarity, the spectral intensity in the range 650–1600 cm^{-1} has been multiplied by a factor of three for SiO_2 and NBS2 glasses.

The spectrum of B_2O_3 glass is dominated by the strong and narrow peak at 806 cm^{-1} attributed to the localised breathing vibration of the oxygen atoms in the boroxol ring, while the broad and weak feature at ~1250 cm^{-1} arises from the vibration of oxygen atoms or $BØ_3$ triangles bridging two boroxol rings.[10–14] The spectrum of the Na-borate glass presents bands at 780 and 806 cm^{-1} and additional weak features at ~490, 920 and 1500 cm^{-1}. According to previous studies,[15–21] the 780 cm^{-1} band can be assigned to the breathing vibration of six-membered borate rings with one $BØ_4^-$ tetrahedral unit, as in tri-, tetra- and pentaborate super structural units. These units are also responsible for the appearance of the weaker bands at 490 and 920 cm^{-1}. The broad feature around 1500 cm^{-1} is due mainly to the stretching vibration of B–O^- bonds of charged triangular borate entities, $BØ_2O^-$. The spectrum of vitreous silica shows its strongest feature as a broad band at ~440 cm^{-1}, this has been assigned to the symmetric stretching vibration of Si–O–Si bridges in six- or five-fold rings of $SiØ_4$ tetrahedra.[22–26] A superimposed peak at ~490 cm^{-1} and a weak feature at ~600 cm^{-1} arise from the ring breathing mode of four- and three-fold silicate rings, respectively, known as defect structures.[22,23] The broad band around 800 cm^{-1} is due to Si–O–Si deformation modes, while the weak features at 1065 and 1200 cm^{-1} have been attributed to the asymmetric stretching vibrational modes of Si–O–Si bridges connecting Q^4 tetrahedra, in agreement with the strong intensity measured in the infrared spectrum of vitreous silica in the same frequency range (Figure 2).[22–26]

The Raman spectra of the low alkali NBS2 glasses maintain the bands characteristic of glassy SiO_2, although the profile of the ~800 cm^{-1} band has changed and an additional band has appeared at 920 cm^{-1}. These findings suggest that the silicate part of the network remains practically unmodified and consists of interconnected Q^4 units. On the other hand, the complex profile at ~800 cm^{-1} originates now from the convolution of modes of three different structures: the bending of Si–O–Si bridges and the breathing mode of boroxol rings and of six-membered borate rings containing one $BØ_4^-$ unit. The new band at 920 cm^{-1} can be attributed to $BØ_4^-$-containing units,[16–21] while stretching vibrational modes of mixed Si–O–B bonds are also expected to be active in this range.[27,28] The broad feature around 1420 cm^{-1} is clearly different from that of the Na-borate glass, and it could be attributed mainly to vibrational modes of neutral $BØ_3$ triangles in different coordination environments.[14,21]

3.2 Infrared spectra

The absorption coefficient spectra, $\alpha(\nu)$, of SiO_2, B_2O_3, $0{\cdot}2Na_2O.0{\cdot}8B_2O_3$ and typical NBS2 glasses are depicted in Figure 2. The spectrum of the pure B_2O_3 glass exhibits a complex profile between 1200 and 1600 cm^{-1} which is attributed to the asymmetric stretching vibrational modes of neutral triangles in boroxol rings and in random configurations.[10,11] The spectrum of the $0{\cdot}2Na_2O.0{\cdot}8B_2O_3$ glass is characterised mainly by the broad profiles situated at 800–1150 and 1150–1550 cm^{-1}. The former profile is associated with the asymmetric stretching modes of tetrahedral units, $BØ_4^-$, and the latter with the asymmetric stretching vibrations of neutral, $BØ_3$, and charged, $BØ_2O^-$, triangular borate units.(9,16–20) The weak band at 710 cm^{-1} is typical of bending modes of borate structural units.(16–20) The infrared spectrum of vitreous SiO_2 exhibits its strongest absorption at 1100 cm^{-1} with a shoulder at ~1210 cm^{-1}. As noted above, these features are weak in the Raman spectrum (Figure 1) due to the nature of the corresponding modes, i.e. asymmetric stretching modes of Si–O–Si bridges. The weaker bands at 800 and 470 cm^{-1} are typical of the bending and rocking motion, respectively, of the Si–O–Si bridges.(22,23)

The infrared spectra of NBS2 glasses show strong envelopes centered at ~1100 and 1400 cm^{-1} (Figure 2). The first feature is characteristic of a SiO_2-like silicate network of neutral Q^4 tetrahedra, while the second one indicates the presence of neutral $BØ_3$ triangles in agreement with the recent NMR study.(7) The weaker bands at 470 and 800 cm^{-1} are quite similar to those in glassy SiO_2, while the band at 920 cm^{-1} could arise from the asymmetric stretching vibration of mixed Si–O–B bridges,(27) without excluding some contribution from the asymmetric stretching vibration of $BØ_4^-$ tetrahedral units.(18–20)

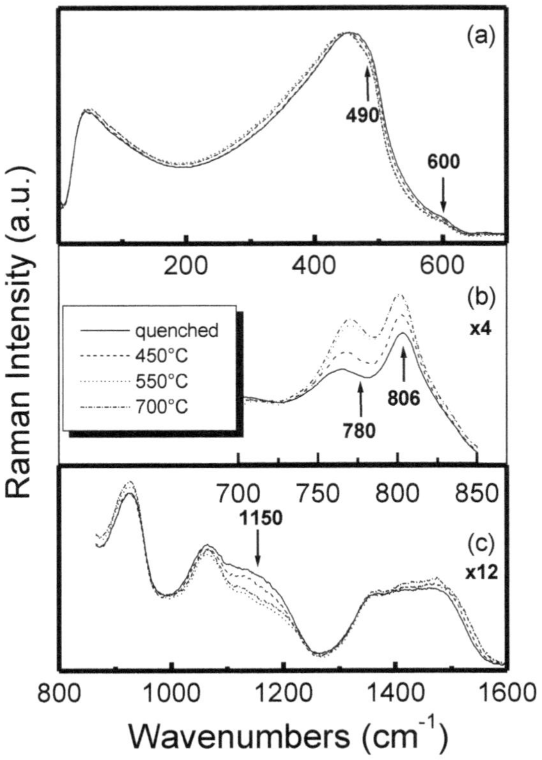

Figure 3. Raman spectra of NBS2 glasses with different melt cooling temperatures (450 to 700°C) and room temperature quenched: (a) 0–700 cm^{-1}, (b) 700-850 cm^{-1} and (c) 800–1600 cm^{-1} range. For clarity, the intensity in (b) was multiplied by a factor of four and in (c) by a factor of twelve relative to the spectral intensity in (a)

4. Discussion

4.1 Raman spectra

As observed in Figure 1, differences in thermal history affect the profile of the Raman spectra of NBS2 samples. To track this effect closely, Figure 3 depicts Raman spectra of NBS2 glasses having the same starting melt temperature (1700°C) but different melt cooling temperatures (450 to 700°C). Due to differences in Raman intensity at various spectral regions, the intensity in the 700–850 cm^{-1} range was multiplied by a factor of four (Figure 3(b)) and in the 800-1700 cm^{-1} range by a factor of twelve (Figure 3(c)) relative to the spectral intensity scale below 700 cm^{-1} (Figure 3(a)). To allow probing the effect of T_{mc}, all spectra were scaled on their most intense band at 440 cm^{-1}. Figure 3 shows that the spectra of NBS2 glasses with different T_{mc} maintain the same characteristic bands, indicating the presence of the same structural units. Nevertheless, differences in relative intensity of certain bands are shown to develop systematically upon increasing T_{mc} suggesting changes in the relative population of structural units.

Spectral variations attributed directly to the silicate part of the network are noted in Figure 3(a). In particular, an increase of T_{mc} results in the decrease of the relative intensity of bands at ~600 and 490 cm^{-1} and the shift of the ~440 cm^{-1} feature to lower frequencies. Considering the origin of these bands, the observed changes suggest the decrease of the relative population of three- and four-membered rings of silicate Q^4 tetrahedra in favour of larger silicate rings as T_{mc} increases. Thus, a more open medium range order silicate structure is preserved when the melt is cooled at a higher T_{mc} temperature.

The bands at ~780 and 806 cm^{-1}, characteristic of borate ring formations, are strengthened significantly with increasing T_{mc} (Figure 3(b)). Besides this overall effect, the 780 cm^{-1} peak due to planar rings with two $BØ_3$ triangles and one $BØ_4^-$ tetrahedron increases slightly in relative intensity compared to the boroxol ring peak at 806 cm^{-1}. The most pronounced variation in the higher frequency region (Figure 3(c)) is the progressive decrease in relative intensity of the ~1150 cm^{-1} shoulder with increasing T_{mc}, implying a decrease in the population of mixed Si–O–B bonds bridging Q^4 silicate tetrahedral with "loose" $BØ_3$ triangles.(24) These Raman changes manifest variations in the borate MRO structure, involving an increasing population of the borate six-membered ring structures with increasing T_{mc}. This finding is consistent with a Raman study of heat treated borosilicate glasses which showed that boroxol rings are favoured as the melt freezes to a glass at higher cooling temperatures.(24) Along the same lines, the recent NMR study(7) concluded that the fraction of borate rings in slow quenched NBS2 glasses is increased compared to that in rapidly-quenched samples.

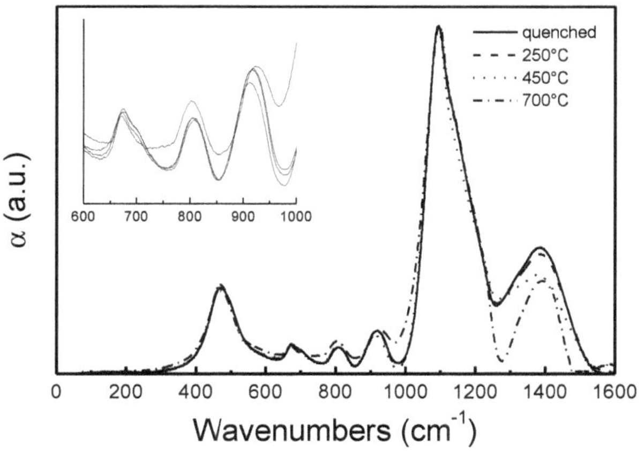

Figure 4. Absorption coefficient spectra, α(ν), of NBS2 glasses with different melt cooling temperatures (450 to 700°C) and room temperature quenched. The inset enlarges the 600–1000 cm^{-1} frequency range region

4.2 Infrared spectra

The absorption coefficient spectra of NBS2 glasses melted at 1700°C and prepared at different melt cooling temperatures are depicted in Figure 4. The main difference in the spectra is seen in the 1250–1500 cm^{-1} range where the B–O stretching modes of $BØ_3$ triangle are active. The bandwidth of this profile decreases significantly with increasing T_{mc} indicating a narrowing in the distribution of the $BØ_3$ bonding environments, in agreement with the Raman results. Thus, while the glass with the highest melt cooling temperature (700°C) shows a well shaped peak at ~1400 cm^{-1}, a shoulder develops at ~1320 cm^{-1} as T_{mc} decreases. Previous infrared studies of borosilicate glasses have attributed bands around 1350 cm^{-1} to vibrations of $BØ_3$ groups linked to either $SiØ_4$ or $BØ_4^-$ units.[15] On these grounds, it is reasonable to suggest that the decrease in intensity at ~1320 cm^{-1} with increasing T_{mc} is due to the decreasing population of $BØ_3$ units participating in non-ring formations such as the B–O–Si mixed bridges. This is in accord with the parallel decreasing intensity of the 680 cm^{-1} band characteristic of the bending vibration of mixed B–O–Si linkages (see inset in Figure 4), as well as with results derived from the Raman spectra.

Additional evidence for a better organisation of borate structural units with increasing T_{mc} is provided by the progressive enhancement of infrared intensity in the 900–1100 cm^{-1} range, where absorption in borate glasses arises mainly from B–O stretching modes of $BØ_4^-$ units in ring formations.[18–20]

5. Conclusions

Borosilicate glasses with composition $0{\cdot}74SiO_2.0{\cdot}21B_2O_3.0{\cdot}04Na_2O.0{\cdot}01Al_2O_3$ (NBS2) were prepared at different melt cooling temperatures, T_{mc}, ranging from 250 to 700°C and their structures was investigated by infrared and Raman spectroscopy. The measured spectra show glass structures consisting of the same short range order (SRO) structural units, i.e. neutral $BØ_3$ and $SiØ_4$ and charged $BØ_4^-$ units, with a small fraction of $AlØ_4^-$, despite the broad variation in T_{mc}. Compared to the SRO structure, the MRO structure of the borosilicate network was found to be sensitive to changes in melt cooling temperature, as manifested by both Raman and infrared spectra. In particular, it was found that the small three- and four-membered rings of $SiØ_4$ units are progressively transformed into larger silicate ring arrangements as T_{mc} increases, followed by a better organisation of the borate SRO units ($BØ_3$ and $BØ_4^-$) into six-membered rings. The progressive evolution of MRO silicate and borate structures with increasing melt cooling temperature happens against the Si–O–B connectivity, but it leads eventually to a better packing of the overall structure as evidenced by the parallel increase in density of NBS2 glasses.

References

1. Griscom, D. L. *Borate glasses: structure, properties and applications*, Edited by L. D. Pye, V. D. Frechette & N. J. Kreidl, Plenum Press, New York, 1978, Pp.11–149.
2. Ehrt, D. *Glass Technol.*, 2000, **41**, 182.
3. Ehrt, D., Reiß, H. & Vogel, W. *Silikattechnik*, 1977, **28**, 359.
4. Johnson, P. A. V., Wright, A. C. & Sinclar, R. N. *J. Non-Cryst. Solids*, 1982, **50**, 281.
5. Bray, P. J., Feller, S. A. & Jellison, G. E. *J. Non-Cryst. Solids*, 1980, **38–39**, 93.
6. Youngman, R. E. & Zwanzinger, J. W. *J. Phys. Chem.*, 1996, **100**, 16720.
7. Möncke, D., Ehrt, D., Eckert, H. & Mertens, V. *Phys. Chem. Glasses*, 2003, **44**, 113.
8. Möncke, D. & Ehrt, D. *Glass Sci. Technol.*, 2002, **75**, 163.
9. Varsamis, C. P., Kamitsos, E. I. & Chryssikos, G. D. *Phys. Rev. B*, 1999, **60**, 3885.
10. Krogh-Moe, J. *Phys. Chem. Glasses*, 1965, **6**, 46.
11. Kristiansen, L. A. & Krogh-Moe, J. *Phys. Chem. Glasses*, 1968, **9**, 96.
12. Galeener, F. L., Lucovsky, G. & Mikkelsen, J. C. *Phys. Rev. B*, 1980, **22**, 3983.
13. Windisch, C. F. & Risen, W. M. *J. Non-Cryst. Solids*, 1982, **48**, 307.
14. Hassan, A. K., Torell, L. M., Börjesson, L. & Doweidar, H. *Phys. Rev. B*, 1992, **45**, 12797.
15. Jellyman P. E. & Procter J. P. *J. Soc. Glass Technol.*, 1959, **39**, 173 T.
16. Konijnendijk, W. L. & Stevels, J. M. *J. Non-Cryst. Solids*, 1975, **18**, 307.
17. Brill, T. W. *Philips Res. Rep. Suppl.*, 1975, No 1, Pp 1–114.
18. Kamitsos E. I., Karakassides M. A. & Chryssikos, G. D. *J. Phys. Chem.*, 1987, **91**, 1073.
19. Kamitsos, E. I., Patsis, A. P., Karakassides, M. A. & Chryssikos, G. D. *J. Non-Cryst. Solids*, 1990, **126**, 52.
20. Kamitsos, E. I. *Phys. Chem. Glasses*, 2003, **44**, 79.
21. Yano, T., Kunimine, N., Shibata, A. & Yamane, M. *J. Non-Cryst. Solids*, 2003, **321**, 137–146; 147–156; 157–168.
22. Galeener, F. L. & Lucovsky G. *Phys. Rev. Lett.*, 1976, **37**, 1474.
23. Galeener F. L. *J. Non-Cryst. Solids*, 1982, **49**, 53; 1985, **71**, 372.
24. Furukawa, T., Fox, K. E. & White, W. B. *J. Chem. Phys.*, 1981, **75**, 3226.
25. McMillan, P. &. Piriou B. *J. Non-Cryst. Solids*, 1982, **53**, 279.
26. Kamitsos, E. I., Patsis, A. P. & Kordas, G. *Phys. Rev. B*, 1993, **48**, 12499.
27. Tenney, A. S. & Wong, J. *J. Chem. Phys.*, 1972, **56**, 5516.
28. Kamitsos, E. I., Kapoutsis, J. A., Jain, H. & Isieh, C. H. *J. Non-Cryst. Solids*, 1994, **171**, 31.

Proc. Fifth Int. Conf. on Borate Glasses, Crystals and Melts Glass Tech.: Eur. J. Glass Sci. Technol. A, October 2006, **47** (5), 138–140

Conductivity and microstructure of silver borate glass/zirconia composites, prepared via a high pressure route

M. Foltyn, M.Wasiucionek, J. E. Garbarczyk, J. L. Nowinski*

Faculty of Physics, Warsaw University of Technology, Koszykowa 75, 00-662 Warszawa, Poland

S. Gierlotka & B. Palosz

Institute of High Pressure Physics, Polish Academy of Sciences, Sokolowska 29/37, 01-142 Warszawa, Poland

Novel dense composites based on $AgI–Ag_2O–B_2O_3$ glass/ZrO_2 powders (~1 or 10 μm grains) were prepared by the application of high pressures (up to ~8 GPa) at 100–200°C. Their electrical properties, microstructures and thermal properties were determined and qualitative evaluation of their mechanical properties was undertaken. The electrical conductivity of the composites was lower by a factor of 2–5 than that of the base glass. Up to ~120°C the conductivity has an Arrhenius dependence on with an activation energy of ~0·3 eV, close to that of the glass. Scanning electron micrographs show that the glassy phase has generally been evenly distributed between the ZrO_2 grains, although some small (~50 nm) drop like features next to the surfaces of zirconia grains were also observed.

Introduction

Silver borate glasses with high AgI contents, have been considered, due to their high ionic conductivity, as good candidates for solid electrolytes in practical devices such as, for example, microbatteries working at ambient temperatures.(1) They have also been of interest from a basic science perspective, in particular with respect to structure–ionic transport relations.(2,3) This interest was considerably enhanced in early 1990s following the successful stabilisation of nanometric crystallites of the α phase of AgI down to room temperature.(4) Important progress has since been achieved in the preparation of AgI-based composites and in understanding their structure and transport properties.(5,6) On the other hand it is well known that the introduction of a foreign phase (a filler) can substantially improve various physical properties (mechanical, electrical, etc.) of a parent homogeneous material. Among the important parameters that determine the effective properties of the composite are the uniform distribution of the filler inside the parent matrix and the smallest possible filler particle size. A promising way to produce homogeneous composites and nanocomposites, free of voids, is by the simultaneous application of high pressures (up to 8 GPa) and elevated temperatures (from 100 to >1000°C) to sandwich structures and/or mixtures of constituents.(7)

In this work we used the high pressure method to synthesise composites based on $AgI–Ag_2O–B_2O_3$ glasses and zirconia powders (1 μm and/or 10 μm average grain size). We examined electrical conductivity of the resulting composites and their microstructures. The composites in this study were based on a $55AgI.30Ag_2O.15B_2O_3$ glass.

Experimental

Silver borate glasses of composition $55AgI.30Ag_2O.15B_2O_3$ were prepared by a standard melt quenching method. Equal volume proportions of glass and zirconia powder (10 μm or 1 μm average grain size) were thoroughly ground and mixed together. The prepared mixture was placed in a graphite tube serving as a furnace and introduced into a special high pressure container. The container was mounted between the anvils of a hydraulic press and heated up to 100–200°C under high pressures of 1·5–8 GPa. Under these conditions the softened glass better penetrates into voids between zirconia grains. The final products were compact, hard cylinders of diameter 5 mm and thickness 2 mm, which exhibited a much higher resistance to fracture than the original glasses.

X-ray diffraction (XRD) analyses of the as-received glasses was undertaken using a Siemens D 5000 diffractometer with Cu K_α radiation. These analyses confirmed the amorphous nature of the samples. XRD patterns of the composites were difficult to analyse, because of the very strong lines from crystalline ZrO_2 grains, which obscured any features characteristic of a

[1] Corresponding author. Email mwas@mech.pw.edu.pl

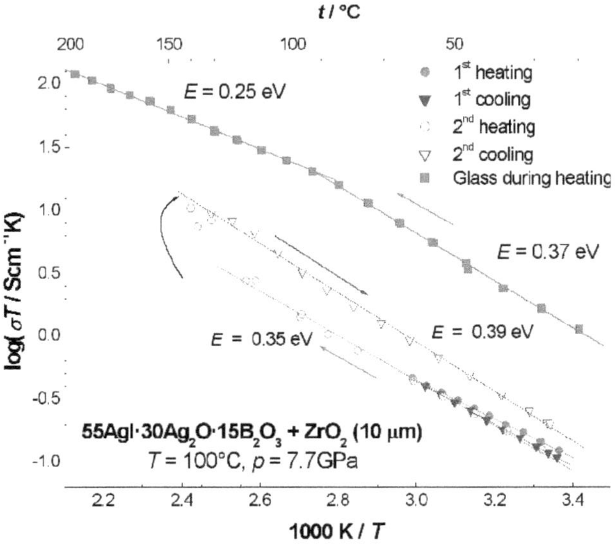

Figure 1. Temperature dependence of electrical conductivity of 55AgI.30Ag$_2$O.15B$_2$O$_3$ glass (□) and of a 55AgI.30Ag$_2$O.15B$_2$O$_3$ glass: 10 μm ZrO$_2$ composite, prepared at p=7·7 *GPa and* T=100°C. *First heating (●), first cooling (▼). Second heating (○) and second cooling (▽)*

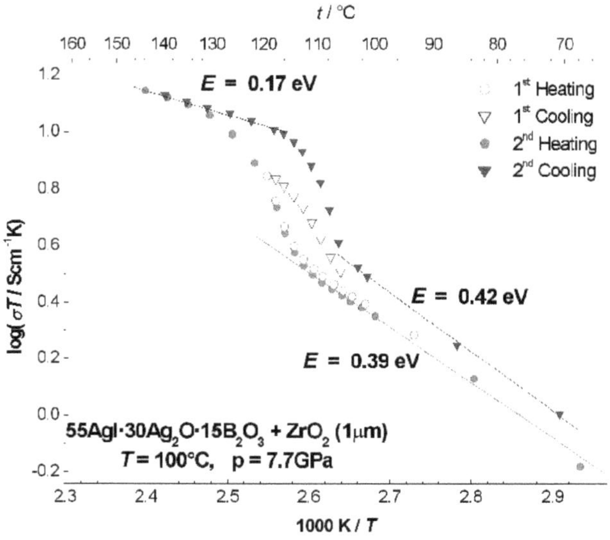

Figure 2. Temperature dependence of electrical conductivity of a 55AgI.30Ag$_2$O.15B$_2$O$_3$ glass: 1 μm ZrO$_2$ composite, prepared at p=7·7 *GPa and* T=100°C. *First heating (○) and cooling (▽). Second heating (●) and cooling (▼)*

glassy phase and/or emerging crystalline phases. The characteristic transition temperatures of the glasses used for preparation of the composites were determined by Differential Scanning Calorimetry (DSC) on a Pyris 1 DSC (Perkin-Elmer) at a 20 K min^{-1} scan rate. Scanning Electron Microscopy (SEM) was carried out using a LEO 1530 SEM. Impedance measurements were carried out using a computer controlled set up based on a Solartron 1260 Impedance/Gain Phase Analyser over a temperature range of 20–300°C, and a frequency range of 10 mHz–10 MHz. For these measurements gold electrodes were sputtered onto opposite polished faces of the samples.

Results

The glass transition temperature T_g of the as-received 55AgI.30Ag$_2$O.15B$_2$O$_3$ glasses as determined by DSC was close to 80°C. This is a typical values for glasses with AgI contents exceeding 50 mol%. The impedance spectra of the as-received 55AgI.30Ag$_2$O.15B$_2$O$_3$ glasses are typical for many ionically conducting glasses; they consist of a single arc at higher frequencies and a steep linear spur at the low frequency end, corresponding to blocking of ionic current at the electrodes. The character of the spectra does not change as the temperature is increased. The electrical conductivity of the glass (squares in Fig.1) has an Arrhenius temperature dependence with an activation energy of 0·37 eV in the 20–80°C temperature range. At higher temperatures the activation energy decreases to 0·25 eV.

The conductivity of composites based on this glass exhibit different temperature dependencies depending on the zirconia particle size. For the composites containing larger particles of ZrO$_2$ (10 μm) several heating–cooling runs were carried out to try and determine the effects of eventual crystallisation within the glassy matrix on the overall conductivity. The first heating and cooling run (full circles in Figure 1) was carried out in the temperature range 20–60°C (in some cases in the 20–80°C range, i.e. up to T_g of the glassy phase). The Arrhenius-like dependence and activation energy of ~0·35 eV indicate that the conductivity of the composite is determined by its amorphous phase. The conductivity values measured during heating and cooling were nearly the same. During the next heating run, this time up to ~150°C, the conductivity also had an Arrhenius dependence, with an activation energy of 0·35 eV (open circles in Figure 1). Close to the maximum temperature the conductivity additionally increased in a non-Arrhenian way. During the cooling run (triangles in Figure 1) temperature dependence of conductivity was Arrhenius-like with an activation energy of 0·39 eV.

Considerably different temperature dependencies were observed for composites with smaller ZrO$_2$ grains (~1 μm) (Figure 2). In this case during the first heating the conductivity had an Arrhenius dependence with an activation energy of 0·39 eV until a temperature of ~115°C. Then the conductivity started to deviate upwards from Arrhenius behaviour. The first heating run was stopped at ~125°C. On cooling (open triangles) the conductivity returned to the previous values, but with some hysteresis. This behaviour was more pronounced in the next heating–cooling run. Again, up to ~115°C the conductivity exhibited

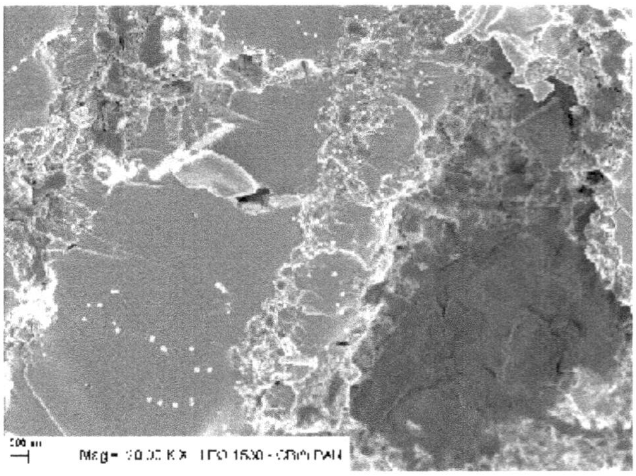

Figure 3. SEM micrograph of a $55AgI.30Ag_2O.15B_2O_3$ glass: 1 μm ZrO_2 composite, prepared at p=7·7 *GPa and* T=200*°C*

Arrhenius behaviour with an activation temperature of 0·39 eV. Then the conductivity underwent a steep increase in the range of 115–125°C and returned to an Arrhenius-like dependence at higher temperatures. On cooling (full triangles in Figure 2) the analogous step-like transition was again observed but shifted some 10 K downwards.

SEM micrographs of these composites (Figure 3) show that the interfaces between their components (i.e. glass and zirconia) are quite ramified. Additionally one can see a number of small (~50 nm size) drop-like inclusions in the vicinity of the interfaces. The nature of these inclusions and composition of the interface regions is as yet unknown, but it seems likely that they contain small AgI crystallites which strongly affect the effective ionic conductivity of the annealed composite.

Discussion

The fact that the conductivities of the composites under study are lower than that of bulk glass (Figure 1) points, at least partly, to the conductivity in these materials having a percolative character, in which a considerable part of the volume (the ZrO_2 grains) does not take part in the conduction.

The pronounced change of conductivity which occurs over a relatively narrow temperature range observed in Figure 2 (but also seen in Figure 1) is similar to the step-like conductivity change seen in bulk AgI when the β–α phase transition takes place. Such a β–α transition may occur also in the composites under study, if there are a number of small crystallites embedded in the glassy matrix of the composites or present at the surfaces of the ZrO_2 grains. The high AgI content (55 mol%) in the glass can lead to the formation of some AgI crystallites either during the heating ramp or, even earlier, during the high pressure fabrication of the composites at temperatures exceeding . Once present, these small AgI crystallites may undergo phase transitions, visible in the variation in conductivity with temperature. The fact that the temperature at which this change takes place is 115–125°C (Figure 2), i.e. some 20 K lower than that of β–α phase transition of pure bulk AgI (147°C) is difficult to explain. A possible explanation for such a substantial temperature shift may be the small size of the AgI crystallites, although other explanations may be possible. To determine the correct explanation for the observed non-Arrhenian dependencies further studies, including DSC analyses, on these and related composites will be carried out.

Conclusions

$55AgI.30Ag_2O.15B_2O_3$ glass/ZrO_2 composites with zirconia grains 1 μm or 10 μm in size, were prepared by a high pressure route. The conductivity of the composites was lower than that of the glass, but the activation energies were similar. The intriguing non-Arrhenian behaviour and the greater than 20 K shift of temperature of the β–α transition of AgI crystallites formed inside the composite was ascribed to the small size of these crystallites.

Acknowledgements

This work was financed by Polish State Committee for Research (Grant No. 4T08E 05223) and, in part, by Centre of Excellence CEPHOMA (Contract No. ENK5-CT-2002-80666).

References

1. Baskaran, N., Govindaraj, G. & Narayanasamy, A. *J. Power Sources*, 1995, **55**, 153.
2. Sidebottom, D. L. *Phys. Rev. B*, 2000, **61**, 14507.
3. Roling, B., Ingram, M. D., Lange, M. & Funke, K. *Phys. Rev. B*, 1997, **56**, 13619.
4. Tatsumisago, M., Shinkuma, Y. & Minami, T. *Nature*, 1991, **354**, 217.
5. Nagai, M. & Nishino, T. *Solid State Ionics*, 1999, **117**, 317.
6. Lee, J.-S., Adams, S. & Maier, J. *Solid State Ionics*, 2000, **136–137**, 1261.
7. Zgirski, M., Garbarczyk, J. E., Gierlotka, S., Palosz, B., Wasiucionek, M. & Nowinski, J. *Solid State Ionics*, 2005, **176**, 2141.

Proc. Fifth Int. Conf. on Borate Glasses, Crystals and Melts *Glass Tech.: Eur. J. Glass Sci. Technol. A*, October 2006, **47** (5), 141–144

Amorphous erbium doped borate thin films obtained by the polymeric precursor method and by sol-gel processing

L. J. Q. Maia,[1,2] *V. R. Mastelaro,*[2] *J. Fick,*[3] *A. C. Hernandes*[2] *& A. Ibanez*[1,*]

[1] *Laboratoire de Cristallographie, CNRS, associé à l'UJF et l'INPG, BP 166, 38042 Grenoble cedex 9, France*
[2] *CCMC-Instituto de Fisica de São Carlos/USP, 400 Av. Trabalhador Sãocarlense, CP 369, 13560-970 São Carlos/SP, Brasil*
[3] *Laboratoire de Spectroscopie Physique, BP 87, 38402 Saint-Martin d'Hères cedex, France*

The synthesis, thermal, structural and optical characterisations of $Y_{0\cdot9}Er_{0\cdot1}Er(BO_3)_4$ amorphous thin films deposited on fused silica substrates by spin coating are reported. The film process starts from solutions prepared by two chemical methods: either the polymeric precursor method or by sol-gel processing. These initial solutions were first polymerised and the resulting powders were analysed by differential scanning calorimetry while the deposits were characterised by x-ray diffraction. The levels of hydroxyl groups in the thin films were evaluated using Fourier transform infrared spectrometry as a function of temperature. The refractive indexes were obtained indirectly by an envelope method using interference fringes of the transmission spectra. Finally, luminescence spectra between 1400 and 1600 nm were obtained using an excitation wavelength of 980 nm, demonstrating the possibility of using these films as planar optical amplifiers.

Introduction

Rare earth doped planar waveguides are an active field of research as there is great interest in optical amplifiers and new laser sources for integrated optical devices. The development of high quality films by techniques such as plasma enhanced chemical vapour deposition (PECVD),[1] flame hydrolysis deposition,[2] liquid phase epitaxy,[3] ion exchange,[4] molecular beam epitaxy,[5] or sol-gel[6] have already enabled the realisation of integrated optical amplifiers and lasers.[1–4] In this field of research, chemical routes, such as the polymeric precursor method (modified Pechini method[7]) and sol-gel processing represent promising low cost processes for the production of good planar waveguides. The high control of chemical purity, the low processing temperature, the possibility of deposition on large area substrates and the easy incorporation of rare earth ions are the main advantages of both the polymeric precursor and sol-gel methods. On the other hand, the inclusion of large amounts of organics in the as deposited coatings is the most important disadvantage for the polymeric precursor method; thermal treatments at high temperatures, under oxygen, have to be used to remove the organics.

Good results have been obtained on erbium doped silica based glasses[1,2] and YAG[3] waveguides. But other systems such as borates $(Y_{1-x}Ln_xAl_3(BO_3)_4$ with Ln=Y, Er, Nd, ...) are also of interest. Indeed, high gains arising from high rare earth concentrations are reported in neodymium doped borate crystals.[8–10] This aluminoborate matrix avoids rare earth aggregation even at high Ln concentrations. Indeed, the rare earth is not a doping element but a constitutive ion of the crystal structure which allows it to maintain a well defined Ln–Ln distance, i.e. greater than 5·6 Å for stoichiometric $LnAl_3(BO_3)_4$ crystals. A solid solution of the rare earth ion associated with yttrium $(Y_{1-x}Ln_xAl_3(BO_3)_4)$ allows us to monitor the average distance between active ions. It is therefore possible to optimise the amount of Ln present to give the highest active ion concentration with limited concentration quenching effects. Finally, it has been shown that amorphous thin films, prepared just below the crystallisation temperature of the crystalline phases, exhibit very interesting optical properties.[6]

In the literature, only the production of $Y_{1-x}Er_xAl_3(BO_3)_4$ single crystals has been reported. The fabrication of bulk glasses is difficult as they require high melting temperatures and high quenching rates resulting in fragile materials. Based on these trends, the present work reports on the synthesis and characterisation of amorphous $Y_{0\cdot9}Er_{0\cdot1}Al_3(BO_3)_4$ (YEAB) thin films. The coating process is based on multilayers annealed at different temperatures. The characterisation methods used include thermal analysis, refractive index determination, evaluation of the levels of hydroxyl groups present and photo-

* Corresponding author. Email alain.ibanez@grenoble.cnrs.fr

luminescence emission. In this paper, we present our first results which confirm that the methods used are a promising way to prepare high quality borate films from solutions.

Experimental

The precursor solutions were prepared by two soft chemical methods: the polymeric precursor method (PPM) and sol-gel (SG) processing.

The polymeric precursor method

Erbium citrate was formed by dissolving erbium nitrate (99·9% purity) in an aqueous solution of citric acid (70–80°C) for 15 min. After homogenisation, yttrium nitrate (99·9% purity) was added to the Er citrate solution, and aluminium nitrate (>98% purity) was added to the Y-Er citrates. After further homogenization (30 min), D-sorbitol (99·6% purity) and boric acid (99·8% purity) were dissolved in water for 1 h and added to the citrate solution. D-sorbitol was added to promote mixed citrate polymerisation by a polyesterification reaction. The molar ratio of citric acid to the metals+boron was 3:1. The citric acid: D-sorbitol mass ratio was set to 3:2. The viscosity of the polymeric solution was adjusted to 0·01 Pas before coating by adjusting the water content (by water evaporation or addition as appropriate).

The sol-gel process

Aluminium acetylacetonate (99% purity) was dissolved at 80°C for 2 h with ethyl alcohol (>99·9% purity), propionic acid (>99% purity) and tri-i-propyl borate (>98% purity). A second solution prepared by the dissolution of erbium and yttrium nitrates (all 99·9% purity) with propionic acid at 80°C for 2 h. The two solutions were mixed together and water was added for the hydrolysis of the boron alkoxide. The molar ratios of the chemicals used were: yttrium nitrate hexahydrate 0·9; erbium pentahydrate 0·1; aluminium acetylacetonate 3; tri-i-propyl borate 4; propionic acid 55; ethyl alcohol 95 and water 5. The viscosity of the SG solution was adjusted before coating by solvent evaporation at 80°C for 70 min.

YEAB thin films were prepared by spin coating onto fused silica substrates. These substrates were cleaned by immersion in a sulfochromic solution, followed by rinsing several times in deionised water. Rotation speeds were fixed at 2500 rpm (PPM) and 2000 rpm (SG); with a spinning time of 5 s in both cases. After deposition, the thin films were dried at 80°C for 15 min on a hot plate to remove the residual solvents. These heat treatments under oxygen were carried out in two stages: an initial annealing at 400°C for 2 h, to pyrolyse the organic materials, followed by annealing at 500°C, 600°C and 700°C for 2 h each to eliminate the residual organic compounds and to promote the film densification. Five, ten and 15 multilayers were deposited on the substrates, each coating being pyrolysed at 400°C and then annealed at 500°C, 600°C and 700°C before the following coating was deposited.

Differential scanning calorimetry (DSC 404 S Netzsch) was used to study the decomposition process of the dried polymeric solution and the crystallisation of the carbon free powders annealed under an oxygen atmosphere. The powders were characterised by thermal analysis from room temperature to 1000°C, using a heating rate of 5°C/min. After the DSC measurements, the crystalline phases were analysed using an x-ray diffractometer with Fe K_α radiation (λ=1·936 Å). Fourier transform infrared spectrometry was used to evaluate the hydroxyl contents of the films as a function of annealing temperature. The spectra were measured using a spectrophotometer in a wavelength range of 3700–2400 cm^{-1}; the silica substrate spectrum was used as a reference. The surface roughness of the thin films was measured by atomic force microscopy using a Digital Nanoscope IIIa equipment operating in tapping mode with an SiN cantilever. Refractive indexes were obtained from transmission interference spectra using the envelope method developed by Manifacier *et al.*[11] This method uses the maxima and minima of the transmission spectra and assumes that we have a non-conducting coating, weak absorption, uniform films and a semi-infinite transparent substrate. The photoluminescence emission was measured in a waveguiding configuration. The films were pumped by a pigtailed laser diode emitting at 980 nm. The films were cleaved using a diamond stylus in order to obtain the end faces necessary for coupling the light in and out.

Results and discussion

DSC measurements were used to observe the glass transition (T_g) and the crystallisation (T_c) temperatures. The powders were annealed at 400°C for 24 h and then at 700°C for 24 h to remove the organic part of the matrix. Between 723°C and 783°C an endothermic peak, which is due to the glass transition, was observed (Figure 1(a)) for the PPM films while for the SG coatings we observed the glass transition between 756°C and 791°C (Figure 1(b)). The powders were amorphous, but in contrast to usual glasses obtained by melt quenching, they had been obtained from dried and heat treated solutions. DSC curves for powders prepared using the two different methods were recorded in the same equipment, and the measurements were repeated. In addition, second samples of both powders were produced and their

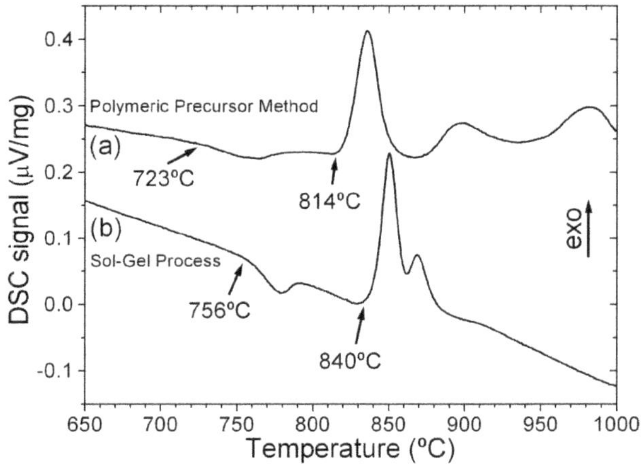

Figure 1. DSC curves of dried $Y_{0\cdot9}Er_{0\cdot1}Al_3(BO_3)_4$ powders obtained by the (a) PPM and (b) SG processes

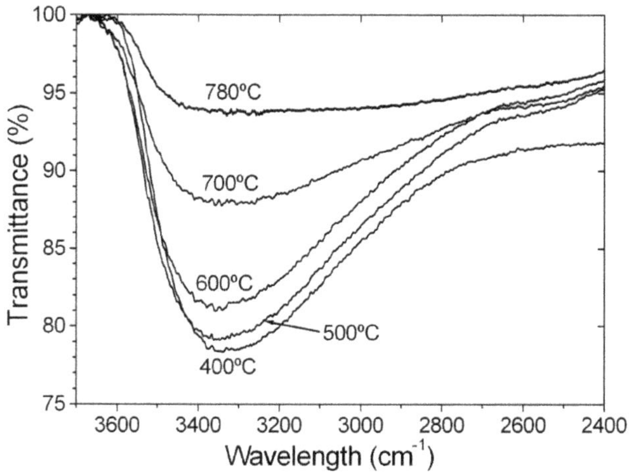

Figure 3. FTIR spectra of $Y_{0\cdot9}Er_{0\cdot1}Al_3(BO_3)_4$ thin films heat treated between 400°C and 780°C for 2 h

DSC curves measured. Each powder exhibited the same T_g slope in all the measurements. Inspection of the T_g slopes (Figure 1(a) and (b)) indicates that the PPM curve is more smooth than the SG one. The different C_p variation leads us to conclude that there are significant structural changes between PPM and SG samples with the two samples containing different amounts of BO_3 and BO_4 groups. Structural characterisation is in progress involving Raman, NMR and IR spectroscopies. Three exothermic peaks were observed between 814°C and 1000°C for PPM and between 840°C and 1000°C for SG films. These peaks are due to the crystallisation of $Al_4B_2O_9$ (T_{c1}=814°C), (Y,Er)BO_3 (T_{c2}=873°C), (Y,Er)$Al_2B_4O_{10\cdot5}$ and (Y,Er)$Al_3(BO_3)_4$ (T_{c3}=955°C); these phases were identified by x-ray diffraction. The temperature difference, $\Delta T=T_{c1}-T_g$, is equal to 92°C for the PPM sample and 84°C for the SG sample, showing that this YEAB glassy composition has good stability against crystallisation. However the PPM sample has a lower thermal stability (T_{c1}=814°C) than the SG one (T_{c1}=840°C). The YEAB thin films prepared at 780°C for 2 h are amorphous, however the annealing time has to be optimised to eliminate residual carbon and possible OH groups.

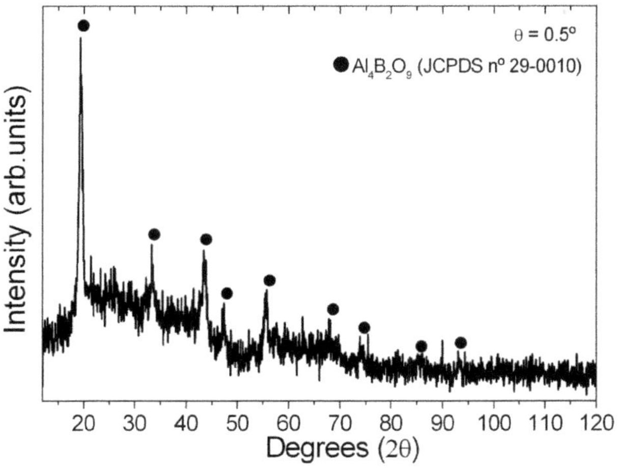

Figure 2. XRD pattern of a $Y_{0\cdot9}Er_{0\cdot1}Al_3(BO_3)_4$ PPM thin film heat treated at 820°C for 2 h

Figure 2 shows a diffraction pattern of a PPM thin film heat treated at 820°C for 2 h. It can be seen that the $Al_4B_2O_9$ phase is the first one to crystallise, in agreement with the DSC results and previous work.(12,13)

Fourier transform infrared spectrometry was used to evaluate the hydroxyl groups (–OH) concentration in the thin films as a function of temperature. The measurements, presented in Figure 3 between 3700 cm^{-1} and 2400 cm^{-1}, are difference spectra between the film+substrate and just the substrate; difference spectra are used to eliminate the hydroxyl contribution from the silica substrates. Our results show that we can strongly reduce the –OH concentration at 700°C (absorption band centred on 3340 cm^{-1}) and much more at 780°C above T_g as shown by DSC (work in progress). The reduction of these –OH groups is an important factor as they cause quenching of the luminescence of the rare earth atoms.

The film roughness, measured by atomic force microscopy over a 100 μm^2 area, varies between 0·4 nm and 0·7 nm RMS. This result confirms the low roughness and the planar surfaces of our films annealed at 780°C. The thin films exhibit a good transmittance between 600°C and 780°C in the ultraviolet, visible and near infrared regions (250–2000 nm range), higher than 90%, but the best results from interference fringes are obtained for thin films annealed just above T_g between 740°C and 780°C. This is due to the greatest coating densification, without crystallisation, corroborating the DSC results [Ref. to be published]. The *m*-line spectroscopy was also involved to determine the refractive indexes, film thicknesses and to verify that the coatings are waveguides. Three TE and TM modes were obtained on 1 μm thick films [Ref.

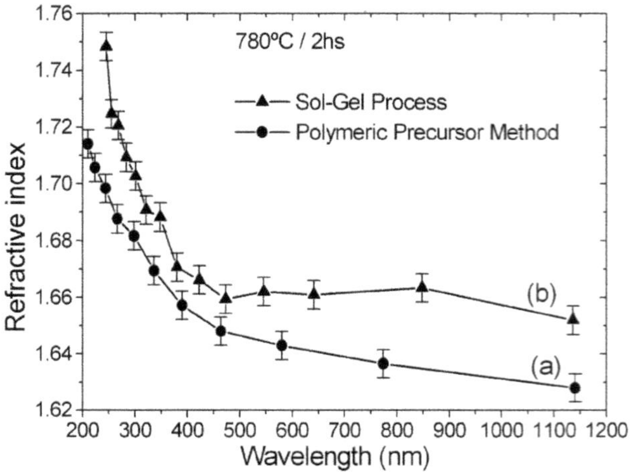

Figure 4. Refractive index as a function of wavelength for $Y_{0·9}Er_{0·1}Al_3(BO_3)_4$ thin films heat treated at 780°C for 2 h: (a) PPM process and (b) SG process

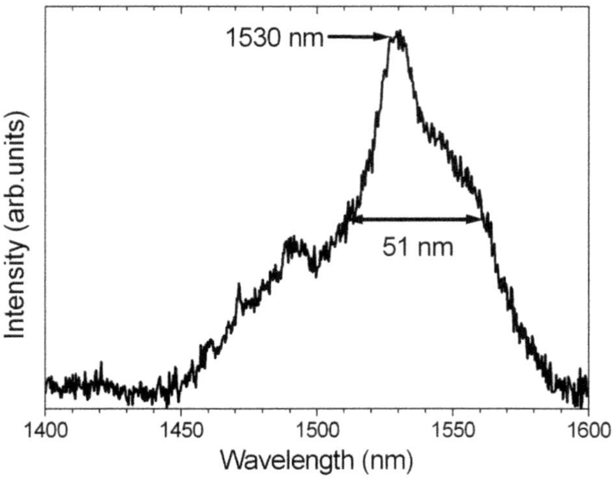

Figure 5. Er^{3+} photoluminescence of a 1·04 μm thick $Y_{0·9}Er_{0·1}Al_3(BO_3)_4$ film annealed at 700°C for 2 h

to be published].

The refractive indexes determined from transmission spectra are shown in Figure 4 for the films prepared from both methods and annealed at 780°C for 2 h. The refractive indices of the SG thin films are significantly higher than those synthesized by PPM, suggesting a denser glassy network for the SG deposits. At the 980 nm excitation wavelength, the index difference between the films and substrate is equal to 0·18 for the PPM route and 0·21 for the SG one. These index differences are important and convenient for the production of efficient waveguides.

Finally, Figure 5 shows the emission spectrum of a 1·04 μm thick film, heated at 700°C. The broad emission is typical for the erbium $4I^{13/2}-4I^{15/2}$ transition in an amorphous matrix. This promising first result shows that Er^{3+} ions are active in these waveguides. Work is now in progress to determine carefully the luminescence efficiency through the optimisation of device preparation, particularly the coupling of light in and light out with optical fibres.

Conclusion

$Y_{0·9}Er_{0·1}Al_3(BO_3)_4$ amorphous thin films were synthesised using the polymeric precursor method and the sol-gel process. The thin films have low roughnesses as measured by atomic force microscopy. Heat treatment at temperatures up to 700°C, are promising for the reduction of hydroxyl groups (–OH) and increasing the film density without crystallisation. The PPM coatings exhibit lower thermal stability than the SG layers, as determined by DSC. The coatings exhibit good optical transparency in the visible and near infrared regions. The waveguide modes were measured by *m*-line spectroscopy. The typical erbium photoluminescence emission was observed in waveguiding geometry. These results confirm the validity of preparing high quality borate thin films via the soft chemistry routes. Work is in progress to optimise the preparation and chemical compositions of the films and to complete optical characterisation of the films.

References

1. Shuto, K., Hattori, K., Ohmori, Y. & Horiguchi, M. *Electron. Lett.*, 1993, **29**, 139.
2. Kitagawa, T., Hattori, K., Shuto, K., Yasu, M., Kobayashi, M. & Horiguchi, M. *Electron. Lett.*, 1992, **28**, 1818.
3. Chartier, I., Ferrand, B., Pelenc, D., Field, S. J., Hanna, D. C., Large, A. C., Shepherd, D. P. & Tropper, A. C. *Opt. Lett.*, 1992, **17**, 810.
4. Feuchter, T., Mwarania, E. K., Wang, J., Reekie, L. & Wilkinson, J. S. *IEEE Photon. Technol. Lett.*, 1992, **4**, 542.
5. Daran, E., Bausa, L. E., Muiioz-Yagile, A. & Fontaine, C. *Appl. Phys. Lett.*, 1996, **62**, 2616.
6. Jenouvrier, P., Fick, J., Audier, M. & Langlet, M. *Opt. Mat.*, 2004, **27**, 131.
7. Neves, P. P., Maia, L. J. Q., Bernardi, M. I. B., Zanatta, A. R., Mastelaro, V. R., Zanetti, S. M. & Leite, E. R. *J. Sol-Gel Sci. Technol.*, 2004, **29**, 89.
8. Chen, X., Luo, Z., Jaque, D., Romero, J. J., Garcia Sole, J., Huang, Y., Jiang, A. & Tu, C. *J. Phys.: Condens. Matter*, 2001, **13**, 1171.
9. Jaque, D., Enguita, O., Garcia Sole, J., Jiang, A. D. & Luo, Z. D. *Appl. Phys. Lett.*, 2000, **76**, 2176.
10. Jaque, D., Enguita, O., Caldino, U. G., Ramirez, M. O., Garcia Solé, J., Zaldo, C., Munoz-Santiuste, J. E., Jiang, A. D. & Luo, Z. D. *J. Appl. Phys.*, 2001, **90**, 561.
11. Manifacier, J. C., Gasiot, J. & Fillard, J. P. *J. Phys. E.*, 1976, **9**, 1002.
12. Beregi, E., Watterich, A., Kovács, L. & Madarász, J. *Vibr. Spectrosc.*, 2000, **22**, 169.
13. Madarász, J., Beregi, E., Sztatisz, J., Földvari, I. & Pokol, G. *J. Thermal Anal. Calorimetry.*, 2001, **64**, 1059.

Investigation of the influence of boron–zirconium frit on the properties of porcelain glazes for sanitaryware

Stoyan P. Djambazov,[1] *Albena P. Yoleva*

Department of Silicate Technology, University of Chemical Technology and Metallurgy, 8 Blvd . Kl. Ohridski, Sofia 1756, Bulgaria

Geray Amzov

Ideal Standard, Sevlievo 5400, Bulgaria

Seven glaze compositions for porcelain, in which 0 to 15 wt% of the feldspar in the base glaze composition is replaced by boron–zirconium frit were prepared. The glaze batches were ground in a ball mill until the residue on a 45 μm sieve was 0·5 wt%. The glaze was applied by spraying to unfired 120×120 mm ceramic tiles prepared from sanitary porcelain slip. Samples for determination of fluidity and coefficient of thermal expansion were made. The samples were fired in industrial tunnel kiln at temperature of 1180°C over a 12 h cycle. With increasing boron–zirconium frit quantity the whiteness and the fluidity of the glazes were raised to a maximum whiteness of 92%. The coefficient of thermal expansion values initially decreased and then increased as the amount of zirconium–boron frit was increased. The thermal shock resistance of the samples to 160°C was assessed. Overall it decided that the best properties were obtained with the glaze composition that contained 5 wt% boron–zirconium frit. No faults were detected in glazes produced from this composition; 2000 kg of this glaze have been produced and used in the glazing of sanitaryware.

Introduction

Developments in the manufacture of sanitaryware are connected with the type and quality of the coating glazes used. The glazes results in a glassy structure on the outside surface of the ceramic body. In addition as a result of chemical reactions occurring at high temperatures, the glaze glasses undergo considerable changes and a very sharp interstitial layer is formed. This layer strongly influences the glaze qualities. Current sanitaryware glazes are multicomponent glasses with varied compositions that contain mostly SiO_2, Al_2O_3 and CaO. Recently there has been an increasing use of zirconium and boron compounds in glazes. In these glazes, B_2O_3 reduces the softening and melting temperatures, lowers the surface tension and improves the mechanical and thermal properties of the glazes. B_2O_3 also influences considerably the aesthetic characteristics of the coating because of the special optical effects created by its inclusion.[1–5]

The aim of the present work is to investigate the influence of boron–zirconium frit on the properties of porcelain glaze for sanitaryware.

Experimental

In order to carry out this investigation the following raw materials were used for preparing the glazes: quartz sand (K2, Senovo, Bulgaria), feldspar (FFF 200M, SP. Minerals, Italy), kaolin (K2, Senovo, Bulgaria), zircon (SZ1, Johnson Mathey, Italy), wollastonite (G38, Ankerpoort, Holand), limestone (Greece), zinc oxide (Bulgaria) and boron–zirconium frit. Seven compositions for sanitary porcelain glazes, in which feldspar was replaced by 0 wt%, 2·5 wt%, 5 wt%, 7·5 wt%, 12·5 wt%, 15 wt%, boron–zirconium frit were prepared. The glaze compositions containing frit in a water suspension with density 1750 kg/m^3 were ground in laboratory ball mill until the residue on a 45 μm sieve was 0·5 wt%. The glazes, with a viscosity of 200°G (Galenkamp), were applied by pulveriser over dried ceramic tiles with sizes 120×120 mm which had been prepared from sanitaryware slip. The whiteness of the studied glaze compositions was determined using a Spectrophotometer Minolta Cm 508 I. The fluidity of the glaze on firing was established using a chute. Several chutes were placed together on unglazed porcelain bodies with slope of 45°. The upper end of the chutes were widened to enable a 12 mm diameter, 3 g tablet of the glazes to be set there. After cooling, the length of glaze that had flowed in the chutes was determined to assess the fluidity of the glaze. The samples were fired in an industrial tunnel kiln at a temperature 1180°C with a 12 h firing cycle. The coefficient of thermal expansion (CTE) in the range 25–400°C was measured by a dilatometer Dil 801 L BAHR. The dilatometer results were used

[1] Corresponding author. Email djambaz@uctm.edu

Table 1. Chemical composition of the B–Zr frit and the glazes

Composition	*Frit*	*0*	*1*	*2*	*3*	*4*	*5*	*6*
Frit (B–Zr), wt%	*0·00*	*0·00*	*2·50*	*5·00*	*7·50*	*10·00*	*12·50*	*15·00*
Oxide, wt%								
K_2O	0·00	2·18	2·01	1·84	1·67	1·51	1·34	1·17
Na_2O	6·60	1·66	1·69	1·71	1·73	1·75	1·78	1·80
CaO	5·04	10·50	10·61	10·71	10·82	10·92	11·03	11·14
BaO	0·45	0·00	0·01	0·02	0·03	0·05	0·06	0·07
MgO	0·11	0·83	0·83	0·83	0·83	0·83	0·84	0·84
ZnO	0·00	2·19	2·19	2·19	2·19	2·19	2·19	2·19
Al_2O_3	6·67	9·24	8·94	8·65	8·35	8·06	7·76	7·47
B_2O_3	14·43	0·00	0·36	0·72	1·07	1·43	1·79	2·15
Fe_2O_3	0·19	0·20	0·20	0·20	0·20	0·21	0·21	0·21
SiO_2	57·48	60·32	60·06	59·80	59·54	59·29	59·03	58·77
ZrO_2	9·10	7·69	7·92	8·15	8·37	8·60	8·83	9·06
TiO_2	0·03	0·02	0·02	0·02	0·02	0·02	0·02	0·02
Loss on ignition		5·18	5·17	5·16	5·15	5·14	5·13	5·12

to define T_g and T_f. The dilatometer program gave the results automatically.

Results

The chemical analyses of the boron–zirconium frit and of the seven studied glaze compositions are presented in Table 1. As the B-Zr frit content increases the SiO_2 and Al_2O_3 content of the glaze decreases. The measured properties of the glazes are given in Table 2. The introduction of B-Zr frit at the expense of feldspar in the glazes leads to an increase in the whiteness, *L*, of the glaze (see Figure 1 and Table 2). The colour characteristics *A*–green-red and *B*–blue-yellow remain in the same region as for the zero composition (see Table 2). The higher whiteness is the result of the increased ZrO_2 content of the glaze.

The most important influence of the B-Zr frit is observed on the fluidity of the glazes (see figure 2). The glaze containing 15 wt% B-Zr frit, which gives a flow length of 43·70 mm±0·01 mm, shows maximum fluidity.

The temperatures, T_g and T_f were defined from the dilatometer analysis by the change of CTE with temperature. The results of the dilatometric measures are presented in Table 2 and Figure 3. The transformation interval is moved toward lower temperatures with the increased B-Zr frit content of the glazes.

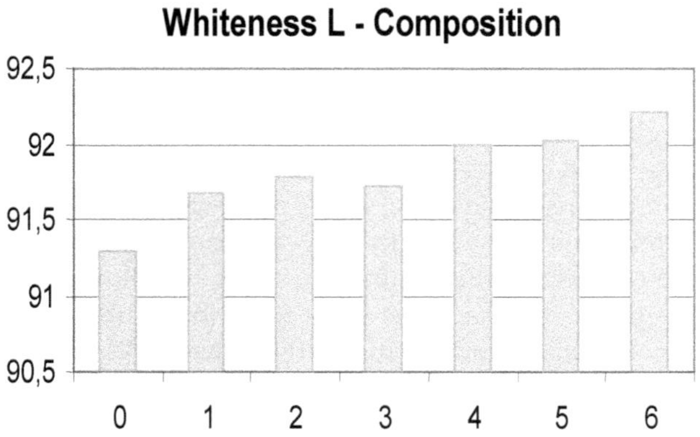

Figure 1. Whiteness of the glazes

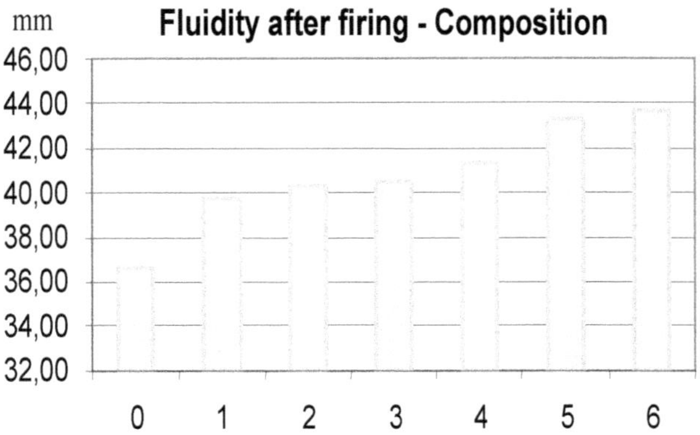

Figure 2. Fluidity of the glazes

The CTE values initial decrease as Zr-B frit is added to the glaze and reaches a minimum for glaze composition 4, increasing again for glaze compositions 5 and 6. The CTE increase observed with larger frit additions is a result of the decreasing SiO_2 content in the glaze composition.

Discussion

Expert evaluation of the glazes indicates that the best quality glazes with fewest defects were obtained with the compositions that contained 2·5 wt%, 5 wt% and 7·5 wt% B-Zr frit. These compositions are worth testing in industrial conditions. These glazes show

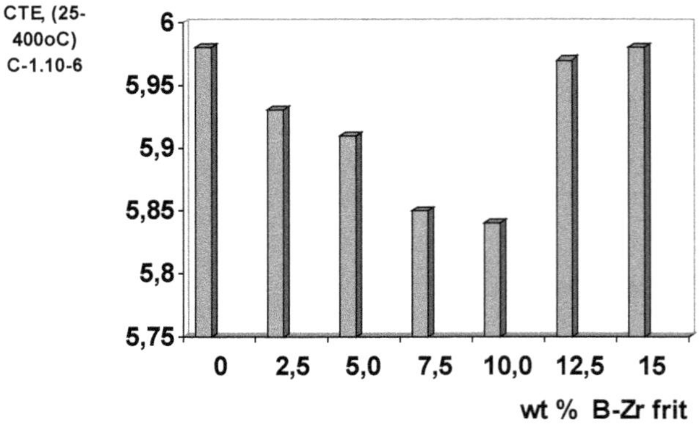

Figure 3. CTE of the glazes

Table 2. Properties of the glazes

Composition	*0*	*1*	*2*	*3*	*4*	*5*	*6*
Frit (B–Zr), wt%	0·00	2·50	5·00	7·50	10·00	12·50	15·00
Whiteness - *L*	91·3	91·67	91·78	91·73	91·99	92·02	92·21
Green - red - *A*	−0·61	−0·73	−0·74	−0·74	−0·79	−0·80	−0·73
Blue - yellow - *B*	2·72	2·67	2·64	2·73	2·70	2·78	2·73
Fluidity after firing, mm	36·70	39·80	40·30	40·50	41·30	43·30	43·70
T_g, °C	710·00	705·00	693·00	689·00	687·00	673·00	659·00
T_t, °C	790·00	785·00	780·00	780·00	779·00	776·00	774·00
CTE, *10-6/K (400°C)	5·98	5·93	5·91	5·85	5·84	5·97	5·98
T Exp., % (400°C)	0·228	0·225	0·224	0·223	0·222	0·228	0·229
Expert valuation	4·5	4·8	5·0	4·7	4·5	4·0	3·5

optimum whiteness and fluidity, and visual defects were not observed. Glaze composition 2, containing 5 wt% B-Zr frit has the highest gloss. The lower expert opinion of the glaze compositions with higher B-Zr frit contents is due to the appearance of pinholes and bubbles in the glaze coatings. The cause of these defects is the lowering of the melting temperature of the glazes during porcelain sanitaryware firing. The glassy coating is formed too early and encapulsates the ceramic body, so that gas can no longer readily escape from the body and it therefore acts as a gas bubble source.

Conclusions

The inclusion of B-Zr frit increases the whiteness of the glaze coatings. Most importantly the B-Zr frit affects the fluidity of the glazes. The transformation interval of the glazes is moved toward low temperatures with increasing B-Zr frit contents. Sanitary porcelain glaze with 5 wt% B-Zr frit has the best properties – namely good finish, brightness and a surface without defects and therefore has the maximal expert valuation of 5. This sanitary porcelain glaze has been tested in production conditions in 2000 kg quantities and it is found that it gives high-quality coatings.

References

1. Stefanov, S. & Batchvarov, S. *Smalti Ceramici*. 1994. Gruppo Editionale Faenca Editrice, Faenca.
2. Djambazov, S., Nedeltchev, N., Dimitriev, Y., Yvanova Y. & Yoleva A. In *Borates glasses, cryastals and melts*. 1997. Edited by A. C. Wright, S. A. Feller & A. C. Hannon. Society of Glass. Technology, Sheffield. Pp1–7.
3. Smith, R. *J. Non-Cryst. Solids*, 1986, **84**, 421.
4. Tozzi, N. *Smalti Ceramici Consideration Teoriche e Practiche*. 1992. Gruppo Editionale Faenca Editirce, Faenca.
5. Barson, T. *Ceram. Eng. Sci. Proc.*, 1997, **18** (2).

Proc. Fifth Int. Conf. on Borate Glasses, Crystals and Melts Glass Tech.: Eur. J. Glass Sci. Technol. A, October 2006, 47 (5), 148–152

The effect of boron on the properties of fibreglass melts

Frederick T. Wallenberger & Robin J. Hicks

PPG Industries, Incorporated, Glass Technology Center, Post Office Box 11310, Pittsburgh, PA 15238-0310, USA USA

Commercial E-glass melts contain 5–7% B_2O_3. It was of interest to determine if the melt viscosity, log3 forming temperature and energy demand of these E-glass compositions can be reduced by reducing the delta between log3 forming and liquidus temperature while still offering crystallisation resistant melts, and if their boron level can be reduced while still meeting ASTM specifications. Trend line design, a new compositional model, showed that the log3 forming temperature of incumbent E-glass melts can be reduced by up to 44°C by reducing the delta between log3 forming and liquidus temperature from 120°C to 65°C. Another new compositional model showed that the log3 forming temperature increases at the same delta temperature (crystallisation resistance) by 12°C for every 1% reduction of the boron level.

1. Introduction

Commercial E-glass melts must meet industry requirements including ASTM specifications,[1] comply with environmental emission standards, have the lowest possible fibre forming temperature (viscosity) to reduce the energy demand and cost, and they must be crystallisation resistant as they pass through the furnace and bushings. Crystallisation resistance of a fibre glass melt depends on the temperature difference (delta) between its log3 forming temperature and its liquidus temperature.

The log3 fibre forming temperature is the temperature of a glass melt with a viscosity of 1000 poise. The liquidus temperature is the highest temperature at which a glass melt, if held sufficiently long, will develop crystals, i.e. at which a solid phase (crystals) and a liquid phase (melt) can form an equilibrium state. The delta temperature is related to the SiO_2/RO (RO=CaO+MgO) ratio of a composition. A specific "delta" between forming and liquidus temperature is therefore required to achieve a fibre glass melt in a commercial furnace that is crystallisation resistant.

Commercial 621 E-glass compositions, which contain 5 to 7% B_2O_3 have delta temperatures ≥120°C and high forming temperatures, while newer boron-free or essentially boron free E-glass melts, which have been designed by regression[2] or trend line design,[3–7] are believed to be crystallisation resistant in commercial furnaces with lower delta temperatures (65 to 85°C). Recently, soda–lime–silica, or container glass compositions have been reformulated to reduce their viscosity (log2 to log4 forming temperatures) as well as their energy demand to enhance their environmental benefit.[8]

2. Compositional design

New experimental 621 E-glass compositions were therefore made in an attempt to reduce their viscosity, log3 forming temperature and energy demand by designing them for a target delta temperature of 65°C. The design of these compositions is based on two new models. Both are empirical, but offer a powerful guide to compositionally reformulate fibreglass compositions. The first model, or trend line design of experiments,[3] reveals the relationship between composition, melt temperature, viscosity as well as crystallisation resistance. The design of optimum boron-free E-glass compositions[3] near the ternary SiO_2–Al_2O_3–CaO eutectic, and the design of optimum boron free and essentially boron free E-glass compositions[4–7] near the quaternary SiO_2–Al_2O_3–CaO–MgO eutectic should be consulted for details.

According to the first model, a reduction of SiO_2 and a corresponding increase in %RO (%CaO+%MgO) reduces the energy demand by reducing the viscosity (forming temperature) as well as both the SiO_2/RO ratio and the delta between the log3 forming and liquidus temperature. Irrespective of boron content, this model can be used to design compositions having the lowest log3 forming temperature in a given system with a desired SiO_2/RO ratio, delta temperature and, therefore, crystallisation resistance.

The SiO_2/RO is not the only, but is the most im-

[1] Corresponding author. Email wallenberger@ppg.com

Table 1. Evolution of commercial E-glass fibres

E-glass	621 (1951–date)	Boron- and essentially boron-free (1984–5)	(1997–date)	(2001–date)
SiO_2, wt%	54·5	59·0	60·01	56·55
B_2O_3, wt%	6·6	-	-	1·30
Al_2O_3, wt%	14·0	12·1	12·99	13·35
CaO, wt%	22·1	22·6	22·13	24·55
MgO, wt%	0·6	3·0	3·11	2·55
TiO_2, wt%	0·5	1·0	0·55	0·55
Na_2O, wt%	0·8	0·9	0·63	0·90
K_2O, wt%	0·2	-	0·14	-
Fe_2O_3, wt%	0·2	0·2	0·25	0·25
F_2, wt%	0·5	-	0·04	-
SiO_2/RO	2·5	2·3	2·3	2·1
Log3 FT, °C	1200	1258	1259	1220
Delta T, °C	136	82	85	65
References	2, 9, 11	11, 13, 14	10, 11, 12	4, 5, 6, 7

Table 2. Commercial E-glass compositions

Oxides	Example 1	Example 2	Example 3
SiO_2, wt%	54·5	54·77	53·91
B_2O_3, wt%	6·6	6·64	6·27
Al_2O_3, wt%	14·0	14·30	14·06
CaO, wt%	22·1	22·64	22·83
MgO, wt%	0·6	0·14	0·55
TiO_2, wt%	0·5	0·57	0·59
Na_2O, wt%	0·8	0·27	0·81
K_2O, wt%	0·2	0·07	0·07
Fe_2O_3, wt%	0·2	0·30	0·33
F_2, wt%	0·5	0·78	0·71
SiO_2/RO	2·5	2·4	2·3
SiO_2/B_2O_3	8·3	8·2	8·6
Log3 FT, °C	1200	1191	1172
Delta T, °C	136	120	133
References	2, 9, 11	2004	2006

portant variable that affects the delta temperature of glass compositions. From a statistical point of view, the next most important ratios are the SiO_2/Al_2O_3 ratio, the Al_2O_3/RO ratio and the SiO_2/B_2O_3 ratio, but it is not possible to cover this topic adequately in the framework of the present paper. Their additive effect may, in fact, cause slight but notable differences between the delta temperatures of a few compositions, which have either been encountered in this study or were reported in the literature.[17] Ratios other than the SiO_2/RO ratio will be discussed elsewhere.

The second design model[7] reveals the effect of percent B_2O_3 on the properties of a melt at the same level of the crystallisation resistance, i.e. at the same delta between log3 forming and liquidus temperature (or at the same SiO_2/RO ratio). This model can lead the way toward the lowest log3 forming temperature (energy use and cost) that would result if the B_2O_3 level (batch cost) were reduced (or increased) at the same delta temperature or SiO_2/RO ratio.

3. Experimental procedures

In addition to the original 621 E-glass composition,[9] a variety of other E-glass compositions with 5 to 7% B_2O_3 are in commercial use today. They were selected as controls for the experimental study that is being reported here. New experimental E-glass compositions were designed to have the lowest log3 forming temperature over a range of B_2O_3 contents at a target delta temperature of 65°C. They were made from reagent grade ingredients and the calculated amount of each reagent grade oxide was corrected for minor and trace impurities. A 1000 g batch sample of the combined ingredients was mixed and then melted in a platinum crucible for four hours at 1485°C in a laboratory furnace. The melts were then poured onto a metal plate and the glasses were analysed for melt properties and oxide composition.

The log3 forming temperatures were determined according to the standard ASTM procedure[15] and against NIST Standard Reference Material 710a. The liquidus temperatures were measured in a gradient temperature furnace according to the standard ASTM procedure[16] against NIST Standard Reference Material 773. The glasses were annealed and their compositions were analysed by x-ray fluorescence (XRF) to confirm their analysed compositions. Boron containing fibreglass compositions may require a confirmatory check by neutron transmission (NTM) analysis to ensure fully reliable results for boron.

XRF analysis requires physical standards for comparison. XRF standards are glasses, which have been statistically designed to bracket each oxide of a target composition and which have then been analysed by wet chemistry for their oxide content. Each type of composition (e.g. boron-free, essentially boron free and high boron E-glass) requires its own XRF standards. The experimental compositions, which are described here, contain 4·45 to 8·37% B_2O_3. The XRF standards, which are used for these glasses are the same as those which are used to analyse commercial 621 E-glass compositions that contain 5 to 7% B_2O_3 and result from the daily production in commercial melt furnaces. An "S4 Pioneer" XRF spectrometer (Bruker Inc., 5465 E. Cheryl Parkway, Madison WI 53711, USA) was used.

4. Results and discussion

Incumbent commercial E-glass compositions have log3 forming temperatures ranging from 1172 to 1200°C and a delta temperatures ranging from 120 to 136°C (Table 2). Twenty new experimental E-glass compositions were made in 2000 (Tables 3 and 4). They contained 4·5 to 8·4% B_2O_3, but had been designed to have the lowest log3 forming temperatures with a delta temperature of 65°C, and with SiO_2/RO ratios ranging from 2·0 to 2·2. Ten additional E-glass compositions, which were found in the patent literature in 2004,[17] will be discussed in the same context (Table 5). The latter were reported to have 5·1 to 6·1% B_2O_3,[17] a delta temperature of 62°C and SiO_2/RO ratios ranging from 2·1 to 2·4. The log3

Table 3. New E-glass compositions made by trend line design

Ex.	B_2O_3	SiO_2	Al_2O_3	CaO	MgO	TiO_2	Na_2O	F_2	Fe_2O_3
1	4·45	53·50	14·01	25·48	0·52	0·53	0·91	0·00	0·37
2	4·45	53·10	14·01	24·00	2·40	0·53	0·91	0·00	0·37
3	4·50	53·50	14·00	24·00	1·50	0·50	0·91	0·50	0·37
4	4·50	53·00	13·50	24·00	2·50	0·50	0·90	0·50	0·37
5	4·50	53·00	13·90	24·00	2·10	0·50	0·90	0·50	0·37
6	4·50	53·00	13·10	24·00	2·90	0·50	0·90	0·50	0·37
7	4·60	52·95	13·90	23·95	2·90	0·50	0·90	0·50	0·37
8	4·60	52·95	13·90	23·95	2·10	0·50	0·90	0·50	0·37
9	4·60	52·95	13·50	23·95	2·50	0·50	0·90	0·50	0·37
10	4·75	53·40	14·01	25·28	0·52	0·53	0·91	0·00	0·37
11	4·75	53·38	14·09	24·82	0·53	0·53	0·91	0·39	0·37
12	5·10	53·80	13·96	24·56	0·52	0·54	0·93	0·00	0·36
13	5·10	53·30	14·10	25·03	0·52	0·53	0·91	0·00	0·37
14	5·10	53·28	14·09	24·57	0·53	0·53	0·91	0·39	0·37
15	6·00	53·27	13·98	23·53	0·68	0·51	0·93	0·48	0·37
16	6·00	53·30	13·98	23·57	0·63	0·51	0·93	0·48	0·36
17	6·00	53·33	13·95	23·60	0·61	0·51	0·93	0·49	0·36
18	6·00	53·33	13·91	23·48	0·73	0·51	0·93	0·48	0·37
19	7·45	51·13	14·11	22·71	2·25	0·04	0·33	0·98	0·36
20	8·37	50·80	14·00	22·59	2·31	0·04	0·32	0·89	0·35

forming, liquidus and delta temperatures of all the experimental compositions will be compared with those of the commercial E-glass compositions.

4.1 Commercial E-glass (target $\Delta T>120°C$)

The compositions and properties of a range of commercial E-glass melts were tabulated in Table 2. The first example represents the original 621 composition[9] that was commercialised in 1951. The data were obtained from the published literature.[2,9,11] The other examples in Table 2 are archive samples from commercial production. They have been analysed by XRF analysis against the same XRF standards as the new experimental E-glass compositions.

It is known that commercial 621 E-glass compositions have diversified over time. Even today they differ slightly from manufacturer to manufacturer. The reasons for the differences are manifold, but they are primarily the result of batch ingredients, which contain "tramp" ingredients in varying amounts that would otherwise not be required for the formulation. TiO_2 and MgO are the most common tramp impurities. Ingredients differ from supplier to supplier and therefore from formulation to formulation. Finally, the fluoride levels may differ from furnace to furnace depending only upon local requirements.

The commercial 621 E-glass compositions, which are listed in Table 2 have high delta temperatures (≥120°C) and therefore high SiO_2/RO ratios (2·3 to 2·5). Finally, while the SiO_2/RO level is the most important parameter for comparing the delta temperatures and crystallisation resistance of glass melts, an important additional parameter would be the SiO_2/B_2O_3 level. The additive effect of this and other oxide ratios on delta temperature and energy demand will be discussed elsewhere.

Table 4. Melt properties of new E-glass compositions (target ΔT=65°C)

Example	Composition %B_2O_3	SiO_2/RO	Log3 FT, °C Predicted	Actual	Actual	ΔT, °C
1	4·45	2·058	1179	1176	−3	50
2	4·45	2·011	1179	1180	+1	67
3	4·50	2·098	1178	1177	−1	55
4	4·50	2·000	1178	1172	−6	69
5	4·50	2·031	1178	1169	−19	51
6	4·50	1·970	1178	1167	−11	57
7	4·60	1·972	1178	1161	−17	45
8	4·60	2·033	1177	1174	−3	57
9	4·60	2·002	1177	1169	−8	66
10	4·75	2·070	1175	1177	+2	52
11	4·75	2·106	1175	1177	+2	59
12	5·10	2·145	1171	1174	+3	82
13	5·10	2·086	1171	1172	+1	58
14	5·10	2·123	1171	1171	0	65
15	6·00	2·200	1160	1161	+1	93
16	6·00	2·202	1160	1158	−2	91
17	6·00	2·203	1160	1157	−3	98
18	6·00	2·203	1160	1157	−3	91
19	7·45	2·048	1143	1142	−1	63
20	8·37	2·040	1133	1133	0	61
Avg.		2·080			−3·4	66·5

4.2 Experimental E-glass (target ΔT=65°C)

Twenty new E-glass compositions were designed in 2000 by the trend line model. They are shown in Tables 3 and 4. They were designed to have a SiO_2/RO ratio ranging from 2·0 to 2·2, i.e. ratios which, in this compositional system, had been predicted to produce a delta temperature of 65°C.

Eleven experimental compositions had between 4 and 5% B_2O_3, seven between 5 and 6% B_2O_3 and two between 7 and 9% B_2O_3. Fourteen compositions had <0·5% fluorine and six were fluorine-free. The individual batch samples (Table 3) were melted and the melt properties (Table 4) were found to confirm the predictions of the trend line design model. Depending on the boron content, the actual log3 forming temperatures ranged from 1133°C (for a melt with 8·37% B_2O_3) to 1180°C (for a melt with 4·45% B_2O_3). As predicted, the actual SiO_2/RO ratio of these compositions ranged from 1·9 to 2·2 with an average of 2·080, and the delta temperatures ranged from 45 to 82°C with an average delta temperature of 66°C. At

Table 5. Additional literature compositions with a target: ΔT of 65°C[17]

Example	Composition %B_2O_3	SiO_2/RO	Log3 FT, °C Predicted	Actual	ΔT, °C Actual	
1	5·14	2·123	1170	1169	−1	54
2	5·15	2·260	1170	1180	+10	61
3	5·47	2·160	1166	1166	0	64
4	5·62	2·114	1164	1163	−1	66
5	5·66	2·078	1164	1163	−1	53
6	5·67	2·061	1164	1169	+5	64
7	5·92	2·362	1161	1163	+2	64
8	5·97	2·144	1160	1161	+1	61
9	6·05	2·141	1159	1171	+12	70
10	6·08	2·143	1159	1156	−3	64
Avg.		2·159			+2·4	62·1

Table 6. Effect of % B_2O_3 at the same delta temperature[7]

B_2O_3 (%)	Log3 FT, °C Change	Lowest log3 FT (°C) predicted for ΔT ~100°C	ΔT ~ 65°C
0·0	0	1272	1232
1·0	−12	1260	1220
2·0	−24	1248	1208
3·0	−36	1236	1196
4·0	−48	1224	1184
5·0	−60	1212	1172
6·0	−72	1200	1160
7·0	−84	1188	1148
8·0	−96	1176	1136
9·0	−108	1164	1124
10·0	−120	1152	1112

an equal boron level, the forming temperature, SiO_2/RO ratio, delta temperature and, implicitly, energy demand are much lower than that of a commercial 621 E-glass compositions.

Ten additional experimental 621 E-glass compositions (Table 5) were found in the patent literature in 2004.[17] These examples further support the above conclusions. The literature compositions contain 5·14 to 6·08% B_2O_3 and zero to 0·29% fluorine. Depending on the boron content, the actual log3 forming temperatures ranged from 1156°C (for a melt with 6·08% B_2O_3) to 1169°C (for a melt with 5·14% B_2O_3). As predicted by the design model, the actual SiO_2/RO ratio of these compositions ranged from 2·1 to 2·4 with an average SiO_2/RO ratio of 2·080.The actual delta temperatures ranged from 53 to 70°C with an average of 62°C. At an equal boron level, the forming temperature, SiO_2/RO ratio, delta temperature and, implicitly, energy demand are about the same as those reported in Table 4, and much lower than that of commercial 621 E-glass melts.

The experimental E-glass compositions listed in Tables 4 and 5 were analysed in the same XRF spectrometer and against the same XRF standards as those which were used for commercial E-glass compositions and are shown in Table 2. Although the results further confirm the validity the design model and of the test results, a statistically designed study would be required to optimise any given composition before it could be entered in a commercial melt furnace.

4.3 E-glass compositions (same target ΔT)

When the data from the trend line design study were evaluated, another new model[7] resulted that predicts the effect of percent B_2O_3 on the log3 forming temperature at the same delta temperature, e.g. at ~100°C or at ~65°C as shown in Table 6.

This model predicts that a 1% change (increase or decrease) of B_2O_3 at the same delta between log3 forming and liquidus temperature (e.g. at ~65 or at ~100°C), would result in a 12°C change (decrease or increase, respectively) of the log3 forming temperature.

5. Conclusions and outlook

The log3 forming temperature of currently commercial 621 E-glass melts ranges from 1200°C to 1172°C. Compositional differences in commercial use can be attributed in part to differences in the boron level (e.g. 6·6% versus 6·3%), and in part to differences in both the SiO_2/RO ratio and delta temperature (e.g. 120 versus 136°C). With the advent of environmentally friendly E-glass compositions ("Advantex" and "1201") it became evident that a delta temperature of 85°C, and even of 65°C, respectively, may actually be sufficient to prevent a commercial E-glass melt from forming crystals in a commercial furnace. It should be noted that the desired delta temperature depend also on furnace design and related factors. These insights were applied to the design of new high boron 621 E-glass melts.

The first option, as shown in Table 7, would be to reduce the delta between log3 forming and liquidus temperature and thereby reduce the log3 forming temperature itself as well as the energy use and cost. For example, a reduction of the delta temperature of commercial E-glass compositions from 120–136°C to 65°C would reduce the log3 forming temperature by 16 to 44°C and would significantly reduce the energy demand.

The second option would be to reduce the delta temperature of commercial 621 E-glass from 120–136°C to 65°C and simultaneously reduce the boron level from 6·3–6·6% to 5·1%. This option would reduce the log3 forming temperature by 1 to 19°C and afford a more moderate energy reduction, and it would reduce the batch cost.

In summary, this paper demonstrates that the energy demand of commercial 621 E-glass melts (6·3–6·6% B_2O_3) can be significantly reduced at the same boron level by reducing the viscosity, the forming and delta temperature and the SiO_2/RO

Table 7. Options for reducing the energy demand of 621 E-glass melts

Composition Status	6·3 to 6·6% B_2O_3 Commercial	6·1% B_2O_3 Predicted	Actual	5·1% B_2O_3 Predicted	Actual
SiO2/RO ratio	2·3–2·5 (avg· 2·4)	2·0–2·2	2·123	2·0–2·2	2·143
Log3 FT., °C	1172–1200	1160	1156	1171	1171
Delta T., °C	120–136	~65	64	~65	65
ASTM E-glass	+	+	+	+	+
Reference	Table 2		Table 5, Ex. 10		Table 4, Ex. 14

ratio. This approach has essentially no effect on the batch cost. Alternatively, this paper shows that the energy demand of commercial 621 E-glass melts can be moderately reduced by reducing the boron level from to 5·1% and remain within ASTM specifications by simultaneously reducing the viscosity, the forming and delta temperature and the SiO_2/RO ratio. This approach therefore reduces the batch cost.

Finally, this paper confirms that trend line design of experiments is a useful conceptual model to find way toward any desired target composition with relatively few experiments, but a statistical study would be required around a target composition, once identified, to provide a commercially robust target composition.

References

1. ASTM Designation D 578-00, Standard specification for glass fiber strands, published May 2000.
2. Aubourg, P. F. & Wolf, W. W., *Advances in Ceramics*, Am. Ceram. Soc., 1986 **18**, 51–63.
3. Wallenberger, F. T., Hicks, R. J. & Bierhals A. T. *J. Non-Cryst. Solids*, 2004,349, 377–387.
4. Wallenberger, F. T., Hicks, R. J. & Bierhals A. T. *Ceram. Trans.*, 2004, **170**, 181–199.
5. Wallenberger, F. T., Hicks, R. J. & Bierhals A. T. *Glastech. Ber. Glass Sci. Tech.*, 2004, **77C**, 170–183.
6. Wallenberger, F. T., Hicks, R. J. & Bierhals A. T. *Am. Ceram. Soc. Bull.*, 2006, **85** (2), 38–41.
7. Wallenberger, F. T., Hicks, R. J. & Bierhals, A. T. *Collection of Papers, 66th Conference on Glass Problems*, American Ceramic Society, Wiley, 2006.
8. Bingham, P. A. & Marshall, M. *Glass Technol.*, 2005, **46** (1), 11–19.
9. Tiede, R. L. & Tooley F. V. *US Patent 2,571,974*, October 9, 1951.
10. Eastes, W. L., Hofmann, D. A. & Wingert, J. W. *US Patent 5,789,325*, August 4, 1998.
11. Wallenberger, F. T. *Advanced Inorganic Fibers*, Kluwer, Dordrecht/Boston/London, 1999.
12. Wallenberger, F. T. *ASM Handbook, Vol. 21, Composites*, ASM International, 2001, 23–26.
13. Sproull, J. F. *US Patent 4,542,106*, September 17, 1985.
14. Das, B., Tucker, B. D. & Watson J. C. *J. Mater. Sci.*, 1991, **26**, 6606–6612.
15. ASTM Designation C 965-96, Standard procedure for measuring glass viscosity, reapproved 2002.
16. ASTM Designation C 829-81, Standard procedure for the liquidus temperature, reapproved 1995.
17. Li, H., Richards C. A. & Cox J. E. *WO 2004/020355 A1*, March 11, 2004.

Adherent antireflection coatings on borosilicate glass for solar collectors

Gundula Helsch, Marta Krzyzak, Gerhard Heide & Günther H. Frischat[1]

Institut für Nichtmetallische Werkstoffe, Technische Universität Clausthal, Zehntnerstrasse 2A, 38678 Clausthal-Zellerfeld, Germany

Borosilicate glass tubes as parts of parabolic-trough solar collectors reflect a total of ≈8% of the incident sun light, thus reducing greatly the efficiency of solar power installations. This paper describes the successful development of a porous SiO_2-containing antireflective coating ~110 nm thick, which was applied by a sol-gel dip coating process and which increased the light transmittance by up to 5%. Previous antireflective coatings did not have sufficient adhesion strength and resistance to wiping on borosilicate glass, so that they do not withstand the rough operating conditions seen in practice. The essential component enhancing the stability of the coating turned out to be H_3PO_4 which is added to the dipping solution. After baking the sol-gel layers for 1 h at 500°C both Na and P concentration peaks develop at both sides of the interface, obviously forming Na_2O phosphosilicate glasses, assuring a high long term stability between glass and layer.

Our sun is an infinite source of energy. The use of this energy is possible mainly by photovoltaic and solar thermal processes, respectively. In the latter case solar collectors are used. In low temperature (~100°C) installations water is heated for domestic uses, at higher temperatures (400 to 1000°C) electrical power is generated in an economically and ecologically advantageous way. One possibility to focus the sunlight is by using parabolic trough solar mirrors with absorber tubes in their focal lines. These are used as parts in power plants with an output of up to 100 MW. Borosilicate glass tubes surround these black absorbers for thermal insulation which due to multiple reflections may reach temperatures up to 400°C. The low thermal expansion of borosilicate glass enables these tubes to withstand such large temperature changes.

The borosilicate glass tubes reflect a total of about 8% of the incident sunlight, thus greatly reducing the efficiency of the solar collectors. A means of overcoming this problem can be achieved by applying a suitable antireflection coating to the glass. Of course, broadband antireflection coatings are already available, e.g. for float glass.[1,2] However, such traditional porous SiO_2 coatings have insufficient adhesion strength and resistance to wiping on borosilicate glass which prove to be decisive disadvantages. The coatings do not withstand the rough operating conditions seen in practice, and the desired effect is no longer present after a short time.

We have recently succeeded in developing and applying a practical and functional antireflection coating which increases the light transmission of these borosilicate glass tubes by up to 5%, thus also increasing the efficiency of the solar collectors.[3] The sol-gel dip coating method was used to apply a coating from a silica sol and the layer was baked at 500°C, displaying a porosity of ~35%. The essential component enhancing adhesion strength and resistance to wiping turned out to be H_3PO_4 which is added to the dipping solution. Schott-Rohrglas GmbH, Mitterteich (Germany) upscaled the process to borosilicate glass tubes of 4 m length which have already been installed in modern solar power stations and are being tested in California (USA).

Experimental

Figure 1 shows a flow scheme of an optimised sol-gel preparation process. Silica sol (Köstrosol 2040, Bad Köstritz, Germany, particle size ≈20 nm, Na_2O content ≈0·1 wt%) and H_3PO_4 (85%) was used together with 2-propanol as solvent. With 1 M HNO_3 the pH value was adjusted to 1–2. The oxide concentrations of the coating solutions amounted to 5%. The technical borosilicate glass Duran (Schott AG, Mainz, Germany, Na_2O content 3·5 wt%) was used as a substrate. For comparison a technical float glass (Pilkington Flachglas AG, Weiherhammer, Germany, Na_2O content 12·8 wt%) was investigated. With a withdrawal speed of 1 mm/s in the dip coating process and a 1 h annealing at 500°C (heating rate 4 K/min) an optimised 110 nm thick coating could be achieved. The advantage offered by the sol-gel

[1] Corresponding author. Email guenther.frischat@tu-clausthal.de

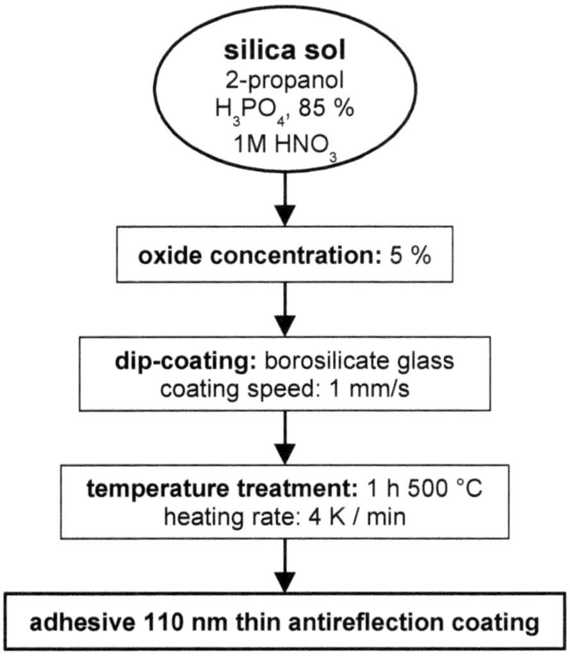

Figure 1. Flow diagram for the preparation of an H_3PO_4 containing porous SiO_2 sol-gel-derived adherent antireflective layer on borosilicate glass

method is the possibility of also flawlessly coating the inside of the glass tubes. Besides P-free layers the molar ratio SiO_2:P_2O_5 in the layers varied between 10:1·7 to 10:2·9. The adhesion strength between layer and substrate was proved by a tape test according to DIN 58196-6[4] and the resistance to wiping by vigorous rubbing with a paper tissue. The reaction profiles between layer and substrate were obtained using secondary neutral mass spectrometry (SNMS, Ions and Neutrals Analyzer, INA 3, Leybold AG, Köln, Germany), operated in the high frequency mode (HFM, Specs GmbH, Berlin, Germany), for details see reference.[5] The transmission of the uncoated and coated glasses was recorded between 300 to 800 nm with a UV-VIS spectrometer (Specord 200, Analytik Jena, Germany).

Results and discussion

Figure 2 displays the UV-VIS transmission spectra of uncoated and coated (on both sides) borosilicate glass samples 5 cm × 2 cm × 0·2 cm in size. At 550 nm wavelength the transmittance of the coated sample is increased to more than 97%, compared to ~92% for the uncoated glass. This is similar to what can be reached by coating technical float glass,[2] however, the differences between the two glasses has to be seen in the fact that only the layer on float glass shows sufficient adhesion strength and resistance to wiping, whereas the antireflection coating on the borosilicate glass has a totally insufficient performance in this respect.

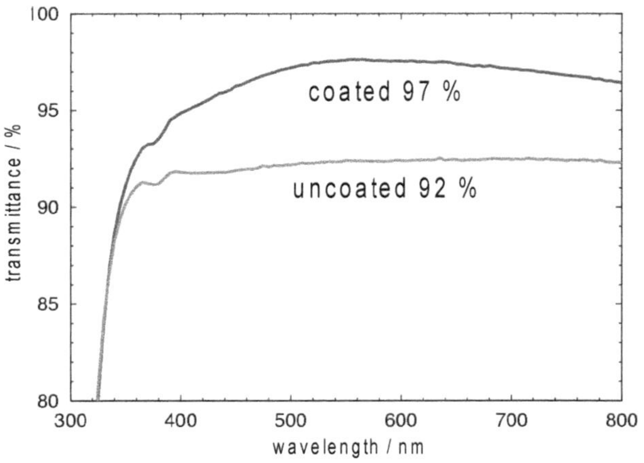

Figure 2. UV-VIS transmission spectra of uncoated and coated (on both sides) borosilicate glass samples

To enhance the adhesion of the antireflective coatings various experiments were performed, e.g. treating the glass substrates differently, changing the baking conditions, and heat treating the already annealed coatings by different solutions in a second step, respectively. Cathro *et al*[6] have already shown that a treatment with a $NaHCO_3$ solution enhances the performance of the antireflective coatings on borosilicate glass. Solutions of either Na_2CO_3, Li_2CO_3, KOH, NH_4HCO_3, $(NH_4)_2CO_3$, or $Ca(OH)_2$ were tried showing some success in the case of the alkali oxide-containing media. However, to circumvent this costly second step it was decided to modify the dipping solution by additions of either Na_2CO_3, H_3BO_3, or H_3PO_4 (85%). With respect to the adhesion strength, all three additions were positive, but the layer with Na_2CO_3 significantly aged with time, thus reducing the antireflective efficiency, and the layer with H_3BO_3 had a reduced transmittance. Only the addition of 85% H_3PO_4 to the dipping solution increased the transmittance, see Figure 2. This coating was stable and showed sufficient adhesion and resistance to

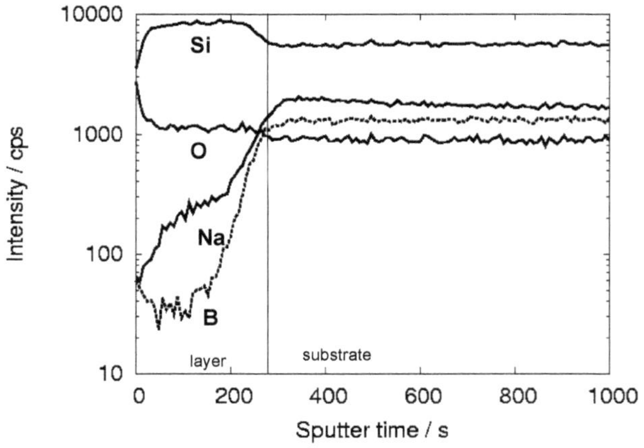

Figure 3. SNMS depth profiles of a P_2O_5 free porous SiO_2 coating on borosilicate glass, annealed for 1 h at 500°C

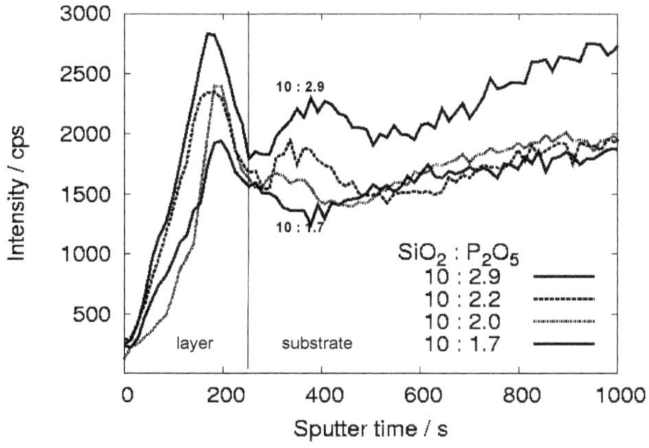

Figure 4. Na depth profiles of porous SiO_2 coatings containing different amounts of P_2O_5 on borosilicate glass, annealed for 1 h at 500°C

wiping. In the cases without H_3PO_4 there was insufficient adhesion and the layers were removed by the wiping process, whereas with H_3PO_4 the layers passed both tests sufficiently.

How can we explain this effect? Figure 3 displays the SNMS in-depth profiles following the reaction between the P-free porous SiO_2 layer and the Duran glass substrate after baking for 1 h at 500°C. The profiles – especially the Na profile – show that there is only a weak reaction near the interface layer/glass and nearly no Na_2O has diffused from the glass into the layer. If one uses float glass as a substrate (not shown here), the content of Na_2O diffused from the glass into the layer is considerably higher, obviously enhancing the stability of the glass/layer adhesion by this reaction. The rapid diffusion of the highly mobile alkali ions also seems to be the reason why the above mentioned solutions, e.g. $NaHCO_3$, Li_2CO_3, and KOH, used to treat the coated borosilicate glass in a second step yield an enhancement in the adhesion strength

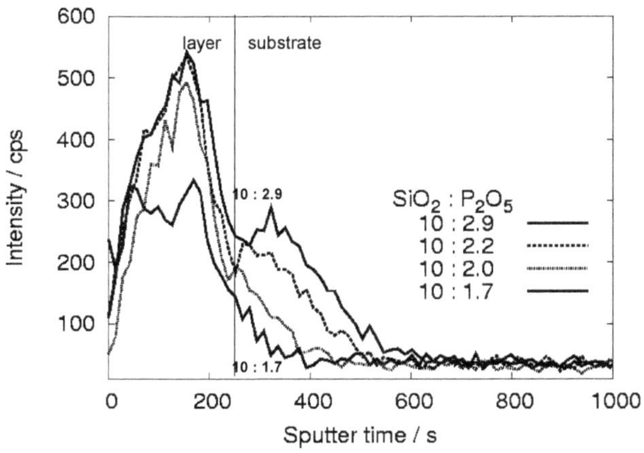

Figure 5. P depth profiles of porous SiO_2 coatings containing different amounts of P_2O_5 on borosilicate glass, annealed for 1 h at 500°C

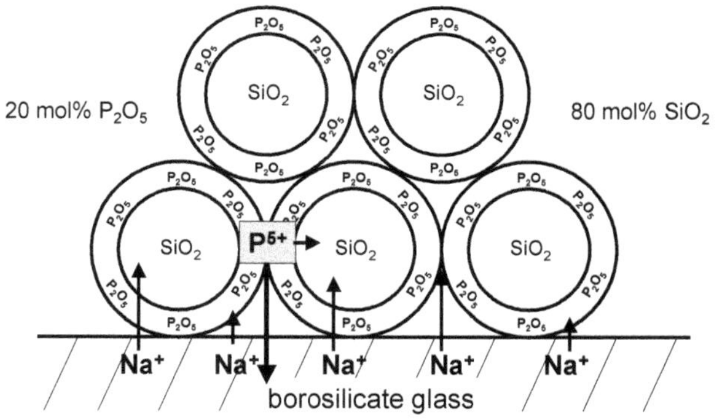

Figure 6. Schematic of possible diffusion processes in the system: porous SiO_2 layer/borosilicate glass

between the layer and the glass.

Figure 4 and 5 show the development of the Na and the P depth profiles both in the layer and in the glass surface as a function of the molar ratio SiO_2:P_2O_5, ranging from 10:1·7 to 10:2·9. Whereas the components Si, O, B, Al do not change, both Na and P display remarkable developments. Although the borosilicate glass contains only ~3·5 wt% Na_2O, Na^+ ions are readily transported by up-hill diffusion from the interior of the glass to its surface and through the interface into the layer, even developing peaks at both sides of the interface. The peaks in Na concentration reach higher maxima the more P_2O_5 the layers contain. The development of the P profiles occurs in conjunction with the Na profiles, i.e. peaks in P concentration are also found at both sides of the interface. Obviously Na_2O phosphosilicate glasses are formed at both sides, assuring high long term stability between layer and glass without affecting the excellent antireflective effect.

What is a possible mechanism explaining the ongoing reactions in the layer, the glass and through the interface? The experimental results tell us that considerable amounts of Na^+ ions diffuse from the interior of the low Na_2O borosilicate glass to the glass surface, across the interface and form a high Na_2O peak in the originally nearly Na_2O free layer, apart from a further (smaller) peak near the glass surface. This Na^+ transport occurs against the Na_2O concentration gradient as an example of an up hill diffusion process. Simultaneously, however, in the opposite direction P^{5+} ions diffuse from the layer through the interface into the glass. Since electroneutrality has to be maintained during this exchange, an interdiffusion process $5Na^+ \leftrightarrow P^{5+}$ seems to be not improbable, with diffusion of the highly charged P^{5+} ions being rate-determining, see Figure 6. As the different in-depth profiles show it cannot be excluded that P^{5+} ions also diffuse into the SiO_2 sol particles of the layer, thus forming some kind of glass with an Na_2O phosphosilicate composition, as may also be the case near the glass adjacent to the interface.

Conclusions

A sol-gel dip coating process was developed to form a porous (~35%) 110 nm thick antireflective coating with high adhesion strength and resistance to wiping on borosilicate glass. Silica sol was used and the essential component which enhanced the stability between layer and glass was H_3PO_4, added to the dipping solutions. Up to 5% increase in light transmittance was achieved. The SNMS in-depth profiles display that after baking for 1 h at 500°C both Na and P concentration peaks develop at both sides of the interface, obviously forming an Na_2O containing phosphosilicate glass, assuring the high long term stability between layer and glass, without reducing the excellent increase in light transmittance.

Acknowledgements

The authors wish to thank Schott-Rohrglas GmbH, Mitterteich, Germany (Dr habil. F.-D. Doenitz, Dr S. Tratzky) for financial and technical support.

References

1. Glaubitt, W., Sporn, D., Hußmann, E., Gombert, A. & Wittwer, V. *Glastech. Ber. Glass Sci. Technol.*, 2000, **73**, 193.
2. Helsch, G., Rädlein, E. & Frischat, G. H. *J. Non-Cryst. Solids*, 2000, **265**, 193.
3. Krzyzak, M., Helsch, G., Heide, G. & Frischat, G. H. *German Offenlegungsschrift DE 10209949 A1*, appl. date 6 March 2002, publ. date 25 Sept. 2003.
4. DIN 58196-6, *Optical coatings, Part 6: Testing of the adhesion with a tape (tape test)*. Deutsches Institut für Normung e.V., Berlin, 1995.
5. Schmitz, R., Frischat, G. H., Paulus, H. & Müller, K.-H. *Fresenius J. Anal. Chem.*, 1997, **358**, 42.
6. Cathro, K. J., Constable, D. C. & Solaga, T. *Solar Energy*, 1984, **32** (5), 573.

Proc. Fifth Int. Conf. on Borate Glasses, Crystals and Melts Glass Tech.: Eur. J. Glass Sci. Technol. A, December 2006, 47 (6), 159–163

Thermal, structural and optical properties of erbium doped oxyfluoride lead borate glasses

W. A. Pisarski[1,*] & J. Pisarska[2]

[1] University of Silesia, Institute of Materials Science, Katowice, Poland
[2] Silesian University of Technology, Department of Materials Science, Katowice, Poland

Thermal, structural and optical properties of Er doped oxyfluoride lead borate glasses in the B_2O_3–PbF_2–PbO–Al_2O_3–WO_3 system have been investigated. Properties of the glasses were analysed as a function of PbF_2 concentration. Two multicomponent oxyfluoride systems were examined, where the B_2O_3:PbO+PbF_2 molar ratio was either 3:1 or 4:5. Structural behaviour has been analysed using Raman and infrared spectroscopy in 1500–400 cm^{-1} spectral region, where characteristic bands related to the B–O vibrations of the borate network are located. Addition of PbO and/or PbF_2 to the borate glass results in the creation of trigonal BO_3 and tetrahedral BO_4 units. Quite intense and long lived near-infrared luminescence of Er^{3+} ions at 1·55 μm related to the main $^4I_{13/2}\rightarrow{}^4I_{15/2}$ laser transition has been observed. Thermal and spectroscopic parameters were measured for the 3:1 and 4:5 systems. The glass transition temperature T_g and thermal stability parameters ΔT and H (the Hruby parameter) decrease, whereas luminescence intensity and the $^4I_{13/2}$ lifetime increase with increasing PbF_2 concentration. Based on analysis of luminescence decay curve it was found that in oxyfluoride lead borate based glasses, with a B_2O_3:PbO+PbF_2, molar ratio 4:5 $^4I_{13/2}$ lifetimes of Er^{3+} ions were longer than in the sample with a 3:1 ratio.

1. Introduction

Oxyfluoride glasses activated by erbium ions have interesting spectroscopic properties in relation to their practical application as laser active media, optical waveguides and infrared-to-visible convertors. In addition, they are used in the preparation of transparent oxyfluoride glass-ceramics for active photonic devices.(1)

On the other hand, Er-doped borate glasses are rather inefficient fluorescent materials. In contrast to the fluoride system,(2) the relatively low energy gaps between the Er excited states high phonon energy (~1300 cm^{-1}) of the host borate glass usually eliminates luminescence from higher energy levels. Addition of a heavy metal oxide such as PbO to the borate matrix increases the radiative parameters of the Er^{3+} ions. Thus, quite strong luminescence from the $^4I1_{3/2}$ level of Er^{3+} ions in lead borate based glasses has been observed.(3) The effect of heavy metal oxides (PbO or Bi_2O_3) on the properties of Er-doped borate based glasses has been extensively studied.(4–7) However, little information is available on the structure and properties of lead borate based glasses in the case of partial or total substitution of PbO by PbF_2.

The research presented here has been focused on novel borate based glasses with a heavy metal fluoride component such as PbF_2. Thermal, structural and optical properties of Er^{3+} ions in the multicomponent oxyfluoride B_2O_3–PbF_2–PbO–Al_2O_3–WO_3 system have been studied. Due to the $^4I_{13/2}\rightarrow{}^4I_{15/2}$ laser transition, trivalent erbium has been used as an active doping ion at around 1·55 μm. Two oxyfluoride lead borate based systems were analysed as a function of PbF_2 concentration, where the B_2O_3:PbO+PbF_2 molar ratio was close to 3:1 or 4:5. The properties of the glass have been examined using differential scanning calorimetry (DSC), Raman, IR and photoluminescence (PL) spectroscopy.

2. Experimental

Two series of multicomponent oxyfluoride lead borate glasses were prepared. System I with a B_2O_3 to PbO+PbF_2 ratio of 3:1 had a molar% composition of $69B_2O_3$–$(23-x)$PbO–$x$$PbF_2$–$6Al_2O_3$–1·5$WO_3$–0·5$Er_2O_3$ (where x=0, 3, 23) and system II with a B_2O_3 to PbO+PbF_2 ratio of 4:5 thus had a molar composition of 39·5B_2O_3–$(49-x)$PbO–$x$$PbF_2$–$9Al_2O_3$–$2WO_3$–0·5$Er_2O_3$ (where x=0, 6, 49).

Anhydrous oxides and lead fluoride (99·99% purity, Aldrich) were used as a starting materials. To prepare the samples, several 10 g batches were mixed homogeneously and heated in atmosphere of dry argon. Glasses were melted at 850°C in Pt crucibles, then poured into preheated copper moulds and annealed below the glass transition temperature. After

[1] Corresponding author. Email wpisarsk@us.edu.pl

this procedure, the samples were slowly cooled to the room temperature. Transparent noncrystalline pink Er-doped oxyfluoride samples were synthesised, whereas the Er-doped oxide samples in system II (B_2O_3:PbO=4:5) were semi-crystalline.

The glass was characterised using a Perkin Elmer differential scanning calorimeter (DSC). The DSC curves were registered with heating rate of 10 K/min. The Raman and IR spectra were stored on a Brücker spectrometer. The spectra were obtained with a standard resolution of 2 cm^{-1}. The photoluminescence of the samples was excited by a Continuum Surelite Optical Parametric Oscillator (OPO), pumped by the third harmonic of a Nd:YAG laser. The luminescence was dispersed by a 1m double grating monochromator and detected with a photomultiplier with S-20 spectral response. The luminescence spectra were recorded using a Stanford SRS 250 boxcar integrator controlled by a computer. Luminescence decay curves were recorded and stored by a Tektronix TDS 3052 oscilloscope. All measurements were carried out at room temperature.

3. Results

The thermal, structural and optical properties of Er-doped oxyfluoride lead borate based glasses were examined for both systems I and II. Figure 1 shows a typical DSC curve of the oxyfluoride lead borate based glass, where PbO was totally substituted by PbF_2. From the DSC curve, characteristic parameters such as glass transition temperature T_g, crystallisation onset T_x and maximum of the crystallisation peak T_p were obtained. These parameters were used to determine the thermal stability parameter, ΔT, and the Hruby parameter, H,[8] from

$$\Delta T = T_x - T_g \quad (1)$$

$$H = (T_x - T_g)/T_g \quad (2)$$

Calculated thermal parameters for the samples studied are given in Table 1.

The Raman and infrared spectra of the oxyfluoride lead borate based glasses are presented in Figures 2 and 3, respectively. Both systems I and II were compared in the case of total substitution PbO by PbF_2. The maxima of the Raman and the infrared bands were estimated based on deconvolution analysis.

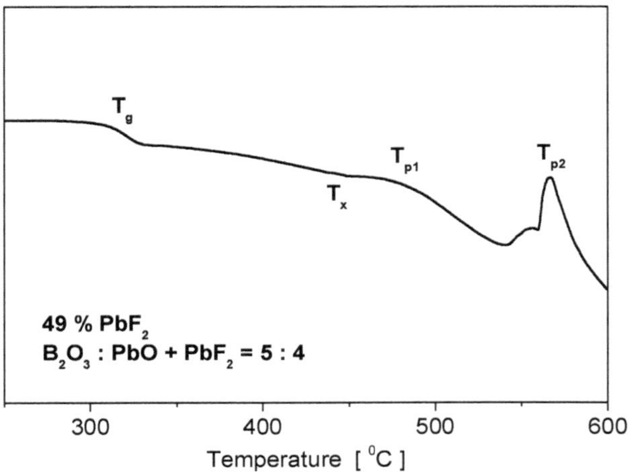

Figure 1. A typical DSC curve of Er doped oxyfluoride lead borate based glass

The deconvolution procedure and the analysis of the Raman and the infrared bands have been described in detail and discussed for borate systems in numerous works.[1,9,10] A detailed analysis of the IR spectra indicates that the main broad infrared asymmetric band between 1200 and 1500 cm^{-1} (Figure 3) has two absorption peaks. Description of the asymmetric infrared profiles requires deconvolution into Gaussian component bands. In this spectral region, two Gaussian component bands were needed for a reasonable fit of each spectrum.

Figure 4 presents the luminescence spectra of Er^{3+} ions in oxyfluoride lead borate glasses. The main infrared emission line at about 1·55 μm is due to the $^4I_{13/2}$–$^4I_{15/2}$ laser transition of erbium. Spectroscopic parameters such as luminescence linewidth and $^4I_{13/2}$ lifetime were obtained for Er-doped oxyfluoride lead borate glasses and the results are given in Table 2.

4. Discussion

4.1 Thermal properties

The thermal behaviour of Er-doped oxyfluoride lead borate based glasses have been examined for two systems with different molar ratios of B_2O_3 to PbO+PbF_2. DSC curves of the oxyfluoride lead borate based glasses were analysed as a function of PbF_2 concentration. A typical DSC curve for an oxyfluoride sample is presented in Figure 1. An additional exothermic peak was observed on the plateau between glass transition temperature T_g and the crystallisation

Table 1. Thermal parameters for Er doped oxyfluoride lead borate based glasses

Borate system	*Batched molar ratio B_2O_3:PbO+PbF_2*	*Batched PbF_2 content [mol%]*	T_g [°C]	T_x [°C]	ΔT [°C]	H
I	3:1	0	475±1	–	–	–
		3	471±1	–	–	–
		23	456±1	–	–	–
II	4:5	0	365±1	–	–	–
		6	353±1	515±1	162±1	0·46
		49	310±1	448±1	138±1	0·44

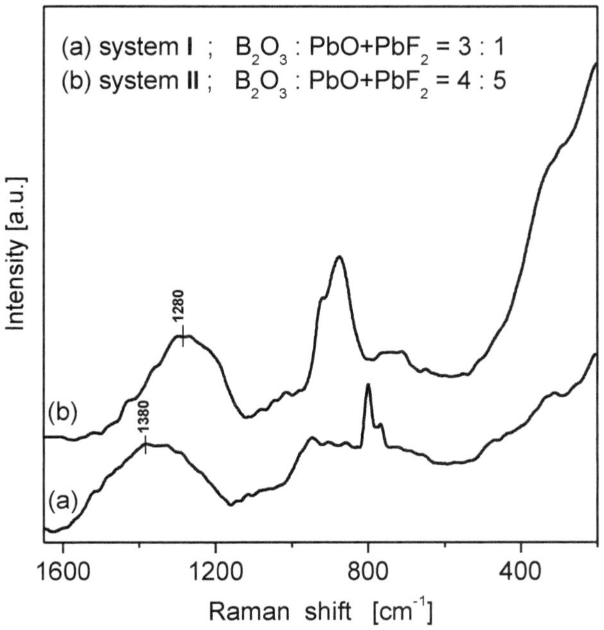

Figure 2. Raman spectra of Er doped oxyfluoride lead borate based glasses

temperatures of the main compounds. A similar effect has been observed for the oxyfluoride lead germanate glass, where, due to the precipitation of a cubic β-PbF_2 phase,[11] an additional DSC peak appeared. Characteristic temperatures and thermal stability parameters have been obtained from the DSC curves (Table 1). Depending on the PbF_2 concentration the glass transition temperature was observed to be in the temperature range 456–475°C or 310–365°C for systems I and II, respectively. In both systems, the glass transition temperature decreases with increasing PbF_2 concentration. The same effect is observed for the thermal stability parameter, which is one of the most important factors for glass fibre drawing. For system II, the calculated ΔT value is close to 162°C for the sample with a low (6 mol%) PbF_2 concentration and decreases with increasing PbF_2 concentration. For the high (49 mol%) PbF_2 concentration the ΔT value is 138°C.

A different situation was found for system I. The glass transition temperature was evaluated, but no obvious crystallisation peak was detected at temperatures below the maximum DSC temperature of 600°C. It has been reported that modifier substitution[12] or increasing PbO/B_2O_3 ratio,[13] strongly affects the thermal behaviour of this type of glass. T_g decreased from 475 to 456°C when PbO was totally substituted by PbF_2. However in all cases T_g was much higher than in system II.

The heavy metal oxide matrix based on PbO/B_2O_3 was successfully matched with PbF_2 in system II to produce homogeneous glass systems with relatively low glass transition temperatures, without crystallisation at temperatures below 500°C. All the highly PbF_2-doped samples in system II have ΔT values exceeding 100°C, which indicates that these oxyfluoride glassy samples are stable against devitrification. Thermal behaviour (especially as evaluated by stability parameters) strongly depends on the B_2O_3 to $PbO+PbF_2$ molar ratio in these glasses.

Table 2. Spectroscopic parameters for Er – doped oxyfluoride lead borate based glasses

Borate system	*Batched molar ratio B_2O_3:PbO+PbF_2*	*Batched PbF_2 content [mol%]*	*λ_p [nm]*	*$\Delta\lambda$ [nm]*	*τ_{meas} [μs]*
I	3:1	0	1527±1	75±1	295±5
		3	1528±1	52±1	380±5
		23	1530±1	56±1	440±5
II	4:5	0	1531±1	97±1	400±5
		6	1532±1	58±1	615±5
		49	1533±1	62±1	820±5

4.2 Structural properties

Generally, vitreous B_2O_3 consists of BO_3 units, which are associated with six-membered boroxol rings; the strong peak at about 805 cm^{-1} observed in Raman spectra is attributed to the totally symmetric vibration (trigonal deformation mode) of the boroxol ring.[14] Addition of a modifier oxide to B_2O_3 causes a conversion of BO_3 into BO_4 units. Raman peaks at about 805 cm^{-1} and near 770 cm^{-1} were also observed in all borate samples (system I), where molar ratio of boron (B_2O_3) to the total lead atoms (PbO and/or PbF_2) is close to 3:1 (Figure 2). The second peak, near 770 cm^{-1}, can be attributed to the formation of six-membered borate rings with one BO_4 unit called pentaborate or triborate groups. Similarly to observations made on the B_2O_3–PbO and B_2O_3–PbO–$PbBr_2$ systems,[15] the variation in the relative intensities of these two peaks with PbO (and/or PbF_2) concentration corresponds to the transformation of BO_3 to BO_4 units. The intensity of the peak at about 805 cm^{-1} considerably decreases with increasing total lead (as PbO and/or PbF_2) concentration. Also increasing lead concentration leads to the formation of two additional broad bands located in the 850–1050 cm^{-1} and 1200–1500 cm^{-1} spectral regions. These bands are associated with BO_4 and BO_3 units, respectively. They are shifted to lower frequencies, when the B_2O_3:PbO+PbF_2 molar ratio in the glass is changed from 3:1 to 4:5 (system II). This is consistent with results obtained by infrared spectroscopy.

Based on literature data,[16–19] the IR spectra in 400–1600 cm^{-1} spectral region consist of four characteristic groups of bands (Figure 3). The first group of bands has been identified as B–O–B and Pb-O-B bending vibrations as well as deformation of borate rings (400–650 cm^{-1}). Due to BO_3 bending a second group of bands occurs at 650–700 cm^{-1}. The third group of bands is associated with the B–O stretching vibration of the tetrahedral BO_4 group (850–1050 cm^{-1}). The

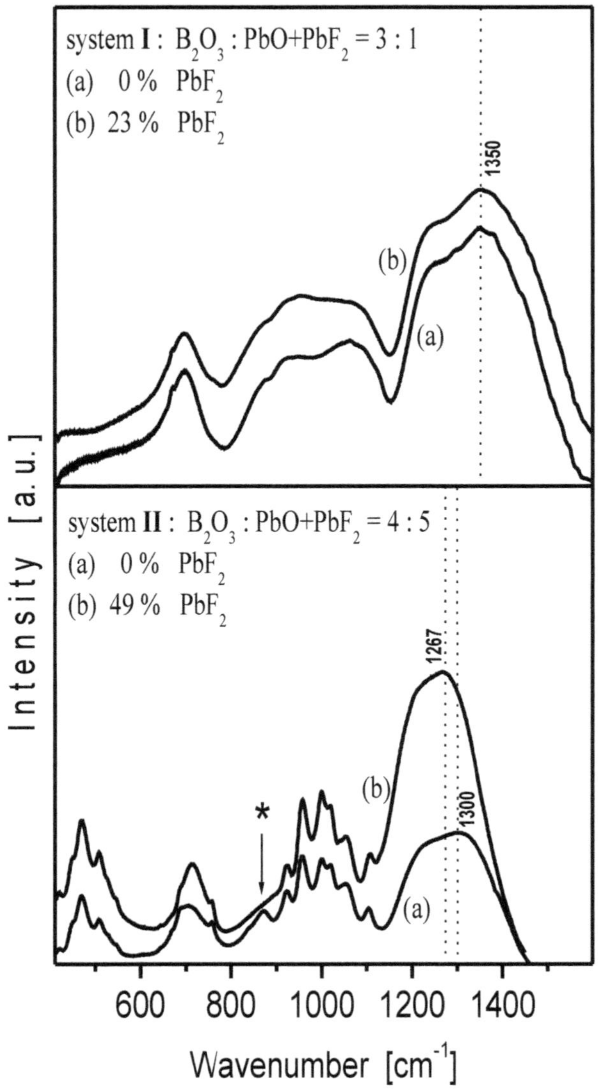

Figure 3. Infrared spectra of Er doped oxyfluoride lead borate based glasses

antisymmetric stretching mode causes bands centred at about 1050 cm^{-1}, whereas the symmetric stretching frequency is located in the 850–900 cm^{-1} region.[19] The fourth intense asymmetric band centered at about 1300 cm^{-1} is due to the antisymmetric B–O stretching vibration of the trigonal BO_3 group. This band is shifted from 1350 cm^{-1} (system I) to 1300 cm^{-1} (system II). A further band shift from 1300 cm^{-1} to 1267 cm^{-1} is observed for system II, when PbO is totally substituted by PbF_2. Addition of larger amounts of heavy metal (PbO and/or PbF_2) to the borate based glass shifts the band to lower frequencies.

It is also clearly seen, that for system II samples are semi-crystalline unlike the system I samples. This is in a good agreement with results obtained from x-ray diffraction analysis, where peaks related to the crystalline $ErBO_3$ phase were identified for the Er-doped sample without PbF_2.[20] These peaks have not been detected by x-ray diffraction in the compositions, where PbO was partially or totally substituted by PbF_2 and results only in the absence of the band located at about 872 cm^{-1} (labelled as * in Figure 3) in the infrared spectrum, which can be identified as being due to the deformation mode of the BO_3 vibration in lanthanide borates. Thus, the observed infrared spectrum for the oxyfluoride sample is similar to the ones obtained for oxide glasses (system II) containing rare earths (except for Er^{3+}) ions.[21] However, we cannot unambiguously exclude the possibility of crystalline phases being present on the nanometric scale. To complete the information about their structure additional experiments, including chemical analysis as well as observation by electron and scanning microscopy, needs to be carried out.

4.3 Optical properties

The influence of PbF_2 additions to the borate based glass on the near infrared emission spectra and luminescence decay curves of the $^4I_{13/2}$ state of Er^{3+} ions were observed. The luminescence linewidth and intensity as well as lifetime are amongst the most important parameters used to compute various spectroscopic properties, necessary for optical device characterisation. In particular, the relatively long lifetime of the $^4I_{13/2}$ metastable level is required in order to obtain large population inversions, which are required for EDFA systems in optical communications. Surprisingly quite strong emission from the $^4I_{13/2}$ level of Er^{3+} ions in lead borate glasses was detected. Unfortunately, the measured $^4I_{13/2}$ lifetime of Er^{3+} ions in lead borate glass (0·44 ms) is one order of magnitude less than that obtained for fluoride glass (4·67 ms) based on InF_3.[3]

Figure 4 presents luminescence spectra recorded at about 1·55 μm, which correspond to the $^4I_{13/2}$–$^4I_{15/2}$ laser transition of Er^{3+} ions in oxyfluoride lead borate based glass. Luminescence linewidth increases with increasing PbO concentration and starts to decrease, when PbO is partially or totally substituted by PbF_2. Luminescence linewidth increases from 75±1 nm to 97±1 nm when the molar ratio B_2O_3:PbO is changed from 3:1 to 4:5. In system II, the lumenscence linewidth decreases from 97±1 nm to 62±1 nm when PbO is completely replaced by PbF_2. The luminescence intensity increases considerably with increasing heavy metal oxide (PbO) and/or fluoride (PbF_2) concentration. Stronger luminescence intensities are observed for highly PbF_2 doped borate samples.

Analysis of the luminescence decay curve shows that the $^4I_{13/2}$ lifetime considerably increases with increasing PbF_2 concentration; the results are given in Table 2. Total substitution of PbO by PbF_2 in borate based glasses results in an increase of the $^4I_{13/2}$ lifetime of Er^{3+} ions from 295±5 μs to 440±5 μs for system I

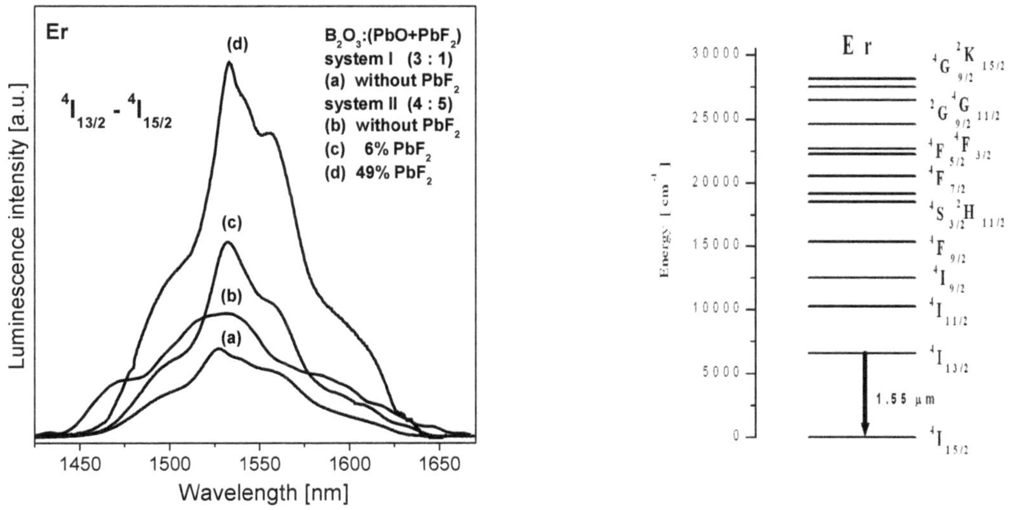

Figure 4. Luminescence spectra of Er^{3+} ions in oxyfluoride lead borate based glasses

and from 400±5 μs to 820±5 μs for system II. Longer $^4I_{13/2}$ luminescence lifetimes are obtained for borate systems containing larger amounts of lead (PbO and/or PbF_2) in the glass composition.

5. Conclusions

Thermal, structural and optical properties of Er-doped oxyfluoride lead borate based glasses in the multicomponent B_2O_3–PbF_2–PbO–Al_2O_3–WO_3 system have been investigated. Properties of the glasses were found to be strongly dependent on the B_2O_3 to PbO+PbF_2 molar ratio. Thermal and spectroscopic parameters have been analysed for two borate systems with the B_2O_3 to PbO+PbF_2 molar ratios close to 3:1 and 4:5, respectively. In both cases as the PbF_2 concentration increases the glass transition temperature decreases, whereas luminescence intensity of $^4I_{13/2}$–$^4I_{15/2}$ laser transition and $^4I_{13/2}$ lifetime of Er^{3+} ions considerably increase. Longer luminescence lifetimes were obtained for the system with smaller amounts of B_2O_3. Stability parameters were calculated only for system with the B_2O_3 to PbO+PbF_2 molar ratio equal to 4:5. Thermal stability decreases with an increasing PbF_2 concentration.

Structural behaviour has been analysed using Raman and infrared spectroscopy. Coexistence of trigonal BO_3 and tetrahedral BO_4 groups has been observed. The characteristic band corresponding to the stretching vibration of BO_3 units was shifted to lower frequencies when PbO was substituted by PbF_2 in oxyfluoride lead borate based glasses.

Acknowledgments

The Committee for Scientific Research supported this work under grant No 4 T08D 017 25.

References

1. Kukkonen, L. L., Reaney, I. M., Furniss, D. & Seddon, A. B. *Phys. Chem. Glasses*, 2001, **42** (3), 265.
2. Pisarski, W. A., Pisarska, J. & Ryba-Romanowski, W. *Chem. Phys. Lett.*, 2003, **380**, 604.
3. Pisarski, W. A. *J. Mol. Struct.*, 2005, **744–747**, 473.
4. Chen, Q., Ferraris, M., Menke, Y., Milanese, D. & Monchiero, E. *J. Non-Cryst. Solids*, 2003, **324**, 1.
5. Chen, Q., Ferraris, M., Milanese, D., Menke, Y., Monchiero, E. & Perrone, G. *J. Non-Cryst. Solids*, 2003, **324**, 12.
6. Yang, J., Dai, S., Zhou, Y., Wen, L., Hu, L. & Jiang, Z. *J. Appl. Phys.*, 2003, **93**, 977.
7. Chen, Y., Huang, Y., Huang, M., Chen, R. & Luo, Z. *Opt. Mater.*, 2004, **25**, 271.
8. Hruby, H. *Czech. J. Phys. B*, 1972, **32**, 1187.
9. Akagi, R., Ohtori, N. & Umesaki, N. *J. Non-Cryst. Solids*, 2001, **293–295**, 471.
10. Yano, T., Kunimine, N., Shibata, S. & Yamane, M. *J. Non-Cryst. Solids*, 2003, **321**, 147.
11. Mortier, M. & Auzel, F. *J. Non-Cryst. Solids*, 1999, **256&257**, 197.
12. Jung, B.-H. & Kim, H.-S. *J. Non-Cryst. Solids*, 2004, **336**, 96.
13. Tawansi, A., Gohar, I. A., Holland, D. & El-Shishtawi, N. A. *J. Phys. D.: Appl. Phys.*, 1988, **21**, 607.
14. Krogh-Moe, J. *Phys. Chem. Glasses*, 1965, **6**, 46.
15. Ushida, H., Iwadate, Y., Hattori, T., Nishiyama, S., Fukushima, K., Ikeda, Y., Yamaguchi, M., Misawa, M., Fukunaga, T., Nakazawa T. & Jitsukawa, S. *J. Alloys Compounds*, 2004, **377**, 167.
16. Harder, U., Reich, P. & Willfahrt, M. *J. Mol. Struct.*, 1995, **349**, 297.
17. Ram, S. & Ram, K. *J. Mater. Sci.*, 1988, **23**, 4541.
18. El-Alaily, N. A. & Mohamed, R. M. *Mater. Sci. Eng. B*, 2003, **98**, 193.
19. Ross, S. D. *Spectrochim. Acta A*, 1972, **28**, 1555.
20. Pisarski, W. A., Goryczka, T., Wodecka-Duś, B., Płońska, M. & Pisarska, J. *Mater. Sci. Eng. B*, 2005, **122**, 94.
21. Pisarski, W. A., Pisarska, J. & Ryba-Romanowski, W. *J. Mol. Struct.*, 2005, **744–747**, 515.

Proc. Fifth Int. Conf. on Borate Glasses, Crystals and Melts *Glass Technol.: Eur. J. Glass Sci. Technol. A*, December 2006, **47** (6), 164–166

Preparation and application of some phosphoborosilicate glasses containing micronutrients as plant fertilisers

M. A. Azooz,[1] *H. A. ElBatal, Kh. M. ElBadry, M. Abd ElMoneim*

Glass Research Department, National Research Center, Dokki, Cairo, Egypt

S. M. ElAshry

Plant Nutrition Laboratory, National Research Center, Dokki, Cairo, Egypt

Some selected phosphoborosilicate glasses containing one or more of the micronutrients: Cu, Mn, Fe, Zn and Mo were prepared from analytical grade chemicals. The grain method was applied to measure the corrosion behaviour of the glass towards the action of different solutions: water, citric acid, and HCl at 95°C for prolonged times of immersion. The weight loss from the grains after each specific time was calculated and compared. The leached solutions were chemically analysed to determine the amount of K^+ *and* P^{5+} *released during the corrosion tests and the data were evaluated. Experimental results reveal that a decrease in* K_2O *content lowers the wt loss%, and that the HCl solution corrodes the glasses more than the citric acid solution or distilled water. Citric acid solution is observed to extract* K^+ *and* P^{5+} *ions to a greater extent than the HCl solution and water. Different corrosion mechanisms are postulated and the various changes in the corrosion data are discussed.*

1. Introduction

All living things grow and reproduce in response to an interaction of dynamic factors in their ambient ecological environment. Maximum yield results only when plants compete successfully for the essentials of life. Essentials for optimum biological efficiency include these positive factors:(1) favourable air and soil temperatures; optimum available soil water and soil air; adequate light as a source of energy for green plants to carry on photosynthesis, and; essential elements for adequate nutrition of plants being available when needed and supplied in balanced proportions.

Plants need 16 elements for normal growth. Supplying trace elements to plants is not always easy and care is required to prevent overdosing, which may damage or kill the crop. To facilitate their application and availability to the plants, trace elements such as copper, iron, manganese, and zinc can now be used in a chelated form. These chelates are 'protected' water soluble complexes of the trace elements with organic substances such as ethylene diamine tetra-acetic acid (EDTA). These organic substances form a temporary protective environment element.(2)

Since plants can take up fertilisers continuously, it is beneficial to provide them with a balance of nutrients as they grow; perhaps the most efficient way to achieve this is to apply a slow release fertiliser, which is designed to release nutrients at the same rate that they are taken up by the plants. Slow release fertilisers contain one or more essential elements. These elements are released or made available to the plant over an extended times.

The idea for the creation, development, and application of agriglass (fertiliser glasses) is not new.(2,3) It is attractive because of several advantages which these materials possess:(1) glass provides almost all of the nutrients required by plants over long or short periods of time, chemical degradation of agriglasses can be controlled by changing their chemical composition and granularity; antibacterial properties of glasses containing Cu, Zn, and Ag have an inhibiting effect on the growth of different kinds of bacteria and fungi, and; a product with valuable characteristics, e.g. easy storage, absence of inert fillers and free from toxic ions.(1)

The first objective of this work was to prepare some phosphoborosilicate glasses compositions containing one or more of the micronutrients Cu, Zn, Mn and Fe. The second objective was to study the corrosion weight loss of the prepared samples by the grain method. The kinetics of glass dissolution can evaluate the rate of assimilation of the nutrients by the plants.(4)

2. Experimental details

2.1. Preparation of agriglass

For the preparation of glassy fertilisers detailed in Table 1 containing different micronutrients the following raw materials were used: silica (SiO_2), purified pulverised quartz was used to obtain silica; and high

[1] Corresponding author. Email address: moenis9@yahoo.com

grade orthoboric acid (H_3BO_3) was used to obtain boron oxide (B_2O_3). Iron was supplied as analar grade ferric oxide (Fe_2O_3). Phosphorus pentoxide (P_2O_5) was added in the form of analar anhydrous ammonium dihydrogen phosphate. Potassium oxide (K_2O), was added in the form of analar anhydrous potassium carbonate. Molybdenum oxide (MoO_3) zinc oxide (ZnO) and copper oxide (CuO) were introduced as such.

The raw materials of each batch were accurately weighed, mixed together, and melted in platinum–2% rhodium crucibles in an electric furnace at temperatures ranging from 1450–1500°C depending on the composition of the glass. To ensure homogenisation, the crucible was rotated at 30 min intervals. The glass melting was continued for 2–3 h depending on the composition of the glass. The melt was poured into stainless steel moulds, and annealed in a muffle furnace at 450°C down to room temperature. For corrosion testing the samples were crushed vertically in an agate mortar and sieved using sieves to mesh sizes of 0·6–0·3 mm.

The chemical corrosion of synthesised agriglasses has been studied using 0·5 g of glass in 50 ml of different kinds of solutions (0·4%HCl, 2% citric acid, and monodistilled water) for 1 h at 95°C in 250 ml polyethylene beakers with covers. 95°C was chosen to accelerate the test to be able in 1 h to evaluate the corrosion weight loss expected for prolonged times at room temperature.

2.2. Determination of potassium and phosphorous in the leachants

2.2.1. Flame photometric determination of potassium

After dilution of the original solution to 1/5 the K-emission was measured in an air–propane flame at 768 nm. A calibration curve was made with standard series of 0–200 mg K/l. The accuracy of the results was found to be ±2%.

2.2.2. Spectrophotometric determination of phosphorous

In the presence of $V^{5+}$and Mo^{6+}, orthophosphates form a yellow coloured phosphovanadomolybdate complex, which shows an optimal absorption at 430 nm. 5 ml of the original solution was pipetted into 50 ml volumetric flask. 10 ml of nitrovanadomolybdate reagent were added and made up to the mark with distilled water. After 1 h the absorbance was measured at 430 nm with a spectrophotometer. Duplicate samples gave ±2% reproducibility.

Table 1. Batch composition of borosilicate glasses

Wt%	*SiO_2*	*B_2O_3*	*P_2O_5*	*K_2O*	*CuO*	*ZnO*	*Fe_2O_3*	*MoO*	*MnO_2*
AG50	48	10	10	20	5	–	7	–	–
AG60	48	10	10	15	5	5	7	–	–
AG70	48	10	10	15	5	–	7	–	5
AG80	48	10	10	15	5	–	7	5	–
AG90	48	10	10	10	5	5	7	–	5

3. Results

The experimental corrosion results are given in Table 2) and the measured K^+ and P^{5+} contents in the leachants are given in Table 3. It is evident that: glass no. AG50 has the higher weight loss values in all leaching solutions than the other glasses in this series; by replacing 5% K_2O with 5% Mn, Zn or Mo oxides the weight loss% values decrease, respectively, in the order Zn>Mn>Mo oxides; the weight loss % value decreases more by replacing 10% K_2O by 5% of both ZnO and MnO, and; the effect of citric acid on the concentration of K^+ and P^{5+} released is greater than the effect of either HCl or H_2O, and the amount of K^+ extracted by citric acid solution is very high for all the glasses tested.

4. Discussion

The chemical durability testing of the studied agriglasses used two methods to evaluate corrosion behaviour: firstly weight loss measurements to provide a simple check of dissolution behaviour, and secondly leachant solution analysis to estimate the amounts of some of the glass components released through corrosion process. Since the same range of grain sizes were used in all the tests there was a constant surface area/specified volume of the attacking solution and thus the weight losses can be directly compared.

Phosphoborosilicate glasses are composed of three glass forming oxides giving connected SiO_4, BO_3, BO_4 and PO_4 groups which produce a complex structure. Therefore the corrosion behaviour of phosphoborosilicate glasses would be expected to be somewhat different than the corrosion of alkali silicate glasses which involves ionic exchange and the release of alkali ions from the glass and their replacement by hydrogen (hydronium) ions from solution. The addition of P_2O_5 and B_2O_3 to a silicate glass can induce phase separation[5,6] which could change the corrosion process.

For the phosphoborosilicate agriglasses studied here some important hydrolytic mechanisms are

Table 2. Weight loss% from the tested glassess in the various leachants (±0·2%)

Glass No.	*Wt loss% Distilled water*	*Wt loss% 0·4% HCl*	*Wt loss% 2% Citric acid*
AG50	3·57	9·18	7·45
AG60	2·34	7·48	5·18
AG70	2·46	7·98	5·46
AG80	2·25	6·66	4·5
AG90	2·14	4·42	4

Table 3. Potassium and phosphorous concentrations in the leachants measured after leach testing (±2%)

Leachants	Conc. of K (ppm)			Conc. of P (ppm)		
	0·4%HCl $\times 10^3$	2%Citric acid $\times 10^3$	DW $\times 10^2$	0·4% HCl $\times 10^3$	2%Citric acid $\times 10^3$	DW $\times 10^3$
AG50	9·8	43	6·14	0·276	4	0·125
AG60	7·3	37	5·46	2·3	1·03	0·15
AG70	7	27	5·10	3	1·9	0·73
AG80	6·1	25	4·66	3·5	2	0·45
AG90	3·9	23	1·10	1·1	1	0·075

suggested beside those known in silicate networks, leading to the formation of silanol by the hydrolysis of Si–O–B, Si–O–P, P–O–P, B–O–B, and P–O–B bonds. Bunker *et al* (1988)[7] suggested that in acidic solution, hydrolysis of Si–O–B and B–O–B bonds (involving either trigonal or tetrahedrally bonded borons) involves the electrophilic attack of protons on bridging oxygens.

The abundant presence of the network modifier oxide content (20% K_2O) decreases the durability of the glass as expected, because the alkali occupies interstitial positions within the network, which creates nonbridging oxygens to provide charge neutrality, thereby decreasing the network connectivity. This open network allows ions to diffuse more easily through the glass structure. Through ion exchange of K^+ and H^+ (as H_3O^+), the surrounding solution becomes basic, causing the main Si–O network to break up. The other multivalent micronutrient ions (Cu^{2+}, Zn^{2+}, Fe^{3+}, Mn^{4+} and Mo^{6+}) possibly exist in their high valence states due to the availability of alkali oxide and their corrosion behaviors will be expected to depend on the ease of breaking their bonds with oxygens and their ease of solubility in the respective attacking solution.

Acidic solutions (HCl and citric acid) attack the studied agriglasses more than water and this seems to be due to their high pH and excellent dissolving power. With glass AG90, which contains four micronutrient oxides (CuO, ZnO, Fe_2O_3 and MnO_3) the corrosion is the least with the acidic solution due possibly to mixed ionic interference during leaching.[8]

Citric acid solution as a leaching or attacking solution is observed to be more corrosive than hydrochloric acid solution as it will chelate with the metals present and make them into soluble complexes. This fact can explain the larger amounts of K^+ and P^{5+} ions that are released after subjecting the glasses to leaching in a citric acid solution rather than an HCl solution or distilled water.

Inspection of the data indicates the total weight loss does not only include the easily released ion exchanged process modifiers but also some of the other ions such as network formers as borate and phosphate phases and the other micronutrient cations. In other word, the corrosion data in the studied glasses comprise both ion exchange and network dissolution processes. Jedlicka *et al*[8] reached the same conclusion.

The studied agriglasses contain three network former ions, namely Si^{4+}, P^{5+}, and B^{3+} and K^+ as the main modifier cation. Also, the compositions include constant percentages of both micronutrient oxides CuO (5%) and Fe_2O_3 (7%), together with varying additions of ZnO, MnO, or MoO_3. The effect of the micronutrient oxides can be understood when the following data are taking into consideration: Zn^{2+} ions are easily dissolved in water and acidic solutions, and; Mo^{5+}, Fe^{3+}, Cu^{2+}, and Mn^{4+} ions are only attacked by aqueous acidic solutions to some extent.

The corrosion tests were carried out at 95°C for 1 h to obtain accelerated data instead of conducting the experiment at room temperature for prolonged times. This type of experiment has been previously accepted to represent the corrosion of the studied glasses for almost 1 year at room temperature.[9,10]

The data indicate that it is possible to formulate agriglass compositions that can give controlled release of micronutrient cations.

5. Conclusions

Agriglasses containing SiO_2, P_2O_5, B_2O_3, K_2O, and the micronutrient oxides CuO, Fe_2O_3, MnO_2, and MoO_3 were prepared. Grain test corrosion measurements of the glasses using HCl, citric acid or distilled water at 95°C for 1 h indicate that HCl solution is more corrosive than citric acid solution and water is the least corrosive medium tested. However, elemental estimations show that citric acid solution extracts K^+ and P^{5+} ions more than HCl or water. Also, it is evident that more K^+ ions are leached out than P^{5+} ions in all media tested. Interpretation of the results is based on the hydrolytic mechanisms of the various network cation-oxygen bonds such as Si–O–Si, Si–O–B, B–O–B, Si–O–P, P–O–P. The presence of the elements such as Zn and Cu should also give the agriglass useful antibacterial properties.

References

1. Follett, R. H., Murphy, L. S., & Donahue, R. L. *Fertilisers and soil amendments*. Prentic-Hall, Inc., Englewood Cliffs, New Jersey, 1981.
2. Nastle, K. *Glastech. Ber.* , 1955, **5**, 24.
3. Gordon, J. R. *J. Am. Ceram. Soc.*, 1975, **12**, 1069.
4. Samuneva, B., Bozajiev, P., Djambas, P., and Rangelova, N. Borate agriglasses: synthesis and application. *Glass Technol.*, 2000, **41** (6), 206–8.
5. McMillan, P. W. *Glass–ceramics*. Second Edition, Academic Press, London, 1979.
6. Doremus, R. H. *Glass Science*. Second Editon, John Wiley& Sons, New York, 1994.
7. Bunker, B. C., Tallant, D. R., Headly, T. J., Turner, G.L. & Kirkpatrick, R. J. The structure of leached sodium borosilicate glasses. *Phys. Chem. Glasses*, 1988, **29** (3), 106–20.
8. Jedlicka, A. B. & Clare, A. G. Chemical durability of commercial silicate glasses I. Material characterisation. *J. Non-Cryst. Solids*, 2001, **281**, 6–24.
9. Bacon, F. R. *Glass Ind.*, 1968, **49** (8), 438.
10. Newton, R. G. *Glass Technol.*, 1985, **26** (1), 21–38.

Proc. Fifth Int. Conf. on Borate Glasses, Crystals and Melts *Glass Technol.: Eur. J. Glass Sci. Technol. A*, December 2006, **47** (6), 167–171

The role of boron in the corrosion of E-glass fibres

Robert L. Jones[1]

Department of Applied Chemistry, Curtin University of Technology, Perth, WA 6845, Australia

Corrosion of E-glass fibres in aqueous solutions of organic acids has been investigated. Of the acids tested, oxalic acid was the most corrosive and causes almost total strength loss after exposures of only a few days. Of particular importance is that oxalic acid is a degradation product of the plastic matrix of glass reinforced plastic materials. The proposed mechanism of corrosion is an ion exchange reaction in which metal ions on the glass surface are replaced by hydrogen ions from the acid. This is enhanced by complex formation of the leached cation with the anion associated with the acid. Oxalic acid is a strong complexing agent for many of the cations found in E-glass fibres, which explains the severity of its action. It also appears that oxalic and other organic acids selectively leach boron to a greater extent than other cations, and as boron is part of the E-glass matrix rather than just an interstitial ion, this would explain the means by which these relatively weak acids can corrode the glass fibres. It is well known that oxalic acid forms strong complexes with boron, and it is likely that other dicarboxylic acids or carboxylic acids with appropriately placed ligand groups (e.g. glyoxylic acid) will do likewise. Complexing acids, such as citric, malic, lactic and tartaric acids, also corrode glass fibres but not nearly to the same extent as oxalic acid, and these acids do not appear to form strong complexes with boron. Also it is unlikely that these acids would be formed from the degradation of the plastic matrix.

1. Introduction

The chemical corrosion of E-glass fibres in acidic environments is now well established.(1–14) E-glass fibres are still the major reinforcing material used in glass reinforced plastic (GRP) composites in Australia, although several reports of GRP failure have been attributed to chemical corrosion.(1–4,10,11,14) The component responsible for stress corrosion failures in GRP materials appears to be the glass fibre rather than the resin, as the latter appears to play a protective role by shielding the fibres from the corrosive environment.(1–4) However, if this is damaged in any way and the fibres are exposed to acidic media, corrosion becomes rapid and may often be catastrophic. The failure of a sulphuric acid elution column in a South African gold mine is an example of this.(11)

Another mechanism of interest is the failure of GRP materials used as insulators on high voltage power transmission lines,(10) where the composite material is under electrical as well as mechanical stress. The mechanism proposed is that partial electrical discharge activity, in an aerobic environment leads to degradation of the bonding resin. This forms low molecular weight dicarboxylic acids and in particular oxalic acid, which then corrode the glass fibre component, resulting in mechanical failure of the insulator. Normally the resin plays a protective role, shielding the glass fibres from the corrosive environment, but in this case the resin contributes directly as the source of the corrosive material.

The accepted mechanism of the corrosion process is an ion exchange reaction in which metal ions associated with the glass surface such as Ca^{2+} and Al^{3+} are replaced by H^+ from the acid medium.(8) The general reaction may be written as

$$\overline{M^{n+}} + nH^+ \rightleftharpoons \overline{nH^+} + M^{n+} \qquad (1)$$

where the bar denotes association of the species with the glass phase. Because protons are smaller in size than the replaced cations, tensile stresses are induced in the surface of the glass and it is suggested that such stresses can become large enough to promote cracking.(7) Originally it was presumed that the corrosive effect was related to the acidic strength of the medium, but later work has shown that the associated anion can also play a significant role if the anion can form complex ions or insoluble species with the cations found in the glass.(9,11–14) This complex ion formation has been invoked to explain the severe corrosive effects of relatively weak organic acids such as oxalic acid in which the strength retention versus concentration curves show a well defined minima.(13) Such minima are observed because at low acid concentration the ion exchange reaction is not favoured on account of the low hydrogen ion concentration, and at high acid concentration the anions present bind to certain metal cations on the glass surface

[1] Email address: L.Jones@curtin.edu.au

Table 1. Percentage composition of E-glass fibres

SiO_2	Al_2O_3	CaO	MgO	Fe_2O_3	Na_2O	K_2O	B_2O_3
52·1	14·86	23·90	1·78	0·39	1·02	1·06	4·29

and prevent their dissolution. This passivation mechanism has been observed with hydrochloric acid solutions, the chloride ion complexing with iron(III) ions in this case,[9,12] and more recently[15] it has been shown that, based on the percentage of cation leached from the glass, iron(III) is selectively leached in unusually high proportions. This paper describes corrosion by hydroxy carboxylic acids such as citric, lactic, malic and tartaric acids, and comparisons are made with recently published data on glyoxylic and malonic acids.[16]

2. Experimental methods

All chemicals used were analytical grade where available or the purest grade available commercially. The material used was an E-glass fibre roving kindly supplied by ACI Fibreglass, Belmont, WA, of chemical composition shown in Table 1.

The acids investigated were malonic acid (Sigma M 1750 98%) 0·5–4 molar; glyoxylic acid monohydrate (Sigma G 4627 97%) 0·5–4 molar; oxalic acid dehydrate (BDH AnalaR 99·5%) 0·2–1·2 molar; citric acid (BDH AnalaR 99·7%0 0·2–1·2 molar; and tartaric acid (BDH AnalaR 99·5%) 0·2–1·2 molar.

Tests were conducted at 20°C for all acids and tests with glyoxylic acid were repeated at 35°C to examine the effect of temperature. Specimens 200 mm long were cut and immersed in 25 ml acid solution for various times up to 24 days, after which the specimens were washed in deionised water, air dried, and then subjected to tensile testing. Control specimens were cut at intervals along the roving and immersed in deionised water to eliminate any handling damage defects. Tensile tests were carried out on the specimens using an Instron 1206 tensiometer. The test length was 100 mm and a crosshead displacement rate of 2 mm min^{-1} was used throughout. Flat rubber-faced grips were used to minimise specimen damage due to clamping. The failure load recorded was the maximum load that the roving could withstand in the tensile test. At least six specimens were tested for each data point and only results showing sharp peaks in the load–deflection curves were recorded. This was necessary to eliminate non-uniform loading, which sometimes occurred in the individual rovings. The mechanical test results are presented in terms of percentage strength retention of corroded samples compared to control specimens. Elemental analyses on the residual solutions were performed using atomic absorption spectroscopy or, when available, inductively coupled plasma optical emission spectroscopy (ICP-OES).

3. Results

3.1. Rate of strength loss

Figure 1 shows the strength loss of E-glass fibres when exposed to various organic acids over several days. Corrosion by oxalic acid is particularly severe and strength loss is complete after a few days. The other acids are much less corrosive and rarely lead to complete strength loss, although the rate of corrosion is rapid in the initial stages of the process. Kinetic models indicate strength loss in oxalic acid follows a second order mechanism, whereas all the other acids follow a first order strength loss curve. As anticipated, increasing the temperature leads to an increase in the rate of corrosion.

3.2. Effect of concentration

The effect of concentration on strength loss is shown Figure 2. Oxalic, malonic and glyoxylic acids show well defined minima over the selected concentration range, whereas citric and tartaric acids do not, although the range may not be sufficiently large in the cases of the last two acids. The data for malic and lactic acids were similar to those for citric and tartaric acids, with no well defined minima for these acids either.

3.3. Rate of ion leaching

The rates of leaching of ions from the glass matrix were determined for all samples tested, with the exception of silicon, which is not leached to any measurable extent. Ions are leached into solution from the glass matrix by an ion exchange mechanism denoted by Equation (1), and all elements tested were found in varying quantities in the leached solution.

The initial rate of dissolution of ions is quite large, and as the dissolution proceeds the rate decreases and eventually equilibrium is reached. This is consistent with the results obtained from strength loss data indicating a direct relationship between strength loss and cation dissolution from the glass matrix.

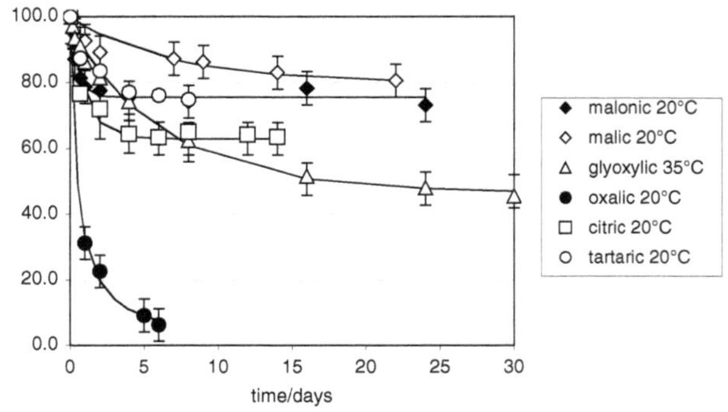

Figure 1. Strength loss over time of E-glass fibres exposed to various organic acids

A feature to note is that in the case of malonic and glyoxylic acids, boron is leached to a much greater extent than the other ions, given its relatively low occurrence in the glass, but this is not the case with the hydroxy organic acids. This is shown in Table 2 for the carboxylic acids at the concentration at which corrosion is at a maximum, together with hydrochloric acid as a comparison. The relative percentage weight was obtained by calculating the total weight of fibers in a typical sample, and relating the amount of each cation leached to its total amount present in the original sample. The calculation was performed for the amount leached after 50% strength loss occurred, or the equilibrium value in the cases of the hydroxy acids, where only approximately 20% strength loss occurred.

4. Discussion

4.1. Rate of strength lost

Strength retention decreases with time for all the acids examined and is particularly severe in the case of oxalic acid. In addition, the rate of strength loss in oxalic acid follows a second order decay whereas the less corrosive acids show a first order decay. This can be attributed to the ability of oxalic acid to form insoluble salts with calcium, and to form strong complexes with aluminium, iron and boron. The other acids are much less destructive, but increasing the temperature leads to increased rates of strength loss. In general, strength loss in all but oxalic acid reaches some limiting value after time, where equilibrium is reached between leaching of the ion and passivation of the glass by complexing with anions. Increasing the temperature speeds up the process and leads to increased rates of strength loss.

4.2. Effect of concentration

Some of the acids in this study show a pessimum concentration at which maximum corrosion occurs. The presence of this pessimum is indicative that something other than hydrogen ion concentration it is important in determining the extent of corrosion. This has been observed before in the case of hydrochloric acid,[9,12] hydrobromic acid,[12] organic acids,[13,16] and more recently in the case of perchloric acid.[17] The explanation offered is related to the

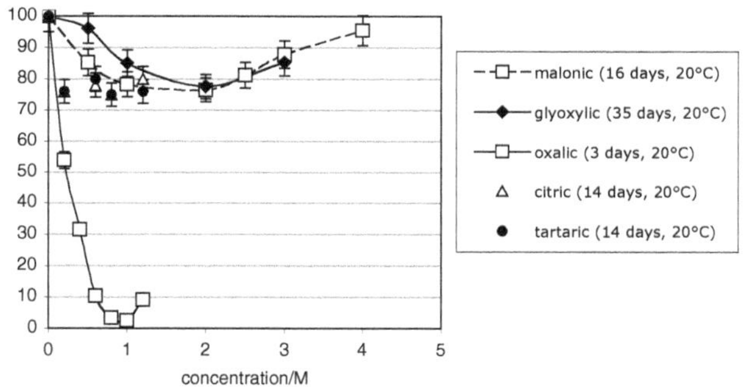

Figure 2. Strength loss of E-glass fibres in organic acids of varying concentrations

complexing nature of the anion associated with the acid. At low acid concentration, the hydrogen ion concentration is too low for effective leaching of the cation from the glass surface, whereas at higher acid concentrations the anion binds to the ion on the glass surface and prevents the dissolution of cations from the glass. Citric, malic, lactic and tartaric acids surprisingly show no such pessimum even though they are known to form complexes with many metal ions. It may be that this study has not explored a wide enough range of concentrations, and further work is in progress on this.

4.3. Rate of leaching

All ions are leached into solution to varying extents, and in general the more abundant ions are leached to a greater extent than the less abundant. The notable exception here in the case of oxalic, malonic and glyoxylic acids is boron, which is leached to a far greater extent than the much more abundant aluminium and calcium ions (Table 2). A possible reason for this is the fact that carboxylic acids, and particularly dicarboxylic acids (oxalic, malonic) and carboxylic acids with adjacent complexing groups (glyoxylic acid), form complexes with boron and this drives the leaching reaction to the right as depicted by

$$\overline{M^+} + H^+ \rightleftharpoons \overline{H^+} + M^+ \qquad (2)$$

$$M^+ + A^- \longrightarrow MA \qquad (3)$$

where as before the bar denotes association with the glass matrix and A^- is the ion in solution as-

Table 2. Relative percentage leaching of cations from E-glass fibres in various acids at 50% strength loss

Acids	Oxalic	Malonic	Glyoxylic	HCl	Citric	Tartaric	Lactic	Malic
Al	73·4	59·1	15·0	34·1	21·9	37·0	57·7	28·5
B	88·7	95·7	50·7	4·0	8·5	20·4	25·6	11·8
Ca	59·2	56·4	14·3	38·8	39·4	64·6	88·0	51·2
Fe	41·2	53·2	17·4	34·7	1·9	3·2	4·2	2·8
K	7·1	6·3	2·7	3·6	0·2	0·5	0·6	0·3
Mg	23·4	33·2	6·0	14·0	1·0	1·8	2·7	1·3
Na	40·0	43·3	17·0	28·0	1·8	3·0	4·6	2·3

sociated with the original acid which complexes the dissolved ion. Oxalic and malonic acids remove cations to approximately the same extent, although the process is much more rapid in the case of oxalic acid. Glyoxylic acid does not appear to complex metal ions very strongly although it does complex boron, and this may account for its corrosive effect. By contrast, hydrochloric acid does not leach boron to any great extent but does leach iron, owing to the iron complexing ability of the chloride ion, and this may be the key to its corrosive properties.[15] In this case it is suggested that the leaching of iron(III) opens a pathway for other ions to be leached and that the kinetics of iron(III) leaching determines the overall rate of leaching of all other ions. A similar mechanism could also explain the corrosion by carboxylic acids, in this case boron being the key atom. This would explain the decreased leaching rate at higher and lower acid concentrations, and the concentration dependence of the corrosion process, as at higher concentrations the anion associated with the acid binds to the glass surface and prevents further leaching. Data from perchloric acid corrosion[17] also show a dependence on concentration and in this particular case the binding of perchlorate and potassium is thought to determine the overall leaching rate. Again, the influence of a minor component of the glass has a major influence on the corrosion process. The passivation effect of perchloric acid has recently been exploited as a protection device,[18] where glass fibres pretreated with concentrated perchloric acid were able to withstand, for a significant time, the corrosive effects of both hydrochloric and oxalic acids, thus substantiating the idea that anions can bind to the glass fibre surface. Citric, lactic, malic and tartaric acids appear not to form strong complexes with boron, although they do form complexes with aluminium and calcium. This may account for the limited strength loss, the absence of a pessimum concentration at which strength loss is at a maximum, and the low percentage of boron leached. However the presence of three carboxyl groups in citric acid and two in tartaric and malic acid makes this result a little surprising. It may be that the carboxyl groups in citric and tartaric acids are too widespread (*c.f.* oxalic and malonic acids) and that the hydroxyl group does not bind sufficiently strongly to form stable complexes with boron.

5. Conclusions

The corrosion of E-glass fibres in carboxylic acids has been studied in an attempt to elucidate the mechanism of the corrosion process. The corrosion of E-glass in oxalic is particularly severe and causes almost total strength loss in a matter of days. Other carboxylic acids also cause strength loss, although not to the same extent. Some of these acids exhibit a pessimum concentration where corrosion is at a maximum, and this is attributed to the ability of the acids to form complexes with ions leached from the glass. At low acid concentrations there are insufficient hydrogen ions for effective leaching of the ions in the glass and at higher acid concentrations the complexing anion binds to the particular ion in the glass surface and prevents the leaching of other ions. The corrosion and hence strength loss of the glass is caused by an ion exchange reaction which leaches ions from the glass surface and replaces them with hydrogen ions. This causes surface tensile stresses, which eventually lead to cracking and consequent weakening of the individual fibres. All non-silicaceous ions are leached from the surface with the more abundant ions being leached to a greater extent. However, in the case of carboxylic acids, there is an unusually high concentration of boron leached into solution and it is thought that this arises from complex formation with the carboxylic acids, and that in turn, this has an overall influence on the leaching of other ions. Citric, malic, lactic and tartaric acids which are known to complex metal ions do not appear to complex boron to any great extent and are not severely corrosive. In the cases of these acids, it may be that the carboxyl groups in these acids are too widespread to form tight complexes, and the hydroxyl group, which is closer, does not have the binding strength of a carbonyl group.

References

1. Hogg, P. J. & Hull, D. Micromechanisms of crack growth in composite materials under corrosive environments. *Metall. Sci. J.*, 1980, **14**, 441–9.
2. Harris, S. J., Noble, B. & Owen, M. J. Metallographic investigation of the damage caused to GRP by the combined action of electrical, mechanical and chemical environments. *J. Mater. Sci.*, 1984, **19** 1596–604.
3. Friedrich, K. Stress corrosion crack propagation in glass fibre reinforced thermoplastic PET. *J. Mater. Sci.*, 1981, **16**, 3292–302.
4. Aveston, J. & Sillwood, J. M. Long term strength of glass-reinforced plastics in dilute sulphuric acid. *J. Mater. Sci.*, 1982, **17**, 3491–8.
5. Jones, F. R., Rock, J. W. & Bailey, J. E. The environmental stress corrosion cracking of glass fibre reinforced laminates and single e-glass filaments. *J. Mater. Sci.*, 1983, **18**, 1059–71.
6. Noble, B., Harris, S. J. & Owen, M. J. Stress corrosion cracking of GRP pultruded rods in acid environments. *J. Mater. Sci.*, 1983, **18**, 1244–54.
7. Garg, A. C. Environmental stress cracking behaviour of glass fibre reinforced epoxy resins. Eng. Frac. Mech., 1983, **17**, 575–7.
8. Metcalfe, A. G. & Schmitz, G. K. Mechanism of stress corrosion in e-glass filaments. *Glass Technol.*, 1972, **13** (1), 5–16.
9. Chandler, H. D. & Jones, R. L. Strength loss in e-glass fibres treated in strong solutions of mineral acids. *J. Mater. Sci.*, 1984, **19**, 3849–54.
10. Chandler, H. D., Jones, R. L. & Reynders, J. P. Stress corrosion failure of composite long rod insulators. *Proc. Fourth Int. Symp. on High Voltage Engineering*. National Technical University of Athens, Athens, 1983, Paper 23.09.
11. Chandler, H. D., Jones, R. L., Eggleston, P. J. & Reynders, J. P. Fractures in GRP – a link with corrosion. Proc. Second Natl. Conf. on Fracture. Edited by R. B. Tait & G. G. Garret, Pergamon Press, Oxford, 1984, p 19.

12. Jones, R. L. & Chandler, H. D. Strength loss in e-glass fibres treated after exposure to hydrochloric, hydrobromic and hydroiodic acids. *J. Mater. Sci.*, 1985, **20**, 3320–4.
13. Jones, R. L. & Chandler, H. D. Strength loss in e-glass fibres treated after exposure to organic acids. *J. Mater. Sci.*,1985, **20**, 3325–8.
14. Jones, R. L. & Chandler, H. D. Chemical corrosion of e-glass fibres in neutral phosphate solutions. *J. Mater. Sci.*, 1986, **21**, 2175–8.
15. Jones, R. L. & Betz, D. The kinetics of corrosion of e-glass fibres in hydrochloric acid. *J. Mater. Sci.*, 2004, **39**, 5633–7.
16. Jones, R. L. Chemical corrosion of E-glass fibres in oxalic and other organic acids. *J. Am. Ceram. Soc.*, 2006, **89** (1), 20–3.
17. Jones, R. L. Corrosion of e-glass fibres in acidic and neutral solutions. *Glass Technol.*, 2002, **43C**, 100–3.
18. Jones, R. L. & Brunt, S. Passivation of e-glass fibres in perchloric acid.

Proc. Fifth Int. Conf. on Borate Glasses, Crystals and Melts Glass Tech.: Eur. J. Glass Sci. Technol. A, December 2006, **47** (6), 172–176

Gas phase deposition and sintering of borosilicate glasses for optical fibres and planar waveguides

Johannes Kirchhof,[1] Sonja Unger, Claudia Aichele, Stephan Grimm & Jan Dellith

Institut für Physikalische Hochtechnologie e.V. Jena, PF 100 239, 07702 Jena, Germany

Borosilicate glasses with the composition $yB_2O_3.(1-y)SiO_2$ (0<y<0·25) have been prepared via gas phase deposition by two different routes: inner tube gas phase oxidation for the fabrication of preforms and optical fibres, and flame hydrolysis deposition for the fabrication of planar waveguide layers. The deposition rate of boron is very different for both cases, which can be understood on the basis of the investigated and calculated chemical equilibria prevailing during deposition. The sintering behaviour of porous layers was studied and interpreted on the basis of a viscous sintering mechanism.

Although borosilicate glass was one of the first materials used for the fabrication of optical fibres[1], such glass plays a rather minor role in today's telecommunication fibres compared with germanium and fluorine doped silica. This lack of importance is due to the infrared absorption tail in borosilicate glasses which gives rise to optical loss at the telecommunication wavelengths around 1·5 μm.[2]

Recently, however, boron has been of increasing interest as a special dopant for optical devices including polarisation maintaining laser and amplifier fibres[3] and birefringence free planar waveguides,[4] along with photosensitive fibres and waveguides for the inscription of Bragg gratings.[5] Silica highly doped with up to 20 mol% B_2O_3 has some outstanding properties compared with the usual high silica glasses: the refractive index is reduced compared with pure silica, high mechanical stresses are built up in fibres or planar waveguides, and a high sensitivity against UV radiation is induced.

The gas phase deposition of borosilicate glasses normally starts with volatile halides such as BBr_3 or BCl_3 together with $SiCl_4$ and can be accomplished by two different routes, oxidation or hydrolysis, see Figure 1. The oxidation process is used in Modified Chemical Vapour Deposition (MoCVD)[6] where the starting gases react with oxygen inside a silica tube, which is heated from the outside by a traversing oxy-hydrogen torch to temperatures of about 1500–1800°C. Several doped layers can be successively deposited. Finally, by collapsing the tube to a solid rod, a fibre preform is produced. Other, less widely used, variants of the oxidation process are inner tube deposition with a microwave plasma[7] or the deposition on a mandrel using a plasma torch[8].

The conventional MoCVD process includes for the formation of each individual layer a fast "in-line" sintering of the glass "soot" which is deposited downstream of the hot zone and immediately afterwards (typically within a few seconds) fused to a glassy film by the moving burner. By using an opposing movement of gas and burner, however, fusion of the soot layer can be prevented and therefore it can be investigated in its initial state[9]. Furthermore, it can be presintered or fully sintered with or without additional doping from the inner gas phase or by solution doping. The latter is very important for the co-doping with oxides which have no volatile precursors, e.g. the rare earths. In solution doping the dopant concentration depends very sensitively on the relative density of the porous layer which needs to to be of the order of 15–25 % of the density of the consolidated layer.[10] A fast sintering process is eminently helpful in preventing phase separation or crystallisation of the co-doped materials.

The hydrolysis route is used by the so-called Outside Vapour Deposition or Vapour Phase Axial Deposition for the preparation of fibre preforms,[2] or the deposition of planar waveguide layers on a substrate, mainly silicon or silica.[11] All these variants use "Flame Hydrolysis Deposition" (FHD), where an oxy-hydrogen torch is directly fed with the raw halide gases. FHD produces initially porous material (as a solid body or a layer). In the preparation of planar waveguides, the initially deposited porous layer is separately sintered to a glassy layer in a furnace at temperatures of about 1100°C and sintering times of about one hour.

Both routes are currently widely used in order to produce optical fibres and planar waveguides, mainly

[1] Corresponding author. Email johannes.kirchhof@ipht-jena.de

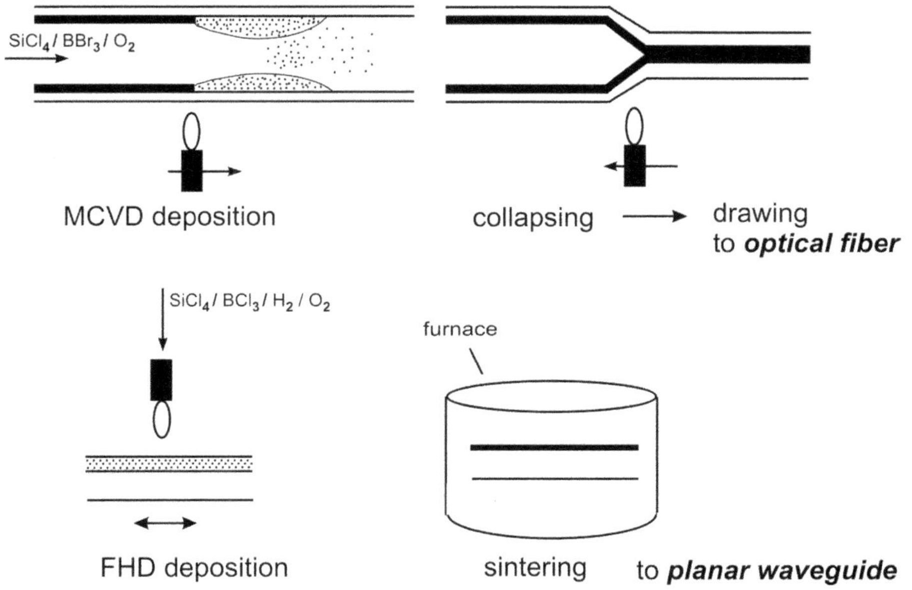

Figure 1. Schematic diagram of the MoCVD process for preparing optical preforms and fibres, and of the FHD process for preparing planar optical waveguide layers

on an empirical basis. Little is known, however, of the mechanism of process steps in terms of the chemistry of deposition or the sintering kinetics. In order to develop a deeper understanding, here we report on the preparation of borosilicate glasses by MoCVD and FHD. We have investigated the efficiency of boron incorporation into the silica glass and the sintering behaviour of porous layers.

Experimental

Boron doped quartz glass samples were prepared by MoCVD from controlled flows of gaseous $SiCl_4$, BBr_3 and O_2. Several layers with a single thickness of about 12 μm were deposited one upon the other on the inner surface of quartz glass carrier tubes with outer and inner diameters of 14 mm and 11·5 mm, respectively, with a traversing burner speed of 10 cm/min at temperatures of 1600°C. Subsequently, the tubes were collapsed to solid rods with a doped core of 5 mm diameter by short term heating to 2200°C during two burner passes. Thin slices have been prepared for the determination of the boron content.

For the FHD preparation, BCl_3 and $SiCl_4$ were fed into the central tube of a standard 4-ring $O_2/H_2/O_2$ torch directed onto a silicon wafer. By an x–y movement of the wafer platform, homogeneous porous layers were deposited on the wafer, which have been separately sintered in a helium atmosphere at 1100°C leading to glassy layers of approximately 25 μm thickness.

In both processes, the exhaust gases were passed through a spectrometer cell, heated to a temperature of 340°C, and the absorbance of the gases was continuously recorded by an infrared spectrometer at wavenumbers between 400 and 4000 cm^{-1}.

The boron content of thin slices of the MoCVD preforms and of the FHD layers was measured using wavelength dispersive x-ray microprobe analysis, and, for comparison, by standard wet chemical analysis for borate glasses. Using these techniques the boron content could be determined with an accuracy of approximately 1 mol%.

In order to study the sintering kinetics, single boron doped layers were deposited in silica tubes by an opposite movement of the gases and the burner, giving low deposition temperatures of <800°C. This produces flocculent deposits, which were subsequently annealed in a pure oxygen atmosphere at different burner temperatures. The applied burner velocity of 2 cm/min led to an effective sintering time of approximately 40 s, owing to the axial temperature profile of the burner. Small rings were cut from the tube and the relative density of the porous layers d_r was determined by microscopic measurements as the ratio of the thickness of a fully densified layer and the thickness of the respective porous layer.

The structure of the initial flocculent deposit was investigated by transmission electron spectroscopy, after washing with carbon tetrachloride. The soot layer in the as deposited stage has a very low relative density, which increases with the boron content from 2% (y=0) to 3–4% (y=0·23), as it consists of a loose agglomeration of spherical particles. The pure SiO_2 particles have small diameters (about 0·02 μm with a wide distribution). With increasing y, the diameters increase and the distribution becomes more homogeneous: 0·02 μm (y=0·09), 0·07 μm (y=0·17), 0·11 μm (y=0·23).

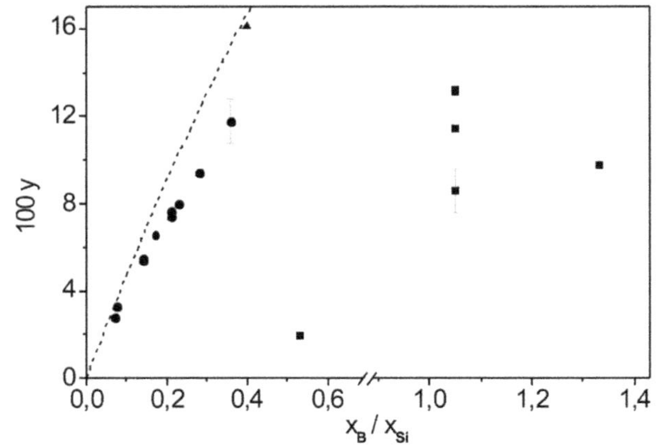

Figure 2. Gas phase deposition: Glass composition y *dependence on the starting gasmixture (•,▲ MoCVD samples, ■ FHD samples)*
x_B*,* x_{Si} *are the molar concentrations of* BBr_3 *(*BCl_3*) and* $SiCl_4$*, respectively, in the starting gas mixture, the broken line gives the relation* $y=x_B/(2x_{Si}+x_B)$*. Typical error bars are given as example*

Results

Boron incorporation via gas phase

In Figure 2, we compare the glass composition $yB_2O_3.(1-y)SiO_2$ with the molar ratio of BBr_3 (or BCl_3) to $SiCl_4$ in the starting gas mixture, x_B/x_{Si}. In the case of oxidation, the molar concentration of boron oxide y in the formed glass corresponds closely with the ratio $x_B/(2x_S+x_B)$, practically independent of other process conditions such as total flow rate or temperature. In the case of hydrolysis however, the boron content of the glass is much lower than expected from the gas phase composition and varies strongly depending on process conditions (fuel gas flow, distance and angle between the burner and the wafer, etc). In both cases, no unreacted halides could be observed

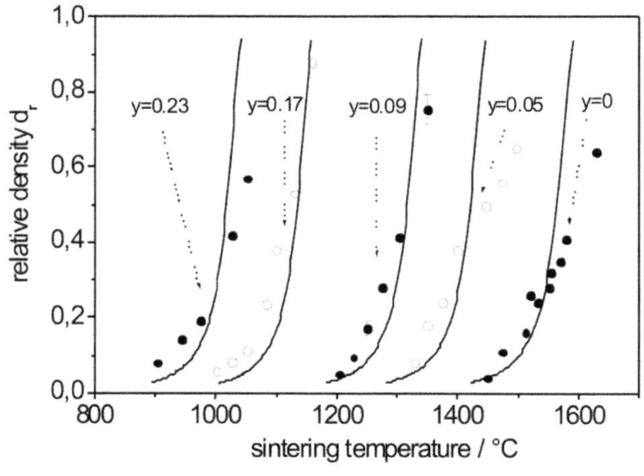

Figure 3. Sintering of porous layers: The dependence of relative density d_r *on temperature and* y*, experimental results (■,•,) in comparison with calculations (solid lines). Typical error bars are given as example*

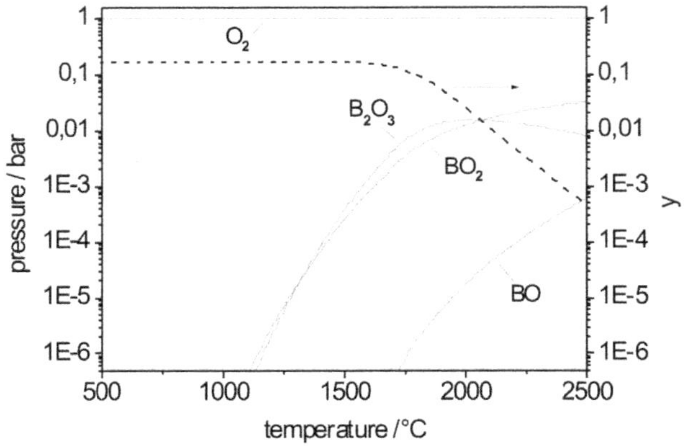

Figure 4. Thermochemical calculations: Glass composition y *and gas phase composition for typical conditions of the oxidation process (*$x_B/x_{Si}=0{\cdot}4$*)*

in the exhaust gases, which shows that the reaction temperatures are high enough and the reaction times are long enough to ensure a complete conversion of all halides. In the MoCVD exhaust, no boron containing gaseous compounds could be detected, but in the FHD exhaust, the IR bands due to gaseous HBO_2 and H_3BO_3 could be qualitatively identified, besides those of the main products HCl and H_2O.[12,13] (In both cases, a certain fraction of the solid soot particles is transported with the gas flow and gives rise to the IR absorption bands associated with $yB_2O_3.(1-y)SiO_2$.[14]) The absence of volatile boron containing compounds in the MoCVD exhaust and their presence in the FHD case respectively, are obviously the reason for the very different incorporation degrees of boron in the two processes.

Sintering behaviour

Annealing of the initial soot layers in the MoCVD tube by the moving burner leads to a progressive consolidation with increasing temperatures. In Figure 3, the development of ρ_r versus the annealing temperature is shown for different y values where T is the maximum temperature of the burner zone. As expected, the sintering temperature is mainly determined by the boron content of the layers and covers a wide temperature range.

Discussion

Boron incorporation via gas phase

The experimental observations can be confirmed by thermochemical calculations. We have calculated the equilibrium gas composition in contact with $yB_2O_3.(1-y)SiO_2$, for the gas system B–O and B–O–H, respectively, at 1 bar total pressure in dependence on temperature, using thermochemical data from Refs 13 and 15. The glass was assumed to be an ideal condensed mixture. To simplify matters, halide species

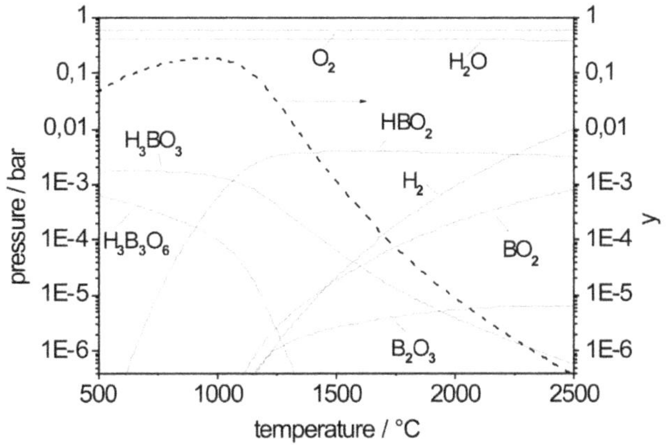

Figure 5. Thermochemical calculations: glass composition and gas phase composition for typical conditions of the hydrolysis process (x_B/x_{Si}=1·0)

(which have a certain influence in the equilibrium composition at very high temperatures) have been omitted. In Figures 4 and 5, some results are shown for typical conditions of the MoCVD and FHD processes, respectively. In the first case, we find only low concentrations of boron compounds in the gas phase, and the glass composition y stays unaffected up to about 1700°C. Only at higher temperatures, are volatile oxides formed and the boron content of the glass is more and more reduced. In the hydrolysis, on the other hand, we have volatile B–O–H compounds, especially HBO_2, and the glass composition y is much lower than the ratio of B to Si. The ratio y reduces with increasing temperatures above 1200°C.

Sintering behaviour

In Ref. 16 it has been shown that the sintering of compact porous bodies made by OVD can be well described by the mechanism of viscous sintering on the basis of a model derived by Scherer.[17] Therefore, we have compared our results with the same theory. The sintering kinetics are determined by a function Φ which depends on the time t

$$d_r=\Phi[(\gamma/l_0)(D_s/D_0)^{1/3}\eta^{-1}t] \quad (1)$$

where γ is the surface tension, l_0 the initial length of the "unit cell" of the porous structure, D_0 the initial density of the porous body, D_s the density of the solid and η the viscosity.[17]

The viscosity of B_2O_3/SiO_2 glasses has been determined by several authors. For our purposes we have re-evaluated the experimental data given in Refs 18–20. In general, for y<0·5 an Arrhenius-like behaviour has been observed

$$\eta(y)=\eta_0(y)\exp(E(y)/RT) \quad (2)$$

We have fitted all reported viscosity data to Equation (2) in order to determine separately the y dependence of the activation energy E and the pre-exponential factor η_0. They could be described by polynomials of third order with sufficient accuracy.

$$\ln(\eta_0[\mathrm{Pa\,s}])=-17{\cdot}08+32{\cdot}52y-4{\cdot}07y^2-37{\cdot}62y^3 \quad (3)$$

$$E/R[\mathrm{K}]=61730-162240y+223460y^2-136780y^3 \quad (4)$$

(For the extrapolation to higher values of y the polynomials are not suitable.)

The surface tension of high silica glasses is expected to be of the order of 0·3–0·4 N/m.[21,22] The value of l_0 however, cannot be exactly determined. Therefore, we have considered l_0 as an adjustable parameter. Using Equations (1)–(4), the relative density was calculated with t=40 s and $(\gamma/l_0)(D_s/D_0)^{1/3}=3\times10^5$ Pa, this value yielded the best agreement with the experiments. The calculated d_r curves are shown in Figure 3 in dependence on the sintering temperature, in each case for the measured value of y. The adjustment leads to a value of about 3 μm for the initial length of the porous unit cell, which is somewhat large but not improbable.

In the first instance, it is obvious that the strong dependence of the sintering temperature on y is well correlated with the calculated curves, i.e. by the viscosity dependence. This confirms the assumption of a viscosity controlled sintering mechanism. Nevertheless, the individual measurements for constant y do not very well correlate with the sintering function. Generally, the experimentally obtained slopes of d_r versus temperature are not as steep as expected from Equation (1). This points to another functional dependence of Φ,[17] possibly, some influence of the thermal gradients across the porous layer during sintering. Another possibility is that Φ, which is derived for free sintering in all directions, is not suited to describing sintering on a rigid substrate. Constrained sintering has been considered in Refs 23 and 24, but a first comparison, indicates that our results cannot be better described with the constained sintering theory. Some further observations (dependences of the sintering kinetics on the layer thickness, the sintering atmosphere and the carrier tube dimensions) cannot be explained here in detail and will be reported elsewhere.

Conclusions

$yB_2O_3.(1-y)SiO_2$ glasses have been prepared by gas phase deposition via oxidation (MoCVD) and hydrolysis (FHD) in the concentration range 0>y>0·23. In the oxidation process, the composition of the glasses is well determined by the relation of boron and silicon in the starting gases. In the hydrolysis process, the boron content of the glasses is much lower than expected from the gas composition, and this is explained by the formation of volatile B–O–H compounds. The sintering kinetics of porous layers deposited by MoCVD were experimentally investigated in terms of the boron content of the layers and the temperature. The kinetics are closely related to the viscosity of the borosilicate glasses.

References

1. MacChesney, J. B., Jaeger, R. E., Pinnow, D. A., Ostermayer, F. W., Rich, T. C. & Van Uitert, L. G. *Appl. Phys. Lett.*, 1973, **23**, 340.
2. Izawa, T. & Sudo, S. *Optical Fibres: Materials and Fabrication*, KTK Scientific Publishers, Tokyo 1987.
3. Farroni, J. A., Manyam, U. H. & Tankala, K. Multimode polarization maintaining double clad fiber, US Patent 2004/0156607.
4. Kilian, A., Kirchhof, J., Kuhlow, B., Przyrembel, G. & Wischmann, W. *J. Lightwave Technol.*, 2000, **18**, 193.
5. Shu, X., Allsop, T., Gwandu, B. & Zhang, L. *IEEE Photon. Technol. Lett.* 2001, **13**, 818.
6. MacChesney, J. B. & DiGiovanni, D. J. *J. Am. Ceram. Soc.*, 1990, **73**, 3537.
7. Geittner, P. & Lydtin, H. *Phil. Tech. Rev.* 1989, **44**, 241.
8. Biriukov, A. S., Dianov, E. M., Golant, K. M., Khrapko, A. V., Koropov, A. V., Perov, A. N., Shakhanov, A. V. & Vasiliev, S. A. *Sov. Lightwave Commun.*, 1993, **3**, 1.
9. Kirchhof, J., Unger, S., Grau, L., Funke, A. & Kleinert, P. *Cryst. Res. Technol.*, 1990, **25**, K29.
10. Kirchhof, J., Unger, S. & Schwuchow, A. *Proc. SPIE*, 2003, **4957**, 1.
11. Kawachi, M. *Proc. Optoelectron.*, 1996, **143**, 257.
12. Ross, S. D. *Inorganic Infrared and Raman Spectra*, McGraw Hill, 1972.
13. JANAF Thermochemical Tables, Third Edition, *J. Phys. Chem. Ref. Data*, Suppl.1, 1985, **14**.
14. Antonova, S. L., D'yakova, V. V. *Fiz. Khim. Stekl.* 1979, **5**, 671.
15. Barin, I. *Thermochemical Data of Pure Substances*, VCH Verlagsgesellschaft mbH, Weinheim 1993.
16. Yan, M. F., MacChesney, J. B., Nagel, S. R. & Rhodes, W. W. *J. Mater. Sci.*, 1980, **15**, 1371.
17. Scherer, G. W. *J. Am. Ceram. Soc.*, 1977, **60**, 236
18. Brückner, R. & Navarro, J. F. *Glastechn. Ber.*, 1966, **39**, 283.
19. Riebling, E. F. *J. Am. Ceram. Soc.*, 1964, 47, 478.
20. Urbain, G., Millon, F. & Cariset, S. *C. R. Acad. Sc. Paris, Serie C*, 1980, **290**, 137.
21. Kingery, W. D. *J. Am. Ceram. Soc.* 1959, **42**, 6.
22. Kirchhof, J. *Cryst. Res. Technol* 1985, **20**, 705.
23. Scherer, G. W. & Garino, T. *J. Am. Ceram. Soc.* 1985, **68**, 216.
24. Garino, T. J. & Bowe, H. K. *J. Am. Ceram. Soc.* 1990, **73**, 251.

Mechanochemical synthesis of silver ion conducting glasses in the system AgI–Ag_2O–B_2O_3

Akitoshi Hayashi,[1] Tomohiko Ishida, Tsutomu Minami & Masahiro Tatsumisago

Department of Applied Chemistry, Osaka Prefecture University, 1-1 Gakuen-cho, Sakai, Osaka 599-8531, Japan

Highly Ag^+ conductive glasses in the system AgI–Ag_2O–B_2O_3 were prepared by mechanochemical synthesis using a high energy planetary ball mill apparatus. The glass forming region for the composition series of xAgI.(100–x)(0·5Ag_2O.0·5B_2O_3) (mol%) was determined. The mechanically milled sample with 60 mol% AgI exhibited the highest conductivity of $1{\cdot}7\times10^{-2}$ S cm^{-1} at room temperature, which is comparable to the conductivity of the corresponding melt quenched glass. The glass transition temperature (T_g) of the milled sample was lower than that of the corresponding melt quenched glasses, but the T_g approached the T_g of the quenched glass with increasing milling time. The local structure, in particular the fraction of four-coordinated BO_4 units, closely related to electrical and thermal properties of the AgI-based borate glasses prepared by mechanical milling as well as by melt quenching.

1. Introduction

Ion conducting solid electrolytes have attracted much attention because these solid materials have potential applications in all-solid-state electrical devices such as batteries and capacitors with high safety and reliability, which could be achieved by replacing conventional liquid electrolytes with inorganic solid electrolytes. Both crystalline and glassy solid electrolytes have been developed. Amorphisation of solid electrolytes is, in general, a useful way of improving ionic conductivity.(1,2) Amongst the ionic conductors, AgI-based oxyhalide glasses are known to exhibit extraordinarily high conductivities in solids, namely 10^{-3}–10^{-1} S cm^{-1} at room temperature.(3–5)

Those superionic glasses are commonly prepared by melt quenching. Mechanochemical synthesis using a high energy ball mill apparatus has been developed as a new preparation procedure for inorganic amorphous materials.(6–8) This technique has several advantages compared to the melt quenching method: the whole process is performed at room temperature and fine product powders are directly obtained without the additional step of pulverising the bulk glass. Fine powders of several Li_2S-based glasses have been prepared by mechanochemistry(9–11) and all-solid-state rechargeable lithium batteries with these glassy powders as a solid electrolyte exhibited excellent cycling performances.(12,13)

Mechanochemical reaction also has a possibility of producing new materials. Machida *et al* have reported that highly Ag^+ conductive new crystals were formed after mechanochemical treatment for the AgI–Ag_3PO_4 system.(14) The AgI-based oxyhalide glasses have been also prepared by mechanical milling in the systems AgI–Ag_2O–P_2O_5,(15) AgI–Ag_2O–MoO_3,(16) and AgI–Ag_2O–V_2O_5.(17) AgI-based borate glasses prepared by melt quenching have a wide glass forming region and high Ag^+ conductivity, and the local structure, in particular the coordination of boron in the glasses, is closely related to the physical properties of borate glass systems.(5) Mechanochemical synthesis of the AgI-based borate glasses has not previously been reported, and the investigation of local structure is important in controlling the properties of these borate glasses.

In the present study, Ag^+ conducting borate glasses in the system AgI–Ag_2O–B_2O_3 were prepared by a mechanical milling technique. The formation process of glassy materials by milling was investigated. Thermal and electrical properties and local structure of the glasses were examined. This paper reports the relationship between properties and local structure of the milled samples as well as the corresponding melt quenched glasses.

2. Experimental

Reagent grade AgI (Aldrich Chem., 99%), Ag_2O (Wako Chem., 99%) and B_2O_3 (Koujundo Chem., 99·9%) powders were used as starting materials for sample preparation. The mechanical milling treatment was carried out for the batches (1 g) of the mixture of the starting materials in an agate pot (volume of 45 ml) with ten agate balls (10 mm in diameter) using a high energy planetary ball mill apparatus (Fritsch Pulverisette 7). The rotation speed of the base disk was fixed at 510 rpm. Melt quenched glasses were prepared for comparison. The mixture of starting materials was heated in a quartz crucible at 700°C for 1 h, and then the molten samples poured out from the crucible were pressed between two iron plates to yield glasses. In the following, 'mechanical milling' or 'mechanically milled' is abbreviated to

[1] Corresponding author. Email: hayashi@chem.osakafu-u.ac.jp

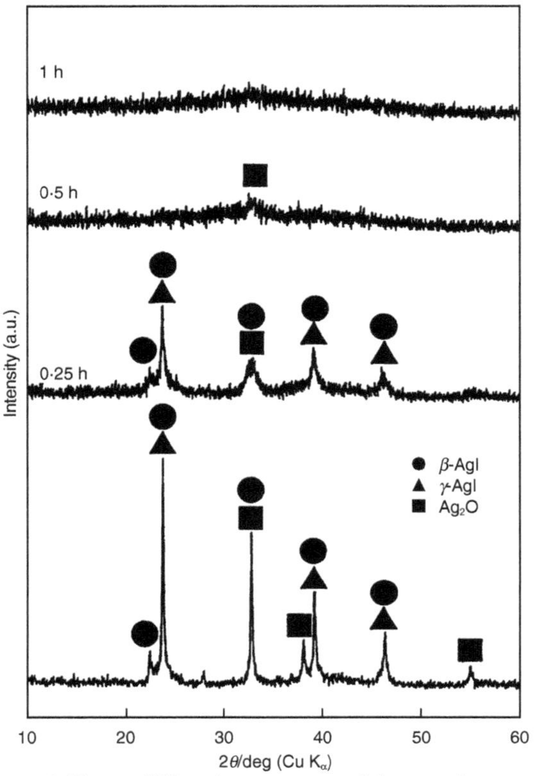

Figure 1. X-ray diffraction patterns of the powder samples with a nominal composition of the 50AgI.50(0·5Ag$_2$O.0·5B$_2$O$_3$) (mol%) prepared by MM. The numbers in the figure denote the milling times

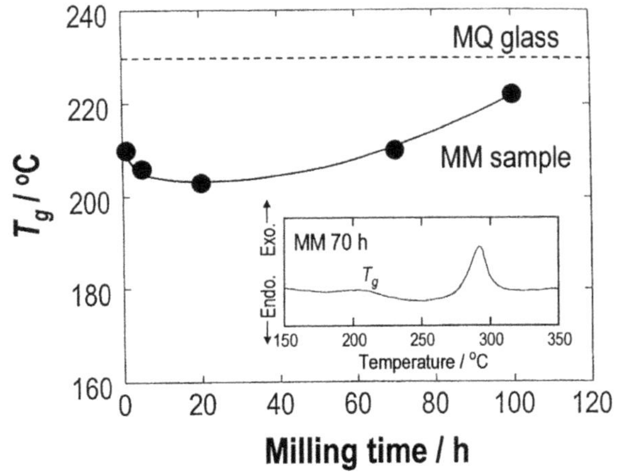

Figure 2. Glass transition temperature (T$_g$) for the 50AgI.50(0·5Ag$_2$O.0·5B$_2$O$_3$) (mol%) samples prepared by MM as a function of the milling time. The T$_g$ of MQ glass (230°C) is also shown in this figure. Inset shows the DSC curve of the amorphous sample prepared by MM for 70 h

'MM', while 'melt quenching' or 'melt quenched' is abbreviated to 'MQ'.

X-ray diffraction measurements (Cu K$_\alpha$) were performed using a XRD diffractometer (Mac Science, M18XHF22-SRA). Differential scanning calorimetry (DSC) of the powder samples was carried out using a DSC apparatus (Perkin Elmer, DSC7) using an Al pan. The heating rate was 10°C / min in the temperature range from 25–500°C under an N$_2$ atmosphere. AC impedance measurements were carried out using a Solartron 1260 impedance analyser. The pelletised samples were obtained by cold pressing glassy powders under a pressure of 3700 kg cm^{-2}. Silver paste was painted onto both faces of the samples and two Au wires were attached to both the faces as a current collector. A frequency range of 10 Hz–8 MHz was used and the measurements were obtained under a dry N$_2$ atmosphere. Solid-state ^{11}B MAS-NMR measurements were carried out at 96·23 MHz on an NMR spectrometer (Varian, UNITY INOVA 300). A single pulse sequence with a $\pi/8$ pulse of 0·7 μs and a recycle pulse delay of 2·0 s were used.

3. Results and discussion

3.1. Formation of glasses by mechanical milling

Initially the amorphisation process of the AgI–Ag$_2$O–B$_2$O$_3$ samples by MM was investigated. Figure 1 shows XRD patterns of the 50AgI.50(0·5Ag$_2$O.0·5B$_2$O$_3$) (mol%) samples prepared by MM of starting materials of AgI, Ag$_2$O, and B$_2$O$_3$ for several hours. The halo pattern gradually becomes dominant with increasing milling time. The peaks due to AgI and Ag$_2$O completely disappear after MM for 1 h, and further MM treatment up to 70 h retained the amorphous state with no formation of new crystalline phases. When just two of three starting materials were used, namely AgI and B$_2$O$_3$, amorphisation did not occur by MM, whereas the mixture of Ag$_2$O and B$_2$O$_3$ became amorphous by MM after 1 h. Our preliminary experiments to clarify the reaction mechanism during MM in the system AgI–Ag$_2$O–B$_2$O$_3$ suggest that firstly Ag$_2$O reacts with B$_2$O$_3$ on the basis of an acid–base reaction, and that the resulting amorphous AgBO$_2$ then reacts with AgI to form the AgI–AgBO$_2$ amorphous materials. FT-IR spectra revealed that the band due to four-coordinated boron units (BO$_4$) increased with increasing the milling time, and the spectrum of the amorphous sample obtained by MM for 1 h was very similar to that of the corresponding MQ glass.

XRD measurements showed that (mol%) xAgI.(100–x)(0·5Ag$_2$O.0·5B$_2$O$_3$) amorphous materials were obtained in the composition range $0 \le x \le 60$ by MM for 20 h. On the other hand, glasses were formed in the composition range $30 \le x \le 60$ by the MQ method. Glasses with AgI contents below 30 mol%, which could not be prepared by the conventional MQ method, can be obtained by MM.

3.2. Thermal and electrical properties of the milled glasses

DSC measurements were carried out to determine glass transition temperatures (T_g) of amorphous samples prepared by MM. Figure 2 shows T_g for

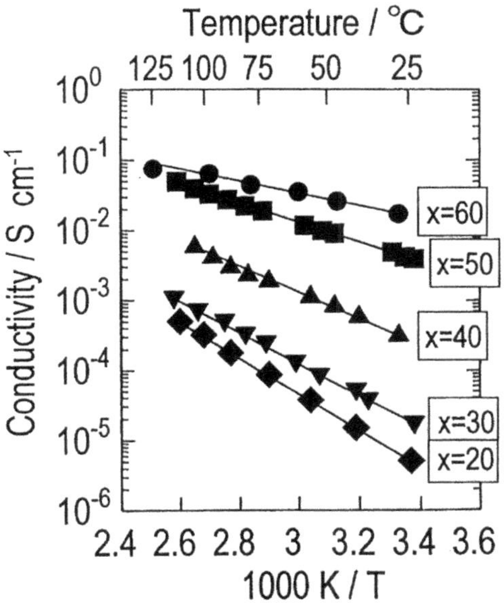

Figure 3. Temperature dependence of electrical conductivity for the xAgI.(100–x)(0·5Ag_2O.0·5B_2O_3) (mol%) glasses prepared by MM

the 50AgI.50(0·5Ag_2O.0·5B_2O_3) (mol%) samples prepared by MM as a function of the milling time. The inset shows the DSC curve of the amorphous sample prepared by MM for 70 h as an example, for which the T_g is observed at 210°C. The MM samples show T_g from 200–220°C, which are lower than T_g of the corresponding MQ glass; the MQ glass has a T_g of 230°C, which almost corresponds to the T_g of the MQ glass at the same composition reported in a previous paper.[18] The T_g of the MM samples tends to approach the T_g of the MQ glass with increasing the milling time.

The phenomena that MM samples have lower T_g than the corresponding MQ glasses have been reported in several systems such as SnO–B_2O_3[19] and AgI–Ag_2O–MoO_3.[16] Tsukushi *et al* have also reported that the T_g of the tri-o-methyl-β-cyclodextrin increases with the milling time, and the saturated T_g of the MM sample is almost the same as the T_g of the MQ glass.[20] Koch *et al* have mentioned that the MM amorphous samples are obtained from a solid state interdiffusion process between particles, not from an equilibrium liquid state,[6,7] suggesting that the difference in T_g between MM and MQ glasses would be based on the degree of homogeneity of the glasses. The increase in homogeneity could bring about the increase in T_g with increasing the milling time as shown in Figure 2.†

Figure 3 shows temperature dependence of electrical conductivity for the xAgI.(100–x)(0·5Ag_2O .0·5B_2O_3) (mol%) glasses prepared by MM. The conductivity monotonically increases and the activation energy for conduction decreases with increasing AgI content. The highest room temperature conductivity of 1·7×10^{-2} Scm^{-1} and the lowest activation energy of 17 kJmol^{-1} are obtained at the composition of 60 mol% AgI. The Ag^+ conducting properties of the MM samples are comparable to those of the corresponding MQ glasses; a compressed powder pellet of the MQ glass with 60 mol% AgI had almost the same conductivity of 1·4×10^{-4} Scm^{-1} and activation energy of 18 kJmol^{-1}.

3.3. Comparison of milled glasses with quenched glasses

Here, we compare the thermal and electrical properties of the MM samples with those of MQ glasses. Figure 4 shows the composition dependence of (a) conductiv-

† Editorial note: despite a request to the authors no compositional analyses have been provided, so it is not possible to ascertain whether any of the observed variations in T_g reported here are due to compositional differences arising from the different preparation routes or are purely a function of milling time.

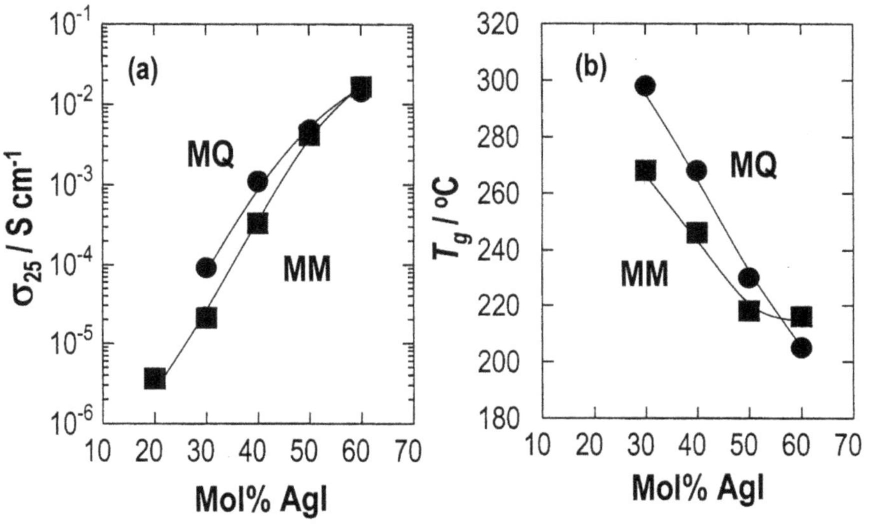

Figure 4. Composition dependence of (a) conductivity at room temperature (σ_{25}) and (b) glass transition temperature (T_g) for the xAgI.(100–x)(0·5Ag_2O.0·5B_2O_3) (mol%) glasses prepared by MM and MQ

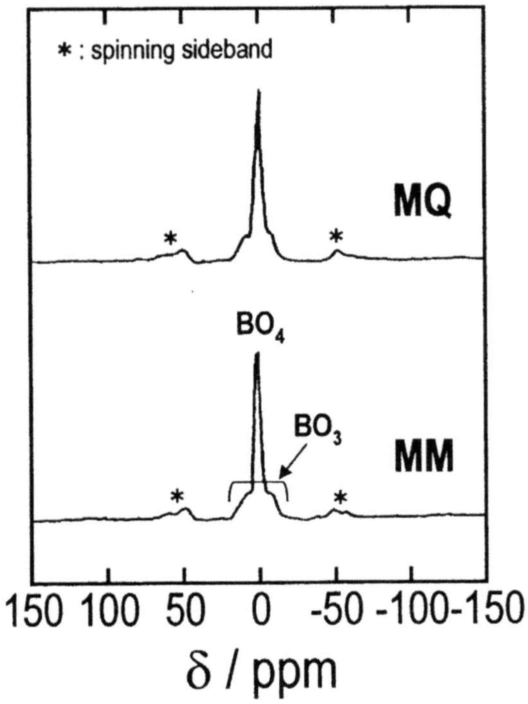

Figure 5. ^{11}B MAS-NMR spectra of the 50AgI.50(0·5Ag_2O.0·5B_2O_3) (mol%) glasses prepared by MM and MQ

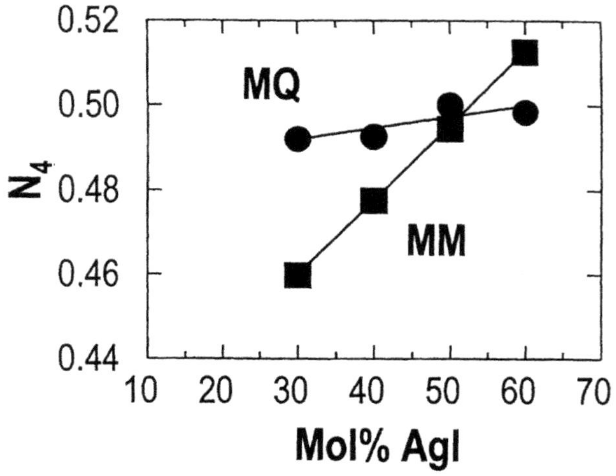

Figure 6. Composition dependence of the fraction of four-coordinated boron (N_4) determined from ^{11}B MAS-NMR studies, for the xAgI.(100–x)(0·5Ag_2O.0·5B_2O_3) (mol%) glasses prepared by MM and MQ

ity at room temperature (σ_{25}) and (b) glass transition temperature (T_g) for the xAgI.(100–x)(0·5Ag_2O.0·5B_2O_3) (mol%) glasses prepared by MM and MQ. In the composition range of x≥50 mol%, both glasses prepared by MM and MQ show almost the same conductivity as shown in Figure 4(a). On the other hand, the conductivities of the MM glasses are lower than those of the MQ glasses at the compositions with the lower AgI contents; the conductivity of the MM glass with x=30 is one order of magnitude lower than that of the corresponding MQ glass. As shown in Figure 4(b), T_g tends to decrease with increasing x content. The T_g of the MM samples is similar to that of the MQ glasses in the composition range of x≥50 mol%, while the T_g of MM samples is lower than those of MQ glasses at the compositions with the lower AgI contents.

The difference in properties such as σ_{25} and T_g of MM and MQ samples at the same composition is probably due to the difference in local structure of the glasses. The local structure around boron atoms in the MM and MQ glasses was investigated by ^{11}B MAS-NMR spectroscopy. Figure 5 shows the ^{11}B MAS-NMR spectra of the 50AgI.50(0·5Ag_2O.0·5B_2O_3) (mol%) glasses prepared by MM and MQ. The sharp peak at 0 ppm is due to four-coordinated boron (BO_4 units) and the broad peak from −20 to +20 ppm is due to three-coordinated boron (BO_3 units). This figure indicates that both BO_4 and BO_3 units are present in both of the glasses. The fraction of BO_4 units (N_4) in the glasses was estimated by the relative area method proposed by Bray *et al*;[21,22] N_4 was determined as the ratio of the area under the sharp peak to the area under the entire curve. Figure 6 shows the composition dependence of N_4 for the xAgI.(100–x)(0·5Ag_2O.0·5B_2O_3) (mol%) glasses prepared by MM and MQ. The N_4 of the MM samples is smaller than that of the MQ glasses for the compositions with x<50 mol%. As mentioned above, conductivity and T_g of the MM samples in this composition range were lower than those of the corresponding MQ glasses, suggesting that the fraction of BO_4 units is closely related to the electrical and thermal properties of the glasses at a given composition. In other words, the lower number of BO_4 units in the MM samples resulted in lower Ag^+ conductivities and lower T_g for relatively low AgI contents.

4. Conclusions

Glasses in the system AgI–Ag_2O–B_2O_3 were prepared by MM for only 1 h at room temperature, and the glass-forming region was determined. The resulting glasses exhibited a maximum conductivity of $1{\cdot}7\times10^{-2}$ S cm^{-1} at room temperature. The T_g of the MM samples was lower than that of the corresponding MQ glasses but the temperatures approached the T_g of the corresponding MQ glass with increasing milling time. The local structure, in particular the fraction of BO_4 units, plays an important role in increasing the conductivity and T_g of Ag^+ conducting borate glasses.

Acknowledgement

This work was supported by the Grant-in-Aid for Scientific Research on Priority Areas (B) and Section (B) from the Ministry of Education, Culture, Sports, Science and Technology of Japan.

References

1. Magistris, A., Chiodelli, G. & Schiraldi, A. *Electrochim. Acta*, 1979, **24**, 203.
2. Minami T. *J. Non-Cryst. Solids*, 1985, **73**, 273.
3. Minami T., Nambu, H. & Tanaka, M. *J. Am. Ceram. Soc.*, 1977, **60**, 467.
4. Minami T. & Machida, N. *Mater. Sci. Eng. B*, 1992, **13**, 203.
5. Minami, T., Ikeda Y. & Tanaka, M. *J. Non-Cryst. Solids*, 1982, **52**, 159.
6. Schwarz, R. B. & Koch, C. C. *Appl. Phys. Lett.*, 1986, **49**, 146.
7. Koch, C. C. *Scr. Mater.*, 1996, **34**, 21.
8. Morimoto, H., Nakai, M., Tatsumisago, M. & Minami, T. *J. Electrochem. Soc.*, 1999, **146**, 3970.
9. Morimoto, H., Yamashita, H., Tatsumisago, M. & Minami, T. *J. Am. Ceram. Soc.*, 1999, **82**, 1352.
10. Tatsumisago, M., Yamashita, H., Hayashi, A., Morimoto, H. & Minami, T. *J. Non-Cryst. Solids*, 2000, **274**, 30.
11. Hayashi, A., Hama, S., Morimoto, H., Tatsumisago, M. & Minami, T. *J. Am. Ceram. Soc.*, 2001, **84**, 477.
12. Komiya, R., Hayashi, A., Morimoto, H., Tatsumisago, M. & Minami, T. *Solid State Ionics*, 2001, **140**, 83.
13. Mizuno, F., Hama, S., Hayashi, A., Tadanaga, K., Minami, T. & Tatsumisago, M. *Chem. Lett.*, 2002, 1244.
14. Machida, N., Nishida, S., Shigematsu, T., Sakai, H., Tatsumisago, M. & Minami, T. *Solid State Ionics*, 2000, **136–137**, 381.
15. Peng, H., Machida, N., Nishida, S. & Shigematsu, T. *J. Non-Cryst. Solids*, 2003, **318**, 112.
16. Dalvi, A., Awasthi, A.M., Bharadwaj, S. & Shahi, K. J. *Phys. Chem. Solids*, 2005, **66**, 783.
17. Dalvi, A. & Shahi, K. *J. Non-Cryst. Solids*, 2004, **341**, 124.
18. Minami, T., Ikeda, Y. & Tanaka, M., *J. Non-Cryst. Solids*, 1982, **52**, 159.
19. Hayashi, A., Konishi, T., Nakai, M., Morimoto, H., Tadanaga, K., Minami, T. & Tatsumisago, M. *J. Ceram. Soc. Jpn.*, 2004, **112**, S713.
20. Tsukushi, I., Yamamuro, O. & Suga, H. *J. Therm. Anal.*, 1991, **37**, 1359.
21. Bray, P. J., Geissberger, A. E., Bucholts, F. & Harris, I. A. *J. Non-Cryst. Solids*, 1982, **52**, 545.
22. Bray, P. J. *J. Non-Cryst. Solids*, 1985, **73**, 19.

Electrochemical properties and local structure of tin borate glasses for lithium secondary batteries

Masahiro Tatsumisago[1] *& Akitoshi Hayashi*

Department of Applied Chemistry, Graduate School of Engineering, Osaka Prefecture University

$SnO–B_2O_3$ glasses were prepared by melt quenching and mechanical milling. $SnO–B_2O_3$ glasses had relatively low glass transition temperatures and worked as high capacity reversible electrode materials in lithium cells with, not only liquid electrolytes, but also solid electrolytes. The thermal and electrochemical properties of the glasses were closely related to the local structure around the boron atoms in the glass network. The relationship between reversible capacity and local structure of the glasses are discussed on the basis of the electrochemical Li insertion mechanism. The electrochemical performance of the SnO-based glassy electrodes depended largely on both the charge–discharge cutoff voltages and the local structure of the glasses.

1. Introduction

Tin oxide based glasses have attracted much attention as low melting glasses and electrode materials for lithium batteries. SnO-based glasses are the most promising candidates for Pb-free low melting sealing glasses; such glasses are desirable from an environmental point of view.(1) Recently, we prepared $SnO–B_2O_3$ and $SnO–P_2O_5$ binary glasses by a melt quenching technique, and investigated the thermal properties and viscosity behaviour of these glasses.(2–5) The glasses had relatively low glass transition temperatures (200–400°C) and their glassy liquids proved to be more fragile than the liquids in alkali borate system. Local structure of these glasses was also analysed by using several spectroscopic techniques such as Raman, NMR, and XPS, and the relationship between glass properties and structure of the glasses was discussed.(3–5)

Application of tin oxide based glasses in lithium rechargeable batteries has been examined following the report by Idota *et al*(6) that SnO-based oxide glasses prepared by melt quenching exhibited large specific capacities as a negative electrode materials for lithium secondary batteries.(6–8) One of the key areas for enhancing the energy density of batteries is the use of electrode materials with high reversible capacities. SnO-based glasses have specific capacities twice as large as carbon materials, which are used as a negative electrodes for commercially available lithium ion secondary batteries. These electrochemically active glasses were prepared by not only melt quenching but also mechanical milling techniques.(9–11) The mechanical milling technique using a high energy ball mill apparatus has attracted much attention as a new preparation procedure for inorganic amorphous materials.(12–14) This technique has several advantages compared to the melt quenching method: (i) the mechanical milling process is basically a room temperature operation and (ii) fine powders are directly obtained without an additional pulverising process being carried out on bulk glasses.

We have succeeded in the preparation of the SnO-based borate and phosphate glassy powders by mechanical milling and have evaluated their electrochemical properties as a negative electrode in lithium cells using a conventional liquid electrolyte and a solid electrolyte.(9–11,15) Nowadays, all-solid-state batteries with solid electrolytes instead of liquid electrolytes are attracting attention because they have high reliability and safety.(16,17) All-solid-state In/$LiCoO_2$ cells with $Li_2S–P_2S_5$ glass-ceramic electrolytes have shown good performance as a rechargeable lithium batteries with high capacity and excellent cyclability at room temperature.(18,19)

The present paper reviews the glass forming region, glass transition temperature, and local structure of tin borate glasses prepared by melt quenching and mechanical milling. Electrochemical performance of the glasses as an electrode was analysed in the cells with both a conventional liquid electrolyte and a high Li^+ conductive, glassy solid electrolyte. The relationship between electrochemical properties and local glass structure was investigated, and the Li^+ insertion mechanism from the viewpoint of local structure is discussed below.

2. Experimental

2.1 Glass preparation

Reagent-grade SnO (Furuuchi Chem., 99·9%) and B_2O_3 (Koujundo Chem., 99·9%) crystalline powders were used as starting materials for sample prepara-

[1] Corresponding author. Email: tatsu@chem.osakafu-u.ac.jp

tion. We used two different methods for synthesis, conventional melt quenching and mechanical milling. In the former case, a mixture of starting materials was heated in a carbon crucible at 900–1200°C for 30 min in a dry N_2-filled glove box. The molten samples were pressed between two iron plates to yield thin disks less than 1 mm thick. The cooling rate was estimated to be 10^2–10^3 K/s. The mechanical milling treatment was carried out for the batches (1–2 g) of the mixed materials of xSnO.(100−x)B_2O_3 (25≤x≤85) in a zirconia pot (volume of 45 ml) with eight zirconia balls (10 mm in diameter) using a high-energy planetary ball mill apparatus (Fritsch Pulverisette 7). The rotation speed was fixed at 370 rpm and all the processes were conducted at room temperature in a dry N_2-filled glove box.

2.2 Structural analysis of glasses

X-ray diffraction (XRD) (Cu K_α) was undertaken using a XRD diffractometer (Shimadzu, XRD-6000). Differential scanning calorimetry (DSC) was carried out using a DSC apparatus (Perkin Elmer, DSC7) on the obtained powder samples which had been sealed in an Al pan in a dry N_2 atmosphere. The heating rate was 10°C / min in the temperature range from 25–500°C under an N_2 atmosphere for all the measurements. Solid state ^{11}B MAS-NMR measurements were carried out at 96·23 MHz on a NMR spectrometer (Varian, UNITY INOVA 300). A single pulse sequence with a $\pi/8$ pulse of 0·7 μs was used. The recycle pulse delay was 2·0 s and the number of scan was 64–320. A cylindrical zirconia spinner was rotated at a speed of about 5 kHz during the measurements. A BPO_4 crystal powder was used as a standard material for ^{11}B NMR measurements. ^{119}Sn Mössbauer spectra were measured with a constant acceleration mode Mössbauer spectrometer using a $CaSnO_3$ source at room temperature. The spectra were obtained by using Wissel spectrometer with a NaI (TI) scintillation counter and recorded on a SEIKO 7700 multi channel analyser.

2.3 Fabrication of electrochemical cells using liquid and solid electrolytes

Electrochemical properties of the glasses used as electrode materials were examined in a simple three electrode cell using a conventional liquid electrolyte. The working electrode was prepared by mixing SnO–B_2O_3 glass powder (70 wt%), acetylene black (20 wt%), and polyvinylidene fluoride (10 wt%). Lithium metal electrodes were used as counter and reference electrodes. A mixture of ethylene carbonate (EC) and diethyl carbonate (DEC) in a 1:1 volumetric ratio containing 1 mol dm^{-3} (1 M) $LiPF_6$ was used as an electrolyte. The charge and discharge measurements were carried out at a constant current density of 1·0 mA cm^{-2} and with a cutoff voltage between 0 and 2·0 V. In the present study, 'charge' corresponds to an insertion of lithium ions to a working electrode, while 'discharge' corresponds to an extraction of lithium ions from the working electrode.

Electrochemical properties for the glasses were also investigated in all-solid-state cells using the highly conductive 95(0·6Li_2S.0·4SiS_2).5Li_3BO_3 (mol%) glassy solid electrolyte. The glass was obtained by melt quenching method[20] and then was ground into fine powder by ball milling. A positive electrode was prepared by mixing of 38 wt% of $LiCoO_2$, 57wt% of powdered glass electrolyte, and 55 wt% of acetylene-black powder. A negative electrode was obtained by mixing the 42 wt% of 50SnO.50B_2O_3 (mol%) glass, 53 wt% glass electrolyte powder and 5 wt% of acetylene-black powder. 100 mg of positive electrode powder, 80 mg of glass electrolyte powder, and 19 mg of negative electrode powder were placed together in an insulating polycarbonate tube (ϕ=10 mm) and then were pressed to obtain three layered solid state cells. The cells obtained were charged and discharged at room temperature under a constant current density of 64 μA cm^{-2} in an Ar atmosphere.

3. Results and discussion

3.1. Glass forming region

Glasses were obtained in the composition range 0≤SnO (mol%)≤75 by melt quenching.[2] The glass forming region is affected by the duration of mechanical milling in an analogous way to how it is affected by quenching rate in the case of melt quenching. First of all, the amorphisation of SnO–B_2O_3 samples by mechanical milling was investigated. Figure 1 shows XRD patterns of the 50SnO.50B_2O_3 (mol%) samples prepared by mechanical milling for several hours. Diffraction peaks due to SnO crystals are observed in the powder mixture before mechanical milling (0 h); the other starting material (B_2O_3) was in an amorphous state. As the milling period increases, the intensity of the SnO peaks decreases and the halo pattern becomes dominant. The peaks due to SnO disappear after milling for 15 h, and further milling treatment retains the amorphous state (no formation of new crystal phases). XRD measurements showed that amorphous xSnO.(100−x)B_2O_3 s materials were obtained in the composition range 0≤x≤80 after mechanical milling for 100 h. Some SnO crystals remained in the 85 mol% SnO composition even after mechanical milling for over 100 h. The glass forming limit for the SnO–B_2O_3 system obtained by melt quenching was 75 mol% SnO. The amorphous material containing 80 mol% SnO, which could not be obtained by melt quenching, was prepared by mechanical milling for 100 h. The SnO-based amorphous

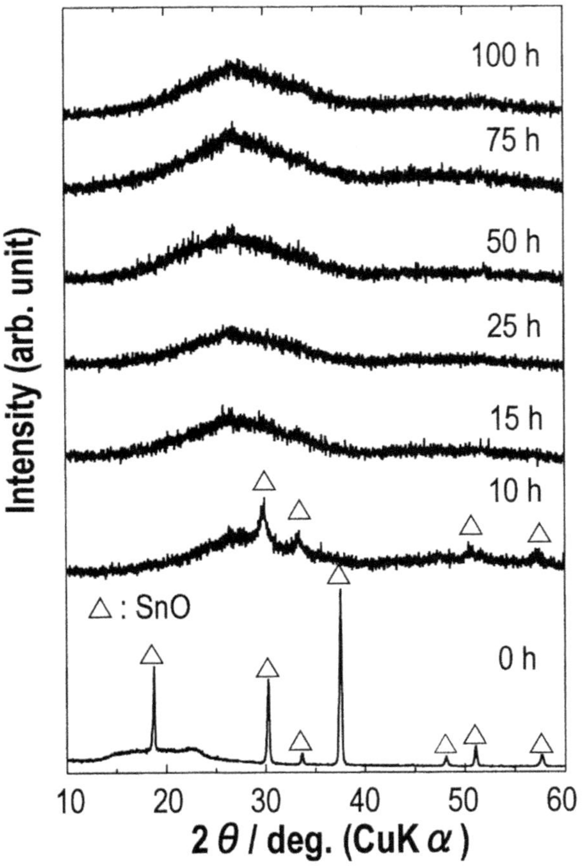

Figure 1. X-ray diffraction patterns of powder samples with a nominal composition of 50SnO.50B_2O_3 (mol%) prepared by mechanical milling for several hours. The numbers in the figure denote the milling times

powders prepared by mechanical milling were a few μm in diameter.[9]

3.2. Thermal properties of the SnO–B_2O_3 glasses

Figure 2 shows the composition dependence of the glass transition temperatures (T_g) for the SnO–B_2O_3 glasses prepared by melt quenching and mechanical milling. The milling period was 75 h for the compositions below 67 mol% SnO, whereas the milling period for 80 mol% SnO was 100 h. In the quenched glasses, the values of T_g increase up to 40 mol% SnO and then decrease with further increases in SnO content. The glasses exhibit relatively low T_gs and have a maximum T_g of 384°C at the composition with 40 mol% SnO. This compositional dependence of T_g is explained from the viewpoint of local structure of the glass, in that the fraction of four-coordinated borons (BO_4) in the glasses is a maximum at the composition with 50 mol% SnO, (for more details see Figure 5 below). The tendency for increasing T_g with increasing four-coordinated borons has been reported in many borate glasses.

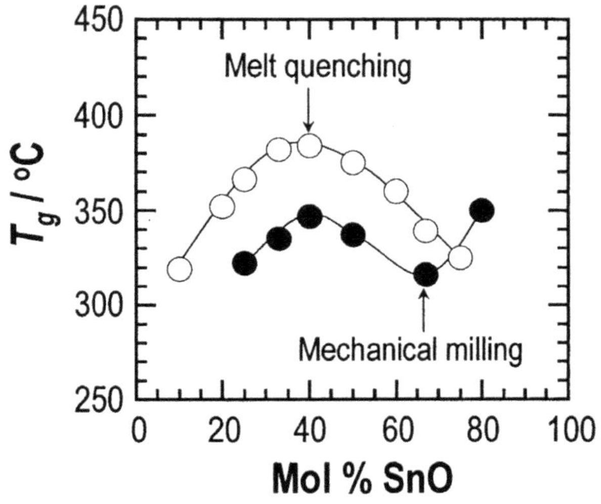

Figure 2. Compositional dependence of the glass transition temperatures for the SnO–B_2O_3 glasses prepared by melt quenching and mechanical milling

Glass transition phenomena were also observed for the amorphous materials obtained by mechanical milling. The compositional dependence of T_g for the milled glasses is similar to that for the quenched glasses except for the compositions with higher SnO contents. The T_g values of the milled glasses are lower than those of the quenched glasses. One possible explanation for the differences in T_g is based on local structure, but ^{11}B MAS-NMR measurements proved that the local structure around boron atoms of the SnO–B_2O_3 milled glasses was similar to that of the quenched glasses.[10,11] Suga *et al* have similarly reported that T_g of milled glasses is lower than that of quenched glasses.[21] They also mentioned that T_g of milled glasses tended to approach T_g of the quenched glasses with an increase in milling time. The differences in T_g would be based on the inhomogeneity of the milled glasses, which arises because the milled glasses are obtained from a solid-state interdiffusion process between particles, not from an equilibrium liquid state. Research designed to explain the relationship is now in progress.

3.3. Electrochemical properties of the SnO–B_2O_3 glasses in lithium cells using liquid electrolytes

Electrochemical measurements were performed in a simple three electrode cell using liquid electrolytes. Figure 3 shows the first charge–discharge curves for the cells using the 50SnO·50B_2O_3 (mol%) glasses prepared by melt quenching and mechanical milling. An abscissa shows the charge–discharge capacities per unit weight of the glasses. Two plateaus at around 1·5 and 0·5 V are observed on the charge curve and one plateau at around 0·5 V is observed on the discharge curve in both cells. The charge–discharge

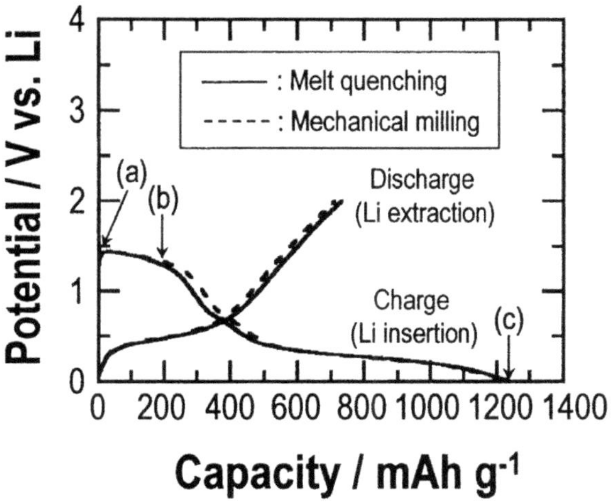

Figure 3. First charge–discharge curves of lithium cells using $50SnO.50B_2O_3$ glassy electrodes prepared by melt quenching and mechanical milling. A mixture of 1 M $LiPF_6$/EC+DEC was used as an electrolyte

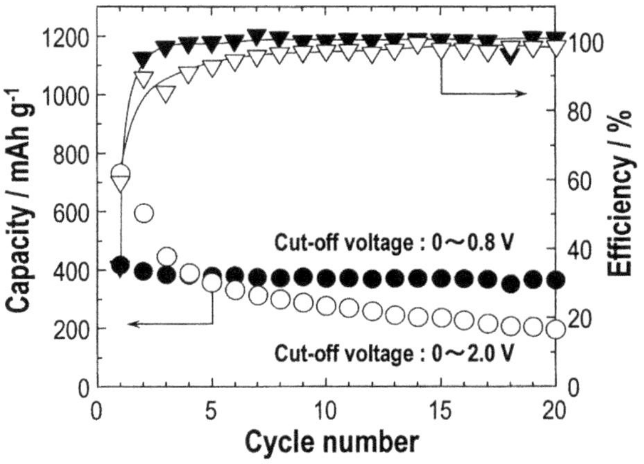

Figure 4. Cycle performances of the lithium cell using the $50SnO.50B_2O_3$ glassy electrodes prepared by melt quenching. A constant current density of 1·0 mA cm^{-2} was used

profiles in both cells are almost the same, suggesting that both the glass prepared by mechanical milling and the glass prepared by melt quenching work as a reversible electrode. The total charge capacity of the $50SnO.50B_2O_3$ quenched glass is 1240 mAh g^{-1}, while the discharge capacity is 733 mAh g^{-1}.

In the charge curve, the first plateau is due to the formation of metallic Sn^0 ($Sn^{2+}\rightarrow Sn^0$) and the Li_2O–B_2O_3 glassy matrix, and the second plateau is due to the formation of a Li–Sn alloy ($Sn^0\rightarrow$Li–Sn).[7,8] The former reaction is known to be irreversible and the formation of metallic Sn is the main reason for the large irreversible capacity on the first cycle. The latter reaction is basically reversible; the formation of Li–Sn alloy ($Sn^0\rightarrow$Li-Sn) occurs on the charge process while the formation of Sn (Li-Sn$\rightarrow Sn^0$) occurs on the discharge process.

The cycling performance of the cell and the effects of a cutoff voltage on the cycling behaviour were investigated. Figure 4 shows the cycle performance for the cells containing $50SnO.50B_2O_3$ glasses prepared by melt quenching. Circles and diamonds denote the discharge capacity and charge–discharge efficiency, respectively. Open and solid marks denote the cutoff voltages of 0–2·0 V and 0–0·8 V, respectively. The capacity gradually decreases during charge–discharge cycling under a cut off voltage 0–2·0 V and the cell reaches the constant capacity of about 200 mAh g^{-1} at the 20th cycle. On the other hand, the cell measured under a cut-off voltage 0–0·8 V maintains higher rechargeable capacity of about 400 mAh g^{-1} for 20 cycles. The charge–discharge efficiency of both cells becomes almost 100% after 20 cycles. The capacity fading during cycling would be caused by a large volume change due to the aggregation of Sn particles, which occurs at over 0·8 V during the discharge process.[7] The cycling performance, therefore, is considerably enhanced by reducing the upper cutoff voltage from 2·0 to 0·8 V. The cycling performance of tin borate glasses is very sensitive to the cut off voltage of the charge–discharge measurements. The mechanically milled SnO–B_2O_3 glasses exhibited almost the same cycling performance as the corresponding melt quenched glasses.[10,11]

Figure 5 shows the composition dependence of the first discharge capacity of the SnO–B_2O_3 glasses prepared by melt quenching. The fraction of four-coordinated boron (N_4), which was determined by ^{11}B solid-state NMR measurements,[4] is also shown in this figure. The capacity (o) increases up to the composition with 50 mol% SnO, then decreases up to 67 mol% SnO, and increases again up to 75 mol% SnO. The maximum discharge capacity is 733 mAh g^{-1} at the composition with 50 mol% SnO. On the other

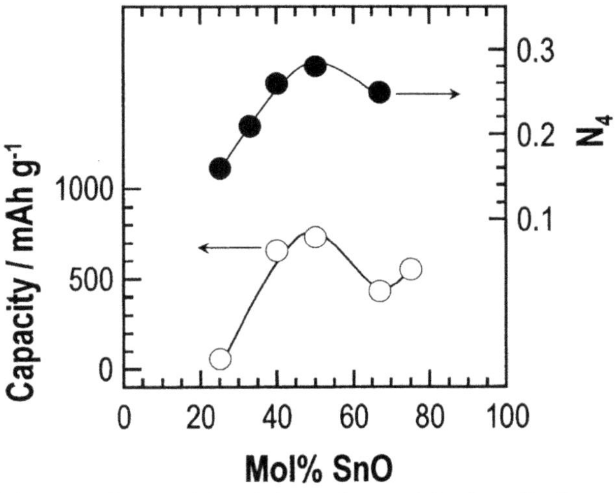

Figure 5. Compositional dependence of the first discharge capacities of lithium cells using SnO–B_2O_3 glassy electrodes prepared by melt quenching. The fraction of four-coordinated boron (N_4) in the glasses, which was determined by ^{11}B solid-state NMR, is also shown in this figure

hand, the value of N_4 (●) increases and then decreases with an increase in SnO content. The maximum N_4 of 0·28 is obtained at the composition with 50 mol% SnO, where the capacity is maximised. It is found that the capacity of the $SnO–B_2O_3$ glasses is closely related to the local structure of the glasses. In general, the cell capacity is expected to increase with an increase in SnO content because Sn works as an active site for lithium insertion. For example, charge–discharge capacities monotonically increased with an increase in SnO content in the $SnO–BPO_4$ system.[2] In the $SnO–B_2O_3$ system, however, charge–discharge capacities have a maximum at the composition with 50 mol% SnO, suggesting that the capacities are affected by not only the SnO content but also glass structure (in particular the fraction of four-coordinated boron) in the system. The reason for showing large capacities at the composition with a large fraction of four-coordinated boron will be discussed in the following section from the point of view of glass structural changes during lithium insertion.

3.4. Lithium insertion mechanism in the $SnO–B_2O_3$ glasses

The investigation of the electrochemical lithium insertion mechanism in the $SnO–B_2O_3$ glasses is helpful in understanding the relationship between capacity and local glass structure as shown in Figure 5. In order to elucidate the lithium insertion mechanism in $50SnO·50B_2O_3$ glass we have investigated structural changes of the glass during the first charge process by several spectroscopic techniques; the electrode samples were obtained after being charged up to 200 $mAh\,g^{-1}$ (the end of the first plateau) as shown in Figure 3(b), and up to 1240 $mAh\,g^{-1}$ (the end of the second plateau) as shown in Figure 3(c).

Figure 6 shows the ^{11}B MAS-NMR spectra for the pristine $50SnO·50B_2O_3$ glass (a) and the glasses charged up to 200 $mAh\,g^{-1}$ (b) and 1240 $mAh\,g^{-1}$ (c). The sharp peak at 0 ppm is due to four-coordinated boron (the tetrahedral $(BO_{4/2})^-$ unit) and the broad peak from –20 to +20 ppm is due to three-coordinated boron (the triangular $BO_{3/2}$ unit).[22] Both three- and four-coordinated borons are present in the pristine $50SnO·50B_2O_3$ glass (a). Little change in the peak position and the shape of the spectrum is observed between spectra (a) and (b). On the other hand, spectrum (c) is drastically changed from spectra (a) and (b); the intensity of the sharp peak at 0 ppm decreases in spectrum (c). The borate glass network in sample (b) is almost the same as that in the pristine glass (a), while the glass network has been rearranged in sample (c) due to a decrease in four-coordinated boron.

Figure 7 shows ^{119}Sn Mössbauer spectra for the pristine $50SnO.50B_2O_3$ glass (a) and the glasses

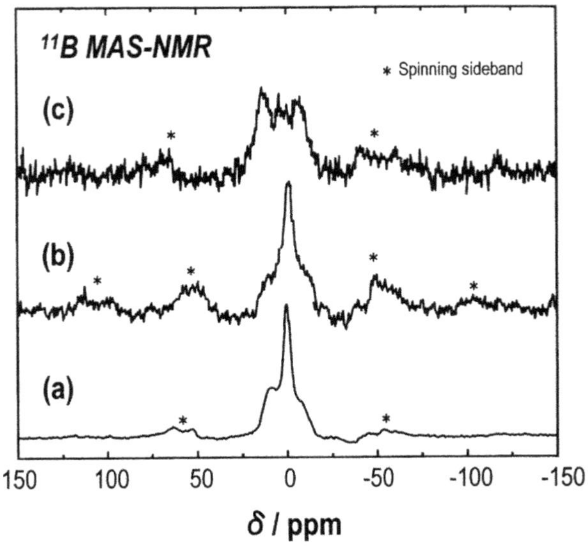

Figure 6. ^{11}B MAS-NMR spectra of the $50SnO.50B_2O_3$ glasses at various levels of charge (a)–(c) as shown on the voltage profile in Figure 3: initial state (a), 200 $mAh\,g^{-1}$ (b), and 1240 $mAh\,g^{-1}$ (c) (the cell voltage reaches 0 V versus Li)

charged up to 200 $mAh\,g^{-1}$ (b) and 1240 $mAh\,g^{-1}$ (c). The spectrum of sample (a) displays a doublet with a shape and peak positions very similar to those of SnO (Sn(II)) crystal, suggesting that the oxidation state of Sn in the pristine glass is only Sn(II).[4] The peaks are broadened but their positions are not changed in sample (b). The peak positions slightly shift to the lower velocity side in sample (c). Courtney *et al* have reported that a broad peak due to metallic Sn is observed in the Mössbauer spectrum of a $SnO–B_2O_3–P_2O_5$ glass at the charging level just after a first plateau.[23] The spectrum of sample (b) suggests that the sample still includes Sn(II) which shows sharp doublet peaks, and thus there is no obvious evidence of metallic Sn, which would only show a weak and broad peak. However, ^7Li NMR spectra indicated that the inserted lithium was present as ionic Li^+ at this stage,[15] suggesting that metallic Sn should be partially formed because Li^+ ions compensate for the negative charge of borate anions instead of Sn^{2+}. Because the first plateau is not completely finished at the charging level of sample (b), a broad peak due to metallic Sn is probably hidden by a sharp peak due to residual Sn(II). Compared with the shape and position of the peaks in the Mössbauer spectra of the lithium inserted glasses,[23] the features in the spectrum of sample (c) are assigned to Li-rich Li–Sn alloy phases such as Li_5Sn_2, $Li_{13}Sn_5$, Li_7Sn_2 and $Li_{22}Sn_5$.

Here, we discuss in detail the rearrangement of borate glass network by lithium insertion on the basis of the occupied volume of discrete borate units in the glasses. The individual volumes of borate units have been obtained from the density data reported

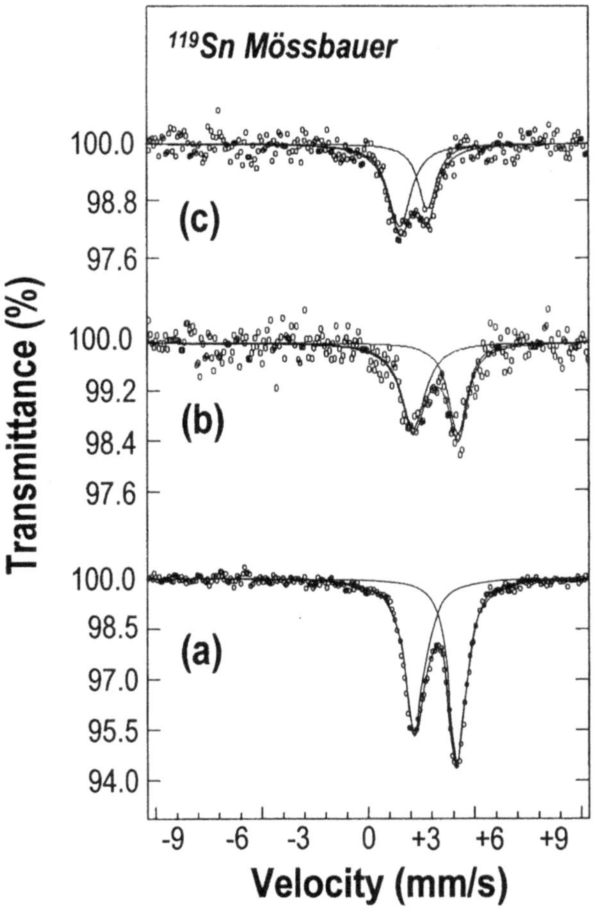

Figure 7. ^{119}Sn Mössbauer spectra for pristine $50SnO.50B_2O_3$ glass (a) and glasses charged up to 200 $mAh\,g^{-1}$ (b) and 1240 $mAh\,g^{-1}$ (c)

for alkali borate glasses.[24–26] In the $Li_2O–B_2O_3$ system, the relative volume of the tetrahedral $(BO_{4/2})^-$ unit compared with the $BO_{3/2}$ unit in the pure B_2O_3 glass is 0·94, which is smaller than the value of 1·28 of the triangular $(BO_{2/2})O^-$ unit. Those relative volumes, however, include empty (free) space in addition to the space originally occupied by the borate unit. In order to consider a structural arrangement of a glass, the concept of 'packing fraction' proposed by Feller *et al*[24–26] is appropriate and useful. The packing fraction is defined as a ratio of the volume calculated from ionic radii to the volume found from the measured density including free space. The calculated packing fraction of the $(BO_{4/2})^-$ and $(BO_{2/2})O^-$ units are 0·65 and 0·39, respectively;[26] i.e. the tetrahedral $(BO_{4/2})^-$ unit has a noticeably higher packing fraction than the planar trianglar $(BO_{2/2})O^-$ unit. These results suggest that the transformation from the $(BO_{4/2})^-$ unit to the $(BO_{2/2})O^-$ unit provides additional free space in the glass network. In fact, the net volume of the trianglar $(BO_{2/2})O^-$ unit calculated from ionic radii is 0·50, which is smaller than the net volume of the tetrahedral $(BO_{4/2})^-$ unit (0·61).

According to previous studies,[7,8] the size of Li–Sn alloy particles gradually increased following lithium insertion into the SnO based glasses, and the oxide glassy matrix played an important role in preventing the aggregation of the Li–Sn alloy particles. In the $SnO–B_2O_3$ glasses including four-coordinated boron, the additional free space generated by the transformation from the tetrahedral unit to the triangular unit can compensate the volume increase of the alloy domains, suggesting that further amounts of lithium could be inserted and Sn active sites would be much more effectively utilised in those glasses. Hence, the unusually large capacity was obtained at the relatively low ratio SnO content of 50 mol% SnO, where the fraction of four-coordinated boron was maximised as shown in Figure 5.

3.5. Electrochemical performance of all-solid-state cells

Figure 8 shows the charge–discharge cycling performance of an all-solid-state cell $50SnO.50B_2O_3$ glass/$LiCoO_2$. The $95(0·6Li_2S.0·4SiS_2).5Li_3BO_3$ (mol%) superionic glass was used as a solid electrolyte.[20] The measurement was performed at room temperature under a constant current density of 64 $\mu A\,cm^{-2}$ between fixed voltage limits of 1·5 and 4·1 V. The abscissa denotes capacities per unit weight of the glass as a negative electrode. The all-solid-state cell using the $SnO–B_2O_3$ glass as a negative electrode works as a lithium secondary battery. Two plateaus around 2·5 and 3·5 V were observed on the first charge process and then one plateau around 3·5 V was observed on the first discharge process. The two

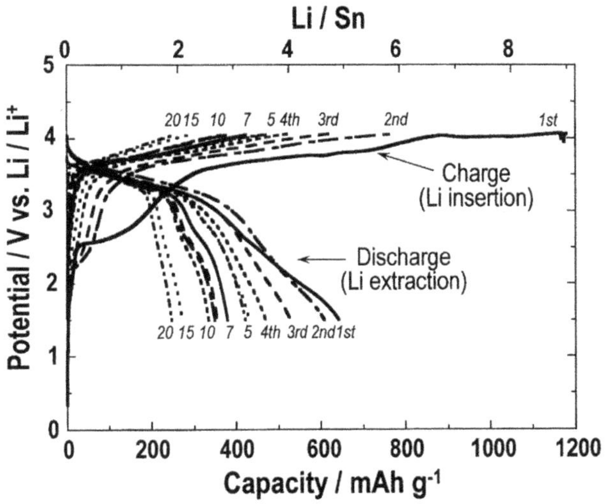

Figure 8. Charge-discharge curves for an all-solid-state $50SnO.50B_2O_3$ glass / $LiCoO_2$ cell. A superionic glass in the system $Li_2S–SiS_2–Li_3BO_3$ is used as a solid electrolyte. The measurement was performed at room temperature under a constant current density of 64 $\mu A\ cm^{-2}$ between fixed voltage limits of 1·5 and 4·1 V. The abscissa denotes capacities per unit weight of the glass

plateaus observed on the charge process correspond to those of 1·5 and 0·5 V observed in the cell using a liquid electrolyte as shown in Figure 3. Total charge and discharge capacities are respectively 1180 and 640 $mA h g^{-1}$ at the first cycle. Both charge and discharge capacities gradually decrease with cycling while the charge–discharge efficiency increases. After 20 cycles the efficiency reaches 100% and the cell maintains capacities over 200 $mA h g^{-1}$.

The observed capacity fading could be due to the large volume change of aggregated Sn particles, which leads to inactive Sn particles because of electrical contact loss. The problem of the degradation of capacity could be overcome by selecting appropriate voltage ranges as shown for a liquid electrolyte cell in Figure 4. The large irreversible capacity is observed during the first cycle in both liquid and solid electrolyte systems; this capacity loss mainly generated by the formation of metallic Sn. Because lithium ions inserted are strongly trapped by borate anions, the Li^+ could not be electrochemically extracted on the discharge process.

4. Conclusions

$SnO–B_2O_3$ glasses were prepared by conventional melt quenching and mechanical milling. The $SnO–B_2O_3$ glasses had relatively low T_g values, which have been related to the local structure of the glasses. The T_g of the milled glasses was lower than that of the corresponding glasses prepared by melt quenching. The glasses worked as high capacity reversible electrode materials in lithium cells with both liquid and solid electrolytes. The charge–discharge capacities of the glasses were closely related to local structure around boron atoms in the glass network. The reversible capacity was maximised at the composition with 50 mol% SnO, where the fraction of tetrahedral boron units was also maximised. The electrochemical Li insertion to the borate glass brought about the transformation of tetrahedral boron units to triangular boron units, which generated additional free space compensating the increase in volume of the Li–Sn alloy domains. The cell using the glass with large amounts of tetrahedral boron units, therefore, exhibited higher reversible capacity. It was found that the electrochemical performance of SnO-based glassy electrodes largely depends on the charge–discharge cutoff voltages and the glass composition.

From the structural point of view, an alternative idea for enhancing electrochemical performance of all-solid-state batteries is using SnS-based sulphide glassy electrodes. We have succeeded in operating all-solid-state cells using $SnS–P_2S_5$ glassy electrodes and the $Li_2S–P_2S_5$ glass-ceramic solid electrolyte; the cells exhibited high reversible capacity of 400 $mA h g^{-1}$ for 50 cycles.[27] A common network former P_2S_5 is used for the electrolyte and electrodes materials in the cells, and this approach would be useful for reducing the interfacial polarisation during Li^+ transfer in all-solid-state cells.

Acknowledgment

We are grateful to Prof. M. Katada for ^{119}Sn Mössbauer experiments and fruitful discussion. This work was supported by the Grant-in-Aid for Scientific Research on Priority Areas (B) and Section (B) from the Ministry of Education, Culture, Sports, Science and Technology of Japan.

References

1. Morinaga, K. & Fujino, S. *J. Non-Cryst. Solids*, 2001, **282**, 118.
2. Nakai, M., Hayashi, A., Morimoto, H., Tatsumisago, M. & Minami, T. *J. Ceram. Soc. Jpn*. 2001, **109**, 1010.
3. Hayashi, A., Konishi, T., Tadanaga, K., Minami, T. & Tatsumisago, M. *J. Non-Cryst. Solids*, 2004, **345–346**, 478.
4. Hayashi, A., Nakai, M., Tatsumisago, M., Minami, T., Himei, Y., Miura, Y. & Katada, M. *J. Non-Cryst. Solids*, 2002, **306**, 227.
5. Hayashi, A., Nakai, M., Tatsumisago, M. & Minami, T. *Comptes Rendus Chimie*, 2002, **5**, 751.
6. Idota, Y., Kubota, T., Matsufuji, A., Maekawa, Y. & Miyasaka, T. *Science*, 1997, **276**, 1395.
7. Courtney, I.A. & Dahn, J.R. *J. Electrochem. Soc.*, 1997, **144**, 2943.
8. Goward, G. R., Nazar, L. F. & Power, W. P. *J. Mater. Chem.*, 2000, **10**, 1241.
9. Morimoto, H., Tatsumisago, M. & Minami, T. *J. Electrochem. Soc.*, 1999, **146**, 3970.
10. Hayashi, A., Nakai, M., Morimoto, H., Minami, T. & Tatsumisago, M. *J. Mater. Sci.*, 2004, **39** 1.
11. Hayashi, A., Konishi, T., Nakai, M., Morimoto, H., Tadanaga, K., Minami, T. & Tatsumisago, M. *J. Ceram. Soc. Jpn*. 2004, **112**, S713.
12. Schwarz, R. B. & Koch, C. C. *Appl. Phys. Lett.*, 1986, **49**, 146.
13. Koch, C. C. *Scr. Mater.*, 1996, **34**, 21.
14. Morimoto, H., Yamashita, H., Tatsumisago, M. & Minami, T. *J. Am. Ceram. Soc.*, 1999, **82**, 1352.
15. Hayashi, A., Nakai, M., Tatsumisago, M., Minami, T. & Katada, M. *J. Electrochem. Soc.*, 2003, **150**, A582.
16. Julien, C. & Nazri, G.-A. *Solid State Batteries: Materials Design and Optimization*. Kluwer Academic Publishers, Boston, 1994, P. 579.
17. Tarascon, J.-M. & Armand, M. *Nature*, 2001, **414**, 359.
18. Mizuno F., Hama, S., Hayashi, A., Tadanaga, K., Minami, T. & Tatsumisago, M. *Chem. Lett.*, 2002, 1244.
19. Tatsumisago, M. *Solid State Ionics*, 2004, 175, 13.
20. Hayashi, A., Komiya, R., Tatsumisago, M. & Minami, T. *Solid State Ionics*, 2002, **152–153** 285.
21. Tsukushi, I., Yamamuro, O. & Suga, H. *J. Therm. Anal.*, 1991, **37**, 1359.
22. Bray, P.J. *J. Non-Cryst. Solids*, 1985, **73**, 19.
23. Courtney, I. A., Dunlap, R. A. & Dahn, J. R. *Electrochim. Acta*, 1999, **45**, 51.
24. Lim, H.P., Karki, A, Feller, S., Kasper, J.E. & Sumcad, G. *J. Non-Cryst. Solids*, 1987, **91**, 324.
25. Chong, B. C. L., Choo, S. H., Feller, S., Teoh, B., Mathews, O., Khaw, E. J., Feil, D., Chong, K. H., Affatigato, M., Bain, D., Hazen, K., & Farooqui, K. *J. Non-Cryst. Solids*, 1989, **109**, 105.
26. Loer, N. P., Mcrae, J. L., Feller, H. A., Betzen, A. R., Kapoor, S., Affatigato, M., & Feller, S. A. *J. Non-Cryst. Solids*, 2001, **293–295**, 669.
27. Hayashi, A., Konishi, T., Tadanaga, K., Minami, T. & Tatsumisago, M. *J. Power Sources*, 2005, **146**, 496.

Author Index

Subject Index

Appendix 1
Conference Participants

Dr Sameh Ahmed
Università di Trento, Dipartimento di Fisica, Via Sommarive 14, I38050 Povo (TN), ITALY. Email: ahmed@science.unitn.it

Professor M. Affatigato
*Coe College, 1220 First Avenue NE, Cedar Rapids, Iowa, USA
Email: maffatig@coe.edu*

Dr Nasser Afify
Università di Trento, Dipartimento di Fisica, Via Sommarive 14, I38050 Povo (TN), ITALY. Email: n_afify@science.unitn.it

Dr Bruce Aitken
*Corning, SP-FR-05 Corning Inc., Corning NY 14831, USA
Email: aitkenbg@corning.com*

Dr Elena Maria Anghel
*Institute of Physical Chemistry "I.G. Murgulescu" of Romanian Academy, Spl. Independentei 202 060021 Bucharest, ROMANIA
Email: manghel@icf.ro*

Mr Graham Appleby
Victoria University of Wellington, PO BOX 600, Wellington, NEW ZEALAND. Email: graham@aenertia.net

Ms Cristina Armellini
Istituto di Fotonica e Nanotecnologie del Consigio Nazionale delle Ricerche - Sezione "ITC CeFSA" di Trento, Via Sommarive 18, I38050 Povo (TN), ITALY. Email: armellin@science.unint.it

Prof. Moenis A. Azooz
*Glass Research Dept., National Research Center, Dokki, Cairo, EGYPT
Email: moenis9@yahoo.com*

Mr Joy Banerjee
*Coe College, 1220 First Avenue NE, Cedar Rapids, Iowa, USA
Email: jbanerje@coe.edu*

Dr Paul Bingham
University of Sheffield, Sir Robert Hadfield Building, Mappin Street, Sheffield, S1 3JD, UK. Email: p.a.bingham@sheffield.ac.uk

Prof. Alexis Clare
*NYSCC Alfred University, 2 Pine Street, Alfred, NY 14802, USA
Email: clare@alfred.edu*

Dr Francesco Congiu
Dipartimento di Fisica, Università di Cagliari, S.P. Monserrato-Sestu, km 0.700, I-09042 Monserrato (Cagliari), ITALY. Email: franco.congiu@dsf.unica.it

Dr Laurent Cormier
*IMPMC-CNRS, 140 rue de Lourmel, 75015 Paris, FRANCE
Email: cormier@lmcp.jussieu.fr*

Prof. Maria Cutroni
Dipartimento di Fisica, Università di Messina, Salita Sperone, 31, 98100 Messina, ITALY. Email: cutroni@unime.it

Prof. G. Dalba
Dipartimento di Fisica, Università di Trento, Via Sommarive, 14-38050 Povo, Trento, ITALY. Email: dalba@science.unitn.it

Prof. Y. B. Dimitriev
*Department of Silicate Technology, University of Chemical Technology & Metallurgy, 8 Kl. Ohridski Blvd., 1756 Sofia, BULGARIA
Email: yanko@uctm.edu*

Dr S. P. Djambazov
*Department of Silicate Technology, University of Chemical Technology &Metallurgy, 8 Kl. Ohridski Blvd., 1756 Sofia, BULGARIA
Email: djam@uctm.edu*

Dr Manfred Dubiel
University of Halle-Wittenberg, Friedemann-Bach-Platz 6, D-06108 Halle, GERMANY. Email: dubiel@physik.uni-halle.de

Dr Olga Dymshits
*State Optical Institute, ul. Babushkina 36-1, 192171 St Petersburg, RUSSIA
Email: vodym@peterlink.ru*

Mr Trenton Edwards
*Coe College, 1220 First Av. NE, 52402 Cedar Rapids, Iowa, USA
Email: tgedward@coe.edu*

Prof. Doris Ehrt
University Jena, Otto-Schott-Institut, Fraunhoferstr. 6, D07743 Jena, GERMANY. Email: doris.ehrt@uni-jena.de

Prof. Steve Feller
*Coe College, 1220 First Ave NE, 52402 Cedar Rapids, Iowa, USA
Email: sfeller@coe.edu*

Dr Guillaume Ferlat
IMPMC, Université P. & M. Curie, Paris Campus Boucicaut, 140 rue de Lourmel, 75015 Paris, FRANCE. Email: ferlat@lmcp.jussieu.fr

Ms Cynthia R. Ferrari Gonçalves
*CNRS BP 166, cedex 09 38042, Grenoble, FRANCE
Email: cynthia.ferrari@grenoble.cnrs.fr*

Prof. Marek Foltyn
Warsaw University of Technology, Koszykowa 75 00-662, Warszawa, POLAND. Email: foltyn@if.pw.edu.pl

Prof. Paolo Fornasini
Università di Trento, Dipartimento di Fisica, Via Sommarive 14, I38050 Povo (TN), ITALY. Email: fornasin@science.unitn.it

Prof. Guenther H. Frischat
*Germany Institut für Nichtmetallische Werkstoffe, TU Clausthal, Zehntnerstrasse 2a, 38678 Clausthal-Zellerfeld, GERMANY
Email: guenther.frischat@tu-clausthal.de*

Prof. Jerzy Garbarczyk
Warsaw University of Technology, Koszykowa 75, 00-662 Warszaw, POLAND. Email: jegar@mech.pw.edu.pl

Dr Olga Golubeva
*Insitute of Silicate Chemistry of the Russian Academy of Sciences, nab. Makarova, d.2, liter B 199034, St.-Petersburg, RUSSIA
Email: olga_isc@mail.ru*

Dr Roberto Graziola
Università di Trento, Dipartimento di Fisica, Via Sommarive 14, I38050 Povo (TN), ITALY. Email: graziola@science@unint.it

Dr Rolly Grisenti
Università di Trento, Dipartimento di Fisica, Via Sommarive 14, I38050 Povo (TN), ITALY. Email: grisenti@science.unint.it

Mr Matt Hall
Coe College, 1220 First Av. NE, Cedar Rapids, Iowa, USA. Email: mbhall@coe.edu

Dr Adrienne Hammerschmidt
University of Münster, Inorganic Chemistry Department, Corrensstr. 30, 48149 Münster, GERMANY. Email: hammers@uni-muenster.de

Dr Katsumi Handa
Synchrotron Radiation Center, Ritsumeikan University, 1-1-1 Nojihigashi, Kusatsu, Shiga 525-8577, JAPAN. Email: handa@se.ritsumei.ac.jp

Prof. Alex Hannon
ISIS Facility, RAL, Chilton, Didcot, Oxon OX11 0QX, UK
Email: a.c.hannon@rl.ac.uk

Mr Richard Haworth
University of Reading, J.J. Thomson Physical Laboratory, Whiteknights, Reading RG6 6AF, UK. Email: r.haworth@rdg.ac.uk

Dr Akitoshi Hayashi
Department of Applied Chemistry, Osaka Prefecture University, 1-1 Gakuen-cho, Sakai Osaka 599-8531, JAPAN
Email: hayashi@chem.osakafu-u.ac.jp

Prof. Zein El-Abidin Heiba
Ain Shams University, Faculty of science,Physics Department,11566 Abassia, Cairo, EGYPT. Email: Zein_KH@gawab.com

Dr Gerhard Herms
Rostock University, Institute of Physics, Universitätsplatz 3, D-18051 Rostock, GERMANY. Email: gerhard.herms@gmx.de

Dr Rafael M. Hovhannisyan
Institute of Materials' Science, 17 Charents Str., 375025 Yerevan, ARMENIA. Email: hovhannisyan@netsys.am

Dr Martun Hovhannisyan
State Engineering University of Armenia, 105 Teryan st., 375009 Yerevan, ARMENIA. Email: martun_h@yahoo.com

Dr Hubert Huppertz
Department Chemie, LMU München Butenandtstr. 5-13, 81377 München, GERMANY. Email: huh@cup.uni-muenchen.de

Dr Alain Ibanez
CNRS BP 166, cedex 09 38042 Grenoble, FRANCE
Email: alain.ibanez@grenoble.cnrs.fr

Ms Junko Ide
Department of Applied Chemistry, Faculty of Science and Engineering Ritsumeikan University, 1-1-1 Nojihigashi, Kusatsu, Shiga 525-8577, JAPAN. Email: ro000998@se.ritsumei.ac.jp

Mr Yuji Ike
Graduate School of Pure and Applied Sciences, University of Tsukuba, Ibaraki, 305-8573 Tsukuba, JAPAN. Email: s-ike@ims.tsukuba.ac.jp

Dr C. K. Jayasankar
Sri Venkateswara University, Department of Physics, Sri Venkateswara University, Tirupati - 517 502 Tirupati, INDIA. Email: ckjaya@yahoo.com

Dr Silvia Larcheri
Istituto di Fotonica e Nanotecnologie del Consigio Nazionale delle Ricerche - Sezione "ITC-CeFSA" di Trento, Via Sommarive 18, I38050 Povo (TN), ITALY. Email: larcheri@science.unitn.it

Dr Robert Lynne Jones
Curtin University of Technology, Department of Applied Chemistry, GPO Box U1987, Perth WA 6845, AUSTRALIA
Email: L.Jones@curtin.edu.au

Dr Satish Khasa
St. Lect. In Physics Govt. College for women, 124001 Rohtak, INDIA
Email: skhasa@rediff.com

Dr Johannes Kirchhof
Institut für Physikalische Hochtechnologie e.V. Jena, PF 100 239 07702, Jena, GERMANY. Email: johannes.kirchhof@ipht-jena.de

Mr Thomas Guido Kloss
Schott AG Jenaer Glas GmbH, 07745 Jena, GERMANY
Email: thomas.kloss@schott.com

Dr Valentin Klyuev
Department of Physics, St.-Petersburg State Technological University of Plant Polymers, Ul. Ivana Chernykh 4, 198095 Saint-Petersburg, RUSSIA
Email: Borpev@mail.ru

Prof. Masao Kodama
Sojo University, 4-22-1 Ikeda, Kumamoto 860-0082, JAPAN
Email: kodama@chem.sojo-u.ac.jp

Prof. Ladislav Koudelka
University of Pardubice, Nam.Cs.legii 565 532 10, Pardubice, CZECH REPUBLIC. Email: ladislav.koudelka@upce.cz

Dr Scott Kroeker
University of Manitoba, Department of Chemistry, 452 Parker Building, Winnipeg, Manitoba, R3T 2N2, CANADA.
Email: Scott_Kroeker@UManitoba.ca

Dr. Raluca Lucacel Ciceo
Babes Bolyai University, Faculty of Physics, St. M. Kogalniceanu nr.1 400084, Cluj-Napoca, ROMANI. Email: rluca@phys.ubbcluj.ro

Dr Andrea Mandanici
Università degli Studi di Messina, Dip.to di Fisica, Salita Sperone 31, Contrada Papardo S.Agata, 98166 Messina, ITALY
Email: andrea.mandanici@unime.it

Ms Tatiana Markova
Department of Physics, St.-Petersburg State Technological University of Plant Polymers, Ul. Ivana Chernykh 4, 198095 Saint-Petersburg, RUSSIA. Email: Markova@land.ru

Dr Doris Moencke
Otto-Schott-Institut, Friedrich-Schiller-University, Fraunhoferstr. 6, Jena, GERMANY. Email: c5modo@rz.uni-jena.de

Prof. Francesca Monti
Dipartimento di Informatica, Università di Verona, Ca' Vignal 2, Strada Le Grazie 15, 37134 Verona, ITALY. Email: francesca.monti@univr.it

Prof. Piercarlo Mustarelli
University of Pavia, Via Taramelli 16, Pavia, ITALY
Email: mustarelli@matsci.unipv.it

Mrs Cristiane Nascimento Santos
CNRS BP 166, cedex 09, 38042 Grenoble, FRANCE
Email: cnsantos@if.sc.usp.br

Ms Elisabetta Negri
Dipartimento di Scienze della Terra Università degli Studi di Pavia, v. Ferrata 1, 27100 Pavia, ITALY. Email: elisabettanegri@manhattan.unipv.it

Dr Daniel Neuville
CNRS-IPGP, 4 place Jussieu, 75005 Paris, FRANCE
Email: neuville@ipgp.jussieu.fr

Dr Jan Nowinski
Warsaw University of Technology, Koszykowa 75, 00-662 Warszawa, POLAND. Email: nowin@if.pw.edu.pl

Prof. Dieter Oelkrug
University of Tuebingen auf des Morgenskelle, Department of Phisical

Chemistry, 72076 Tuebingen, GERMANY
dieter.oelkrug@ipc.uni-tuebingen.de

Prof. Norikazu Ohtori
Graduate School of Science and Technology, Niigata University, Ikarashi 2-no cho, Niigata 950-2181, JAPAN.
Email: ohtori@chem.sc.niigata-u.ac.jp

Mr Greg Ongie
Coe College 1220 First Av. NE Cedar Rapids, Iowa, USA
Email: GJONGIE@coe.edu

Dr. Boris Pevzner
Department of Physics, St.-Petersburg State Technological University of Plant Polymers, Ul. Ivana Chernykh 4, 198095 Saint-Petersburg, RUSSIA. Email: Borpev@mail.ru

Dr Joanna Pisarska
Silesian University of Technology,, Krasinskiego 8, 40-019 Katowice, POLAND. Email: gabrys@polsl.katowice.pl

Dr Wojciech Pisarski
University of Silesia, Bankowa 12, 40-007 Katowice, POLAND
Email: wpisarsk@us.edu.pl

Mr Steffen Pütz
University of Münster, Inorganic Chemistry Department, Corrensstrasse 36, 48149 Münster, GERMANY. Email: putzs@uni-muenster.de

Mr Lauro J. Queiroz Maia
CNRS BP 166, cedex 09, 38042 Grenoble, FRANCE
Email: cnsantos@if.sc.usp.br

Dr Anna Raimondo
Dipartimento di Fisica, Università di Messina, Salita Sperone 31, Messina, ITALY. Email: anna.raimondo@unime.it

Prof. Christian Rüssel
Otto-Schott-Institut Fraunhoferstr. 6, 07743 Jena, GERMANY
Email: ccr@rz.uni-jena.de

Prof. Witold Ryba-Romanowski
Institute of Low Temperature and Structure Research, Okolna 2, 50-422 Wroclaw, POLAND. Email: W.Ryba-Romanowski@int.pan.wroc.pl

Dr Francesco Rocca
Istituto di Fotonica e Nanotecnologie del CNR, Sezione CeFSA di Trento, Via Sommarive 18, 38050 Povo, Trento, ITALY
Email: rocca@science.unitn.it

Dr Andrea Sanson
Istituto di Fotonica e Nanotecnologie del CNR, Sezione CeFSA di Trento, Via Sommarive 18, -38050 Povo, Trento, ITALY. Email: sanson@science.unitn.it

Prof. A. A. A. Shama
Physics Department, Faculty of Science, Ain Shams University, 11566 ABBASSIA, Cairo, EGYPT

Mr Alexandr Shashkin
State Optical Institute, ul. Babushkina 36 – 1, 192171 St Petersburg, RUSSIA. Email: shashkin@goi.ru

Dr Mikhail Shepilov
State Optical Institute, ul. Babushkina 36-1, 192171 St Petersburg, RUSSIA. Email: shep@peterlink.ru; shep@goi.ru

Prof. Simion Simon
Babes-Bolyai University, Faculty of Physics, M. Kogalniceanu Str. 1, 400084 Cluj-Napoca, ROMANIA. Email: simons@phys.ubbcluj.ro

Prof. Roger Sinclair
Reading University Physics, Whiteknights, Reading, UK
Email: R.N.Sinclair@rdg.ac.uk

Prof. Steve Singleton
Coe College, 1220 First Ave NE, 52402 Cedar Rapids, Iowa, USA
Email: ssinglet@coe.edu

Dr Ondrej Sipr
Institute of Physics, Academy of Sciences CR, Cukrovarnicka 10, 162 53 Praha 6, CZECH REPUBLIC. Email: sipr@fzu.cz

Prof. Akira Takada
Asahi Glass,1150 Hazawa-cho, Kanagawa-ku, Yokohama-shi, JAPAN
Email: akira-takada@agc.co.jp

Dr Masahiro Tatsumisago
Department of Applied Chemistry, Osaka Prefecture University, Sakai, Osaka 599-8531, JAPAN. Email: tatsu@chem.osakafu-u.ac.jp

Dr Norimasa Umesaki
Japan Synchrotron Radiation Research Institute JASRI/Spring-8 Kouto, Mikazuki, Sayo Hyogo 679-5198, JAPAN. Email: umesaki@spring8.or.jp

Dr. Marco Vaccari
Università di Trento, Dipartimento di Fisica, Via Sommarive 14, I38050 Povo (TN), ITALY. Email: vaccari@science.unitn.it

Dr Rene Vacher
Universite Montpellier, LCVN Case 069 - Cedex 5, F-34095 Montpellier, FRANCE. Email: rene.vacher@lcvn.univ-montp2.fr

Dr Christos-Platon Varsamis
National Hellenic Research Foundation, Vass. Constantinou 48 Ave, 11635 Athens, GREECE. Email: cvars@eie.gr

Dr Natalia M. Vedishcheva
Insitute of Silicate Chemistry of the Russian Academy of Sciences, Ul. Odoevskogo 24, Korp. 2, St. Petersburg, RUSSIA. Email: natalia@mail.rcom.ru

Mr Juergen Walker
Alfred University, 2 Pine Street/BMH, Alfred, NY 14802, USA
Email: juergen.walker@gmx.net

Dr Frederick T. Wallenberger
PPG Industries - Glass Technology Center, PO BOX 11310, Pittsburgh - PA 15238-0310, USA. Email: wallenberger@ppg.com

Dr Marek Wasiucionek
Warsaw University of Technology, Koszykowa 75, 00-662 Warszawa, POLAND. Email: mwas@mech.pw.edu.pl

Prof. Adrian C. Wright
University of Reading, J.J. Thomson Physical Laboratory, Whiteknights, Reading RG6 6AF, UK. Email: a.c.wright@reading.ac.uk

Mr Guang Yang
ISL, University of Sheffield, Department of Engineering Materials, Mappin Street, Sheffield S1 3JD, UK. Email: g.yang@shef.ac.uk

Dr Randall Youngman
Corning Incorporated, SP-AR-02-4, Corning, NY 14831, USA
Email: youngmanre@corning.com

Dr Alexandr Zhilin
State Optical Institute, ul. Babushkina 36-1, 192171 St Petersburg, RUSSIA
Email: nitiom@goi.ru

Prof. Josef Zwanziger
Dalhousie University, NS B3H 4J3 Halifax, CANADA
Email: jzwanzig@dal.ca

www.ingramcontent.com/pod-product-compliance
Ingram Content Group UK Ltd.
Pitfield, Milton Keynes, MK11 3LW, UK
UKHW050624170926
15229UKWH00009B/36

9 780900 682612